AF835008

Index Theory and Price Statistics

Peter von der Lippe

Index Theory and Price Statistics

PETER LANG
Frankfurt am Main · Berlin · Bern · Bruxelles · New York · Oxford · Wien

Bibliographic Information published by the Deutsche Nationalbibliothek
The Deutsche Nationalbibliothek lists this publication in the Deutsche Nationalbibliografie; detailed bibliographic data is available in the internet at <http://www.d-nb.de>.

ISBN 978-3-631-56317-5

© Peter Lang GmbH
Internationaler Verlag der Wissenschaften
Frankfurt am Main 2007
All rights reserved.

All parts of this publication are protected by copyright. Any utilisation outside the strict limits of the copyright law, without the permission of the publisher, is forbidden and liable to prosecution. This applies in particular to reproductions, translations, microfilming, and storage and processing in electronic retrieval systems.

www.peterlang.de

Affectionately dedicated to
my wife Yanqian

献给爱妻燕倩

Preface

This book grew out of some preparatory texts written for participants of various courses on "Index Theory and Price Statistics" organized by the TES Institute (Training of European Statisticians) of Eurostat (the Statistical Office of the European Union). Courses of this kind were designed for the staff of National Statistical Institutes (NSIs) as well as research institutes, central banks and ministries of a great number of countries (in one case also non-European Mediterranean countries). They took place in Düsseldorf and Luxembourg (both repeatedly) as well as in Brussels and Ljubljana. Later, after the liquidation of TES, there were similar courses for Balkan NSIs in Skopje and Belgrade. My subject as course leader was primarily index theory, and in the later courses I also took over the price statistics part, initially as a rule presented by representatives of the German NSI and Eurostat.

The book evidently owes a lot to the many TES-co-trainers from official statistics and also to quite a few participants. For the sake of brevity I have to refrain from naming them here.

So this book has the same genesis as my book "Chain Indices, A Study in Index Theory", published in 2001 in a publication series of the German Federal Statistical Office. It is, however, more of a text book, whereas the chain index book should be seen rather as a pamphlet against chain indices. In a sense, chapter 7 of the present book summarizes a good deal of my criticism of chain indices in the aforesaid book, which unfortunately has not yet found much (benevolent) attention. Given that Alfred Marshall's (1842 – 1924) pleading in favour of chain indices remained more or less unheard for about a hundred years until it became all the rage thanks to the SNA 1993, I am pretty sure that it is most unlikely that I will outlive the chain index era. Yet I still hope for the best, and that my work was not in vain.

This should not be misunderstood. Chaining or not, the present book is first and foremost simply a "normal" textbook, mostly dealing with non-controversial issues. It does so, however, with the intention of giving a comprehensible and comprehensive introduction for users of statistics.

Parts of the book have also been the subject of lectures for graduate students at the University of Essen, so I had some feedback not only from the practice of official statistics, but also from students. Some of them later became assistants of mine and cooperated with me in order to take their doctor's degree. They made very useful contributions to this book. In this context I would like to mention Dr. Andreas Kladroba, Michael Westermann, Jens Mehrhoff, and my present assistants Janina Küter and Claus Christian Breuer. The latter is on the right track to become an expert in the economic theory of index numbers (section 2.4 of the present book), but he also left a mark in other parts of the book and was a great help to me in some work on applied price statistics, such as an expertise for the German Constitutional Court. Janina Küter also went through the whole text of the book and contributed in particular to topics on official price statistics. She was particularly helpful in that she mildly (and politely) insisted that I should not be content with the incomplete collection of downloads we had offered on the Internet for quite a while, but to finish up with a fine-looking book after all. So I only recently filled the gaps (especially sections 2.1,

4.2 and 4.3) and once more checked everything in order to prepare it for a proper publication.

In this phase, in which there was more time consuming work left to be done than expected (and in which Janina Küter's talent in organizing a working party was again indispensable), I was lucky to have met two students who both made most valuable contributions and proved really a blessing. Jinny Meyer has done a very good (and laborious) job in the final editing and formatting of the whole text on the computer. She has also paid attention to the content and language and she worked out the references. Matthew Harrison – the only native speaker – checked the language, not only with language proficiency but also with thoroughness and a good understanding of what the text is about. I am fully aware of the fact that his work is essential for a hopefully good acceptance of the book, so I do feel much gratitude.

It is a pleasure and honour for me to mention two prominent persons who both made a number of helpful comments on a draft version of parts of this book, viz. Bert Balk, CBS and Erasmus University in the Netherlands and Hans Albert Leifer, head of the Economics Statistics Division of the Deutsche Bundesbank.

I am also grateful to the publisher, in particular Dr. Ühlein from the Peter Lang book publisher, who demonstrated much patience and forbearance with the author, as well as competence in producing nice books.

Finally, I would like to mention my wife, Dr. Yanqian von der Lippe-Fan, to whom the book is dedicated (which is only a modest way of expressing my gratitude and affection). She not only directly, but also indirectly, supported me in the time-consuming creation of this book. A brief remark on what is meant by "indirectly" might be appropriate. Conditions to accomplish the arduous task of unfolding some mathematical proofs in full length were particularly favourable for me in China - and they were indeed nowhere better than there - far away from my everyday life in Essen. The deplorable fact that I am still so lazy and perhaps not gifted enough to learn the terribly difficult mother tongue of my wife (shame on me!) meant that there were evenings in which I had plenty of time to think over difficult things. Quite a few parts of the book were written in Shanghai, while the rest of the family was sitting in another room, talking in Chinese. This applies for example to section 8.4, where I worked out in detail the derivation of the EKS parities from minimizing a distance. I could not find the proof in any other textbook. There are some other intricate parts of the book likewise "born" in Shanghai, but the vast majority of the book is much easier reading, which is fully compatible with every day life and does not require sitting alone in a room in a remote country.

So I hope that the reader will find the book altogether useful and suitable as an introduction into a perhaps unexpectedly interesting part of statistics.

Essen, May 2007 Peter von der Lippe

Brief contents

Table of contents (complete)

List of figures and tables

Chapter 1

General introduction and elementary price index theory

The objective of this book is to integrate a comprehensible presentation of the theory of index numbers and practical problems of price statistics in statistical offices.[1] This textbook owes its existence to TES-Courses[2] on "Price Index Theory and Price Statistics". The theory of index numbers is dominated by mathematical considerations, and it focuses on the justification (foundation), and assessment of index functions (formulas) against the background of formal aspects. The practice of price statistics on the other hand, has to cope with different problems, like the regular compilation of prices as well as "weights", adjustments for quality changes and the like. For each group, index theoreticians on one hand and official price statisticians on the other, it is not easy to understand the problems of the other. Moreover the two groups appear to diverge rather than to converge.[3] Therefore we aim to build a bridge between mathematical theory and practice in statistical offices and ministries. The book attempts to help those who have to deal with the production and interpretation of price statistics to understand index numbers better. It therefore tries to provide a balanced introduction with respect to both, index theory and practice of official statistics. This also implies a kind of presentation which throughout the book is rather moderately mathematical.

1.1 Fundamental principles of price statistics and price indices

In this section some most elementary and introductory aspects of index formulas will be discussed explicitly with reference to problems inspired by the practice of official statistics and the way the results of price statistics are used and interpreted in politics. After giving a short overview we give a first (to be refined) definition of index functions and introduce some fundamental methodological problems of price statistics, i.e. organization of data collection and index compilation.

a) The scope and structure of the book

In **fig. 1.1.1** an overview of the structure of this book is given. Index theory primarily if not almost exclusively deals with the formula of the (price) index to be compiled. It

[1] For this reason we also try to provide a balanced introduction with respect to the variety of problems, opinions and "approaches" in index theory and to examine the usefulness of the results *for the purposes of official price statistics.*

[2] TES stands for "Training of European Statisticians" and was an institute providing courses for official statisticians in Europe. In the courses "Price Index Theory and Price Statistics", held repeatedly, the author was course leader and he presented the theoretical part of the topic, while lecturers of the German Statistical Office and Eurostat dealt with the practice of price statistics and index compilation.

[3] Mathematical theory of index functions seems to grow in sophistication and complexity and hence becomes less accessible to "ordinary" economists and statisticians. On the other hand, for mathematicians familiar with index theory, there is little or no attraction in topics related to the practice of "official statistics".

may be questioned, however, if the formula itself really should occupy the center of interest. Perhaps some of the problems of getting the right data to enter the formula have an even greater effect on the result of index calculation than just the formula has. In index theory we find a variety of "approaches" to find and evaluate formulas, in particular a group of more "formal" ones (of which the axiomatic approach is arguably the most prominent) and the so-called "economic theory" (which is perhaps more "theory" than "economic") approach to index numbers.

It is perhaps a fruitless effort to count all the different index formulas that have been invented and proposed by somebody, as being the "best" or "ultimate" formula, but none of them rightly deserves absolute priority over the remainder.[4] There are hundreds of formulas and the set of formulas we might think of being useful in some respect or other is unbounded. Thus theory can at best offer some tools to enable users to decide on which formula to apply.

The axiomatic approach is such a device for a systematic assessment of index formulas. However, the criteria (desirable properties of an index number) used in this kind of evaluation, that is the "axioms" or "tests", are open to debate, often in conflict with one another and we cannot expect the axiomatic theory to "prove" in such a situation which criteria should come first and which should come second and therefore could be sacrificed.

To comment briefly on **fig. 1.1.1** we start with the index-theory part.

A fundamental distinction applying to inter-**temporal** comparisons only is the division into

a) **direct** binary (two-situation) comparisons (unfortunately often called "fixed base"[5] indices) as opposed to

b) the **chain index** approach (see **sec. 2.5** and **chap. 7**),

depending on whether simply the end-points of an interval in time are compared ("directly") or the interval is subdivided into a number of sub-intervals and the comparison is made indirectly by linking the results for the sub-intervals.

In addition to "tests" and "axioms" (for example the factor reversal test) as "finders of formulae" (I. Fisher[6]) we can also construct formulas on the basis of certain "models", for example the model of an additive decomposition of the value change (Stuvel), of a "factorial experiment" (Banerjee), or of continuous functions in the variable "time" (Divisia[7]). This will be introduced in more detail in **sec. 2.1**.

---

[4] This applies also to the famous "ideal" index of Irving Fisher. In addition to a criticism of the chain index approach it is one of our main concerns to show that the never ending enthusiasm for this formula is delusive.

[5] For reasons to be explained later, we prefer the terms "direct" (instead of "fixed base") and "chain" indices.

[6] Fisher's approach was to say "mechanistic" in examining (or rather trying out) all sorts of means and weights as elements of possible index formulas. By contrast the modern "axiomatic approach" is focusing on "systems" of axioms and "characterizing" (to be explained later) rather than deriving (or justifying) formulas.

[7] His approach is related to the chain index idea.

Figure 1.1.1: General scope and structure of the book

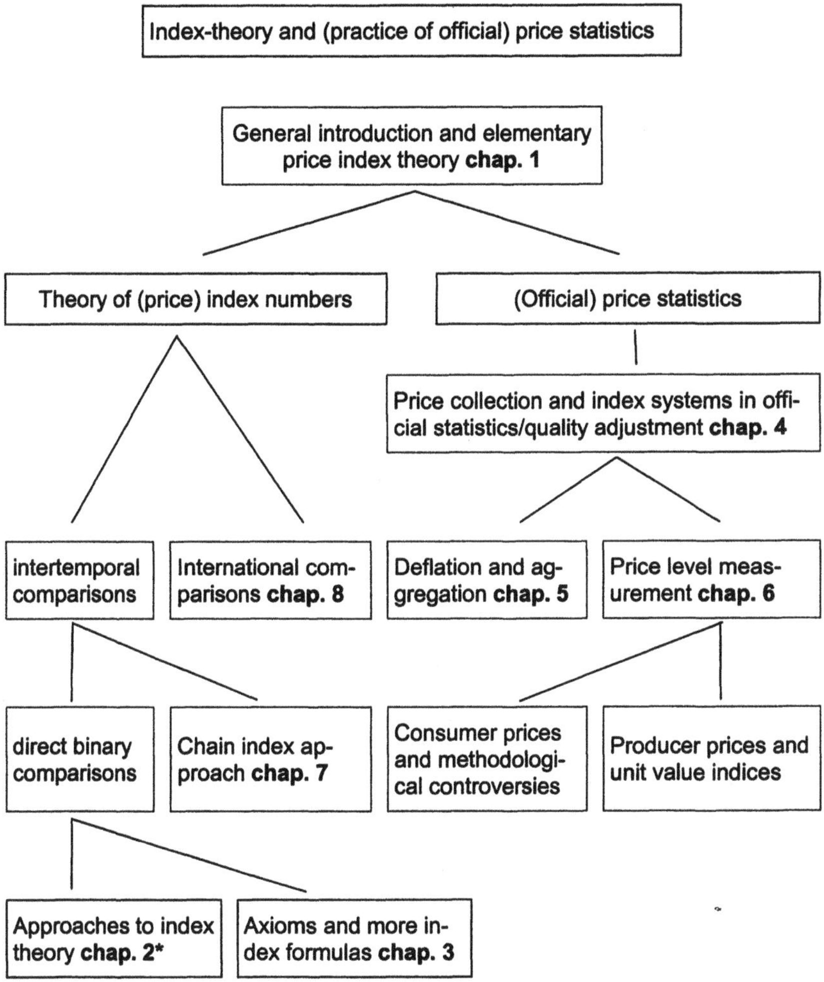

*e.g. formal index theory (focusing on mathematical properties of index functions) and economic theory of index numbers (aiming at a microeconomic foundation of index formulas).

To turn now to the price statistics part it should be noted that another fundamental distinction refers to the use made of price indices, viz. tools for price level measurement (inflation measurement in particular) on the one hand and for deflation of aggregates (in the National Accounts framework in particular) on the other.[8]

For the most part price statistics and price index compilation face a conflict between the principles of

- "representativity" (or "representativeness"), and of
- "pure comparison" (i.e. pure price - or pure quantity comparison respectively).

This conflict is also part of the choice between true (price/quantity) indices on the one hand and (price/quantity) indices on the basis of "unit values" on the other.[9]

In the price statistics part of this book great emphasis is put on the presentation of international (and more distinctly European) recommendations and harmonization projects of official statistics, for example the System of National Accounts (SNA93[10]), or the ongoing publication of "manuals" such as the **PPI manual** of the IMF, or the **CPI manual** of the ILO.

Finally some **technical notes** should be added.

Each chapter of the book attempts to be more or less self contained and is divided into a number of *sections*, numbered at two levels. Equations are indicated by three numbers "x.y.z", the first two refer to the section (x.y) and the third to the order of occurrence within that section. Tables, figures and examples are indicated in the same manner. *Matrices* are indicated by bold uppercase symbols and *vectors* by bold face lowercase symbols.

b) The concept of a price index

The word index is also used in a much wider sense of an *absolute* figure. There are "indices" to represent a summary measure of some more or less complex magnitude (or "construct") like welfare, social status, the business cycle etc. in *absolute* terms.[11] An index in the sense of what follows, however, expresses a *comparative* statement comparing a situation with a *base*-situation. We will discuss only *price* indices. The formal difference to other indices, like indices of wages, new orders, production and so on is in general deemed small.

---

[8] To solve certain practical price statistics problems, such as the deflation of value aggregates or consistency in making international price comparisons again gives rise to theoretical considerations. We will thus refer to index theory considerations in these more practical parts of the book too.

[9] See part e of this section for a definition of "unit values". As the unit value index type commonly occurs in the case of foreign trade and wage statistics unit values will be discussed in depth in **sec. 6.4**.

[10] SNA revision 1993.

[11] This is usually done by adding "scores" for various aspects of the construct in question to a weighted or unweighted sum of such scores. Some authors call "index" also descriptive statistics other than price indices, like measures of concentration (disparity) or so.

We call "indices"[12] only *relative* figures designed to measure a change (or difference) with respect to a complex (an aggregate) of *several* related variables. It is a measure of aggregative comparison showing in a single figure how a *set* of n related variables has changed (over time) or differs (from place to place). This leads to the following working definition:

> An index P_{0t} is a function P: $\mathbb{R}^{kn} \rightarrow \mathbb{R}$ mapping k = 4 real valued vectors with 4n dimensions[13]
>
> (1.1.1) $$P_{0t} = P(\mathbf{p}_0, \mathbf{q}_0, \mathbf{p}_t, \mathbf{q}_t)$$
>
> into a one dimensional positive real number for comparative purposes.

Note that this definition does not apply to chain indices nor to the so called "true cost of living index" resulting from the "economic theory" approach. The function P( ) should satisfy certain functional equations (axioms) and should have a meaningful interpretation.

In the case of a *binary temporal* comparison, to which index theory originally was exclusively devoted (and also the greater part of this book) it is common practice to use the following notation: P_{0t} (the price *index*) denotes the level of prices at time (period) t, the *current* period (or reporting period), compared to (relative to) time 0, the period known as the *base* period (or simply "base") because it is the basis of comparison in the sense that the level of prices at period 0 is 1 or 100%. Hence the result P_{0t} = 1.25 means a rise of prices between 0 and t amounting to 25% (per cent of the level of the base period)[14]. Note that the subscripts [or labels] 0 and t have a definite order, interchanging is not permissable since P_{t0} and P_{0t} are definitely different index numbers.

The word **base** is ambiguous because it may refer to the period to which

1. we compare the current state (**reference base**), or to which
2. the weights refer (**weight base**).

Some comments on the key words of the working definition should be added.

- By variables we may understand prices and quantities of commodities bought or sold, wages, industrial production, all (or selected) inputs, or the output of the manufacturing industries etc. The only limitation is that the variables should be positive and measurable on a metric scale. Otherwise there will be no real numbers to be combined to an "index".

---

[12] To express the plural of "index" there was originally in English no other term in use than "index numbers". But it has become more and more common in the last years to say "indices" or "indexes" instead of index numbers. Some writers distinguish an "index" (in the sense of "index function" or "index formula") and "index number" which is a specific value of this function.

[13] With two price vectors and two quantity vectors for n commoditites we have 4n dimensions (k = 4). In the case of chain indices we have k > 4.

[14] In practice it may be more common to state the result as P_{0t} = 120 but multiplying with 100 is only an operation to improve comprehension by users who are not so familiar with statistics. This is mathematically of minor significance and we will define index numbers without the multiplication by 100, that is we prefer the quotation 1.2 instead of 120.

- The set (or aggregate) of prices can be consumer prices in the case of a consumer price index as opposed to prices of raw materials or wholesale prices. The set is defined only qualitatively, i.e. consumer prices refer to a collection of commodities which is said to be representative of the "cost of living", whereas wholesale or retail prices belong to a set defined somewhat differently (e.g. there are very few services in a retail price index whereas there will be quite a few of them like bus fares, housing, bank or postal services, theatre or cinema performances in a consumer price index although there may be many overlaps between a "retail price index" and a "consumer price index" or a "cost of living index").[15]
- As to the number of comparisons there is not much theory related to the case of a true multi-situation instead of only two-situation ("binary") comparison. Of primary concern in index theory is to compare
  1. periods[16] in time (**intertemporal** comparisons), or
  2. locations, i.e. **interspatial** (or "place to place", that is in practice international and interregional) comparison.

Much less common is to consider explicitly **structural** comparisons[17] in index theory, like the "parity ratio" between prices received by farmers and prices paid by farmers or the well known "terms of trade", i.e. the ratio between export- and import prices.

Another example of a "structural" comparison is the comparison of subdivisions of the Consumer Price Index (CPI) which is usually stuctured into major categories like food and beverages, housing, clothing, personal care etc., or the structural comparison between CPIs *for different sub-populations.*[18]

c) Objectives and methodological principles of price statistics, the concept of deflation

Price statistics deals with the collection of *individual* prices in *absolute* terms (i.e. expressed in $, £ or so) and with synthetic expressions of "price levels" referring to *aggregates* of goods and services in *relative* terms of indices for the relevant aggregates. For some purposes absolute prices are interesting results as such but in most parts they are used as raw material only to compile price indices. Moreover, if we

---

[15] The type of variables and the definition of the "set" constitutes the type (or name) of the index number in question, e.g. whether we are dealing with a production index, an index of new orders, a price index or more specifically a consumer-, retail-, agriculture-sales-price index, or agriculture-input-price index etc..

[16] Intervalls like months, years etc. or (in the case of stocks) points in time (like Dec. 31st 2002).

[17] For example the comparison of index numbers of different type or among various sectoral indices (sub-indexes), each of them being part of the same global (overall) index.

[18] Different types of index numbers may not only refer to different subcollections of goods and services, they may also - in the case of a consumer price index - intend to measure differences in the change of well being of different groups of the population ("sub-populations" defined by types of households).

want price statistics to deliver information on absolute prices in contrast to relative prices only the system of price quotations ought to be much more detailed and comprising many more reports on prices.

Purposes price indices have to serve (see also **fig. 1.1.2**) are twofold[19]

a) to describe the "price **level**" of an aggregate, i.e. for a set of commodities, and

b) to "**deflate**" an amount of money representing a purchase or sale of commodities (goods and services) or simply an expenditure (e.g. payment of taxes).

To define "deflation" consider an aggregate at **current** (period t) prices,

$$(1.1.2) \qquad V_t = \sum_i p_{it} q_{it}$$

which is called a **value,** V_t (or "nominal" aggregate). The same (with respect to the selection of commodities and their quantities)[20] aggregate valued at **constant** prices of the base period 0

$$(1.1.3) \qquad Q_t = \sum_i p_{i0} q_{it}$$

is called a **volume** (or "real" aggregate). The objective of "**deflating**" an aggregate V_t, or a **"value index"** (value ratio)

$$(1.1.4) \qquad V_{0t} = V_t / V_0$$

is to estimate the volume Q_t which a value represents, given the value V_t, or to derive a quantity index Q_{0t} (showing the change of volume $Q_{0t} = Q_t/Q_0$)[21] from a value index V_{0t}. The aforementioned definition of an index as well as the two major uses of index calculations (see **fig. 1.1.2**) suggest that data collections on prices and index compilations have to meet certain general requirements, standards or fundamental methodological principles of price statistics and price indexes, viz.

a) the selection of "representative" prices, and

b) the principle of "pure price comparison".

Price statistics and index calculations always imply a selection (sampling) from the universe of prices and commodities. Care should be taken to arrive at a selection that is both representative, or in the case of a consumer price index, *typical* for the effective, or most *recent*, consumption pattern of specified types of households[22] and at the same time also sufficiently *comparable* (if possible *constant*).[23]

---

[19] The same formal requirements generally have to be met in both applications of index theory (price level measurement and deflation) but there are also situations in which a conflict may arise.

[20] It is important to make sure that both, the "nominal" and the "real" aggregate refer to the same commodities such that (ideally) the difference between them is attributable to the difference in prices only.

[21] The definition of "quantity" and "volume" will be made more precise in connection with the problem of aggregation (see part e of this section).

[22] mostly defined by income and family composition categories.

[23] The selection should not undergo frequent changes in the course of time (or differ much at both locations to be compared) in order to enable the result to be interpreted in terms of causes and effects ("pure" comparison).

As conditions inevitably change, it is easily seen that both principles, **representativity** and **pure price comparison**, are hard to reconcile, (so that a compromise is needed). Moreover, as change takes place more rapidly the task becomes even more difficult.

From the general objectives of price indices it follows:

To understand the two requirements it should be recalled that the *object of observation* in price statistics is *not* the mere *price but* the whole *act of buying* or selling in which the price is *one* element but not the only one and perhaps not even the most decisive one. Other factors are the quantity and quality of the commodity, the shop (outlet) in which the sale takes place, a bonus granted or services rendered in connection with the sale if applicable, arrangements made concerning delivery, availability of spare parts, insurance etc. Such "side conditions", or "**price determining characteristics**" (**PDC**) are elements that have to be constantly observed and taken into account in price statistics in order to avoid comparisons that are clearly unfair and misleading.

The concept of PDCs makes it easy to define the two principles:

The **representative principle (R)** consists in chosing those PDCs that are typical, often met in every-day purchases and that may be looked upon as being the "average" specification of the purchase in question.[24] As the number of purchases every day is immense, it is completely impossible to observe them all in any systematic manner and the contracts would also differ greatly if we would go into more details describing and classifying them. Thus a statistic of *all* contracts and prices would be not only prohibitively costly, but also of very limited use. Since a selection has to be made it seems reasonable to concentrate on "typical" contracts.

Pure price comparison (P) requires conditions under which any two price observations are made to be as similar as possible except for time or location, i.e. except for just that aspect, that the comparison intends to demonstrate. The reason is: the rise or decline of a price should not be attributable to factors other than the time or location of observation. Comparisons are called "pure" only if different prices can *not* be "explained" by factors (like different quality, type of shop etc.) other than time or place of observation.[25] In practice really "pure" comparison can hardly be made:

- For example in international comparisons it is often simply not possible to find the same good or only an "equivalent" good in the other country.[26]
- P requires that "*adjustments*" have to be made on observed prices in the case of *quality change*, in order to distinguish between a rise of price due to quality improvement and a "genuine" rise of the price (it is only the latter that price statistics should attempt to measure).

[24] The type of selection procedure (e.g. random sample or other) is left open in this context.

[25] Note that "pure" is defined negatively as *not* reflecting other factors, *not* being affected by structural change etc.

[26] That is the reason why e.g. expenditures for housing, accomodation or other services are often excluded from international price comparisons.

Figure 1.1.2: Uses of price statistics and price indices

a) **Purposes of price indices** in intertemporal comparisons

- analytical use: the measurement of "price levels" of certain markets
- deflation: the estimation of the underlying quantity of an aggregate

b) **Methodological principles** of price indices and price statistics

All price determining characteristics (PDC) of contracts have to be taken into account

representativeness (R)	pure price comparison (P)
select contracts governed by the most common and typical PDCs	PDCs prevailing in both situations to be compared as similar as possible

Moreover, in practice it is by no means easy to comply with *both* principles, P and R *simultaneously*, because if one of them is adhered to most strictly the other almost automatically will be violated. If applied absolutely the two principles will be contradictory. This can easily be seen by some examples:

- How should we deal with the emergence of a new product B, gradually replacing A? P requires carrying on in observing the price of A, while in order to satisfy R it would be better to change over to B, at least once B has become the more typical product than A. It should be added that for some other reasons (on the basis of the "economic" index theory) it is nowadays often vigorously claimed to include new products as soon as possible.
- To satisfy P product specification should be most detailed and concrete in order to ensure that prices of the same products are reported. However, R requires a more flexible description to enable a switch to new (more representative) products if necessary.

Hence as a consequence

> in practice always in one way or another some *compromise* between R and P has to be found which tries to comply with *both* conflicting principles acceptably. No solution can meet both of them to full satisfaction. This conflict also applies to index formulas.

There is for example an index concept, like direct Laspeyres indices, in which much attention is given to the idea of "pure" comparison at the expense of the representativeness of the basket to which the index refers. On the other hand, to want an index taking also the most recent changes of the consumption pattern into account would make a "chain index" preferable. But this necessarily implies an "impure" comparison in the sense that what is supposed to be a measure of a change in *prices* is to some extent also reflecting change in quantities and structure (mix of products, composition of the "basket"). Thus the result has some ambiguity, the avoidance of which is in essence the rationale of the principle of pure price comparison.

Interestingly pure price comparison is incontested in the case of a *single commodity*. A lot of attention is given to all sorts of problems that may impair comparabilty, like quality change or availability of similarly specified products in two different countries. There is a vast amount of discussion on how to handle these types of problems. But surprisingly when applied to a collection, "bundle" or *"basket" of commodities*, the same principle seems to be forgotten or even turned into its opposite. To incorporate the most recent representative basket in index calculation appears to dominate all other considerations.

It is generally accepted that each individual commodity should be comparable over time or space but it is also widely accepted that this can well be neglected in the case of a "composite" commodity, that is in the case of indices.[27]

A reasonable compromise between R and P respectively is - in our view - neither a chain index (focusing on R at the expense of P) nor keeping weights of a Laspeyres index unchanged for quite a long period in time (that is pursuing exclusively P and neglecting R unduly), but rather a Laspeyres index in which weights are reviewed and updated in intervals of say five years or so.

In the chain index "ideology" the fundamental principle of pure price comparison is habitually and deliberately ignored (or at least treated as less relevant). Even worse, it should be noted that the obsession with the updating of weights entails massive problems with the interpretation of the results and consistency requirements as regards aggregation and deflation.

d) Simple comparisons (a single commodity)

In what follows we try to show that there are three ways of describing the change in an individual price of commodity i say:

1. ***price relatives*** (≈ growth *factors*) with reference to a base period price p_{i0} (eq. 1.1.5), the aggregation of which leads to most of what are known as the "traditional" index formulas, for example the formulas of Laspeyres, Paasche, Fisher etc., or relatives (***links***) with reference to the price $p_{i,t-1}$ of the preceding period (**eq. 1.1.6**), or

[27] Obviously this remark refers to chain indices. They are vigorously recommended in SNA 93. On the other hand the authors of SNA 93 rejected so called "unit value" indices which "measure the change in the average value of units that are not homogeneous and may therefore be affected by changes in the mix of items as well as by changes in their prices. Unit value indices therefore cannot be expected to provide good measures of average price changes over time" (SNA 93, *no.* 16.13). *It did not occur to* them that the same criticism would apply to chain indices as well.

2. ***log-changes*** (≈ growth *rates* referring to a logarithmic mean as the reference value)[28] leading to "log-change indices" (like for example the index formulas of Varita and Törnqvist), and
3. ***differentials*** with respect to time, $dp_i(t)/dt$ terms, involved in the Divisia index.

Relatives (related to a fixed base)

Consider a (discrete) time series of prices i.e. a sequence of prices recorded for the same commodity in successive equally spaced time intervals, $p_{i\tau}$ (τ = 0, 1, ..., t)[29]. The price relative, a^i_{0t} or simply a_{0t} to compare prices of commodity i (i = 1,2,...,n) between two periods (e.g. months or years), 0 and t is defined as[30]

$$(1.1.5)\qquad a_{0t} = a^i_{0t} = \frac{p_{it}}{p_{i0}}.$$

A price relative is a pure number, free of any unit of measurement. The word "relative" makes clear that the price is not measured in absolute terms (in this respect we lose information) but relative to the price of the same commodity at the base period (which mostly is a past period) and in a way that is more appropriate for comparisons with other prices that tend to be on a different level (in this respect additional information is gained).

In the same manner, a quantity relative b_{0t} and a value relative c_{0t} (or more precise b^i_{0t}, c^i_{0t}) are defined as

$$(1.1.5a)\qquad b_{0t} = \frac{q_t}{q_0} \text{ and } \qquad (1.1.5b)\qquad c_{0t} = \frac{v_t}{v_0} = \frac{p_t q_t}{p_0 q_0} = a_{0t} b_{0t},$$

where q denotes the quantity and v the value of the commodity i (note that the sum of values over all i = 1, ..., n commodities will be called $V = \Sigma v$).

Price, quantity and value relatives have the six properties (also called axioms) easy to prove and listed in **tab. 1.1.1**. All six axioms are trivially fulfilled when applied to relatives but applied to index numbers, that is to $n \geq 2$ commodities

- at least some of them may pose problems, and will not be satisfied, and
- there are some axiomatic requirements to add, like mean value property or proportionality which make sense only if the number of commodities is $n \geq 2$.

The first three axioms (identity, dimensionality and commensurability) are minimum requirements to be met also by any index function in order to avoid ambiguity. They are clearly essentials for any intertemporal comparison. By

---

[28] See **sec. 3.4** for more details.

[29] or simply $p_0, p_1, ..., p_t$.

[30] For convenience, a price relative (like an index) is also often expressed in percentage form requiring the multiplication by 100. Following the practice in defining indices we refrain from multiplying by 100 in defining relatives. Relatives are also called *simple* index numbers or *elementary* index numbers in contrast to index numbers (sometimes also called *aggregative* index numbers).

1. **identity** a clear meaning is given to $a_{0t} = 1$ as opposed to $a_{0t} \neq 1$. Otherwise a relative would lack a reference point. Identity provides a normalization condition to judge the direction of change. Axiom
2. **dimensionality** makes sure that the calculation is independent of the currency in which the prices are expressed. Dimensionality is also called **"homogenity of degree 0 in the prices"**, and finally by axiom
3. **commensurability** independence is also given with respect to the arbitrary choice of units (e.g. quarts, gallons, pounds, ounces etc.) to which the price quotation refers (it is of course something different if the price of a quart or of a gallon is referred to).

Most index formulas satisfy these minimum requirements as relatives always do (if not the index formula would be almost useless). The remaining three axioms are often *not* met, however, when applied to indices instead of relatives. For example relatives (but mostly not index numbers) necessarily satisfy the **time reversal** test and the **factor reversal** test (axioms 4 and 5). The latter (axiom 5) requires that the value relative c_{0t} of a commodity i should be decomposable *exactly* into two factors, a price component (a_{0t}) and a quantity component (b_{0t}) respectively, just in the same way as the absolute monetary expression[31] p_tq_t can be partitioned into a price (p_t) and a quantity factor (q_t) as absolute figures without any residual (mixed or structural) component. As multiplying a price by a quantity gives a value (expenditure or turnover), the same should be true for price-, quantity- and value-changes (as measured by relatives).

Finally, **transitivity** (or chainability) means that the relative change as measured for a long interval from 0 to t should be the same as the corresponding change measured over a sequence of sub-intervals and "linked" (multiplied) together. This should be true for *any*[32] subdivisions of the interval into two or more adjacent sub-intervals, like from 0 to s and from s to t. This principle sounds reasonable and is automatically fulfilled in the case of relatives, but applied to indices it is responsible for much debate and much confusion. Transitivity may be regarded as a kind of consistency in temporal aggregation (i.e. over intervals in time).

A requirement not satisfied by simple relatives but in general by indices is the determinate-test by which an index should not be zero or infinity if zero prices are introduced (the same applies mutatis mutandis to quantity indices). A base price p_{i0} amounting to zero renders the price relative infinite but most index formulas made up of sums, like $\sum p_{i0}q_i^*$ where prices are multiplied by some quantities q* (like $q^* = q_{i0}$ as an example) will remain determinate.

[31] To be more distinct this is true only in the case of a "commodity flow" (CF), as opposed to a non comomodity flow (NCF) representing only an amount of money without an underlying quantity-structure (see **chap. 5** for more details).

[32] It will be shown in **chap. 7** that it is important to underline this point with much emphasis. There are many misunderstandings with respect to "chain indices" that simply result from not giving much attention to the word "any" here. For different subdivisions of the same time interval for example a chain index usually will *not* yield the same result.

Table 1.1.1: Axioms satisfied by price and quantity relatives (fixed base)

no	Axiom	definition and interpretation	
1	**Identity**	$a_{00} = a_{tt} = 1$	uniqueness of the reference point
2	**Dimensionality** (price dimensionality)	$\frac{\lambda p_t}{\lambda p_0} = \frac{p_t}{p_0}$	a_{0t} is independent of the currency in which the prices are expressed [b)]
3	**Commensurability** [a)]	$\frac{\lambda p_t}{\lambda p_0} = \frac{p_t}{p_0}$	independence of the unit of quantity to which the price of commodity i refers [b)]
4	**Time reversal test**	$a_{t0} = \frac{1}{a_{0t}}$	consistency of relatives with different base periods
5	**Factor reversal test**	$c_{0t} = a_{0t} b_{0t}$	the value change is decomposable in price change and quantity change
6	**Transitivity**	$a_{0t} = a_{0s} a_{st}$	for *all* three periods 0, s and t

a) if only one commodity is involved the mathematical representation of 2 and 3 can not be distinguished.

b) in both periods, 0 and t.

Links (chain base) and Log changes

Relatives providing a comparison with a fixed base value (p_0 in the definition of a price relative a_{0t} and q_0 in b_{0t}), normally the starting value in a time series, are also called fixed base relatives. An alternative is the chain based comparison, using "links" to be multiplied (chained). "Chaining" means to represent a_{0t} as product of link relatives, like links of a chain. A link (link relative) is a growth factor, defined by

(1.1.6) $$\ell_t = \frac{p_t}{p_{t-1}} = a_{t-1,t},$$

which is a price relative in which the base is always the previous period.[33] It can easily be seen that the axioms of tab. 1.1.1 will also hold in the case of link relatives and that

(1.1.7) $$a_{0t} = \ell_1 \ell_2 \ldots \ell_{t-1} \ell_t.$$

[33] Since the base period is always just one period apart from the reference period the notation can be simplified such that only one subscipt t is sufficient.

The link relatives, or growth factor ℓ_t of i is closely related to the growth rate of the i-th price $r_t = \frac{p_t - p_{t-1}}{p_{t-1}} = \frac{\Delta p_t}{p_{t-1}} = \frac{p_t}{p_{t-1}} - 1 = \ell_t - 1$, and the appropriate average describing the average rate of increase over an interval is $\bar{r}_{0t} = \sqrt[t]{\ell_1 ... \ell_t} - 1 = \sqrt[t]{a_{0t}} - 1$, the geometric mean of links, using the t-th root of the product of t successive links. Note that links are a device to handle a problem of aggregation over time (the multiplication in a_{0t} and in $\bar{r}_{0t}$ takes place over time-intervals). The geometric mean is also used in averaging over n commodities (instead of n intervals in time). In this context the meaning of the geometric mean is less evident, however.

The terms

(1.1.8) $Da_{0t} = \ln(p_t/p_0)$ or

(1.1.8a) $D\ell_t = \ln(p_t/p_{t-1})$

are called **log-changes**. There is a close relation between Da_{0t} and a_{0t} and there also exists an appropriate mean of terms like Da_{0t}, the so-called **logarithmic mean** of, for example, the two positive numbers, x and y ($x \neq y$)

(1.1.9) $L(x,y) = \frac{y - x}{\ln(y/x)} = L(y,x).$

There are some index formulas based on log changes and L, like the Törnqvist index or Vartia indices, but their role in official statistics was rather small until now.

Alternatives other than links or log changes to the fixed base approach are rarely proposed. Some writers also examined (price) relatives based on moving averages of prices (or quantities respectively) in the denominator such as $\frac{p_t}{\frac{1}{3}(p_{t-1} + p_t + p_{t+1})}$ or similar expressions.

A characteristic of all aforesaid instruments to measure changes in prices, quantities or volumes introduced till now is that they are relative figures. At first glance it appears to be equally reasonable to measure change in absolute terms (in terms of differences), like $\Delta p_t = p_t - p_{t-1}$ and Δq_t, Δv_t defined correspondingly. In general relative expressions (as opposed to absolute figures) are preferred to absolute figures in price statistics, mainly for two reasons

- being pure numbers they are independent of the currency unit in which prices are expressed and independent of the physical unit to which the price quotation refers;
- in the case of n = 1 commodity a price level can be expressed in absolute terms, as the word "level" suggests an absolute measurement; however in the

case of n ≥ 2 goods involved a "level" of prices[34] can only be conceived as a relative figure.

Both points need further clarification:

Firstly being pure numbers does *not* imply, however, that relatives are meaningfully additive across different goods and services. A sum of relatives or links in general has no interpretation. Reference to a common unit of measurement, like the common monetary unit (currency) in case of summing over prices or using pure numbers with no dimension like relatives, may be a *necessary* condition for calculating a sum but it cannot be a *sufficient* condition to arrive at a meaningful result.

Secondly it should be realized that there is a *marked* difference between the simple case of one-commodity only (relatives) and situations in which n ≥ 2 commodities are involved (indices) in that the use of absolute figures is definitely less reasonable in the many commodities case.

e) Aggregative comparisons (two or more commodities) and unit values

Aggregation problems are the main reason for a need for a special index theory. With the transition from n = 1 commodity (the case of relatives) to n ≥ 2 commodities (the case of indices) the situation changes fundamentally in that

- both "price" and "quantity" have to be redefined if applied to a "bundle" (or "basket") of commodities instead of a single commodity;
- aggregation (across commodities) will destroy many useful properties of relatives or links mentioned above,[35]
- a structural problem, unknown to the single commodity case, is the question to what extent a price index measures the increasing *level* of prices and to what extent it also reflects a change in the *structure* of prices or in the consumption pattern.[36]

Such problems give rise to a special statistical theory of price index numbers not being simply a part of the theory of means. We may even say that index theory in general is that part of statistics that deals with the measurement of "a complex that is made of individual measurements for which no common physical unit exists" as Ragnar Frisch has put it (quoted by Allen, 1975, p. 5). A key problem then is the elimination of structural change in aggregates in order to provide a pure price comparison. As price index formulas will yield different results only in the case of such a change,

---

[34] This applies to an even greater extent to a level of "quantity".

[35] indices fail some tests like time reversal, or factor reversal test etc. which simple relatives are satisfying by definition.

[36] The same applies to quantities of course. In the single commodity case there is no point in separating and isolating these two aspects and in defining a "pure" price-level-component in contrast to a mixed (price/quantity) component or a "structural" component. The *structure* of prices is an issue that always caused a lot of confusion in index theory. The change of structure entails a non-zero variance of price relatives. On the other hand if the variance vanishes virtually all index formulas would yield the same result. A problem of selecting the appropriate index function only arises if there is some amount of dispersion, i.e. if there is some *structural* change involved.

different index formulas can also be viewed as different ways of dealing with structural change by some kind of *standardization.*[37] In what follows the need for such a standardization will be demonstrated with a simple numerical example.

Example 1.1.1

Imagine an economy with only two industries A and B, and wages of $10 and $16 paid at base period:

situation in base period			
industry	wage	hours	payment
A	10	50	500
B	16	50	800
sum*	13	100	1300

* or average

It would be highly misleading to compare simply the average wage per hour presently paid with the average wage formerly paid at the base year. Assume two alternative (presented for demonstrative purposes) situations in t.

situation in t			
industry	wage	hours	payment
A	15	90	1350
B	24	10	240
total	15.9	100	1590

alternative situation in t			
industry	wage	hours	payment
A	15	10	150
B	24	90	2160
total	23.1	100	2310

The average wage (price) per hour changed from 1300/100=13$ to 15.9 $ (i.e. growth of only 22.3%) or in the alternative situation from $13 to $23.1 (by 77.69%), although both industries experienced exactly the same rise of 50% in wages. Hence the average wage obviously does not reflect the rise of wages (at an equal rate of 50% in both branches) correctly. The reason is that the structure of employment has changed in favor of the low-earning industry or (alternatively) of the high-earning industry. ♦

From this it follows that an **index of wages** (which is formally nothing else than a price index) should provide weights for the wages, derived from the *same* employment structure in both situations 0 and t in order to show the true ("pure") rise of the wage-level[38] which is undoubtedly 50%. Any wage index not showing a rise of 50%

[37] Standardization in statistics means making aggregates (or averages) comparable by using the same set of "weights".

[38] In this case it is irrelevant whether the Laspeyres- or the Paasche type of wage index is used since the growth rates of wages are the same in both industries (+ 50%).

does not appear reasonable.[39] There are two reasons for using indices instead of simple averages to compare aggregates:

1. A sum or average might not be meaningful,[40] and it
2. might be affected by a change in the *structure* (for example the structure of the manpower employed in the case of wages).

Structural change therefore has to be eliminated by some method of weighting (for example by weighting with a constant base year structure, like in a Laspeyres type wage index). Indices differ from averages (i.e. so called "unit values") primarily by their invariance to a possible structural change and thus their ability to provide a "pure" comparison.

If there were no such change it would be legitimate to measure a "price level" simply in absolute terms of some sums or averages like

- value aggregates $V_t = \sum p_{it} q_{it}$ or
- the "unit value" as indicator of the "average price" (provided the sum $\sum q_{it}$ is defined)

$$\tilde{p}_t = \frac{\sum p_{it} q_{it}}{\sum q_{it}} = \sum p_{it} \frac{q_{it}}{\sum q_{it}} \quad (1.1.10)$$

But in general such measures are inappropriate because they are sensitive to a change in structure. The value of, say, automobile-output, can rise ($V_t > V_0$) only due to an increase of prices and not of quantities and values as well as unit values can rise ($\tilde{p}_t > \tilde{p}_0$) only because of a switch from low to high priced models even though the total number of automobiles (quantity) as well as the volume (quantity valued at constant prices) produced remains unchanged.

1.2 Unweighted indices

This section starts with some historical notes, showing how index formulas emerged, and what kind of debates took place at the early stage of index theory. The topic of unweighted index formulas re-gained much attention only recently due to some controversy about how to average price quotations on the lowest level of aggregation.[41]

---

[39] This tacitly requires a measure which satisfies "*proportionality*": if all prices (wages) rise by 50% the index should display a rise by 50%, too.

[40] This applies in particular to prices referring to *different commodities*: an "average-price" of a bottle of milk and a car is nonsense.

[41] A vivid discussion of this issue is owed to the famous "Boskin report" dealt within **sec. 1.3**.

a) Dutot's index and Drobisch's unit value index

The French author Dutot[42], one of the first to use an index, had to answer the question whether Louis XV, with revenue of 100 million livres in 1735, was better off than Louis XII had been, who had 7.65 million livres in 1515. He decided that Louis XV was considerably worse off due to an enormous inflation in the 220 years between 1515 and 1735. Dutot selected a *large* number of commodities (e.g. a goat, a chicken, a rabbit, a pigeon, a rick of hay), obviously mostly food, but also, which is interesting enough and has not been imitated for a long time since, services (a day's labor for man and for woman) and he devised perhaps the historically first formula in index theory which is a ratio of mean prices

$$(1.2.1) \qquad P_{0t}^{D} = \frac{\bar{p}_t}{\bar{p}_0} = \frac{\frac{1}{n}\Sigma p_{it}}{\frac{1}{n}\Sigma p_{i0}} = \frac{\Sigma p_{it}}{\Sigma p_{i0}} \quad \text{(price index of Dutot 1738).}$$

There are two main objections to be made to this formula:

1. The index does not satisfy the **commensurability** axiom: a change in the unit of measurement of *some* commodities (e.g. introducing prices referring to gallons [4 quarts] instead of quarts) does not leave the index unchanged which will be demonstrated in **ex. 1.2.1**.
2. The formula does not depend explicitly on quantities consumed and the unweighted arithmetic mean is not the only way to define an average price. An alternative method is to construct so called "unit values" $\tilde{p}$ (= average expenditures) as a type of "average prices".

M. W. Drobisch proposed the following index formula[43] based on unit values

$$(1.2.2) \qquad P_{0t}^{UD} = \frac{\tilde{p}_t}{\tilde{p}_0} = \frac{\Sigma p_{it}q_{it}/\Sigma q_{it}}{\Sigma p_{i0}q_{i0}/\Sigma q_{i0}} \quad \text{(unit value index of Drobisch 1871).}$$

Both formulas have in common that they are ratios of absolute figures, that is of average prices expressed in €, $, £ or the like. Such indices are necessarily liable to violate the commensurability requirement. The sums $\Sigma p_t q_t$ and $\Sigma p_0 q_0$ are not affected by a change in the physical units of quantities, but the sums Σq_t and Σq_0 are and so are the sums Σp_t and Σp_0 in Dutot's index. P^{UD} and P^{D} have this defect in common, however, as **fig. 1.2.1** shows, P^{UD} has three additional shortcomings making P^{UD} definitely even less acceptable than Dutot's formula:

- P^{UD} does not meet **the mean value condition**, that is its value can be greater (lower) than the greatest (lowest) individual price relative.
- The sums $\Sigma_i q_{it}$ and $\Sigma_i q_{i0}$ are not defined for the universe of goods of a whole economy,[44]

[42] "Reflexions politiques sur les finances et le commerce", The Hague 1738.

[43] There are some other formulas Drobisch invented. The unit value index of Drobisch (the ratio of two unit values at the scale of the whole economy) has to be distinguished from unit value indices (of wages or prices in foreign trade) discussed in **sec. 6.4**.

- P^{UD} can indicate a change of the price level although all prices remain unchanged, simply because quantity changed (in level or in structure). Hence P^{UD} violates the **identity** axiom.

Interpreted in terms of the economic theory of index numbers (**sec. 2.4**) the P^{UD}-index is no less problematic. Balk (1995)[45] has shown that this formula only makes sense in the case of a "simple sum utility function" $U = (\mathbf{p}_t'\mathbf{q}_t / \mathbf{p}_0'\mathbf{q}_0)(\mathbf{i}'\mathbf{q}_0 / \mathbf{i}'\mathbf{q}_t)$ where $\mathbf{i}' = [11 \ldots 1]$.

Moreover the arguments carried out in **sec.1.1e** concerning the inapplicability of an average wage (instead of a price index of wages) apply to P^{UD} too. By contrast to an index as average of relatives (see part b of this section), a relative (ratio) of averages is *not* a useful formula of an index.

Figure 1.2.1: Indices as ratios of average prices (as absolute figures)

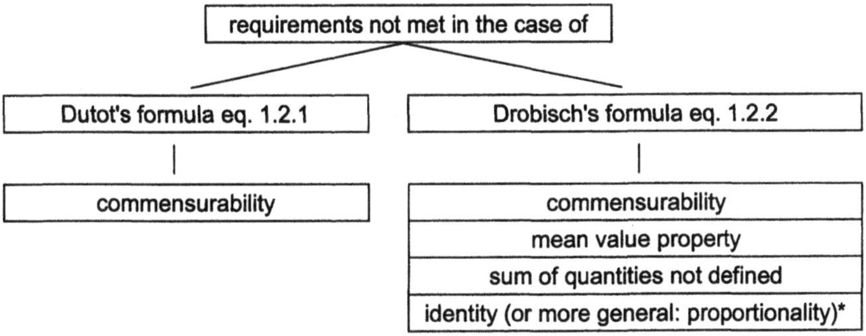

* identity is a special case of proportionality.

Example 1.2.1

Consider the following prices and quantities of two commodities, A and B

i	p_{i0}	p_{it}	q_{i0}	q_{it}		q_{it}^*
A	10	15	5	8		10
B	30	35	2	4		2
Σ	40	50	7	12		12

44 This is true, simply due to lacking a common unit of measurement. Only in some instances, esp. in foreign trade statistics unit values are calculated in official statistics for certain groups of homogeneous goods.

45 Bert M. Balk, On the use of unit value indices as consumer price subindices, Report. Presented at the Joint ECE/ILO Meeting on Consumer Price Indices, 20-24 November, Geneva, at the International Seminar "Improving the Quality of Price Indices", 18-20 December, Florence, and at the Fourth Meeting of the International Working Group on Price Indices, 22-24 April 1998, Washington DC.

Index of Dutot: $P_{0t}^{D}=\frac{\bar{p}_t}{\bar{p}_0}=\frac{25}{20}=\frac{\sum p_t}{\sum p_0}=\frac{50}{40}=1.25$. Now asssume prices p_{A0} and p_{At} refer to quarts and price statistics changes to a quotation in terms of gallons.[46] With prices of A on the basis of gallons we get $P_{0t}^{D}=\frac{60+35}{40+30}=\frac{95}{70}=1.357$ indicating a rise of prices by more than 25%. The unit value index P^{UD} amounts to

$$P_{0t}^{UD}=\frac{(120+140)/12}{(50+60)/7}=\frac{260/12}{110/7}=\frac{21.67}{15.71}=1.3788.$$

Assume now prices remained unchanged, so that p_{At} = 10 and p_{Bt} = 30. A reasonable index should be unity but $P_{0t}^{UD}=\frac{200/12}{110/7}=\frac{16.67}{15.71}=1.0606$, indicating a rise by 6%. Since

$$\text{(1.2.2a)} \qquad P_{0t}^{UD}=\frac{\Sigma p_{it}\frac{q_{it}}{\Sigma q_{it}}}{\Sigma p_{i0}\frac{q_{i0}}{\Sigma q_{i0}}}=\frac{\Sigma p_{it}\alpha_{it}}{\Sigma p_{i0}\alpha_{i0}}$$

weights α, reflecting the structure of quantities, play an essential part in P^{UD}. As weights α_{it} differ from α_{i0} such that commodity B, which is cheaper than A, will get a weight α_{it} greater than α_{i0} the result is $P^{UD} < 1$. Taking quantities q* (variation of the example) we obtain: $P_{0t}^{UD}=\frac{(100+60)/12}{110/7}=\frac{13.33}{15.71}=0.8485$. ♦

An analysis of P^{UD} also shows why the ratio of "average wages" (virtually unit values with quantities q being numbers of employees) in ex. 1.1.1 is inacceptable. Since in this example we have $\Sigma q_0=\Sigma q_t$ = 100 persons, P^{UD} equals the simple value ratio (index) that can be decomposed into a price and a structural component as follows:

$$\text{(1.2.3)} \qquad V_{0t}=\frac{\Sigma p_t q_t}{\Sigma p_0 q_0}=\frac{\Sigma p_t q_0}{\Sigma p_0 q_0}+\frac{\Sigma p_t(q_t-q_0)}{\Sigma p_0 q_0}=P_{0t}+S_{0t}.$$

The price component or the pure wage movement is $P_{0t} = \Sigma p_t q_0/\Sigma p_0 q_0 = 1.5$ (which is by the way the Laspeyres-price index P_{0t}^{L}), in accordance with the 50% rise of wages in both branches, and the structural component, i.e. the expression $S_{0t} = \Sigma p_t(q_t - q_0)/\Sigma p_0 q_{0t}$ can be positive or negative. From this it follows that – unlike Dutot's index - ratios of values or unit values may violate proportionality, a property requiring that the index is 1.5 (150%) if prices (wages) rise in unison by 50% in all branches.[47]

[46] Prices and quantities of B remain unchanged.

[47] It can easily be verified that if all prices are rising yet the value index and Drobisch's index may nonetheless be displaying a decline.

b) Carli's index formula

The discussion of Dutot's and Drobisch's formula suggests the following conclusion:

In order to satisfy commensurability and to eliminate the structural component a price index should be a mean of ratios (relatives) instead of a ratio of means.

It should be noted that a mean of ratios and a ratio of means are something very different. The former is commensurable, the latter is not. The reason is that price relatives are invariant under a change of physical units (which changes numerator and denominator in proportion) while sums of prices are not.

The first index formula, or at least one of the earliest price index formulas that really solved the aggregation and commensurability problem was developed by the Italian Carli[48] in 1764, who used the unweighted arithmetic mean of price relatives:

(1.2.4) $P_{0t}^{C} = \frac{1}{n}\sum_i \frac{p_{it}}{p_{i0}} = \frac{1}{n}\sum a_{0t}^{i}$ (price index of Carli 1764).

Carlis formula does not only satisfy the commensurability axiom but also the condition of any kind of mean to fall between the smallest and the largest price relative:

(1.2.5) $a_{0t}^{min} \leq P_{0t}^{C} \leq a_{0t}^{max}$.

Much of the discussion in the more recent development of index theory was devoted to two problems closely related to the Carli formula. Instead of using the *unweighted arithmetic* mean of price relatives it was discussed

- whether means other than the arithmetic should be used, and
- whether weights should be introduced.[49]

It turns out that Carli's formula is - very much in contrast to Dutot's or Drobisch's formula - already very close to well established formulas, especially to the formula of Laspeyres which has played an important part up to today. But as a means to aggregate prices in official price statistics it has of late been discarded (in the first place obviously because of not satisfying the time reversal test, a requirement which however appears not particularly well reasoned).

c) Commensurability and time reversal test, choice of the type of average in aggregating price quotations

A price index P_{0t} as opposed to an individual price should measure a price *level* and should have "no dimension" in two respects: it should remain unchanged

[48] Gian Rinaldo Carli (1720 - 1795) was a professor of astronomy at Padua. Later he moved to Milan where he established a considerable reputation as antiquary, historian and economist. He tried to show to what extent the value of money had declined in Italy since the discovery of America. Carli took three commodities only, grain, wine and oil in 0 = 1500 and t = 1750.

[49] The first who explored systematically the effects of various choices concerning the type of mean and of weighting schemes was Irving Fisher (see **sec. 2.2** which is devoted to his approach).

- if the unit of the currency in which prices are expressed is changed (the price dimensionality axiom),[50] and if
- the conventional unit of quantity, to which the price quotations refers is changed (i.e. the commensurability axiom).[51]

Note that a change of the currency unit only affects prices, while a change of the unit of "quantity" to which price quotations refer will affect both prices and quantities, and that there are a lot more axioms which could be regarded as "invariance" requirements. Invariance in a sense is also given in the case of **time reversibility** (or the "time reversal test") defined by

$$(1.2.6) \qquad P_{t0} = \frac{1}{P_{0t}},$$

making a price index invariant to a change of the base period. Interchanging 0 and t results in the reciprocal index. To "prove" the importance and meaningfulness of this axiom (or "test") it is often stated that

- it is equally justifiable to calculate a change forward from year 0 to a later year, t ($0 \rightarrow t$) as the other way (i.e. backward, $0 \leftarrow t$), one way should be a function (the reciprocal) of the other, such that strictly speaking there is only one (unequivocal) result (as one follows from the other), and in a similar way
- it is said to be most desirable to have only one parity P_{AB} in the case of two countries, say A and B (because P_{BA} is simply the reciprocal of P_{AB}), and finally
- as time reversibility always holds for relatives (the eqation $a_{t0} = 1/a_{0t}$ is satisfied by definition),[52] i.e. in the case of a single commodity it might appear simply "logical" and "consistent" to have this fulfilled in the case of an index too.

It should be noted, however, that this reasoning may well be questioned and that the rationale of this time reversibility property is far less convincing than it might appear on the face of it. Worse even: unfortunately many of the best index theoreticians are setting great store in this property (at the expense of other ones no less, if not even much more important). If a selection is made where time reversibility is regarded as irremissible many useful formulas are ruled out (for example the formulas of Laspeyres and Paasche).[53] This is not only unfortunate, it is also unwarranted:

---

[50] These considerations do not apply, however, to the concept of price level in international price comparison: in this context the function P, called "parity" has a dimension which is the relation of two currency units.

[51] That means a move from for example prices in terms of kilograms to those in pounds should not influence the result. The concept of commensurability is not restricted to indices in which quantities explicitly appear as "weights". It also applies to indices that are composed of prices only, like e.g. Dutot's index in which quantities appear only implicitly, as point of reference for prices.

[52] Although price relatives necessarily comply with time reversibility, this need not hold for *index* numbers as means of $n \geq 2$ relatives.

[53] Moreover some definite principles of index construction and weighting for example are postulated *once we adopt this position: this explains for example the emphasis put on considerations of* "symmetry" and the like.

The emphasis placed on time reversibility (and the esteem this property enjoys) as well as the idea that the formulas of Laspeyres and Paasche are "equally justified" (prompting the taking of an average of both formulas) rests on the tacit assumption that both periods, 0 and t have the same logical status. This is root-and-branch erroneous, and so are most of the far reaching consequences drawn from this proposition (i.e. the equivalence of 0 and t). Second thoughts reveal that 0 and t are *not* on the same footing. This can easily be seen as 0 is kept constant for some periods at least, while t denotes a *sequence* of periods, 1, 2,...

For an unweighted mean of price relatives to satisfy time reversibility the range of possible formulas reduces to unity. As shown in **fig. 1.2.2** there is one only type of mean remaining, viz. the geometric mean.

Figure 1.2.2: Behaviour of unweighted indices by type of mean

unweighted mean of relatives

arithmetic (PA)	geometric (PG)	harmonic (PH)
(1.2.7) $PA_{t0} > (PA_{0t})^{-1}$	(1.2.8) $PG_{t0} = (PG_{0t})^{-1}$	(1.2.9) $PH_{t0} < (PH_{0t})^{-1}$
Carli's index (1.2.4) $P_{0t}^{C} = \sum(p_{it}/p_{i0})/n$ $P_{t0}^{C} > 1/P_{0t}^{C}$	Jevons' index (1.2.10) $P_{0t}^{JV} = \sqrt[n]{\prod \frac{p_{it}}{p_{i0}}}$	unnamed index (1.2.11) $PH_{0t} = \frac{n}{\sum \frac{p_{i0}}{p_{it}}}$

These relations also hold for weighted means since they refer to the *kind of mean*. Thus for to the Laspeyres index (a PA-type) eq. 7 applies whereas for the Paasche index (PH-type) eq. 9 applies.

The unweighted geometric mean of price relatives is known as price index of **Jevons** 1863 which so far has however a very limited influence on the practice of price statistics.[54] Irving Fisher studied systematically the obvious but also the more hidden properties of various index formulas. He discovered in connection with his time reversal test the following relation between the "arithmetic forward" mean and the "harmonic backward" mean, as he called the formulas (both means being unweighted):

[54] The unweighted geometric mean of Jevons has some useful properties (it does not only meet the time reversal test but also the "circular test" in the sense of Irving Fisher) and it is nowadays sometimes used for special index numbers, such as an index of exchange-rates of currencies or share price indices, or in the aggregation of price quotations referring to the same good (see below). Köves 1983, p.32-36 studied the properties of P^{JV} in connection with a quantity index analogously constructed, to be called Q^{JV} in some detail, but in general both indices, P^{JV} and Q^{JV} are of historical and theoretical interest only.

$$(1.2.12)\quad PA_{0t} = \frac{1}{PH_{t0}},$$

a relation which holds for any number n of commodities and which is simply the definition of the harmonic mean.[55] In Fisher's terminology the arithmetic mean is the "*time antithesis*" of the harmonic mean and vice versa. **Tab. 1.2.1** once more summarizes the properties of some unweighted means. It should be noted in passing that a table like this is the usual way to show that time reversability (T) and commensurablity (C) are *independent*: It is possible to satisfy C and violate T (as P^C does) or vice versa (like P^D). It is also possible that both requirements are satisfied. Thus C and T are *consistent*.

Table 1.2.1: Some properties of unweighted means as indices[a)]

	commensurablity (C)	
time reversal test (T)	satisfied	violated
satisfied	Jevons P^{JV}	Dutot P^D
violated	Carli[b)] P^C	Drobisch P^{UD}

a) including Drobisch's unit value index; all index functions listed violate the factor reversal test

b) the same is true for an unweighted *harmonic* mean.

Given that nowadays great store is set in the time reversal condition it is not surprising that in international standards for an unweighted aggregation of price quotations* the formulas P^{JV} and P^D are recommended while P^C (Carli) is banned.

*referring for example to the same commodity in different outlets

In order to justify the rejection of Carli's formula for aggregation in the train of index compilation (on the "lowest level of aggregation" as it is usually put) some suitable numerical demonstrations are quite popular. For example Haschka 1999 gave the following presentation: Consider two goods with prices in 0 and t as follows

I	p_{i0}	p_{it}	price relative
1	20	10	0.5
2	10	20	2
	$\bar{p}_0$ = 15	$\bar{p}_t$ = 15	

price index
Dutot $P^D = 15/15 = 1$
Jevons $P^{JV} = \sqrt{0.5 \cdot 2} = 1$
Carli $P^C = (2+0.5)/2 = 1.25$

Both Dutot's as well as Jevons' index show no change (unity) which appears justified as the "market situation" (in terms of the sum or average of prices) is unchanged. Carli's index stands out negatively as clearly overstating the rise in prices if there is

55 Since generally PA > PH holds (except for all relatives being equal, a case in which PA = PH) equation 9 easily follows from eq. 7 (or 7 from 9) due to eq. 12 and 4. The relations will hold whenever the variance of the price relatives is greater than zero (i.e. not all relatives being equal).

any. The demonstration owes much of its appeal, however, to the fact that the p_0 are not too different. Assume for example prices of the second commodity as follows: p_{20} = 100 and p_{2t} = 200. This would not change P^J but it would give a Dutot index of P^D = 1.75 (since $\bar{p}_0$ = 60 and $\bar{p}_t$ = 105) which is plausible since what people experience is likely to depend on the *absolute* figures (in contrast to P^{JV} and P^C these absolute prices influence P^D). Given that P^D = 1.75 the result P^C = 1.25 is no longer ridiculously high. It is rather P^{JV} the index that might be questioned now: that a rise 100 → 200 will be perceived as offset by a decline 20 → 10 is doubtful. P^C and P^D differ due to the structure of prices in period 0, because

(1.2.13) $$P_{0t}^D = \sum\left(\frac{p_{i0}}{\sum p_{i0}}\right)\frac{p_{it}}{p_{i0}} \neq P_{0t}^C = \sum\left(\frac{1}{n}\right)\frac{p_{it}}{p_{i0}}.$$

Since P^D is being a weighted arithmetic mean of price relatives, it can easily be seen that $P^D = 1.75 > P^C = 1.25$ is reflective of the high weight assigned to the second as compared to the first commodity (since $p_{20} = 100 > p_{10} = 20$ which yields a weight of 100/120 = 5/6 for good 2).[56] We can also think of examples in which the above quoted "market situation" (on the basis of average prices) is unchanged and yet P^J does not remain unchanged:

i	p_{i0}	p_{it}	price relatives
1	200	210	1.05
2	20	10	0.5
	$\bar{p}_0$ = 110	$\bar{p}_t$ = 110	

price index
Dutot P^D = 1
Jevons P^{JV} = 0.72457
Carli P^C = 0.775

To conclude: numerical examples do not "prove" the superiority of a type of unweighted index over another. Nor should a decision be made based on one single test (like the time reversal test). We can always avail of other tests, and for example easily argue against Jevons' formula from the point of view that both indices, P^D and P^C, are "additive" (linear in the prices p_t) while P^{JV} is the only one of the three which fails this test.

The question of which kind of mean should be used had been of considerable interest in the early days of price index computation because at that time a consumer price index usually consisted of a couple of commodities only.[57] Nonetheless, nowadays, when a list of several hundred commodities is standard, the choice of the appropriate type of average is still viewed as an issue of some importance.

[56] It is clear that $P^C \geq P^J$ as a consequence of the type of average (arithmetic vs. geometric). The relationship between P^C and P^D depends on the correlation between the level of prices and the amount of change. If prices of (at time 0) low-priced goods rise more (less) than those of high-priced goods, that is if the correlation is positive (negative), we will get $P^C < P^D$ (or $P^C > P^D$ respectively).

[57] Weights are also (implicitly, indirectly) introduced into the form of the index by the selection of commodities. In this way the Carli-index has a bias towards food (combining exclusively series representing food) and a zero-weight of non-food commodities and also services.

In our days the inadequacy of unweighted means as measures of price levels or deflators isn't questioned. Such formulas are inadequate since they do not take quantities explicitly into account, so that there is nothing that would relate them to the purpose of "deflation" (aiming at a sort of total "quantity").

It is however possible to provide a theoretical underpinning to unweighted formulas as will be shown subsequently.

d) Stochastic and aggregative approach to index theory

To understand the issue of unweighted vs. weighted indices better it might be useful to briefly comment on a controversy in the early theory of price index numbers, which nowadays has somewhat fallen into oblivion, and which deals with the problem of how to define the complex and not directly observable general price level or the value of money (purchasing power of money), one the reciprocal of the other.

One group of thinking aimed at a broad *objective*, statistically representatively defined general level of prices without specific reference to a group of consumers. The other school referred to the aggregative expenditure of a specified group of consumers with the aim to define a *subjective* value of money (or real income). Ragnar Frisch called the first the stochastic[58] approach and the other the aggregative approach. To set out the main differences:

	(old*) stochastic approach	aggregative approach*
notion of the price level	objective, inflation as the result of monetary factors (equally influencing all prices)	subjective, i.e. defined with reference to the expenditures ("baskets") of specific consuming units (e.g. households)
preferred type of index	**unweighted means of price relatives** (index conceived as [arithmetic, geometric etc] mean of a distribution of relatives)	derived from comparing expenditures (aggregates) in period t with those in the base period 0, consumer price index as a **ratio of expenditures** (of households)

*as opposed to new (see sec. 2.3), some authors distinguish *unweighted* (= old) and *weighted* (= new)

The economic justification of the stochastic (or statistical) approach to index numbers was the hypothesis that non-specific monetary factors, such as changes in the quantity of money, are responsible for changes in the value of money. The central concept in the stochastic approach is the unweighted[59] **mean of the** frequency **distribution** of price relatives (whatever this concept may mean economically) for example the

[58] It will be shown in **sec. 2.3**, that the term "stochastic index theory" is used nowadays in quite a different way.

[59] Given n items, the prices of which have been observed, then each of them will get a weight of 1/n.

geometric mean,[60] whilst the aggregative approach typically provides a "changing cost of a basket" - interpretation with explicit reference to the expenditures of a particular subject consuming (e.g. household).[61] These expenditures may permit defining "weights", denoting the different importance (in the perception of representative households) of various goods and services. The introduction of weights can also be viewed as a reconciliation of the two approaches since certain weighted means are in line with both approaches:

> The formulas of Laspeyres and Paasche for example, can be interpreted in *both* ways, as (weighted) means of price relatives (in line with the stochastic approach) and as ratios of expenditures (as required in the aggregative approach). It should be noticed that *none* of the interpretations applies to the "ideal" index of Fisher or to all sorts of chain indices.

Some of the controversy "stochastic vs. aggregative" depicted above is turning up again in the distinction between formal and economic theory of index numbers. The controversial issue here is whether index theory should be confined to observable variables, and treat prices and quantities as unrelated, or rather take into account the (nonobservable) relationship (explained using "indifference curves" [the preference, or utility function] in microeconomic theory) between prices and (as a response to prices) quantities.

1.3 Index formulas of Laspeyres and Paasche

In this section an intuitive introduction into the two perhaps most frequently used index formulas, the index of Laspeyres (denoted by P_{0t}^{L}) and the index formula of Paasche, P_{0t}^{P} will be given along with a short presentation of some fundamental mathematical relations between the two formulas, and of a recent controversy about the (fixed basket) concept of Laspeyres.

a) Price indices, dual interpretation

Both index formulas P_{0t}^{L} and P_{0t}^{P} can be interpreted in an intuitive and straightforward way, as

---

60 Edgeworth was in favor of the median (due to its "robust" nature) and the geometric mean of price relatives, the latter mainly because it satisfies the time reversal condition. Advocates of the *geometric mean*, as Edgeworth in general argued that the distribution of price relatives is skewed with an expectation of unity, a minimum value of zero and maximum values unlimited. Variations in price relatives were viewed as following a "geometric law". The notion of a *frequency distribution of prices*, however, is not without difficulties. To begin with, what are the "frequencies" in this case? Moreover, if there were such a frequency distribution in the "ordinary" meaning of statistics it would not make sense to discuss questions like: should we prefer an unweighted mean to a weighted mean? Or which variables should serve as weights?

61 or revenues of a subject producing.

- comparing **expenditures**[62] at two points in time, and being
- **weighted means** of price relatives.[63]

Perhaps the major shortcoming of Carli's formula P_{0t}^{C} (when used as Consumer Price Index for example) was the lack of "weights" expressing the different "significance" or "importance" of different goods consumed. Hence there is the need for explicitly taking into account weights reflecting the consumption pattern of the type of household to which the index refers. It is interesting to note that by introducing appropriate weights the index formula can also gain a "rising costs of a comparative budget" interpretation (that is raising costs of living if the quantities consumed of each commodity are equal in both[64] periods, 0 and t).[65]

Denoting prices and quantities of n commodities by p and q, the actual expenditures (or costs of living) are $\Sigma p_{i0}q_{i0}$ and $\Sigma p_{it}q_{it}$ at the current period (t) and the base period (0) respectively. We may recall the definition of the value index (or value relative, or value ratio) as follows:

$$(1.3.1) \qquad V_{0t} = \frac{\Sigma p_{it}q_{it}}{\Sigma p_{i0}q_{i0}} = \frac{\Sigma p_t q_t}{\Sigma p_0 q_0} = \frac{\mathbf{p}_t'\mathbf{q}_t}{\mathbf{p}_0'\mathbf{q}_0} \qquad \text{(value index).}$$

Note that, in contrast to a price index, which is devoted to measure the pure price component of the cost of living, the value index V_{0t} is affected by both, price changes and changes in quantities consumed. Furthermore eq. 1 cannot be regarded as a weighted average of price relatives, since in $V_{0t} = \Sigma \frac{p_{it}}{p_{i0}} w_i^*$ the "weights" w_i^* = $p_{i0}q_{it}/\Sigma p_{i0}q_{i0}$ do not add up to 1. The changing-costs-of-a-basket-interpretation requires the same budget (which is the vector of quantities **q**) in both situations 0 and t respectively

$$(1.3.2) \qquad P_{0t} = \frac{\Sigma p_{it}q_i}{\Sigma p_{i0}q_i} = \frac{\Sigma p_t q}{\Sigma p_0 q} = \frac{\mathbf{p}_t'\mathbf{q}}{\mathbf{p}_0'\mathbf{q}}.$$

Note that eq. 2 in contrast to eq. 1 has a weighted mean interpretation since an algebraic equivalent form is

---

[62] The term "expenditure" applies of course to the example of a consumer price index.

[63] This does not mean that the two index formulas are equally plausible. There are of course interesting differences between the Laspeyres- and the Paasche formula respectiveley, concerning the type of expenditures compared and the type of weights used and therefore concerning the purposes for which they are most suitable.

[64] This is done in order to eliminate the structural component [change of quantities] in the expenditures compared.

[65] Note that reference is made to the *quantities* not the level of utility in both periods, 0 and t. A "changing cost" or "ratio-of-aggregates" definition of a price index is also in line with the aggregative (see **sec. 1.2**) and economic (see **sec. 2.4**) approach as opposed to the idea of a simple (unweighted) mean of relatives preferred by the (old) stochastic index theory (see **sec. 2.3**).

(1.3.2a) $$P_{0t} = \Sigma \frac{p_t}{p_0} \cdot \left(\frac{p_0 q}{\Sigma p_0 q} \right) = \Sigma a_{0t}^i w_i$$ (fixed budget index)[66]

and $\Sigma w_i = 1$. In the "fixed budget" - form (eq. 2a) the quantitiy vector **q** (that is a set of quantities q_i) has to be *the same* at both periods to be compared. To which point of time **q** is related appears arbitrary or at least of minor importance *at first sight*. We may basically think of two ways of defining quantities:

- selection of base year quantities q_{i0} leading to P_{0t}^L, or
- taking the current year quantities q_{it} as weights, leading to P_{0t}^P.

The first solution gives the famous *base* weighted price index formula of **Laspeyres** (1864)[67]

(1.3.3) $$P_{0t}^L = \frac{\Sigma p_t q_0}{\Sigma p_0 q_0}$$

in its aggregative form[68] and the second (current quantities for weighting the prices) leads to the price index formula of Paasche P_{0t}^P (1874)[69]:

(1.3.4) $$P_{0t}^P = \frac{\sum p_t q_t}{\sum p_0 q_t}$$

again in its aggregative form. There are of course many other meaningful definitions of **q** in eq. 2 possible, but the point is: In both formulas, P_{0t}^L and P_{0t}^P the same quantities (either base or current quantities) are used in the numerator and in the denominator in order to avoid the ambiguity of a cost of living (or more general: value) index.

Both price indices, Laspeyres and Paasche, can be interpreted in terms of "changing costs of a budget (basket)". It should be noted, however, that there is not a complete symmetry[70]:
period 0 denotes a *single*, **constant** (for the "life time" of an index) period while
period t is a *variable* period referring to **many** different periods, t = 1, 2 .... .

As to the weighted-mean interpretation, both formulas can be written either as arithmetic or as harmonic means of relatives. It is more common to express P_{0t}^L as the weighted arithmetic mean of price relatives, the weights being relative expenditures in the base period:

---

66 Note that Carli's unweighted (or more precisely: equally weighted) index is simply the special case where w_i= 1/n. The effect of weighting depends on the variance of weights, e.g. to what extent the weights w_i differ from their mean, which is 1/n.

67 Etienne Laspeyres (1864 - 1913) was professor at Giessen, Germany and he was also the first to introduce a quantity index.

68 or "ratio-of-aggregates" formula.

69 Hermann Paasche, German economist and politician (1851-1925); he also was for some years vice president of the German Reichstag (Parliament).

70 A symmetric interpretation of the Laspeyres and Paasche formula is also very popular in the case of the so-called "economic theory" of index numbers.

(1.3.3a) $P_{0t}^{L} = \sum \frac{p_t}{p_0} \left(\frac{p_0 q_0}{\sum p_0 q_0} \right)$ Laspeyres index mean-of-relatives-form.

The Paasche price index is more frequently represented as a weighted harmonic mean of price relatives, weights being expenditure shares in the current period t, $w_i = p_{it} q_{it} / \sum p_{it} q_{it}$

(1.3.4a) $\left(P_{0t}^{P}\right)^{-1} = \sum \frac{p_{i0}}{p_{it}} \frac{p_{it} q_{it}}{\sum p_{it} q_{it}}$ Paasche index mean-of-relatives-form.[71]

Note that equations 3a and 4a are mathematically equivalent to equations 3 and 4 respectively.

But there is a considerable difference concerning the

- interpretation, and the
- applicability of the formula in practice

since quantities q_{i0} by which prices are weighted in eq. 3 or shares of expenditures (i.e. the ratios $p_{i0} q_{i0} / \sum p_{i0} q_{i0}$) by which price-relatives are weighted in eq. 3a are more readily available than quantities q_{it} and ratios $p_{it} q_{it} / \sum p_{it} q_{it}$ in the case of eq. 4 and 4a. It may be interesting to see that the value index V_{0t} as defined by eq. 1 can be expressed as a weighted mean of value relatives (not price relatives) in both ways, that is as an arithmetic mean with base weights $p_{i0} q_{i0} / \sum p_{i0} q_{i0}$ (Laspeyres type), and as a harmonic mean with actual weights of the Paasche type (i.e. expenditure shares $p_{it} q_{it} / \sum p_{it} q_{it}$).

b) Price indices and quantity indices

"Index numbers come in pairs in economic theory, one of price and the other a matching one of quantity. In economic practice they tend to be found paired off in this way ... Such a pair may be designed to account for the variation in a value aggregate, as when movements in aggregate expenditure of consumers are analysed into the two components of changes in prices and in real consumption" (Allen 1975, p. 1).

The idea of measuring volume changes by quantity indices arose only much later than the price level measurement by price indices. In general three ways of deriving a quantity index (denoted by Q_{0t}) should be distinguished as indicated in **fig. 1.3.1**

- a direct method by *interchanging prices and quantities* (more exactly: price and quantity vectors) in the aggregative form of a price index,
- indirectly *based on* the idea of the "*factor reversal test*" which requires the product of P_{0t} and Q_{0t} to equal the value ratio (or value index) V_{0t} and

[71] It can easily be seen that P^L can be expressed as a weighted harmonic mean too and P^P as a weighted arithmetic mean using expenditure shares as weights, in which prices and quantities do not refer to the same period (see sec. 2.2).

- a third less common method in which both indices are derived *independently* or both from a system of interrelated equations; the resulting formulas of the two indices, P_{0t} and Q_{0t} then need not look similar.[72]

Figure 1.3.1: Price and quantity indices

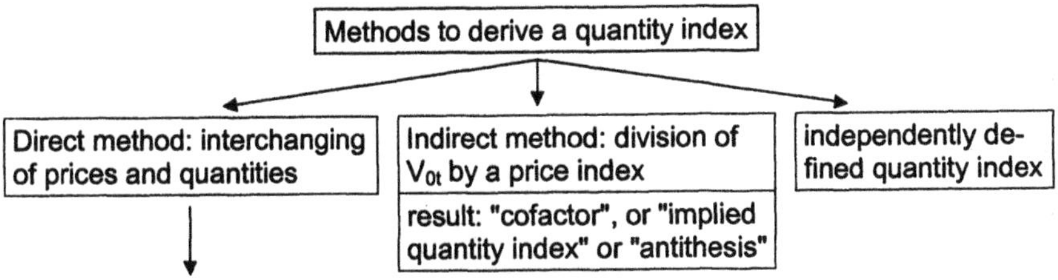

(direct) indices of the Laspeyres and Paasche type

index	Laspeyres index	Paasche index
price	$P^L_{0t} = \frac{\sum p_t q_0}{\sum p_0 q_0} = \frac{\mathbf{p}_t' \mathbf{q}_0}{\mathbf{p}_0' \mathbf{q}_0}$	$P^P_{0t} = \frac{\sum p_t q_t}{\sum p_0 q_t} = \frac{\mathbf{p}_t' \mathbf{q}_t}{\mathbf{p}_0' \mathbf{q}_t}$
quantity	$Q^L_{0t} = \frac{\sum q_t p_0}{\sum q_0 p_0} = \frac{\mathbf{q}_t' \mathbf{p}_0}{\mathbf{q}_0' \mathbf{p}_0}$	$Q^P_{0t} = \frac{\sum q_t p_t}{\sum q_0 p_t} = \frac{\mathbf{q}_t' \mathbf{p}_t}{\mathbf{q}_0' \mathbf{p}_t}$

The formulas of Laspeyres and Paasche are related to the value index as follows

(1.3.6) $V_{0t} = P^L_{0t} Q^P_{0t} = P^P_{0t} Q^L_{0t}$

showing that Q^P_{0t} is the cofactor (or "factor antithesis") of P^L_{0t} and so is Q^L_{0t} of P^P_{0t}.

The first method (interchanging of vectors) expressed in matrix (vector) notation consists of arriving at $Q = f(\mathbf{q}_0, \mathbf{p}_0, \mathbf{q}_t, \mathbf{p}_t)$ from $P = f(\mathbf{p}_0, \mathbf{q}_0, \mathbf{p}_t, \mathbf{q}_t)$, which means that the vector $\mathbf{q}_0$ takes the place of $\mathbf{p}_0$, and $\mathbf{p}_0$ the place of $\mathbf{q}_0$ and so on in the same function f. The results gained on the basis of this interchanging (direct) method in the case of the Laspeyres formula and the Paasche formula is given in the table which is part of fig. 1.3.1. It is justified to call Q^L_{0t} and Q^P_{0t} quantity indices because both indices also have a weighted mean of relatives interpretation (in this case quantity relatives instead of price relatives) completely analoguous to the mean of (price) relatives interpretation of price index numbers. For example in the case of the Laspeyres formula the expression

(1.3.5) $Q^L_{0t} = \sum b^i_{0t} \; w_i$

is a mean of quantity relatives $b^i_{0t} = q_{it}/q_{i0}$ with weights $w_i = p_{i0}q_{i0}/\sum p_{i0}q_{i0}$[73]. It should be added that the direct method to define a quantity index sometimes fails in

[72] An example for this method is Stuvel's pair of indices, which can be viewed, however, also as an example for the second method.

the sense that the result can hardly be demonstrated to be a quantity index (this is not unlikely in cases in which the price index is developed by "crossing" index formulas [see sec. 2.2]).

The decomposion of the value ratio (value index) V_{0t} according to eq. 1.3.6 is the basis for making use of P^P_{0t} for purposes of deflation (using P^P_{0t} as a "deflator").

Example 1.3.1

Prices and quantities of four commodities (i = 1,...., 4) are given as follows:

	base period (0)		current per. (t)					
i	p_{i0}	q_{i0}	p_{it}	q_{it}	$p_{i0}\,q_{i0}$	$p_{i0}\,q_{it}$	$p_{it}\,q_{i0}$	$p_{it}\,q_{it}$
1	10	16	15	12	160	120	240	180
2	16	15	18	14	240	224	270	252
3	20	10	24	12	200	240	240	288
4	15	20	21	15	300	225	420	315
				sums	900	809	1170	1035

Since the relevant sums are given it is now easy to apply the aggregate formulas:

$P^L_{0t} = \frac{1170}{900} = 1.3$	$P^P_{0t} = \frac{1035}{809} = 1.279$	$Q^L_{0t} = \frac{809}{900} = 0.899$	$Q^P_{0t} = \frac{1035}{1170} = 0.885$

Note that the product $P^P_{0t}Q^L_{0t}$ equals the product $P^L_{0t}Q^P_{0t}$ which is (by eq. 6) the value ratio $V_{0t} = 1035/900 = 1.15$. Hence the rise in value by 15% can be decomposed into

- a rise in prices of 30% (P^L_{0t}=1.3), and a decline in volume of 11.5% (Q^P_{0t}-1= 0.885-1 = - 0.115), or alternatively
- a rise in prices of 27.9% (P^P_{0t}=1.279), and a decline in volume of 10.1% (Q^L_{0t}-1= 0.899-1 = -0.101).

Furthermore note that $P^L_{0t} > P^P_{0t}$ and $Q^L_{0t} > Q^P_{0t}$, which in general is more likely to occur than $P^L_{0t} < P^P_{0t}$ and $Q^L_{0t} < Q^P_{0t}$ respectively. It can also easily be verified that the results in applying the means-of-relatives formulas and those obtained from the aggregative formulas are identical[74]:

[73] The equivalent interpretation for Paasche's quantity index as the harmonic mean of quantity relatives weighted with actual (current period) expenditure shares can easily be verified.

[74] This will be shown for the two Laspeyres indices subsequently (the correspondig demonstration for the Paasche formulas is left to the reader).

i	$a_i = p_{it}/p_{io}$	$b_i = q_{it}/q_{io}$	$p_{io}q_{io}$	w_i*	$a_i w_i$	$b_i w_i$
1	1.5	0.75	160	0.1778	0.2667	0.1333
2	1.125	0.9333	240	0.2667	0.3	0.2489
3	1.2	1.2	200	0.2222	0.2667	0.2667
4	1.4	0.75	300	0.3333	0.4667	0.25
		sum	900	1.0	1.3	0.8989

* $w_i = p_{io}q_{io}/\sum p_{io}q_{io}$ ♦

c) Asymmetry in the interpretation of the Laspeyres and Paasche formula

As to the ease of interpretation it is certainly an advantage that both index formulas, P_{0t}^{L} and P_{0t}^{P} offer two ways of looking at them[75]:

- in the aggregative form (eq. 3 and 4) they can be regarded as ratios of expenditures[76] (actually incurred or imputed), or as a comparison of "standardized" expenditures in the sense of referring to the same quantities as weights in both periods to be compared;
- and secondly they can be viewed as means of price relatives (eq. 3a and 4a), or in other words as a **typical** (representative) **price relative**.

In both cases the "Laspeyres principle" consists in taking constant (base period) weights,[77] while the "Paasche principle" on the other hand consists in using variable weights related to the current period. Both formulas, Laspeyres and Paasche are reasonable solutions to settle the ambiguity problem of a value index in that both formulas use the same quantities in the numerator and in the denominator of the ratio-of-aggregates formula. But (as already mentioned in sec. 1.2) the impression of symmetry is suggestive only if we confine the comparison to two periods at a time. In this case everything concerning P_{0t}^{L} and P_{0t}^{P} appears to be completely symmetrical. It can often be read in index literature that both formulas, P_{0t}^{L} and P_{0t}^{P} are conceptually equivalent to one another,[78] or that both formulas are based logically on the same footing. This is not true for a number of reasons[79] and it can easily be seen once a series of successive periods t = 0, 1, 2,..., a time series or, a "run" of index numbers,

---

[75] This is a great advantage of the two index formulas, because there are other formulas, sometimes preferred to them, like Fisher's "ideal" index, or "chain indices" to which *neither* of the two interpretations applies.

[76] When the price index formula applies to consumer prices as supposed throughout this section.

[77] i.e. quantities to be multiplied by prices (aggregative form, eq. 3), or expenditure shares to be multiplied by price relatives (eq. 3a).

[78] This is one of the points repeatedly emphasized in the book of Mudgett 1951, a book which apparently shaped or at least influenced in no small measure the American way of thinking about index numbers, and which in particular advocated the chain index approach.

[79] As already noted a decisive point is that period 0 is fixed, whereas the period t is not. For a certain while at least there is only *one* constant base period, 0 but a *multitude* of current periods, t as time goes on.

as Allen called it, that is $P_{01}^L, P_{02}^L, P_{03}^L$ ... as opposed to $P_{01}^P, P_{02}^P, P_{03}^P$ ..., is considered there is a remarkable difference between the two formulas, as shown in **tab. 1.3.1**, in particular with respect to:

- data requirements,
- an interpretation in terms of representativity and pure price comparison, and
- the underlying concept of measuring a price movement (rise or decline of prices) which may be more direct by looking at the changing costs of a fixed budget or more indirect by comparing values with volumes.

Generally speaking weights should be capable of measuring the relative "importance" of items. In the case of relative expenditures this is performed in a way strongly appealing to common sense: a commodity i is more important than another commodity j, say, if the amount of money spent by an average consumer on buying i is greater than the amount spent on j. This argument applies to both periods, to the base period 0 and to the current period t as well, or: to the historical budget as well as to the current (or most recent) budget. In the view of most people the current budget appears to be preferable to the historical budget, however. This is an issue to be discussed in detail in connection with chain index numbers. Constant weights on the other hand, as applied in the Laspeyres philosophy, are justified on the following grounds:

1. Since the weights do not change, the variations in the Laspeyres price index (or quantity index) from year to year can be attributed exclusively to variations in the prices (or in quantities respectively). A mean using constant weights represents a pure change of prices (in P_{0t}^L), or a pure change in quantities (in the case of Q_{0t}^L). Thus the Laspeyres principle appears to be more in line with "pure" price (quantity) comparison, whereas Paasche indices are superior as regards the representativity criterion.
2. Constant percentage weights[80] permit operation with the same constant basket for all successive years in which the same base period is maintained. Thus Laspeyres' formula is less demanding from the point of view of data-collection (price- and quantity-reports) than Paasche's formula, requiring an update of the structure of weights in every period.
3. The Laspeyres-approach is to infer rising prices directly from rising costs of a fixed budget while Paasche's approach to conceptualize price movement is more indirect: prices move by the amount to which actual costs diverge from those costs (expenditures) that were to be paid if prices remained constant.[81]

[80] Expenditure shares in the case of a consumer price index.

[81] Of course an analoguous distinction can be made with respect to how "quantity movement" is measured.

Table 1.3.1: The interpretation of the Laspeyres and Paasche price index and quantity index formula (ratio of aggregate forms)

	Price-index formula	
	Laspeyres	Paasche[1)]
numerator	**imputed** expenditures, i.e. expenditures as they *were*, if quantities were kept constant	**empirical**, i.e. observed actual expenditures referring to actual quantities (*not* to constant quantities)
denominator	**empirical**, i.e. actual base period expenditures (being constant)	**imputed** expenditures as they *were*, if prices were kept constant; measures of volume (as substitute for quantity)
time series	time series interpretation possible because only the numerator varies	both numerator and denominator vary, indices in a "run" are not comparable
price movement	**directly** measured: rising (descending) prices inferred from rising (decreasing) costs of a fixed budget[2)]	**indirectly**[3)] measured: rising prices because actual costs are higher as they were when prices remained constant
	Quantity-index formula	
concept of quantity movement	**direct**: quantities (volume) increased to the extent to which expenditure valued at constant prices has increased (i.e. rising volume)	**indirect**[3)]: quantities are rising if value at current prices rises more than value at constant prices (for a fixed budget i.e. more than volume).

1) It is assumed that the Paasche price (quantity) index is used to measure price (quantity) *level* movement, but the real merits of the Paasche formula can be seen in the case of deflation (**sec. 5.2**).

2) a quantitatively fixed budget

3) indirect approach means: by comparing value and volume (both referring to actual consumption quantities).

The last point may be clarified by looking at numerator and denominator in the sequence of the two price index formulas in their aggregative form: in the case of Laspeyres price indices

$$P_{01}^{L}, P_{02}^{L}, P_{03}^{L} \text{ ... we have } \frac{\sum p_1 q_0}{\sum p_0 q_0}, \frac{\sum p_2 q_0}{\sum p_0 q_0}, \frac{\sum p_3 q_0}{\sum p_0 q_0}, \ldots$$

such that each index differs from other indices *only* with respect to prices in the numerator, quantities being the same throughout. The situation is fundamentally different, however, with successive Paasche-index numbers, that is

$$P_{01}^{P}, P_{02}^{P}, P_{03}^{P} \text{ ... where } \frac{\sum p_1 q_1}{\sum p_0 q_1}, \frac{\sum p_2 q_2}{\sum p_0 q_2}, \frac{\sum p_3 q_3}{\sum p_0 q_3}$$

is a series in which not only prices, but prices and quantities are different in each element of the sequence of current period weighted Paasche indices. The same consideration applies to series of successive Laspeyres- and Paasche quantity indices. In the series of Laspeyres quantity indices Q_{01}^L, Q_{02}^L, Q_{03}^L ..., only the numerators $\sum p_0q_1, \sum p_0q_2, \sum p_0q_3,...$ are different with respect to quantites.[82] To sum up:

> The **Laspeyres** indices (of prices or quantities) provide a result fully in line with the spirit of the principle of **pure comparison**, successive values are influenced only by the variable in question, that is prices or quantities respectively. Paasche indices, however, are always affected by both variables.

As to the interpretation of the sequences P_{01}^L, P_{02}^L, P_{03}^L ... and P_{01}^P, P_{02}^P, P_{03}^P ... as listed above we recall that the aggregates primarily of interest for the interpretation are those expenditures which are imputed (tab. 1.3.1), that is the numerator of P^L and the denominator of P^P

- The P^L numerators $\sum p_1q_0, \sum p_2q_0, \sum p_3q_0$ are reflective of a pure price movement (not disturbed by quantity movements).[83] The Laspeyres-formula therefore provides a direct approach to measure price movements;
- The P^P denominator, $\sum p_0q_1, \sum p_0q_2, \sum p_0q_3$ on the other hand traces a quantity movement (a series of expenditures at constant prices or "real expenditures" [as an estimate for the quantity development]) rather than a price movement. The Paasche-formula is therefore more an indirect approach in comparing the value movement $\sum p_1q_1, \sum p_2q_2, \sum p_3q_3,...$ with the movement of volume, $\sum p_0q_1, \sum p_0q_2, \sum p_0q_3,...$

Hence the underlying concept of (price) inflation is

- direct (Laspeyres): observable due to rising cost of a fixed budget or
- indirect (Paasche): a situation which makes nominal (at current prices) expenditures rise faster than real expenditures (or expenditure at constant prices as a proxy for quantities).

In describing a specific price level the Laspeyres type of index therefore has a clear advantage due to its simple and straightforward interpretation. The Laspeyres approach is in keeping with the principle of pure price (volume) comparison. This does not apply to the same extent to the Paasche formula[84], nor to entirely different

---

[82] Hence it is the change in quantity and only in quantity that is responsible for the result of the Laspeyres quantity index.

[83] Likewise the numerators of Q^L describe a pure quantity movement, because Q^L is influenced solely by the numerator, which is Σq_tp_0, an expenditure at constant prices (a volume), reflecting changes in quantities only, while the denominator Σp_0q_0 remains constant. Again Q^P measures quantity movement *indirectly* in the sense that the value movement Σp_tq_t (numerator), influenced by *both* prices and quantity, differs from Σp_tq_0.

[84] The Paasche index using a *variable* set of weights over the years is not only more demanding (since statistics of quantities [budgets] are not usually available every year) as far as the practical calculation (the application by a statistical office) is concerned but also less suitable for a pure-price-(or pure-quantity) movement-interpretation.

index devices, like e.g. the chain indices. The merits of P^P_{0t} lie more in utilizing them for purposes of deflation.

Finally mention should be given to the following "antithetic" relationship known already to Irving Fisher and easily to be verified

(1.3.7) $P^L_{t0} = \frac{1}{P^P_{0t}}$ and (1.3.8) $P^P_{t0} = \frac{1}{P^L_{0t}}$.

Hence Paasche and Laspeyres price indices are "time antithesis" of each other. The same is true for quantity indices

(1.3.7a) $Q^L_{t0} = \frac{1}{Q^P_{0t}}$ and (1.3.8a) $Q^P_{t0} = \frac{1}{Q^L_{0t}}$.

This also explains why neither P^L nor P^P (or Q^L and Q^P) satisfies time reversibility, and that the index $P^F_{0t} = \sqrt{P^L_{0t}\,P^P_{0t}}$, known as Fisher's "ideal" index is able to pass this test, since

(1.3.9) $P^F_{0t}P^F_{t0} = \sqrt{P^L_{0t}\,P^P_{0t}}\sqrt{P^L_{t0}\,P^P_{t0}} = \sqrt{P^L_{0t}\,P^P_{0t}}\sqrt{\left(1/P^P_{0t}\right)\left(1/P^L_{0t}\right)} = 1$.

As already mentioned, Fisher called the Laspeyres formula and the Paasche formula "time antithesis" and "factor antithesis" of each other and in his view his "ideal index", P^F was superior to both in that P^F meets both tests, the time reversal as well as the factor reversal test.

d) Theorem of Ladislaus von Bortkiewicz concerning the relationship between Paasche and Laspeyres formulas

An extremely valuable equation in comparing the Laspeyres and the Paasche price index has been derived by Ladislaus von Bortkiewicz[85]. A generalization of the following theorem will be presented in **sec. 3.2**. Denoting price relatives and quantity relatives by a^i_{0t} and b^i_{0t} respectively we obtain $c^i_{0t} = a^i_{0t} \cdot b^i_{0t}$ for the value ratio (relative) of an individual commodity and by using weights $w_i = \frac{p_{i0}q_{i0}}{\sum p_{i0}q_{i0}}$ as base period expenditure shares (Laspeyres weights) it is easy to verify that the means of relatives take the following forms

(1.3.10) $P^L_{0t} = \sum_i a^i_{0t} w_i$	(1.3.11) $Q^L_{0t} = \sum_i b^i_{0t} w_i$	(1.3.12) $V_{0t} = \sum_i a^i_{0t} b^i_{0t}\, w_i$

The covariance C between (weighted) price and quantity relatives is given by

(1.3.12a) $C = \sum_i (a^i_{0t} - P^L_{0t})(b^i_{0t} - Q^L_{0t})\, w_i = V_{0t} - P^L_{0t}Q^L_{0t} = Q^L_{0t}(P^P_{0t} - P^L_{0t}) = P^L_{0t}(Q^P_{0t} - Q^L_{0t})$,

[85] To many statisticians von Bortkiewicz (1868 - 1931) is firstly known for the example of the Poisson distributed event of dying from a horse kick in the Prussian cavalry. Bortkiewicz, born into a Polish family, lived at first in Russia, and later for most of his life in Germany, where he was a professor in Berlin.

or using r_{ab}, the correlation coeffient, and V_a, V_b denoting the coefficients of variation, we get

(1.3.13) $$C = V_{0t} - P^L_{0t}Q^L_{0t} = r_{ab}(s_a s_b) = r_{ab}P^L Q^L V_a V_b$$

which is known as theorem of Bortkiewicz. It is often stated as follows

(1.3.14) $$\frac{V_{0t}}{P^L_{0t}Q^L_{0t}} = 1 + r_{ab}V_aV_b.$$

The interpretation usually given to this finding reads as follows:

- If price and quantity relatives correlate negatively[86] ($C < 0$, prices and quantities move in opposite direction) the difference $P^P_{0t} - P^L_{0t} < 0$, that is $P^L_{0t} > P^P_{0t}$, and
- on the other hand we get $P^L_{0t} < P^P_{0t}$ (which implies $Q^L_{0t} < Q^P_{0t}$) whenever the covariance C is positive (prices and quantities tend to move in the same direction between 0 and t).

Hence the direction of the divergence between P^L and P^P depends on the covariance, but the extent of the divergence also depends on the variances, s^2_a and s^2_b of the price and quantity relatives, a and b. An interesting implication of the formula of Bortkiewicz is the following statement: equality of Laspeyres- and Paasche price index, that is $P^P_{0t} = P^L_{0t}$, is to be expected:

1. in the case of the proportionality test: all prices rise at the same rate λ ($p_{it} = \lambda p_{i0}$) or all quantities rise at the same rate λ (or remain constant [$\lambda = 1$][87]), that means: the variance of price ratios and/or of quantity ratios is zero,
2. when the covariance C between a_{0t} and b_{0t} vanishes.

In **ex. 1.3.2** the covariance C equals zero. From eq. 12 we also may conclude:

(1.3.15) $$\frac{P^L_{0t}}{P^P_{0t}} = \frac{Q^L_{0t}}{Q^P_{0t}} = 1 - \frac{C}{V_{0t}} = \frac{P^L_{0t}Q^L_{0t}}{V_{0t}}$$ and

(1.3.15a) $$\frac{P^P_{0t}}{P^L_{0t}} = \frac{Q^P_{0t}}{Q^L_{0t}} = 1 + r_{ab}V_aV_b,$$

which means that the amount by which the Laspeyres formula (usually) exceeds the Paasche formula is the same in both cases, the price index and the quantity index. Some other useful relations are compiled in **tab. 1.3.2**.

The so called "Laspeyres-effect", that is the situation in which $P^L > P^P$ (and consequently also $Q^L > Q^P$) can easily be made plausible now: it occurs when the price of commodity i rises (that is $a^i_{0t} > 1$) and the quantity tends to be reduced ($q_{it} < q_{i0}$ hence $b^i_{0t} < 1$) and vice versa. Since $P^L_{0t} = \sum \frac{p_t}{p_0}\frac{p_0 q_0}{\sum p_0 q_0}$ whilst $P^P_{0t} = \sum \frac{p_t}{p_0}\frac{p_0 q_t}{\sum p_0 q_t}$, a price

---

[86] This is supposed to be the normal (or at least more frequent) situation.

[87] The so called *identity* test ($\lambda = 1$) is a special case of proportionality.

relative which is greater than unity (that is $a^i_{0t} > 1$) will have a lower weight (since $p_0q_t < p_0q_0$) in the Paasche formula compared with the Laspeyres formula.[88] On the other hand: if $a^i_{0t} < 1$ (falling price of commodity i) a higher weight will be assigned to it in P^P compared with P^L.

Table 1.3.2: Relations between Laspeyres- and Paasche formulas

price index	quantity index
$P^P = P^L \frac{V}{V-C}$	$Q^P = Q^L \frac{V}{V-C}$

Product	Paasche	Laspeyres
$P_{0t}Q_{0t}$	$P^PQ^P = V^2/(V-C)$ $< V$ if $C < 0$, $> V$ if $C > 0$	$P^LQ^L = V-C$ $> V$ if $C < 0$, $< V$ if $C > 0$
$P_{0t}P_{t0}$	$P^P_{0t}P^P_{t0} = P^P_{0t}/P^L_{0t}$ < 1 if $C < 0$, > 1 if $C > 0$	$P^L_{0t}P^L_{t0} = P^L_{0t}/P^P_{0t}$ > 1 if $C < 0$, < 1 if $C > 0$

Two additional comments

1. on the covariance C

A situation in which we expect the covariance to be negative (that is $P^L > P^P$, $Q^L > Q^P$) is typical in the case of a demand-oriented market where buyers dominate and sellers have to face a given demand curve with a negative slope (prices and quantities move in opposite directions: as prices rise [fall] buyers will buy less [more] and they are always ready for rational substitution). The market for consumer goods, esp. those that are not absolutely vital and not to be substituted by other goods, is the leading example for the situation $C < 0$.

On the other hand a positive covariance ($C > 0$) is likely to occur when the market is dominated by suppliers and consumers are unable to react to rising prices by reducing the consumption or by rational substitution. This has been demonstrated to occur in the case of rising prices for lodging and energy (e.g. fuel consumed in cars) for example.

88 In other words: whenever prices and quantities change in opposite directions P^L gives a higher weight (lower weight) to rising (decreasing) prices than P^P does, hence $P^L > P^P$.

2. on international comparisons

The relationship between Laspeyers- and Paasche- indices has also been demonstrated in international comparisons. The finding of Bortkiewicz is known there as "*Gerschenkron-effect*". The purchasing power parity (PPP) usually will be lower when consumers assimilate their consumption pattern to the foreign country (Paasche approach, migrator model), and conversely they will be higher whenever they maintain their accustomed consumption structure of their native country through living abroad (Laspeyres approach, tourist model). Accomodation to the new, foreign consumption habits makes the covariance negative.

Example 1.3.2

In the following numerical example the covariance is zero and therefore the results obtained for P_{0t}^{L} and P_{0t}^{P} will be equal although not all price and/or quantity relatives are equal:

	Prices			Quantities		
i	p_{i0}	p_{it}	p_{it}/p_{i0}	q_{i0}	q_{it}	q_{it}/q_{i0}
1	10	14	1.4	40	48	1.2
2	20	28	1.4	30	24	0.8
3	30	48	1.6	20	16	0.8
4	40	64	1.6	10	12	1.2

The weights to be attached to the relatives $p_{i0}q_{i0}/\sum p_{i0}q_{i0}$ are 0.2 for commodities 1 and 4, and 0.3 for commodities 2 and 3. The reader should verify that P_{0t}^{L} = P_{0t}^{P} = 1.5 (in fact an average between 1.4 and 1.6), Q_{0t}^{L} = Q_{0t}^{P} = 0.96 (which is 1.2 • 0.4 + 0.8 • 0.6), and that the value index is $V_{0t} = 1.44 = 1.5 \cdot 0.96$ which is compatible with the rise of total expenditures from $\sum p_{i0}q_{i0} = 2000$ to $\sum p_{it}q_{it} = 2880$. ♦

e) Critique of the Laspeyres index and the principle of "pure price comparison" by the US-Senate Advisory Commission (Boskin Commission, BC)

The Laspeyres approach to price level measurement has often been criticized because of its fixed basket. There is a widespread belief that a "fixed basket" index overstates inflation (is "biased" upwards) and that a chain index, or any other index giving a higher weight to the more recent consumption pattern will do a better job and will result in a lower inflation rate. Furthermore advocates of the "*economic theory*" (or true cost of living, COLI) approach (see **sec. 2.4**) also heavily criticize the Laspeyres principle[89] because in their view the point of reference should not be the

[89] Both lines of reasoning, the chain index and the economic theory (COLI) approach have in common a certain down-weighting of the (analytical) usefulness of making "pure" price comparisons using a

same quantity of products in period t and 0 but rather the same utility. Such ideas were vigorously advanced by an Advisory Commission (also known as the Boskin Commission [BC] because it was headed by Michael Boskin) to the US-Senate Commission of Finance[90] to explore a possible upwards bias of the US Consumer Price Index (CPI, then a traditional fixed basket Laspeyres index). The BC published an interim report in 1995 and a final report in 1996 titled "Toward a More Accurate Measure of the Cost of Living"[91] - both widely made public in the internet - from which subsequently some quotations are taken. The position of the BC may be summarized in two points:

1. the BC called for the **abolition of the** traditional **fixed basket** U.S.-CPI in favor of a "True Cost of Living Index" (COLI) as defined in the "economic index theory" framework. The traditional P^L approach was found inadequate for a dynamic and flexible economy like the U.S. economy, and the BC
2. tried to establish that the U.S.- CPI methodology considerably **overstates** the inflation rate by about 1.5 (interim report) or 1.1 (final report) percentage points anually, due to five types of **"biases"** or "errors" (i.e. the substitution- (or high level aggregation-), outlet-, new product-, quality- and formula-bias (or low level aggregation bias)).

The BC insisted on the need to revise and change weights more frequently, preferably every year, and readily endorsed all popular beliefs on the damage done by the "growing irrelevance" of baskets based on past consumption patterns. Interestingly the BC, though heavily supporting the idea of frequent updating, did not recommend (at least not explicitly) chain indices to be the only solution, but a so called (direct) superlative index instead, that is an index also making use of current expenditure data, but still representing a direct index rather than a chain index.

Most of the "biases" directly contradict the Laspeyres idea, according to which

- the "basket" (selection of goods and specification of their weights), and
- the PDCs[92] (sample of price reporting outlets, quality etc)

should be kept constant if possible (otherwise corrections or adjustments [for quality change for example] should be made). Hence what is a virtue in the context of the Laspeyres index (like constancy for the sake of comparability) is called "bias" by the

---

constant basket. They vigorously attack the "deficiencies" of relying on weights being several years out of date and criticize Laspeyres indices for making unrealistic assumptions.

[90] It is worth noticing that the installation of the BC should be seen against the background of making extensively use of indexation and thus having a problem with budget and debt management in the USA, calling for a cut in expenditures for social purposes in these days. Hence there was most obviously a certain political interest in getting a lower rate of inflation. It is interesting (and depressing) to note that given this political background there were many theoreticians and statisticians who readily agreed in a unanimous critique of the Laspeyres formula as overstating inflation and who worked out estimates (then sometimes called "guestimates") of the amount of bias in the US-CPI. On the other hand, those defending Laspeyres' index formula were few, if any.

[91] The title was misleading because the concern was a *new* conceptual basis of measurement, rather than how to observe and measure more accurately price movement on the basis of a *given* generally accepted way of measuring inflation.

[92] Price determining characteristics see **sec. 1.1**.

BC. Surprisingly enough there is, however, the "bias" of imperfect adjustment for quality change in both, the Boskin Report (BC) as well as the Laspeyres context. Yet one may ask[93]

> When there is no need for comparability over more than just two adjacent periods, t-1 and t respectively, because the price index has to contain at any time the newest products and most recent quality variants, and nothing is to be held constant, why is there any need to make adjustments for quality changes relating to the "same" good for several years?

It seems contradictory and inconclusive to require, as the BC did,

- the use of the most recent basket and to include as soon as possible new products, new outlets etc., thus making prices incomparable,
- and to request adjustments for quality changes (in order to make prices comparable by measuring [correctly] the impact of quality change on prices).

But the contradiction dissolves when the motivation for such adjustments[94] is considered: the traditional Laspeyres index makes adjustments in order to ensure price comparability (based on *comparable* purchases), for the COLI-framework (of the "economic index theory") on the other hand it is essential to infer gains or losses in utility. With focusing on utility levels[95] we should account for all sorts of substitutive reactions by which households save money without losing utility (by changing their consumption habits, switching to new products, or to new outlets and so on).[96] **Tab. 1.3.3** gives some examples of this kind of "utility" reasoning.

Needless to say that the estimates (in these days triggered by the Boskin Report) of the amount of unmeasured gains from quality improvements and new goods differed enormously.

Fig.1.3.2 once more shows that the motivation for making adjustments for quality change in the COLI-type of index theory on the one hand and the traditional Laspeyres thinking in terms of "pure" price movement (underlying the US-CPI in these days) on the other hand is fundamentally different.

Most economists are ready to view a COLI as theoretically superior to a price index and thus being the preferred instrument for measuring inflation. But the alledged

---

[93] The question also becomes relevant in the "chain index" context. I asked this question to many "chainers" but interestingly I never got an answer.

[94] There are also substantial differences with respect to the amount of "adjustments". While the BLS made adjustments for minor changes in different variants of the same good (e.g. higher capacity of a refrigerator) the BC requested accounting for quality improvement of a totally different new good as compared with the old one (like utility gain of a PC as compared with an electric typewriter). The BC said for example that PCs with word processing and spell-checking capability represent a quantum jump as compared with traditional electric typewriters, and an estimate of such a gain in utility should in some way or other enter the CPI formula and thereby reduce inflation.

[95] As will be shown, the underlying gauge of this kind of inflation measurement is "cost of a 'unit' utility".

[96] The result of this type of thinking is a remarkably one-sided view of progress. It is not by coincidence that the BC regarded everything "new" in the first place as something "better" and increasing wealth and purchasing power, that therefore should be accounted for in a CPI.

advantages tend to diminish once some implications of the notion of "utility" as the basis of the COLI-approach are considered.[97] The critique of the COLI-approach may be summarized as follows:

> 1. The distinction between inflation and welfare measurement becomes blurred, questionable imputations of gains or losses in utility are instigated, 2. the notion of "good" becomes boundless (comparing all sorts of notional "advantages"), and finally 3. we move away from statistics of observable phenomena to speculations about levels of utility or a "fair" amount of income necessary for a "compensation".

The examples of **tab. 1.3.3** show that the utility approach triggers speculations about what might offer a roughly equal level of satisfaction. With a little imagination we will always find some additional sources of utility. The US-CPI had been criticized by the BC for example for not taking into account "the benefit to consumers of being able to keep track more easily of children, spouses, or of aged parents" in reporting prices of cellular telephones, as if these kind of benefits should be seen as equivalent to offering telephones at a lower price.

Table 1.3.3: Some examples for BC suggestions for improvement of the CPI

	price quotations in tradional CPI	Boskin Commision's proposal
medical care*	expensive surgical operations in case of heart attacks or ulcer	treating heart attacks or ulcer with generic drugs
entertainment	visits to cinemas and theatres buying old-fashioned bound-book versions of encyclopedias	rent of videos, seeing a movie at home, CD-ROM encyclopedias, surfing the internet, going to libraries

*Interestingly substitutions to maintain a constant level of *utility* may occur across borders of a commodity classification: for example services (surgical operations) are substituted by goods (drugs).

The COLI approach also tends to soften the concept of goods and make distinctions of classifications fade away. The substitution of surgical operations by generic drugs is a substitution across categories of a classification (services for goods).[98] An index has to operate on the basis of some product classification and depends on observations related to such entities.

By the notion of being "at least as well off", however, we move beyond a basket of market goods and "private consumption". Utility results from many factors, not all of them traded on markets, and purchased and consumed individually. It has been claimed for example that "greater variety and convenience" in purchases of food and

[97] This will be done in more detail in **sec. 2.4**.

[98] To find a solution for *sub*-indexes of a CPI theoretically equally well established as the *overall*-CPI in terms of the microeconomic theory of a utility maximizing household is another problem of the utility approach.

beverage should be accounted for by corrections (reductions) of prices (of which goods in particular?),[99] as if "convenience" or "variety" had a "price" of its own.

By now it does not come as a surprise that there was a discussion in the BC about the inclusion or exclusion of higher rates of crimes, air and water pollution, colder winters, the appearance of AIDS, traffic congestion, and of course also the impact of various sorts of societal and moral changes (declining family loyalty, job insecurity, political instability etc) as possible elements of a "true" measurement of inflation.

Figure 1.3.2: Conceptual differences in price level measurement

Price index should measure according to	the Laspeyres's formula	the Boskin Commision (BC)
the general concept	the change of **expenditure** needed to **buy a** constant (with respect to kind and quantities of goods) **"basket"**	change of average **costs of** a unit of **"utility"** U (on the basis of observed purchases and imputed effects on U)
objective of measurement	"pure" comparison of prices of a well defined set of commodities	changing (due to prices) cost of living at a constant level of U (or wellbeing)
why adjustment for quality change	in order to ensure comparability of prices over time (to distinguish a genuine increase from a rise of the price due to quality improvement)	to infer gains in utility (U) incorporated in new goods (making the maintaining of a given level of U cheaper: more U for the same price)

The problems with such ideas inspired by thinking in terms of utility are

- to lose contact with observable prices for observable goods and with the scope of a price index, and more specific a consumer price index, and
- to be in need of many subjective judgements, and an aggregation of a principally unbounded set of beneficial or detrimental factors influencing utility.

There are many factors affecting utility (well-being) of households, *not* related to the costs of living, and much less to individual *private* purchases. It is doubtful how such factors might be related to a "true" actual rate of inflation and how to obtain an accurate measure of the increase of prices when higher expenses may be offset by

99 Assume "variety" means not only apples but also pears. Should we make a correction (or deduction to be clear) of the price of apples, or pears, or both, in order to account for "variety"?

more satisfaction or utility-gains in one domain should be balanced with losses in another field.

Many economists welcome the idea that theory should provide the basis of measuring inflation. Thus the **COLI concept** appears superior to the rather simple idea of a "pure" price comparison of the traditional Laspeyres concept, apparently a sort of "measurement without theory". Likewise it is an intriguing idea that a **chain index** might better fit to a dynamic economy. However, it is our firm belief that the alleged advantages or improvements (compared to Laspeyres) of the two concepts (COLI and chain indices) are misconceived.

Digression: A linear approximation to the Jevons index[100]

Due to an inequality of Cramer[101] relating to the "power mean" (eq. 2.2.17) the indices of Carli (P^C), Jevons (P^{JV}) and the "unnamed" harmonic index PH (eq. 1.2.11 and **fig. 1.2.2**) are related as follows: $P_{0t}^C \geq P_{0t}^{JV} \geq PH_{0t}$. It therefore gives rise to the idea to approximate P^{JV} by taking the geometric mean of two indices, one consistently "biased" upwards (P^C) and the other downwards (PH).[102] The resulting index

(1.3.16) $$P_{0t}^{CSWD} = \sqrt{P_{0t}^C PH_{0t}} = \frac{\sqrt{\sum p_{it}/p_{io}}}{\sqrt{\sum p_{io}/p_{it}}}$$

has been proposed by Carruthers, Sellwood and Ward (1980) as well as Dalén (1992) and will henceforth be referred to as CSWD-index.[103] The similarity of the empirical results of P^{CSWD} and P^{JV} can be shown by applying multiple first order Taylor expansions to P^{CSWD} around $p_{it}/p_{io} = 1$.

The following "hybrid" formula

(1.3.17) $$P_{0t}^{HYB} = \frac{\sum \sqrt{p_{it}/p_{io}}}{\sum \sqrt{p_{io}/p_{it}}} = \frac{\sum p_{it}\sqrt{1/p_{io}p_{it}}}{\sum p_{io}\sqrt{1/p_{io}p_{it}}} = \sum \frac{p_{it}}{p_{io}} \frac{\sqrt{p_{io}/p_{it}}}{\sum \sqrt{p_{io}/p_{it}}}$$

approximates P^{CSWD} and therefore also the Jevons index P^{JV} (see below).

Both indices P^{CSWD} and P^{HYB} meet commensurability *and* time reversibility just like P^{JV} (see **fig. 1.2.1**), however, P^{HYB} is unlike P^{CSWD} linear i.e. it can be expressed as

[100] Jens Mehrhoff (Deutsche Bundesbank) has contributed this note on his "hybrid index", a linear unweighted index providing an approximation to the well known Jevons index (see eq. 1.2.10). This paper represents the author's personal opinion and does not necessarily reflect the view of the Deutsche Bundesbank or its staff.

[101] See footnote 28 on page 61.

[102] The same idea can be found in Vartia's 'system of six index formulas' (see **sec. 3.4** for details) which now – in the absence of weights - reduces to a system of three indices, with upper and lower biases in the sense of Fisher's 'five tines fork'.

[103] It apparently fell into oblivion that Fisher (1922) was the first to propose this index which he listed as his formula 101. So it became rediscovered by CSWD before long. Fisher already remarked that the CSWD index and Jevons' index are two indices which are empirically very close to each other.

P^{HYB} = $\mathbf{p}_t'\mathbf{q}^*/\mathbf{p}_0'\mathbf{q}^*$ where the vector $\mathbf{q}^*$ is given by $\mathbf{q}^{*\prime} = \left[\sqrt{\frac{1}{p_{10}} \cdot \frac{1}{p_{1t}}} \quad \ldots \quad \sqrt{\frac{1}{p_{n0}} \cdot \frac{1}{p_{nt}}}\right] = \left[\sqrt{q_1^C \cdot q_1^H} \quad \ldots \quad \sqrt{q_n^C \cdot q_n^H}\right]$. In the case of P^C = $\mathbf{p}_t'\mathbf{q}^C/\mathbf{p}_0'\mathbf{q}^C$ and PH = $\mathbf{p}_t'\mathbf{q}^H/\mathbf{p}_0'\mathbf{q}^H$ the weights are represented by the "quantities"[104] $q_i^C = 1/p_{i0}$ and $q_i^H = 1/p_{it}$ respectively. This explains the name "hybrid" as the weights constitute a mixture of Carli- and harmonic-"quantities".

P^{HYB} may be preferred to P^{CSWD} in the first place because it is linear and therefore can directly be compared to other linear indices such as Laspeyres and Paasche using a theorem of L. v. Bortkiewicz (see **fig. 3.2.2**). Furthermore P^{HYB} possesses a "ratio-of-expenditures" (for fictitious quantities $\mathbf{q}^*$, second expression in eq. 17) as well as a mean-of-relatives interpretation (third expression). The latter interpretation gives rise to the following idea (of H. A. Leifer [see the preface]): If each price relative is getting the same weight 1/n instead of $p_{i0}q_{i0}/\Sigma p_{i0}q_{i0}$ we have $P_{0t}^C = \sum \frac{p_{it}}{p_{i0}} \frac{1}{n}$ instead of P_{0t}^L. Likewise replacing $p_{it}q_{it}/\Sigma p_{it}q_{it}$ in P_{0t}^P by 1/n yields PH_{0t}. So P_{0t}^{CSWD} may be viewed as taking the part of P_{0t}^F in the unweighted case.

To see why P^{CSWD} can be approximated by P^{HYB} assume $p_{it}/p_{i0} \approx p_{jt}/p_{j0} \quad \forall i \neq j$ and perform squaring and rooting which gives:

$$(1.3.18) \qquad P_{0t}^{HYB} = \frac{\sqrt{\left(\sum \sqrt{p_{it}/p_{i0}}\right)^2}}{\sqrt{\left(\sum \sqrt{p_{i0}/p_{it}}\right)^2}} = \frac{\sqrt{\sum_i p_{it}/p_{i0} + \sum_{i \neq j} \sqrt{(p_{it}/p_{i0})(p_{jt}/p_{j0})}}}{\sqrt{\sum_i p_{i0}/p_{it} + \sum_{i \neq j} \sqrt{(p_{i0}/p_{it})(p_{j0}/p_{jt})}}}, \text{ and}$$

$$(1.3.19) \qquad P_{0t}^{HYB} \approx \frac{\sqrt{\sum p_{it}/p_{i0} + (n-1)\sum p_{it}/p_{i0}}}{\sqrt{\sum p_{i0}/p_{it} + (n-1)\sum p_{i0}/p_{it}}} = \frac{\sqrt{n}}{\sqrt{n}} \frac{\sqrt{\sum p_{it}/p_{i0}}}{\sqrt{\sum p_{i0}/p_{it}}} = P_{0t}^{CSWD} \approx P_{0t}^{JV}.$$

The assumption of all price relatives being nearly equal to each other seems to be very restrictive. It should be noted, however, that the approximation is exact in the case of two goods only (see eq. 2.2.4 below). The same applies to the CSWD and Jevons index, no matter how different the price relatives are. It can be shown that even with more than two goods and a rather high variance of price relatives the approximation holds to an extremely high degree which has been checked with simulations.[105]

Finally it is noteworthy that P^{HYB} approximates P^{JV} to the second order following the approach of Dalén (1992). While P^{JV} passes all six axioms stated by Dalén, both the P^{CSWD} and P^{HYB} index fail the permutation test.

---

[104] It was not uncommon in the early days of index theory to consider fictitious "quantities" as reciprocal prices and thereby introduce a sort of weight in an unweighted index formula. These p^{-1} - terms are in general viewed as those quantities which can be bought with exactly 1 € (or 1 $).

[105] More results and descriptions of methodology are available on request from the author (Mehrhoff).

Chapter 2

Approaches in index theory

In this chapter an attempt is made to present an overview of theories or "approaches" in index theory along with some index formulas (some of them, however, failed in gaining importance in the practice of official price statistics).[1] There is no end to the search for more general index functions or in attempts to derive classes of index functions on the basis of some few fundamental principles or some new ways of looking at the index problem. Thus the chapter can necessarily present only fragments of more or less theoretical nature.

2.1 Outline of index theories or "approaches"

Index theory has a wide variety of "approaches" to offer. For the most part index theory, and hence also the rivalling approaches, have to do with the assessment of index formulas. It is perhaps a fruitless effort to count all the different index formulas that have been invented and proposed by anybody, as being the "best" or "ultimate" formula. There are hundreds of formulas and the set of formulas we might think of being useful in some respect is unbounded. Introducing this chapter, therefore, the best we can do is to try to give a systematic, fair, and balanced overview. Reviewing textbooks on index theory and price statistics, a salient feature of most of them is that often some "approaches" are exempted from presentation, or that at least the place devoted to some topics may appear to be too small as compared to the relevance of the topic.[2] We therefore aim at an overview which is as complete as possible.

a) Formal and economic theory of index numbers

A fundamental distinction is between the so called "formal"[3] approaches (in general not claiming to be seen as "theories") as opposed to the "economic theory of index numbers" claiming to give formulas a microeconomic theoretic foundation (or rather justification). In **fig. 2.1.1** a tentative structuring of the "formal" approach is given.

---

[1] It may nonetheless be worthwhile to give some thought to the ideas underlying the formulas, to their formal properties, potential interpretations, and relationships to other formulas they reveal. In many cases it is more the way to derive a new formula than the formula itself which is interesting and which throws some light on more general principles of index theory, thereby possibly improving our general understanding of index problems.

[2] Some textbooks in this field are deliberately incomplete, perhaps because of the dominating desire of the authors to convince the reader of their specific approach. We can hardly expect that an author adhering to the "economic theory" of index numbers will pay as much attention to the axiomatic approach as he does to the economic theory and vice versa. Although the present author is not neutral concerning the approaches presented here (which becomes obvious in chapter 7) we try to be fair to each school in index theory.

[3] Also named: axiomatic, statistical, mathematical, mechanistic or mechanical theory. In what follows we consider "axiomatic" as part of "formal" which is not, however, common practice. So one may also use "axiomatic" and "formal" as synonyms.

To solve certain problems, like deflating value aggregates or making international price comparisons consistently, will require in general a specific reasoning, different from the standard problem of a binary comparison between two price levels. To this standard problem refers the distinction, usually made between an axiomatic and a constructive (see part b) approach respectively, within the "formal" approach.

As index formulas abound, index theory can at best offer some tools to enable users to decide on which formula to apply. The axiomatic approach explicitly aims at systematically providing criteria for this evaluation problem in terms of "axioms", of which some examples (e.g. identity, commensurability, time reversibility etc.) are already introduced here in **tab. 1.1.1** in connection with the discussion of properties of price relatives. The constructive approach is the attempt to find an index formula resulting from a model (for example of an additive as opposed to a multiplicative decomposition of value-dynamics) or satisfying certain properties like the factor reversibility.

This leads over to the economic theory of index numbers. In **sec. 1.2** part d the (old) stochastic index approach as opposed to the aggregative approach has been introduced. The latter explicitly made reference to the expenditures (aggregates) of households in measuring consumer prices, thereby deriving "weights", or a "basket" while the former approach favored unweighted index formulas without any quantities and specific households involved. In a sense the economic theory of index numbers is in the tradition of the aggregative approach in that it relates prices and (as a response to prices) quantities. In the case of the Consumer Price Index (CPI) or "True Cost of Living Index" (COLI) this means that recourse is made to non-observable "indifference curves" and "utility levels".[4]

The formal theory of index numbers on the other hand is restricted to observable variables, treating prices and quantities as exogenous[5] and as unrelated arguments in the index function

(1.1.1) $P_{0t} = P(\mathbf{p}_0, \mathbf{q}_0, \mathbf{p}_t, \mathbf{q}_t)$.

This does not mean, however, that questions of "meaning", "sense" and the like are generally irrelevant for the point of view of the formal theory. Such considerations are of a certain importance, more indirectly however, because they inspire the definition of "axioms", that is some minimum quality standards (desirable properties) for an index, making them meaningful. Such axioms are open to debate, however, they often are in conflict with one another and we cannot expect the axiomatic theory to "prove" in such a situation which criteria should come first and which should come second and therefore could be sacrificed. From this it follows that it depends on the emphasis given to conflicting axioms which index formula will be preferred. In a collection of axioms it is difficult if not impossible to establish axioms deserving absolute priority over the remainder, and therefore it is unlikely that controversies about index problems (such as chain indices or direct indices, see **ch. 7**) will ever be resolved using axioms alone. Moreover it is not even clear that axioms really focus on the

[4] The concept "utility function" is the central category in the COLI-theory. Quantities are regarded as determined by prices and demand functions (derived from utility functions).

[5] Formal theory does not take into account the existing economic relationship between prices and quantities. For economists this is the major shortcoming of the formal theory since the economic interpretation of the price index formula is beyond the scope of this type of theory. The focus is exclusively on formal properties of the index function (eq. 1.1.1).

relevant problems rather than concentrating on those features of formulas that have an impact on aspects of mathematical elegance only.[6]

As no index approach is superior in all respects an attempt should be made to examine as many "approaches" as possible in order to gain a wide understanding of index problems.

Figure 2.1.1: Structure of formal index theories

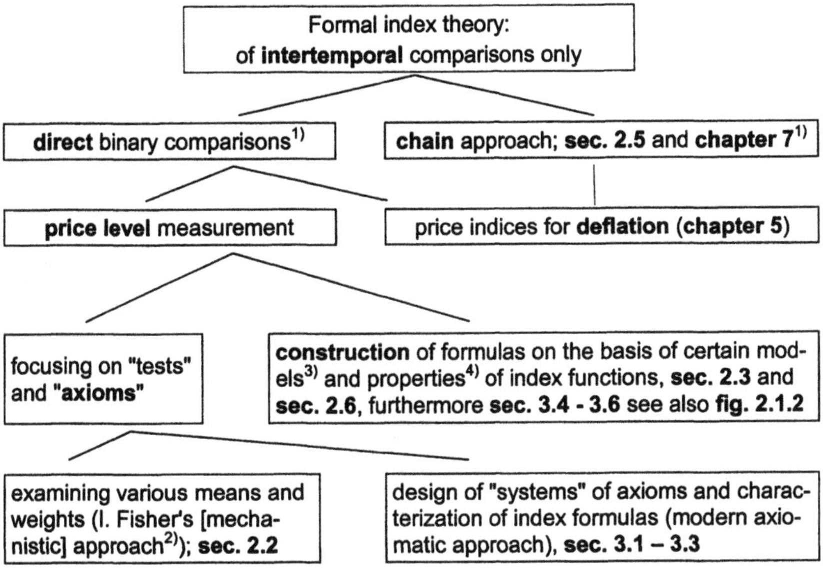

1) for reasons to be explained later, we distinguish "direct" (instead of "fixed base") and chain indices.

2) or old axiomatic approach

3) as an example: the additive decomposition of value change (Stuvel) or the "factorial approach" (Banerjee).

4) e.g. search for formulas that satisfy the factor reversal test, or formulas constructed as means of "log changes".

b) Constructive approaches

Index construction often aims to find an index satisfying certain criteria (e.g. the factor reversal test), or being optimal in some sense, or a more and more general formula, containing other formulas as special cases. The solution of an exercise of this type is not always a new index formula, sometimes it is "only" a new interpretation of

[6] Furthermore axioms are not self-evident, and criteria like understandability, comparability of results, ease to provide empirical data, "acceptance" by non-statisticians and the like are generally seen as mathematically not manageable and therefore readily ignored.

known formulas, gained as a by-product in following a new "approach", a new look at index theory altogether. Ideas that may in fact constitute a new and fertile approach to index theory are

- the additive model in defining a price and quantity component of value change which is applied in the closely related approaches of Stuvel and Banerjee (see **sec. 2.6**),
- the construction of "log-change" index numbers, new schemes of weights and the discovery of hitherto unknown relationships among index formulas (**sec. 3.4** to **3.6**),
- some other ideas, like applying the least square principle to obtain formulas able to *approximately* fulfil certain criteria, or the construction of measures for the "structure" of prices in contrast to the "level" of prices and the idea of index numbers emerging as coefficients in some econometric models (**sec. 2.3**).

Of course the approaches listed above differ greatly with respect to their significance, scientific fertility and the like. The status of some of these "approaches" is open to discussion. But each of them will at least add a bit to a better understanding of index numbers in general. Thus they are worth being explored to some extent at least.

In **fig. 2.1.2** an attempt is made to define a system into which most of the issues, theories, formulas etc. may fit. The scheme is limited to what might be called "constructive" approaches, the common goal of which is to find a general formula of an index function as a result of conceptual considerations, based on the premise that the *observable* (manifest) value change V_{0t} (or value difference ΔV) is to be decomposed into *two non-observable* (hypothetic or latent) *components*, the first, P accounting for the price movement and the second, Q measuring the impact of quantity change on value change.

Much of our criticism against the chain index as well as the economic theory approach has been mentioned already in **sec. 1.3**, in particular (as the COLI approach is concerned) in our exhibition of the Boskin Report (sec. 1.3 part e).

There are three different ways of describing changes in individual prices:

price relatives	***log-changes***	***differentials*** with respect
(≈ growth factors)	(≈ growth rates)	to time $dp_i(t)/dt$
→ traditional indices	→ log-change indices	→ Divisia index

Apart from these constructive contributions to index theory there are many other contributions that do not fit into this scheme. Take for example the idea to decompose V_{0t} (or ΔV) into three or more components. In addition to P and Q we might think of the structural component or the "change in quality" component or some other component.

Figure 2.1.2: Some[1)] "constructive" approaches in index theory

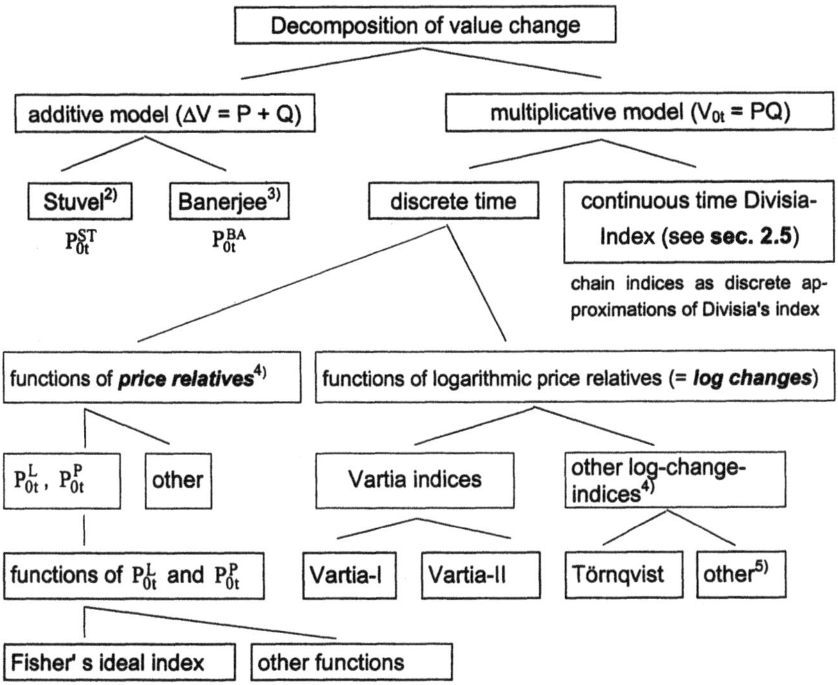

1) not integrated in this chart: Theil's approach and the so called "(new) stochastic approach"
2) shading of boxes indicates that this index is an "ideal index" (i.e. satisfying the factor reversal test).
3) factorial approach.
4) these indices are not derived from a model of decomposing the value change.
5) as an example: Cobb-Douglas index (transitivity holds) or refined Törnqvist approaches in order to come closer to factor reversal (Theil, Sato).

Another example is the stochastic approach, in which the focus is not on the index formula and its properties taken in isolation, nor on the relationship between the formula and its elements (prices, quantities), but on an econometric model to "explain" the generation of prices, quantities etc.

As aforementioned one might safely assume that not much of what is discussed in this and the following chapter was hitherto put into effect in the practice of official price statistics.[7] The general remarks on **fig. 2.1.2** might give some indication why this is so. Many new indices combine remarkable advantages with serious disadvan-

[7] This also seems true for the above mentioned "economic theory" approach, which is also devoted to models explaining the generation of price and quantity relations in a somewhat different manner, however, than the stochastic approach.

tages. To list some examples, without giving a complete account of all negative properties:

- indices derived from an additive model and being functions of the traditional Laspeyres- and Paasche formula, like Stuvel's formula, P_{0t}^{ST} or Banerjee's formula, P_{0t}^{BA} will in general *not* be *linear homogeneous* in the prices p_{it},
- *all* so called *log-change indices* violate the *monotonicity* axiom,[8] which is a serious defect since some of them (or generalizations of them) are used for purposes of international comparisons,[9]
- a log-change index like the Cobb-Douglas index (CD-index, or P_{0t}^{CD}) is, despite being transitive, not suitable for international comparisons nor for deflation.[10]

The greatest shortcoming from which perhaps most of the indices mentioned in **fig. 2.1.2** (in particular the log-change indices) suffer is, however, the lack of understandability and intuitive meaning of the formula[11] though an attempt is made in **sec. 3.5** to show that once the combination of log-changes and the log mean is well understood it turns out that the concept of Vartia indices is quite straightforward and not so difficult to understand as often believed.

2.2 Irving Fisher's approach and reversal tests

This section is devoted to Irving Fisher's[12] contribution to index theory. In his seminal book "The Making of Index Numbers" (1922) he was systematically in search of an "ideal" index function trying various means and weighting schemes and also examining means of index formulas already found and means of weighting schemes with respect to so-called "tests".[13] Fisher was not the first to think of "tests" as criteria in evaluating index functions but probably the first[14] to use this kind of reasoning on a large scale and in search of new formulas.

[8] The Vartia II quantity index resulting from deflation with the Vartia II price index does not necessarily satisfy monotonicity in the quantities.

[9] Furthermore some of them (like Vartia I) do not meet proportionality, other indices (like Vartia II) pass the factor reversal test but are not consistent in aggregation.

[10] Because the CD-index does not depend on quantities. There is no quantity index that can reasonably be related to the CD-price index.

[11] To recall only one of the aspects involved in "understandability": none of the proposed formulas has an explicit "budget"-interpretation (rising costs of a "basket" due to rising prices).

[12] Irving Fisher, American economist (1867 - 1947). His kind of approach is sometimes called "mechanistic", formal, or "test approach".

[13] This taking of means of formulas or weights was called "crossing" by Fisher.

[14] Certain criteria, in particular "reversal tests" are "finders of formulae". Fisher himself made reference to Pierson to whom he owed the idea of the time reversal test. Other authors well known for having worked with tests in these days were A. Wald and H. Westergaard.

a) Fisher's systematic search for formulas

Pal Köves (1983) introduced Fisher's famous system of index functions with reference to the so called "family tree of index formulae"[15] in which three "generations" of index formulas are distinguished (see **fig. 2.2.1**):

1. first generation indices: simple unweighted means of prices and price relatives,[16] like indices of Dutot, Carli, or Jevons (see **sec. 1.2**);
2. second generation indices, gained from combining all kinds of averages with all kinds of weights (see below), to which belong the formulas of Laspeyres, or Paasche for example,
3. third generation indices derived from crossing of second generation formulas with their "time antithesis"; or from crossing weights. Examples for this are Fisher's ideal index, or the Edgeworth - Marshall index.

Fisher's types of weights and means

Fisher started with *six* types of (unweighted) *means* to derive index numbers, among them traditional measures of central tendency, like 2, 3, 4 (in **fig. 2.2.1**), but also median and mode, and a kind of mean which lies somewhat apart, the "aggregative" index as Fisher called it (like the Dutot index $\Sigma p_t / \Sigma p_0$). He then introduced *four* methods of *weighting* price relatives:

I	base weights: $p_0 q_0$ (Laspeyres)	III	pure price movement: $p_t q_0$ "hybrid"
II	real expenditures: $p_0 q_t$ "hybrid"	IV	current weights: $p_t q_t$ (Paasche)

and he combined them with the abovementioned six types of means. The result is a collection of "second generation" index-formulas such as P^L, P^P, P^{AH}, P^{PA}, P^{HB}, and P^{HH} (**fig. 2.2.1**).

Fisher did not make an attempt to interpret most of the index formulas he arrived at in a purely mechanic manner. Some index formulas, however, though seldom used deserve some further discussion:

Firstly the index

$$P_{0t}^{PA} = \Sigma \frac{p_t}{p_0} \frac{p_t q_t}{\Sigma p_t q_t} = \frac{\Sigma p_t \tilde{q}_t}{\Sigma p_t q_t} \geq P_{0t}^{P} \text{ where } \tilde{q}_t = q_t \frac{p_t}{p_0} \tag{2.2.1}$$

because it originally was proposed by R. H. I. Palgrave, 1886, and

$$P_{0t}^{HB} = \frac{\Sigma p_0 q_0}{\Sigma (p_0^2 q_0 / p_t)} \leq P_{0t}^{L}, \tag{2.2.2}$$

[15] Evidently the "family tree" is incomplete as none of the indices of **sec. 2.5, 2.6** and **3.3** etc. is included.

[16] At the beginning we only had *price* indices. Only in a later stage also quantity indices, direct or derived as co-factors (factor antithesis) of first generation price indices were introduced.

a "no-name" index we might call the "harmonic base" index.

Secondly in addition to P_{0t}^{HB} there are some more awkward formulas, especially P_{0t}^{AH}, and P_{0t}^{HH} in **fig. 2.2.1** to which we refer later to demonstrate some properties of index formulas.[17]

Thirdly the attention given to the group of geometric-mean-indices greatly increased recently. In Fisher's thinking they apparently did not play an important role.

Figure 2.2.1: Family tree of index formulas (according to Köves)

First generation indices

Fisher's system of six types of means				
additive structure			geometric block	positional parameters
1	2	3	4	5, 6
aggregative	arithmetic	harmonic	geometric mean	median and mode
Dutot P_{0t}^{D}	Carli P_{0t}^{C}	P_{0t}^{HM} see below	Jevons P_{0t}^{JV}	

Second generation indices

means	Fisher's four types of weights			
	I: base $p_0 q_0$	II: hybrid $p_0 q_t$	III: hybrid $p_t q_0$	IV: current $p_t q_t$
arithmetic	P_{0t}^{L} Laspeyres	P_{0t}^{P} Paasche	P_{0t}^{AH} see below	P_{0t}^{PA} Palgrave
harmonic	P_{0t}^{HB} see above	P_{0t}^{HH} see below	P_{0t}^{L} Laspeyres	P_{0t}^{P} Paasche
geometric	less well known formulas, see text below and table 2.2.1			

[17] Hence there are formulas, the use of which is foremost to serve for mathematical demonstration purposes, mainly with respect to the time reversal test and factor reversal test but also to the monotonicity axiom.

Figure 2.2.1 continued

Note that some of the combinations are identical due to inherent relationships between the arithmetic and the harmonic mean: both index formulas, P^L and P^P can be expressed in two ways, using pure and hybrid weights.

Third generation indices see tab. 2.2.1

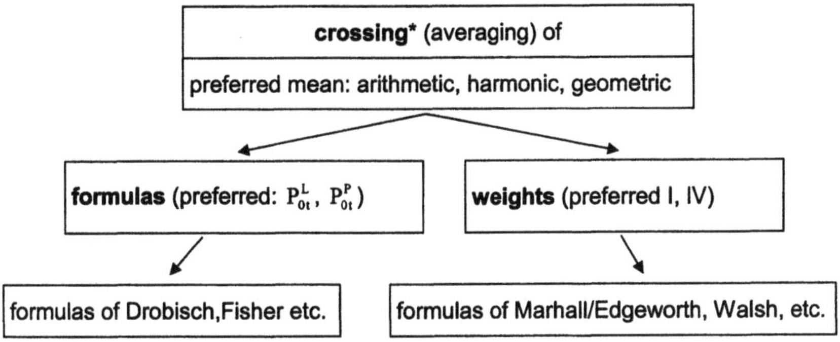

* "Crossing" in a more specific use refers to averaging an index P with its "antithesis". If this is done using a *geometric* mean Fisher called it "rectifying".

A formula of this kind of indices is given by

(2.2.3) $$P_{0t}^{CD} = \prod \left(\frac{p_t}{p_0} \right)^{\alpha_i} = \prod \left(a_{0t}^i \right)^{\alpha_i}$$ (CD = Cobb-Douglas)[18],

where the α_i-coefficients are any real valued arbitrary weights not further specified, except for $\alpha_i \geq 0$ and $\Sigma\alpha_i = 1$. Three of the best known members of the "geometric bloc" (Köves) or "family of log-change-indices" are presented in **tab. 2.2.1**. These formulas will be dealt with in **sec. 3.4**. Note that only type I and type IV weights are used. This is because type II or type III are uncommon to consider in combination with the geometric mean.

An interesting feature of index theory is that some index formulas are rediscovered from time to time, probably because of a widespread unawareness of the history of index numbers. Thus, for example, the formula known as "logarithmic Laspeyres" due to the utilization of Laspeyres type weights $w_0 = p_0q_0/\Sigma\, p_0q_0$ was also proposed by the Swiss economist W.A. Jöhr in the 1970's.

[18] The name refers to the well known Cobb-Douglas (production) function.

Table 2.2.1: Some "log-change-indices" derived from Fisher's scheme to systematize index formulas, geometric means and types of weights

weights I: $w_{i0} = p_0 q_0 / \Sigma p_0 q_0$	weights IV: $w_{it} = p_t q_t / \Sigma p_t q_t$
logarithmic Laspeyres index	logarithmic Paasche* index
$DP_{0t}^{L} = \prod \left(\frac{p_{it}}{p_{i0}} \right)^{w_{i0}}$	$DP_{0t}^{P} = \prod \left(\frac{p_{it}}{p_{i0}} \right)^{w_{it}}$

geometric mean of both indices: Törnqvist index
$P_{0t}^{T} = \sqrt{DP_{0t}^{L} DP_{0t}^{P}} = \prod (p_t / p_0)^{\overline{w}}$ where $\overline{w} = \frac{1}{2}(w_0 + w_t)$ or $\ln(P_{0t}^{T}) = \frac{1}{2}\left[\ln(DP_{0t}^{L}) + \ln(DP_{0t}^{P})\right]$ There are striking similarities between P^F and P^L, P^P on the one hand and P^T and DP^L, DP^P on the other hand.

*Obviously the construction of DP_{0t}^{P} suggests the name "logarithmic Palgrave" rather than "logarithmic Paasche". Yet it is consistent to refer to Paasche in this context as shown in **sec. 3.4**.

Crossing weights and formulas (creation of third generation indices)

We now turn to third generation formulas gained by crossing weights and formulas.[19] Some well known formulas resulting from crossing of *formulas* are the arithmetic and the geometric mean of the Paasche and Laspeyres index known as formulas of Drobisch[20] (or Sidgwick) and as Fisher's "ideal index". They are both presented in **tab. 2.2.2** together with formulas resulting from "crossing *weights*", an exercise performed mostly on the basis of Laspeyres and Paasche weights, that is, using weighting schemes I and IV respectively.

Consider two factors, x_1, x_2 only. It is easy to see that the geometric mean of these values, that is $\overline{x}_G = \sqrt{x_1 \cdot x_2}$, will be the geometric mean of the arithmetic and the harmonic mean, thus

(2.2.4) $\qquad \overline{x}_G = \sqrt{\overline{x} \cdot \overline{x}_H}$

which is the reason for the relation between P^F, P^{HPL} and P^{DR}. Note further that the

[19] Fisher used the terms "crossing" and "rectifying" of formulas also averaging an index with its "time-" and "factor antithesis". See chapter 7 and 8 of Fisher's book.

[20] Note that this is another, more famous formula than the unit-value formula (P^{UD}, **sec. 1.2**) of the same author.

harmonic mean of quantities q_H is definded as $q_{iH} = \frac{2(q_{i0}q_{it})}{q_{i0} + q_{it}}$ and that the Marshall – Edgeworth index is the only index of additive structure in **tab. 2.2.2** to which **both** interpretations (crossing of weights and of formulas) apply.[21]

Table 2.2.2: Some well known formulas derived from crossing[1)] formulas and weights

arithmetic mean	geometric mean	harmonic mean
	crossing Laspeyres and Paasche index formula	
(2.2.5) Drobisch 1871 (or Sidgwick1883) $P_{0t}^{DR} = \frac{1}{2}\left(P_{0t}^{L} + P_{0t}^{P}\right)$	(2.2.6) Irving Fisher 1922 $P_{0t}^{F} = \sqrt{P_{0t}^{L} \cdot P_{0t}^{P}}$ well known as "ideal index" of Fisher	(2.2.7) no-name index[2)] (not in use), to be called P^{HPL} $P_{0t}^{HPL} = \frac{\left(P_{0t}^{F}\right)^2}{P_{0t}^{DR}}$
	crossing Laspeyres and Paasche weights	
(2.2.8) Marshall and Edgeworth 1887 $P_{0t}^{ME} = \frac{\sum p_t(q_0 + q_t)}{\sum p_0(q_0 + q_t)}$	(2.2.9) Walsh 1901 $P_{0t}^{W} = \frac{\sum p_t\sqrt{q_0 q_t}}{\sum p_0\sqrt{q_0 q_t}}$	(2.2.10) Geary-Khamis[3)] $P_{0t}^{GK} = \frac{\sum p_t \frac{q_0 q_t}{q_0 + q_t}}{\sum p_0 \frac{q_0 q_t}{q_0 + q_t}}$

1) Because the Paasche formula is the time and factor "antithesis" of Laspeyres' formula and vice versa, it is (in this case) also correct to say, second generation indices are derived from crossing a formula with its antithesis instead of crossing two formulas.

2) The name, given here for convenience of presentation, should be "Harmonic-Paasche-Laspeyres" (HPL).

3) Result of applying the method of Geary and Khamis for multinational comparisons to m = 2 countries only, A and B. The periods 0 and t are taking in eq. 10 the place of the countries A and B.

P^{ME} makes use of arithmetic mean of weights, $\frac{1}{2}(q_{i0} + q_{it})$ and it also takes the form of a weighted mean of P^L and P^P,

(2.2.8a) $$P_{0t}^{ME} = \frac{P_{0t}^{L} + V_{0t}}{1 + Q_{0t}^{L}} = \left(\frac{1}{1 + Q_{0t}^{L}}\right)P_{0t}^{L} + \left(\frac{Q_{0t}^{L}}{1 + Q_{0t}^{L}}\right)P_{0t}^{P}.$$

Fisher conjectured that in general a "compromise" (result of crossing) should be nearer to the truth than each of the two formulas crossed taken in isolation. The idea

[21] A similar relation is true for the "geometric" Törnqvist index.

of "crossing" (or finding a "compromise" formula) played an important part in Fisher's thinking. For Fisher this position, perhaps founded on some philiosophical grounds[22], was unquestioned and fascinating. He went even further and proposed a **double crossing** of formulas, both of which were derived by crossing. There seemed to be no doubt for him that a successive crossing would carry us nearer and nearer to an ideal. One of those "doubly rectified" formulas of Fisher's was the following most remarkable monster:

$$(2.2.11)\quad \frac{\sum p_t q_t}{\sum p_0 q_0}\left[\frac{\sum p_0 q_0 \frac{q_0}{q_t}\sum p_t q_0\left(\frac{q_0}{q_t}+\frac{p_t}{p_0}\right)\sum p_t q_t \frac{p_t}{p_0}\left(\sum p_0 q_t\right)^2}{\sum p_0 q_0 \sum p_t q_0 \sum p_t q_t \frac{q_t}{q_0}\sum p_0 q_t\left(\frac{q_t}{q_0}+\frac{p_0}{p_t}\right)}\right]^{1/4},$$

a formula that received a ranking of "excellent" in Fisher's seven grades quality scale[23] (in which the indices of Laspeyres and Paasche only received "very good") but

- this "monster" formula has no obvious economic interpretation,
- the "message" of the formula can hardly be understood by anybody, and
- it is not even clear why this index should be viewed as a price rather than quantity index.

The procedure of crossing formulas apparently has only been applied to weighted indices, not to unweighted indices. The reason presumably will be that those authors who proposed compromises by crossing were already aware of the limitations of unweighted indices, especially of their uselessness for deflation.

To evaluate second and third generation formulas a first criterion should be the kind of *interpretation* a formula permits (see **fig. 2.2.2**), in contrast to Fisher, who judged formulas mainly on the basis of his *reversal tests,* we make use of the criterion of interpretability in terms of weighted means and ratios of aggregates. With respect to *his* (Fisher's) criteria, the "reversal tests", his ideal index is in fact superior to most or even to all indices introduced till now. There are however many criteria other than reversibility, like for example "aggregation properties" (see **sec. 5.2c**) that might be chosen to evaluate index functions as well. And notably, according to such properties the "ideal" index no longer seems to be a "supreme", “superlative” or unsurpassed formula.

In Fisher's index theory the crossing of formulas and weights was regarded as a most pertinent method to derive an "ideal" index formula. Yet, as will be shown in what follows, it should be noted

- his system of means and weights is incomplete, likewise his system of
- criteria (tests, axioms) to judge the quality of formulas is inconsistent, and
- the underlying philosophy of Fisher's approach is highly questionable.

---

[22] Of course as a general principle "compromise" is not tenable. A compromise between two equally fallacious positions is no more trustworthy than just either of them.

[23] See part e of this section.

We begin the discussion with the first of these remarks.[24] There were mainly two kinds of means that were generally known in Fisher's days, and hence candidates to be used for crossing formulas and weighting schemes: the arithmetic and the geometric mean. Both means now can be shown to be only special cases of some more general formulas.

b) Generalization of means[25]

The unweighted arithmetic mean used in crossing formulas can be regarded as a special case of the weighted linear combination[26] (or weighted arithmetic mean, or **convex combination**)

(2.2.12) $$P_{0t}^{CC}(\alpha) = \alpha P^L + (1-\alpha)P^P \qquad (0 \leq \alpha \leq 1)$$

of two indices, P^L and P^P.

In an analogous manner the usual concept of an unweighted geometric mean is to be replaced by the more generalized *weighted geometric mean* P_{0t}^{GM}

(2.2.13) $$P_{0t}^{GM}(\alpha) = \left(P_{0t}^L\right)^{\alpha} \cdot \left(P_{0t}^P\right)^{(1-\alpha)}$$

known also as the generalized Fisher index. Depending on the parameter α we get

arithmetic: $P_{0t}^{CC}(\alpha = 0) = P_{0t}^P$, $P_{0t}^{CC}(\alpha = \frac{1}{2}) = P_{0t}^{DR}$, $P_{0t}^{CC}(\alpha = (1 + Q^L)^{-1}) = P_{0t}^{ME}$, $P_{0t}^{CC}(\alpha = 1) = P_{0t}^L$

geometric: $P_{0t}^{GM}(\alpha = 0) = P_{0t}^P$, $P_{0t}^{GM}(\alpha = \frac{1}{2}) = P_{0t}^F$, $P_{0t}^{GM}(\alpha = 1) = P_{0t}^L$.

Hence many traditional formulas, not only crossed ones, can be viewed as special cases of the class of P_{0t}^{CC} - indices or P_{0t}^{GM} - indices.

It follows from above that[27] $P_{0t}^P \leq P_{0t}^F \leq P_{0t}^L$, or $P_{0t}^P \geq P_{0t}^F \geq P_{0t}^L$.

There also exists a generalized (weighted) *harmonic* crossing of formulas

(2.2.14) $$P_{0t}^H(\alpha) = \frac{P_{0t}^L P_{0t}^P}{\alpha P_{0t}^L + (1-\alpha)P_{0t}^P},$$

---

[24] Discussion in this section then will be continued with the third remark in part e, and **sec. 3.3** deals with the second problem. Some of Fisher's "tests" always were controversial; and his system was contradictory (inconsistent) and hence could not be fulfilled.

[25] The following considerations refer to some issues in the theory of means that were probably not yet known at the time, when Fisher wrote his "Making of Index Numbers". After this short digression we will return to Fisher's contribution to index theory.

[26] or general weighted mean.

[27] Likewise P^{DR} (Drobisch), and P^{HPL} (harmonic mean) will always lie in the interval bounded by P^P and P^L (where normally P^P will be the lower bound and P^L the upper bound).

such that $P_{0t}^{H}(\alpha = \frac{1}{2}) = P_{0t}^{HPL}$ and harmonically crossed weights $q_i(\alpha) = q_{i0}q_{it} / (\alpha q_{i0} + (1-\alpha)q_{it})$ with the special case $\alpha = 1/2 \rightarrow P^{GK}$ (cp. **tab. 2.2.2**).

Figure 2.2.2: Interpretations of second and third generation indices

second generation	Laspeyres (P^L), Paasche (P^P), Palgrave (P^{PA}), log. Lasp. (DP^L)
third generation	Drobisch (P^{DR}), Fisher (P^F), Marshall-Edgeworth (P^{ME}), Walsh (P^W), Geary-Khamis (P^{GK})

Admissible interpretations

mean of individual price relatives	ratio of aggregates (expenditures, revenues)[1]	mean [2] of index formulas P^L and P^P
applies to *all* second generation indices[3]	P^L, P^P: yes; with reference to q_0, q_t;	P^L: $\alpha = 1$, P^P: $\alpha = 0$, P^{PA}, DP^L no, P^{PA} farfetched [4]
third generation index only P^W, P^{GK}	yes (average quantities): P^{ME}, P^W, P^{GK}, no: P^{DR}, P^F	P^{DR}, P^F yes (α = ½), P^{ME} yes; P^W, P^{GK} no

1) that is, numerator and denominator can be expressed (regarded) as sums of expenditures (products of prices and certain quantities q*, not necessarily the same in numerator and denominator).

2) this means: the index function can be expressed as a special case of the general (weighted) arithmetic mean $\alpha P^L + (1-\alpha)P^P$ or of the general (weighted) geometric mean $(P^L)^{\alpha} + (P^P)^{1-\alpha}$.

3) interpretation possible by kind of construction of this type of indices (as they are derived as means).

4) quantities in the numerator, $q_t(p_t/p_0)$ might be viewed as "adjusted" quantities q_t; the situation is similar in other cases, like P^{AH}, P^{HB} and P^{HH} in fig. 2.2.1.

Note that $P_{t0}^{H}(\alpha) = \alpha P^L + (1-\alpha)P^P = P_{0t}^{CC}(\alpha)$, a relationship which explains why P^{DR} and P^{HPL}, being indices of type $P^{CC}(\alpha)$, and $P^H(\alpha)$ respectively (where α = ½), both violate the time reversal test and are "time antithesis" of one another. The equivalent relation in the case of a weighted geometric mean is

(2.2.13a) $$P_{t0}^{GM}(\alpha) = \frac{1}{P_{0t}^{GM}(\alpha)},$$

by virtue of which the (generalized) Fisher index complies with time reversibility (for *all* α). Note that from eq. 2.2.13 follows

(2.2.15) $$P_{0t}^{GM}(\alpha) \cdot Q_{0t}^{GM}(\alpha) = \left(P_{0t}^{L}Q_{0t}^{L}\right)^{\alpha} \cdot \left(P_{0t}^{P}Q_{0t}^{P}\right)^{(1-\alpha)},$$

the result of which differs from V_{0t} if $\alpha \neq$ ½. Hence

All indices of the the generalized Fisher index type ($0 < \alpha < 1$, $\alpha \neq 0$ and $\alpha \neq 1$) will satisfy the time reversal test, but fail the factor reversal test, except for the "ordinary" Fisher index (e.g. the case $\alpha = ½$).

As the index P^{ME} is also to be viewed as a weighted arithmetic mean of P^L and P^P, a direct comparison between P^{ME} and P^{DR} (provided $P^L > P^P$) is possible and we may conclude

$Q^L < 1 \rightarrow P^{ME}$ closer to P^L than P^{DR}, hence $P^{ME} > P^{DR}$

$Q^L > 1 \rightarrow P^{ME}$ closer to P^P than P^{DR}, hence $P^{ME} < P^{DR}$.

(the converse would be true if $P^L > P^P$).

In a purely formal manner Fisher's "ideal index" P^F as unweighted **geometric** mean of P^L and P^P can also be expressed as **arithmetic** mean (Köves 1983, p.79) as follows

(2.2.16) $$P^F = \frac{P^F - P^P}{P^L - P^P} P^L + \frac{P^L - P^F}{P^L - P^P} P^P,$$

and by analogy Q^F is a weighted arithmetic mean of Q^L and Q^P with weights being $(Q^F-Q^P)/(Q^L-Q^P)$ and $(Q^L-Q^F)/(Q^L-Q^P)$, again adding up to unity.

It can be shown that not only crossing of formulas but also crossing of weights q_0 and q_t leads to formulas with values that lie in the intervall $[P^P, P^L]$.

The concept of **a power mean (or moment mean)** $\bar{x}_p(r)$ of degree r with weights (relative frequencies) $h_1, h_2, ..., h_m$ in statistics is defined as follows:

(2.2.17) $$\bar{x}_P(r) = \left[h_1 x_1^r + h_2 x_2^r + ... + h_m x_m^r\right]^{1/r}.$$

It immediately follows that we get some well known formulas as special cases: $r \rightarrow 0$ gives the geometric mean, $r = 1$ the arithmetic mean $\bar{x}$, $r = -1$ the harmonic mean $\bar{x}_H$, and $r = 2$ is the quadratic mean $\bar{x}_Q$.[28] An index derived as a weighted quadratic mean of price relatives is given by

(2.2.18) $$P_{0t}^{QM} = \sqrt{\sum \frac{p_t^2}{p_0^2} \frac{p_0 q_0}{\sum p_0 q_0}} = \sqrt{\frac{\sum \left(p_t^2 / p_0\right) q_0}{\sum p_0 q_0}},$$

or with weights $p_0^2 q_0^2 / \sum p_0^2 q_0^2$ we get[29] $P_{0t}^{QM2} = \sqrt{\frac{\sum p_t^2 q_0^2}{\sum p_0^2 q_0^2}}$.

An even wider class of index formulas can be derived by introducing *products of power means*, like $\bar{x}_p(r) \cdot \bar{x}_p(k)$ especially if $k = -r$. Thus we get the following general formula:

---

[28] According to a corrolary first proven by Cramer, the power mean is a monotonous function in the parameter r such that harmonic ≤ arithmetic ≤ quadratic.

[29] This is a formula having some resemblance to an index proposed by Young and to be introduced later.

(2.2.19) $$P_{0t}^{PP}(r)=\left[\sum_i s_{i0}\left(\frac{p_{it}}{p_{i0}}\right)^{r/2}\right]^{1/r}\cdot\left[\sum_i s_{it}\left(\frac{p_{it}}{p_{i0}}\right)^{-r/2}\right]^{-1/r},$$

where $s_{i0}=\frac{p_{i0}q_{i0}}{\sum p_{i0}q_{i0}}$ and $s_{it}=\frac{p_{it}q_{it}}{\sum p_{it}q_{it}}$ are expenditure shares. Again Fisher's ideal index is a special case (r = 2). Fisher's index has in fact so many relations to other index formulas and so many interesting properties, that Fisher himself called the formula "ideal index". Another special case is r = 1

(2.2.20) $$P_{0t}^{PP}(r=1)=V_{0t}/Q_{0t}^{W} \text{ where } Q_{0t}^{W}=\frac{\sum q_t\sqrt{p_0p_t}}{\sum q_0\sqrt{p_0p_t}},$$

where P^{PP} is the Walsh type cofactor price index (or factor antithesis of the quantity index Q^W).[30]

Finally two types of mean should be mentioned, both not yet known to Fisher. The function

(2.2.21) $$Ex(x)=Ex(x_1,...,x_n)=\ln\left[\tfrac{1}{n}\sum\exp(x_i)\right]$$

is known as the (unweighted) **exponential mean**. With weights h_i introduced, this reads as follows

(2.2.21a) $$Ex(x)=Ex(x_1,...,x_n)=\ln\left[\sum\exp(x_i)h_i\right] \text{ where } \sum h_i=1.$$

The crossing of two indices, P_1 and P_2 yields P^{EX} = Ex(P_1, P_2) = $\ln\left[\frac{1}{2}\left(e^{P_1}+e^{P_2}\right)\right]$, which in fact is a mean because exp(P^{EX}) is the unweighted arithmetic mean of exp(P_1) and exp(P_2) and the exponential function is strictly monotonous in real arguments.

The **logarithmic mean** of two variables, x_1 and x_2 (introduced already in **sec.1.1**) is given by

(2.2.22) $$L(x_1,x_2)=\frac{x_1-x_2}{\ln(x_1/x_2)}=\frac{x_2-x_1}{\ln(x_2/x_1)}$$

and will play an important part in combination with index functions of the "geometric block" or so called "log change indices" (see **sec. 3.4, 3.5**).

To summarize, it would be easy to add another 134 index functions to Fisher's system of index functions by utilzing some of the above-mentioned new formulas of means, not yet known to Fisher or by introducing some alternative weighting schemes, like $p_0^2q_0^2/\sum p_0^2q_0^2$ (used to derive P_{0t}^{QM2} above for example) but it is of course doubtful whether any progress is to be expected from this kind of approach to index problems.

[30] Other cases (e.g. the cases r = -1 and r = -2) are unable to yield formulas of much use (because "prices" $p^{3/2}$ and the like occur in such formulas).

c) Fisher's reversal tests, "crossing" and "rectifying" of formulas

Fisher introduced three reversal tests,[31] but the first of them (the commodity reversal test) is almost "trivial", so that it is nowadays rarely mentioned at all. The second and third are the two "great reversal tests" (as Fisher himself put it), the time - , and the factor reversal test. The three tests correspond to the three elements of the formula of an index, viz. commodities (i = 1, 2,..., n), points in time (or spatial units), like the periods 0 and t, and variables (prices and quantities). The idea and meaning of these tests is as follows:

1. According to the commodity reversal test (also known as symmetry axiom) the order (sequence) of the commodities ought to make no difference for the result of the index calculation.
2. The so called **time reversal test**, already introduced in tab. 1.1.1 and **sec. 1.2.c**, requires that $P_{t0} = (P_{0t})^{-1}$, or in words: Interchanging base time (0) and comparison (reference) time (t) should lead to the inverse result. To require this "axiom" appears more reasonable in interspatial (e.g. international) rather than in intertemporal comparisons: The **country reversal test** between any two countries, A and B then is defined by $P_{BA} = 1/P_{AB}$ (the first subscript denoting the "base" country). The reason for this test is to avoid ambiguity in establishing the "parity", i.e. the relative purchasing power.[32]
3. By the **factor reversal test** the product of a price and a quantity index should equal the value index V_{0t} (expenditure ratio) $P_{0t}Q_{0t} = V_{0t}$. It was primarily[33] an argument of symmetry between prices and quantities on one hand, and relatives and indices on the other that led Fisher to emphasize this test. The formula of Q_{0t} should be gained by interchanging price vectors and quantity vectors in the formula of P_{0t} (that is Q_{0t} is the **direct** quantity index, otherwise we speak of the **product test** which is a weak factor reversal test).[34] In other words: the factor reversal test is met when the **cofactor** of the price index P_{0t} (or **"factor antithesis"** of P_{0t}) equals the direct quantity index.

Reasons given for the tests: Fisher himself did not elaborate very detailed arguments in favor of his time reversal and factor reversal test. As aforementioned, the reversal tests were mainly motivated by the anaology to individual price relatives (see

---

[31] For Fisher these reversal tests were deemed most important, and at least the "factor reversal test" was most probably his own invention. Fisher also used his reversal tests as "finders of formulae". At least some of the tests are poorly motivated, and the "philosophy" behind them is controversial.

[32] Moreover the rationale of this test seems to be clear: after reaching the conclusion that A is k times as much than B, it would be hard to understand why B should not be one k-th of A.

[33] Or better: again, as Fisher's idea of *"fairness"* (as he himself put it), or *symmetrical* treatment (of periods 0 and t) was in the first place the motivation for his time reversiblity, so the symmetry between prices and quantities appears attractive too. Moreover Fisher was influenced to a great deal by his desire to see analogies as perfect as possible between indices and simple (price) relatives. We will see that most of this kind of reasoning is not tenable, though it appears convincing at first sight.

[34] The so called *"value index preserving test"* is another weakening of factor reversibility.

tab. 1.1.1) and an argument of "fairness".[35] That this type of reasoning is not tenable can be seen by the following fact: Fisher thought his three reversal tests correspond to the three "elements", commodities, periods and variables (prices, quantities), and therefore they should have the same "logical" status. That this is erroneous becomes evident once we realise that the tests are enormously different as far as the difficulty to pass them is concerned: while the commodity reversal test is almost trivial, there are only very few index functions that comply with the factor reversal test. Time reversal appears to have a middle position in this respect.

With regard to two countries, A and B, there is apparently no reason to prefer one of them to the other so that it is reasonable or even desirable to treat them symmetrically in the sense of country reversibility. But with regard to points in time, that is in the intertemporal framework, it should be recognized that

- history takes one direction only: nothing can be made undone or redone once more, and
- 0 and t are not just **two** points in time but rather one **single, fixed** period 0 and a "period" t which is a **variable**, a multitude of points in time.[36]

Hence there is no need for "symmetry". We have many similarities but also many dissimilarities between intertemporal and interregional price comparisons (see **sec. 8.1**). Thus an argument that is convincing for interregional price comparison may not be valid in the case of intertemporal comparison and vice versa.

According to quite a few authors, the factor reversal test is supposed to be of considerable importance for "deflation" of aggregates (see **chap. 5**), and it is said that the idea of this criterion is to decompose unambiguously an increase in value into two components, a price and a quantity component with no residual term left. However, we don't know why this should not apply to the weaker product test as well[37]:

The motivation to require the much stronger, and highly restrictive factor reversal test instead of the product test is rarely if ever spelled out in detail. It seems to be the desire to do both inflation measurement and deflation with the help of the same price index.

The **product test** does not require that P and Q have to be formulas of the same structure. The product test is met if both indices, P and Q, comply with the axioms required for a price index or quantity index respectively and their product PQ equals the value ratio (index) V_{0t}. The formulas P and Q can be different, provided they both

[35] As Fisher put it: "Index numbers to be fair ought to work both ways - both ways as regards ... the two times to be compared or as regard the two sets of associated elements for which index numbers may be calculated - that is, prices and quantities" (Fisher 1922, p. 62).

[36] As already stated it is therefore not correct to say that both the Laspeyres (base weight) and Paasche (current weight) approach rest "on just as solid a logical foundation as the other ", like Mudgett 1951, p. 21 expressed the idea, which can also be found in many other books and articles.

[37] An index formula unable to pass the product test is for example the *Törnqvist* index. This is a serious defect of a formula which is one of the most favored formulas in SNA and in the "economic theory of index numbers".

fulfill certain axioms.[38] A pair of Laspeyres and Paasche indices, for example, satisfies the product test (despite violating the factor reversal test), since the product $P_{0t}^{L}Q_{0t}^{P}$ as well as $P_{0t}^{P}Q_{0t}^{L}$ is equal to V_{0t}.[39]

Reversal tests as "finders of formulae": Fisher used the notion of "time antithesis" P* of an index P, that is P* = T(P) and "factor antithesis" P** of the quantity index Q (that is P** = F(Q) which is a price index)[40] and the concept of a double antithesis (see fig. 2.2.3).[41] Another idea of Fisher was to cross (average) a price index P (a corresponding result applies to a quantity index Q) to one of its three antithesis in order to find a new price index formula. A special relationship exists in the case of a geometric mean for any P in that

(2.2.26) $\tilde{P} = \sqrt{P \cdot T(P)}$ meets the time reversal test, $\tilde{P}_{0t} = 1/\tilde{P}_{t0}$, and

(2.2.27) $\tilde{P} = \sqrt{P \cdot F(Q)}$ along with (2.2.27a) $\tilde{Q} = \sqrt{Q \cdot F(P)}$

is factor reversible: $\tilde{P}\tilde{Q} = V$. The special case of crossing with a *geometric* mean is called *rectification*. In this sense P_{0t}^{F} is a rectification of the Laspeyres- and Paasche formula.

The following relationships hold for the antithesis relation:

1. they are *reflexive* T(T(P)) = P, and F(F(P)) = P, and
2. the *symmetry* between time - and factor antithesis relation T(F(P)) = F(T(P))

Note that by using equation 2.2.27 it is always possible to construct a pair of index formulas $\tilde{P}$ and $\tilde{Q}$ conforming with factor reversibility. Often, however, it is difficult enough to find a meaningful interpretation for such index formulas, like $\tilde{P}$ and $\tilde{Q}$. For example the index

(2.2.28) $$Q_{0t}^{GK} = \frac{\Sigma q_t \left(\frac{p_0 p_t}{p_0 + p_t} \right)}{\Sigma q_0 \left(\frac{p_0 p_t}{p_0 + p_t} \right)}$$ is the direct quantity index of the Geary - Khamis

type and the geometric mean of its factor antithesis and P_{0t}^{GK} is the following price index (according to eq. 2.2.27)

---

[38] We may define the factor reversal test also as the condition requiring direct index and implicit (co-factor) index to be equal.

[39] The product test criterion facilitates greatly the finding of an appropriate quantity index Q, because Q can also be defined implicitly (i.e. by the division V/P) as opposed to a explicit or direct definition (i.e. an index derived by the same underlying reasoning as applied in case of the price index, or simply by interchanging prices and quantities in the formula of the price index).

[40] Accordingly $F(P_{0t})$, the factor antithesis of P_{0t}, is a quantity index.

[41] In this sense notably the Laspeyres formula is the "time antithesis" and also the "factor antithesis" of the Paasche formula and vice versa.

$$\tilde{P}_{0t}^{(GK)} = \sqrt{\frac{\Sigma p_t q_t}{\Sigma p_0 q_0}\left(\frac{\Sigma p_t\left(\frac{q_0 q_t}{q_0+q_t}\right)\Sigma q_0\left(\frac{p_0 p_t}{p_0+p_t}\right)}{\Sigma p_0\left(\frac{q_0 q_t}{q_0+q_t}\right)\Sigma q_t\left(\frac{p_0 p_t}{p_0+p_t}\right)}\right)} = \sqrt{\frac{\Sigma p_t q_t}{\Sigma p_0 q_0}(A)},$$

and the formula $\tilde{Q}$ according to eq. 2.2.27a is simply $\tilde{Q}_{0t}^{(GK)} = \sqrt{\frac{\Sigma p_t q_t}{\Sigma p_0 q_0}\left(A^{-1}\right)}$.

Hence the interesting finding of equations. 2.2.27 and 2.2.27a is in pratice of limited use only.

Figure 2.2.3: Crossing P with an antithesis as finder of formulas

crossing = average of P and an antithesis of P	→ special case: **rectifying**

time antithesis P = T(P)	factor antithesis P = F(Q)	double: F(T(Q)) =T(F(Q))	take the (unweighted) **geometric** mean

time (2.2.23) $P_{0t}^{*} = T(P_{0t}) = 1/P_{t0}$	$P_{0t}^{*} = T(P_{0t})$ is the time antithesis of P_{0t}	P^L is $T(P^P)$ since $P_{0t}^{L} = 1/P_{t0}^{P}$ and vice versa $P^P = T(P^L)$ as $P_{t0}^{P} = 1/P_{0t}^{L}$
factor (2.2.24) $P_{0t}^{**} = F(Q_{0t}) = V_{0t}/Q_{0t}$	the price index $P_{0t}^{**} = F(Q_{0t})$ is the factor antithesis of Q_{0t}	due to $V_{0t} = P_{0t}^{L}Q_{0t}^{P} = P_{0t}^{P}Q_{0t}^{L}$ we have: $P^L = F(Q^P)$ and vice versa $P^P = F(Q^L)$
double (2.2.25) $P_{0t}^{*} = F[T(Q_{0t})] = T[F(Q_{0t})]$	P_{0t}^{*} is the double antithesis of Q_{0t}	P^L is the double antithesis of Q^L, and P^P is the double antithesis of Q^P

We look now at some well-known **"second generation formulas"** to check whether they pass the time (or country) reversal test and the factor reversal test. The result is given in **fig. 2.2.4.** It is well known that the Laspeyres and Paasche formula are antithetic in both respects, and it easily follows that Fisher's index meets both tests since[42]

42 In addition **fig. 2.2.4** once more reveals some interesting relationships among second generation indices like those mentioned in **fig. 2.2.1**, for example $Q^{AH} = Q^P$, $Q^{HH} = Q^L$, and $Q^{HB} = 1/Q^{PA}$ relating

$$P_{0t}^{F} = \sqrt{P_{0t}^{L}\,P_{0t}^{P}} = \frac{1}{P_{t0}^{F}} \text{ and } V_{0t} = P_{0t}^{F}Q_{0t}^{F} = P_{0t}^{F}\sqrt{Q_{0t}^{L}\,Q_{0t}^{P}} = \underbrace{\sqrt{P_{0t}^{L}\,Q_{0t}^{P}}}_{\sqrt{V_{0t}}}\,\underbrace{\sqrt{P_{0t}^{P}\,Q_{0t}^{L}}}_{\sqrt{V_{0t}}}\,.$$

Figure 2.2.4: Time reversal test and factor reversal test of second generation indices (additive block only)

weighting schemes: I = p_0q_0; II: p_0q_t; III: p_tq_0 and IV: p_tq_t

arithmetic means (A-) and weights (I → IV)			
A - I: P_{0t}^{L}	A - II: P_{0t}^{P}	A - III: P_{0t}^{AH}	A - IV: P_{0t}^{PA} *
↓	↓	↓	↓

time antithesis			
harmonic means (H-) and weights (IV → I)			
H - IV: P_{0t}^{P}	H - III: P_{0t}^{L}	H - II: P_{0t}^{HH}	H - I: P_{0t}^{HB}
↓	↓	↓	↓

factor antithesis		
$P_{0t}^{L} \rightarrow Q_{0t}^{P} = Q_{0t}^{AH}$	$P_{0t}^{P} \rightarrow Q_{0t}^{L} = Q_{0t}^{HH}$	formulas not useful

Hence the following relationships between arithmetic and harmonic indices hold

price indices			
A-I: P^L	A -II: P^P	P^{AH}	P^{PA}
P^{HB}	P^{HH}	H-III: P^L	H-IV: P^P

quantity indices			
A-I: Q^L		A-III: Q^P	
	H-II: Q^L		H-IV: Q^P

* PA = Palgrave

As for the **third generation indices**, we should start with the following general observation: Unlike crossing weights the crossing of formulas does not lead to time reversible formulas (with the prominent exception of Fisher's index, however).

Palgraves index with the "harmonic base" index in the case of the direct quantity index (the same relation holds for the cofactor - quantity indices).

Index numbers with weights obtained by crossing of weights (e.g. quantity weights q_{i0} and q_{it} in a price index) such as the index formulas of Edgeworth-Marshall (P^{ME}), Walsh (P^{W}) or Geary Khamis (P^{GK}) always satisfy the *time* reversal test (since the weights are the same whatever the base period may be, 0 or t respectively), but not necessarily the *factor* reversal test.[43]

For the crossing of formulas take the formula of Drobisch as an example:

$P_{0t}^{DR} = \frac{1}{2}\left(P_{0t}^{L} + P_{0t}^{P}\right)$ does not meet the time reversal, and the factor reversal condition either. We get, not surprisingly $P_{t0}^{DR} = \frac{1}{P_{0t}^{HPL}} \neq \frac{1}{P_{t0}^{DR}}$. Thus the harmonic P^{HPL} - index and the arithmetic Drobisch index are a pair of time antithetic indices like the arithmetic Laspeyres and the harmonic Paasche index.

The cofactor (antithetic) quantity index of the Geary - Khamis type s given by

$$(2.2.28a) \qquad \Sigma p_t q_t \Sigma p_0\left(\frac{q_0 q_t}{q_0 + q_t}\right) \Big/ \Sigma p_0 q_0 \Sigma p_t\left(\frac{q_0 q_t}{q_0 + q_t}\right),$$

which differs from the direct quantity index (eq. 2.2.28) such that factor reversibility is violated. In the same manner the indices of Walsh and Marshall - Edgeworth confirm with the time reversal but not with the factor reversal test: the direct quantity index and the cofactor (antithetic) quantity index are different. The product

$$P_{0t}^{W} \cdot Q_{0t}^{W} = \frac{\sum p_t \sqrt{q_0 q_t}}{\sum p_0 \sqrt{q_0 q_t}} \cdot \frac{\sum q_t \sqrt{p_0 p_t}}{\sum q_0 \sqrt{p_0 p_t}}$$

does not equal the value index. The same is true for the product

$$P_{0t}^{ME} Q_{0t}^{ME} = V_{0t} \frac{\left(1 + \frac{1}{Q^P}\right)\left(1 + \frac{1}{P^P}\right)}{\left(1 + Q^L\right)\left(1 + P^L\right)}.$$

It may be interesting to use the theorem of L. v. Bortkiewicz (**sec. 1.3** eq. 1.3.12) to describe the extent to which the indices of Laspeyres and Paasche fail the reversal tests. We get

$$P_{0t}^{L} P_{t0}^{L} = \frac{P^L Q^L}{V} = \frac{P^L}{P^P} = 1 - \frac{C}{V} \text{ and } P_{0t}^{P} P_{t0}^{P} = \frac{P^P Q^P}{V} = \frac{P^P}{P^L}.$$

We also see that

$$\frac{P^{DR} Q^{DR}}{V} - 1 = \frac{1}{2}\left[\frac{1}{2}\left(\frac{P^L Q^L}{V} - 1\right) + \frac{1}{2}\left(\frac{P^P Q^P}{V} - 1\right)\right] = \frac{1}{2}\left[\frac{1}{2}\left(-\frac{C}{V}\right) + \frac{1}{2}\left(\frac{C}{V - C}\right)\right],$$

such that in fact the relative deviations of the indices L and P are averaged. The deviation will not vanish in general when an average is taken, though it will be reduced *substantially* in amount. The following numerical example will illustrate these results.

[43] The same is true for index numbers using any constant weights (the same in numerator and denominator of course) *not depending on periods 0 and t* (like Lowe's prices index, or the Cobb Douglas index).

Table 2.2.3: Factor reversal test in the case of **third** generation indices
TR = time reversal test, FR = factor reversal test

name of index	(direct) quantity index	cofactor quantity index	TR	FR
		Crossing of **formulas** P^L and P^P		
Drobisch	$Q^{DR} = \frac{1}{2}\left(Q^L + Q^P\right)$	$Q^{HPL} = \left(Q^F\right)^2 / Q^{DR}$	no	no
Fisher	$Q^F = \sqrt{Q^L Q^P}$	$Q^F = \sqrt{Q^L Q^P}$	yes	yes
harmonic (HPL)	$Q^{HPL} = \left(Q^F\right)^2 / Q^{DR}$	$Q^{DR} = \frac{1}{2}\left(Q^L + Q^P\right)$	no	no
		Crossing of **weights***		
Marhall - Edgeworth	$\frac{1}{1+P^L}Q^L + \frac{P^L}{1+P^L}Q^P$	$\frac{V}{V+P^L} + \frac{P^L}{V+P^L}Q^P Q^L$	yes	no
Walsh	$\frac{\sum q_t \sqrt{p_0 p_t}}{\sum q_0 \sqrt{p_0 p_t}}$	$\frac{\sum p_t q_t}{\sum p_0 q_0} \frac{\sum p_0 \sqrt{q_0 q_t}}{\sum p_t \sqrt{q_0 q_t}}$	yes	no

*as for the Geary - Khamis formula see equations 2.2.28 and 28a

Example 2.2.1

Assume the following indices are given: P^L_{0t}=1.5, P^P_{0t} = 1.2 and V_{0t} = 1.75. Calculate the indices gained from crossing the Laspeyres and Paasche indices and verify the relations valid for factor and time reversibility. Consider the *factor reversal test* first:

It follows $Q^P_{0t} = V_{0t} / P^L_{0t}$= 7/6 = 1.167 and $Q^L_{0t} = V_{0t} / P^P_{0t}$ = 1.4583.

Thus the "crossed" indices are

index	price P_{0t}	quantity Q_{0t}	product $P_{0t}Q_{0t}$, deviation
Laspeyres	1.5	1.4583	2.1875 (125% of V_{0t} = 1.75)
Paasche	1.2	7/6	1.4 (80% of V_{0t} =1.75)
Drobisch	1.35	1.3125	1.771875 > V_{0t} = 1.75
Fisher	$\sqrt{1.8}$ = 1.3416	1.3044	1.75 = V_{0t}
harmonic	1.8/1.35=1.333	1.29696	1.7284 < V_{0t} = 1.75

P^L/P^P = 1.25 hence P^L/P^P - 1 = +0.25 and C = -0.25V = -0.4375. In the same manner we get P^P/P^L -1 = - 0.2 or using the result for the covariance C/(V-C) = - 1/5 = P^P/P^L -1. Drobisch's index fails the factor reversal test by only 1.25% because $\frac{P^{DR}Q^{DR}}{V} - 1$ = +0.0125. This is half of the mean relative deviation of P^LQ^L (+ 0.25) and P^PQ^P (- 0.2) since (0.25 - 0.2)/2 = 0.025 and half of this is 0.0125. Furthermore $P^{HPL}Q^{HPL}/V$ = 1.7284/1.75 = 0.98765 = V/ $P^{DR}Q^{DR}$ = 1.75/1.771875 such that the pair of HPL-indices has a relative deviation of -1.23% only since 0.98765 - 1 = - 0.01235.

Now consider the *time reversal test*: From the time-antithesis-relationship we obtain $P^L_{t0} = 1/P^P_{0t}$ = 5/6 and $P^P_{t0} = 1/P^L_{0t}$ = 2/3. Hence the crossed formulas yield the following results

Index	price index P_{0t}	price index P_{t0}	product $P_{0t}P_{t0}$
Laspeyres	1.5	5/6	1.25 > 1
Paasche	1.2	2/3	0.8 < 1
Drobisch	1.35	0.75	1.0125 > 1
Fisher	$\sqrt{1.8}$ = 1.3416	$\sqrt{5/9}$ = 0.74536	1
Harmonic	1.8/1.35=1.333	0.74074	0.98765 < 1

Note that the sign and the amount of the relative deviation from unity now is exactly the same as in the case of the factor reversal test (deviation from the value ratio). The results are fully in line with our expectations with the properties of the arithmetic and harmonic mean in mind. ♦

In Fisher's theory the amount to which a pair of indices fails the time reversal test or the factor reversal test is an important criterion.[44] From this point of view indices like "Drobisch" or "HPL" are *much* better (relative deviations of only + 1.25% or - 1.24%) than Paasche and Laspeyres, both violating these tests to a greater extent (+ 25% and - 20%).

d) Digression: a weak variant of the time reversal test

It is obviously rather restrictive to require an index P_{t0} to be the inverse index P_{0t}. It appears sufficient to postulate:

(2.2.29) if $P_{0t} > 1$ then $P_{t0} < 1$ and if $P_{0t} < 1$ then $P_{t0} > 1$.

This requirement seems to be reasonable and not too ambitious: it is only desired that an increase in the direction $0 \to t$ should correspond to a decline in the opposite direction $t \to 0$ and vice versa. In the case of the Laspeyres price index this is equivalent to

if $P^L_{0t} > 1$ then $P^P_{0t} > 1$ and if $P^L_{0t} < 1$ then $P^P_{0t} < 1$

a property which does not seem to be too demanding but rather easy to ensure because $P^L_{t0} = 1/P^P_{0t}$ and $P^P_{t0} = 1/P^L_{0t}$. The following numerical example will show, however, that this is not necessarily the case.

Example 2.2.2

Assume the following prices and quantities

i	p_{i0}	p_{it}	q_{i0}	q_{it}
1	12	15	80	20
2	20	18	10	80

Calculate the following indices P^C (Carli), P^L, P^P, P^{DR} (Drobisch), each in both directions, that is $0 \to t$ and $t \to 0$.

The results are as follows:

[44] We found a similar reasoning (arguments we disagree with to be more precise) in writings of Mudgett and Banerjee.

formula	P_{0t} direction 0 → t	P_{t0} direction t → 0
Carli	P^{C}_{0t} = (1.25 + 0.9)/2 = 1.075 > 1	P^{C}_{t0} = 0.9555 < 1
Laspeyres	P^{L}_{0t} = 1380/1160 = 1.1897 > 1	$P^{L}_{t0} = 1/P^{P}_{0t}$ = 1.0575 > 1
Paasche	P^{P}_{0t} = 1740/1840 = 0.9457 < 1	$P^{P}_{t0} = 1/P^{L}_{0t}$ = 0.8406 < 1
Drobisch	P^{DR}_{0t} = 1.0677 > 1	P^{DR}_{0t} = 0.9490 < 1

Thus both the Laspeyres- as well as the Paasche formula may fail this weak time reversal test, whereas Carli's index and the index of Drobisch (or Sidgwick) will necessarily pass this test (though both indices do *not* satisfy the time reversal test), since

$P^{DR}_{t0} = \frac{P^{L}_{0t} + P^{P}_{0t}}{2P^{L}_{0t}P^{P}_{0t}} = \frac{P^{DR}_{0t}}{P^{L}_{0t}P^{P}_{0t}} = \frac{1}{P^{HPL}_{0t}}$ (where P^{HPL} is the harmonic mean index which is 1.0537 and necessarily smaller than the arithmetic mean index P^{DR} = 1.0677) or in the example

$$P^{DR}_{t0} = \frac{P^{DR}_{0t}}{P^{L}_{0t}P^{P}_{0t}} = \frac{1.0677}{1.125} \text{ which is} < 1 \text{ because}$$

$$P^{L}_{0t}P^{P}_{0t} > \tfrac{1}{2}\left(P^{L}_{0t} + P^{P}_{0t}\right) > 1 \text{ or } P^{L}_{0t}P^{P}_{0t} < \tfrac{1}{2}\left(P^{L}_{0t} + P^{P}_{0t}\right) < 1 .$$ ♦

e) Fisher's philosophy in evaluating formulas by reversal tests

Apart from the index function P^{F}_{0t} named in honour of Fisher[45], as a rule much emphasis was given to his "test" (or axiomatic) approach. To understand this approach better, it is useful to recall that Fisher published *two* books on index theory

- in 1911 "The Purchasing Power of Money" and
- in 1922 the famous book "The Making of Index Numbers".

In his first book he set out a system of in total eight tests, of which the two reversal tests constitute an important part only. The full system of Fisher's tests (axioms) will be discussed in **sec. 3.3**. In the course of time till his second book was published he not only changed his opinion on some of his tests (especially the circular test) but also his center of interest. While he initially focused on the algebraic properties of simple (mostly unweighted) formulas, his chief concern in his "Making of Index Num-

[45] It is perhaps a less well known fact that the formula was not created by Fisher himself who qualified it, however, as "ideal" (including quotation marks). In order to get a fair idea of Fisher's contribution to index theory it seems to be better to comment on his ideas on several occasions rather than to try an overall evaluation at this point of development. See **sec. 3.3** for his complete axiomatic system.

bers" was rather the search for new formulas and the behavior of formulas when applied to data. His focus changed to examine numerical results the formulas rendered when applied to data. This is nowadays often called a **"quantum theory"** of indices as opposed to a "pure" axiomatic theory of indices. At the same time two of his tests, the time reversal test and the factor reversal test gained a dominant position in his thinking. He evaluated all 134 formulas he either knew of or found by his systematic search with the help of these two tests and in doing so he performed numerous calculations with numerical data to reach some definite conclusions.[46]

In an attempt to provide a summary evaluation of all index formulas known to him he introduced (unfortunately) the concept of "*bias*", by which he meant the degree to which a formula violates the time and/or factor reversal test. A formula which violates the two reversal tests in any direction[47] depending on the data entering the formula is called "*erratic*". If the error (bias) is substantial or occurs in one direction only irrespective of the data, Fisher called the formula "*freakish*". As to the source of bias: Fisher made a distinction between a bias due to the type of mean and a bias resulting from the type of weight (I through IV) used. The problem was the combination of both sources which can well tend in opposite directions. Fisher conjectured that[48]

- harmonic means and weights I (p_0q_0) and II (p_0q_t) tend to produce a **downward** bias, while
- arithmetic means, especially if combined with weights III (q_tq_0) and IV (p_tq_t) are likely to be biased **upwards.**

The combination of two sources of bias caused some confusion. As an example, P_{0t}^{L} is an arithmetic mean (hence upwards biased) of weights type I (which should produce a downward bias). Another result Fisher arrived at with the help of the loads of numerical examples he used, and which he erroneously believed to be representative, was a classification of all his 134 index formulas according to a ***seven point scale*** ranging from worthless to superlative:

1. worthless, 2. weak, 3. correct, 4. good, 5. very good, 6. excellent and 7. superlative

examples: 5: Laspeyres and Paasche, 6: equation 2.2.11, 7: the two crossed indices P^F and P^{ME}

The basis of this classification was the extent to which the numerical result the formula in question differed from his ideal index P_{0t}^{F}. But clearly this deviation is influenced by both the formula as well as the data and it was repeatedly objectioned that the deviation can hardly be attributed to the formula alone.

46 In his times, when computers did not yet exist, a tremendous amount of calculating had to be performed. But in his own view and of some of his critics the ranking of index functions he arrived at was not fully satisfactory, with both respects, the method used and the results attained.

47 Such as the formulas of Laspeyres and Paasche depending on the sign of the correlation among price relatives and quantity relatives.

48 According to Köves (1983, p. 53) Fisher's statements concerning the weight bias are inconclusive. As a rule, in cases of negative correlation weights of type I will have a downward bias and those of type IV will have an upward bias respectively. General conclusions are difficult to be drawn, however, especially with respect to weights II and III.

Finally most of the criticism of Fisher's work of course concentrated on his test approach in general, on certain tests in particular,[49] and on his expectation to possibly obtain a definite "test winner" in the sense of a formula which (most unlikely)[50] combines all desirable properties we might think of. Furthermore Fisher's approach has often been criticized in that it attaches too much weight to formal aspects at the expense of criteria like availability of data (to enter the formula), understandability and ease of economic interpretation of the formula, aggregative properties and accounting for interdependence of prices and quantities. Nonetheless Fisher's "Making of Index Numbers" has of course been a milestone in the development of index theory.

f) The meaning and significance of Fisher's circular test

Circularity[51] (or the so called circular test, or "chain test") is highly restrictive and introduces an element of inconsistency when combined with other criteria. Thus it always has been the object of much controversy. Let 0, 1, 2 be some (not necessarily adjacent) periods in time, then the following relation should hold

(2.2.30) $P_{01} P_{12} = P_{02}$, and in connection with identity

(2.2.30a) $P_{01} P_{12} P_{20} = P_{00} = 1$,

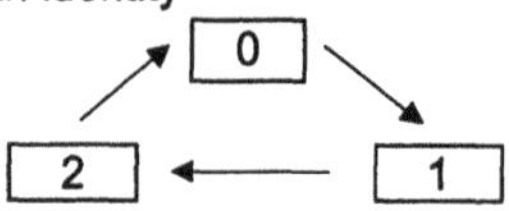

which explains the name "circular" test. Circularity should of course also apply to more than three periods, for example

(2.2.30b) $P_{01} P_{12} P_{23} = P_{03}$.

The name "circular" or "chain" illustrates one of the applications: by virtue to this criterion in intertemporal comparisons the separate year to year index numbers (or "links") can be joined together by successive multiplications like the links of a chain. The name chain index is misleading, however, since this type of index is not in line with the spirit of circularity[52].

Circularity has interesting interpretations in intercountry applications where it - like time reversibility – is gaining more and more importance as being a fundamental condition of consistent comparisons: all *direct* comparisons, eg. of the parity between country A and B (P_{AB}), or of volumes (Q_{AB}) will yield the same result as the *indirect*

---

[49] This applies in particular to the circular test (the position Fisher himself held with respect to this test was always found somehow contradictory), and with good reasons to Fisher's decision to rely heavily on his two reversal tests.

[50] It is often maintained that this inherent difficulty of a test- or axiomatic approach, makes this kind of approach to index numbers altogether useless. This is erroneous, however, a choice should be made (and cannot be made on formal grounds only) with respect to which criteria should be essential, and which less important or only desirable (see **sec. 3.1** for such a tentative ranking of axioms).

[51] The notion of "circularity" (or "transitivity") is also of some importance in the chapters 7 and 8.

[52] More precisely: chain indices are *not* transitive.

comparisons via any other country C, D or so and also indirectly with several links like C **and** D.

Fisher was initially a fan of the circular test but later he proposed to discard this test.[53] He even went further and rigoriously argued against it and called the test "theoretically a mistaken one" such that the fulfillment of the circular test "should really be taken as proof that the formula which fulfills it is erroneous".

The crucial point in the system of tests of Fisher is that he could not find any index formula that satisfies both the circular test and the factor reversal test simultaneously. Thus he already suspected (without proof) that adding the circular test for example to the factor reversal test (or vice versa) introduces an element of inconsistency. He conjectured that in order to meet the transitivity an index should have weights **neither** depending on 0 nor on t, but be constant throughout the whole interval of successive periods under consideration.[54]

Fisher was not right, however, in thinking that circularity can occur only if weights are kept constant. Variable weights are possible, but only if rather unusual functions, hardly to be viewed as "index" functions at all, are considered. A pair of indices which satisfies both the circular test *and* the factor reversal test is for example P^{RR} and Q^{RR} (see **sec. 3.2**). But the functions meet the factor reversal test only in a rather unorthodox way: price- and quantity index being identical $P_{0t}^{RR} = Q_{0t}^{RR}$, it is impossible to say what qualifies P_{0t}^{RR} to be a price index, and Q_{0t}^{RR} to be a quantity index. Hence P_{0t}^{RR} is more or less a "nonsense" index only.

The price **index of Lowe**[55] uses quantities q_i in both the numerator and denominator of the aggregate formula that are constant, not depending on the periods 0 or t (or any other in a series of index numbers P_{01}, P_{02} ...). The index P_{0t}^{LW} is defined as

(2.2.31) $$P_{0t}^{LW} = \frac{\sum p_t q}{\sum p_0 q}$$ (Lowe's price index).

The problem with Lowe's index is, however, that P^{LW} does **not** satisfy **commensurability** and therefore does not comply with the minimum requirements for an index. Furthermore there is no reasonable Lowe-quantity index and Lowe's approach is of no use for the deflation task. The question of the factor reversal test or of the product test is void of meaning in this case.[56]

---

[53] Often it is erroneously said that he did so when he noticed that his "ideal" index satisfied all of his tests except for the circular test. However, it was in fact mainly this test that introduced an element of inconsistency into his system of "tests" or "axioms".

[54] Lowe's index (eq. 2.2.31), or the Cobb-Douglas index (eq. 2.2.3) would be an example for this. Ironically just for such indices, not dependent on quantities, q_0 and q_t, the idea of factor reversibility could not apply.

[55] Joseph Lowe, perhaps the father of a "fixed-basket-thinking" proposed this formula in 1822. Fixed weights q (unequal q_0 or q_t) were also used by Scrope in 1833.

[56] Lowe's method of weighting P_{01} with quantities, neither q_1 nor q_0 has to be rejected according to Mudgett 1951, p. 20 on "logical grounds" (probably rather on economic grounds): "Exact weights for expressing the importance of prices must be either q_1 or q_0 because these q's are the only ones which are causally related to these p's".

g) A critique of circularity and time reversibility (Pfouts)

Pfouts 1966 vigoriously criticised Fisher's circular and time reversal test and Fisher's ideal index altogether. This will again be discussed in **sec. 5.4**. He has shown that circularity is tantamount to the requirement that a certain matrix **P** of index numbers has to be *singular*. **P** is defined as follows (in the case of T+1 = 4 rows and colums, t = 0,1,...,T)

$$P = \begin{bmatrix} P_{00} & P_{01} & P_{02} & P_{03} \\ P_{10} & P_{11} & P_{12} & P_{13} \\ P_{20} & P_{21} & P_{22} & P_{23} \\ P_{30} & P_{31} & P_{32} & P_{33} \end{bmatrix}.$$

Pfouts now claims that **P** should **not** be identically singular, because if **P** is singular (that is the determinant $|\mathbf{P}| = 0$) the following consequences are intolerable in his view

- a linear relationship must exist between index numbers having the same base and worse
- the same relationship applies to **all** base periods.

Thus singularity of **P** imposes "artificial restrictions" on index numbers "that typically would not be present in the empirical data", and this would rob index numbers of desirable flexibility. It is therefore desirable to have a *nonsingular* rather than singular matrix **P**. This rules out:

1. the circular test, which implies for any two periods, say 2 and 4 independence of the base because we then obtain $P_{24} = \frac{P_{04}}{P_{02}} = \frac{P_{14}}{P_{12}} = \frac{P_{34}}{P_{32}}$ and so on, and
2. the time reversal test (being implied by circularity in combination with identity).

Fisher's tests, however, tacitly assume **P** being singular. This can easily be seen since in the case of T = 2 we obtain:

$$P = \begin{bmatrix} P_{00} & P_{01} & P_{02} \\ P_{10} & P_{11} & P_{12} \\ P_{20} & P_{21} & P_{22} \end{bmatrix} = \begin{bmatrix} 1 & P_{01} & P_{01}P_{12} \\ 1/P_{01} & 1 & P_{12} \\ 1/P_{01}P_{12} & 1/P_{12} & 1 \end{bmatrix}$$

and the determinant $|\mathbf{P}|$ in fact vanishes: multiplication of row 1 by $1/P_{01}$ gives row 2, or multiplication of row 3 by P_{12} gives row 2. Hence **P** is singular ($|\mathbf{P}| = 0$).

A consequence is that a single additional value, P_{23}, is sufficient to calculate a fourth row and column, although we do not even have to know what index formula is being used, because the multiplication of **P** by a column vector **c** yields a column vector **p** which is

$$\mathbf{Pc}=\begin{bmatrix}1 & P_{01} & P_{02}\\ 1/P_{01} & 1 & P_{12}\\ 1/P_{01}P_{12} & 1/P_{12} & 1\end{bmatrix}\begin{bmatrix}0\\0\\P_{23}\end{bmatrix}=\begin{bmatrix}P_{02}P_{23}\\P_{12}P_{23}\\P_{23}\end{bmatrix}=\begin{bmatrix}P_{03}\\P_{13}\\P_{23}\end{bmatrix}=\mathbf{p}.$$

Thus it is simple to calculate an additional, fourth column (with elements P_{03}, P_{13}, P_{23} and 1) and a fourth *row* (with elements $1/P_{03}$, $1/P_{13}$, $1/P_{23}$ and 1) with *one* single observation, P_{23} only. There are actually only two independent observations, P_{01} an P_{12} assembled in the 3 × 3 matrix **P**. This rigid relationship appears to be too restrictive, at least in the case of intertemporal comparison.[57] It can easily be verified that time reversibility implies all second order principal minors of **P** are identically singular, hence determinants like $\begin{vmatrix}P_{00} & P_{01}\\ P_{10} & P_{11}\end{vmatrix}$ or $\begin{vmatrix}P_{33} & P_{35}\\ P_{53} & P_{55}\end{vmatrix}$ and so on will all vanish.

But Pfouts also rejects the time reversal test for other reasons. Since he primarily examines Laspeyres-type index functions, weighted arithmetic means of price relatives, a_{0t}^{i}, he points out this test also implies a peculiar condition on the weighting functions w to be admissible

$$\text{(2.2.32)}\qquad \sum_i w_{i0}a_{0t}^{i}\sum_i w_{it}\frac{1}{a_{0t}^{i}}=1,$$

where w_0 and w_t are weights applying to base period 0 and t respectively.

h) Some weak versions of the circular test, the base invariance test and base test

It very soon became apparent that circularity is difficult to attain. Hence attempts to weaken the condition of circularity have a long tradition. Such a test, called **base (period) invariance test** was already known to Edgeworth (1896) and Fisher (1911). This test is a weak version of the circular test and requires a ratio (function) f of two indices

$$\text{(2.2.33)}\qquad \frac{P_{02}}{P_{01}}=\frac{P(\mathbf{p}_0,\mathbf{q}_0,\mathbf{p}_2,\mathbf{q}_2)}{P(\mathbf{p}_0,\mathbf{q}_0,\mathbf{p}_1,\mathbf{q}_1)}=f(\mathbf{p}_1,\mathbf{q}_1,\mathbf{p}_2,\mathbf{q}_2)=f$$

not depending on prices and quantities of period 0, but only on prices and quantities of the two periods, 1 and 2 to be compared. Obviously the conclusion

if the circular test is met	⟶	then also the base invariance test

is valid, but the converse is not true. If circularity is met, we simply have f = P_{12}, which does not depend on prices and quantities of periods other than 1 and 2. However the test is weaker than circularity because f( ) need not have the same functional form as P( ) has. On the other hand, in this test the tight restrictions of the circular test are relaxed only *slightly*. It is difficult to find an example of an index function satisfying the base invariance test but not at the same time the circular test. Both indices, the

[57] In multinational price comparisons transitivity has considerable importance, however, see chap.8.

Laspeyres and Paasche index, are not capable of satisfying even this weakened version of the circular test, since

$$\frac{P_{02}^{L}}{P_{01}^{L}} = \frac{\sum p_2 q_0}{\sum p_1 q_0} = f_L(...) \text{ and } \frac{P_{02}^{P}}{P_{01}^{P}} = \frac{\sum p_2 q_2}{\sum p_0 q_2} \frac{\sum p_0 q_1}{\sum p_1 q_1} = f_P(...)$$

the function f_L does not depend on $\mathbf{q}_1$ and $\mathbf{q}_2$ but on $\mathbf{q}_0$ instead, and f_P on the other hand depends on $\mathbf{p}_1$, $\mathbf{p}_2$, $\mathbf{q}_1$ and $\mathbf{q}_2$ as required, but also on $\mathbf{p}_0$ and $\mathbf{q}_0$ not in agreement with the test. It also follows that Fisher's index does *not* meet this test, since $f_F = \sqrt{f_L f_P}$.

The base invariance test is weakened further by the so called **base test** (Eichhorn/Voeller 1983, p. 432, Olt 1996, p. 47) according to which one of the following equations should hold

(2.2.34) either: $P(p_0, q_0, p_1, q_1) f(p_1, q_1, p_2, q_2) = P(p_0, q_0, p_2, q_2)$

or: $f(p_0, q_0, p_1, q_1) P(p_1, q_1, p_2, q_2) = P(p_0, q_0, p_2, q_2)$.

2.3 The stochastic approach in price index theory

The name "stochastic approach" is misleading in that this type of analysis does not really deal with sampling problems (sampling error due to the fact that only a part of the universe of prices has been observed). A distinction should be made between the traditional (19th century) notion of "stochastic" in the context of index theory and the revival of the stochastic approach in our days, i.e. in more recent publications, like articles of Clements and Izan (1987) or the monograph of Selvanathan and Prasada Rao (1995). In what follows we will first describe the focus and some underlying ideas of this approach, then demonstrate some applications, and conclude with a criticism of this sort of index theory.

a) Objectives of the stochastic approach

The stochastic approach as well as the axiomatic approach is regarded as "atomistic" or "formal" in the sense of treating prices and quantities as independent variables and not accounting for (economic) relations that may exist among them. Whilst the axiomatic approach concentrates on (hidden) mathematical properties of the index formula and its implications when applied to some stylized conditions (axioms),

- the old stochastic approach (OSA, see **fig. 2.3.1**) as opposed to the so called "aggregative" approach is known for its advocating an unweighted index[58] as a measure of the not directly observable general price level; while

[58] This issue (weighted vs. unweighted) was a controversy in the early theory of price index numbers, which nowadays has somewhat fallen into oblivion. One group of thinking aimed at a broad objective, a statistically defined general level of prices (as the reciprocal of the purchasing power of money) without specific reference to a group of consumers. The other school referred to the aggregative ex-

- at the heart of the new stochastic approach (NSA) is the idea that index functions are to be derived as parameters (regression coefficients) in regression equations and that in the light of empircal data the "goodness of fit" of the equation is a justification for the index formula in question (hence data should influence decisions on conceptual issues such as the choice of an index formula).

The name "stochastic" or "statistic" (or sometimes also "mechanic") approach is used in this context because both OSA and NSA focus on the assessment of index functions in terms of distributions of price relatives, the variability of the results of index calculations, the coefficient of determination of a regression and similar statistical concepts. The central point in the OSA is the mean of the distribution of price relatives. No recourse is made to "consumers", and much less to their preferences, whilst the aggregative approach typically provides a "changing cost of a basket" - interpretation with explicit reference to a particular subject consuming (e.g. household).[59] The problem with OSA always was that it was far from clear how to define the "frequencies" in the distribution of price relatives $x_i = p_{it}/p_{i0}$. Does for example $f(x_i = 1.2) = 0.1$ mean that in 10% of all commodities the price went up by 20%? This would mean that the result would depend on the degree of detail the commodity classification offers. If we would make distinctions among 100,000 commodities instead of 10,000, the frequency belonging to x_i = 1.2 would possibly be 7% or 9% or so instead of 10%. If, on the other hand, frequencies are related to expenditure shares, what is then the difference between OSA and the aggregative approach?

Edgeworth followed in his pioneer work in the 1880s the stochastic approach, favoring the median[60] and the *unweighted* geometric mean of price relatives.[61] The economic justification of OSA was the hypothesis that mainly monetary factors, such as changes in the quantity of money, are responsible for changes in the value of money, and they would influence, according to Edgeworth, each price to the same extent.[62]

J.M. Keynes became a supporter of the aggregative view,[63] favoring a *weighted* mean of price relatives. In this view an index should show the changing (in practice mostly rising) costs of buying a certain, well defined constant basket (or a fixed budget) of goods and services.

---

penditure of a specified group of consumers defining a subjective value of money (or real income). Ragnar Frisch called the first the stochastic and the other the aggregative approach.

[59] Note that introducing weights denoting the different importance of various goods and services in the concept of the distribution can also be viewed as reconciliation of stochastic and aggregative approach, and that the Laspeyres formula, as well as the Paasche formula can be interpreted in both ways (Fisher's index or all sorts of chain indices as an example are unable of either interpretation).

[60] It is interesting that this idea only recently re-gained some significance in the context of the measurement of the so-called "core inflation" (**sec. 6.2**).

[61] The reason was that it satisfies the (time) reversal condition, an aspect which (unfortunately) even today is found most important by lots of index theoreticians.

[62] To summarize: the price level was conceived as "objective", inflation as a result of (nonspecific) monetary factors as opposed to the "subjective" notion of the aggregative approach,with its reference to the expenditures of a specific consuming unit.

[63] "... purchasing power must always be defined with reference to a particular set of individuals in a given situation" Keynes said in "A Treatise of Money" Vol. I, London 1930, p. 54 quoted by Allen, p. 7.

Figure 2.3.1: The place of the stochastic approach in index theory

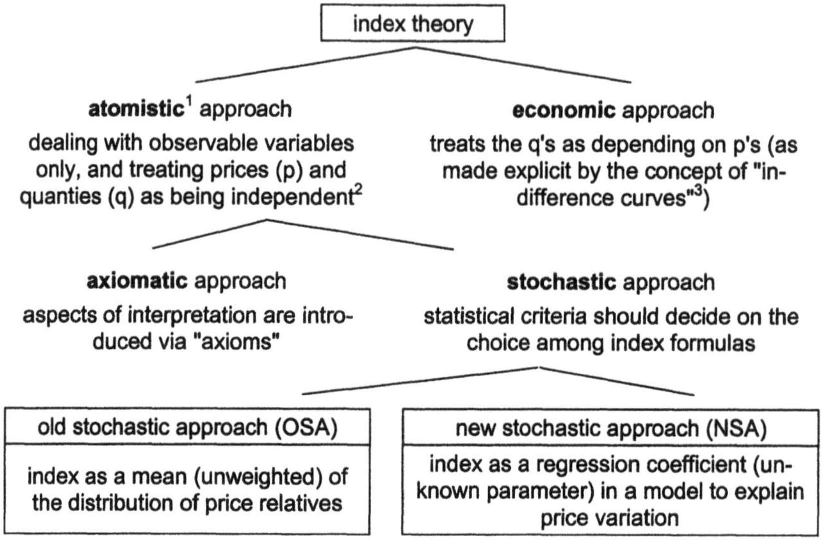

1) also called "formal" or "mechanic" (mechanistic) and the like

2) variations in q in response to variations in p are captured indirectly in the NSA since in some regression models expenditure shares are involved

3) or preference orderings, a purely theoretical and nonobservable concept of economic theory

The common conviction and expectation of those who favor the new stochastic approach (NSA) is that it should be possible and useful

1. to show how different index formulas can be derived as regression coefficients in the context of different stochastic models (hoping that this will lead to a better understanding of index formulas);
2. to provide an "estimation error"[64] associated with the point estimate "index", the source of "error" being the variability of price relatives[65], and finally
3. that this "sampling error" might give an indication of the appropriateness of the index function in question.

As to no. 2 it may in fact appear to be somewhat unsatisfactory that index compilations end up only with a single figure without providing any measure of its reliability which of course is a function of the dispersion in the individual price relatives that enter into the index formula. There certainly is a need for a kind of "level of confi-

[64] i.e. the variance of the estimate "index" from sample data.

[65] or of course of quantity relatives if the method is applied to quantity indices.

dence"[66] to be attached to the value of an index and this level will certainly be different when, as an example, the formula of Laspeyres instead of Paasche is applied to the same data. The question

- whether the variation between price relatives can be viewed as a source of a "sampling" error in the familar sense of this concept, and even if so whether
- the resulting goodness of fit of a regression function really is a reliable guide for determining the best index formula,

casts some doubts as to the the validity of the aforesaid elements no. 2 and 3 of the NSA concept. To sum up

The centerpiece of the new stochastic apoproach (NSA) is to regard the standard errors of the estimates of various index formulas as caused by the dispersion in the (sample) price data and the index formula in question. This error is seen as a characteristic of the index formula, and can be taken as an indication for the "reliability" or appropriateness of the index function. It is our view, however, that the underlying conception of the relationship between data and index formulas is a misconception. It is inadmissible to interpret the results of an index calculation when the formula is applied to these data in terms of appropriateness of the formula.

Before entering into a criticism of the rationale of the NSA in evaluating index formulas we should briefly present some results of this stochastic modelling of index functions.

b) Applications: some simple regression equation models for index formulas

The basic assumption in a model for an unweighted average of price relatives like Carli's index is that "each observed price relative is a reading on the index of overall price changes 'contaminated' by the random term" (Selvanathan and Prasada Rao (1995), p. 48), such that

(2.3.1) $p_{0t}^{i} = \theta_t + \varepsilon_{it}$,

where $p_{0t}^{i} = p_{it}/p_{i0}$ and ε_{it} is an error term with zero-expectation $E(\varepsilon_{it}) = 0, \forall t$, a common variance $V(\varepsilon_{it}) = \sigma_t^2$, and uncorrelated over the n commodities, that is, it possesses a zero-covariance $C(\varepsilon_{it}\varepsilon_{jt}) = 0$ for any two commodities i and j respectively. The parameter θ_t is interpreted as the common trend in all prices (general price level in t as compared to 0, due to monetary expansion for example).

It is clear that *Carli's* index (unweighted arithmetic mean of price relatives) will be the least squares (LS) estimator of θ_t in the model of eq. 1, that is $\hat{\theta}_t = \sum p_{0t}^{i}/n = P_{0t}^{C}$.

By the same token we get *Jevons'* index (unweighted geometric mean of price relatives)

[66] It is maintained that the traditional index theory is only able to provide point estimates whilst the NSA is superior in delivering an interval estimate in addition to the point estimate.

(2.3.2) $\quad Dp^i_{0t} = \ln(p^i_{0t}) = \theta_t + \varepsilon_{it}$,

where θ_t is again the common trend in all prices (not depending on i), and eq. 2.3.1 applies again to the error term ε_{it} (reflecting also aspects related to the commodity i not only to t). The LS estimator now is given by $\hat{\theta}_t = \ln\{(\prod p^i_{0t})^{1/n}\} = \ln(P^{JV}_{0t})$, and according to Clements and Izan (1987) the sampling variance of this best linear unbiased estimator $\hat{\theta}_t$ is

(2.3.2a) $\quad V(\hat{\theta}_t) = V(\varepsilon_{it})/n = \sigma^2_t/n$

where σ^2_t can be estimated unbiasedly as follows

(2.3.2b) $\quad \hat{\sigma}^2_t = \frac{1}{n-1}\sum(Dp^i_{0t} - \hat{\theta}_t)^2$.

We now quote a finding of Clements and Izan (C&I)[67] and Selvanathan and Prasada Rao (S&P), according to which a simple modification concerning the error term in eq. 2.3.2 (replacing ε_{it} by $u_{it} = \varepsilon_{it}/\sqrt{\overline{w}_{it}}$) will lead to the *Törnqvist* index $P^T_{0t} = \prod_i (p^i_{0t})^{\overline{w}_{it}}$ where i = 1, ..., n and $\overline{w}_{it}$ is the arithmetic average of two expenditure shares of commodity i (see below), that is $\overline{w}_{it} = (w_{i0} + w_{it})/2$. Consider

(2.3.3) $Dp^i_{0t} = \ln(p^i_{0t}) = \theta_t + u_{it}$,

where $u_{it} = \varepsilon_{it}/\sqrt{\overline{w}_{it}}$, and such that $E(u_{it}) = 0$ for all i and all t, and $V(u_{it}) = \sigma^2_t/\overline{w}_{it}$ regarding the assumptions concerning the error term ε_{it} (i.e. expectation $E(\varepsilon_{it}) = 0$, a common variance σ^2_t). Note that strictly speaking $\overline{w}_{it}$ cannot be treated as a constant since it depends on prices p_{it} (and quantities) being random variables. The generalized least squares (GLS) estimator of θ_t is said to be

(2.3.3a) $\quad \hat{\theta}_t = \ln(P^T_{0t})$ with an estimated sampling variance amounting to

(2.3.3b) $\quad \hat{\sigma}^2_\theta = \frac{1}{n-1}\sum \overline{w}_{it}(Dp^i_{0t} - \hat{\theta}_t)^2$.

Hence $\ln(P^T_{0t})$ can be derived from a model which is linear in the log changes (logarithms of price relatives), and the confidence interval of $\ln(P^T_{0t})$ should be calculated using eq. 2.3.3b.

It should be added that in addition to the derivation of P^T in terms of a regression model with an error term u_{it}, we can also motivate P^T as the overall mean of logarithmic price reatives $Dp_i = \ln(p_{it}/p_{i0})$ resulting from a sampling procedure such that the probability of i to be drawn is $\overline{w}_{it}$.

It may be noted in passing that C&I also consider the situation in which we have logarithmic links $Dp^i_{t-1,t} = \ln(p_{it}/p_{i,t-1})$ instead of the log changes Dp^i_{0t} as defined

[67] Clements and Izan (1987).

above. In this case the deriviation corresponding to eq. 2.3.3 would result in the Divisia index according to C&I.

Figure 2.3.2: Some "Budget share weighted average models" (BSW models)

General structure of the model $y_{it} = \theta x_{it} + u_{it}$, where u_{it} is a function of ε_{it} and ε_{it} fulfills the standard assumptions concerning $E(\varepsilon_{it}), V(\varepsilon_{it}) = \sigma_t^2$, and $C(\varepsilon_{it}\varepsilon_{jt})$, the other symbols are defined in the text

index formula[1)]	y-variable	x-variable	error term u	$(n-1)\hat{\sigma}^2_{\hat{\theta}}$ [2)]
Laspeyres	$p^i_{0t}\sqrt{w_{i0}}$	$\sqrt{w_{i0}}$	$\varepsilon_{it}\sqrt{w_{i0}}$	$\sum w_{i0}(p^i_{0t} - \hat{\theta}_t)^2$
Paasche	$p^i_{0t}\sqrt{w^\bullet_{it}}$	$\sqrt{w^\bullet_{it}}$	$\varepsilon_{it}\sqrt{w^\bullet_{it}}$	$\sum w^\bullet_{it}(p^i_{0t} - \hat{\theta}_t)^2$
Törnqvist $\ln(P^T)$	Dp^i_{0t}		$\varepsilon_{it}/\sqrt{\overline{w}_{it}}$	$\sum \overline{w}_{it}(p^i_{0t} - \hat{\theta}_t)^2$
Jevons $\ln(P^{JV})$	Dp^i_{0t}		ε_{it} [3)]	$\frac{1}{n}\sum(Dp^i_{0t} - \hat{\theta}_t)^2$

1) parameter $\hat{\theta}$

2) $\hat{\sigma}^2_{\hat{\theta}}$ denotes the estimated sampling variance of the regression coefficient $\hat{\theta}$.

3) variance $V(\varepsilon) = \sigma^2$

The model of eq. 2.3.3 already represents so called budget share weighted average models (BSW) according to S&R.[68] Some well known traditional index formulas may be derived from such BSW models, for example the Laspeyres and the Paasche formula. The following budget (expenditure) shares should be defined

- $w_{i0} = p_{i0}q_{i0} / \sum p_{i0}q_{i0}$, and correspondingly $w_{it} = p_{it}q_{it} / \sum p_{it}q_{it}$,
- $w^\bullet_{it} = p_{i0}q_{it} / \sum p_{i0}q_{it}$ as a "hybrid" share, and
- $\overline{w}_{it}$ as an average budget share (as defined above).

Fig. 2.3.2 shows how we may think of special BSW-models depending on what is assumed for x, y and u in a general regression equation which reads as follows

68 Selvanathan and Prasada Rao 1995, they also studied so called "expenditure based regression models" (EBR)).

(2.3.4) $y_{it} = \theta_t x_{it} + u_{it}$.

The sampling variance in the case of $\ln(P_{0t}^{JV})$ can also be expressed as $\sum \tilde{w}_i \left(Dp_{0t}^i - \hat{\theta}_t\right)^2$ with weights $\tilde{w}_i = 1/n$ for all i = 1, ..., n.

Note that the sampling variance as the basis for choosing among index formulas is in some cases related to P_{0t}^*, in other cases referring to the logarithm of the indexes,that is to $\ln(P_{0t}^{**})$, which makes comparisons difficult. The criterion on which we will focus therefore is the length of the (transformed) confidence interval.

C&I also considered models which allow for systematic changes (captured by the parameter ψ_i) depending on the type of commodity (i) such that we get[69]

(2.3.4a) $y_{it} = \theta_t + \psi_i + u_{it}$ instead of for example $y_{it} = \theta_t + u_{it}$.

In order to get a better understanding of the strengths and weaknesses of the NSA it may now be appropriate to proceed by presenting some calculations with a numerical example.

Example 2.3.1

The following data for n = 5 commodities are given

i	p_{i0}	p_{it}	q_{i0}	q_{it}	w_{i0}	w_{it}	w_{it}^*	$\overline{w}_{it}$	p_{0t}^i
1	10	12	40	35	0.200	0.1750	0.1392	0.1875	1.200
2	15	20	30	25	0.225	0.2083	0.1491	0.2167	1.333
3	20	15	10	32	0.100	0.2000	0.2545	0.1500	0.750
4	30	25	20	30	0.300	0.3125	0.3579	0.3063	0.833
5	20	20	17.5	12.5	0.175	0.1042	0.0994	0.1396	1.000

The definitions of the expenditure shares w_{i0}, w_{it}, w_{it}^*, and $\overline{w}_{it}$ are given in the text above. The price relatives are $p_{0t}^i = p_{it}/p_{i0}$. Calculate the following indices: P_{0t}^L, P_{0t}^P and P_{0t}^T with their respective (estimated) sampling variances.

We get the following results: $P_{0t}^L = \frac{2080}{2000} = 1.04$ further $P_{0t}^P = \frac{2400}{2515} = 0.9543$ and

$P_{0t}^T = 0.99753, (\ln P_{0t}^T = -0.00247)$.

[69] With this model C&I again claim to come close to the Divisia index (obviously seen as some kind of ideal).

To get the error of the estimate P^L_{0t} (a regression coefficient) we have to calculate $(n-1)\hat{\sigma}^2_\varepsilon$ as given in fig. 2.3.2 by[70] $\sum w_{i0}(p^i_{0t} - \hat{\theta}_t)^2 = 0.2\ (1.2\text{-}1.04)^2 + 0.225\ (1.333\text{-}1.04)^2 + ...+ 0.175\ (1\text{-}1.04)^2 = 0.045983$, such that we have $\hat{\sigma}^2_\theta$ after dividing this result by $n - 1 = 4$ (giving 0.01150). And then following eq.2.3.2a we get

(2.3.5) $\hat{\sigma}^2_{PL} = \hat{\sigma}^2_U = \hat{\sigma}^2_\varepsilon/(n-1)$ such that $\hat{\sigma}_{PL} = 0.1072$, a term permitting the calculation of a 95% confidence interval (t = 2.78) with the lower and upper bounds 0.7419 and 1.3381 respectively giving a length of the interval of 0.5962, which is possibly the best (and only) measure to compare the goodness of fit across widely different models.

index $\hat{\theta}_t$	$\hat{\sigma}_U$	bounds of confidence interval*	length of conf. interval
Laspeyres $P^L_{0t} = 1.04$	0.1072	0.7419; 1.3381	0.5962
Paasche $P^P_{0t} = 0.9543$	0.1071	0.6565; 1.2521	0.5956
logarithm of Törnqvist $\ln(P^T_{0t}) = -0.00247$	0.1058	-0.2967; 0.2917	
retransformed $P^T_{0t} = 0.9977$		0.7433; 1.3388	0.5955
logarithmic Jevons $\ln(P^{JV}_{0t}) = 0.0$	0.1072	-0.2981; 0.2981	
retransformed $P^{JV}_{0t} = 1$		0.7422; 1.347	0.6051

*lower and upper bounds, where $t_{n-1,0.95} = 2.78$; hence $\hat{\theta}_t \pm t \cdot \hat{\sigma}_\theta$.

In a similar manner we get the error (standard deviation of the sampling distribution) of the estimate P^P_{0t} (Paasche) and $\ln(P^T_{0t})$ (logarithmic Törnqvist index) from which we can get after retransformation P^T_{0t} (and for comparison Jevons' index too).

As to the point estimates we have

$P^P_{0t} = 0.954 < P^T_{0t} = 0.998 < P^J_{0t} = 1 < P^L_{0t} = 1.04$,

and as for the length of the confidence interval we get: the smallest is 0.5955 (P^T_{0t}), with only a tiny difference from the second index, P^P_{0t}. ♦

[70] This is simply the variance of the price relatives weighted with w_{i0} (budget shares of the base period).

From a practical point of view the result is most disappointing:

> Interestingly the point estimates (indicating a change from - 4.6% to + 4%) differ a lot more than just those figures from which we expect some guidance in assessing the "reliability" of the formulas to be compared, that is the lengths of the confidence intervals.

In other words, the indices among which a choice has to be made are more different than expected, showing that it *does* matter which formula is taken. On the other hand the standard errors which intend to measure the appropriateness of a formula in the NSA framework show little difference. Thus the NSA may be of little, if any, help as there is almost no difference between the standard errors.

c) Some questions concerning the general programme of the stochastic approach

The starting point of our criticism is the vaguenesses concerning terms like "random", or "sampling error" and the like which on the other hand are central in the NSA. Secondly we form the conjecture that the increase in our understanding of index formulas will be astoundingly small once indices are conceived as parameters of regression functions, and thirdly the NSA is grounded on an interpretation of the relationship between index functions, data and estimates which is erroneous.

Variation and sampling

It is true that whenever the prices of n commodities change disproportionately such that we get a non-zero variance of price relatives, different index formulas will produce different estimates of the same "general" change of prices. On the other hand if all prices would change exactly equiproportionally[71] there would be no change in the structure of prices and of course no index problem could arise. We cannot take it for granted, however, that

- the dispersion observed in the sample can be viewed as produced by a random variable (the price relative itself or another dependent variable[72] in a regression function due to some sort of an error term) or being the result of a (random) sampling procedure, and that
- a sampling error calculated on this basis and interpreted as arising from having observed only a part of the universe of prices indicates different degrees of "validity" of index formulas, which is, however, primarily a conceptual problem rather than a data problem.

The problem of viewing the i-th price relative as "random" is that the i-th commodity is in general not an independent replication of the j-th commodity. Furthermore

[71] To assume this is the same as if there were only one single commodity, a situation in which no aggregation problem would exist, and therefore no index problem either.

[72] As shown in **fig. 2.3.2** most of the dependent variables are functions of the (logarithms of) price relatives and expenditure shares.

genuine random samples of commodities are the exception if not non-existent in the practice of price statistics. Note that it is *not* the impact of inaccuracy or sampling error in the data[73] but the effect of the variability in price relatives on the result of the index calculation which is the major concern of the stochastic approach.

In fact C&I themselves made a distinction between their "errors" resulting from the dispersion of price relatives (in the sample *as well as* in the universe) and two "other sources of errors in conventionally employed price indexes", one of them being[74] "the error caused by partial sampling of the individual prices". Such sources "are analytically distinct from the problem studied in this article, whereby the source of the sampling error in the estimate of inflation is the disproportionate movement of the individual prices" (C&I, 1987, p. 340).

In our view it is not clear how a genuine "sampling error" can emerge when it is doubtful that a sample has been drawn at all, let alone that this "error" should be able to provide a basis for ranking formulas which primarily reflect *conceptual* differences.

In C&I we find the example with three commodities showing log changes Dp_{0t}^{i} as follows:

	Dp_{0t}^{1}	Dp_{0t}^{2}	Dp_{0t}^{3}
case 1	0.10	0.00	0.05
case 2	0.05	0.05	0.05

Of course we have a dispersion in case 1 as opposed to case 2 where all three logarthmic price relatives are identical. So in case 1 different index formulas will not yield the same result (unlike case 2 where whatever formula is applied, the index will yield exp(0.05) = 1.0513, that is a rise of 5.13%). It is true that conventional index calculation fails to take into account that the cases are different, but it rather presents one single figure as a result in both cases. However, does the dispersion in case 1 give rise to a *sampling* error,[75] and does it allow us to decide which formula is better than which other formula on the basis of different numerical results?

New insights in the rationale of index formulas? Lack of (conceptual) justification of regression equations

For Selvanathan and Prasada Rao (1995) the attraction of this approach lies in that it provides a new interpretation of some well-known index formulas that can be derived from various regression models. The "interpretation" is given by demonstrating in which kind of regression equation the index formula occurs, and which kind of as-

[73] If this were the case the focus of this stochastic apoproach should be on conceiving better sampling procedures rather than on using the sampling error for making choices among index formulas.

[74] The other one being the "substition bias", which is irrelevant in this present context.

[75] There is certainly reason to be more confident with the point estimate in case 2 as compared to case 1, but it is doubtful that an "error" in this situation can legitimately be viewed as a "sampling error".

sumptions are made concerning the error term (random variable). It is doubtful however that the inherent strengths or weaknesses of a particular index *formula* will appropriately be revealed by viewing an index as a regression coefficient.

What can we learn for example about the rationale of the price index of Laspeyres when P_{0t}^L is recognized as the point estimate $\hat{\theta}_t$ in the following equation

(2.3.6) $$p_{0t}^i \sqrt{w_{i0}} = \theta_t \sqrt{w_{i0}} + u_{it},$$

where $u_{it} = \varepsilon_{it} \sqrt{w_{i0}}$ and thus $V(u_{it}) = V(\varepsilon_{it}) w_{i0}$ holds for the variances. What does it mean that the error term varies in proportion to the square root of the budget share $\sqrt{w_{i0}}$?

It should be noted that we encounter the opposite assumption too, i.e. an error term u_{it} of i in t depending *inversely* on the budget share[76] in the case of the Törnqvist index where it is assumed that $u_{it} = \varepsilon_{it} / \sqrt{\overline{w}_{it}}$ and thus $V(u_{it}) = V(\varepsilon_{it}) / \overline{w}_{it}$.

Given equation 2.3.6: can we interpret $\sqrt{w_{i0}}$ in one sense or another as "causing" or "explaining" the dependent variable $p_{0t}^i \sqrt{w_{i0}}$? How can we make plausible why possibly only slightly modified data will make a modell $\sqrt{w_{it}^*}$ (x-variable) $\rightarrow p_{0t}^i \sqrt{w_{it}^*}$ (y-variable), better fitting the data? In the NSA each index formula is viewed as parameter of a model. However, an evaluation of a formula should not be completed by inventing a regression model to which the fomula fits. The model should also be discussed with respect to its being realistic, economically meaningful and the like.[77]

When the use of the stochastic approach is limited to demonstrating in which model which index formula appears as a parameter, without trying to give a meaningful interpretation to a specific model (in contrast to other models) we cannot expect new insights in the interpretation of index formulas. Moreover without such an interpretation the inference from the sampling error of Θ to the appropriateness of the index formula Θ does not appear reasoned.

---

[76] C&I justify this assumption as follows: "Thus the variability of the relative price of a good having a large budget share, such as food, will be lower than of a commodity with a smaller share, such as cigarettes. This is a plausible specification, since there is less scope for a relative price to change as the commodity in question grows in importance in the budget, (implicitly assuming that volatility is necessarily great/small when the budget share of a commodity is great/small)". Now if this is plausible the opposite assumption made in the case of Laspeyres and Paasche should be particularly implausible.

[77] There are, however, not many examples in NSA known to us in which much care and sophistication has been given to this task. It is fairly generally true that if there is not much that can be said about the rationale of the assumptions of a model, we also can't learn much from the model. Moreover as some of them appear even somewhat farfetched there is not much to be gained from this kind of approach.

Figure 2.3.3: Main ideas of the new stochastic approach

Data		index **function**		**results** of calculations
n different price relatives	→	different index formulas Θ_1, Θ_2... applied	→	different formulas will yield different results*)
variability of price relatives gives rise to a disturbance term	→	index function as regression coefficient Θ (explanation?)	→	the standard error (S.E.) of the estimate of Θ (criterion for model-fit)

Main conclusions and rules of NSA
1. understand "index" by identifying it as a parameter in a regession equation
2. take standard error as an indication of the "reliability" or appropriateness of the index function Θ (prefer Θ_1 to Θ_2 when the S.E. of Θ_1 is smaller than of Θ_2)

*) It is the essence of the NSA method that the direction of this last arrow can be inverted in that the variability of the results permits an assessment of the index formulas.

The underlying methodology of the NSA (fig. 2.3.3)

Looking at fig. 2.3.3 we may conclude that the central idea of the NSA is that the direction of the first row of arrows can be inverted: as different index formulas applied to the same data cause a variability in the results, which will in turn result in a better or poorer fit of the model, so we can conclude in a reverse direction: a smaller or greater "error" in estimating the regression coefficients Θ_1, Θ_2 ... (representing different index functions) will be a proof of superiority or inferiority of the formula in question. This kind of reasoning is fundamentally wrong:

Evidently different index formulas (Θ_1, Θ_2, ...) will produce different results even when applied to the same data unless all price relatives are equal. But this variability of results does not in itself permit an evaluation of the index functions which is basically a conceptual matter. Strengths or weaknesses of different index formulas (even if viewed as parameters of different regression functions) will be revealed by discussing the properties of the formula Θ rather than by measuring the goodness of fit of the regression function in which Θ is a coefficient.

There is a causal relation between data D (variation in price relatives) and formulas F on the one hand and variation in the results of index calculations R when different index formulas are used. Arrows ($D \rightarrow F \rightarrow R$) indicate this in **fig. 2.3.3.** To invert the direction of the arrow $F \rightarrow R$ to $F \leftarrow R$ does not make sense. There are essentially two questions provoked by this kind of reasoning and they are to our knowledge unanswered:

1. Is the standard error a measure of the appropriateness of the index formula?

2. To what extent this error reflects characteristics of the index formula and to what extent it is affected by the data?

Relevance for the practice of official statistics

Another question of some relevance for the practice of official statistics is how the results of the stochastic approach should be used to decide when a choice has to be made between different index formulas. How to decide when the empirical results are inconclusive or even contradictory? Should we decide to use Törnqvist's index (P^T) to deflate consumption aggregates and to use Stuvels index (P^{ST}), for example, to deflate investment aggregates because P^T had a smaller standard error than P^{ST} when the P^T- model was applied to the consumption price data and correspondingly the P^{ST}- model produced a smaller standard error as compared with the P^T- model when applied to investment data? Would this be satisfactory or even feasible, taking into account that the situation concerning the empirical data may change from time to time? By the same token: it would be fully in line with the NSA reasoning to measure inflation for example in the period 1996 to 2000 using the Laspeyres formula P^L then in the interval 2000 to 2003 the Paasche formula P^P and so on. There can be no doubt that such a procedure would be found intolerable by most of the users of official price statistics.

2.4 Elements of the economic theory of index numbers

> The economic theory approach "... is a useful approach but it must not be followed very far".[78]
>
> "Many price statisticians find the economic approach to the determination of an appropriate functional form for the consumer price index to be overly formalistic and intuitively implausible".[79]
>
> "Indeed many people ... are quite uncomfortable with the idea that the introduction of Viagra should reduce the CPI".[80]

The so called "economic theory of index numbers" claims to provide a microeconomic foundation of index functions based on an optimising behaviour of economic agents (households, producers etc.).[81] The "formal" index theories (axiomatic and

[78] Allen, 1975, p. 44.

[79] W. Erwin Diewert, The Consumer Price Index and Index Number Purpose (1999) Paper presented at the Fifth Meeting of the International Working Group on Price Indices (The Ottawa Group) Reykjavik, Aug. 25 - 27, 1999, revised Sept. 1999, p. 18.

[80] At What Price (Schultze Report), Washington 2002, p. 67.

[81] The standard argument against other approaches is that they suffer from a general lack of theoretical foundation. Note, however, that nobody has ever asked for a microeconomic foundation of the covariance for example and that the way the so called "economic theory of index numbers" gives a theoretic underpinning is far from convincing and no less, if not more formal than the index theories (or "approaches") which it purports to be able to improve and replace. We are going to show that the COLI approach loses much of its attraction once the assumptions are made explicit and some "practical" questions are considered.

stochastic approach) have been criticised for being focused primarily on merely formal aspects, treating prices and quantities as separate independent variables and disregarding microeconomic relations between them. Hence, it is claimed that in order to derive and justify index functions one has to take into account the typical functional relations between prices and quantities as explained by microeconomic models, such as

- the utility function an individual household faces in the case of an "economic theory" consumer price index (CPI), the so called "true cost of living index" (COLI), and
- if the theory is applied to (output and input) prices of a producer, i.e. to a (theoretic) producer price index (PPI), the role of the utility function is played by the production function (representing a given technology).

There is also an "economic theory" approach in the case of an output or "production index", a quantity rather than a price index.[82] To cover both utility as well as production functions, Diewert coined the term "aggregator" function.

In general the "economic" approach offers an additional solution to the problem of choice of the index formula. For some authors[83] the outstanding result of this economic approach is to recommend certain symmetrically weighted index formulas ("symmetry" refers to using quantities q_0 as well as q_t in a symmetric fashion) like P_{0t}^{F} (Fisher) P_{0t}^{T} (Törnqvist) or P_{0t}^{W} (Walsh) as "superlative"[84] indices (see below part e of this section). For them, the fact that these formulas also turned out to be preferable from the point of view of *a number of* quite different index-theories[85] gives them (additional) validity.[86]

a) Definition of the "true cost of living index" (COLI), introduction of the utility concept

It is a widespread view, promoted especially by the Boskin Commission (1996/7) that a COLI would be the best measure of inflation possible and can be approximated by

---

[82] There is an aggregation problem in all these applications of the economic index theory, consumer prices, producer prices, and production index (aggregation over economic agents, not aggregation in the sense of "aggregator" function) that has to be solved in order to derive demand functions, or supply functions respectively, since index numbers do not refer to individual households or producers but to an economy as a whole.

[83] For example Erwin Diewert, perhaps the most prominent and influential author in economic index theory.

[84] Note that the class of "superlative" indices (SI) is a sub-set of the symmetrically weighted (SW) indices. There is no SI which is not SW, but there exist SW indices that are not superlative (SI), for example the formula of Marshall and Edgeworth P^{ME}.

[85] This is true for the axiomatic and the stochastic approach as well as economic reasoning.

[86] That different approaches starting out with different concepts nonetheless end up in recommending the same formulas for indices is clearly a strong case for this position, if this in actual fact were the case. However, there are good reasons *not* to recommend P^{F} from the point of view of certain axioms (in particular those related to aggregation properties) or P^{T} in the framework of the stochastic approach.

a "superlative" index. Therefore official statistics should aim at such an index instead of a fixed weighted Laspeyres index. Interestingly some years later (2002) the Schultze Panel[87] adopted a much less enthusiastic position as regards the COLI.

The "true cost of living index" (COLI), Konüs-index[88], or "constant utility index" (CU-index) is defined as the ratio of the minimum expenditures required to attain a particular indifference curve (the same utility level) under two different price regimes.

Note that the minimum costs are not observable but defined as the solution of a constrained maxima/minima problem.

Note further that we have only *one* utility level so far under consideration. No comparison between different levels is involved and thus, no cardinalization of utility is required.

Assume a private household "possessing" a utility function $U_0 = f(q_{10},q_{20})$, having the possibilities to choose between two commodities, say 1 and 2. The utility function (preference order, indifference or preference map) assigns a certain utility level (welfare) U_0 to infinitely many combinations of quantities q_1 and q_2 at base time 0. It is graphically represented by the well-known "indifference curve" shown in **fig. 2.4.1**. Each specific combination of quantities q_{10} and q_{20}, determined by the base-period's prices p_{10}, p_{20} and the base period's outlay or costs C turns out to be a point on this curve.

In general it will be assumed that the function $U = f(\mathbf{q})$ has the following properties

1. a continuous[89] non-decreasing and concave[90] function in n well defined commodities (or "items") kept constant over time such that we have a positive

   and diminishing marginal utility $\frac{\partial U}{\partial q_i} = U_i^{'} > 0$ and $\frac{\partial^2 U}{\partial q_i^2} = U_{ii}^{''} < 0$, for all i = 1,..., n

   and $\frac{\partial^2 U}{\partial q_i \partial q_j} = U_{ij}^{''} > 0$ (in general more of good j [another good] increases the marginal utility of good i, such as more of i [the same good] entails a decrease of the marginal utility of i),
2. all goods are perfectly divisible and homogenous, and (most important)
3. all intertemporal aspects of the consumer's allocation are ignored (this applies to savings, consumer durables), as well as problems of the allocation of time (to work and leisure etc), and the same is true for a possible change of tastes, qualities, or the availability of the goods, the supply of public goods and envi-

[87] Charles L. Schultze, Chr. Mackie (Eds.), At What Price? National Research Council, Washington 2002.

[88] It is not uncommon to make reference to the Russian economist A. A. Konüs as having been the first to introduce this index approach in 1924.

[89] continuity is a more or less technical assumption.

[90] The function g(q) is concave if and only if $g(\lambda q_1+(1-\lambda)q_2) \geq \lambda g(q_1) + (1-\lambda)g(q_2)$ where $0 < \lambda < 1$.

ronmental conditions of all sorts (in all these cases possible changes, though relevant for the consumer's preferences, are not considered).

Figure 2.4.1: Indifference curve and cost minimization of a household in the micro-economic consumption theory

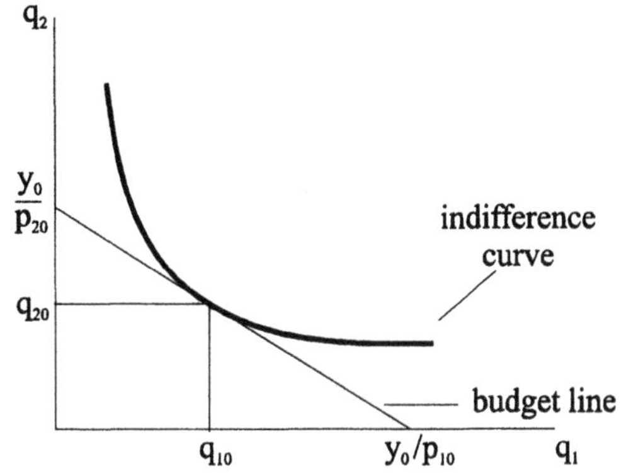

From above it follows that the indifference curve has a negative slope – dq_2/dq_1 given by the "rate of substitution" (of less q_2 by more q_1 or vice versa). The assumption of maximising the utility U under the constraint of balancing the "budget", i.e. the total expenditure (y_0) or costs (C_0)[91] or equivalently (the dual problem to the primal one, i.e. utility maximi-zing) of minimizing C_0 to attain a given utility level has a well known unique solution:

$$\left|\frac{dq_2}{dq_1}\right| = \frac{U_1'}{U_2'} = \frac{\partial U/\partial q_1}{\partial U/\partial q_2} = \frac{p_1}{p_2}$$

the ratio of marginal utilities is equal to the ratio of prices

or graphically: the tangential point in **fig. 2.4.1** determines uniquely the quantities q_{10} and q_{20}. Strictly speaking, the cost minimization $y = C(a, b) = C(U_a, \mathbf{p}_b) = \min\{\Sigma p_b q: f(q) = U_a\} = \Sigma p_b q$ means that the consumer (household)

[91] Thus the model of consumer choice reads as follows: max $f(q_{10}, q_{20})$ given $C_0 = p_{10}q_{10} + p_{20}q_{20}$ (the expenditure function, or "budget line" where the C_0 are kept constant and spent completely to the purchase of the quantities q_{10} and q_{20} (utility maximization) or written in its dual (equivalent) form (as done above) minimize $C_0 = p_{10}q_{10} + p_{20}q_{20}$ (by choosing quantities q_{10} and q_{20}) given $U(q_{10}, q_{20})$ and a given utility level U_0.

- has well defined preferences over combinations of n items (vector **q**) represented by a utility function f
- is facing the price vector $\mathbf{p}_b$ and
- is choosing a vector **q** such that his (absolute) costs (budget) y, or the cost C(a, b) to attain the utility level U_a are the smallest possible facing prices $\mathbf{p}_b$.

Note that this cost function $C(a,b) = C(U_a,\mathbf{p}_b)$ *defines* the true cost of living (price) index and is based on cost minimization. It is in general not an empirically observed expenditure but has to be inferred from the utility function and optimization behaviour. Nothing is said about how to estimate C(a, b) empirically.

In principle the COLI can refer to any reference utility level U; it is natural (and a well established convention), however, to choose two specific utility levels and hence quantity vectors, viz. U_0 and U_t respectively and to assume that a household

- facing prices $\mathbf{p}_0$ will as a rule minimize his costs in order to attain the level U_0 while
- in period t (prices $\mathbf{p}_t$) the household will minimise the costs of the utility level U_t.

Hence, the minimum costs C(0, 0) and C(t, t) are in principle observable *if* in fact the household is engaged in cost minimization while all other combinations such as $C(0, t) = C(U_0,\mathbf{p}_t)$ or C(t, 0), defined correspondingly are unobservable. Using those minimum expenditures (costs) the COLI- or "constant utility" (CU) –index (P^{CU} index) is defined as the ratio of the minimum expenditures, or "costs" C required to attain the same utility level U (or in particular U_0 or U_t respectively[92]) given price vectors $\mathbf{p}_t$ and $\mathbf{p}_0$:

$$(2.4.1) \quad P_{0t}^{CU}(U_0) = \frac{C(U_0,\mathbf{p}_t)}{C(U_0,\mathbf{p}_0)} = \frac{C(0,t)}{C(0,0)}, \text{ and } (2.4.2) \quad P_{0t}^{CU}(U_t) = \frac{C(U_t,\mathbf{p}_t)}{C(U_t,\mathbf{p}_0)} = \frac{C(t,t)}{C(t,0)}.$$

It should be noticed that

- this is a *definition* only, it does not provide a formula by which one might calculate the index in practical applications with empirical data, since the utility function is in general unknown, nor can it be assured that the *observed* quantities, a household h = 1, 2, ..., H consumes, will in actual fact represent an *optimal* (theoretical) position of h as required in the definition,
- as the (theoretical) quantities results from the utility maximization (or cost minimization) problem, the price index function is no longer given by

  (1.1.1) $\quad P_{0t} = P(\mathbf{p}_0,\mathbf{q}_0,\mathbf{p}_t,\mathbf{q}_t)$, but rather by

  (2.4.3) $\quad P_{0t}^{CU}(U_0) = P(\mathbf{p}_0,\mathbf{p}_t,U_0;U = f(\mathbf{q}))$ and in a similar manner

  (2.4.3a) $\quad P_{0t}^{CU}(U_t) = P(\mathbf{p}_0,\mathbf{p}_t,U_t;U = f(\mathbf{q}))$;

92 Some authors speak of the Laspeyres-Konüs and the Paasche Konüs index respectively.

eqs. 2.4.1/2 as well as 2.4.3/3a are often interpreted in the sense of the "pure price comparison", that is since the price vector is the only thing which is different in the numerator (pt) as compared with the denominator (p0),

- the CU-index does not only depend on price vectors but also on the utility function U(**q**) and the utility level in question, which means that the two indices in eq. 2.4.1 and 2.4.2 are not necessarily equal: it *does* matter which particular indifference curve is chosen from the indifference map,
- the definition in these equations refers to a *single household* facing the same utility function in both situations, 0 and t, and consuming on the *same* utility level, either U_0 or U_t, and we do not yet consider a comparison of different utility levels (more than one indifference curve involved),[93] or an aggregation over households or areas (regions) of an economy. We have not yet considered the *many household case* (see below part f of this section),
- a utility function is only determined up to a monotonic transformation: it is therefore no more justified to call a man's utility today equal to (or greater or smaller than) yesterday's utility than to make assumptions over the "amount" of utilities of two different men at the same time. Thus the economic theory of index numbers cannot provide intertemporal or interspatial *utility* measurements (resulting in "levels" of utility) and its statements and conclusions are to be understood as implications of *hypothetical* considerations only.

Recall that the COLI (as a *price* index) does not attempt to compare situations in which a different level of utility or even a different utility function in two periods, say 0 and t is involved. However, this no longer applies once we consider *the implicit quantity index* resulting from the COLI (P^{CU} index) as a price index, or in other words, the index that is derived by dividing the value index V_{0t} by the P^{CU}- index.

To summarize

> The Laspeyres price index compares expenditures at different price regimes referring to the **same quantities** whilst the constant utility (CU)-index P^{CU} compares expenditures referring to the **same level of utility** (thus allowing for substitution of q_{i0} (i = 1, ..., n) by other quantities q_{it} provided that both sets of quantities will offer the same utility level).

It should also be noted that the utility level matters in the sense that in general the CU-index does not only depend on the consumer's preference *function* (the *shape* of the indifference map) but also on the choice of the utility *level* (i.e. the *location* of the indifference curve selected) serving as reference point, or "basis" (that is either U_0 or U_t). Hence, in general,[94] it *does* matter which initial level of utility is chosen out of the

[93] In practice, there is a sequence of indifference curves U_0, U_1, ... shifting away from the origin as (income and) the utility level rises. This makes it necessary to distinguish substitution and income effects. There is no duality between an "economic theory" - price index (COLI) on the one hand and an "economic theory" - quantity index Q on the other hand. The latter can be defined with reference to the "product test" (that is as the "cofactor" of COLI, regarded as "price" index), and is sometimes called also "index of real consumption" or index of "real income". Q measures that rise of quantities induced by a change of prices that is necessary to maintain the same level of satisfaction.

[94] Unless the preferences are "homothetic" (see below).

sequence of indifference curves: the COLI depends on the utility function as well as the specific level of utility chosen as a reference level (say U_0 or U_t).[95] A rich man's COLI will be different from the COLI of a poor man.[96]

It can be argued that a similar situation applies to traditional price indices too, where a decision has to be made whether weights q_0 or q_t (or both) should be used, i.e. whether a Laspeyres or a Paasche approach should be preferred. By the way, a situation equivalent to homotheticity in the "basket approach" would mean $q_{it} = \lambda q_{i0}$ for all i (where λ may also equal unity).

Though this is clearly a limitation, it appears all too easy to understand[97]: if utility determines the hypothetical quantities on which the COLI is based, all influences which affect the level of satisfaction have to be considered, for example

- the level of income,
- demographic and environmental variables, and
- all other "side conditions" relevant for the level of satisfaction to be attained.[98]

This is entirely in line with, or rather a simple consequence of, the underlying concept of measurement, which is decidedly "subjective": the P^{CU} is the amount of income necessary to leave somebody *as well off as before* the price change took place. Quite often, but not necessarily, this amount will be greater in the case of a rich consumer than a poor consumer.

The more general formulation of the cost minimization problem in the COLI framework runs as follows: let **q** be a nonnegative consumption vector and $\mathbf{e} = [e_1,...,e_m]$ be a vector of quantities of (all!) "other" m variables that matter, called "environmental" variables (or demographic variables, or "public goods" variables) in addition to the consumption quantities of n goods in **q**.[99] Furthermore let f(**q**, **e**) denote the consumer's utility or preference function so that u = f(**q**, **e**) is the level of utility that the consumer achieves. Then the cost function

(2.4.4) $$C(u,\mathbf{p},\overline{\mathbf{q}},\overline{\mathbf{e}}) \equiv \min\{\mathbf{p}'\mathbf{q} : f(\mathbf{q},\overline{\mathbf{e}}) \geq f(\overline{\mathbf{q}},\overline{\mathbf{e}})\}$$

---

[95] In other words: P^{CU} depends not only on the consumer's preferences but also on the initial level of utility the consumer is starting from. The same rise or decline of prices will yield a different result depending on which (base) level of utility is chosen.

[96] Given that the utility function is not homothetic, then the same isolated increase in one price does not necessarily give rise to a higher increase in the COLI of a wealthy consumer (*higher* indifference curve as the common level of utility in *both* periods) as compared to the less wealthy consumer (lower indifference curve). The situation may well be the opposite, that is the rich man will suffer less from inflation than the poor man.

[97] Whatever the consequences for this conceptual framework regarding monetary policy may be.

[98] Here in particular all characteristics of the kind of civilisation in which the household is living are involved. An obvious criticism of the COLI approach is of course that utility depends not only on the quantities consumed of certain individual goods obtainable at the market.

[99] In the literature e is quite often called y, however, to avoid confusion we prefer e since we used y also to denote the budget (maximum possible expenditure) of the household. One of the problems with the COLI approach is that in practice it can be difficult to draw a clear line between the m goods in e (held constant in the conditional COLI) and the n goods in q (which are allowed to vary).

gives the minimum commodity cost $\mathbf{p}'\mathbf{q} = \Sigma pq$ required to achieve the reference utility level $f(\mathbf{q},\bar{\mathbf{e}})$ given that the e-vector is fixed at $\bar{\mathbf{e}}$. Note that the notation reveals that minimization of costs refers to **q** (in order to attain at least $\bar{\mathbf{q}}$), not to **e**, which is fixed as $\bar{\mathbf{e}}$. Since the reference utility level u is a function of $\bar{\mathbf{q}}$ and $\bar{\mathbf{e}}$ we may also write $C(\mathbf{p},\bar{\mathbf{q}},\bar{\mathbf{e}})$, and because the costs C (costs of constant quantities $\bar{\mathbf{e}}$ excluded) in eq. 2.4.1, 2 and 4 denote an inner product Σpq, they are linearly homogeneous in the elements of the nonnegative p-vector.

Finally the (due to the $\bar{\mathbf{e}}$ quantities kept constant) **"conditional" COLI price index** under the price regimes $\mathbf{p}_0$ and $\mathbf{p}_t$ is defined as $P_{0t}^{CU} = C(\bar{u},\mathbf{p}_t)/C(\bar{u},\mathbf{p}_0)$ where the "quantity" (or utility level) $\bar{u}$ in the numerator and denominator refer to the same environmental vector $\bar{\mathbf{e}}$ and to quantities of consumer goods **q**, which as a rule are *not* the same in the numerator and denominator, though they generate (in combination with $\bar{\mathbf{e}}$) *the same utility* level $\bar{u}$.

b) Upper and lower bounds for the COLI provided by the Laspeyres and Paasche price index formula

In view of the principle difficulties with a preference (utility) function (indifference curves) the economic theory of index numbers developed three major lines of research which proved to be more or less helpful in enabling the theory to be applied empirically, viz.

1. attempts to estimate empirical utility functions using econometric methods which, however, turned out prohibitively difficult and almost impossible,
2. to establish observable bounds, for example P_{0t}^{L} and P_{0t}^{P}, between which every COLI should fall whatever the underlying utility function may be, and
3. to study fairly general functions and derive the index functions (functional forms) that fit to them (or which are "exact" with respect to the function in question), i. e the approach searching for so-called "superlative index functions" (see part e of this section).

To turn to 2: research soon concentrated on inferring "bounds" for price index formulas actually used in practical price statistics. A well known result is the following statement:

The Laspeyres index P^L provides an upper bound to the theoretic index *if* the base time level of utility, that is U_0, is referred to. In the same manner the Paasche price index is a lower bound in the case of U_t, such that

(2.4.5) $P_{0t}^{CU}(U_0) \le P_{0t}^{L}$ (the COLI is bounded from *above* by Laspeyres), and

(2.4.6) $P_{0t}^{P} \le P_{0t}^{CU}(U_t)$ (bounded from *below* by Paasche), provided that indices P^L and P^P refer to a single utility maximizing household "possessing" indifference curves with utility levels U_0 and U_t respectively.[100] Recall that assuming

- a household facing prices $\mathbf{p}_0$ will as a rule minimise the costs to attain the level U_0
- while in period t (prices $\mathbf{p}_t$) the household will minimise the costs of the utility level U_t rather than U_0

amounts assuming that the (*theoretical*) *minimum* costs C(0,0) and C(t,t) are in principle observable, such that the denominator of P^L (that is C(0,0)) and the numerator of P^P (the minimum costs C(t,t)) are treated as empirically given.[101] Under such conditions the explanation of the findings in the issue of upper and lower bounds is quite simple:

P^L measures the expenditure to be incurred if the consumer buys the same quantities as before. In this case he is *by definition* (same quantities) as well off as before. However, by substituting products in a rational manner the consumer should be able to obtain a higher level of utility or the same level at less cost. Therefore, P^L provides an upper bound to the true cost of living index P^{CU} (but only if we refer to U_0, not to U_t).[102]

The proof (concerning the *upper bound*) is built on the following reasoning: Consider first the case[103] in which both prices rise with the same rate $\lambda = P_{0t}^L$ and suppose the income were to be increased λ-fold such that the amount $y_t = P_{0t}^L \sum p_0 q_0 = \lambda \sum p_0 q_0$ is available to the consumer at time t, thus enabling him/her to purchase exactly the same quantities q_0 as bought in the base period[104] and thus (by definition) resulting in at least the same level of satisfaction (or utility) U_0 as in the base period; it should be possible, however, to attain a higher level of utility with the same expenditure (or attain the same level at lower costs) due to rational substitution.

The consumption which (providing at least the utility U_0) could be purchased by spending y_t is given by the intersections of the new (parallel) budget line A'B' (parallel to AB) and the original indifference curve U_0, that is by the points P_1 and P_2 respectively (in **fig. 2.4.2**). It is possible, however, for a consumer disposing of an amount y_t of income to obtain a higher level of utility U_t. The quantities q_{1t} and q_{2t} are given by the point P_t in which the new budget line A'B' is tangent to the "higher" indifference

[100] Note that the upper bound (P^L) and the lower bound (P^P) of a CU-index refer to different levels of utility. The two inequalities can be combined into one, however, in the case of "homothetic preferences" (see below).

[101] Of course this is a rather "daring" though quite common assumption.

[102] This is due to the fact that the household maintains his level of utility but does *not* make use of the possibility to achieve the same level at lower costs by virtue of a rational substitution.

[103] It is in principle easy to generalize the result by considering an increase of prices with different rates. The assumption of a *uniform* price rise by λ amounts to considering the income effect in isolation (not disturbed by the substitution effect).

[104] Later such a λ-fold compensation for a price increase to ensure the purchase of the same quantities of goods will be called a *Slutzky*-type (as opposed to a *Hicks* type) of income-escalation.

curve U_t. It follows that the true cost of living index $P_{0t}^{CU}(U_0)$ cannot exceed P_{0t}^{L} since an increase of income proportional to P_{0t}^{L} would enable the consumer to be *at least* as well off as before.

The nature of the "results" of eqs. 2.4.5 and 6 is often misunderstood as providing some kind of new insight, or throwing light on the "correct" index formula. It is important to note that they are simply implications of the *definition* of the "cost function" (built on the assumption the household is acting in a cost minimizing way):

$C(U_0,\mathbf{p}_t) = C(0,t)$ is *defined* as the (non-observable) *minimum* cost to attain utility level U_0, given a price vector $\mathbf{p}_t$. Thus, we necessarily get $C(U_0,\mathbf{p}_t) \leq \Sigma p_t q_0$, and since C( ) is a *minimum* expenditure, no other [empirical] expenditure can be less than C. It also follows using $C(U_0,p_0) = C(0,0) = \sum p_0 q_0$

$$\text{(2.4.5a)} \quad P_{0t}^{CU}(U_0) = \frac{C(U_0,\mathbf{p}_t)}{C(U_0,\mathbf{p}_0)} = \frac{C(0,t)}{C(0,0)} = \frac{C(0,t)}{\Sigma p_0 q_0} \leq \frac{\Sigma p_t q_0}{\Sigma p_0 q_0} = P_{0t}^{L}.$$

Hence the upper bound is only an implication of the definition of *minimum* cost C and the assumption of cost minimising behaviour.

As, in general, there is nothing known about how observable (empirical) expenditures, like $\sum p_t q_0$, could be related to (theoretic) minimum costs, like C(t,0), it is impossible to find a bound lower than P^L to P^{CU} in default of a specified utility function.

In the same manner the Paasche bound is derived from assuming cost minimizing behaviour at time t so that $C(U_t,p_t) = \sum p_t q_t$, whereas $C(U_t,p_0) \leq \sum p_0 q_t$. It follows that

$$\text{(2.4.6a)} \quad \mathbf{P}_{0t}^{CU}(U_t) = \frac{C(U_t,\mathbf{p}_t)}{C(U_t,\mathbf{p}_0)} = \frac{\Sigma \mathbf{p}_t \mathbf{q}_t}{C(U_t,\mathbf{p}_0)} \geq \mathbf{P}_{0t}^{P} = \frac{\Sigma \mathbf{p}_t \mathbf{q}_t}{\Sigma \mathbf{p}_0 \mathbf{q}_t}.$$

P^L (as well as P^P) are the best (closest) bounds we can get. We cannot improve on them unless we make specific assumptions about the functional forms of the aggregator (utility) function.

c) "Homothetic" preferences and the "unit cost" function

It has been shown that in principle there are two ways to combine the two inequalities by which

(2.4.5) $P_{0t}^{CU}(U_0)$ is bounded from *above* by the observable Laspeyres index P_{0t}^{L} while

(2.4.6) $P_{0t}^{CU}(U_t)$ is bounded from *below* by the observable Paasche index P_{0t}^{P} into one single inequality,[105] viz.

- either to assume an intermediate utility level U* (as a kind of average of U_0 and U_t respectively) to which all indices involved (i.e. the indices $P^{CU}(U^*)$, P^L and P^P) apply, and therefore an intermediate consumption vector **q*** defined as the straight line joining the vectors $\mathbf{q}_0$ and $\mathbf{q}_t$ respectively, that is

[105] Both ways have already been demonstrated by A. A. Konüs.

(2.4.7) $\mathbf{q}^* = \lambda \mathbf{q}_o + (1-\lambda)\mathbf{q}_t$ where $0 < \lambda < 1$,

- or to assume "homothetic preferences" (or equivalently a utility function which is linear homogeneous in the quantities $q_1,...,q_n$ [or a monotonous transformation of such a utility function]).

A utility function, that is a system of indifference curves (IC) $f(q_1,...,q_n)$ is called **homothetic** (or **linear homogeneous** in the quantities such that $f(\lambda q_1,..., \lambda q_n) = \lambda f(q_1,...,q_n)$ if each IC is a uniform enlargement, or contraction of each other[106]. When total expenditure z increases λ-fold resulting in λz, then homotheticity means that all optimal quantities will also rise λ-fold to λq_1, λq_2, ... (which is clearly not a realistic assumption).[107]

A trivial case permitting a combination of the two inequalities into one single inequality is of course to assume the *same* utility level at time 0 and t (such that $U_0 = U_t$). It follows directly from the convex shape of the IC that[108] $P_{0t}^P \le P_{0t}^{UC}(U_t) = P_{0t}^{UC}(U_0) \le P_{0t}^L$, or more precisely, since under such conditions we get C(0,t) = C(t,t) and C(0,0) = C(t,0)

(2.4.8) $$P_{0t}^P \le P_{0t}^{UC}(U_t) = \frac{C(t,t)}{C(t,0)} = P_{0t}^{UC}(U_0) = \frac{C(0,t)}{C(0,0)} \le P_{0t}^L = V_{0t} = \frac{\Sigma p_t q_t}{\Sigma p_0 q_0} = \frac{C(t,t)}{C(0,0)}.$$

The consequence of "homotheticity" is that the absolute, or "normal" cost function $C(U = \bar{u}, \mathbf{p}) = C(\bar{u}, \mathbf{p})$ can be factorised in the utility level $\bar{u}$ and the unit cost function $c(\mathbf{p})$ as follows

(2.4.9) $C(\bar{u}, \mathbf{p}) = \bar{u}\, c(\mathbf{p})$, so that $c(\mathbf{p}) = C(\bar{u} = 1, \mathbf{p})$.

Consequently the COLI referring to the utility level $\bar{u}$ (and the same homothetic utility function in both periods, 0 and t) $P_{0t}^{CU} = C(\bar{u}, p_t)/C(\bar{u}, p_0)$ can be defined in terms of unit costs

(2.4.9a) $$P_{0t}^{CU} = C(\bar{u}, \mathbf{p}_t)/C(\bar{u}, \mathbf{p}_0) = \bar{u} \cdot C(1, \mathbf{p}_t)/\bar{u} \cdot C(1, \mathbf{p}_0) = c(\mathbf{p}_t)/c(\mathbf{p}_0),$$

without making reference to the utility level $\bar{u}$.

Furthermore time reversibility is given in the case of homotheticity. The COLI based on a non-homothetic utility function does not necessarily pass the time reversal test.

---

106 To be more specific: a utility function $f(\mathbf{q})$ is called neoclassical (nc) if it is linear homogeneous, positive and concave over the set $\mathbf{q}$ and a function F is called homothetic (h) if an increasing continuous function exists such that g(F) is neoclassical. It will not be too wrong to use both terms, nc and h as synonyms, and define h as linear homogeneous in the first place.

107 The assumption of homotheticity, however, is well known to be bluntly unrealistic because it amounts to requiring an income elasticity (with respect to the demand of all goods) to be equal to 1 or equivalently all income-consumption curves (Engel curves) to be rays from the origin.

108 assuming a convex shape of the IC; alternatively, if the IC were concave $P^L < P^P$ would result.

Figure 2.4.2: Income escalation at a rate of P_{0t}^{L} and a constant structure of prices

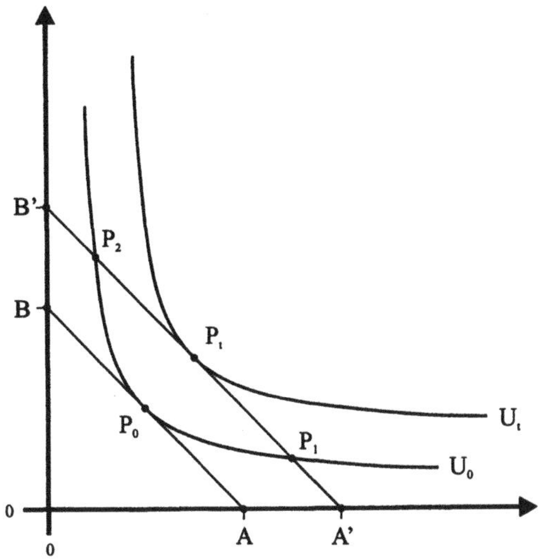

d) Example: the logarithmic Laspeyres index as an "exact"[109] index function

Before going any further it might be helpful to look at an example to clarify and illustrate the examination so far.

Example 2.4.1

The following example may show that compensation according to P^L may lead to a considerable overcompensation. Assume the utility function[110] $U = \sqrt{q_{1t} q_{2t}}$, for the two periods t=0 and t=1, or more general

(2.4.10) $U = q_1^{\alpha 1} \cdot q_2^{\alpha 2} \cdot \ldots \cdot q_n^{\alpha n} = \prod_i q_i^{\alpha i}$,

which is obviously homothetic $f(\lambda \mathbf{q}) = \lambda f(\mathbf{q}),\ \lambda \neq 0$ provided that the sum over the α coefficients is unity, as otherwise $\prod \lambda^{\alpha i} = \lambda$ would not hold. Assume the following cost-minimal quantities given the prices $\mathbf{p}_0$ and $\mathbf{p}_t$ (see points N_0 and N_t respectively in **fig. 2.4.3**):

[109] "Exact" means that an index function (e.g. the log-Laspeyres) corresponds to (one or more) specified utility functions such as the "Cobb-Douglas" utility function; in the case of the following example it corresponds to the log-Laspeyres index.

[110] Note that this specific utility function is "homothetic" theoretical indices $P_{0t}(U_0)$ and $P_{0t}(U_t)$ coincide.

	prices		quantities	
commodity	0	t	0	t
1	8	2	1	4
2	2	8	4	1

Calculate P^L, P^P, and P^{CU} and show that the suitable COLI index formula is the logarithmic Laspeyres index.

In both periods the household enjoys the same utility $U = U_0 = \sqrt{4} = 2$, such that substitution (reduction of $q_{20} = 4$ to $q_{2t} = 1$) appears to be not unacceptable. Costs (total consumption expenditures) amount to $C_0 = 8 \cdot 1 + 2 \cdot 4 = 16$ in 0 (N_0) and $C_t = 16$ in t (for N_t) respectively such that we get

COLI (P^{CU})	Laspeyres	Paasche
$P_{0t}^{CU} = 16/16 = 1$	$P_{0t}^{L} = 34/16 = 2.125$	$P_{0t}^{P} = 16/34 = 0.4706$

P^{CU} lies between P_{0t}^P and P_{0t}^L, and the result of $P_{0t}^{CU} = 1$ seems to be reasonable as an increase in p_1 is perfectly offset by a decrease in p_2. The budget line (restraint) in period 0, that is the straight line AB, is given by $q_{20} = 8 - 4q_{10}$ and shows all those combinations of quantities q_{10}, q_{20}, that result in expenditures of $8q_{10} + 2q_{20} = 16$. If the compensation were granted according to $P_{0t}^L = 2.125$ the utility could be more than redoubled allowing a consumption of $q_{1t} = 8.5$ and $q_{2t} = 2.125$ rather than 4 and 1 (see **fig. 2.4.4**). In this case costs amount to 34 instead of 16. The slope of the budget line $q_2 = 4.25 - 0{,}25q_1$ as well as the indifference curve $U_t = \sqrt{q_{1t}q_{2t}} = 4.25$ in $q_1 = 8.5$, i.e. in the point N_t^*, is given by $dq_2/dq_1 = -(4.25)^2/q_1^2 = -1/4$. ♦

For the "Cobb-Douglas" (CD) function 2.4.10 with coefficients α_i denoting expenditure shares rather than arbitrary constants (as usual in the case of CD), or more specificly

(2.4.10a) $$U = q_1^{\alpha 1} q_2^{\alpha 2} \quad (\alpha_1 + \alpha_2 = 1,\ 0 < \alpha_i < 1)$$

we get the marginal rate of substitution $-dq_1/dq_2$ (of good 1 to be substituted by 2)

(2.4.11) $$\frac{\partial U/\partial q_2}{\partial U/\partial q_1} = \left|\frac{dq_1}{dq_2}\right| = \frac{\alpha_2 q_1^{\alpha 1} q_2^{\alpha 2 - 1}}{\alpha_1 q_1^{\alpha 1 - 1} q_2^{\alpha 2}} = \frac{\alpha_2 q_1}{\alpha_1 q_2},$$

which is equal to the price ratio at the optimum

(2.4.12) $$\frac{\alpha_2 q_1}{\alpha_1 q_2} = \frac{p_2}{p_1} \text{ such that } \frac{p_1 q_1}{p_2 q_2} = \frac{\alpha_1}{\alpha_2}, \text{ and } \frac{p_1}{\alpha_1} = \frac{p_2}{\alpha_2} q_1 q_2 .$$

The α_i coefficients thus represent expenditure shares, and $\alpha_1 = \alpha_2 = 1/2$ in the example (N_0, N_t and N_t^*).

Figure 2.4.3: Consumer equilibriums N_0 and N_t

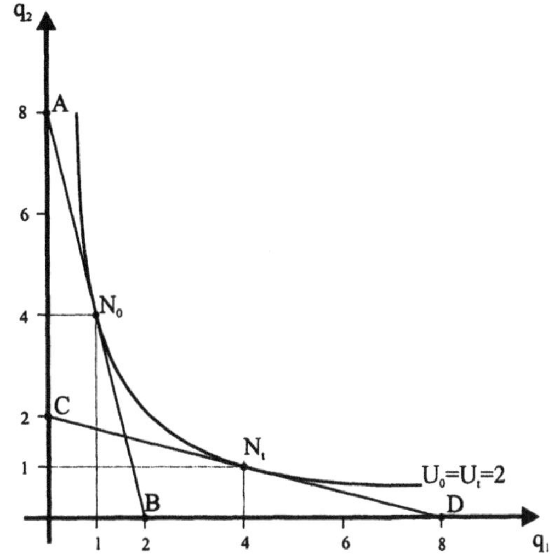

Eq. 12 is also derived from taking the total differential

$$dU = \frac{\partial U}{\partial q_1} \cdot dq_1 + \frac{\partial U}{\partial q_1} \cdot dq_2 = 0.$$

Using eqs. 10a and 12 in combination with $C = p_1q_1 + p_2q_2$ gives

(2.4.13) $$C = \left(\frac{p_1}{\alpha_1}\right)^{\alpha 1} \left(\frac{p_2}{\alpha_2}\right)^{\alpha 2} q_1^{\alpha 1} q_1^{\alpha 1} = \left(\frac{p_1}{\alpha_1}\right)^{\alpha 1} \left(\frac{p_2}{\alpha_2}\right)^{\alpha 2} U = C(U,p),$$ such that

(2.4.13a) $$c(p) = C(U=1,p) = \left(\frac{p_1}{\alpha_1}\right)^{\alpha 1} \left(\frac{p_2}{\alpha_2}\right)^{\alpha 2}$$ is the unit cost function.

With $p_{10} = 8$, $p_{20} = 2$, $q_{10} = 1$ and $q_{20} = 4$ we have total costs $C = 2\sqrt{p_1p_2q_1q_2} = 16$. Hence, the index function which "corresponds" to the utility function U of eq. 10a (the CD function), or which is "exact" for the CD utility function, is the log.-Laspeyres index

(2.4.14) $$P_{0t}^{CU} = \left(\frac{p_{1t}}{p_{10}}\right)^{\alpha 1} \left(\frac{p_{2t}}{p_{20}}\right)^{\alpha 2} \frac{U_t}{U_0} = \left(\frac{p_{1t}}{p_{10}}\right)^{\alpha 1} \left(\frac{p_{2t}}{p_{20}}\right)^{\alpha 2} = \frac{c(p_t)}{c(p_0)},$$

or in general $DP_{0t}^{L}=\prod_i \frac{p_{it}}{p_{i0}}^{\alpha_{i0}}$ such that $\ln\left(DP_{0t}^{L}\right)=\sum_i \ln\left(\frac{p_{it}}{p_{i0}}\right)\alpha_{i0}$, which explains the name "log. Laspeyres" since $P_{0t}^{L}=\sum_i\left(\frac{p_{it}}{p_{i0}}\right)\alpha_{i0}$.

Figure 2.4.4: "Compensation" according to P^L = 2.125

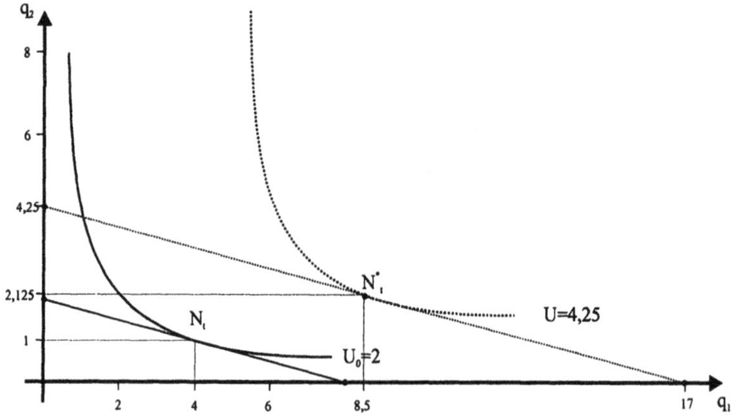

The more general result reads as follows: the "geometric" (or "log-change") indices (see **sec. 3.4** and **3.5**) $\prod_i\left(\frac{p_{it}}{p_{i0}}\right)^{\alpha_i}$ fit to (are "exact for") the Cobb-Douglas function $\prod_i q_i^{\alpha_i}$ where the α_i coefficients denote some kind of symmetric average m (arithmetic, or logarithmic mean etc.) of expenditure share so that $\alpha_i = m(s_{i0}, s_{it}), m(s_i, s_i) = s_i$. This covers log-change indices like Vartia II, Walsh II, Törnqvist as well as the extreme cases log. Laspeyres ($\alpha_i = s_{i0}$) and log. Paasche index ($\alpha_i = s_{it}$).

On the basis of the COLI-theory described so far, it may be useful to recall some concepts of textbooks in microeconomics and to explore their relationship to some theorems needed for proofs in the COLI framework. This is done in **box 2.4.1**.

We arrive from 1 to 2 (or 1 → 2 for short) by solving for q (q*), 2 → 3 by substitution of optimal q*'s in u( ) or y (giving C), 3 → 2 by differentiation of C with respect to p (Shephard's lemma) or of u(q*) with respect to prices (p) and the costs (y, or C), which is Roy's identity.

Box 2.4.1: Some microeconomic concepts used in COLI – theory

Utility function u(**q**), a specific utility level $\bar{u}$, the optimum quantities (labelled **q***) are gained from solving the optimization problem, costs denoted by y (providing a restraint) or C (in the case of minimum costs)

Level	Primal problem	Dual problem
1	max u(**q**) subject to $\sum p_i q_i \leq y$ (budget)[1]	min $\sum p_i q_i = y$ subject to $u(q) = \bar{u}$
2	solution: vector optimal quantities **q*** in terms of prices **p** and the budget y **q*** = g(**p**, y) is Marshall's demand equation (ordinary demand function)	$\mathbf{q}^* = h(\mathbf{p}, \bar{u})$ is known as "compensating" (or Hicksian) demand curve (a constant [given] utility attained at minimum costs)
3	substitution of **q*** in u(**q**) gives the indirect utility function, i.e. utility determined by prices and costs instead of quantities, so that direct u(**q**), indirect u(**q***) and $u(\mathbf{q}^*) = u(\mathbf{p}, y) = \bar{u}$ [2]	upon substitution of optimal q_i^* for the q_i in $y = \sum p_i q_i$ we get C = $\sum p_i q_i^*$, the cost function C = C(u, **p**) and unit cost function $c(\mathbf{p}) = C(u = 1, \mathbf{p})$ homotheticity means $C(\bar{u}, \mathbf{p}) = \bar{u} \cdot c(\mathbf{p})$

1) costs are henceforth denoted by y or also by C (minimum costs)

2) for the indirect utility function the cost function is the inverse function and vice versa

Roy's theorem: optimal demand quantities are gained from the indirect utility function u(q*) by differentiation with respect to prices (p) and the costs (y) giving

$$q_i^* = \frac{\partial u / \partial p_i}{\partial u / \partial y}.$$

Shephard's lemma relating optimal quantities to the cost function, using the derivatives of C with respect to prices

(2.4.) $q_i^* = \partial C(p,u) / \partial p_i$,

or in the homothetic case $C(p, \bar{u}) = \sum p_i q_i^* = \bar{u} \cdot c(p)$. We then get $\frac{q_i^*}{\sum p_i q_i^*} = \frac{\partial c(p) / \partial p_i}{c(p)}$.

This is Shephard's lemma in the homothetic case.

Otherwise $\frac{\partial C(p) / \partial p_i}{C(p)}$ for the right hand side RHS. Or: for small changes in prices the effect of the cost of living of a price increase is equal to the amount of the good (q*) purchased. This lemma relates the cost function (defining the COLI) to the observable quantities of the demand function (if households are engaged in utility maximization).

Box 2.4.1 (continued)

Wold's identity relates the marginal utility[111] to prices and costs C

$$\frac{p_i}{C} = \frac{p_i}{\sum_j p_j q_j} = \frac{\partial u(q)/\partial q_i}{\sum q_j (\partial u(q)/\partial q_j)}$$, or $\frac{\partial u(q)/\partial q_i}{u(q)}$ for the RHS in the homothetic case.

This also leads to the well known relation between the marginal utilities and prices for any pair of commodities.

We are also able now to express the COLI in terms of unit costs, viz. $P_{0t}^{CU} = c(\mathbf{p}_t)/c(\mathbf{p}_0)$ and the corresponding economic theory quantity index $Q_{0t} = u(\mathbf{q}_t)/u(\mathbf{q}_0)$ relating the utility levels so that the product $P_{0t}^{CU} Q_{0t} = \frac{u(\mathbf{q}_t)\cdot c(\mathbf{p}_t)}{u(\mathbf{q}_0)\cdot c(\mathbf{p}_0)} = \frac{C(u_t, \mathbf{p}_t)}{C(u_0, \mathbf{p}_0)} = V_{0t}$ gives the value ratio provided that the household is engaged in utility maximization in both periods, 0 and t respectively.

Finally it may be useful to verify Shephard's lemma for **ex. 2.4.1**

> In this example: eq. 2.4.13 represents the cost function C, and the derivative of C with respect to p_1 (in the case of prices and utility of period 0), that is $\partial C/\partial p_1$ is given
>
> by $q_1^* = \alpha_1 p_1^{\alpha_1 - 1}\left(\frac{1}{\alpha_1}\right)^{\alpha_1}\left(\frac{p_2}{\alpha_2}\right)^{\alpha_2}\cdot u$ or $q_1^* = \frac{1}{2}\frac{1}{\sqrt{p_1}}\sqrt{2}\sqrt{4}\cdot 2$
>
> using u = 2, p_2 = 2 and $\alpha_1 = \alpha_2 = 1/2$. This yields q_1^*= 1 with $p_1 = 8$. In the same manner we get from $\partial C/\partial p_2$ for the second commodity q_2^* = 4.

From eq. 13 it follows that the indirect utility function u = u(**p**, C) as the inverse cost function is given by $U = \bar{u} = C\left(\frac{\alpha_1}{p_1}\right)^{\alpha 1}\left(\frac{\alpha_2}{p_2}\right)^{\alpha 2}$ or numerically in the base period

$2 = 16\cdot\sqrt{\frac{1/2}{8}\cdot\frac{1/2}{2}}$ and $2 = 16\cdot\sqrt{\frac{1/2}{2}\cdot\frac{1/2}{8}}$ in period t.

End of box 2.4.1

e) The notion of "exact" and "superlative" index functions (cost of living indices)

In the 70's and 80's of the last century the "economic - theory approach" experienced a considerable upgrading, owing to the theory of "exact" and "superlative" index functions developed by W. Erwin Diewert. Until then the standard argument against the COLI approach used to be as follows: to identify an index function equivalent to the CU-index (COLI) we usually have to specify the precise functional form of the utility

[111] derivative of the utility with respect to quantities.

function (indifference curves), as done in ex. 2.4.1 using the Cobb-Douglas function as an example. However since there is no known procedure to gain empirical data for utility (preference) functions, practical price statistics can hardly make much use of this highly theoretical COLI approach. Thanks to Diewert's theory we may conclude that under certain quite general conditions (often not made explicit by those who refer to this theory, however) some indices using observed data only, like P_{0t}^{F} (Fisher), or P_{0t}^{T} (Törnqvist), will approximate the COLI for whatever underlying utility function, so that there is no need to know empirical indifference maps in order to compile a true COLI.

1. The meaning of "second order approximation"

In order to understand better the theory of superlative index formulas it seems useful to get an idea of what is meant by "second order approximation" at the outset. The expression

$$f(x) = f(0) + f'(0) \cdot x + \frac{f''(0)}{2!}x^2 + \ldots$$

is called the Taylor (or polynomial) expansion of f(x) around the point $x_0 = 0$ or more general: for any points $x, x_0 \in [a, b]$ of the domain of definition [a, b] of f(x) we have

$$f(x) = f(x_0) + \frac{f'(x_0)}{1!}(x - x_0) + \frac{f''(x_0)}{2!}(x - x_0)^2 + \ldots + \frac{f^{(n)}(x_0)}{n!}(x - x_0)^n + R$$

with the residual R. To give an example, in the case of $f(x) = a_0 + a_1 x + a_2 x^2 + \ldots$ at the point x = 0 we have f(0) = a_0, furthermore

$$f'(x) = \frac{df(x)}{dx} = a_1 + 2a_2 x + \ldots \rightarrow f'(0) = a_1$$

$$f''(x) = 2a_2 + 3 \cdot 2 \cdot a_3 x + \ldots \rightarrow f''(0) = 2a_2 \text{ etc.}$$

$\hat{f}(x) = f(0) + \frac{f'(0)}{2!}x = a_0 + a_1 x$ is called the first order approximation of the function f(x) at the point x = 0 (while $\hat{\hat{f}}(x) = \hat{f}(x) + \frac{f''(0)}{2!}x^2$ is the second order or "quadratic" approximation to f(x)). The straight line $\hat{f}(x)$ or $p_2(x) = a_0 + a_1 x$ in **fig. 2.4.5** is the tangent of f(x) such that f(x) and y are locally (that is, in x = 0) already quite similar. The divergence (distance) of the function f(x) from this tangent is itself again a function in x, viz. g(x) so that

$$g(x) := f(x) - (f(0) + f'(0) \cdot x) = \frac{f''(0)}{2!}x^2 + \frac{f'''(0)}{3!}x^3 + \ldots$$

which shows that f(x) will "bend" upwards (relative to the tangent) if f"(0) > 0 (or f(x) is said to be "locally convex", or "concave up" at the point x = 0 where f(x) has a local minimum), or it tends downwards if f"(0) < 0 (concave down, local maximum, negative second derivative).

A distinction is made between approximations of different order depending on how successively additional derivatives are taken into account of an increasing order. Obviously

- equality of f(x) and the approximating function p(x) at the point $x = x_0$, that is $f(x_0) = p(x_0)$, still allows for possibly very different curves (indicated in **fig. 2.4.5**) while
- equality of the slope (tangent) of f(x) and p(x) in x_0, that is, $f'(x_0)=p'(x_0)$, is already able to provide a much better approximation of the function f(x) by p(x),

and accordingly we may expect an even better approximation once the third order term $\frac{f''(0)}{3!}x^3$ is taken into account as well, and so on.

In the case of our example the quadratic (second order) approximation (SOA) of f(x) is given by $\hat{\hat{f}}(x)=f(x_0)+\frac{f'(x_0)}{1!}(x-x_0)+\frac{f''(x_0)}{2!}(x-x_0)^2$ which in turn is the parabola $p(x)=a_0+a_1(x-x_0)+a_2(x-x_0)^2$, or $p(x)=a_0+a_1x+a_2x^2$ at $x_0 = 0$, since $a_0=f(x_0)$, $a_1=f'(x_0)$, and $a_2=\frac{1}{2}f''(x_0)$, and SOA requires $p(x_0)=f(x_0)$, $p'(x_0)= f'(x_0)$ and $p''(x_0)=f''(x_0)$, that is equality of derivatives up to the second order. In the same fashion the n-th order approximation would require equality of derivatives up to the n-th order around the point x_0.

We are now ready to examine generalizations as regards approximations (of different orders) for functions in two variables, x and y, for example

$F(t) = f(x_0 + ht, y_0 + kt)$ where $0 \le t \le 1$.

$F(0)+\frac{1}{1!}F'(0)+\frac{1}{2!}F''(0)t+\ldots$ where $F(0) = f(x_0,y_0)$,

$F'(t)=hf_x(t)+kf_y(t)$, and

$F''(t)=h^2f_{xx}(t)+2hkf_{xy}(t)+k^2f_{yy}(t)$ etc., and the first, second, etc. derivatives are

$f_x=\partial F/\partial x$, $f_y=\partial F/\partial y$ and $f_{xy}=\partial^2F/\partial x\partial y$ etc. The vector of first order derivatives of the function F with respect to **x** evaluated at $\mathbf{x}_0$ denoted

$$\nabla_x F(x_0)=\begin{bmatrix}\partial F(x_{10},x_{20},\ldots,x_{n0})/\partial x_1\\ \vdots \\ \partial F(x_{10},x_{20},\ldots,x_{n0})/\partial x_n\end{bmatrix}$$, or in the case of two x-variables only, x and y,

$$\nabla F=\begin{bmatrix}\partial F/\partial x\\ \partial F/\partial y\end{bmatrix}=\begin{bmatrix}f_x\\ f_y\end{bmatrix}$$ is called the gradient (vector) of F. Correspondingly the second order derivatives form a (Hessian) matrix $\mathbf{H}=\begin{bmatrix}f_{xx} & f_{xy}\\ f_{yx} & f_{yy}\end{bmatrix}$.

In the theory of "superlative" indices a cost - or aggregator-function is called "flexible" if it can provide a second order approximation to an arbitrary twice continuously differentiable cost - or aggregator function. A price (quantity) index is said to be superlative if it is exact for a flexible cost (aggregator) function.

Figure 2.4.5: Equal values of functions in $x = x_0$ and first order approximation (tangent)

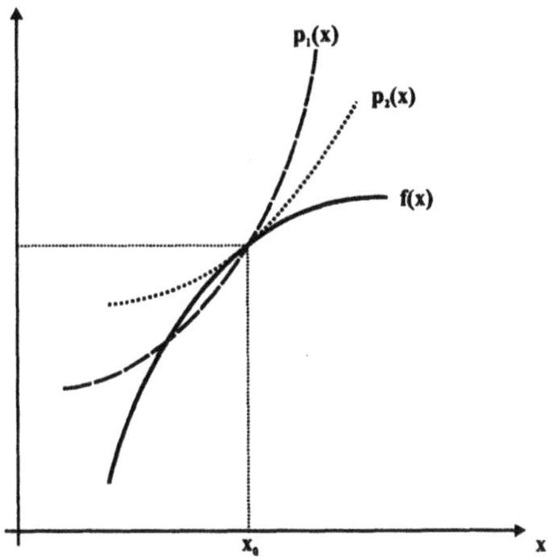

2. Fisher's ideal index is exact for a homogeneous quadratic aggregator function

The aggregator function

(2.4.15) $U(q_1,...,q_n) = \sqrt{\sum\sum a_{ik} q_i q_k}$, $i,k = 1,...,n$ where $a_{ik} = a_{ki}$ or equivalently

(2.4.15a) $U(\mathbf{q}) = f_2(\mathbf{q}) = (\mathbf{q}'\mathbf{A}\mathbf{q})^{1/2}$ with the symmetric matrix **A** = **A'**

is called the "homogeneous quadratic" aggregator function. Obviously this is the special case r = 2 of the more general "quadratic mean of order r" aggregator function

(2.4.16) $U(q_1,...,q_n) = f_r(\mathbf{q}) = \sqrt[r]{\sum\sum a_{ik} q_i^{r/2} q_k^{r/2}}$,

which is homothetic[112] and a special case of the famous CES-function in microeconomic theory.

Another special case r =1 of $f_r(\mathbf{q})$ is $f_1(\mathbf{q}) = \Sigma\Sigma a_{ij} q_i^{1/2} q_j^{1/2}$, a utility function related to the likewise (like Fisher and Törnqvist) superlative index of Walsh $P_{0t}^W = \frac{\Sigma p_t \sqrt{q_0 q_t}}{\Sigma p_0 \sqrt{q_0 q_t}}$.

The i-th element of the gradient of $f_2(\mathbf{q}_t) = U(\mathbf{q}_t)$ is given by the following marginal utility

[112] This applies of course also to the special case r = 2 of eq. 15.

(2.4.17) $$\frac{\partial f_2(\mathbf{q}_t)}{\partial q_{it}} = \frac{\Sigma_j a_{ij} q_{jt}}{\sqrt{\Sigma_i \Sigma_j a_{ij} q_{jt} q_{jt}}} = \frac{\Sigma_j a_{ij} q_{jt}}{U(\mathbf{q}_t)}$$

(the same applies to the gradient of $f_2(\mathbf{q}_0)$). The Fisher quantity index may be written as $Q^F_{0t} = \left(\frac{\sum p_{i0} q_{it}}{\sum p_{i0} q_{i0}}\right)^{1/2} \left(\frac{\sum p_{it} q_{i0}}{\sum p_{it} q_{it}}\right)^{-1/2} = \left(\frac{\sum p_{i0} q_{it}}{U(\mathbf{q}_0)}\right)^{1/2} \left(\frac{U(\mathbf{q}_t)}{\sum p_{it} q_{i0}}\right)^{1/2}$.

Using Wold's identity (see **box 2.4.1**) for homothetic aggregator functions, we may relate the prices p_{i0} and p_{it} to marginal utilities and we can then substitute eq. 2.4.17 for the marginal utilities, giving the following ratio of utility levels

(2.4.18) $Q^F_{0t} = \frac{U(\mathbf{q}_t)}{U(\mathbf{q}_0)} = \frac{f_2(\mathbf{q}_t)}{f_2(\mathbf{q}_0)}$,

and since the cost functions in the homothetic case are related to the utility levels $U(\mathbf{q}_t)$ and $U(\mathbf{q}_0)$ and unit cost functions $c(\mathbf{p})$ as follows

$V_{0t} = \frac{C(U(\mathbf{q}_t), \mathbf{p}_t)}{C(U(\mathbf{q}_0), \mathbf{p}_0)} = \frac{U(\mathbf{q}_t) \cdot c(\mathbf{p}_t)}{U(\mathbf{q}_0) \cdot c(\mathbf{p}_0)} = Q^F_{0t} P^F_{0t}$, we can see that

(2.4.19) $P^F_{0t} = \frac{c(p_t)}{c(p_0)}$,

so that the Fisher Price index is exact for the homogeneous quadratic aggregator function.

Table 2.4.1: Some flexible functions and the corresponding superlative indices

function	aggregator function $U(q_1,\dots,q_n) = U(\mathbf{q})$ 1)	Exact quantity index[1]	cost (or unit cost) function $C(U,\mathbf{p})$ or $c(\mathbf{p})$ 2)	exact price index 2)
homogeneous quadratic 3)	$f_2(\mathbf{q}) = \sqrt{\sum\sum a_{ik} q_i q_k}$ or $= (\mathbf{q}'\mathbf{A}\mathbf{q})^{1/2}$	Q^F_{0t}	$C(u,\mathbf{p}) = (\mathbf{p}'\mathbf{B}\mathbf{p})^{1/2}u$ → unit cost ($u = 1$): $c(\mathbf{p}) = c_2(\mathbf{p}) = (\mathbf{p}'\mathbf{B}\mathbf{p})^{1/2}$	P^F_{0t}
homogeneous quadratic of order r	$f_r(\mathbf{q}) = \sqrt[r]{\sum\sum a_{ik} q_i^{r/2} q_k^{r/2}}$	$r = 1$ $\rightarrow Q^W_{0t}$ 4)	$c_r(\mathbf{p}) = \left(\sum\sum b_{ik} p_i^{r/2} p_k^{r/2}\right)^{1/r}$	$r = 1$: P^W (Walsh) and P^F 4)
homogeneous translog function 5)	$f_r(\mathbf{q})$ when $r \rightarrow 0$ (eq. 2.4.22)	Törnqvist Q^T_{0t}	$\ln[c_0(\mathbf{p})]$ according to eq. 2.4.23/23a	P^T_{0t}

Table 2.4.1 (continued)

1) $Q_{0t} = U(\mathbf{q}_t)/U(\mathbf{q}_0)$

2) $P_{0t} = c(\mathbf{p}_t)/\ c(\mathbf{p}_0)$

3) The more general aggregator function $a_0 + \mathbf{a}'\mathbf{q} + \frac{1}{2}\mathbf{q}'\mathbf{A}\mathbf{q}$ leads to complicated formulas so that it is common to assume $a_0 = 0$ and $\mathbf{a} = \mathbf{0}_n$ (zero vector with n components)

4) Walsh index, such that P_{0t}^{W} is exact for the corresponding unit cost function when r = 1; and of course we get Q_{0t}^{F} in the case of r = 2.

5) the limit of $f_r(\mathbf{q})$ as r tends to 0 is defined by eq. 2.4.22, and likewise the limit ($r \to 0$) of $f_r(\mathbf{q})$ is $c_r(\mathbf{p})$ according to eq. 2.4.23; homotheticity is not required, the assumption is made, however, that the underlying utility level is given by $u^* = \sqrt{u_0 u_t}$.

This can also be shown by directly using the unit cost function corresponding to the aggregator function 2.4.15 which is given by

$$(2.4.20) \qquad C(u,\mathbf{p})=(\mathbf{p}'\mathbf{B}\mathbf{p})^{1/2}u$$

where **B** is the inverse matrix **A** so that $\mathbf{B} = \mathbf{A}^{-1}$, and therefore of course also symmetric (**B** = **B'**) and **p'** = $[p_1 \ldots p_n]$ is the price (row) vector.

The (general) unit cost function c(**p**) = C(1,**p**) related to $f_r(\mathbf{q})$ can be written as

$$(2.4.21) \qquad c_r(\mathbf{p}) = \left(\sum_i \sum_j b_{ij} p_i^{r/2} p_j^{r/2} \right)^{1/r}$$

of which eq. 20 (and thus $c(\mathbf{p})=(\mathbf{p}'\mathbf{B}\mathbf{p})^{1/2}$) is the special case r = 2, that is $c_2(\mathbf{p})$. As usual, cost-minimal quantities are gained by differentiating c(**p**) with respect to the prices. It may be easier to ignore for the moment the square root over the quadratic form **p'Bp** and examine only period 0.

The gradient of $Q = \mathbf{p}_0'\mathbf{B}\mathbf{p}_0$ as a row vector is then $(\nabla_p Q)' = \left[\frac{\partial Q}{\partial p_{10}} \ \frac{\partial Q}{\partial p_{20}} \right]$, which may easily be verified for the case of only n = 2 commodities.

$$Q = [p_{10} p_{20}] \begin{bmatrix} b_{11} b_{12} \\ b_{12} b_{22} \end{bmatrix} \begin{bmatrix} p_{10} \\ p_{20} \end{bmatrix} = p_{10}^2 b_{11} + 2p_{10}p_{20}b_{12} + p_{20}^2 b_{22}$$ yields the following gradient

$$(\nabla_p Q)' = \left[\frac{\partial Q}{\partial p_{10}} \ \frac{\partial Q}{\partial p_{20}} \right] = [2p_{10}b_{10} + 2p_{20}b_{12} \quad 2p_{10}b_{12} + 2p_{20}b_{22}] = 2\mathbf{p}_0'\mathbf{B} = [p_{10} \quad p_{20}] \begin{bmatrix} b_{11} & b_{12} \\ b_{12} & b_{22} \end{bmatrix},$$

or as a column vector $\nabla_p Q = \begin{bmatrix} \partial Q/\partial p_{10} \\ \partial Q/\partial p_{20} \end{bmatrix} = 2\mathbf{B}\mathbf{p}_0$.

Accounting for the square root now and utilising the chain rule we get

$\mathbf{q}_0 = \nabla_p C(U_0 = u_0, \mathbf{p}_0) = (\mathbf{p}_0'\mathbf{B}\mathbf{p}_0)^{-1/2}\mathbf{B}\mathbf{p}_0 u_0$ and correspondingly

$\mathbf{q}_t = \nabla_p C(U_t = u_t, \mathbf{p}_t) = (\mathbf{p}_t'\mathbf{B}\mathbf{p}_t)^{-1/2}\mathbf{B}\mathbf{p}_t u_t$.

Upon inserting these expressions for $\mathbf{q}_0$ and $\mathbf{q}_t$ in the terms $\mathbf{p}_0'\mathbf{q}_0 = \sum p_0 q_0$ and $\mathbf{p}_t'\mathbf{q}_t = \sum p_t q_t$ of the Fisher index formula and assuming $u_t = u_0$ for the utility levels we get

$$P_{0t}^F = \left[\frac{\mathbf{p}_t'\mathbf{q}_0(\mathbf{p}_t'\mathbf{B}\mathbf{p}_t)^{1/2}u_t}{\mathbf{p}_0'\mathbf{q}_t(\mathbf{p}_0'\mathbf{B}\mathbf{p}_0)^{1/2}u_0}\right]^{1/2} = \left[\frac{(\mathbf{p}_0'\mathbf{B}\mathbf{p}_0)^{-1/2}(\mathbf{p}_t'\mathbf{B}\mathbf{p}_t)^{1/2}}{(\mathbf{p}_t'\mathbf{B}\mathbf{p}_t)^{-1/2}(\mathbf{p}_0'\mathbf{B}\mathbf{p}_0)^{1/2}}\right]^{1/2}$$

(taking into account that $\mathbf{p}_t'\mathbf{B}\mathbf{p}_0 = \mathbf{p}_0'\mathbf{B}\mathbf{p}_t$) and we see that in fact

$$P_{0t}^F = \left[\frac{\mathbf{p}_t'\mathbf{B}\mathbf{p}_t}{\mathbf{p}_0'\mathbf{B}\mathbf{p}_0}\right]^{1/2} = \frac{C(U_t, \mathbf{p}_t)}{C(U_0, \mathbf{p}_0)},$$

so that P_{0t}^F is indeed exact for the $c_2(\mathbf{p})$ function of eq. 2.4.21.

3. Other superlative index formulas

Another set of flexible forms is, according to Diewert, the so called translog aggregator function

(2.4.22) $$\ln f_0(\mathbf{q}) = a_0 + \sum_i a_i \ln q_i + \frac{1}{2}\sum_i\sum_j a_{ij}(\ln q_i)(\ln q_j)$$

and the corresponding unit cost function

(2.4.23) $$\ln c_0(\mathbf{p}) = a_0 + \sum_i a_i \ln p_i + \frac{1}{2}\sum_i\sum_j a_{ij}(\ln p_i)(\ln p_j)$$ or especially

(2.4.23a) $$\gamma = \boldsymbol{\pi}'\mathbf{A}\boldsymbol{\pi}$$

for which the Törnqvist index is exact and where i, j = 1,..., n, $\sum_i a_i = 1$ and $\sum_j a_{ij} = 0$ for all i, and $a_{ij} = a_{ji}$ (symmetry of **A**). In eq. 2.4.23a the logarithm of costs $\gamma = \ln(c)$ is a quadratic form of the logarithms of prices $\pi = \ln(p)$.

Starting with aggregator functions of the type $f_r(\mathbf{q})$ of eq. 2.4.16 or $\left(\sum_i\sum_j a_{ij} q_i^{r/2} q_j^{r/2}\right)^{1/r}$ we may go a step further by assuming aggregator functions built as products of such power means giving rise to general formulas with expenditure shares s_{i0} and s_{it} ($s_{ik} = p_{ik}q_{ik}/\sum p_{ik}q_{ik}$ of period k = 0, t) for example[113]

(2.4.24) $$P_{0t}^{QMr} = \left[\sum s_{i0}\left(\frac{p_{it}}{p_{i0}}\right)^{r/2}\right]^{1/r}\left[\sum s_{it}\left(\frac{p_{it}}{p_{i0}}\right)^{r/2}\right]^{-1/r}$$

which is the so-called "quadratic mean of order r price index". We also find the fairly general formula

(2.4.25) $$P_{r,k} = \left(\sum_i s_{ik}(p_{it}/p_{io})^r\right)^{1/r}$$ for $r \neq 0$ and k = 0,t and

[113] The index functions of Fisher and Walsh are obviously special cases.

(2.4.25a) $P_{r,k} = \prod_i (p_{it}/p_{io})^{s_{ik}}$ for r = 0.

Evidently special cases are $P_{1,0} = P_{0t}^L$ (Laspeyres, r = 1 and k = 0) $P_{-1,1} = P_{0t}^P$ (Paasche), so that $P_{0t}^F = \sqrt{P_{1,0}P_{-1,1}}$, furthermore $P_{0,0} = DP_{0t}^L$ (log. Laspeyres index) and $P_{0,1} = DP_{0t}^P$ (log. Paasche).

4. Laspeyres and Paasche; relations between superlative index formulas and some criticism

The formulas of Laspeyres and Paasche are exact only for the fixed coefficients (or Leontief) aggregator functions f (**q**) = $\min_i(q_i/b_i)$ where **b'** = [b_1, ..., b_n] is a vector of positive constants. They are also exact for a linear aggregator function b(**q**) = **a'q**. Note that an index formula can be exact for more than one single type of aggregator function.

For quite a few results the assumption of homotheticity is essential though well known to be unrealistic. The case of non-homotheticity can be accommodated, however, once a utility level u* as a value of the constant [for both periods, 0 and t] utility function somewhere between u_0 and u_t has been chosen.

The results of the theory of exact and superlative indices are in principle derived using the Taylor series expansion technique, that is, by evaluating the first and second order derivatives of functions around the "***equal price and quantity point***" (i.e. assuming $p_t \approx p_0$ and $q_t \approx q_0$).[114]

Diewert has also shown that superlative price indices such as P^T, P^F and P^W approximate one another to the second order (i.e. have the same first and second derivatives) around any given price-quantity combination,[115] so that the choice between them is unlikely to matter much in practice. Recently this result has been challenged by R. J. Hill (2002), however, who stated that it applies only if the choice of r in f_r(**q**) or c_r(**p**) is restricted to the interval $0 \leq r \leq 2$, on which the focus is usually placed. For a wider interval we may well see superlative indices that differ between one another by as much as the factor two or even four.

It has become evident that only such index formulas are eligible as superlative in which quantities of both periods are taken into account in a symmetric fashion. Conspicuously there is no asymmetric, or even one-sided superlative index,[116] moreover both quantities have to be accounted for in a particular manner, for example using a

[114] Diewert also made use of this technique in order to show that deviations of chained direct Fisher indices from transitivity for any three points in time 0, 1, and 2, are small only (provided that $p_0 \approx p_1 \approx p_2$ and the same applies to quantities, an assumption which seems to be fairly restrictive in the case of quite long chains).

[115] On the other hand Paasche and Laspeyres indices approximate the superlative indices only to the first order. Another result of the approximation theory is that the Fisher index, though not strictly aggregatively consistent (see **sec. 5.2**), may still be "additive" *approximately*.

[116] If only one quantity (e.g. the q_0's) is taken into account the index would be unable to reflect substitution.

geometric, rather than an arithmetic mean. Thus, as Ralph Turvey, a critic of the COLI approach has pointed out, the ME-Index (see **tab. 2.4.2**) is not superlative, even though P^{ME} makes use of both quantities and has (in contrast to P^F or P^T) favorable aggregation quantities.

On the other hand the theory is compatible with many more formulas. If r is specialized to r = 1 in the case of $U(q_1,...,q_n) = \sqrt[r]{\Sigma\Sigma a_{ik} q_i^{r/2} q_k^{r/2}}, \quad i,k = 1,...,n$, we get the formula of Walsh, but for other values of r we would likewise gain "superlative" index formulas, which are then "no-name" superlative index formulas.

It should also be noted that flexible forms may be fairly general but yet not free of certain tacit assumptions. In the case of r = 2 we have the homogeneous quadratic function of eq. 15a for which Fisher's index is exact. This amounts to

Table 2.4.2: Relationships* between symmetry, kind of mean and the notion "superlative"

Superlative indices as a subset of symmetric indices

	symmetric	not symmetric
superlative	**Törnqvist, Fisher, Walsh**	impossible
not superlative	Edgeworth-Marshall P^{ME}, Drobisch, Vartia I and II	Laspeyres, Paasche, logarithm. Laspeyres and *many* more

geometric mean of		arithmetic mean of	
formulas	quantities	formulas	quantities
$\sqrt{P_{0t}^L P_{0t}^P}$	$\sqrt{q_{i0} q_{it}}$	$(P_{0t}^L + P_{0t}^P)/2$	$(q_{i0} + q_{it})/2$
Fisher	Walsh	Drobisch	Edgeworth Marshall
superlative		not superlative	

*The question arises how these relationships should be interpreted from the point of view of (micro) economic theory. Furthermore it should be noted that interestingly no superlative index is aggregatively consistent.

$(U_0)^2 = \mathbf{q}_0'\mathbf{A}\mathbf{q}_0 = [q_{10} q_{20}] \begin{bmatrix} a_{11} & a_{12} \\ a_{21} & a_{22} \end{bmatrix} \begin{bmatrix} q_{10} \\ q_{20} \end{bmatrix}$ if n = 2 (in period t by analogy $U_t = \sqrt{\mathbf{q}_t'\mathbf{A}\mathbf{q}_t}$) and due to the required symmetry of **A** ($a_{12} = a_{21}$) to

$U_0 = \sqrt{a_{11}q_{10}^2 + 2a_{12}q_{10}q_{20} + a_{22}q_{20}^2}$, which is rather specific even though the parameters a_{ij} are not specified numerically. The term a_{12} implies that utility does not only depend on the two quantities taken in isolation (q_{10} and q_{20} respectively) but also on the combination of both commodities in a particular manner.

Finally, it should be borne in mind that the COLI approach rests on the assumption that the utility function remains constant over the interval under consideration. This may be viewed as no less restrictive and "static" a concept as the familiar constant basket paradigm approach.[117]

f) Conditional COLIs and the many households case (social COLI)

1. Conditional COLI

We have already made mention of the problem that utility does not only depend on the amount of goods and services bought individually at free markets. An unconditional COLI would also respond to changes in the supply of non-market goods[118] and all general conditions of well-being. Such goods are for example air quality, less commuting time, more public safety, political stability, progress in curing diseases such as AIDS, or in the case of market goods a greater variety of goods from which a choice can be made. More or better (resp. less and worse) in this respect does not have much to do with deflation, or inflation. Thus, it is widely seen as unacceptable if the COLI were to react to such changes.[119]

A *conditional* as opposed to an unconditional COLI is not subject to such non-price changes, and therefore seems to be a better choice. However, even a conditional COLI may reflect changes in well-being with no price movement involved, and which therefore are not accounted for in a COGI, for example the appearance of new goods and services enhancing the price-utility-ratio and thus giving rise to an estimate of surplus-utility which may offset a price increase.

A conditional COLI is a special sub-index of a more comprehensive, all encompassing unconditional index[120] (see **fig. 2.4.6**). Parts of the domain of the total COLI are irrelevant for the comparison because the quantities (and thus their contribution to the total utility) are kept constant. In contrast to a partial index, separability (see below) of the utility function is not required. To define the appropriate domain of a COLI, and to make a clear distinction between goods covered and goods "conditioned" (that is not covered, deliberately left out due to constancy), is one of the most

---

[117] A criticism again according to R. Turvey. For him it is in particular not warranted to take the difference between the Laspeyres index P^L and a "superlative" index, for example Fisher's P^F, as a measure of the "substitution effect". This would be admissible only if all factors influencing the consumption pattern of the (!) household remained constant.

[118] especially goods provided by governments free of charge.

[119] see the quotation concerning Viagra at the beginning of this section.

[120] A conditional COLI is a COLI with restricted coverage (domain). In a way the traditional COGI may also be interpreted as a conditional COLI (in the sense of restricted to market goods, exclusive of certain non-market goods).

intricate practical problems to tackle in the case of implementing a COLI approach for an official CPI. The difficulties may become apparent when we consider medical care as a part of the CPI-domain.

"Utility" in this field may generally be viewed as consisting in "extended and better quality of life". This may arise, however, from

- new drugs, enhanced medical services due to progress in diagnosis, surgery and the like which may be called "embodied", or materialised improvements[121] as opposed to
- "disembodied" medical progress consisting, for example of a healthier lifestyle (discontinuation of smoking etc.), new results in medical research, or goods which make life more enjoyable even though they do not really cure disease.

The latter may also contribute to more utility or an improved price-utility ratio, and it is generally accepted that a conditional COLI should not comprise those elements. There are situations too in which a border is difficult to draw between improvements and changes of taste (as for example in the case of discontinuation of smoking). If technology is treated as an "environmental variable" to be held constant in a conditional COLI, quality improvement in medical care which comes through new technology would also not be reflected.[122] Hence, both COLIs, the unconditional as well as the conditional, pose difficult problems of interpretation inherent in the notion of "utility" and in consequence of the need to define the domain of definition of an index.

Figure. 2.4.6: Types of sub-indices

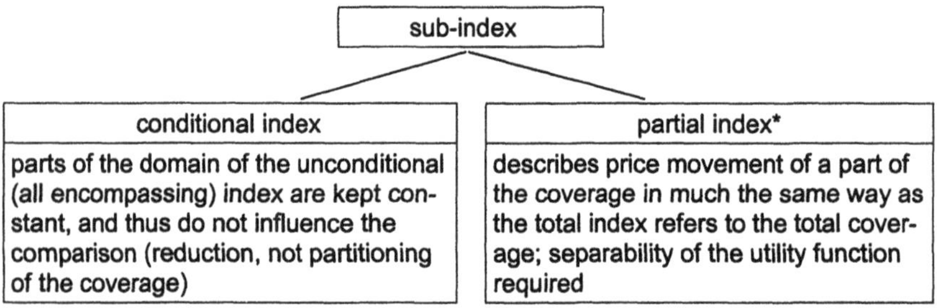

* see part h1 of this section

[121] They use to be reflected by quality adjusted prices of specific goods and services.

[122] In this case not only disembodied progress but also such sort of progress would not give rise to an implicit utility-increase, not even indirectly via quality adjustment. In general problems with quality adjustment abound in the case of medical care. How to account for advanced techniques in the treatment of heart attacks for example? A correct utility based adjustment would require to assign an estimated market value (expressed in $) to possibly more satisfaction due to more life expectancy, and or to less pain etc.

2. Aggregation over households

As already pointed out, another difficult problem is aggregation over households. Obviously a CPI has got to reflect a price movement referring to the inflation experience of (ideally) *all* private households at each income level and of various sizes (number of persons). The switch from the single-household to the many-household case now entails lots of aggregation problems and this is arguably one of the major objections critics have raised against the COLI approach. For many of these problems, a unique and satisfactory solution has not, to date, been found.

Given that for a specified good or service each household has to pay the same price (an assumption which is well known to be unrealistic), the problem of aggregation over households boils down to establishing the suitable weights for individual households in order to account for their relative importance. Generally two variants of an aggregated COLI are distinguished

- the *plutocratic* COLI (or the "social" COLI as it was called originally by Pollak, one of the first to introduce this idea), in which the weight assigned to household h is determined by its consumption expenditure (costs) , as opposed to
- the *democratic* COLI where each household is given the same weight 1/H.

Both variants can be conceived in the Laspeyres or in the Paasche version, depending on the period (either 0 or t) to which the utility levels and functions as well as the vector **e** of "quantities" of the conditioned environmental goods and services refers. Thus for example

$$(2.4.26)\quad P_{0t}^{pL} = \frac{\sum_h C_h(\mathbf{p}_t, U_0(\mathbf{q}_{h0}), \mathbf{e}_{h0})}{\sum_h C_h(\mathbf{p}_0, U_0(\mathbf{q}_{h0}), \mathbf{e}_{h0})} \le \frac{\sum_i \left(p_{it} \sum_h q_{hi0} \right)}{\sum_i \left(p_{i0} \sum_h q_{hi0} \right)} = P_{0t}^{L},\quad h = 1,\dots,H$$

denotes the plutocratic (p) all-households COLI of the Laspeyres (L) type based on the households' (h) utility function and level as well as endowment with environmental and public goods referring to period 0 being kept constant. As the individual COLI of household h is defined as $P_{0t}^{CU}(U_0, h) = \frac{C_h(\mathbf{p}_t, U_0(\mathbf{q}_{h0}), \mathbf{e}_{h0})}{C_h(\mathbf{p}_0, U_0(\mathbf{q}_{h0}), \mathbf{e}_{h0})}$ we get

$$(2.4.26a)\quad P_{0t}^{pL} = \sum_h P_{0t}^{CU}(U_0, h) \frac{C_h(\mathbf{p}_0, U_0(\mathbf{q}_{h0}), \mathbf{e}_{h0})}{\sum_h C_h(\mathbf{p}_0, U_0(\mathbf{q}_{h0}), \mathbf{e}_{h0})} = \sum_h P_{0t}^{CU}(U_0, h) \cdot \gamma_h$$

where the weights γ_i are determined by expenditure (therefore "plutocratic) shares of the base period, viz. $\gamma_h = C_h(\mathbf{p}_0, U_0, \mathbf{e}_0) / \sum_h C_h(\mathbf{p}_0, U_0, \mathbf{e}_0)$ as opposed to equal weights

$$(2.4.27)\quad P_{0t}^{dL} = \sum_h P_{0t}^{CU}(U_0, h) \cdot \frac{1}{H} = \frac{1}{H} \sum_h P_{0t}^{CU}(U_0, h) = \bar{P}_{0t}^{CU}(U_0)$$

in the democratic (d) social COLI of the Laspeyres (L) type. In a similar vein the expression

(2.4.28) $$P_{0t}^{dP} = \frac{1}{H}\sum_h P_{0t}^{CU}(U_t, h) = \bar{P}_{0t}^{CU}(U_t)$$

denotes the democratic Paasche version of a social COLI, and

(2.4.29) $$P_{0t}^{pP} = \frac{\sum_h C_h(\mathbf{p}_t, U_t(\mathbf{q}_{ht}), \mathbf{e}_{ht})}{\sum_h C_h(\mathbf{p}_0, U_t(\mathbf{q}_{ht}), \mathbf{e}_{ht})} \geq \frac{\sum_i \left(p_{it}\sum_h q_{hit}\right)}{\sum_i \left(p_{i0}\sum_h q_{hit}\right)} = P_{0t}^{P}, \quad h = 1,\dots,H$$

its plutocratic variant, again bounded from below by the "usual" Paasche index when the usual assumptions of utility maximization and constancy of preferences and certain levels are made.[123]

The theory of the many-households-COLI now unfolds in much the same way as in the single household case. Most of the results carry through, in that (Laspeyres and Paasche) bounds are established, or results are derived for intermediate utility levels $U^* = (1-\lambda)\,U_0 + \lambda\, U_t$ (or a reference utility *vector* **u*** with H components). Consequently a sort of balanced average of the Laspeyres and the Paasche form (such as the Fisher geometric average)[124] is recommended as a well reasoned point estimator of the unknown social COLI.

It is noteworthy to point out that we don't generally need to tackle the intricate if not insoluble problem of aggregation over utility functions or even to accept the restriction that all households will possess the same preference order[125]. The proposals regarding a sort of "social" COLI usually only make the assumption that aggregation (summation) takes place over costs or individual COLIs rather than utility functions or utility levels.

This clearly makes the COLI approach more manageable from a practical point of view but on the other hand it also deprives the COLI to a certain extent of its theoretic (microeconomic) appeal. We cannot possibly interpret a social COLI in terms of a collective utility maximising or cost minimizing exercise in which an indefinite and unexplored multiplicity of decision makers acting on the basis of an equally vast and unknown variety of utility functions is involved. Hence, the COLI loses much of its justification (consumption as generated by the behaviour of utility maximising households) if compared to the allegedly arbitrary COGI approach.

Moreover, for a number of considerations (for example in eq. 26) the assumption has to be made that quantities q_i of a given commodity i referring to various households h = 1, 2, ..., H can be (and will be) summated over the H households, which is clearly unrealistic and not in line with the practical compilation of price indices in official statistics.

---

[123] Bounds for the democratic variants are obviously not so readily available.

[124] The "superiority" of P^F is generally claimed on two grounds: symmetry and time reversibility.

[125] which would result in the awkward situation of having only one type of car for each income-class on offer, because all preferences are identical, such that those who have the same income and hence can afford the same car will of course also like the same car most. There is no point in producing additional cars, and the same applies to apparel, restaurant meals, books, and perhaps everything else.

Due to the vector $\mathbf{e}_0$ or $\mathbf{e}_t$ respectively in the utility function, both non-existent in the usual P^L and P^P index, the bounds established for the conditional COLI (plutocratic version) will not in general coincide with empirical indices such as P^L or P^P. Moreover a social COLI is *not* a COLI showing how the society as a whole may optimise welfare in view of changing prices (this could also be done for example by redistributing income) but rather an average impact on cost of living given that each household may react in its specific way.

Most of the well-known statistical (observed) price indices may be interpreted in terms of the plutocratic concept. Take for example the Walsh index

$$P_{0t}^{W} = \Sigma p_t \sqrt{q_0 q_t} / \Sigma p_0 \sqrt{q_0 q_t} = \Sigma \frac{p_t}{p_0} \sqrt{p_0 q_0 p_0 q_t} / \Sigma \sqrt{p_0 q_0 p_0 q_t}$$

which may also be written as

$$(2.4.30) \qquad P_{0t}^{W} = \sum_i r_i \frac{\sqrt{v_{00} v_{0t}}}{\sum_i \sqrt{v_{00} v_{0t}}} = \sum_i r_i \frac{\sum_h \sqrt{v_{ih00} v_{ih0t}}}{\sum_h \sum_i \sqrt{v_{ih00} v_{ih0t}}},$$

where the $r_i = p_{it}/p_{i0}$ denote price relatives (i = 1,...,n) applicable to all households, and $v_{km} = p_k q_m$ denotes a value related to commodity i. For household h we have

$$(2.4.31) \qquad P_{0t}^{Wh} = \sum_i r_i \frac{\sqrt{v_{ih00} v_{ih0t}}}{\sum_i \sqrt{v_{ih00} v_{ih0t}}} = \sum r_i \frac{\sqrt{v_{ih00} v_{ih0t}}}{V_h},$$

and we therefore get for the all-households Walsh index

$$(2.4.30a) \qquad P_{0t}^{W} = \sum_h P_{0t}^{Wh} \frac{V_h}{\sum_h V_h} = \sum_h P_{0t}^{Wh} \Psi_h .$$

The expenditure shares $\Psi_h = V_h / \sum V_h$ may be interpreted as the relative (related to all households) sums (over i = 1, ..., n) of geometric mean expenditures $\sqrt{v_{ih00} v_{ih0t}} = p_{i0}\sqrt{q_{ih0} q_{iht}}$ when a mean is taken of expenditures for q_{i0} and q_{it} and prices p_{i0} prevail in both periods 0 and t. Thus, the usual Walsh index may be viewed as the arithmetic mean of household-specific Walsh indices P_{0t}^{Wh} with weights Ψ_h assigned to household h proportional to h's expenditures V_h, that is, with weights established in a plutocratic manner. In the same way other statistical indices, such as P_{0t}^{L} or P_{0t}^{P}, may be regarded as a plutocratic aggregation over households.[126] Making reference to the individual households frequently gives rise to terribly complicated formulas, however. Moreover the assumptions implicitly made in some of those aggregated COLI formulas may well be somewhat farfetched at times. We found for example a democratic COLI to compare the price levels faced by H_1 households in country 1 to the level relevant for H_2 households in country 2 as follows[127]

[126] In the case of $q_{hit} = q_{hi0}$ for all commodities i and all households h=1, ... H the term V_h in eq. 30a would yield $V_h = \Sigma p_{ih0} q_{ih0} = \Sigma p_{ih0} q_{iht}$ giving P^L (in this case identical with P^P) as a plutocratically weighted index.

[127] W.E. Diewert, Index Numbers, in: The New Palgrave Dictionary of Economics, Vol. 2, p. 774, London 1987.

$\prod_{h=1}^{H_1}\prod_{k=1}^{H_2}\left[P_{12}^{T}\left(p_1, p_2, q_{1h}, q_{2k}\right)\right]^{1/H_1H_2}$, utilizing price vectors as well as quantities in both countries combined in H_1H_2 different Törnqvist indices. In particular we here also need quantities of each product consumed by each household h = 1, ..., H_1 in country 1 and at the same time the respective quantities (of the same goods) consumed by each household of the second country. Hence we would need H_1 vectors of n-components (comprising n quantities for n goods) $\mathbf{q}_{11},...\mathbf{q}_{1,H1}$ and H_2 vectors $\mathbf{q}_{21},...,\mathbf{q}_{2,H2}$ (of the same n components), and compile H_1H_2 different Törnqvist (P^T) indices resulting from combining each of the H_1 consumption patterns of country 1 with each of the H_2 patterns of country 2, and finally to take a geometric mean of all those P^T indices. It is unlikely that we will ever be able to establish an estimate of such an index.[128]

We now turn to the deflation problem in the framework of a social (aggregated over households) COLI theory, that is the relation between a plutocratic, or democratic social COLI and the corresponding quantity index Q (a quantity index such that the product of Q and the COLI gives the observable value ratio) and the properties and underlying assumptions of this implicit index "quantity" index Q.

g) The quantity index corresponding to the COLI as a price index

Recall that the COLI (as a *price* index) does not attempt to compare situations in which a different level of utility or even a different utility function at two periods, say 0 and t, is involved. However, this no longer applies once we consider *the implicit quantity index* Q resulting from dividing the value index V_{0t} by the COLI (P^{CU}) as a price index. The Q index also requires a cardinalization of the utility concept (a money-metric utility for each household), that is, the "amount" of utility should be measurable on a metric scale. In other words, a Q-index of q = 1.2 states that the level U_t is 20 per cent higher than U_0.

A set of "economic" quantity indices (Q) can be defined as implicit (or "cofactor") quantity indices upon dividing the value ratio V_{0t} by a suitable COLI-price index. The Q indices of the Laspeyres and Paasche type may be viewed as special forms of the Allen quantity index (where the specific type of price index is left open). The focus of the theory is again on establishing nonparametric (not depending on parameters of the aggregator function) bounds of the economic quantity indices (primarily Q^L and Q^P as bounds) and on axioms satisfied by those indices. It is for example generally true that the implicit quantity index is proportional in the quantities $(\mathbf{q}_t = \lambda\mathbf{q}_0 \rightarrow Q = \lambda)$. Another group of economic quantity indices (such as the so-called Malmquist index) is built on the notion of the "deflation" or "distance" function corresponding to the aggregator (e.g. utility) function F($\mathbf{q}$) denoted by $d(F(\mathbf{q}), \mathbf{q}_t) = d(q, t)$, which is the factor by which the quantities $\mathbf{q}_t$ have to be divided in order to get quantities $\mathbf{q}$ able to "deliver" a utility amounting to F($\mathbf{q}$). The Malmqvist Q-index is defined as ratio $d(q,t)/d(q,0)$.

---

[128] Moreover the superiority of this index is based on the assumption that household h in country 1 and household k in country 2 are possessing translog cost functions *with identical coefficients* and thereby identical preferences over the same set of commodities.

The different quantity indices coincide in the case of homotheticity. Furthermore certain general and superlative price indices such as P_{0t}^{QMr} in eq. 24 and their corresponding indirect quantity indices $\tilde{Q}_{0t}^{QMr} = V_{0t}/P_{0t}^{QMr}$ will form a pair of superlative indices (quadratic mean of order r indices).[129] Likewise

$$\text{(2.4.24a)} \quad Q_{0t}^{QMr} = \left[\sum s_{io}\left(\frac{q_{it}}{q_{io}}\right)^{r/2}\right]^{1/r}\left[\sum s_{it}\left(\frac{q_{it}}{q_{io}}\right)^{r/2}\right]^{-1/r}$$

and $\tilde{P}_{0t}^{QMr}$ is another pair of superlative indices ($P, \tilde{Q}$ just as $\tilde{P}, Q$). Obviously the pair P^F, Q^F is a special case (r = 2). Again various superlative quantity indices approximate each other to the second order. For the proof it is not necessary to assume utility maximizing behaviour.[130]

Given this abundance of quite sophisticated mathematical considerations it may perhaps be overlooked that the nature of an "economic" quantity index is far from being clear and transparent. The interpretation of "quantity" and its relation to a kind of (aggregated!) "utility" as well as the equivalence of various kinds of direct and indirect (via price indices) conceptualisations of quantities or deflated aggregates is all but evident. Conspicuously the relevant literature of the COLI theory is dominated by the price index interpretation of the COLI while the implicit or explicit quantity index receives less attention.

Digression: Banerjee's economic-theory indices

In **sec. 2.6** two groups of new indices of Banerjee, denoted by P^{BA1}, and P^{BA2} and Q^{BA1}, Q^{BA2} are introduced. The indices $P_{0t}^{BA2} = \frac{P_{0t}^{P}(P_{0t}^{L}+1)}{P_{0t}^{P}+1}$, and $Q_{0t}^{BA2} = \frac{Q_{0t}^{P}(Q_{0t}^{L}+1)}{Q_{0t}^{P}+1}$ are derived from "economic theory" considerations. Banerjee's idea was simply to reverse the argument:

- if two situations (vectors of quantities $\mathbf{q}_0$, $\mathbf{q}_t$) are equivalent (on the same indifference curve) a true quantity index should be unity, and "therefore" he concludes (which should be viewed with great suspicion, however):
- "if the quantity index ... is equal to unity, it would imply that the standards of living ... in the two situations are the same".

Banerjee calls Q_{0t}^{BA2} the "true index of quantity" or "economic quantity index", an index which is interestingly derived without explicit reference to any utility function or

---

129 Note that the Törnqvist Q and P indices are defined as limiting cases of QM indices as r tends to 0.

130 It is assumed, however, that the derivatives are evaluated at the equal quantities (or prices) point, that is, the variation of quantities (prices) is not too great going from period 0 to period t. Interestingly the fact that the theory of superlative indices time and again refers to this point of equal prices or quantities is also taken as a justification for chain indices in official statistics since successive links of a chain index will not differ much as regards prices and quantities such that various superlative links will produce quite similar results.

production function whatsoever, (which makes it difficult to have faith in such an approach).

h) Some extensions and some more criticisms of the COLI reasoning

In the following we try to give a brief account of some additional considerations as well as additional arguments against the COLI approach to be found in the relevant literature.

1. Existence and theoretical justification of sub-indices, separability of a utility function

In order to justify the existence of subindices of the overall COLI, assumptions like homotheticity and separability (as mentioned already in **fig. 2.4.6**) are necessary. A utility function $F(\mathbf{q}, y)$ is said to be homogeneously separable if

(2.4.32) $F(\mathbf{q}, y) = F^*[f(\mathbf{q}), y]$ where $f(\lambda q) = \lambda f(q)$

that is, the micro aggregator function f is linearly homogeneous and F* is increasing in its first argument $\mathbf{q}$. Furthermore F has the structure given by eq. 30.

Applied to the partitioning of a set of goods (and services) into two subsets, $\mathbf{q}_1$ and $\mathbf{q}_2$, eq.32 states

(2.4.32 a) $F(\mathbf{q}) = F(\mathbf{q}_1, \mathbf{q}_2) = F[f(\mathbf{q}_1), \mathbf{q}_2]$

where $f(\mathbf{q}_1)$ is called the "category utility function" for the category (or the subset) $\mathbf{q}_1$, or more restrictivly, the "symmetric" concept of separability requires

(2.4.32b) $F(\mathbf{q}) = F[f_1(\mathbf{q}_1), f_2(\mathbf{q}_2)]$,

where not only are goods in $\mathbf{q}_1$ separable from those in $\mathbf{q}_2$ but also vice versa. For the sake of distinctiveness it should be added that this requires the marginal rates of substitution among goods of group 1 to be invariant to the amount of any good in the group 2 and vice versa. Note that this assumption of separability needed for establishing a COLI approach to the construction of sub-indices is clearly highly restrictive and – even worse – certainly not realistic.[131]

Separability is also important for justifying a reduction of the domain of a COLI.[132] To give an example: time offered for work as employee or devoted to work in the household is usually omitted from the COLI, which is then in a sense "only" a sub-COLI.[133] In order to rationalise for such omissions, separability has to assumed.

---

[131] The problem is to find a conceptual (microeconomic) foundation of sub-indices of "food" or "shelter (accommodation)" for example, where the assumption has to be made that, among other things, households make decisions on the basis of "separable" utility functions, that is, by planning expenditures in one field irrespective of what they need, desire or may be glad of having in other fields.

[132] Generally speaking, separability "provides the justification for ignoring 'other goods' in constructing subindexes "(R. A. Pollak).

[133] see below part 7 for problems of incorporating the household production function in a more general COLI.

The problem with sub-indices becomes even more intricate once we not only presuppose separability for subsets $\mathbf{q}_1$, $\mathbf{q}_2$ as regards the separate utility functions but also explore the nature of the utility function F[...] underlying the combination of sub-COLIs to a total COLI. According to R. A. Pollak 1975, "the complete cost of living index cannot be constructed from the partial indexes unless preferences are of a very special form".[134] It is not clear how partial indices should be combined to a total index. Recall that both types of indices, the total index as well as each of the partial indices should have a COLI interpretation, that is, we want each index to be derived from an expenditure minimization and the result of the minimization to be invariant upon a partitioning of the set of goods into subsets for example $\mathbf{q} = [\mathbf{q}_1\ \mathbf{q}_2]$.

This also demonstrates the fundamental difference between true sub-indices (requiring separability) and conditional indices. Assume $\mathbf{q}_2$ denotes the environmental and public goods to be held constant. The complete index should reflect changes in the supply of $\mathbf{q}_2$ much like changes regarding $\mathbf{q}_1$ and we simply fade out this effect on the partial index of $\mathbf{q}_1$ (of the private goods) by assuming constant supply conditions of $\mathbf{q}_2$. If, however, the objective is to construct a sub-index (rather than a conditional index) of $\mathbf{q}_1$ which has a COLI interpretation of its own,[135], variations in the provision of $\mathbf{q}_2$ should not have an effect (just like the constancy of $\mathbf{q}_2$ – evidently – has no effect) and the index should not reflect changes in the prices of the second group of goods.

If this is true, then the question of course arises as to what the role of an overall utility function F[...] in eq. 32 b can be: a preference order which usually also allows for substitutions *across* commodity groups?

To sum up, it is one of the criticisms of the COLI approach that it does not easily integrate a special microeconomic aggregation theory. To aggregate over sub-indices may be easy if we simply take weighted averages and do not emphasise the need for consistent COLI interpretations of all more or less aggregated (sub) indices.

Another problem in this context is worth mentioning. As "utility" refers to a much broader concept than what is identifiable as an observable "good", a COLI may well to a certain extent also disregard the need to make distinctions between of goods and services. This poses problems for the definition of partial indices along distinctions made by a classification. From a utility point of view there is no problem in substitutions across groups of goods or in substituting goods for services and vice versa, for example treating heart attacks or ulcers with drugs (goods) instead of surgical operations (services), or using books from a public library (service) instead of buying books (goods).[136]

---

[134] Subindexes in the Cost of Living Index, International Economic Review Vol. 16, No 1, Febr. 1975, p. 135 (139).

[135] Or which complies with certain axiomatic properties like for example monotonicity (in the prices of period t, i.e. assume a higher price vector $\mathbf{p}^*$ in COLI (P^*) as compared to COLI (P) with prices $\mathbf{p}$ entails a higher price index $\mathbf{p}^* > \mathbf{p} \rightarrow P^* > P$), or linear homogeneity or other axioms.

[136] Moreover, the satisfaction enjoyed from consumption more often than not cannot be attributed to a specific good but to an "ensemble" of circumstances. How to account for example for increase of "variety", "safety", or "convenience" etc. by adjusting prices (which price?)? The examples mentioned above are taken from the "Boskin Report".

2. Axiomatic properties of "economic" indices

Several attempts have been made to formulate economic theory counterparts of some "tests" or "axioms" introduced in the axiomatic theory. Thus, there is an economic theory version of the "time reversal test" (ET) or of the "circular test" (EC). However these tests are *not* equivalent to what is familiar in the *statistical* (axiomatic) price index theory (**sec. 3.1** etc.), where the focus is on which type of index function satisfies the test. Now we ask whether or not the test holds for a certain utility function. A first result is that *any* homothetic utility function will also meet the ET and EC test. It comes as no surprise that in the case of non-homotheticity we may think of a symmetric function F combining two utility levels, u_0 and u_t respectively, $F = F(u_0,u_t) = F(u_t,u_0)$, the modified COLI $C(\mathbf{p}_t,F(u_0,u_t))/C(\mathbf{p}_0,F(u_0,u_t))$ will satisfy the ET and EC test.[137] Due to these theorems in which such symmetric mean utility levels occur, it also comes as no surprise that authors following the "economic theory" line mostly prefer index functions which make use of q_0 and q_t in a symmetric manner like the Fisher, Walsh, or Törnqvist index.

3. Application of the economic theory (COLI) reasoning to the PPI (producer price index)

The so called "economic (theory) approach" to index numbers can also be used to define output price and quantity indices for producing units in much the same way as done in the case of the consuming household (or the "COLI-theory"). Hence we may for example think of a "true" output- (TO), intermediate input- (TI), and value added (the latter being the difference between TO and TI) *price*-index, as shown in **tab. 2.4.2**.

In all these cases the common feature of this "microeconomic" index approach is

- the index is generated in the way of some sort of optimisation
- quantity as well as price indices are *derived* from behavioural equations and aggregator functions such as production and utility functions, and we have to assume
- that all determinants of the optimisation task except for the relevant price vectors $\mathbf{p}_0$ and $\mathbf{p}_t$ are constant,

such that finally the observed quantities will in general differ from the "*theoretic*" ones and the index will not be defined empirically without adding some further - often quite restrictive, or unrealistic - assumptions (for example establishments producing one single homogeneous good and acting in a competitive price taking market). We will not go into the details of these PPI index considerations.

---

137 Again we find that to refer to an intermediate utility level will remedy against shortcomings of non-homotheticity (F is some kind of a symmetric mean utility level). An asymmetric choice of the utility reference level, using u_0 for example in both the numerator as well as the denominator like in a Laspeyres type COLI, will fail the ET-test unless u_0 refers to a homothetic utility function. Circularity may be fulfilled if F is such that $F(u_0,u_1) = F(u_1,u_2) = F(u_0,u_2) = u^*$ where u^* is a constant utility level, "fixed reference level irrespective of the actual utility levels" (Bossert and Pfingsten 1987, p. 283) unless U is a homothetic utility function.

Table 2.4.3: Economic theory ("true") price indices on the producer price level (PPIs)

	True Cost of Living Index (COLI)	Theoretical Output (TO) Price Index	Theoretical intermediate input (TI) Price Index
Optimization	minimize costs of quantities of consumer goods	maximize revenue by determining quantities of outputs	minimize costs of bought-in intermediate goods (inputs)
Subject to (given)	constant utility (standard of living), preference function (U)	constant technology[1] (T) and inputs (quantities and prices)	technology, fixed primary inputs (labor and capital) and given output
Price regimes[2]	of consumer goods	output goods	intermediate inputs
Optimizing unit[3]	household[3]	establishment[3]	establishment[3]
Upper bound	$P_{0t}^{L} \geq P_{0t}^{CU}(U_0)$	$P_{0t}^{P} \geq P_{0t}^{TO}(T_t)$	similar to COLI
Lower bound	$P_{ot}^{P} \leq P_{0t}^{CU}(U_t)$	$P_{0t}^{L} \leq P_{0t}^{TO}(T_0)$	similar t COLI

1) T is given by a production function (= production possibilities set [or curve])

2) Under different price regimes (price vectors) of ...

3) acting in a competitive price taking market

4. The nature of the COLI: genuine "findings" or only implications of assumptions

It was already contended that the results of eqs. 2.4.5 and 6 (concerning the upper bound (P^L) and the lower bound (P^P) of a CU-index) are mere implications of the *definition* of the "cost function" and a consequence of cost minimizing and facing a given utility function: C(t,0) is *defined* as the (non-observable) *minimum* cost to attain utility level U_0, given a price vector $\mathbf{p}_t$, so that C(t,0) = $C(\mathbf{p}_t, U_0) \leq \sum p_t q_0$ and likewise C(0,t) is *by definition* $C(\mathbf{p}_0, U_t) \leq \sum p_0 q_t$.[138] It comes therefore as no surprise that $P_{0t}^{CU}(U_0) \leq \sum p_t q_0 / \sum p_0 q_0 = P_{0t}^{L}$, and $P_{0t}^{CU}(U_t) \geq P_{0t}^{P}$.

It is also anything but surprising that in general - unless certain conditions (i.e. homotheticity) prevail - the same rise or decline of prices will yield a different result in terms of what is needed to maintain the level of well-being. And this *of course* depends on which level of utility the comparison is based. This is a consequence of the well known notion of a diminishing "marginal utility", according to which (a change of) utility depends on the quantities consumed and thus the level of utility attained.

[138] Where $U_0 = U(\mathbf{q}_0)$ and $U_t = U(\mathbf{q}_t)$.

5. The legacy of the Boskin report and official price statistics in the USA

Among economists, especially in the USA, it is a widespread belief that the "true cost of living index" (COLI) is the best measure of inflation and official statistics therefore should aim at a "superlative" index (like Fisher or Törnqvist) as a target concept instead of a fixed weighted Laspeyres index P_{0t}^{L}. In particular the Boskin Commission[139] 1996/97 (see **sec. 1.3**) adopted this position in its critique of the US Consumer Price Index (CPI). In the European Union as well as in the so called "Schultze panel" succeeding the Boskin Commission in 2001 there was much less enthusiasm for the COLI concept and the stance concerning P_{0t}^{L} was much more tolerant. The overarching recommendation was to substitute a COLI for a fixed weight formula of Laspeyres as hitherto used. In its obsession for the economic index theory the Commission had no doubt that a COLI should be the target concept of inflation measurement.[140]

K. Abraham, then BLS[141] commissioner responsible for the CPI admitted that this "seems basically right to us". Objections were made only as far as certain practicalities of implementation and compilation were concerned.[142] In addition to difficulties of disposing of the most recent weights in order to calculate a superlative index, the BLS rightly worked out some problems of estimating "utility" in dealing with quality adjustment. This applied in particular to what the Commission called "intangible aspects of quality change" such as improved safety of home power tools, improved quality of stereo sound and TV pictures, increased mobility or - expressly noted as to be accounted for - the increased freshness and timeliness of fruits and vegetables, or improvements in bypass operations, ulcer treatments, or cataract surgeries.[143] For the BLS the problem was not to ignore or recognise the welfare impact of certain quality improvements but to find a measure for the gains in utility and to make deductions on prices (choosing the correct commodities from which deductions should be made if double counting is to be avoided). On the other hand the BLS rightly emphasised aspects like understandability by the general public and the need to avoid frequent changes in fundamental features of index compilation, "subjective" and "judgmental" methods. Again it did not come to their mind that this might have a bearing on the theoretical basis of the (subjective) COLI.

139 A Commission appointed by the U.S. Senate Finance Committee.

140 In its criticism of the Laspeyres principle as well as in its advocating the economic theory approach the Boskin Commission, however, was not uncontested. In this section some of the arguments challenging the Commission are mentioned.

141 Bureau of Labor Statistics. The BLS did not call the conceptual centrepiece of the Boskin report into question, it only emphasised practical problems.

142 "if the BLS staff or other technical experts knew how to produce a true cost-of-living index on a monthly production schedule, that would be what we would produce" (Abraham et al., 1997, p. 27).

143 Interestingly doubts concerning the notion "utility" came up in BLS's comment only in connection with corrections of quality change in reporting prices for individual commodities. It did not occur to them difficulties in assessing "utility" might cast doubts also on the usefulness of COLI reasoning (for which "utility" is central) altogether.

6. Difficulties with implicit utility comparisons and quantifying the "substitution bias", utility or income to be compared?

The "economic theory index" COLI is often interpreted as answering the question: how much would it cost in today's prices to make the consumer as well off (in terms of "utility") as he was yesterday under the regime of yesterday's prices. There is, however, also a problem in stating equality of utility, whatever its value may be. To achieve such a comparative statement without giving up much of the theory, it is often assumed that the indifference map has remained unchanged. This, however, in turn entails a number of things to remain constant: quality and taste change, new goods on the market, disappearing of old goods etc. are assumed away.[144] Moreover, to contend that two situations represent the same absolute level (or "amount") of utility is inadmissible and invalid because a utility function is only determined up to a monotonic transformation.[145]

By the same token[146], it is also erroneous to treat the difference between such superlative indices like P_{0t}^F and P_{0t}^T and the Laspeyres type CPI, that is, P_{0t}^L - P_{0t}^F, or P_{0t}^L - P_{0t}^T, as correctly reflecting "substitution" and measuring the "substitution bias". Such an interpretation would only be valid and "legitimate in the extremely unlikely circumstance that all other factors determining the pattern of consumption had remained unchanged over the interval covered by the index" (Turvey). Such factors which in actual fact do change are among other things "changes in disposable income distribution and levels; advertising, fashion, magazine articles, pop stars, TV, weather, the fact that households which bought a durable good ... in the first period usually won't want to repeat the purchase in the second period; and that people and their children are older, some have died and new households have been formed."

All sorts of explicit or implicit[147] comparisons between theoretical indices with empirical ones are flawed because the latter will hardly not reflect of these kind of influences. And for all those reasons there is a strong desire not to measure utility but rather income needed to achieve a certain, not specified "level" of utility. And it also should be borne in mind, that the statements and conclusions of the economic theory of index numbers are to be understood as implications of *hypothetical* considerations.

---

[144] Assumptions sometimes made when theory makes allowance for changes of this kind are hardly tenable in real situations. So for example consumers allegedly already know of not yet available goods, they account for quality variations in advance (prior to their appearance!), or a change of taste occurs but does not influence the utility function.

[145] This means: If $U(q_1, q_2, ..., q_n)$ is a "direct" utility function assigning a number u_0 (denoting "amount of utility") to a combination of quantities $q_1, q_2, ..., q_n$, then any increasing monotonic transformation of U is also a direct utility function. Thus, the absolute level of "utility" is as a rule not defined.

[146] An argument repeatedly stressed by Ralph Turvey (see below).

[147] As in the case of equating the observable Fisher-index to the unobservable COLI.

7. Some more technical problems and limitations of the COLI approach

In his paper[148] "The Consumer Price Index and Index Number Purpose" (1999) W. Erwin Diewert enumerates some "criticisms of the cost of living approach". Without going into details we simply quote in what follows the relevant statements used by Diewert in order to characterize the points of criticism:

a) Problems regarding the price vectors

1. we sometimes have in actual fact only unit values rather than true prices,
2. prices are not constant across households, a problem which Diewert induced to some mathematically complicated additions to his theoretic exhibition of the COLI
3. prices are not independent of quantities purchased;

b) Problems concerning the households and the constancy of their utility functions

4. preferences of households and environmental variables are not constant,
5. the assumption that the household has well defined preferences over all possible commodities is unrealistic (even in the case of a single household we can not presuppose that the consumer is completely informed over all [!!] consumption possibilities, independent and stable in his preferences),[149]
6. household composition is not constant over time,
7. the COLI ignores the problem posed by household production (COLI theory views the household only as consuming unit, a more realistic theory should, however, also allow for the allocation of time to market production, work within the household and leisure which also has an impact on utility, consideration which again gave rise to some mathematical refinement of the COLI definitions and relevant theorems);

c) Some more technical aspects

8. goods sometimes cannot be purchased in fractional units as assumed in an optimisation model but rather in integral amounts only,
9. the COLI approach has difficulties in dealing adequately with temporary non-availability as, for example, in the case of seasonal goods.

---

148 Paper presented at the Fifth Meeting of the Ottawa Group, Reykjavik, Aug. 25 - 27, 1999, revised Sept. 1999.

149 Both problems number 4 as well as 5 refer to the intricate domain issue of what sort of prices should affect the COLI as opposed to those conditions which will be kept constant and where consequently a price change will not influence the index.

8. A fundamental criticism of Ralph Turvey

One of the few economists not in line with the majority of COLI fans among economists is Ralph Turvey.[150] His criticism may be summarised in three points.

1. Firstly he is not very impressed by Diewert's theory of superlative indices because it is derived "under strict and unrealistic assumptions", and for him it is difficult to accept that indices like the Fisher, Törnqvist and Walsh index are superlative while the index of Edgeworth[151] (P_{0t}^{ME}) is not, although P_{0t}^{ME} combines both quantities, q_0 and q_t, no less than the other indices. The ME-index deserves rather to be preferred to the other ones because of its quite favorable aggregation properties.
2. Turvey's main objection against the COLI theory is that the reasoning is a good example of a *comparative static* and *purely theoretic* approach in economics which does not lend itself to empirical validation as it is built on:
   - highly unrealistic ideas for example that a household is striving for a level of satisfaction and is engaged in utility maximization on the basis of a utility function which belongs to another period (or a function which is determined by parameters relating to another period);[152] furthermore, and not surprisingly, the theory is also remarkably vague as far as the length of the interval under consideration (from 0 to t) is concerned;[153]
   - conspicuously more concrete descriptions of more or less realistic utility functions are avoided (the focus is rather on functions which are as general as possible and it is no coincidence that no mention is given to how to discriminate between more and less realistic functions), nor is there any theory about how changes in the preferences evolve over time.
3. In addition to such mainly theoretical points, Turvey also raises some questions of practical application where a satisfactory solution is still lacking: the treatment of durable goods, the use of which "extends beyond the period in which they were purchased",[154] the theoretical underpinning of adjusting for quality changes using hedonic regression,[155] computing elementary aggregate indices, the treatment of new goods, seasonally unavailable goods, the treatment of income-dependent prices, missing observations, discounts and other sales in-

[150] We quote some notes made available in the Internet by Ralph Turvey.

[151] or Marshall and Edgeworth in our notation.

[152] For example in $P^{CU}(U_t)$ in period 0 a valuation of goods and services in terms of utility takes place within the framework of preferences and tastes and other side conditions of period t in the future.

[153] This point refers to the existence of price vectors, the availability of the same commodities at two different points in time, and of course also to the interpretation of a time series of COLIs.

[154] This aspect as well as the aforesaid changes in the preferences calls for a multi-period extension of the COLI theory (involving more than two periods to be compared).

[155] According to Turvey COLI theory has not much to offer in this respect. The assumption of a preference function over characteristics on the one hand is farfetched, and in practice quality adjustment can well do without such assumptions. For more details see **sec. 4.2**.

ducements, brokerage fees, insurance,[156] complicated tariff systems in utilities, traffic etc., or finally the design of the item and outlet sampling. In all those instances the COLI theory proved unsatisfactory, or is otherwise of little or no use from the practical point of view.

In summary, at least for the time being,[157] the practice of price statistics may - according to Turvey - safely ignore COLI considerations, and continue to recourse to acceptable un-theoretic solutions.

9. The treatment of new goods

Both approaches, the COLI as well as the COGI approach favor a timely incorporation of new goods in the basket of a CPI. The strategy may produce lower inflation rates in that the typical rapid price decrease of new goods after their introduction into the market will be captured whereas the traditional procedure of accounting for new goods only at the start of new base periods may impart an upwards bias. There are, however, at least two issues specific to the COLI approach, viz.

- the conceptual problem (at least in theory) of establishing a preference order inclusive of a new good (first available in t for example) which is also applicable to a period (say in 0) in which it does not yet exist, and
- the estimation of the utility surplus (offsetting to a certain extent inflation) that the new good represents compared to the consumption pattern valid prior to its appearance.[158]

The utility gain as an inflation-reducing device is unknown to the COGI approach but central to inflation in COLI terms. In other words, to treat increasing utility as equivalent to the lowering of prices is uncommon in the COGI framework, but a necessary consequence of an index approach in which utility is a pivotal element. Two problems are inseparably connected to this kind of reasoning,

- there are not only serious doubts about the empirical estimates but also regarding the theoretical reasonableness of such imputations, moreover
- virtual price reductions have to be compared to a hypothetical base-period price "valid" for a period when the new good was not yet existent.

In micro-economic theory the latter is usually referred to as estimating the "reservation price". This price of a hitherto not available good is the – as a rule extraordinarily high – price at which demand reduces to zero.[159] In view of the insurmountable econometric difficulties involved in the estimation of the reservation price as well as

---

[156] Decisions like taking net or rather gross premiums as determining prices and/or weights.

[157] Turvey acknowledges, however, that continued research along the lines of the highly sophisticated conceptual apparatus of the COLI theory may turn out ultimately as rewarding but under present conditions it is of limited value for the practice of official statistics.

[158] Interestingly, to our knowledge nobody has yet proposed the notion of a utility gain provided by the disappearance of an old good.

[159] It sounds odd, but the estimation is based of course on the existence of a demand function for the good in question at a period in which the good was not yet available or even known to the consumers.

the monetary value of the utility surplus, this treatment of new goods (though from the COLI theory point of view warranted) is usually not recommended, not even by those who advocate the COLI approach in most or all of its other aspects. This may be regrettable as everybody is of course fully aware of the fact that in our days a great deal of improvements which should be contrasted against inflation materialises in new goods and increased consumption possibilities.

10. Final remark: consequences of coupling "inflation" and "compensation"

Criticism of a Laspeyres CPI P_{0t}^{L} is usually concentrated on the importance of accounting for substitution. P_{0t}^{L} is said to provide an upper bound of the COLI and thus lead to "overcompensation". Although there may be no disagreement that it is acceptable, if not imperative that households ameliorate their situation by consuming relatively less (more) of those goods of which cost relatively more (less), this is obviously not a value-free assertion. The justification for this kind of reasoning (i.e. substitution lessens inflation and thus gives rise to a smaller compensation) is not often spelled out in detail. It should be noticed that with this linking of "inflation" and justifiable "compensation"

1. statisticians may well have to face a lack of understanding of their figures in the general public, and even worse
2. they needlessly make a measurement issue dependent on problems of equity that are to be decided politically, and they thereby place statistics in an uncomfortable position.[160]

The implications of both are substantial. As Kathleen Abraham et al. pointed out, the substitution issue is "a case in which what seems obvious to the average economist is not at all obvious to the average citizen" and they quote a statement of an "individual", presumably an average citizen, as follows:

> "... your suggestion that switching [from butter to margarine] reduces inflation is flatly wrong. Were price increases ... to lead people to live in cardboard boxes and eat gravel, that logic would lead to the statistical conclusion that there is no inflation at all." (Abraham et al., 1997, p. 28).

It is clear which kind of answer an economist would give: eating gravel would occur only on a lower level indifference curve and therefore is not the type of substitution an economist has in mind. But the question remains to be answered: how an economist can decide from observation whether a substitution on a constant level of utility has taken place (which should not justify compensation) or a substitution in which the consumer has to put up with a lower level?

[160] I found that in particular Robert A. Pollack made this point very clear. He urges caution with the COLI approach primarily because of such political implications. In order to remove statistics from the political firing line it may be better - especially for official statistics and when indexation is widely used - not to claim to be able to solve the kind of problems politicians often do not dare to decide, viz. which is the "correct" amount of compensation for inflation, or how to settle budget problems by - if need be - reducing certain social programmes?

2.5 Chain indices and Divisia's approach, general introduction

The controversy about advantages or disadvantages of the chain index design is owed in no small measure to ambiguities in the relevant concepts, in particular of the notion "chain index" itself.[161] It is thus useful to start with some terminological clarifications. A more detailed look at the properties of chain indices will be given in chapter 7.

a) Some really necessary terminological distinctions

A distinction between "chain-" and "fixed based" indices is often made. From this we might conclude that a chain index has a different kind of "*base*" than the "ordinary" index formula of Laspeyres or Paasche. Another distinction widely used is between a "fixed weighted" index versus "chain-weighted" or simply "chain" index. This is again misleading. It is for example consistent with calling $P_{0t}^{P} = \frac{\sum p_t q_t}{\sum p_0 q_t}$ a "fixed weighted" Paasche index[162] but the weights q_t vary, as t proceeds from 0 to 1, to 2 etc. weights are q_1, q_2 etc.[163]

Figure 2.5.1: Terminological distinctions referring to chain indices

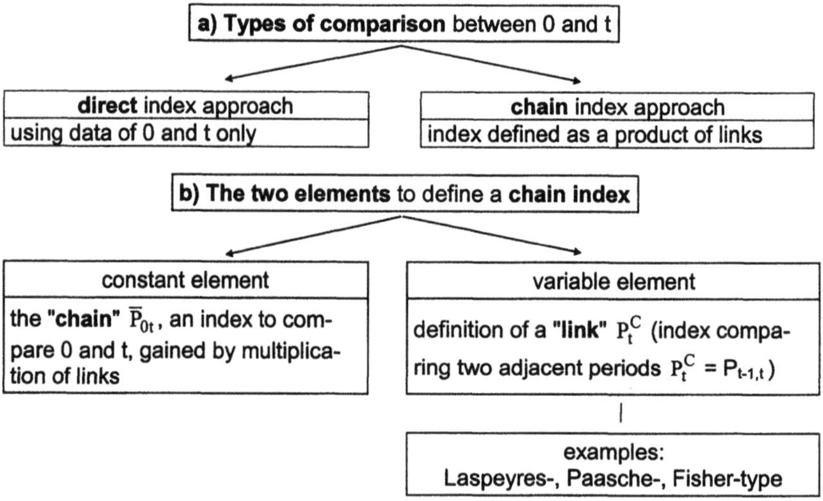

---

[161] Another source of confusion is that axioms as evaluation-criteria, known in the context of direct indices, do not exist in the case of chain indices.

[162] This will be called *direct* Paasche index later, as opposed to the Paasche *chain* index.

[163] There is no point in calling P^P "fixed" weighted by contrast to chain weighted, since the weights of P^P are updated no less frequently than the weights of a chain index.

More importantly, the terms "fixed base" or "fixed weighted" are conjuring on emotions, more precisely: they are degrading[164] of traditional indices such as P_{0t}^{L} or P_{0t}^{P} and they aid and abet misunderstandings concerning the chain indices approach, as if they have solved the problem of choosing a suitable base period, or as if they were the only indices incorporating the most recent weights.[165]

Therefore we will henceforth distinguish between a ***direct*** and a ***chain*** approach to compare the two periods, 0 and t, because what really matters is that the chain principle (much like Divisia indices) consists in explicitly taking into account the *intermediate* periods 1, 2,..., t-1, not just the two end points 0 and t, and thus a chain index compares two periods indirectly rather than directly. To make clear that the results will generally be different, we introduce the symbol $\overline{P}_{0t}$ to denote a chain index in contrast to P_{0t}, which is the corresponding direct index.

A chain index is essentially a specific type of temporal aggregation (over intervals in time) and description of a time series rather than a comparison of two states taken in isolation; it provides a measure of the *cumulated effect* of successive steps (and the shape of the path) from 0 to 1, 1 to 2, ... , t-1 to t.

In the "direct approach" on the other hand, the result P_{0t} is not influenced by what happens to prices and quantities in the intermediate points in time. As a consequence we have some striking features of the chain approach worth noting just at the outset of this section:

There are necessarily always at least **three sources of variation**[166] responsible for the result we get in the case of a chain index,

1. the difference in prices in t compared to 0,
2. the change weights (quantities) have undergone (in response to a change in prices) in the interval from 0 to t, and
3. prices and quantities in the intermediate points in time, that is in 1, 2,. .., t-1.

The **first source** of variation refers to what a price index (as opposed to a quantity index for example) primarily has got to measure. In essence the idea of "pure" is to make index calculation depend on this source only.

The **second source** of variation accounts for the change of quantities ($q_0 \rightarrow q_t$) in response to changes in prices. In view of the idea of a "substitution bias" (in the "economic theory" approach) or of some *direct* index functions characterized by making

[164] As everything changes, and perhaps in these days more rapidly than before, the term "fixed" appears to refer to old-fashioned types of indices not suitable to modern times.

[165] The common prejudice of course is that such weights are the "best" weights and chain indices thereby measure a change of prices more "accurately" and more in line with the relevant consumption pattern than all other indices.

[166] A fourth source is the ever changing "domain of definition" of the index function, that is, the often praised ease with which the selection of goods can be changed from one period to another (not only in the case of rebasing).

use of weights related to *both* periods, 0 and t, an index should also reflect this (second) aspect.[167]

Interestingly there is much discussion about why a chain index is better than a direct index like P_{0t}^{L} for example, but there is little or no reason given for their superiority over those direct indices which also take current weights into account, like P_{0t}^{P}, or "superlative" indices like P_{0t}^{F} (Fisher), P_{0t}^{T} (Törnqvist), and P_{0t}^{W} (Walsh) for example.

The **third source** of variation is clearly something specific for chain indices. Unlike all other index types (except for the Divisia index), a chain index accounts for the *path* connecting 0 with t and is thus reflected by the specific shape of the time series of prices and quantities, that is its own "history". However this seems to be an unintended by-product of chaining rather than deliberately intended or required on theoretical grounds.

Moreover, the operation of chaining (multiplying) is not motivated by some causal model explaining how present prices evolve from past prices but by aiming at consistent temporal aggregation (i.e. over time). Interestingly "path dependence"[168] is just the opposite of this consistency (that is of "chainability" or "transitivity"), such that what multiplication should achieve (i.e. consistent aggregation over time) ironically just cannot be achieved using chain indices.[169]

To define the notion of "chain index" carefully and to avoid much of the confusion met in this field we should introduce the following *two* "elements" of the definition of chain price[170] indices (see **fig. 2.5.1** part b):

1. A constant element is the **"chain"** $\overline{P}_{0t}$, which is always a product of "links" P_{t}^{C}, each of which is a direct index comparing t with the preceding period t-1

   (2.5.1) $\overline{P}_{0t} = P_{1}^{C}P_{2}^{C}...P_{t}^{C} = \prod_{\tau=1}^{\tau=t} P_{\tau}^{C}$, and

2. in defining the **link**[171] $P_{t}^{C} = P_{t-1,t}$ there are numerous solutions we might think of[172], giving rise to Laspeyres- , Paasche- Fisher- and other chain index numbers (depending on the type of link P_{t}^{LC}, P_{t}^{PC}, P_{t}^{FC} etc. that are multiplied

---

[167] Hence on the face of it chain indices and some direct "superlative" indices have the first and the second source of variation in common. The difference between them lies primarily in the third source of variation reflecting the impact of past prices and quantities on present (or most recent) ones in the course of time. There is no explicit "economic theory" of the type of behavior resulting in the path dependence typical for chain indices, and by which chain indices differ from some direct "superlative" indices.

[168] typical for chain indices and the Divisia index.

[169] Conspiciously none of the few transitive indices has variable weights.

[170] The definition also applies mutatis mutandis to quantity indices.

[171] Since a link always compares the reference period t with the *preceding* base period t-1 there is no need for two subscripts. It is sufficient to use only one subscript, t.

[172] This is, so to say, the variable element of the definition of a chain index.

["chainlinked"] to get the chain $\overline{P}_{0t}$, or in particular $\overline{P}_{0t}^{LC}$, $\overline{P}_{0t}^{PC}$ or $\overline{P}_{0t}^{FC}$). The Laspeyres link,[173] as an example is defined as follows

(2.5.2) $$P_t^{LC} = P_{t-1,t}^{L} = \frac{\sum p_t q_{t-1}}{\sum p_{t-1} q_{t-1}},$$

such that $\overline{P}_{0t}^{LC} = P_1^{LC} \cdot \ldots \cdot P_t^{LC}$ is the Laspeyres chain.

There are two important reasons to insist on a distinction between *two* elements of the definition:

- A closer look at arguments put forward in favor of chain indices (see **sec. 7.1**) reveals that most of them result from calling "chain index" what in fact is "only" the link, and from comparing a link $P_t^C = P_{t-1,t}$ with P_{0t} instead of correctly comparing $\overline{P}_{0t}$ with P_{0t}.
- There is some discussion about which formula of the link to prefer, completely ignoring that the properties of the chain in general cannot be concluded from properties of the links.

Note that *only the link* comparing a period t with t-1 *is an index* in the usual manner of being capable of satisfying or violating certain "axioms". A chain is something different. It is *not* an index and can violate axioms, though its links may well satisfy them all.

Due to the multiplication the chain $\overline{P}_{0t}$ is in general a function of the price and quantity vectors $\mathbf{p}_0, \mathbf{q}_0, \mathbf{p}_1, \mathbf{q}_1, \mathbf{p}_2, \mathbf{q}_2, \ldots, \mathbf{p}_{t-1}, \mathbf{q}_{t-1}, \mathbf{p}_t, \mathbf{q}_t$, and not only of the first and last pair of vectors. Hence it should be most unlikely that applying the chain index approach and the traditional direct approach to the same data will yield the same results.

Note also that the existence of a product representation of an index as such is not sufficient to characterize a chain index.[174] To see this, compare the Laspeyres *chain* index $\overline{P}_{03}^{LC}$ which is

(2.5.3) $$\overline{P}_{03}^{LC} = \left(\sum \frac{p_1}{p_0} \frac{p_0 q_0}{\sum p_0 q_0}\right)\left(\sum \frac{p_2}{p_1} \frac{p_1 q_1}{\sum p_1 q_1}\right)\left(\sum \frac{p_3}{p_2} \frac{p_2 q_2}{\sum p_2 q_2}\right)$$

with the *direct* Laspeyres index P_{03}^{L} given by

(2.5.4) $$P_{03}^{L} = \left(\sum \frac{p_1}{p_0} \frac{p_0 q_0}{\sum p_0 q_0}\right)\left(\sum \frac{p_2}{p_1} \frac{p_1 q_0}{\sum p_1 q_0}\right)\left(\sum \frac{p_3}{p_2} \frac{p_2 q_0}{\sum p_2 q_0}\right) = \sum \frac{p_3}{p_0} \frac{p_0 q_0}{\sum p_0 q_0}.$$

---

173 The Paasche link obviously is $\Sigma p_t q_t / \Sigma p_{t-1} q_t$.

174 It is important to recognize that multiplication with respect to index numbers can well have different meanings. The products in eq. 2.5.3 and 4 are not only made of different factors, they also have a different relevance for the respective type of index.

Note that the factors on the right hand side (RHS) of eq. 4 are *not* the "ordinary" indices, P^L_{01}, P^L_{12} and P^L_{23},[175] but a sequence of *rebased* Laspeyres indices

$$P_{01} = \frac{P_{01}}{P_{00}} = \frac{\sum p_1 q_0}{\sum p_{01} q_0},\ P_{12(0)} = \frac{P_{02}}{P_{01}} = \frac{\sum p_2 q_0}{\sum p_1 q_0},\ \text{and}\ P_{23(0)} = \frac{P_{03}}{P_{02}} = \frac{\sum p_3 q_0}{\sum p_2 q_0}.$$

The difference between eq. 3 and 4 is in the first place that in P^L_{03} *only* prices and not at the same time also quantities are updated (which appears less reasonable) whereas in $\overline{P}^{LC}_{03}$ *both* prices and quantities are updated continuously (ostensibly economically more meaningful).[176] With quantities reacting to prices, however, it is not astonishing that $\overline{P}^{LC}_{03}$ will in general differ from P^L_{03}, which is known as *drift* of the chain index[177]. There are some points worth stressing.

1. Calculation, accuracy

   Note that P^L_{03} can be calculated in two ways, by multiplication *and* directly. In the case of $\overline{P}^{LC}_{03}$ there is no such choice, however. Multiplication is the *only* way of compiling a chain index. The implication of this is significant: $\overline{P}_{0t}$ depends on $\overline{P}_{0,t-1}$, $\overline{P}_{0,t-2}$, and so on, while each element of the sequence P_{0t} is an independent estimate.

2. Transitivity

   Many writers conclude from a chain index being defined as a product that the chain index is transitive (in the sense of Fisher's chain test), or "chainable" *by definition* since

   (2.5.5) $\overline{P}_{0t} = \overline{P}_{0k}\overline{P}_{kt}$

   holds.[178] In contrast, the idea of the chain test (or "chainability" or "transitivity") is that the result ($\overline{P}_{0t}$) should be the same for *any* k, irrespective of how the interval (0, t) is partitioned into sub-intervals. However, in general this is *not* true in the case of chain indices. Not only is the indirectly gained result $\overline{P}^{LC}_{06}$ different from the direct one, P^L_{06}, the case of 1-period-links, $\overline{P}_{06} = P_{01}P_{12}...P_{56}$ will in gen-

---

[175] The reason is that P^L is *not* transitive.

[176] With quantities remaining constant there would obviously be no difference between P^L and P^{LC} which makes some people believe that chain indices are the more general approach. If this could be seen this way (which we doubt) a direct index P^L appears only as a badly justified (all quantities held constant) special case. It is of limited use, however, to compare two index functions on this basis. Moreover, notice that eq. 4 shows that weights are not really constant but up-dated at least as far as prices are concerned.

[177] The term "drift" does not mean that the incorrect chain index is drifting away from the correct direct index. We may of course as well think of the direct index drifting away from the (correct) chain index.

[178] According to Banerjee 1975 for example "the circular test is ... satisfied by the chain index" (p. 55). It is often (erroneously) said in literature that this is so "by definition" or "by construction".

eral not yield the same result as for example the chaining of 2-period-links $P_{02}P_{24}P_{46}$.[179]

The name "chain index" is misleading: multiplication ("chaining") may give rise to the impression chainability is met. But this is not true. Different subdivisions of the same interval in time will yield different results, and these results will in general also differ from a direct comparison.

A correct understanding of transitivity (circularity) requires that *each* chaining over the *same* interval in time should yield the same result, that is, indirect comparisons of *all* types should be consistent with direct comparison and also consistent among themselves. This is not given, since the result of annual chaining

(2.5.3a) $\overline{P}_{04}^{LC} = \frac{\sum p_1 q_0}{\sum p_0 q_0} \frac{\sum p_2 q_1}{\sum p_1 q_1} \frac{\sum p_3 q_2}{\sum p_2 q_2} \frac{\sum p_4 q_3}{\sum p_3 q_3}$ and bi-annual chaining

(2.5.3b) $\overline{P}_{04}^{LC*} = P_{02}^{LC} P_{24}^{LC} = \frac{\sum p_2 q_0}{\sum p_0 q_0} \frac{\sum p_4 q_2}{\sum p_2 q_2}$

will in general be different. In contrast to $\overline{P}_{04}^{LC}$, in $\overline{P}_{04}^{LC*}$ there are no prices and quantities of periods 1 and 3 for example. Not only is $\overline{P}_{04}^{LC}$ not equal to P_{04}^{L}, but $\overline{P}_{04}^{LC}$ is also not invariant upon different partitions of the interval from 0 to 4, such that $\overline{P}_{04}^{LC} \neq \overline{P}_{04}^{LC*}$.

3. Reflecting changes in intermediate periods (mystification of multiplication)

   For some "chainers" the idea is attractive that a (better) comparison over a long interval in time is achieved by multiplying comparisons over short sub-intervals (see **sec. 7.1**).[180] The reason[181] given is that a chain index takes all data of a time series into account (there is some appeal in this argument by leaving open precisely what is meant by "taking into account"), in contrast to the traditional direct approach, amounting "to establishing a trend by a kind of end-point method" (Craig 1969, S. 147). However

---

[179] Moreover, it can be shown that the idea of chainability as a requirement is necessarily in conflict with other important aspects of index construction, like additivity and adaptability of weights.

[180] It can be shown that multiplication of links is justified by (implicitly) assuming that prices will change in proportion when change is measured using *different* weights, such that the weight structure will not matter. On the other hand, the change of weights on an ongoing basis (most often viewed as a major advantage of chain indices) is justified on just the opposite grounds: using different (more recent) weights will make a difference, and will result in a more "relevant" measurement of the change in prices. Chain indices obviously try to reconcile two ideas clearly contradictory in themselves: making long term comparisons by chaining and at the same time providing a continuous updating of weights.

[181] As with many other arguments this one has yet to make its meaning clear (see **sec. 7.1**). Chaining seems to create some additional information (the exact nature of which is not specified, however) or to make better use of data (by letting more data enter the formula). The problem with this kind of reasoning is that in statistics more phenomena affecting a result will not necessarily give a more "informative" result. In fact, in statistics more often than not the very opposite is true, and not infrequently any effort is made to reduce the effect of irrelevant and incomparable data and to construct a (standardized) measure in which such influences are eliminated or neutralized, rather than to make calculations sensitive to a greater number of influences.

> Multplication is not a unique, defining feature of chain indices. Nor are there any desirable properties of chain indices to be concluded from multiplication as such. On the contrary: to study figures resulting from a complicated mix of influences such as unit values, is in general of rather limited value.

Multiplication is only a device of *temporal aggregation* by integrating over subintervals. It does not contribute additional information to indices but rather disturbs their axiomatic properties. Interestingly, the SNA 93 discards "unit value indices" (see **sec. 7.2**) for their violation of the principle of pure comparisons with the correct argument.

> Unit value indices are "affected by changes in the mix of items as well as by changes in their prices. Unit value indices therefore cannot be expected to provide good measures of average price change over time" (para 16.13),

but the very same SNA obviously (and amazingly) did not realize that the same type of objections can also be raised against chain indices, which the SNA advocated with great vigour.

4. Changes in the domain of definition (of one link compared with another)

   In eq. 4 it is requisite that summation takes place over the same commodities i = 1, 2, ..., n in each factor of the product. However we may well define

   (2.5.3c) $$\overline{P}_{03}^{LC} = \frac{\sum_i p_{1i}q_{0i}}{\sum_i p_{0i}q_{0i}} \frac{\sum_k p_{2k}q_{1k}}{\sum_k p_{1k}q_{1k}} \frac{\sum_m p_{3m}q_{2m}}{\sum_m p_{2m}q_{2m}}$$

   It is even maintained that one of the major advantages of chain indices is that they enable every new development to be accounted for as rapidly as possible.

b) Weights in the chain approach

A distinction is made by many authors between fixed based indices and variably based indices (see **fig. 2.5.2**). This does not make clear how to classify the chain approach with respect to its type of weights. It should be taken into account that $\overline{P}_{0t}^{LC}$, or any other type of chain index, implicitly has many weighting structures, not only one. The link, not the chain, has a single weighting scheme only. As to the chain, it would be better characterized by saying that it has multiple or more precisely even *cumulative* weights. We therefore prefer the typology of **fig. 2.5.3** while the typology of **fig. 2.5.2** is obscuring, for the following reason: It implicitly uses the term "base" with two different meanings and not only seems to refer to a single weight-base in both cases (chain based and fixed based), but also to a choice between a constant period base and the previous period base (the latter applies to links only).

It is useful (as done already in **sec. 1.1**) to distinguish between two types of "base", the ***reference base (RB)***, that is, the period to which the comparison is related (the period set 100), and the ***weight base (WB)***, that is, the period to which the weights belong, respectively.

Note that each index has one, and only one RB, but an index may have several WBs, for example weights being an average of several weights.[182] It should also be remembered

- that weights related to the base period 0 and weights related to t are *not* logically equivalent, in the sense that for example P_{0t}^{L} (WB: 0) and P_{0t}^{P} (WB: t) are equally "logical", since 0 is kept constant for at least a certain time, thus *one* single period, whilst t is variable (t = 1,2, ...), denoting *several* periods, and
- there is a strange tendency to prefer weights related "symmetrically" to both periods 0 and t, as if this "impartiality" were an advantage as such.

Figure 2.5.2: Traditional classification of weighting schemes*

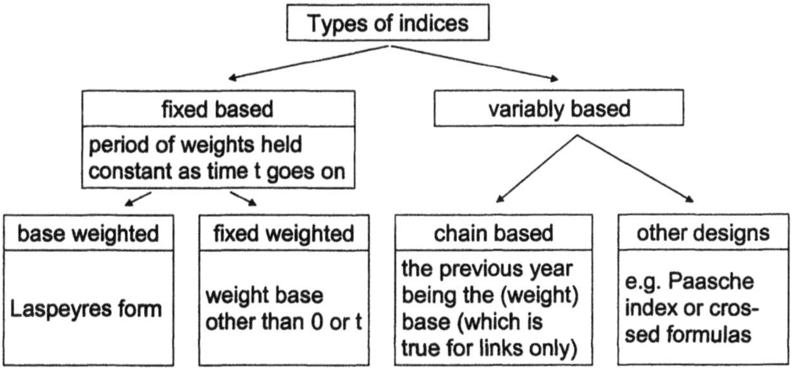

* Other possibilities exist. The weight base can be a period other than 0 or t in the interval (0,t) or it can be periodic (e.g. each starting point of a business cycle).

As stated repeatedly here, a closer look at the nature of 0 and t shows that the idea of a logical "equivalence" of the Laspeyres and Paasche formula[183] and of some alleged advantage of a "symmetric" construction over a single weight is not tenable.

[182] Examples for this quoted in fig. 2.5.3 are the Edgeworth - Marshall -, the Törnqvist - or the Walsh index.

[183] One of the messages Mudgett 1951 set out repeatedly is that the Laspeyres and Paasche formulas are equally good or bad they are logically on "the same footing", "one rests on just as solid a logical foundation as the other" (p. 21), "neither can be considered preferable to the other for logical reasons" (p. 28). Mudgett advocated chain indices with much vigor, predominantly because he rejected fixed weights: "It is my view, however, that fixed weights are the greatest weakness of modern indexes, and one purpose of the following pages is to show that a basic analysis of the problem of index number measurements leads infallably to this conclusion" (p. vi). Mudgett certainly has done much to support the obsession with most recent weights and the idea of logical superiority of the chain approach to the direct approach. It is our firm belief that fixed weights are anything but a weakness. Their strength is to allow "pure" comparisons. *Hence our point is in all statements just quoted that it is the very opposite of Mudgett's standpoint that is true.*

Both periods, RB and WB, coincide in the case of the direct Laspeyres index. Equality of RB and WB is just the main characteristic of the Laspeyres approach: P_{0t}^{L} (WB = RB) is said to be a *base weighted* index[184]. It is often maintained that to change the RB is a purely "formal" maneuver where no problem whatsoever arises. But this is true only for (simple) price relatives, where - interestingly - no problem of choosing a WB and aggregating over commodities is given.

It is common practice to give the following interpretation to "2000 = 100": the results obtained for 2001, 2002, ... are expressed "in terms of", or "in units of" the level of 2000. In view of this interpretation there should be some relation between RB and WB; to change the RB should no longer be viewed as a purely "formal" maneuver, providing a strong case for the direct Laspeyres index where WB = RB.

Figure 2.5.3: Alternative principles to define weights

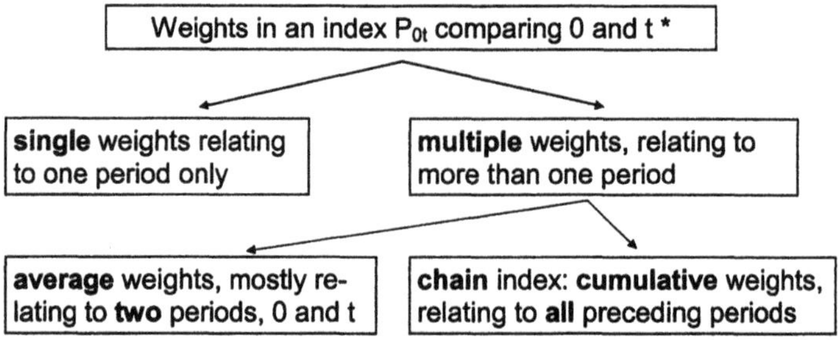

* This scheme compares **direct** indices with **chains**, not direct indices with **links** (as **fig. 7.1.2** does). A direct index will in general be single-weighted, like P^{L} or P^{P}, or average weighted, like P^{ME}, P^{T} or P^{W}, whereas a chain index *always* has cumulative weights.

Turning now to *fixed weighted* indices, take for example a direct Laspeyres index P_{0t}^{L}, rebased to the reference base s (≠ 0, t) by the operation $P_{0t}^{L}/P_{0s}^{L} = \frac{\sum p_t q_0}{\sum p_s q_0} = P_{st(0)}^{L}$ with a constant WB other than s. This also allows a more general notation for all single weighted index formula. Thus we have $P_{0t}^{L} = P_{0t(0)}$ and $P_{0t}^{P} = P_{0t(t)}$, and we get

(2.5.4a) $P_{03}^{L} = P_{01(0)} P_{12(0)} P_{23(0)}$, with factors having a moving RB but a single WB (in brackets) remaining constant as opposed to

(2.5.4b) $\overline{P}_{03}^{LC} = P_{01(0)} P_{12(1)} P_{23(2)}$,

where both RB and WB are moving. In the same manner, in $\overline{P}_{03}^{PC} = P_{01(1)} P_{12(2)} P_{23(3)}$ the links are changing with respect to both RB and WB. It should thus be clear that $\overline{P}_{03}^{LC}$

[184] By contrast a *fixed weighted* index - to be denoted by $P_{0t}(s)$ - also has a constant WB (which is 0), but a WB being neither 0 (the RB) nor t but s.

has three weighting schemes, not only one. We often find comparisons that might be called "unfair", in trying to give the false impression of $\overline{P}_{0t}$ having only one single WB, just like P_{0t} in fact has. It is also far from clear why $\overline{P}_{0t}$ should have an economic-index-theory interpretation (or "true cost of living", COLI interpretation).[185]

In chapter 7 we present merits and (more so, it appears to us) demerits of chain indices on the basis of this definition which hopefully will help to prevent some misjudgements of chaining.

c) Divisia index and its relation to chain indices

The key element of Francois Divisia's index approach in 1925 was to introduce time as a *continuous* variable. It is assumed that two functions, $p_i(t)$ and $q_i(t)$ exist for each commodity (i = 1, ... , n) at *any* point in time. Since in practice conditions like these are unlikely to be found, the significance of Divisia's methodology for official statistics is most limited and likely to remain so in future (a mindset the euthor often met at the University of Essen). On the other hand we will probably never run short of economists who are keen in showing off and enthusing over everything that has the appearance of complicated mathematics[186] and thus are well prepared to mistake Divisa's "integral index" for something of theoretical perfection, in the sense that all index formulas used in practice should approximate it. In a similar vein "chainers" are fond of making reference to Divisia in the false belief that this would give chain indices some reputation as chain indices allegedly simply "translate" the Divisia index into the reality of price observations at discrete points in time.[187]

Let $P(\tau)$ denote the price level function varying continuously over time and let $Q(\tau)$ denote the quantity level function defined analogously. Both $P(\tau)$ and $Q(\tau)$ are assumed to be functions of n continuous price functions $p_i(\tau)$ or quantity functions $q_i(\tau)$ respectively for each individual commodity i = 1, ... , n at each *point* τ in time as mentioned above. It is a mere matter of definition that a value function $V(\tau)$ exists as follows

$$(2.5.6) \qquad V(\tau) = \sum_{i=1}^{n} p_i(\tau) q_i(\tau),$$

denoting an *absolute* value at τ, a point in time (τ continuous). $V(\tau)$ is not a value index measuring value at time τ = t relative to τ = 0, but a value *level*, just like $p_i(\tau)$ and $q_i(\tau)$ denote absolute individual levels, and $P(\tau)$ and $Q(\tau)$ absolute aggregative

---

185 To our knowledge there is no economic theory index referring to a *sequence* of price and quantity vectors.

186 You have a good chance of encountering this type of conceited and yet rather mediocre econometrician at the University of Essen.

187 In truth there are infinitely many discrete approximations (among them even the direct Laspeyres index) to Divisia's continuous time index. Moreover a closer look at Divisia's index reveals that there is no reason to regard the index as a theoretical "ideal" and that the relationship between Divisia's approach and chain indices is not such that the latter is capable of providing a theoretical foundation of chaining.

levels of prices and quantities. As τ is just a point in time, there is no problem in postulating functions $P(\tau)$ and $Q(\tau)$ to have the property that their product is

(2.5.7) $\quad V(\tau) = P(\tau)\,Q(\tau)$.

Unlike the functions $p_i(\tau)$, $q_i(\tau)$ and $V(\tau)$, the levels $P(\tau)$ and $Q(\tau)$ are unobservable. They will eventually lead to a "price index" and "quantity index" respectively. Eq. 7 only *defines* the "levels" $P(\tau)$ and $Q(\tau)$ implicitly by stipulating a relationship between $V(\tau)$, $P(\tau)$ and $Q(\tau)$. It is worth mentioning that:

- in eq. 7 no instruction is given, how $P(\tau)$ and Q, and the "indices" derived from them should be calculated in practice, and
- in contrary to a frequently expressed belief it is not true that "factor reversibility"[188] somehow enters into eq. 7 as an "assumption".

The elegance of Divisia's approach rests on the idea of imagining a point in time τ, or a sufficiently small interval, like $\tau + \Delta\tau$, which simply is so short that no third (structural) component will exist.[189] Hence factor reversibility is an implication of time (τ) being a continuous variable,[190] but no longer valid when a discrete time approximation to the continuous time approach is made.

The next "logical" step is to consider differential changes of $V(\tau)$ according to eq. 6

(2.5.8) $\quad dV(\tau) = \sum_i q_i(\tau)\,dp_i(\tau) + \sum_i p_i(\tau)\,dq_i(\tau)$.

Dividing both sides by V(t), as given in eq. 6, leads to

$$\frac{dV(\tau)}{V(\tau)} = \frac{\sum q_i(\tau)\,dp_i(\tau)}{\sum q_i(\tau)p_i(\tau)} + \frac{\sum p_i(\tau)\,dq_i(\tau)}{\sum p_i(\tau)q_i(\tau)}, \text{ and}$$

(2.5.9) $$\frac{dV(\tau)/d\tau}{V(\tau)} = \frac{\sum q_i(\tau)\,dp_i(\tau)/d\tau}{\sum q_i(\tau)p_i(\tau)} + \frac{\sum p_i(\tau)\,dq_i(\tau)/d\tau}{\sum p_i(\tau)q_i(\tau)}.$$

The next idea is to do the same in the case of eq. 7 (using the product rule of differentiation)

(2.5.8a) $$\frac{dV(\tau)}{d\tau} = Q\frac{dP(\tau)}{d\tau} + P\frac{dQ(\tau)}{d\tau} \text{ or}$$

(2.5.9a) $$\frac{dV(\tau)/d\tau}{V(\tau)} = \frac{dP(\tau)/d\tau}{P(\tau)} + \frac{dQ(\tau)/d\tau}{Q(\tau)}.$$

---

[188] Eq. 7 only postulates the existence of two functions, P and Q, such that their product equals V. It turns out that they are specified in such a way that one function is derived by interchanging arguments of the other function. Hence we may say that factor reversibility holds *by definition* for points in time or infinitesimally small intervals but not necessarily for a long interval from 0 to t.

[189] In other words, the interval is so short that there is no room for a structural component (consisting in *correlated changes* of P and Q) to be effective and observable.

[190] It should be noted that there is no specific assumption in eq. 6 and 7 apart from τ being continuous, a fact responsible for the widespread belief that Divisia'a approach is something "natural", or indisputable.

This eq. simply states the well known fact that the (continuous time) growth rate (or logartihmic derivative) of $V(\tau)$, a *product* of $P(\tau)$ and $Q(\tau)$, is the *sum* of the growth rates (rates of change) of the two factors, $P(\tau)$ and $Q(\tau)$ respectively.

An obvious way to find an index P or Q is to relate the right hand sides of eq. 9 and 9a, and to identify the first (second) term as growth rate of the unknown price level (or quantity level)

$$(2.5.10) \quad \frac{dP(\tau)/d\tau}{P(\tau)} = \frac{d \ln P(\tau)}{d\tau} = \frac{\sum_i q_i(\tau)}{\sum_i q_i(\tau) p_i(\tau)} dp_i(\tau)/d\tau,$$

$$= \sum w_i(\tau) \frac{dp_i(\tau)/d\tau}{p_i(\tau)} = \sum w_i(\tau) \frac{d \ln p_i(\tau)}{d\tau}.$$

where $w_i(\tau) = p_i(\tau) q_i(\tau) / \sum p_i(\tau) q_i(\tau)$. Thus:

The logarithmic derivative (continuous time growth rate) of the unknown price level $P(\tau)$ is the weighted average of individual price levels $p_i(\tau)$, where weights $w_i(\tau)$ are expenditure shares at point τ (and hence of course changing with time) and summation takes place over i = 1, ..., n commodities.

In the same manner, the growth rate of the function $Q(\tau)$ is a weighted arithmetic mean of growth rates of n functions $q_i(\tau)$

$$(2.5.11) \quad \frac{dQ(\tau)/d\tau}{Q(\tau)} = \frac{d \ln Q(\tau)}{d\tau} = \sum w_i(\tau) \frac{d \ln q_i(\tau)}{d\tau}.$$

Thus the two functions $P(\tau)$ and $Q(\tau)$ are defined

- using their relation to $V(\tau)$ (eq. 7), and
- via their continuous time growth rates derived from differential equations (10 and 11).

Calling $\frac{dP(\tau)/d\tau}{P(\tau)}$ the growth rate of the *price* level (and $\frac{dQ(\tau)/d\tau}{Q(\tau)}$ the growth rate of the *quantity* level respectively) is justified on the following grounds (sometimes called Divisia's axiom)[191]:

Assume that quantities in eq. 9 don't change, such that $dq_i(\tau) = 0$ for all I, then

- the change of the quantity level (that is of $Q(\tau)$) according to eq. 11 should also be zero, or in other words
- the change of volume (in eq. 9) should equal the change of prices.

This applies mutatis mutandis to the assumption of no change of prices $dp_i(\tau) = 0$.

It is this consideration that allows us to separate the two differentials (in prices and in quantities) and to identify them as the growth rate of price and quantity level respectively. Interestingly, apart from the length of the time interval in question, this

191 Vogt 1979, p. 50.

idea is no different from the notion of P_{0t}^{L} and pure price comparison, according to which P_{0t}^{L} measures how costs (revenue or more general: value) changed when quantities had remained constant. Change in prices (quantities kept constant) should only result in $V_{0t} = P_{0t}$ and $Q_{0t} = 1$.

However, to assume only this is not enough to derive P_{0t}^{L} and this again shows that such highly general considerations will not suffice to justify any index construction. Assume *both* types of variables will change ($dp_i(\tau) \neq 0$ *and* $dp_i(\tau) \neq 0$) and then of course Divisia and Laspeyres will offer widely different solutions.

Note that, in contrast to other index constructions, Divisia's $Q(\tau)$ and $P(\tau)$ are in fact conceived as some absolute (though not directly observable) figures and that growth of $Q(\tau)$ is in fact to be derived from growth of $P(\tau)$ and vice versa by interchanging arguments.[192]

Another way of looking at Divisia's method is indicated in eq. 10 and 11 and mentioned above already. In essence

Divisia's method consists in constructing a (continuous time) growth rate of the (absolute) price level $P(\tau)$, or quantity level $Q(\tau)$ respectively as a weighted sum of n growth rates; weights $w_i(\tau)$ being shares in total value and varying continuously with time τ.

It became fashionable to call all kinds of aggregates, the growth rates of which are gained from averaging growth rates using variable weights, "Divisia". In this sense there is for example a discussion of the use of "Divisia monetary aggregates", expected to be superior to simple (unweighted) sums of money stock components.

The next step in arriving at Divisia's index is usually explained as follows: To get the price index, solve ("integrate") the differential equation (eq. 10) for P, yielding

$$\text{(2.5.12)} \quad P(t) = P(0)\exp\left(\int_0^t \frac{\sum q_i(\tau) \dfrac{dp_i(\tau)}{d\tau}}{\sum q_i(\tau)p_i(\tau)} d\tau \right) = P(0)\exp\left(\int_0^t \sum w_i(\tau) \frac{d\ln p_i(\tau)}{d\tau} d\tau \right),$$

and thus the price index is given by

$$\text{(2.5.12a)} \quad P_{0t}^{\text{Div}} = \frac{P(t)}{P(0)}.$$

Correspondingly, Divisia's quantity index is

$$\text{(2.5.13)} \quad Q_{0t}^{\text{Div}} = \frac{Q(t)}{Q(0)} = \exp\left(\int_0^t \frac{\sum p_i(\tau)\, dq_i(\tau)}{\sum p_i(\tau) q_i(\tau)} \right).$$

---

192 The incorrect common interpretation of this is that the factor reversal condition is satisfied (or imposed on) by the pair of Divisia indices (whereas in our view factor reversibility in the growth rates is simply a consequence of continuous time; such that there is no place for a "mixed term" on the right hand side of eq. 9).

The name "integral index" stems from the fact that the pair of Divisia indices are derived by solving (integrating) differential equations. The problem, however, is that the integration suffers from *lack of path invariance:* the solutions (integral functions) of the "line integrals" in eq. 12 and 13 depend on the path connecting 0 and t. In contrast the integration

$$(2.5.14)\quad V_{0t} = \frac{V(t)}{V(0)} = \exp\left(\int_0^t \frac{dV(\tau)/d\tau}{V(\tau)} d\tau\right) = \exp\left(\int_0^t \frac{dV(\tau)}{V(\tau)}\right) = \exp[f(0,t)]$$

depends on the endpoints 0 and t only, not on the (shape of the) path connecting them. Thus

$$(2.5.14a)\quad V_{0t} = V_{0k} V_{kt} = \exp\left(\int_0^k \frac{dV}{V} + \int_k^t \frac{dV}{V}\right) = \exp\left(\int_0^t \frac{dV}{V}\right) \text{ or } f(0,t) = f(0,k) + f(k,t).$$

In contrast to f( ), the corresponding functions in eq. 12 and 13 are *not* path invariant.[193] Both approaches, Divisia's and chaining, have path dependence in common, which is often viewed as a major disadvantage as ensues from it (Hulten 1973, p. 1017f):

- "that for at least one of the paths the index does not return to its initial value" (violation of identity),
- that it (the index) "can become arbitrarily large (or small)",[194] and that
- "a multiplicity of index values may be associated with any given point in the set of variables to be indexed" (that is, in the comparison of, for example, p_t with p_0).[195]

All these properties (e.g. violation of identity, reaction to cyclical movement in prices) are well known to be possessed by chain indices as well, another reason for "chainers" to believe in an inherent relationship between the two types of indices (Divisia and chain indices). In particuar it should be noted that not only chain indices but also certain direct indices can be derived

- under specific assumptions concerning the functions $p_i(\tau)$ and $q_i(\tau)$, or
- from various types of discrete time approximations to the continuous time Divisia index

and therefore it may be seen as having an interpretation in terms of Divisia as well.

In some textbooks, for example, it is shown that the direct Laspeyres index can be derived from some specific (though farfetched) conditions (Banerjee 1975, p. 126).

---

[193] Divisia himself was already aware of this drawback and he also proposed the compilation of chain indices as a discrete approximation.

[194] Alluding to a phenomenon called "bouncing" in the case of chain indices. See **sec 7.2**, part d.

[195] This consequence of path dependence gave rise to many investigations in index formulas resulting from taking other paths than the empirical path, and in conditions under which the Divisia index becomes path independent (or "non cycling"). An interesting approach, though highly complex and theoretical (and hence not having received much attention), in this respect is the so called "natural index" of A. Vogt with a straight line path (Vogt 1978b, 1979). See also Usher 1974.

Consider in τ the same (or for all i proportional) quantites as at 0, or assume a (most unlikely) path of quantities such that $q_i(\tau) = \lambda q_i(0) = \lambda q_0$. This gives

(2.5.15) $$\frac{dP(\tau)}{P(\tau)} = \frac{\sum q(\tau)dp(\tau)}{\sum q(\tau)p(\tau)} = \frac{\sum \lambda q_0 dp(\tau)}{\sum \lambda q_0 p(\tau)} = \frac{d\sum q_0 p(\tau)}{\sum q_0 p(\tau)},$$

where the subscript i denoting the commodity is deleted for convenience. Integration of the price differential $\int_0^t \frac{dP(\tau)}{P(\tau)} d\tau$ subject to constant or proportional quantities[196] leads to

(2.5.16) $$\ln P(t) = \ln\left[\sum q_0 p_t\right] + C \text{ where } p_t = p(\tau = t)$$

with an arbitrary constant C, the value of which can be determined by assuming prices p_0 to enter P(0), the price level in base period 0, as do prices p_t with P(t). This means

(2.5.16a) $$\ln\frac{P(t)}{P(0)} = \ln P(t) - \ln P(0) = \ln\frac{\sum p_t q_0}{\sum p_0 q_0}$$

such that we finally arrive at the familiar Laspeyres index $\frac{P(t)}{P(0)} = \frac{\sum p_t q_0}{\sum p_0 q_0} = P_{0t}^L$. Of course P_{0t}^L is (as a result of constant quantities) path invariant as opposed to P_{0t}^{Div}.

There are also attempts to show that other direct indices, like Paasche[197] (P_{0t}^P) or Fisher[198] (P_{0t}^F), might also be regarded as special cases of the Divisia index.

d) Discrete time approximations and weights in Divisia's approach

Divisia's approach differs from other ways of finding index functions in at least three aspects:

- the focus is on (infinitely small) subintervals of the interval from 0 to t, not only on the end points,
- to describe the movement of individual prices, differentials $\frac{d\ln p_i(\tau)}{d\tau}$ are considered, that is, *increments* instead of ratios, like $\frac{p_{it}}{p_{i0}}$ or log changes $\ln\left(\frac{p_{it}}{p_{i0}}\right) \approx \frac{p_{it}}{p_{i0}} - 1$, and finally

---

196 Or: in which the individual price changes are weighted with constant base period weights.

197 The method is similar to the way in which the Laspeyres index was derived: assume $q(\tau) = \lambda q(t)$.

198 making some peculiar and much less reasonably motivated assumptions, however.

- weights $w_i(\tau)$ used in aggregating over commodities (individual price changes) in order to get a price index P (or to get its continuous time growth rate $\frac{dP(\tau)/d\tau}{P(\tau)}$) are not only taken from either or both points 0 and t but they are related to *all* points τ in time in the interval $0 \leq \tau \leq t$.[199]

This makes clear that a certain resemblance between Divisia's index and chain indices exists. The problem is, however, that there are *various* different discrete approximations to Divisia's index. Aside from path dependence this is one of the main shortcomings of Divisia's approach and thus subject to much criticism.

In what follows we will refer to two approximations only, Laspeyres and Paasche, that is to $\overline{P}_{0t}^{LC}$ and $\overline{P}_{0t}^{PC}$. Substituting forward differences $\Delta V_t = V_{t+1} - V_t = \Sigma p_{t+1} q_{t+1} - \Sigma p_t q_t$ for the differential dV (and correspondingly Δp_{it} and Δq_{it} for dp and dq) leads to

(2.5.17) $\Delta V_t = \sum_i q_{it} \Delta p_{it} + \sum_i p_{it} \Delta q_{it} + \sum_i \Delta p_{it} \Delta q_{it}$,

an equation equivalent to eq. 2.5.8, but with a mixed element $\sum_i \Delta p_{it} \Delta q_{it}$ (assumed to tend to zero). It is reasonable therefore to define P_t and Q_t in such a way that

(2.5.18) $$\frac{\Delta P_t}{P_t} = \frac{\sum q_t \Delta p_t}{\sum q_t p_t} \text{ and } \frac{\Delta Q_t}{Q_t} = \frac{\sum p_t \Delta q_t}{\sum q_t p_t}$$

omitting subscripts i for convenience. Using eq. 17 this gives

(2.5.19) $$\frac{\Delta V_t}{V_t} = \frac{\Delta V_t}{\sum q_t p_t} = \frac{\Delta P_t}{P_t} + \frac{\Delta Q_t}{Q_t} + \frac{\sum \Delta p_t \Delta q_t}{\sum q_t p_t}, \text{ or}$$

(2.5.19a) $$\frac{\Delta V_t}{V_t} = \left(\frac{P_{t+1}}{P_t} - 1\right) + \left(\frac{Q_{t+1}}{Q_t} - 1\right) + R_t,$$

where the residual term $R_t = \frac{\sum \Delta p_t \Delta q_t}{\sum q_t p_t}$ will tend to zero and will henceforth be neglected.

Thus a growth factor of a "price level" defined on the basis of this kind of reasoning is

(2.5.20) $$\frac{P_{t+1}}{P_t} = \frac{\sum q_t p_{t+1}}{\sum q_t p_t} = P_{t+1}^{LC},$$

which is the Laspeyres link, and the corresponding *index* (comparing period t with 0) is then

(2.5.21) $$\frac{P_{t+1}}{P_0} = \frac{P_1}{P_0} \frac{P_2}{P_1} \cdots \frac{P_{t+1}}{P_t} = P_1^{LC} P_2^{LC} \ldots P_{t+1}^{LC} = \overline{P}_{0,t+1}^{LC}, \text{ and analogously}$$

---

199 In the view of "chainers" this aspect of updating weights in particular makes Divisia's methodology attractive.

(2.5.22) $\frac{Q_{t+1}}{Q_t} = Q_{t+1}^{LC} = \frac{\sum q_{t+1} p_t}{\sum q_t p_t}$ and (2.5.22a) $\frac{Q_{t+1}}{Q_0} = \overline{Q}_{0,t+1}^{LC}$.

In a similar manner, we may derive Paasche chain indices $\overline{P}_{0t}^{PC}$ (and $\overline{Q}_{0t}^{PC}$) by using *backward* differences $\Delta^* P_t = P_t - P_{t-1}$ and $\Delta^* p_{it} = p_{it} - p_{i,t-1}$ respectively[200]

(2.5.23) $\frac{\Delta^* P_t}{P_t} = 1 - \frac{\sum q_t p_{t-1}}{\sum q_t p_t} = 1 - \frac{P_{t-1}}{P_t} = 1 - \left(\frac{P_t}{P_{t-1}}\right)^{-1} = 1 - \left(P_t^{PC}\right)^{-1}$.

On the basis of such considerations, chain indices are often regarded as being practical (discrete time) approximations of Divisia's index, implicitly claiming chain indices as having some desirable properties in common with Divisia's index, such as factor reversibility. However, as soon as we move from continuous time to discrete time the factor reversal test will no longer be passed.[201] This also applies to links, let alone to chains (products of links). Not surprisingly we again find the "antithetic" relation between Paasche and Laspeyres links to be $V_t / V_{t-1} = P_t^{LC} Q_t^{PC} = P_t^{PC} Q_t^{LC}$.

In what follows we quote the position taken by Diewert, with which we agree (Diewert and Nakamura 1993). In taking infinitesimally small intervals not only do many problems of index construction disappear, such as lack of factor reversibility,[202] but unfortunately also the object of observation itself,

> ".... the smaller we make the unit of time or space within which production or consumption takes place, the less actual production or consumption there will be to observe, and comparisons between these tiny units will become meaningless". (p. 3).
>
> "The problem with this approach is that economic data are almost never available as continuous time variables ... Hence for empirical purpose it is necessary to approximate the continuous time Divisia price and quantity indexes by discrete time data. Since there are many ways of performing these approximations, the Divisia approach does not seem to lead to a definite result". (p. 23). More important still, since the approximations "can differ *considerably* (in amount), the Divisia approach does not lead to a practical resolution of the price measurement problem" (p. 43).[203]

"Chainers" are not the only people who believe to have found theoretical support in Divisia, a situation as such is conducive in casting some doubt on the usefulness of the approach altogether.

To sum up:

---

[200] See Allen, 1976, p. 181, Banerjee 1975, p.127. The analogous derivation of Fisher's ideal index given in these textbooks needs assumptions, however, for which an interpretation is difficult to find.

[201] As with "links vs. chains", a relationship valid in the continuous time approach (for differentials) need not hold also in the discrete time framework (i.e. for differences instead of differentials).

[202] It is often conjectured that Divisia's solution of the factor reversal test is a most elegant one (see for example Neubauer, 1995), but it should not be overlooked that the solution loses much of its elegance when discrete approximations are made.

[203] Emphasis and text in brackets added.

The approach is interesting from the point of view of index theory[204] (and on the part of the economic theory where we habitually have enthusiasm for Divisia in abundance), and thus as is more often than not the case in such situations, its significance is likely to be overrated. However, as regards the needs of practical price statistics, or the desire for a theoretical underpinning of chain indices, disappointment is likely to set in after a while.

2.6 Additive models, Stuvel's and Banerjee's index formulas

Georges Stuvel developed a pair of index formulas, P_{0t}^{ST} and Q_{0t}^{ST}, with some favorable properties and based on what he called "additive analysis" of value *differences* (as opposed to the value *ratio* = value index). Another example of an additive model is the approach of Kali S. Banerjee, in which index numbers are conceived in analogy with factorial experiments.[205] We first show how Stuvel found his formulas along with some alternative ways of deriving these formulas, and discuss their interesting properties and possible generalizations. Thereafter we will proceed to discuss the index theory of Banerjee.

a) Additive decomposition of the value difference

It is useful to make a distinction between a multiplicative analysis and an additive analysis of value decomposition on the one hand and between an analysis on the microlevel (that is, at the level of the i-th individual commodity) and on the macrolevel (in terms of aggregates relating to n commodities) on the other. The objective of the **multiplicative** analysis is to find factors whose product is the value ratio and the objective of the **additive** analysis is to decompose the value difference into a sum of components. The fundamental concepts of Stuvel's approach are summarized in **fig. 2.6.1**.

From this it follows that by equation (*m*) in the multiplicative approach the decomposition is uniquely determined in the single commodity case (microlevel). However, on the macrolevel (in eq. *M*) only V is determined uniquely, P and Q are not. The values of P and Q will depend on which indices are chosen for their measurement. In contrast, in the additive analysis the decomposition is not uniquely determined, even for single commodities (on the microlevel). There are two equations possible (*a1* and *a2*) to measure the price and quantity component respectively, and no good reason can be found to prefer one to the other.

Hence by "additive approach" is meant that the difference of values should be decomposed into a sum of two components, A_i and B_i respectively (or A and B on the macrolevel). The first term A_i (or A) is the price component and the second term, B_i (or B) represents the impact of quantity (volume) rise or decline on the increase or

204 This is owing to the impression that Divisia's index seems to rest on little or no restrictive asumptions.

205 It will be shown, however, that Banerjee's enthusiasm about the factorial approach as being able to provide a unified and general approach to all index number problems, comprising also the so called "economic theory" of index numbers, is (in our view at least) not justified.

decline of value. As the analysis is not even uniquely determined on the microlevel, the following *two* equations with relatives $p_i^* = p_{it} / p_{i0}$ and $q_i^* = q_{it} / q_{i0}$ will hold

$$(a1) \quad v_{it} - v_{i0} = (v_{it} - v_{i0}p_i^*) + (v_{i0}p_i^* - v_{i0}) = v_{i0}p_i^*(q_i^* - 1) + v_{i0}(p_i^* - 1)$$

$$(a2) \quad v_{it} - v_{i0} = (v_{it} - v_{i0}q_i^*) + (v_{i0}q_i^* - v_{i0}) = v_{i0}q_i^*(p_i^* - 1) + v_{i0}(q_i^* - 1).$$

Figure 2.6.1: Stuvel's approach

	Microlevel	Macrolevel
multiplicative approach	(*m*) $\frac{v_{it}}{v_{i0}} = \frac{p_{it}}{p_{i0}} \frac{q_{it}}{q_{i0}}$	(*M*) $V_{0t} = \frac{\Sigma v_{it}}{\Sigma v_{i0}} = P_{0t}Q_{0t}$
additive approach	(*a*) $\Delta v_i = v_{it} - v_{i0} = A_i + B_i$	(*A*) $\Delta V = \Sigma v_{it} + \Sigma v_{i0} = \Sigma A_i + \Sigma B_i = A + B$

The terms A_i and B_i, (or A and B on the macrolevel) are supposed to measure the price and quantity component respectively. The **two** decompositions (equations) in additive analysis are as follows:

Microlevel	Macrolevel
(*a1*) $\Delta v_i = v_{i0}p_i^*(q_i^* - 1) + v_{i0}(p_i^* - 1)$	(*A1*) $\Delta V = V_0P(Q - 1) + V_0(P - 1)$
(*a2*) $\Delta v_i = v_{i0}q_i^*(p_i^* - 1) + v_{i0}(q_i^* - 1)$	(*A2*) $\Delta V = V_0Q(P - 1) + V_0(Q - 1)$

where p_i^* and q_i^* denote the price relative and quantity relative respectively.

Summation over all n commodities and division by $V_0 = \Sigma v_{i0} = \Sigma p_0 q_0$ yields equations A1* and A2* (omitting the subscripts 0 and t for convenience of presentation). The corresponding macroequations are

$$(A1^*) \quad V - 1 = (V - P^L) + (P^L - 1) = P^L(Q^P - 1) + (P^L - 1)$$

$$(A2^*) \quad V - 1 = (V - Q^L) + (Q^L - 1) = Q^L(P^P - 1) + (Q^L - 1),$$

where $V - 1 = V_{0t} - 1 = \Delta V/V_0$

and the task is to define a pair of unknown indices, P and Q, in (A1) and (A2) such that they comply with (*A1**) and (*A2**).

Note that all equations presented so far are nothing but simple identities. To define the two components, A and B, or the indices P and Q, a decision has to be made as to which equation should be used, (*A1**) or (*A2**), or both of them. Stuvel's way of defining A and B and thereby of finding two new index formulas (P and Q) is to give both equations equal weights.[206]

The details of Stuvel's steps in deriving his new index formulas are somewhat complicated. His derivation ends up with a quadratic equation in P from which he gets P_{0t}^{ST} and a quadratic equation in Q giving Q_{0t}^{ST}. Moreover, one of the equations relating the terms A and B to P and Q (the unknown index numbers becoming P_{0t}^{ST} and Q_{0t}^{ST})[207] implies factor reversibility. The final result is the quadratic equation[208]

(2.6.1) $$P^2 + \frac{A-B}{V_0}P - V_{0t} = 0,$$

the nonnegative solution of which renders Stuvel's price index:

(2.6.2) $$P = \frac{B-A}{2V_0} + \sqrt{\left(\frac{B-A}{2V_0}\right)^2 + V_{0t}} = \frac{P_{0t}^L - Q_{0t}^L}{2} + \sqrt{\left(\frac{P_{0t}^L - Q_{0t}^L}{2}\right)^2 + V_{0t}} = P_{0t}^{ST}.$$

The corresponding quadratic equation in Q reads as follows:

$Q^2 + \frac{B-A}{V_0}Q - V_{0t} = 0$. One of the two roots of this equation is given by

(2.6.3) $$Q = \frac{A-B}{2V_0} + \sqrt{\left(\frac{A-B}{2V_0}\right)^2 + V_{0t}} = \frac{Q_{0t}^L - P_{0t}^L}{2} + \sqrt{\left(\frac{Q_{0t}^L - P_{0t}^L}{2}\right)^2 + V_{0t}} = Q_{0t}^{ST}$$

This is the formula for a quantity index Stuvel was searching for. Again, the other root ($-\sqrt{\ }$) is irrelevant because the result will be negative.

b) Properties of Stuvel's formulas

What follows is an attempt to list all properties and characteristics of Stuvel's indices as far as they are known to us and reported in index literature. The outstanding property of Stuvel's pair of index functions, P_{0t}^{ST} and Q_{0t}^{ST}, is of course their being "ideal" (ability to pass the factor reversal test), like for example Fisher's index and only very

---

[206] In his view it is legitimate to take a simple (unweighted) mean of the two equations resulting from additive analysis, because each of the two equations is onesided. For Stuvel, this evenhanded ("*symmetric*", "fair" or "unbiased") *treatment of both equations* (a1 and a2, or A1* and A2* respectively) is a major advantage of his indices which also made him believe that his formulas are to be preferred to the two extremes of onesidedness in additive analysis, the indices of Laspeyres and Paasche (it turned out that these traditional formulas make use of only one of the two equations of the additive model; therefore Stuvel's critcism of onesidedness).

[207] $A + B = V_0(PQ - 1) = V_t - V_0$ and $A - B = V_0(Q - P)$.

[208] See the preceding footnote for the relationship between A and B and P and Q.

few other index formulas. This can easily be verified taking into account that $\frac{Q_{0t}^L - P_{0t}^L}{2} = -\frac{P_{0t}^L - Q_{0t}^L}{2}$ and thus $P_{0t}^{ST} Q_{0t}^{ST} = V_{0t}$.

Furthermore, in Stuvel's understanding – as mentioned above – the "symmetry" with respect to the two types of additive decomposition is another element of superiority of his pair of index funtions over the "heavily biased" formulas of Laspeyres and Paasche which simply represent the two extreme solutions.

axioms, "tests" and other criteria **satisfied**
all fundamental criteria, like dimensionality, comensurability, (strict) monotonicity, proportionality, **factor reversal** test, time reversal test, consistency in aggregation and equality test.
axioms **violated** and some additional remarks
linear homogeneity[1], no interpretation as means of price relatives and in terms of rising costs of a budget (basket)[2].

1) this is true also for the generalized Stuvel index except for the (limiting) cases of the pairs P^L, Q^P and P^P, Q^L.

2) the Stuvel indices are generally difficult to interpret economically.

It is easy to see that linear homogeneity is violated (though only slightly as demonstrated in the following **ex. 2.6.1**) due to the following inequality

(2.6.4) $$P_{0t}^{ST}(\lambda) = \frac{\lambda P_{0t}^L - Q_{0t}^L}{2} + \sqrt{\left(\frac{\lambda P_{0t}^L - Q_{0t}^L}{2}\right)^2 + \lambda V_{0t}} \neq \lambda P_{0t}^{ST} \text{ because}$$

$$\frac{\lambda P_{0t}^L - Q_{0t}^L}{2} \neq \lambda \frac{P_{0t}^L - Q_{0t}^L}{2}.$$

Example 2.6.1

To show that Stuvel's index meets proportionality but not linear homogeneity we start with making the assumption that all prices will rise by 50% relative to the base period prices.

i	p_{i0}	p_{it}	q_{i0}	q_{it}
1	12	18.0	10	8
2	15	22.5	8	6
3	20	30.0	5	4

Hence $P^L_{0t} = 1.5$ and taking quantities as given above we get $Q^L_{0t} = 266/340 = 0.7824$ and V_{0t} = 399/340 = 1.1735. Stuvel's indices therefore are

$$P^{ST}_{0t} = \frac{1.5-0.7824}{2} + \sqrt{\left(\frac{1.5-0.7824}{2}\right)^2 + 1.1735} = P^L_{0t} = 1.5 \text{ and } Q^{ST}_{0t} = Q^L_{0t},$$

which shows that proportionality is satisfied. Next we attempt to verify that linear homogeneity is violated. Assume prices p_{it} being unanimously 50% higher than before, that is 27, 33.75 and 45 (instead of 18, 22.5 and 30). The consequence is that P^L_{0t} and V_{0t} will also be 50% higher but Q^L_{0t} remains unchanged. The new result for Stuvel's index is

$$P^{ST}_{0t} = \frac{2.25-0.7824}{2} + \sqrt{\left(\frac{2.25-0.7824}{2}\right)^2 + 1.7603} = 2.24636 < P^L_{0t} = 2.25 \text{ (as required}$$

by linear homogeneity). Thus P^{ST}_{0t} now is not exactly 50% (but only 49.76%) higher than before. Q^{ST}_{0t} remains 0.7852. ♦

c) Alternative ways of deriving Stuvel's index formulas

We are now able to add three[209] considerations conducive to new insights into Stuvel's formulas and thereby not only show how one might arrive at these formulas but also indicate how one might give them some new meaningful interpretations:

1. Distances to the Laspeyres formulas

We may say *firstly* that Stuvel's approach aimed at finding a pair of index functions, P and Q, such that two conditions are satisfied simultaneously viz.

(2.6.5) factor reversal test	$V_{0t} = PQ$ **and**
(2.6.6) distance condition[210]	$P - P^L_{0t} = Q - Q^L_{0t}$

We arrive at P^{ST}_{0t} and Q^{ST}_{0t} when this system is solved for P and Q. There is an obvious way of arriving at a more general system of equations of which eqs. 2.6.5 and 6 are only special cases (see **fig. 2.6.2**). Note that we find a *difference* in eq. 2.6.6 while a similar condition for *ratios* leads to Fisher's indices: $P/P^L_{0t} = Q/Q^L_{0t}$ holds for $P = P^F_{0t}$, $Q = Q^F_{0t}$.

---

[209] There are two other derivations of Stuvel's formulas, the first shown within the framework of Banerjee's analysis (this section, see below), and the second in **sec. 3.6**.

[210] This condition is implicitly postulated in Stuvel's equation $(A-B)/V_0 = Q^L - P^L = Q - P$ in his original analysis. The equation results from the idea of having P and Q conform with the criteria of additive analysis (or: to ensure an additive *macro*analysis consistent with the additive decomposition in *micro*-analysis).

Stuvel himself already found this interpretation: Whilst the pair P^L_{0t} and Q^L_{0t} fails the factor reversal test, another pair of indices, P and Q should be found, such that the price index P is equally away from P^L_{0t} as the quantity index Q is away from Q^L_{0t} and P and Q are factor reversible.

2. Corrections applied to Laspeyres indices to make indices factor reversible

According to a *second* interpretation[211] (Olt 1996, 154), P^{ST}_{0t} and Q^{ST}_{0t} (and some other index formulas as well) can be found by making "corrections" of P^L_{0t} and Q^L_{0t} such that

1. *multiplicative* corrections: $P^g_{0t} := g\,P^L_{0t}$, $Q^g_{0t} := g\,Q^L_{0t}$ and $P^g_{0t}\,Q^g_{0t} = V_{0t}$

   (P^g_{0t} and Q^g_{0t} satisfy the factor reversal test)

2. *additive* corrections: $P^s_{0t} := s + P^L_{0t}$ and $Q^s_{0t} := s + Q^L_{0t}$ and $P^s_{0t}\,Q^s_{0t} = V_{0t}$.

The solution of the first exercise is $g = \sqrt{P^P_{0t}/P^L_{0t}}$ (the term g is obviously a function of the covariance in Bortkiewicz's theorem) yielding Fisher's indices: $P^g_{0t} = P^F_{0t}$ and $Q^g_{0t} = Q^F_{0t}$ (see the first uniqueness theorem UT-1 in **sec. 3.1**).

The second exercise results in $P^s_{0t} = P^{ST}_{0t}$ and $Q^s_{0t} = Q^{ST}_{0t}$, since the "problem" is simply a reformulation of eqs. 2.6.5 and 6, the latter requires $P - P^L_{0t} = Q - Q^L_{0t}$ or, equivalently $s = P^s_{0t} - P^L_{0t} = Q^s_{0t} - Q^L_{0t}$.

3. Accounting for the covariance C between price and quantity relatives

A *third* interpretation is less appealing and somewhat complicated. It starts with the fact that $P^L Q^L > V_{0t}$ (or $< V_{0t}$) depending on the sign of the covariance C between price and quantity relatives ($> V_{0t}$ if $C < 0$ and $< V_{0t}$ if $C > 0$) since according to Bortkiewicz's theorem $C = Q^L(P^P - P^L) = P^L(Q^P - Q^L) = V - P^L Q^L$.

---

[211] It can easily be seen that this is in fact only another way of expressing the idea of the first interpretation.

Figure 2.6.2: Alternative ways[1] of deriving and interpreting P_{0t}^{ST} and Q_{0t}^{ST}

first condition: Find a pair of indices P (price index) and Q (quantity index) such that they pass factor reversal test $V_{0t} = P_{0t}Q_{0t}$ (equation2.6.5) **and**

second condition (**special**)	**second** condition (**general**)
give both types of additive decompostion (*A1*, *A2*) the same weight ½,	give additive decompostion (*A1*, *A2*) weights w and 1-w respectively
or equivalently: $P - P_{0t}^{L} = Q - Q_{0t}^{L}$	equivalently: $w(P - P_{0t}^{L}) = (1-w)(Q - Q_{0t}^{L})$
P should be equally away[2] from P_{0t}^{L} as Q is away from $Q_{0t}^{L} \rightarrow$ ordinary Stuvel indices.	$(0 \leq w \leq 1)$ P_{0t}^{ST} and Q_{0t}^{ST} is the special case of w = ½ → **Generalized Stuvel** indices

1) several other interpretations are presented in the text below

2) difference, additive "correction"

In writing (*) $P = \frac{1}{2}(P^L - Q^L + R)$ and $(**) Q = \frac{1}{2}(Q^L - P^L + R)$

where R is a "residual" we get an expression for 4V = 4PQ, namely $4V = R^2 - (P^L - Q^L)^2$, from which it is easy to arrive at R. Substituting $\sqrt{(P^L - Q^L)^2 + 4V}$ for R in eq. (*) we obtain

$$P = \frac{P^L - Q^L}{2} + \frac{1}{2}R = \frac{P^L - Q^L}{2} + \sqrt{\frac{1}{4}(P^L - Q^L)^2 + V}$$ which is simply P_{0t}^{ST}.

In the same manner, we get the equation for Q_{0t}^{ST} from eq. (**). The relationship between the equations (*) and (**) and the theorem of Bortkiewicz can be seen when in the equation of R the term V is replaced by $Q^L P^P$, giving

$$R = \sqrt{(P^L)^2 + 2P^L(2Q^P - Q^L) + (Q^L)^2} = \sqrt{(P^L + Q^L)^2 + 4C}.$$

We now may distinguish the following cases:

covariance C[1]	correction R	P^P, P^L	Stuvel's formulas
zero (C = 0)	$\sqrt{(P^L + Q^L)^2} = P^L + Q^L$	$P^L = P^P$, $Q^L = Q^P$, and $V = P^L Q^L$	$P = P^L = P^{ST}$ and also[2] $Q = Q^L = Q^{ST}$
negative (C < 0)	$R = \sqrt{(P^L + Q^L)^2 + 4C} < P^L + Q^L$	$P^P < P^L$, $Q^P < Q^L$,	$P^{ST} < P^L$ and $Q^{ST} < Q^L$
positive (C > 0)	$R > P^L + Q^L$ [3]	$P^P > P^L$, $Q^P > Q^L$ and $V > P^L Q^L$	$P^{ST} > P^L$ and $Q^{ST} > Q^L$

Note that the amount of the positive or negative gap between P^L and P^{ST} and between Q^L and Q^{ST} is the same (= first interpretation).

1) between price and quantity relatives

2) due to eq. **

3) from $R > (P^L + Q^L)$ follows with eq. (**) that $Q = Q^{ST} > Q^L$.

We will return to this kind of analysis in **sec. 3.6** where another (closely related) interpretation of Stuvel's indices is given.

d) Generalization of Stuvel's index formulas

A more general solution in dealing with the duality of additive decompositons is to introduce weights w and 1-w ($0 \leq w \leq 1$) to be attached to the two fundamental equations of the additive approach, eq. *A1** and *A2** respectively. This is a rather "natural" method and Stuvel's solution is comprised in it as the special case w = 1/2.

Multiplication of A1* and A2* with weights w and 1-w (where $0 \leq w \leq 1$) results in

(*A1***) $w(V - 1) = w(V - P^L) + w(P^L - 1)$
(*A2***) $(1-w)(V - 1) = (1-w)(V - Q^L) + (1-w)(Q^L - 1)$

In the case of of w = 1- w= ½ we get:

(*A1**) $(V - 1) = (V - P^L) + (P^L - 1)$
(*A2**) $(V - 1) = (V - Q^L) + (Q^L - 1)$

Upon summation of both equations *A1*** and *A2*** we get

$$V - 1 = [w(V-P^L) + (1-w)(Q^L-1)] + [w(P^L-1) + (1-w)(V-Q^L)].$$

It is remarkable that in both parts of the equation the indices P^L and Q^L appear in the same combination

$$w(V-P^L) + (1-w)(Q^L-1) = [wV - (1-w)] - [wP^L - (1-w)Q^L] = [wV - (1-w)] - S \text{ and}$$
$$w(P^L-1) + (1-w)(V-Q^L) = [(1-w)V - w] + [wP^L - (1-w)Q^L] = [(1-w)V - w] + S.$$

Thus any pair of index formulas P and Q which satisfies the condition

$$S = wP^L - (1-w)Q^L = wP - (1-w)Q$$

will be consistent with the additive analysis.

The generalized Stuvel indices, P = $P_{0t}^{ST}(w)$ and Q = $Q_{0t}^{ST}(w)$, can be found by solving the following system of equations:

(2.6.5) $PQ = V_{0t}$

(2.6.6) $wP - (1-w)Q = wP^L - (1-w)Q^L$, equivalently: $w(P-P^L) = (1-w)(Q-Q^L)$.

Figure 2.6.3: Index functions as variants of the generalized Stuvel indices

w	C	$P_{0t}^{ST}(w)$	$Q_{0t}^{ST}(w)$
1/2	1	Stuvel $P_{0t}^{ST}(1/2) = P_{0t}^{ST}$	Stuvel $Q_{0t}^{ST}(1/2) = Q_{0t}^{ST}$
1	0	Laspeyres P_{0t}^{L}	Paasche Q_{0t}^{P}
0	∞	Paasche P_{0t}^{P}	Laspeyres Q_{0t}^{L}
$Q^L/(P^L+Q^L)$	P^L/Q^L (*)	Fisher P_{0t}^{F}	Fisher Q_{0t}^{F}
$Q^L/(P^P+Q^L)$	P^P/Q^L	D+ $\sqrt{D^2 + (P_{0t}^{P})^2}$ where $D = \frac{1}{2}(P_{0t}^{L} - P_{0t}^{P})$	D+ $\sqrt{D^2 + (Q_{0t}^{L})^2}$ where $D = \frac{1}{2}Q_{0t}^{L}(1 - P_{0t}^{L} / P_{0t}^{P})$
$Q^P/(P^L+Q^P)$	P^L/Q^P	D+ $\sqrt{D^2 + (P_{0t}^{L})^2}$ where $D = \frac{1}{2}P_{0t}^{L}(-Q_{0t}^{L} / Q_{0t}^{P})$	D+ $\sqrt{D^2 + (Q_{0t}^{P})^2}$ where $D = \frac{1}{2}(Q_{0t}^{L} - Q_{0t}^{P})$

(*) Setting $w = Q^P/(P^P+Q^P)$ and $c = P^P/Q^P$ will yield the same result, i.e. Fisher's index.

The solution found is:

(2.6.7) $$P_{0t}^{ST}(w) = \frac{P_{0t}^{L} - \frac{1-w}{w}Q_{0t}^{L}}{2} + \sqrt{\left(\frac{P_{0t}^{L} - \frac{1-w}{w}Q_{0t}^{L}}{2}\right)^2 + \frac{1-w}{w}V_{0t}}\ ,\text{ and}$$

(2.6.8) $$Q_{0t}^{ST}(w) = \frac{Q_{0t}^{L} - \frac{w}{1-w}P_{0t}^{L}}{2} + \sqrt{\left(\frac{Q_{0t}^{L} - \frac{w}{1-w}P_{0t}^{L}}{2}\right)^2 + \frac{w}{1-w}V_{0t}}\ .$$

The next interesting feature of this approach is that some well known index formulas can be derived by choosing appropriate values for w or $c = \frac{1-w}{w}$. Some formulas resulting from certain w,c - parameter constellations are listed in **fig. 2.6.3**. In some textbooks P^{ME} and Q^{ME} are also mentioned as special variants of the generalized Stuvel indices.

e) Stuvel's approach and interpretations of price index formulas

Laspeyres

The choice w = 1 in eq. 8 and 9 leads to $Q = Q^P$ and $P = P^L$. This is equivalent to the case of looking at eq. *A1** only while ignoring eq. *A2** altogether

(*A1**) $V - 1 = \frac{\Delta V}{V_0} = (P^L - 1) + (Q^P - 1)P^L = (P - 1) + (Q - 1)P$

The microfoundation of this equation is

(*a1*) $v_{i0}p_i^*(q_i^* - 1) + v_{i0}(p_i^* - 1)$ and the general conception is:

value change = pure* price change + quantity change valued at *new* prices

* at **constant** (base period) **quantities** (P-1).

Paasche

Analoguously we may say: putting w = 0 into eq. 8 gives $P = P^P$ and consequently $Q = Q^L$; this is equivalent to focusing on eq. *A2** only

(*A2**) $V - 1 = (Q^L - 1) + (P^P - 1)Q^L = (Q - 1) + (P - 1)Q$

The microfoundation of this equation is

(*a2*) $v_{i0}q_i^*(p_i^* - 1) + v_{i0}(q_i^* - 1)$ and the general concept of this approach is

value change = pure* quantity change + price change at *new* quantities

* at **constant** (base period) **prices** (Q-1).

Fisher

Upon setting $c = \frac{1-w}{w} = \frac{P^L}{Q^L}$ in eq. 8 we obtain $P^2 = \frac{1-w}{w} V = \frac{P^L}{Q^L} Q^L P^P$ and therefore $P = \sqrt{P^L P^P} = P^F$.

f) Banerjee's factorial approach in index theory

1. Prices and quantities as factors in an experiment (analogy with factorial experiments)

The analysis of variance (ANOVA) applied to the value difference $\Delta V = V_t - V_0$ (additive model) is known as the "factorial approach" to index theory. It may be useful to recall at first the ANOVA method in the simplest case, in which only two "factors", A and B, are involved. Each of the factors is dichotomous, that is, A has only two levels, a_1 and a_2, and it is convenient to set $a_1 = 1$ (existence of the factor or "treatment" A) and $a_2 = 0$ (nonexistence). The levels (of factor B) $b_1 = 1$ and $b_2 = 0$ are defined analogously. An observation y_{ij} is "explained" by

(2.6.9) $$y_{ij} = \mu + \alpha_i + \beta_j + \gamma_{ij} + \varepsilon_{ij} \qquad i,j = 0,t$$

where μ represents an overall "level" (the general effect) and ε_{ij} some (random) disturbance. The coefficients α_i and β_i measure the isolated effect of the i-th level of factor A and the j-th level of factor B respectively. The term γ_{ij} is known as "interaction", measuring the extent to which y is affected by the *combined* influence (which is not necessarily simply the sum) of A and B. Forthwith the factors A and B are identified as the price-(P) and quantity-(Q) factors, observed at two points in time, and for which "absolute indices" (denoted by an asterix) are defined.[212] The ensuing (traditional) indices (as relative terms) then are given as follows

(2.6.9a) $$P_{0t} = \frac{P_t^*}{P_0^*} \text{ and } Q_{0t} = \frac{Q_t^*}{Q_0^*}.$$

This leads to the following system of equations (notations)

	price 0	Price t
quantity 0	$Y_{00} = \sum p_0 q_0 = P_0^* Q_0^*$	$Y_{t0} = \sum p_t q_0 = P_t^* Q_0^*$
quantity t	$Y_{0t} = \sum p_0 q_t = P_0^* Q_t^*$	$Y_{tt} = \sum p_t q_t = P_t^* Q_t^*$

The capital letter Y in contrast to y in eq. 10 is introduced to indicate that the terms refer to aggregates with n commodities involved and no longer to a single commodity.

[212] One of the key ideas of Banerjee's factorial approach is to assign what may be called "absolute indices" to "levels" of prices and quantities and to define the required (relative) price index P_{0t} and quantity index Q_{0t} respectively in terms of these absolute indices. In this respect we have a marked similarity to Divisia's approach. However, unlike Divisia's method, such absolute levels are defined now for *discrete* points in time only.

2. The system of equations to derive formulas for price and quantity indices

Minimizing $\sum_{i,j}\varepsilon_{ij}^2$ under the constraint $\sum_i \alpha_i = \sum_j \beta_j = \sum_i \gamma_{ij} = \sum_j \gamma_{ij} = 0$ leads to the Least Squares (LS) estimation equations as follows

(2.6.10a) $4\hat{\mu} = (\Sigma p_0 q_0 + \Sigma p_t q_0 + \Sigma p_0 q_t + \Sigma p_t q_t) = + P_0^* Q_0^* + P_t^* Q_0^* + P_0^* Q_t^* + P_t^* Q_t^*$

(2.6.10b) $2\hat{\alpha} = [(\Sigma p_t q_0 - \Sigma p_0 q_0) + (\Sigma p_t q_t - \Sigma p_0 q_t)] = - P_0^* Q_0^* + P_t^* Q_0^* - P_0^* Q_t^* + P_t^* Q_t^*$

(2.6.10c) $2\hat{\beta} = [(\Sigma p_0 q_t - \Sigma p_0 q_0) + (\Sigma p_t q_t - \Sigma p_t q_0)] = - P_0^* Q_0^* - P_t^* Q_0^* + P_0^* Q_t^* + P_t^* Q_t^*$

(2.6.10d) $2\hat{\gamma} = [(\Sigma p_t q_t - \Sigma p_t q_0) - (\Sigma p_0 q_t - \Sigma p_0 q_0)] = + P_0^* Q_0^* - P_t^* Q_0^* - P_0^* Q_t^* + P_t^* Q_t^*$

The meaning of the four equations in terms of factors and their effects can be demonstrated, taking eq. 2.6.10b as an example:

the effect of prices changed $p_0 \to p_t$ taken in isolation and assuming	
constant quantities of period 0	constant quantities of period t
$\Sigma p_t q_0 - \Sigma p_0 q_0 = \hat{\alpha}_0$	$\Sigma p_t q_t - \Sigma p_0 q_t = \hat{\alpha}_t$.

A reasonable estimate of $\hat{\alpha}$ now appears to be $\hat{\alpha} = (\hat{\alpha}_0 + \hat{\alpha}_t)/2$ *
giving $\hat{\alpha} = \frac{1}{2}[(\sum p_t q_0 - \sum p_0 q_0) + (\sum p_t q_t - \sum p_0 q_t)]$, that is, $\hat{\alpha}$ as defined in eq. 2.6.10b.

* the analogous estimate of $\hat{\beta}$ is $\hat{\beta} = (\hat{\beta}_0 + \hat{\beta}_t)/2$ in eq. 2.6.10c.

Assuming nonexistence of interaction ($\gamma = 0$) is equivalent to $\hat{\alpha}_0 = \hat{\alpha}_t$ and $\hat{\beta}_0 = \hat{\beta}_t$, or $[(\Sigma p_t q_t - \Sigma p_0 q_t) - (\Sigma p_t q_0 - \Sigma p_0 q_0)] = 0$ amounting in fact to equation 2.6.10d. After some algebra not worth describing here Banerjee ended up with system of equations (resulting from a re-arrangement of equations 10a through 10d)

(2.6.11a) $4\hat{\mu} = (P_t^* + P_0^*)(Q_t^* + Q_0^*) = V_0(P_{0t} + 1)(Q_{0t} + 1)$

(2.6.11b) $2\hat{\alpha} = (P_t^* - P_0^*)(Q_t^* + Q_0^*) = V_0(P_{0t} - 1)(Q_{0t} + 1)$

(2.6.11c) $2\hat{\beta} = (P_t^* + P_0^*)(Q_t^* - Q_0^*) = V_0(P_{0t} + 1)(Q_{0t} - 1)$

(2.6.11d) $2\hat{\gamma} = (P_t^* - P_0^*)(Q_t^* - Q_0^*) = V_0(P_{0t} - 1)(Q_{0t} - 1)$

where $V_0 = \Sigma p_0 q_0 = P_0^* Q_0^* = Y_{00}$.

Two equations, viz. 11b and 11c of this system of equations are already known from Stuvel's analysis. It thus comes as no surprise that one of Banerjee's solutions, which is the solution using eqs. 11b and 11c, turns out to be exactly Stuvel's pair of indices[213].

Figure 2.6.4: Banerjee's factorial approach

The system of equations

	Equation	derived from condition for
(a)	$(P_{0t}+1)(Q_{0t}+1)=V_{0t}+P_{0t}^{L}+Q_{0t}^{L}+1$	grand mean (μ)
(b)	$(P_{0t}-1)(Q_{0t}+1)=V_{0t}+P_{0t}^{L}-Q_{0t}^{L}-1$	factor price (α)
(c)	$(P_{0t}+1)(Q_{0t}-1)=V_{0t}-P_{0t}^{L}+Q_{0t}^{L}-1$	factor quantity (β)
(d)	$(P_{0t}-1)(Q_{0t}-1)=V_{0t}-P_{0t}^{L}-Q_{0t}^{L}+1$	interaction (γ)

Solution of the equations

Equations	Price index = P_t^{*}/P_0^{*}	Quantity index = Q_t^{*}/Q_0^{*}
a and b	index of ME* $P_{0t}^{ME}=\frac{P_{0t}^{L}+V_{0t}}{1+Q_{0t}^{L}}$	index of Laspeyres Q_{0t}^{L}
a and c	index of Laspeyres P_{0t}^{L}	index of ME* $Q_{0t}^{ME}=\frac{Q_{0t}^{L}+V_{0t}}{1+P_{0t}^{L}}$
a and d	no meaningful result	
b and c	index of Stuvel P_{0t}^{ST}	index of Stuvel Q_{0t}^{ST}
b and d	index of Laspeyres P_{0t}^{L}	**new formula** $Q_{0t}^{BA1}=Q_{0t}^{L}\frac{P_{0t}^{P}-1}{P_{0t}^{L}-1}$
c and d	**new formula** $P_{0t}^{BA1}=P_{0t}^{L}\frac{Q_{0t}^{P}-1}{Q_{0t}^{L}-1}$	index of Laspeyres Q_{0t}^{L}

* ME= Marshall Edgeworth

Hence formulas of Stuvel, Laspeyres and Marshall-Edgeworth, as well as two new formulas appear as special cases of the factorial approach.

The next step is to combine the two systems of equation 11 and 10 such that

$$(\bullet)\quad 4\hat{\mu}=\Sigma p_0q_0+\Sigma p_tq_0+\Sigma p_0q_t+\Sigma p_tq_t=V_0\left(V_{0t}+P_{0t}^{L}+Q_{0t}^{L}+1\right)$$

and so on.[214] The result is the system of equations a through d in **fig. 2.6.4**.

[213] This gives rise to an additional interpretation of Stuvel's formulas.

Since the equations are independent it should be possible to take any two of these equations to arrive at the same pair of indices, P_{0t} and Q_{0t}. This, however, is not the case. The fact that no pair P, Q exists that can satisfy simultaneously all four equations indicates already "that the underlying idea of the factorial approach is erroneous" (Stuvel 1989, p. 102). Moreover, combining eqs. a and d results in two identical quadratic equations

$$P^2 - P(P^L + Q^L) - V = 0 \quad \text{and} \quad Q^2 - Q(P^L + Q^L) - V = 0,$$

such that it is impossible to decide which of the solution of this equation is P and which is Q. Banerjee on the other hand has seen in this a confirmation of his theory because the equations 4a and 4d are a pair of equations which "recognizes the existence of only the interaction, and not of any of the main effects, prices or quantities" (Banerjee 1961, p. 596).

Furthermore the two quadratic equations cannot yield positive results for P and Q. This can easily be seen since for that purpose the root $\frac{1}{2}\sqrt{(P^L + Q^L)^2 - 4V}$ should be greater than, but due to the fact that V_{0t} is always positive this is impossible.

Apart from the pair a and d there is no pair of equations which produces necessarily unacceptable results. The new formulas derived from Banerjee's analysis take superscripts BA in **fig. 2.6.4** in which all results are listed that will be obtained in solving the system of equations for P and Q. A numerical example (**ex. 2.6.2**) will demonstrate, however, that the two index functions, P_{0t}^{BA1} and Q_{0t}^{BA1}, can well produce highly absurd results.

Since there are two price and two quantity indices proposed by Banerjee these indices will subsequently be called P^{BA1}, P^{BA2} and Q^{BA1}, Q^{BA2} respectively. The second pair of indices is derived from (Banerjee's quite specific) considerations concerning the "true cost of living" index (or "economic" index, or "utility" index) and P^{BA2} is defined as

$$\text{(2.6.12)} \quad P_{0t}^{BA2} = \frac{P_{0t}^{P}\left(P_{0t}^{L} + 1\right)}{P_{0t}^{P} + 1}.$$

P_{0t}^{BA2} is also the factor antithesis of the quantity index of Marshall-Edgeworth since

$\frac{V_{0t}}{Q_{0t}^{ME}} = V_{0t}\frac{1 + P_{0t}^{L}}{V_{0t} + Q_{0t}^{L}} = P_{0t}^{BA2}$ was already known to Irving Fisher.[215] The corresponding

quantity index is given by

[214] The term in brackets on the RHS of eq. (•) is the RHS of eq. a in fig. 2.6.4 and the LHS of this eq. a is given by eq. 10a.

[215] According to Olt (1996), p. 74 this formula was listed under number 2154 in Fisher's "Making of Index Numbers" and the formula violates linear homogeneity.

(2.6.12a) $Q_{0t}^{BA2} = \frac{Q_{0t}^{P}\left(Q_{0t}^{L}+1\right)}{Q_{0t}^{P}+1}$

which is the factor antithesis of the Marshall-Edgeworth price index.

g) Some comments on Banerjee's approach

It has been said that nonexistence of a single solution of the whole system of equations for $P_{0t} = P_t^* / P_0^*$ and for $Q_{0t} = Q_t^* / Q_0^*$ indicates that the approach as a whole might be "erroneous" (Stuvel) or, to say the least, not really convincing. This view is supported by looking at

- the appropriateness of the underlying analogy to factorial experiments and
- the plausibility of the solutions presented in **fig. 2.6.4**.

According to Stuvel 1989, p. 103, it is doubtful that there is a clear correspondence between a factorial experiment and the situation to be described by virtue of a price index. Stuvel in particular finds

> "it hard to conceive of the n associated price and volume changes of the n single commodities in the aggregate ... as replications of each other".

It ought to be noticed that it is the same assumption that one commodity is to be viewed as a "replication" of the other that is inherent in the "stochastic approach". In both approaches, Banerjee's as well as the stochastic approach, it is essential that commodities can be regarded as identical to each other, differing only with respect to a random variable ε, accounting for all remaining influences and being distributed with zero mean and constant variance.

A summary table of Banerjee's new index formulas might be useful to facilitate discussion of properties:

index of	Figure 2.6.4	economic theory
prices	$P_{0t}^{BA1} = P_{0t}^{L}\frac{Q_{0t}^{P}-1}{Q_{0t}^{L}-1}$	$P_{0t}^{BA2} = \frac{P_{0t}^{P}\left(P_{0t}^{L}+1\right)}{P_{0t}^{P}+1}$
quantities	$Q_{0t}^{BA1} = Q_{0t}^{L}\frac{P_{0t}^{P}-1}{P_{0t}^{L}-1}$	$Q_{0t}^{BA2} = \frac{Q_{0t}^{P}\left(Q_{0t}^{L}+1\right)}{Q_{0t}^{P}+1}$

The new indices P^{BA1} and Q^{BA1} meet linear homogeneity, since $\lambda P_{0t}^{L}\frac{Q_{0t}^{P}-1}{Q_{0t}^{L}-1} = \lambda P_{0t}^{BA1}$ (and the same is true for Q^{BA1}). However, this is not true for the economic indices P^{BA2} and Q^{BA2} because $\frac{\lambda P_{0t}^{P}\left(\lambda P_{0t}^{L}+1\right)}{\lambda P_{0t}^{P}+1} \neq \lambda P_{0t}^{BA2}$ (and Q^{BA2} correspondingly).

It can be shown that both Banerjee-2 indices are bounded by the respective Laspeyres- and Paasche-indices P^L, Q^L and P^P, Q^P.

Given that $P_{0t}^L = (P_{0t}^P + \Delta) > P_{0t}^P, \Delta > 0$ we obtain $\left(P_{0t}^P + 1\right) P_{0t}^{BA2} = P_{0t}^P\left(P_{0t}^P + 1 + \Delta\right)$ and thus $P_{0t}^{BA2} = P_{0t}^P + \frac{\Delta P_{0t}^P}{P_{0t}^P + 1} > P_{0t}^P$ and in just the same manner we get $P_{0t}^{BA2} < P_{0t}^L$ in this situation ($P_{0t}^P < P_{0t}^L$), and $P_{0t}^L < P_{0t}^{BA2} < P_{0t}^P$ in the case $P_{0t}^L < P_{0t}^P$.

Laspeyres- and Paasche-bounds do *not* apply, however, to the Banerjee-1 indices, and as the following example shows, these indices can even yield completely absurd results.

Example 2.6.2

The following data relating to three commodities A, B and C permit calculation of indices as well as sums of squares necessary for the analysis of variance

	p_0	q_0	p_t	q_t
A	10	18	15	15
B	16	10	14	12
C	20	5	18	10

The sums calculated with the data above are $\sum p_0 q_0 = 440, \sum p_t q_0 = 500$, $\sum p_0 q_t = 734$ and $\sum p_t q_t = 573$. From this the requested indices follow easily:

$P_{0t}^L = \frac{500}{440} = 1.136, P_{0t}^P = \frac{573}{734} = 0.781, Q_{0t}^L = \frac{734}{440} = 1.668, Q_{0t}^P = \frac{573}{500} = 1.146$,
$V_{0t} = 573/440$.

Hence we get $P_{0t}^{ME} = \frac{1073}{1174} = 0.914$, $Q_{0t}^{ME} = \frac{1307}{940} = 1.390$ and the Stuvel indices yielding $P_{0t}^{ST} = 0.9058$ and $Q_{0t}^{ST} = 1.4377$ further $P_{0t}^{BA1} = 0.2483$ (miles away from P^L!!) and finally even more awkward: $Q_{0t}^{BA1} = -2.6833$. ♦

The result obtained for P_{0t}^{BA1} shows that this index can only be biased badly. The formula for P_{0t}^{BA1} reveals that the factor $f = \left(Q^P - 1\right)/\left(Q^L - 1\right)$ can produce a P^{BA1} being substantially lower than P^L, actually only one fifth (0.2185) of P^L. This occured because Q^P is only 68.7% of Q^L resulting in f = 0.2185. The smallest price relative in the example is 0.875 but $P_{0t}^{BA1} = 0.248$. The negative result for Q_{0t}^{BA1} reveals, how this formula of Banerjee's borders on the absurd. Yet this complete absurdity is not unusual but rather to be expected in all situations in which $P^P < 1$ and $P^L > 1$. The BA-2 formulas on the other hand, $P_{0t}^{BA2} = 0.9366$ and $Q_{0t}^{BA2} = 1.4248$, appear much more reasonable.

Chapter 3

Axioms and more index formulas

In the framework of the so called axiomatic approach index functions[1] are explored with reference to a set of axioms, some of them introduced already in preceding chapters. The primary concern of this chapter is to present some more details concerning the interpretation of axioms, relationships among various axioms, and selections among them made by index theoreticians in order to form their "systems of axioms" as well as to present some additional index formulas and "approaches" in index theory.[2]

3.1 The axiomatic approach, some theorems and fundamental axioms

a) Uses of "axioms" and axiomatic systems

An axiom is a functional equation, which is an equation describing a relation which holds for a function, like the index function. An example of a functional equation would be

(3.1.1) $\varphi(y,x) = [\varphi(x,y)]^{-1}$

which states that upon interchanging the arguments x and y in the function φ (a function of x and y, $\varphi(x,y)$) the result is the inverse value of $\varphi(x,y)$ since $\varphi(x,y) = [\varphi(y,x)]^{-1}$. Eq. 3.1.1 applies for example to $\varphi(\mathbf{p}_t,\mathbf{p}_0) = \Sigma p_{it}/\Sigma p_{i0}$ which is Dutot's price index P_{0t}^{D}, as $\varphi(\mathbf{p}_0,\mathbf{p}_t) = \Sigma p_{i0}/\Sigma p_{it} = [\varphi(\mathbf{p}_t,\mathbf{p}_0)]^{-1} = P_{t0}^{D}$, however, it would not apply for example to the exponential mean P_{0t}^{EX} (see below eq. 3.2.15), again an unweighted index function which, however, does satisfy the time reversal test $P_{t0}^{EX} \neq 1/P_{0t}^{EX}$ (as opposed to Dutot's index, where $P_{t0}^{D} = 1/P_{0t}^{D}$ holds). In the context of "axioms" we look for functional equations which describe properties of a function an index function should satisfy.

Some authors make a distinction between "axioms" and "tests" (**fig. 3.1.1**) in order to single out those requirements which are so fundamental that they are deemed necessary to *define* "index"[3] as opposed to functions not worth being called "index".

---

[1] Throughout this book we are (as index theory in general does) primarily dealing with *price* index functions but mutatis mutandis also with quantity index functions. Furthermore, the axiomatic theory is more or less limited to binary temporal comparisons, that is, thinking in terms of axioms, is known primarily in *temporal* comparisons of only *two* situations.

[2] This applies in particular to **sec. 3.4** through **sec. 3.6**.

[3] In other words' axioms are those properties of a function which are elements of the definition of an index. If a function complies with them it is called "index", if not the function should not be regarded as an index whatever useful properties the function may have. The same property of an index function may be called "axiom" by one author, or "test" by another.

The paramount aim of axiomatic considerations is to find criteria to assess index formulas, and thus help to make a choice from among index formulas without recourse to lengthy numerical examples. In principle we are at liberty to think of any property of a function to serve as an axiom with respect to index functions, or to invent some "new" postulates and call them "axioms". But it is of course desirable that

- it is possible to give some motivation or justification for the axiom proposed in demonstrating why an index function satisfying the requirement is "meaningful" and otherwise meaningless;
- an exact mathematical formulation can be found so that the idea can be expressed in a functional equation.[4]

Figure 3.1.1: Properties of an index function

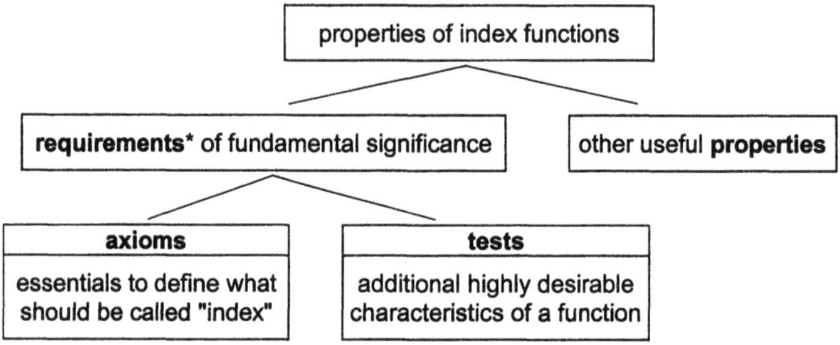

* functional equation an index function should satisfy

It would be unsatisfactory to dispose of one axiom only, as this would widen or narrow down unduly the class of admissable index functions.[5] Once two or more axioms are introduced some additional considerations are called for, however. It is necessary

- to make sure that no axiom is in contradiction with another, because in this case the condition to meet all axioms would prevent *any* function φ from existing;
- that in principle at least one reasonable function φ exists that is able to fulfill *all* axioms required simultaneously and
- to avoid axioms which are simply implications of other axioms, and finally
- to give at least some interpretation of all axioms in examining the implications of each of the axioms, not only taken in isolation but also considered in combination with other axioms.

---

[4] This turns out to be a serious limitation of axiomatic reasoning. A property like "understandability" of an index formula, for example, is no doubt highly desirable but all attempts to find an exact concept of "understandability" were not yet successful. Thus understandability is likely to be played down.

[5] It is of course possible to define axioms that can be fulfilled by only one function. See **sec. 5.2** for an example (structural consistency of volumes, satisfied only by a Paasche price index as "deflator").

Figure 3.1.2: Types of axiomatic approaches

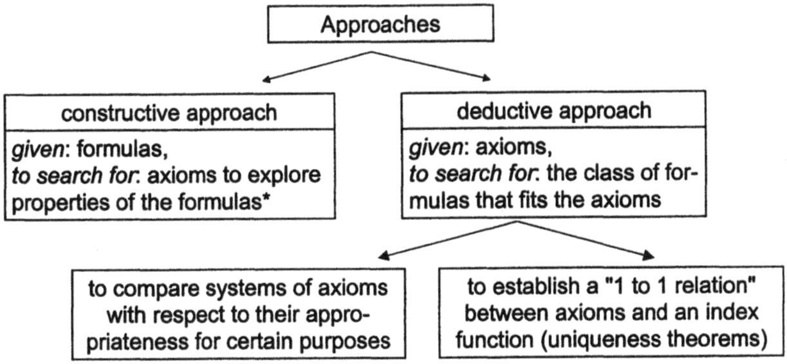

*To search systematically for new formulas was the way in which e.g. Fisher has used his "tests".

The first two characteristics are called "consistency" and the third refers to the "independence" of a system of axioms. They apply to all reasonable axiomatic systems (**sec. 3.3**), and subsequently in part b of this section it will be shown how to prove consistency and independence.

With a system of axioms we aim to find a class of functions able to meet the whole set (or "system") of axioms. An even more ambitious use of axioms is made when it is possible to find the only (the "***unique***") function or class of functions that satisfies the system of axioms, that is, to find a relation connecting unambiguously axioms, say A_1, A_2, A_3, with the index function I_1, axioms A_2, A_3, A_4, A_5 with the index I_2 and so on. In this case the particular index function in question is said to be „***characterized***" by the set of axioms in question.

With respect to how use is made of axioms we may distinguish between

- the ***constructive*** use of axioms[6]: if axioms or systems of axioms are invented to evaluate some given (known) index formulas, or to select the most appropriate formula; while
- in a "***deductive***" approach an attempt is made to establish and compare whole axiomatic systems, to deduce the set of "admissable" index functions, and to "characterize" a formula completely by certain axioms (to deduce ***uniqueness theorems***).

Axioms are powerful tools to describe the properties of index formulas and to better understand the "behaviour" of a formula by referring to some fundamental and general a priori aspects found reasonable or indispensable in view of what an index should measure. An axiomatic system is usually a collection of certain *minimum* standards an index should obey. The following final remark should be borne in mind:

6 The constructive approach (as exercised by I. Fisher in particular) begins with the axioms and makes use of them for the assessment of given formulas or for the development of new formulas. It is not primarily concerned with questions of consistency and independence of the axiomatic system. The focus rather lies on the interpretation of axioms and on arriving at a better understanding of formulas.

Chapters 2 and 3 are exclusively concerned with binary (or "bilateral") price indices for the intertemporal comparison between two periods, 0 and t. Thus the underlying definition of a price index to which we refer throughout is

(1.1.1) $P_{0t} = P(\mathbf{p}_0, \mathbf{q}_0, \mathbf{p}_t, \mathbf{q}_t)$.

The axioms, or tests, examined in what follows can therefore be called "binary-" or " bilateral tests".[7] It should be noted that the considerations are generally *not applicable to chain indices* (or more concrete: they are applicable to indivdual links but not to chains derived from such links), nor does eq. 1.1.1 apply to the "economic index theory".

As an example of a test explicitly taking into account a multisituation (multilateral) comparison we might address the *circular test* (or *circularity, transitivity*) in Fisher's system of tests (axioms). To find more details on typical multilateral (in this case interspatial rather than intertemporal) comparisons see chapter 8.

b) How to prove consistency and independence of a system of axioms?

A system of axioms as a set of "reasonable" index formulas should contain only those formulas that meet *all* axioms of the system. A system of axioms is a tool to decide which functions to admit and which to reject. To serve this purpose, a system of, say, k axioms ought to be independent and consistent:

- ***Consistency*** means: there is no conflict between axioms (they are non-contradictory), which can be proved by showing that there exists at least one formula which satisfies all of the k axioms.[8] The system of "tests" devised by Irving Fisher was inconsistent. This could be *presumed* because even Fisher was unable to find a formula that could pass all of his tests. However, this could only be a hint that inconsistency might exist, not a proof of inconsistency, since we may always argue that a more intensive search might be successful.

  In contrast to inconsistency, it is much easier to prove consistency. To do so, only a single example of an index complying with *all* k axioms suffices to demonstrate that the system can be *fulfilled*, it is not contradictory and hence consistent.

- ***Independence*** means: any k-1 of the k axioms can be satisfied by a function unable to fulfill the remaining axiom. It is sufficient to give examples to demonstrate this. Thus at least one example of an index function has to be found that violates axiom 1 but satisfies the remaining axioms, 2, 3 and so on, or that violates axiom 2 but satisfies axiom 1, 3 and so on. Independence guarantees that there is no redundancy in the system so that no axiom can be shown to be an implication of other axioms in the system.

---

[7] This limitation applies also to axiomatic systems in **sec. 3.3** and to formulas in **sec. 3.4** to **sec. 3.6**.

[8] The two axiomatic systems of Eichhorn and Voeller, as an example (to be introduced in **sec. 3.3**) are proved to be consistent.

c) Quantum theory of index formulas

In contrast to an axiomatic theory which focuses on mathematical aspects only without using data and in which only two categories (to pass or to fail a test) are distinguished, a so called "quantum theory" deals with the amount or *degree of deviation*, "*bias*" or violation of an axiom an index formula will produce when applied to data. The aim of such a theory is to

- give an estimate for the difference of any two or more index formulas under very general conditions (concerning the data that enter the formula), or
- if this cannot be performed, an attempt should be made to establish at least some inequalities, or to determine *upper and lower bounds* of an index function.[9]

This type of analysis has already been applied by Irving Fisher, who introduced a so called *five tines fork* (see **sec. 3.4** for more details and for his concept of "bias") by which a distinction was made between various degrees of bias in two directions, as opposed to a (*seven* levels) scale designed to measure some kind of overall "quality of an index function", implicitly relying to a certain degree on the biasedness:

Irving Fisher's "**five** tines fork"

biased upwards	(2 +) uppermost, (1 +) mid-upper
neutral	(0) middle (unbiased)
biased downwards	(1 -) mid-lower, (2 -) lowermost

Irving Fisher's **seven** grades quality scale (see **sec. 2.2**)

1. worthless, 2. weak, 3. correct, 4. good, 5. very good, 6. excellent and 7. superlative

The conclusions Fisher has drawn from calculating a vast amount of numerical examples have been criticized, however, with respect to two questions: how the "unbiased" situation is defined and whether the inequalities or deviations found in empirical applications are "typical" and still valid if we could derive them from sufficiently general conditions concerning the data?

---

[9] Certain variants of the "mean value tests" can also be regarded as making reference to certain bounds, such as the lowest and highest price relative (supposed to be bounding a reasonable price index), or the price index formula of Paasche (usually the lower bound) and of Laspeyres (upper bound) respectively.

Some of Fisher's statements were conclusions drawn from the results of an example which was proved to not correctly display the general situation. This applies in particular to Fisher's contention that P^L and P^P will not differ greatly from one another and that both indices are approximately equal to (results not deviating much from) his "ideal" index P^F. As mentioned before, it turned out that the data Fisher used to calculate the index numbers happened to be *not* that representative as Fisher himself erroneously believed they were. Furthermore, in many cases Fisher was not yet able to establish the relevant formulas to explain the amount of difference between any two index formulas he compared. Instead, he often could only refer to his numerical examples (like for example of the type of Bortkiewicz's theorem).

d) A tentative list and grouping of axioms

There always will be a lot of well-reasoned and well-intended debate as to which axiom is more important to be satisfied and which less. Furthermore, there are not infrequently desirable properties of a formula to be received only at the expense of some disadvantages in other respects. We may for example argue in favor of a formula satisfying the factor reversal test, even though this means that some desirable properties are sacrificed.[10] However, we can of course make another decision as to the relative importance of certain axioms and we have thereby implicitly made another choice concerning the preferred index formula(s). Once a system of axioms has been established, each axiom is regarded as equally justified and necessary as each other.[11] There is usually no hierarchy among axioms. However, to decide on which axiomatic system should be preferred to another, it is in general desirable to classify axioms according to their "importance" for the specific purpose the index function should serve.[12]

Most of the controversies in index theory concerning the superiority or inferiority of certain index formulas are directly related to the different significance authors attribute to the same axiom.

As will be shown below our (most positive) assessment of traditional formulas like Laspeyres and Paasche as opposed to Fisher's "ideal index" (vigorously advocated

---

[10] Indeed the factor reversal test is for example definitely inconsistent with circularity. Therefore, doubt might arise on what justifies such a restrictive criterion as factor reversibility and why the weak version of the test (the so-called product test) should not suffice.

[11] However there are enormous differences as to the restrictiveness of axioms. In the list of **tab. 3.1.1** we have for example the "positivity", which is trivial, while "factor reversibility", for example, is most demanding. This should be kept in mind when counting of the following (not uncommon) kind takes place: formula P_1 is better than P_2 because the number of axioms fulfilled is greater in the case of P_1 than P_2. The *dominance principle* instead of the simple *number* of axioms satisfied had therefore been proposed: P_1 is "better" than P_2 because P_1 satisfies exactly the same axioms as P_2 plus in addition the axioms x, y etc.

[12] It will be shown for example in **sec. 8.1** that the relative importance certain axioms like transitivity etc depends on whether intertemporal or international comparison is intended.

by other authors) is a consequence of much less emphasis we are willing to give to axioms like time reversiblity or the so-called "quantity reversal test"[13] than others do.

Hence some idea on more or less "importance" of certain properties of an index function seems to be useful, although this cannot be an issue to be settled by a mathematical proof. Before trying to give some hints hopefully useful to contemplate aspects of "importance" and "significance" etc. applied to axioms, an attempt should be made to give as comprehensive a list of axioms/tests as possible. The following list and rough classification of axioms presented in **tab. 3.1.1** is based on the draft version of the CPI and PPI Manual of the ILO and IMF disseminated in the Internet and articles of W. E. Diewert. The complete list of Fisher's tests (denoted by **F** in tab. 3.1.1) to which we refer in **sec. 3.3** is integrated.

Table 3.1.1: List of "tests"/axioms of price index functions version $P(\mathbf{p}_0, \mathbf{q}_0, \mathbf{p}_0, \mathbf{q}_0)$

(**B.**∗,**T**∗ refer to Diewert's list of 20 or 21 tests, and **F**∗ to Fisher's system of tests)

	Name of test	Comment
	Group B.1: Basic tests	
T1	Positivity	P_{0t} = P(...) and all constituent vectors are positive
T2	Continuity	P(...) is a continuous function of its vectors
F1	Determinate (determinateness) test or weak continuity axiom	if any scalar argument in P(...) tends to zero, then P tends to a unique positive real number
T3 **F2**	Identity $P_{00} = 1$ (or: Constant prices test)	$P(\mathbf{p}_0, \mathbf{q}_0, \mathbf{p}_t, \mathbf{q}_t) = 1$ if for all i = 1, 2,..., n commodities $p_{it} = p_{i0}$ [1)]
T4	Fixed basket test (or: Constant quantities test)	$P(\mathbf{p}_0, \mathbf{q}_0, \mathbf{p}_t, \mathbf{q}_0) = V_{0t}$
	Group B.2: Homogeneity tests	
T5 **F3**	proportionality (strict version) in current prices $P(\mathbf{p}_0, \mathbf{q}_0, \lambda\mathbf{p}_0, \mathbf{q}_t) = \lambda$	if all prices move in proportion, so does the index, or: if all period t prices change λ-fold then the value of P is also changed by λ ($\lambda \in \mathbb{R}$)[2)]

[13] This is "test" T12 in tab. 3.1.1. Especially Diewert sets great store by these properties and he therefore rejects the traditional formulas. T12 for example rules out all formulas other than those in which quantities of 0 and t enter the formula in a symmetric manner (interestingly all so-called superlative indices are of precisely this type, i.e. complying with T12).

F3a	weak version $P(\mathbf{p}_0, \mathbf{q}_0, \lambda\mathbf{p}_0, \mathbf{q}_0) = \lambda$	F3a is equal to F3 provided quantities do not change $\mathbf{q}_t = \mathbf{q}_0$
T6	Inverse proportionality in $\mathbf{p}_0$	$P(\lambda\mathbf{p}_0, \mathbf{q}_0, \mathbf{p}_0, \mathbf{q}_t) = 1/\lambda$
T7	Invariance to proportional changes in current quantities	$P(\mathbf{p}_0, \mathbf{q}_0, \mathbf{p}_0, \lambda\mathbf{q}_t) = P(\mathbf{p}_0, \mathbf{q}_0, \mathbf{p}_0, \mathbf{q}_t)$
T8	Invariance to proportional changes in base quantities	$P(\mathbf{p}_0, \lambda\mathbf{q}_0, \mathbf{p}_0, \mathbf{q}_t) = P(\mathbf{p}_0, \mathbf{q}_0, \mathbf{p}_0, \mathbf{q}_t)$

Note that Diewert's definition of "proportionality" resembles the notion of linear homogeneity:

(T5)	proportionality in current prices $P(\mathbf{p}_0,\mathbf{q}_0,\lambda\mathbf{p}_t,\mathbf{q}_t) = \lambda\, P(\mathbf{p}_0,\mathbf{q}_0,\mathbf{p}_t,\mathbf{q}_t)$	if all current period prices are multiplied by $\lambda > 0$ the new price index is λ times the old price index[3)]
(T6)	Inverse proportionality in prices $\mathbf{p}_0$	$P(\lambda\mathbf{p}_0, \mathbf{q}_0, \mathbf{p}_t, \mathbf{q}_t) = 1/\lambda\, P(.\mathbf{p}_0, \mathbf{q}_0, \mathbf{p}_t, \mathbf{q}_t)$

Group B.3: Invariance and symmetry tests

T9	Commodity reversal test	invariance upon changes in the ordering of com.
T10 **F4**	Invariance to changes in the units of measurement = commensurability	independence of the quantities to which price quotations refer (i.e. units of measurement)[4)]
T11 **F5**	Time/country reversal $P(\mathbf{p}_0, \mathbf{q}_0, \mathbf{p}_t, \mathbf{q}_t) = P(\mathbf{p}_t, \mathbf{q}_t, \mathbf{p}_0, \mathbf{q}_0)^{-1}$	interchanging $(\mathbf{q}_0,\mathbf{p}_0) \leftrightarrow (\mathbf{q}_t,\mathbf{p}_t)$, i.e. reversing the direction of comparison yields $P_{t0} = 1/P_{0t}$
T12	Quantity reversal test (quantities of both periods must enter the index formula symmetrically)	$P(\mathbf{p}_0, \mathbf{q}_t, \mathbf{p}_t, \mathbf{q}_0) = P(\mathbf{p}_0, \mathbf{q}_0, \mathbf{p}_t, \mathbf{q}_t)$, that is the index remains invariant upon interchanging of quantity vectors
T13	Price reversal test PRT[5]	quantity index remains invariant upon interchanging of price vectors

Group B.4: Mean value tests		
T14	Mean value test for prices (often simply called: Mean value test)	P_{0t} lies between minimum and maximum price relative
T15	Mean value test for quantities	implicit Q_{0t} (= V_{0t}/P_{0t}) lies between minimum and maximum quantity relative
T16	Paasche + Laspeyres bounding test	$P^P \leq P_{0t} \leq P^L$ or $P^L \leq P_{0t} \leq P^P$

Group B.5: Monotonicity[6] tests		
T17	Monotonicity in current prices	if any p_{it} increases ($p_{it} > p_{i0}$) P_{0t} increases ($P_{0t} > 1$)
T18	Monotonicity in base period prices	if any p_{i0} increases P_{0t} must decrease
T19	Monotonicity in current quantities	implicit Q_{0t} must increase if any q_{it} increases
T20	Monotonicity in base quantities	implicit Q_{0t} must decrease if any q_{i0} increases

Other tests, additivity (aggregative)[8] properties		
T21 **F6**	Factor reversal $P_{0t}Q_{0t} = V_{0t}$	$Q_{0t} = P(\mathbf{q}_0,\mathbf{p}_0,\mathbf{q}_t,\mathbf{p}_t)$ if $P_{0t} = P(\mathbf{p}_0,\mathbf{q}_0,\mathbf{p}_t,\mathbf{q}_t)$ that is Q_{0t} is derived from P_{0t} by interchanging prices and quantities[7]
F7	Circular test (see **sec. 3.2**)	also called transitivity test or chain test
F8	Withdrawal-and-entry test [9] and equality test (see **sec. 5.2**)	Index should remain invariant if a price relative or sub-index is added or removed which is exactly equal to the overall index
	Aggregative consistency of the index formula (see **sec. 5.2**)	Aggregation of relatives to subindices and subindices to the overall index follow same function[10]
	Structural consistency of volumes (deflated values), SCV in **sec. 5.2**	Using P_{0t} as deflator should result in volumes that satisfy the same definitional equations values do

1) or: $P(\mathbf{p}_0,\mathbf{p}_t) = 1$ for an index **not** depending on quantities.
2) note that identity is obviously the special case $\lambda = 1$.
3) or: the price index function is (positively) homogenous of degree one in the components of the current period price vector $\mathbf{p}_t$.
4) We first referred to the commensurability test/axiom in connection with Dutot's index (**sec. 1.2**).
5) According to Diewert the indices of Laspeyres (P^L) and Paasche fail this test while they are able to pass the differently defined PR-Test in **sec. 3.2**.
6) What is defined here is strictly speaking weak monotonicity as opposed to strict monotonicity
7) otherwise product test (= weak factor reversal test).
8) The purpose of these tests is to make sure that the overall-index can be compiled from sub-indices or be decomposed into sub-indices without difficulties and that aggregation and deflation will yield reasonable results.
9) rarely mentioned at all; to be discussed (along with the equality test) in **sec. 5.2/4**.
10) with weights adjusted appropriately to the aggregation problem in question.

e) First considerations concerning the importance of axioms/ test and the time reversal test

The requirements F1, F2 and F4 of Fisher's system are fundamental for index functions in general and they are all met by P^L (Laspeyres), P^P (Paasche) and all second and third generation formulas as well. Their common objective is to avoid ambiguity.[14] In this context it is of paramount importance to have

- a kind of "***neutral***" starting ***point*** (what does $P_{0t} = 1$ mean?) from which a move up or down is defined (identity [F2] serves this purpose),
- the next prerequisite of measurement appears to be to avoid ambiguities as to the ***unit of measurement*** (Fisher's tests F1 and F4 belong to this group, but also (price) dimensionality not mentioned by Fisher).

The next step then will be to give some thought to how to establish correctly the direction and amount of movement away from the neutral point. As to the ***direction***, the idea of monotonicity comes into play and as far as the ***amount*** is concerned a most fundamental idea is proportionality (F3). We are going to continue this kind of reasoning in part i of this section.

It is of course possible to argue that, for example, the **time reversal test** (F5, T11) is again motivated by the desire to avoid ambiguity. In this kind of reasoning the fact that both P^L and P^P fail this test ("a fundamental test") is "a severe limitation. That the *conceptually* arbitrary decision to measure forward in time gives different results to the reciprocal of the measure backwards, does not give faith in either measure".[15] In our view this is not tenable since

[14] In a sense this applies also to the **principle of pure price comparison** (not called "axiom", however, because of difficulties in finding an exact mathematical description of this somewhat broader idea). See sec. 8.2 in my book on chain indices for an attempt to find an operational definition of "pure" in this context.

[15] See CPI/PPI Manual chapter "The axiomatic and the stochastic approach" paragraph 5.

- much in contrast to the interspatial situation (where the time reversal test in the shape of country reversibility rightly has some relevance) in intertemporal comparisons the direction of comparison is *not* arbitrary (it is not unjustified to prefer a forward movement to moving backwards as it might be a matter of taste whether we compare the countries A and B on the basis of A = 100 or B = 100);
- even in those cases in which the decision might be arbitrary (as in comparing A with B) we cannot find good reasons why an inverse relation must hold, since it definitely makes a difference to express a situation in A in terms of B or the other way round.

> Note that Fisher (poorly) motivated his reversal tests with "fairness" or by analogy to simple relatives. He also could not yet incorporate in his system some axioms nowadays deemed most useful, like monotonicity, nor was he aware of the importance of aggregation properties in a time in which National Accounts were yet to be invented.

Hence it is interesting to refer to Fisher but his way should not be followed too far.

f) Comments on the meaning and significance of some basic axioms I: determinateness, commensurability and dimensionality

Swamy 1965 found that in particular the factor reversal test and the ***determinateness*** test are suspect. His view is supported by some arguments derived from the economic theory of index numbers. He wrote: "It is not clear at all why this test should receive any attention, because zero, near zero, infinite, and near infinite prices are rare things. Even if some prices were to vanish, the price index must show a zero value because the consumer, under the usual assumption of non-satiety, will derive infinite utility" (p. 621).

Commensurability can be expressed as follows

$$\text{(3.1.2)}\qquad P(\mathbf{L}\mathbf{p}_0,\mathbf{L}^{-1}\mathbf{q}_0,\mathbf{L}\mathbf{p}_t,\mathbf{L}^{-1}\mathbf{q}_t)=P(\mathbf{p}_0,\mathbf{q}_0,\mathbf{p}_t,\mathbf{q}_t)$$

where **L** is a n × n diagonal matrix with elements $\lambda_1,\dots,\lambda_n$, such that

$$\mathbf{L}=\begin{bmatrix}\lambda_1 & 0 & \dots & 0\\ 0 & \lambda_2 & \dots & 0\\ \dots & \dots & \dots & \dots\\ 0 & 0 & \dots & \lambda_n\end{bmatrix}\quad\text{and}\quad \mathbf{L}^{-1}=\begin{bmatrix}1/\lambda_1 & 0 & \dots & 0\\ 0 & 1/\lambda_2 & \dots & 0\\ \dots & \dots & \dots & \dots\\ 0 & 0 & \dots & 1/\lambda_n\end{bmatrix}$$

Commensurability is a very powerful property. When commensurability is satisfied the index function can be expressed in price relatives. Consider a matrix **L** as follows

$$\mathbf{L}=\begin{bmatrix}1/p_{10} & 0 & \dots & 0\\ 0 & 1/p_{20} & \dots & 0\\ \dots & \dots & \dots & \dots\\ 0 & 0 & \dots & 1/p_{n0}\end{bmatrix}$$ with main diagonal elements $1/p_{i0}$.

Then we obtain

$\mathbf{Lp}_0 = \mathbf{1}$, where $\mathbf{1}' = [1\ 1\ \ldots\ 1]$ and

$\mathbf{Lp}_t = \mathbf{a}$, the vector of price relatives $\mathbf{a}' = [p_{1t}/p_{10}\ \ p_{2t}/p_{20}\ \ \ldots\ \ p_{nt}/p_{n0}]$. Furthermore

$\mathbf{L}^{-1}\mathbf{q}_0 = \mathbf{v}_0$ the vector of base period *values* $\mathbf{v}_0' = [p_{10}\,q_{10}\ \ p_{20}\,q_{20}\ \ \ldots\ \ p_{n0}\,q_{n0}]$ and

$\mathbf{L}^{-1}\mathbf{q}_t = \mathbf{v}_t$ the vector of *volumes* $\mathbf{v}_t' = [p_{10}\,q_{1t}\ \ p_{20}\,q_{2t}\ \ \ldots\ \ p_{n0}\,q_{nt}]$.

When commensurability holds, the index function can be expressed as follows:

(3.1.3) $\quad P(\mathbf{p}_0,\mathbf{q}_0,\mathbf{p}_t,\mathbf{q}_t) = P(\mathbf{a},\mathbf{v}_0,\mathbf{v}_t)$,

that is as a *function* of only 3n (instead of 4n) variables, the price relatives a_i, the reference period volumes v_{it} (valued at base period prices) and the base period values v_{i0}.

By (***price***) ***dimensionality*** or homogeneity of degree 0 in prices (not explicitly listed in **tab. 3.1.1** but already introduced in **sec. 1.1**d, **tab. 1.1.1** where it is applied to price relatives, a trivial property) the index function is not affected by a change in the currency unit in which the prices are expressed[16]:

(3.1.4) $\quad P(\lambda\mathbf{p}_0,\mathbf{q}_0,\lambda\mathbf{p}_t,\mathbf{q}_t) = P(\mathbf{p}_0,\mathbf{q}_0,\mathbf{p}_t,\mathbf{q}_t)$ $\qquad$ (price) dimensionality.

In combination with commensurability quantity dimensionality is implied:

if price dimensionality and commensurability is met	→	then also quantity dimensionality

but the converse is not true. Dutot's index for example satisfies quantity dimensionality and price dimensionality but not commensurability.

Quantity dimensionality (also "weak commensurability ") is defined as follows

(3.1.5) $\quad P(\lambda\mathbf{p}_0,\lambda^{-1}\mathbf{q}_0,\lambda\mathbf{p}_t,\lambda^{-1}\mathbf{q}_t) = P(\mathbf{p}_0,\mathbf{q}_0,\mathbf{p}_t,\mathbf{q}_t)$,

and is much weaker a requirement than commensurability (sometimes also called "dimensional invariance"), because it is the special case of all $\lambda_1,\ldots,\lambda_n$ being equal[17] ($\lambda_i = \lambda\ \forall\ i = 1, 2, \ldots, n$), such that $\mathbf{L}$ reduces to the scalar matrix $\mathbf{L} = \lambda\mathbf{I}$

g) Comments on the meaning and significance of some basic axioms II: identity, a first look at monotonicity, and proportionality

The idea of the "***identity*** test" has been introduced by Laspeyres. As already explained above

[16] Index numbers for intertemporal comparison have no dimension, those for the comparison of countries with different currencies refer to a ratio of currencies.

[17] Quantity dimensionality is just the special case in which each commodity is subject to the *same* change in the unity of measurement. Quantity dimensionality means that if, for example, *all* prices in *both* periods no longer refer to kilogramms but to pounds, 500 gramms, instead ($\lambda = \frac{1}{2}$) the index should not be affected. This is true for Dutot's index and for all other index formulas that depend on sums of prices only, although they do not satisfy commensurability.

identity: if *no* price changes the price index function should be 1 (unity).

There is no axiom stating "the price index function should be unity if and only if ..." because we may think of infinitely many constellations of some prices rising and other prices declining and exactly offsetting the rise.

The most frequently quoted example of an index violating (strict) identity is the value ratio (or value index) V_{0t} due to different quantities in numerator and denominator (in fact they are likely to differ since we may assume in most practical applications that different quantities $q_{it} \neq q_{i0}$ will emerge in response to different prices in 0 and t). Often a distinction is made between strict and weak identity in the following manner:

(3.1.6) strict identity $P(\mathbf{p}_0, \mathbf{q}_0, \mathbf{p}_0, \mathbf{q}_t) = 1,$ if $\mathbf{p}_t = \mathbf{p}_0$

(3.1.6a) weak identity $P(\mathbf{p}_0, \mathbf{q}_0, \mathbf{p}_0, \mathbf{q}_0) = 1,$ where $\mathbf{p}_t = \mathbf{p}_0$ *and* $\mathbf{q}_t = \mathbf{q}_0$.

The value index V_{0t} satisfies weak identity, but not strict identity. Weak identity is less demanding in that it is assumed that not only prices remain unchanged but also quantities.

As usual the weak version is gained from the strict one by adding further assumptions (like $\mathbf{q}_t = \mathbf{q}_0$). Obviously strict identity implies weak identity but the converse is not true (as could be learnt from the example of V_{0t}).

A statement to be regarded in a certain sense as the "opposite" of identity is:

if any *one* single price taken in isolation* is rising (or declining), the index function should *not* be 1 but indicate a rise or decline.

* hence also the case of *all* prices or *some* prices is covered.

This is guaranteed by the (weak) ***monotonicity*** (in current period prices) axiom. Thus identity and monotonicity are two very fundamental characteristics required for any index to be meaningful in the following sense: the index should not indicate a change (or no change) if prices in fact remain constant (are changing). On the other hand it should indicate a change if prices are in fact changing.

In order to prove that identity and monotonicity indeed represent two independent and different properties it should be demonstrated that at least one example of an index function exists that fits to field (1, 2) and to field (2, 1) respectively:

index	prices remain constant $p_{it} = p_{i0}$	prices going up/down
constant $P_{0t} = 1$	(1,1) identity	(1,2) $\Sigma p_0 q_0 / \Sigma p_0 q_0 = 1 =$ const.
indicates a change*	(2,1) $V_{0t} = \Sigma p_t q_t / \Sigma p_0 q_0$	(2,2) monotonicity

* in the correct direction

The simplest function for the case (1, 2) is of course an index function without any prices p_{it}.[18] In the same manner V_{0t} would also fit to case (1, 2): consider the case where a rise of prices may be exactly offset by a decline in quantities, such that $\Sigma p_t q_t = \Sigma p_0 q_0$ and thus[19] $P_{0t} = V_{0t} = 1$. The value index V_{0t} can of course also remain constant in spite of changing prices because of a change in quantities in the opposite direction (field (2, 1)).

> Identity can also be related to *time reversibility*: If prices rise in the interval from 0 to 1 such that $\mathbf{p}_1 > \mathbf{p}_0$ but return (decline) thereafter to their initial level, such that $\mathbf{p}_2 = \mathbf{p}_0$ then the price index should be unity ($P_{02} = 1$). It is often pointed out as an example of a serious defect of chain indices, that such indices need not result in unity but may well indicate increase ($P_{02} > 1$) or decrease ($P_{02} < 1$) under such conditions, depending on what happened in the intermediate period 1.

According to ***proportionality*** an index should be λ if all prices change at the same rate λ such that $p_{it} / p_{i0} = \lambda$ for all i = 1, 2,..., n. This requirement is intuitively appealing since it refers to an extremely simple situation. Its rationale is that an index P( ) should *at least* reflect a *uniform* change correctly: if P( ) fails to do so, it is more likely that P( ) will also reflect a *mixed* change incorrectly, as generally occurs in practice. Hence proportionality gives a hint to a correct measurement in an extremely simple, or "idealized" case, in which no difficulty exists as to what the result of measurement should be.

Again a distinction can be made between a strict and a weak version:

(3.1.7) strict proportionality $P(\mathbf{p}_0, \mathbf{q}_0, \lambda\mathbf{p}_0, \mathbf{q}_t) = \lambda$, where $\lambda \in \mathbb{R}$, and $\mathbf{p}_t = \lambda\mathbf{p}_0$

(3.1.7a) weak proportionality $P(\mathbf{p}_0, \mathbf{q}_0, \lambda\mathbf{p}_0, \mathbf{q}_0) = \lambda$, where $\lambda \in \mathbb{R}$, $\mathbf{p}_t = \lambda\mathbf{p}_0$, $\mathbf{q}_t = \mathbf{q}_0$

Even a badly biased index such as Drobisch's unit value index (eq. 1.2.2) is able to meet weak proportionality but not strong proportionality.

Proportionality implies identity but not vice versa:

if an index satisfies proportionality	→	then also identity

The first part of the assertion is simple to verifiy: identity is simply the special case of $\lambda = 1$ in eq. 7 or 7a respectively. Thus no index function exists which satisfies proportionality but not at the same time identity as well. To prove the second part ("the converse is not true") a simple example is sufficient. The function

---

[18] The function $\Sigma p_0 q_0 / \Sigma p_0 q_0$ probably is not a good example for the case (1, 2); a better choice might be mod (p_{it}/p_{i0}), that is the modal (most frequent) price relative taken as the price index. It meets identity and at the same time will not comply with monotonicity in the current prices, because if n-1 price relatives are unity and only one is > 1 or < 1 this function would of course remain unity and thereby violate monotonicity.

[19] Because for some p_{it} on the LHS we have $p_{it} > p_{i0}$ and on the other hand for some $q_{it} < q_{i0}$. This would by the way not contradict monotonicity because violation of monotonicity is not given if $P_{0t} = 1$ as a consequence of *several* prices changing, but only if $P_{0t} = 1$ in the case of an *isolated* change (other prices remain constant).

(3.1.8) $P_{0t}^{q1} = \frac{1}{n}\sum\left(\frac{p_{it}}{p_{i0}}\right)^2$ or in the weighted form $P_{0t}^{q2} = \sum\left(\frac{p_{it}}{p_{i0}}\right)^2 \frac{p_{i0}q_{i0}}{\sum p_{i0}q_{i0}}$

fulfills identity but not proportionality,[20] whereas $P_{0t}^{QM1} = \sqrt{P_{0t}^{q1}}$ and $P_{0t}^{QM2} = \sqrt{P_{0t}^{q2}} = P_{0t}^{QM}$, which is simply the quadratic mean of price relatives, agrees with both tests.

The relationship between proportionality and identity also shows that Fisher's system (**tab. 3.1.1**) is *redundant*: There is no need to introduce the identity test if proportionality is already required. The two tests are clearly *not* independent.[21]

h) Logical relations among the most fundamental axioms

The identity (and more general proportionality), commensurability, and determinateness tests are indeed most elementary and fundamental. Their significance can also be derived from the fact that virtually all reasonable index formulas are able to pass these tests, which is not true for tests like time- oder even factor reversibility.

Since P_{0t}^{L} and P_{0t}^{P} use the same quantities, either q_{i0}, or q_{it} in both the numerator and the denominator, it is immediately clear that they fulfill identity and proportionality. Furthermore, since they add products of prices and quantities (instead of merely prices or quantities, that is, sums like Σp_{it}, Σq_{i0} or so) in numerator and denominator, they also meet the determinateness and commensurability test. The results are valid for all (third generation) indices that are derived by crossing formulas and weights as well. Thus

P^L, P^P, P^F, P^{DR}, P^{HPL}, P^W, P^{ME} and P^{GK} **all** satisfy the following tests: 1. identity, 2. determinateness, 3. commensurability, and 4. proportionality

Fig. 3.1.3 is an attempt to show how the most fundamental axioms quite naturally follow from simple deliberations about the reasonable reaction an index should display to certain simplified processes (assumptions). Assume that all prices change *somehow* and differently, then we cannot deduce logically what numerical results an acceptable index funtion should yield.

If, however, *all* prices have undergone the same change such that $\mathbf{p}_t = \lambda \mathbf{p}_0$ holds for the vectors, the n-commodities situation then reduces to a one-commodity situation and it would be illogical to have an index $P_{0t} \neq \lambda$. On the other hand, the situation is less clear if we consider only *some* prices changing at the rate λ while other prices change more ($> \lambda$) or less ($< \lambda$). It appears reasonable to require that in such a more complicated case the price index should represent some kind of average rate. This quite naturally brings us to the "mean value test" (T14 in tab. 3.1.1) which thus is again a criterion reflecting some simple but convincing common sense reasoning.

[20] because $\lambda = 2$ results in a new index four times the old one.

[21] More about the logical status of Fisher's system of tests see **sec. 3.3**. Another violation of independence (with regard to time reversibility F5) is the fact that the circular test (F7) in combination with identity (F2) implies F5.

Figure 3.1.3: Relations among some axiomatic properties

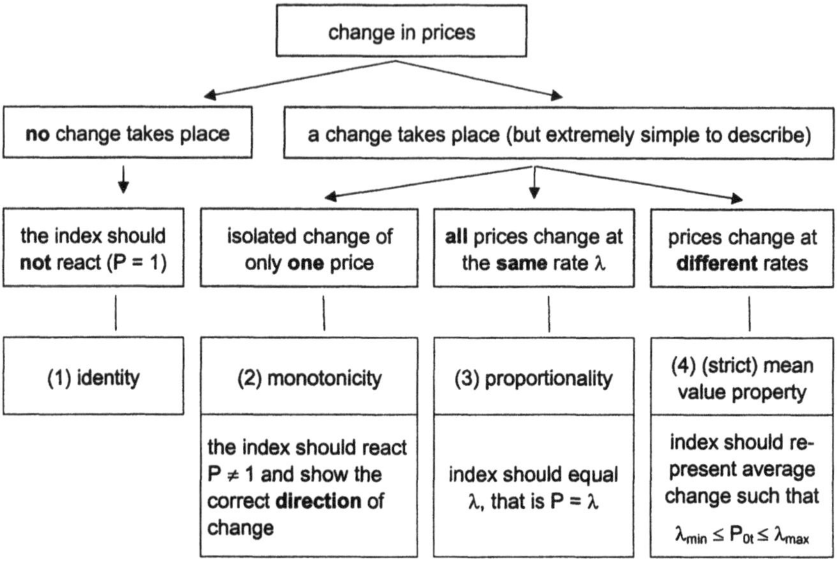

Relations among the properties are for example: (3) → (1), and (4) → (3).

i) A tentative ranking of axioms in terms of importance

We already made note of the idea that some reasoned views regarding the relative "importance" of certain properties of an index function are indispensible or at least desirable, though arguably this may never be decided in a mathematical manner.

In part e of this section we already gave some hints as to what might be considered as most fundamental: we found that certain most general prerequisites of measurement should come first in a ranking in terms of importance. As a minimum, there should exist a point of origin, or a "neutrality" notion [the meaning of "no change"], a benchmark against which to measure "change", and no less important, a clear concept of the direction and amount of movement away from this origin. We now try to elaborate this point a bit further. The tentative ranking of axioms presented in **fig. 3.1.4** distinguishes axioms of three types, namely axioms that are

a) expressing some logical ***prerequisites of measurement*** in general (some properties *any* reliable measuring rod must have),

b) necessary (or at least desirable) with respect to the ***purposes of analysis*** (regarding for example the intended economic interpretations) for which an index is compiled: the axioms of this group refer to the correct reflection of the movement of n individual prices by a suitable summary measure, and they strongly appeal to an intuitive understanding of what an index intends to measure;

c) finally a group of "axioms" is introduced by ***analogy to relatives***, they are based on the belief that index numbers should behave as if they were indvidual relatives (scalars instead of vector functions). In a sense they may be called invariance[22] axioms, or axioms to ensure consistency of operations with index numbers.

Since most parts of the axioms introduced are known and discussed in some detail already, in what follows some brief comments should be sufficient.

Figure 3.1.4: Tentative classification of axioms and their uses

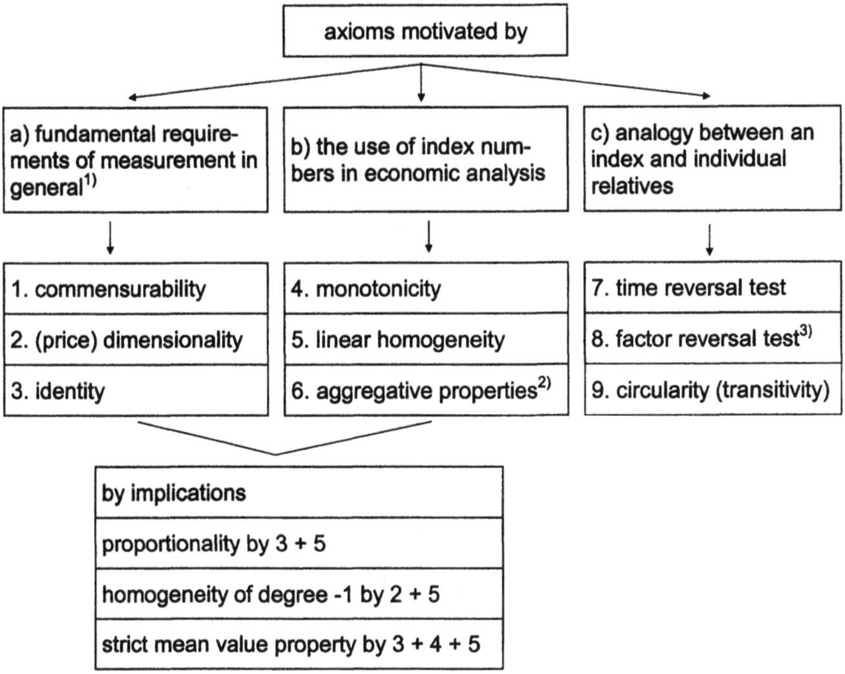

1) these may be called invariance axioms

2) see **sec. 5.2**

3) unlike the factor reversal test, the product test is deemed necessary

[22] As abovementioned time reversibility (T11) is grouped under "B.3 invariance and symmetry tests" in **tab.3.1.1**. Interestingly, the circular test is not mentioned there. This test has always been the object of much controversy. The idea is to make consistent intertemporal comparisons over an interval in time irrespective of how the interval is subdivided into sub-intervals. If and only if transitivity holds can aggregation over time be made consistently by successive multiplication of separate year to year index numbers. They are joined together like the links of a chain. However, this test should not be confused with the chain index principle (see chap. 7).

Ad a: Prerequisites of measurement

Requirements belonging to the first group are those, according to which an index should be independent of, or *invariant* under a change in the units of measurement, that is

1. quantities to which the price quotations refer (*commensurability*), and
2. the currency unit used to express the prices (price *dimensionality*); by which an index should be a relative figure in the case of intertemporal comparisons, not depending on a unit of measurement. This implies in addition
3. a neutral element, or definition of a *reference point* (unity or 100%), from which to measure unambiguously a change (deviation): by the *identity* requirement P = 1 is made a well defined situation from which P > 1 and P < 1 reflects a rise or decline respectively without ambiguity.

Note that so far nothing is said about the ability of the index function to differentiate correctly among different directions and "degrees" of changes in prices.

Ad b: Correct reflection of price movement

According to the second group of requirements it is highly desirable that a price index P indicates correctly the *direction* and *amount* of change (or no change) if prices change (or remain constant). It is indispensable that an index also indicates the amount of change properly, at least in cases in which a change of extremely simple nature takes place, like for instance

- an isolated change where all other prices remain constant (*monotonicity*), or
- a uniform change (*proportionality*).

This appears to be a minimum standard for an index to be used in analysing the movement of a price level, since if the task is to detect and measure a movement it is of course fundamental to have a clear idea

- of some kind of a starting point and the *direction of movement*,
- the *amount of change* in simple situations where no doubt arises as to what the amount to be reflected should be, and it is also necessary
- to separate the type of movement we want to measure from irrelevant changes, like changes in units of measurement (in a wider sense this does not only cover the invariance properties above but also the idea of *"pure" price movement* where the effect of changes in variables other than prices should be eliminated).

Furthermore, being a measure of *aggregate* movement, an index and calculations derived by using indices should be meaningfully decomposable.

Whenever an "*explanation*" of a price movement, for example in the case of consumer prices, is intended to show to what extent certain prices contributed to the overall price movements, certain aggregation properties of the *index function* are highly desirable.

Whenever indices are used to *deflate* aggregates it should be possible to decompose volumes derived from values in aggregative (i.e. measured using indices) prices and quantities. A result of deflation violating definitional equations (that hold by defintion) in aggregates is to be avoided. Hence aggregative properties (see **sec. 5.2** for more details) refer to

- the index function as well as
- to volumes as the results of deflation.

Certain minimum standards with respect to both aspects of aggregation are therefore most important for economic applications (see **sec. 5.2**).

Ad c: Analogy to relatives

It is not legitimate to postulate properties an index should have simply by analogy to individual price relatives. As mentioned above, axioms like time reversibility and circularity, known from the single-commodity case, might also be viewed as aspects of invariance in a certain sense.[23] But there are reasons *not* to insist on such criteria to be met by indices in case of inter-*temporal* comparisons: No symmetry exists between 0 and t (there is no point in interchanging 0 and t), nor is there any reason why a relation between P_{0s}, P_{st} and P_{0t} should exist by definition.

In our view group a of axioms should be considered more important than group b, and group b in turn more important than group c. In principle many other classifications of axioms might be conceived and different views on the relevance of certain axioms may be adopted. Any attempt to classify axioms is inevitably affected by opinions on what makes axioms more or less justified, more or less dispensable and the like. Widely divergent views on this issue are possible without calling into question the general usefulness of the axiomatic approach.

j) First example of uniqueness theorems and an inconsistency theorem

To conclude this section it is useful to have a look at two prevailing types of theorems and proofs in index theory,

- the *uniqueness theorem* (**UT**), which states that there is one and only one index function able to satisfy the entirety of certain conditions (the difficulty consists in the exclusion of all other conceivable (possibly not yet known) functions unless the conditions are such that the function in question directly follows from the assumptions),[24] and
- the *inconsistency* (or non-existence) *theorem* (**IT**) proving certain conditions to be contradictory and thus impossible to be satisfied simultaneously.

[23] Time reversibility refers to the irrelevance of the choice of the base, which is found desirable by some authors, and circularity can be regarded as a type of consistency (in aggregating over periods in time), in the sense that a "direct" calculation P_{0t} is consistent with *any* "indirect" calculation, like $P_{0s}P_{st}$.

[24] This applies in the case of the theorem subsequently presented for demonstration purposes.

As mentioned repeatedly a UT is a tool to "characterize" an index. The "message" of the formula can be captured completely by the conditions of the UT. This is not to say that this is the *only* "characterization", as the same formula can well be described by a number of UTs. This is true especially for Fisher's "ideal" index, which owes much of its renown to the fact that it not only satisfies many (if not in fact most) known axioms, but also emerges as the result from most exercises in constructing UTs.

The following uniqueness theorem (**UT - 1**) is easy to verify[25]:

A pair of Fisher indices P_{0t}^F, Q_{0t}^F is the only pair of indices that satisfies the product test (or factor reversal test) $P_{0t}^{*}Q_{0t}^{*} = V_{0t}$ *and* $P_{0t}^{*}/P_{0t}^{L} = Q_{0t}^{*}/Q_{0t}^{L}$

In **sec. 3.6** we present an interpretation of this property of P_{0t}^F and Q_{0t}^F. It can be shown that Fisher's indices can be obtained from Laspeyres' indices by "adding" a part of the structural component (SC for short) to the price index and the other "half" (strictly speaking, in both cases it is $\sqrt{SC}$ that is allocated to P_{0t} and Q_{0t} respectively) to the quantity index.

The ***proof of theorem UT-1*** is straightforward. Substitution of $Q_{0t}^{*} = V_{0t}/P_{0t}^{*}$ in $P_{0t}^{*}/P_{0t}^{L} = Q_{0t}^{*}/Q_{0t}^{L}$ gives $(P_{0t}^{*})^2 = V_{0t}P_{0t}^{L}/Q_{0t}^{L} = P_{0t}^{P}P_{0t}^{L}$ such that $P_{0t}^{*} = P_{0t}^{F}$.

On the other hand, a most complicated uniqueness theorem is addressed in Diewert's well known statement according to which Fisher's index (or the indices P^F and Q^F) is the only type of index that satisfies all of the 20 tests of tab. 3.1.1 and in addition the factor reversal test (T21). In particular this formula is the only one able to meet the "four minimal tests" T1, T11, T12 and T21.[26]

As Diewert pointed out, all other formulas fall short of this impressive number 20: Walsh's formula P_{0t}^{W} falls short of 4 tests (T13, 16, 19, and 20), P_{0t}^{L} and P_{0t}^{P} of only 3 tests (T11, 12 and 13, conspicously only reversal tests), and Törnqvist's formula fails 9 out of the 20 tests.[27] It will make of course a great difference as far as the performance of P_{0t}^{F} (as compared with P_{0t}^{L} and P_{0t}^{P}) is concerned if we would replace some of the 20 tests by other tests, for example substitute reversal tests with aggregation properties.

Finally, we give an example of an ***inconsistency theorem***, the proof of which (taken from Balk[28]) is most impressive for its mathematical elegance:

Functions, P_{0t} and Q_{0t} which satisfy simultaneously 1. the identity (strict or weak) axiom, 2. the circular test, and 3. the product test do not exist.

[25] Balk 1995 (Theorem 4).

[26] The result is deprived of most of its appeal, however, once we call some tests into question. Interestingly, apart from T1, which is trivial, virtually all tests (T11, T21 and not least T12) are doubtful.

[27] See PPI and CPI Manual for more details.

[28] Balk 1995, p. 76 refers to Frisch (1930) and Wald (1937) as having proved it already.

It is interesting to demonstrate how the ***proof*** can be established. This is done by showing that admissible assumptions, like $\mathbf{p}_0 = \mathbf{p}_2, \mathbf{p}_1 = \mathbf{p}_3$, and $\mathbf{q}_0 = \mathbf{q}_1, \mathbf{q}_2 = \mathbf{q}_3$ will produce an equation which does not necessarily need to be true, but can be violated.

By virtue of identity we get with vectors as above

(a1, a2) $\quad P_{02} = P(\mathbf{p}_0, \mathbf{q}_0, \mathbf{p}_0, \mathbf{q}_2) = P_{13} = P(\mathbf{p}_1, \mathbf{q}_1, \mathbf{p}_1, \mathbf{q}_3) = 1$ and

(b1, b2) $\quad Q_{01} = P(\mathbf{p}_0, \mathbf{q}_0, \mathbf{p}_1, \mathbf{q}_0) = Q_{23} = P(\mathbf{p}_2, \mathbf{q}_2, \mathbf{p}_3, \mathbf{q}_2) = 1$.

If the product test is passed we obtain from (a)

(c1, c2) $\quad Q_{02} = \mathbf{p}_2'\mathbf{q}_2 / \mathbf{p}_0'\mathbf{q}_0 = V_{02}$ and $Q_{13} = \mathbf{p}_3'\mathbf{q}_3 / \mathbf{p}_1'\mathbf{q}_1 = V_{13}$ and from (b)

(d1, d2) $\quad P_{01} = \mathbf{p}_1'\mathbf{q}_1 / \mathbf{p}_0'\mathbf{q}_0 = V_{01}$ and $P_{23} = \mathbf{p}_3'\mathbf{q}_3 / \mathbf{p}_2'\mathbf{q}_2 = V_{23}$

and if circularity holds we expect

$P_{03} = P_{01}P_{13} = V_{01}$ (by a2 and d1)	$P_{03} = P_{02}P_{23} = V_{23}$ (by a1 and d2)

or virtually the same result

$P_{03} = \frac{V_{03}}{Q_{01}Q_{13}} = \frac{V_{01}V_{13}}{Q_{01}Q_{13}} = \frac{V_{01}}{Q_{01}} = V_{01}$ (by b1 and c2)	$P_{03} = \frac{V_{03}}{Q_{02}Q_{23}} = \frac{V_{02}V_{23}}{Q_{02}Q_{23}} = V_{23}$ (by b2 and c1)

To have a consistent result, $V_{01} = \frac{\mathbf{p}_1'\mathbf{q}_1}{\mathbf{p}_0'\mathbf{q}_0} = V_{23} = \frac{\mathbf{p}_3'\mathbf{q}_3}{\mathbf{p}_2'\mathbf{q}_2}$ must hold, but in

(e) $$V_{01} = \frac{\mathbf{p}_3'\mathbf{q}_0}{\mathbf{p}_0'\mathbf{q}_0} = V_{23} = \frac{\mathbf{p}_3'\mathbf{q}_3}{\mathbf{p}_0'\mathbf{q}_3}$$

the vectors $\mathbf{q}_0$ and $\mathbf{q}_3$ can well be different and chosen in a way that (e) will *not* hold.

3.2 Fundamental axioms and their interpretation

Though a number of the properties of index functions we are going to deal with in what follows, that is monotonicity, additivity, linear homogeneity, proportionality, and mean value property, have been mentioned and even explained in some detail above already, it appears useful to recall and comment on the definitions, and to give a more comprehensive account of the confusingly many closely inter-related properties of index functions and the nature of the relationship between them.

a) The meaning of strict and weak monotonicity

The presentation of this section starts with monotonicity, a fundamental requirement to ensure that an index function will behave like an average, and indicate the direction of change correctly. Monotonicity is also a more general concept than "additivity", by virtue of which indices are easily decomposable into "subindices" and suitable for deflation.[29]

Strict monotonicity applied to an index function P(...) is defined as follows:

(3.2.1)	$P(\mathbf{p}_0,\mathbf{q}_0,\mathbf{p}_t^*,\mathbf{q}_t) > P(\mathbf{p}_0,\mathbf{q}_0,\mathbf{p}_t,\mathbf{q}_t)$, if $\mathbf{p}_t^* \geq \mathbf{p}_t$ and
(3.2.2)	$P(\mathbf{p}_0^*,\mathbf{q}_0,\mathbf{p}_t,\mathbf{q}_t) < P(\mathbf{p}_0,\mathbf{q}_0,\mathbf{p}_t,\mathbf{q}_t)$, if $\mathbf{p}_0^* \geq \mathbf{p}_0$.

By $\mathbf{p}_t^* \geq \mathbf{p}_t$ is meant that at least one element of the nonnegative vector $\mathbf{p}_t^*$, say p_{it}^*, is greater than the corresponding element of $\mathbf{p}_t$, such that $p_{it}^* > p_{it}$, while to other commodities $p_{jt}^* = p_{jt}$ applies (and $\mathbf{p}_0^* \geq \mathbf{p}_0$ is defined analogously).

An index for which strict monotonicity holds will satisfy the following *two* conditions[30]:

1. by eq. 1: when the prices p* (at least one of them) in the **current period** t are higher compared to the prices p, the index has to reflect this rise, and correspondingly
2. by eq. 2: when **base period** (0) prices p_{i0}^* are assumed to be higher [lower] than p_{i0}, the index should be lower [higher].

Thus a price index has to indicate an increase (P > 1) whenever any of the prices (at least one of them, say the i-th) in the *current* period p_{it}^* is greater than p_{it} and if at least one of the *base* period prices is lower, that is if $p_{i0}^* < p_{i0}$, everything else being constant.

In contrast to strict monotonicity the so called *weak monotonicity* is defined by[31]

(3.2.1a)	$P(\mathbf{p}_0,\mathbf{q}_0,\mathbf{p}_t,\mathbf{q}_t) > P(\mathbf{p}_0,\mathbf{q}_0,\mathbf{p}_0,\mathbf{q}_t)$ if $\mathbf{p}_t \geq \mathbf{p}_0$ and
(3.2.2a)	$P(\mathbf{p}_0,\mathbf{q}_0,\mathbf{p}_t,\mathbf{q}_t) < P(\mathbf{p}_0,\mathbf{q}_0,\mathbf{p}_0,\mathbf{q}_t)$ if $\mathbf{p}_t \leq \mathbf{p}_0$.

Strict monotonicity implies weak monotonicity

if strict monotonicity is met ⟶ then also weak monotonicity

but the converse is not true.

Equation (or to be more specific: inequality) 1a follows from eq. 1 by substituting $\mathbf{p}_t = \mathbf{p}_0$ on the righthand side (RHS) and $\mathbf{p}_t^* = \mathbf{p}_t$ on the lefthand side (LHS) of eq. 1 (in this order). In the same manner we get eq. 2a from eq. 2 by substituting in eq. 2 as follows, firstly $\mathbf{p}_t = \mathbf{p}_0$ (right) and then $\mathbf{p}_0 = \mathbf{p}_0^*$ (left). **Ex. 3.2.1** shows that the func-

[29] Not surprisingly, such considerations, while neglected in the older index theory, have now gained importance in our days due to the appearance of National Accounts.

[30] In the next part of this section it will be shown that the two conditions are independent.

[31] In contrast to eq. 1 and 2 the two following inequalities are in fact only one.

tion $P_{max} = P(...) = \max\left(\frac{p_{it}}{p_{i0}}\right)$ and Palgrave's index satisfy weak monotonicity but not strict monotonicity.[32]

Figure 3.2.1: Strict and weak monotonicity

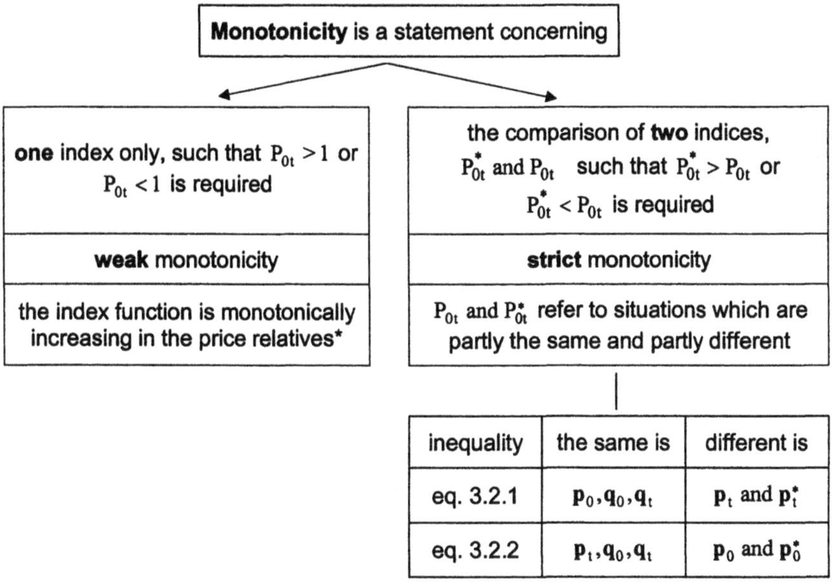

This simply means that we should get P > 1 (or P < 1) when the price relatives p_t/p_0 show a rise (or decrease) irrespective of whether the rise (decline) is due to rising prices p_t or lowered prices p_0 (or lowered prices p_t and increased prices p_0). That is the reason why there is only one condition comparing p_t with p_0, whereas strict monotonicity needs two (comparing p with p in period t or in 0).

In other words: weak monotonicity only requires a price index showing an increase whenever any of the prices in the current period are raised, or any of the prices in the base period are lowered. Strict monotonicity on the other hand requires a price index P^*_{0t} to show a greater increase than another price index P_{0t} whenever any of the price relatives in P^*_{0t} shows a greater increase than the price relatives in P_{0t}.

[32] In the same manner, the function $\min(p_{it}/p_{i0})$ satisfies weak monotonicity but not strict monotonicity.

Example 3.2.1

Assume three commodities with prices and quantities given as follows:

i	p_{i0}	p_{it}	q_{i0}	q_{it}
1	12	18	10	8
2	15	15	8	6
3	20	20	5	5

p_{it}^*
18
16
20

Calculate P_{max}, P_{0t}^L and P_{0t}^P and explain "weak" as opposed to "strict" monotonicity.

Disregarding prices p* we get $P_{max} = \max(p_t/p_0) = 18/12 = 1.5 > 1$, hence *weak* monotonicity is met, requiring only: when the price of at least one commoditiy is rising (in this case the price of commodity 1), all other prices remaining constant, the index should be $P > 1$. This is true also for $P_{0t}^L = 400/340 = 1.176 > 1$ and $P_{0t}^P = 334/286 = 1.168 > 1$.
To see whether all three index functions also satisfy *strict* monotonicity this result should be compared with the situation in which p_{it} is replaced by p_{it}^*.

Since $\mathbf{p}_t^* > \mathbf{p}_t$, in order to satisfy strict monotonicity the index should be greater than before. The results are: $P_{max} = \max(1.5, 1.067, 1) = 1.5$ as before, hence there is no increase in P_{max} although $p_{2t}^* > p_{2t}$. Thus P_{max} satisfies weak but not strict monotonicity. P_{0t}^L and P_{0t}^P on the other hand are both strictly monotonous since we now get $P_{0t}^L(p^*) = 408/340 > 400/340 = P_{0t}^L$ and $P_{0t}^P(p^*) = 340/286 = 1.189 > 1.168 = 334/286 = P_{0t}^P$. ♦

Note that - as aforementioned - the *two* inequalities (eq. 1a and 2a) correspond to only *one* (eq. 1) in the case of strict monotonicity. Furthermore both inequalitites, 1a and 2a, defining weak monotonicity are independent (see **ex. 3.2.2**, and **tab. 3.2.1**)

Table 3.2.1: Two variants of weak monotonicity

prices decline	prices rise ($p_t > p_0$), eq. 1a	
($p_t > p_0$), eq. 2a	yes	no
yes	Palgrave's index	$P_{min} = \min\left(\frac{p_{it}}{p_{i0}}\right)$
no	$P_{max} = \max\left(\frac{p_{it}}{p_{i0}}\right)$	median of price relatives

The implications of monotonicity will be shown in more detail when special cases of monotonicity, such as additivity or multiplicativity, will be discussed (see part b of this section). An important relationship should be mentioned now already:

Index functions that can be conceived as means (averages) of price relatives are always monotonically increasing (decreasing) when the price relatives rise (decrease), that is, they are by implication monotonous in the *weak* sense.

Monotonicity is a most fundamental characteristic of all reasonable means (or averages in general). If a single price rises (declines), everything else being constant, there should be a reaction of the index formula also showing a change in the proper direction, that is an increase or a decrease in the price level. If there were no such reaction, the function would not be reasonable. The median of price relatives is an example of an index that might violate this condition.

According to **tab. 3.2.2** the ***two parts*** (eqs. 3.2.1 and 3.2.2) of the definition ***of strict monotonicity*** are independent since there are index functions that satisfy[33]

- the first part (strict monotonicity in current period prices, eq. 1) but not necessarily also the second part (strict monotonicity in base period prices, eq. 2), for example

  $$P_{0t}^{HB} = \left[\sum \frac{p_0}{p_t} \frac{p_0 q_0}{\sum p_0 q_0}\right]^{-1}$$ (harmonic mean with base year budget[34]),

  and functions that satisfy

- the second part but not necessarily also the first part, for example

  $$P_{0t}^{PA} = \sum \frac{p_t}{p_0} \frac{p_t q_t}{\sum p_t q_t}$$ Palgrave's index.

That P_{0t}^{PA} meets strict monotonicity with respect to base period prices can easily be seen since at least one of the terms p_t/p_0 will become smaller while nonetheless the weights $p_t q_t/\sum p_t q_t$ are not affected. But monotonicity with respect to current period prices cannot be deduced from this consideration, because assuming prices $p_t^* > p_t$ will influence price relatives as well as weights.

Table 3.2.2: Two conditions of strict monotonicity

eq. 2 base period prices	eq. 1 current period prices: yes	eq. 1 current period prices: no
yes	P^L, P^P etc.	Palgrave's index P^{PA}
no	harmon. Lasp. P^{HB}	median of price relatives

Numerical examples to show this need to be somewhat farfetched, however, since in all more or less "normal" situations both indices will satisfy both conditions of monotonicity. In what follows a simple example is given for the violation of eq. 1 in the case of P^{PA}.

---

[33] See Olt 1996, p. 129 ff. for a proof.

[34] Later also called P^{HL} (harmonic Laspeyres).

Example 3.2.2

The following numerical example will demonstrate the second assertion (with respect to the violation of strict monotonicity in current period prices in the case of P_{0t}^{PA}):

	p_0	p_t	p_t^*	q_t
A	80	30	31	4
B	20	25	25	8

We obtain $P_{0t}^{PA}(\mathbf{p}_0,\mathbf{p}_t,\mathbf{q}_t)=\frac{30}{80}\cdot 0.375+\frac{25}{20}\cdot 0.625$ =0.92188, but with prices p* we get

$P_{0t}^{PA}(\mathbf{p}_0,\mathbf{p}_t^*,\mathbf{q}_t)=\frac{31}{80}\cdot\frac{124}{324}+\frac{25}{20}\cdot\frac{200}{324}$ = 0.91991, which is *not* greater but smaller than the first result (0.921875), although $\mathbf{p}_t^* > \mathbf{p}_t$. Strict monotonicity with respect to current (t) period prices requires, however, that under such circumstances $P(\mathbf{p}_t^*) > P(\mathbf{p}_t)$. Hence Palgrave's index can violate strict monotonicity in the sense of eq. 3.2.1. ♦

b) Additivity and multiplicativity as special cases of strict monotonicity

Monotonicity only requires that the index function reflects an increase or a decrease whenever prices increase or decrease, that is the *direction* of change has to be indicated correctly. But nothing is said about the *amount* of increase or decrease the index should be able to reflect.

1. Additivity as a type of monotonicity

An interesting subset of the set of monotonous index functions is known as "additive" index functions. The advantage of an index being additive (in current period prices) is that the difference of two indices due to different prices can be expressed as a function of the extent to which the individual prices differ. By additivity of the index function (note that the term "additivity" is used in index theory in many different ways, as will be demonstrated in much more detail in **sec. 5.2** below) the following property of the function P(...) is meant:

Assuming nonnegative price vectors, $\mathbf{p}_t^*$ and $\mathbf{p}_0^*$, which are defined as sums of two price vectors, then the function P(...) is additive if

(3.2.3) $P(\mathbf{p}_0,\mathbf{p}_t^*)=P(\mathbf{p}_0,\mathbf{p}_t)+P(\mathbf{p}_0,\mathbf{p}_t^+)$ = A + B where $\mathbf{p}_t^*=\mathbf{p}_t+\mathbf{p}_t^+$, and

(3.2.4) $\frac{1}{P(\mathbf{p}_0^*,\mathbf{p}_t)}=\frac{1}{P(\mathbf{p}_0,\mathbf{p}_t)}+\frac{1}{P(\mathbf{p}_0^+,\mathbf{p}_t)}=\frac{1}{C}+\frac{1}{D}$ where $\mathbf{p}_0^*=\mathbf{p}_0+\mathbf{p}_0^+$.

The reader should verify (using **ex. 3.2.3**) that P^L, P^P, and P^D (Dutot) all satisfy both conditions, while Fisher's ideal index P^F is *not* additive, although both components, P^L and P^P, are.

Moreover the two conditions, eq. 3 and eq. 4, are independent:

	eq. 3 satisfied	eq. 3 violated
eq. 4 satisfied	P^L, P^P, Dutot P^D	unweighted harmonic mean of P^L and P^P
eq. 4 violated	Carli's index P^C index of Drobisch $½(P^L + P^P)$	Fisher's ideal index P^F

Eq. 3 (4) applied to P^L and P^P leads to the arithmetic (harmonic) mean of P^L and P^P, the reciprocal of the latter is $½(1/P^L + 1/P^P)$.

Example 3.2.3

Assume three commodities with prices and quantities given as follows:

i	p_{i0}	p_{it}	q_{i0}	q_{it}
1	12	18	10	8
2	15	12	8	6
3	20	24	6	5

p^*_{i0}	p^*_{it}
15	15
16	16
22	25

Calculate P^L, P^P, P^D (Dutot), P^C (Carli), the unweighted harmonic mean index, and Fisher's "ideal" index P^F using

1) prices a) p_{i0} and p_{it} and b) p_{i0} and p^*_{it} to examine eq. 3.2.3, and

2) prices a) p_{i0} and p_{it} and b) p^*_{i0} and p_{it} to examine eq. 3.2.4.

ad 1: the vector $\mathbf{p}_t^+$ as column vector has now the following form: $\mathbf{p}_t^+ = \begin{bmatrix} -3 \\ 4 \\ 1 \end{bmatrix}$ (the sum of the elements amounts to 2; → Dutot's index below), and index calculations result in

formula	$P(\mathbf{p}_0,\mathbf{p}_t)$	$P(\mathbf{p}_0,\mathbf{p}_t^*)$	$P(\mathbf{p}_0,\mathbf{p}_t^+)$
Laspeyres	420/360	428/360	8/360
Paasche	336/286	341/286	5/286
Dutot	54/47	56/47	2/47
Carli [a)]	3.5/3	3.5667/3	0.0667/3
harmonic mean [b)]	12/11	240/203	36/237
Fisher [c)]	1.1707388	1.1905971	0.0197104

a) Carli's index is in this case the same as the Laspeyres index because the value shares of all commodities at base time are identically 120/360 = 1/3.

b) The sum 12/11 + 36/237 = 1.2428 differs from 240/203 = 1.18227.

c) The sum of 1.1707 and 0.0197 is 1.1904492 differs (slightly) from 1.1905971.

The example shows that the index functions of Laspeyres, Paasche and Dutot meet additivity in the sense of both equations 3 and 4, while Fisher's index is clearly not additive, i.e. unable to meet either condition.

ad 2: the vector $\mathbf{p}_0^+$ as row vector has now the following form: $\mathbf{p}_0^+ = [3 \quad 1 \quad 2]$

formula	$1/P(\mathbf{p}_0,\mathbf{p}_t)$	$1/P(\mathbf{p}_0^*,\mathbf{p}_t)$	$1/P(\mathbf{p}_0^+,\mathbf{p}_t)$
Laspeyres	360/420	410/420	50/420
Paasche	286/336	326/336	40/336
Dutot	47/54	53/54	6/54
Carli[a)]	3/3.5	3/3.04091 [a)]	3/30
harmonic mean	8.25/9 = 11/12	9.25/9	1/9
Drobisch [b)]	0.85437	0.9732028	0.119048
Fisher [c)]	0.85416	0.9732097	0.119047

a) = 0.986547, but the sum of 3/3.5 = 0.857143 and 3/30 = 0.1 is 0.957143 instead of 0.986547.

b) 0.85437 + 0.119048 = 0.9734414

c) The sum of 0.85416 and 0.11905 is 0.9732091, and (slightly) different from 0.9732097. ♦

Not being additive has been regarded as a serious defect of Fisher's index by Pfouts 1966. For him the notion of additivity has an intuitive appeal:

> The difference between two indices reflects a difference between two price vectors. In the absence of additivity no inference from price indices to the difference in prices is possible.

Weak variant of additivity: To make additivity more plausible, Pfouts presents yet another argument: Additivity "states if all prices are increased by the same amount , the index of the new price should equal to the old index number plus the index number of the constant amount" (p.176). This requirement might be regarded as a weaker version of additivity: the vector $\mathbf{p}_t^+$ being a vector of constants is $\mathbf{p}_t^+ = \begin{bmatrix} b \\ \dots \\ b \end{bmatrix}$ and $\mathbf{p}_0^+$ is defined correspondingly. The indices of Laspeyres and Paasche calculated for the vector of differences $\mathbf{p}_t^+$ will be simply $b\sum q_0/\sum p_0 q_0$ in the case of P_{0t}^L (and $b\sum q_t/\sum p_0 q_t$ in the case of P_{0t}^P), but there is still no reason why P_{0t}^F should satisfy only one of the two eqs. 3 and 4, let alone both of them, under these relaxed conditions.

2. Multiplicativity as type of monotonicity

Multiplicativity of the index function is defined as follows (see Eichhorn/Voeller1976, p. 14):

(3.2.5) $$P(\mathbf{p}_0^*,\mathbf{p}_t^*) = P(\mathbf{K}\mathbf{p}_0,\mathbf{L}\mathbf{p}_t) = P(\mathbf{p}_0,\mathbf{p}_t)\cdot\phi(\kappa_1,\dots,\kappa_n,\lambda_1,\dots,\lambda_n)$$

where **K** and **L** are diagonal matrices $\mathbf{K}=\begin{bmatrix}\kappa_1 & & 0\\ & \ldots & \\ 0 & & \kappa_n\end{bmatrix}$ and $\mathbf{L}=\begin{bmatrix}\lambda_1 & & 0\\ & \ldots & \\ 0 & & \lambda_n\end{bmatrix}$

and ϕ is a function depending on the real numbers $\kappa_1, \kappa_2, \ldots, \kappa_n, \lambda_1, \lambda_2, \ldots, \lambda_n$ only such that ϕ is a positive real number. An example of an index function satisfying multiplicativity is the Cobb-Douglas-Index (eq. 2.2.3) because

$$\left(\frac{\lambda_1 p_{1t}}{\kappa_1 p_{10}}\right)^{\alpha_1}\ldots\left(\frac{\lambda_n p_{nt}}{\kappa_n p_{n0}}\right)^{\alpha_n}=\left(\frac{p_{1t}}{p_{10}}\right)^{\alpha_1}\ldots\left(\frac{p_{nt}}{p_{n0}}\right)^{\alpha_n}\phi \text{ where } \phi=\left(\frac{\lambda_1}{\kappa_1}\right)^{\alpha_1}\ldots\left(\frac{\lambda_n}{\kappa_n}\right)^{\alpha_n}$$

is dependent on the constants $\kappa_1, \kappa_2, \ldots, \kappa_n, \lambda_1, \lambda_2, \ldots, \lambda_n$ only.

The logarithmic Laspeyres (DP^L) index, for example, with weights representing base period expenditure shares (instead of arbitrary constants $\alpha_1, \ldots, \alpha_2$), can violate eq. 5 because alternative assumptions concerning base period prices ($\mathbf{K} \neq \mathbf{I}$) will entail different (expenditure share) weights, hence

Laspeyres type	formula	property
normal	$\sum_i \frac{p_{it}}{p_{i0}} s_{i0}$	additive in both respects: in current (eq. 3.2.3) and base period prices (eq. 3.2.4)
logarithmic	$\prod_i \left(\frac{p_{it}}{p_{i0}}\right)^{s_{i0}}$	multiplicative in current period prices only (eq. 3.2.5)

3. Theorem on additive indices

Aczel and Eichhorn 1974 (see also Eichhorn/Voeller 1976, p.18 - 21) have proved that an index function P( ) that satisfies the conditions of additivity (eq. 3.2.3 and 3.2.4) neccessarily

1. also meets the following condition

   (3.2.6) $$P\left(\mathbf{p}_0,\left(\frac{M}{N}\right)\mathbf{p}_t\right)=\frac{1}{N}P(\mathbf{p}_0,M\mathbf{p}_t)=\frac{M}{N}P(\mathbf{p}_0,\mathbf{p}_t)$$

   which implies linear homogeneity, homogeneity of degree -1 and (price) dimensionality (the converse is not true),

2. will be P( ) = 0 if all current prices decrease to zero and otherwise P( ) > 0, and finally (and most importantly), such a function P( )

3. must have the following form

   (3.2.7) $$P^A=\frac{\mathbf{a}'\mathbf{p}_t}{\mathbf{b}'\mathbf{p}_0},$$

which is an ***additive index***, the ratio of two inner products ($\mathbf{a}$ and $\mathbf{p}_t$ need not contain the same number of commodities as $\mathbf{b}$ and $\mathbf{p}_0$), like the Laspeyres form ($\mathbf{a} = \mathbf{b} = \mathbf{q}_0$) or the Paasche form ($\mathbf{a} = \mathbf{b} = \mathbf{q}_t$), but also the value index V_{0t}, in which $\mathbf{a} = \mathbf{q}_t$ and $\mathbf{b} = \mathbf{q}_0$, or the index of Walsh, in which $\mathbf{a} = \mathbf{b}$ contains elements $\sqrt{q_{i0} q_{it}}$.

The following two statements are equivalent (or **1 ↔ 2**): index P( ) is additive in the sense of being constructed according to eq. 3.2.7 and index P( ) satisfies eq. 3.2.3 and 3.2.4 (defining additivity)

In other words: additivity is *characterized* by eq. 3.2.7 or by eq. 3.2.3 and 3.2.4. This explains also why for example Fisher's ideal index P^F does not fulfil the conditions of additivity. The same is true for for the quadratic mean index P_{0t}^{QM}, defined as

$$(3.2.8) \qquad P_{0t}^{QM} = \sqrt{\sum \left(\frac{p_t}{p_0}\right)^2 \frac{p_0 q_0}{\sum p_0 p_0}}\,.$$

Also the index of Drobisch $P_{0t}^{DR} = \frac{1}{2}\left(P_{0t}^{L} + P_{0t}^{P}\right) = \frac{1}{2} \cdot \frac{\sum p_t q_0 \sum p_0 q_t + \sum p_t q_t \sum p_0 q_0}{\sum p_0 q_0 \sum p_0 q_t}$ cannot be transformed into the form $\mathbf{a}'\mathbf{p}_t / \mathbf{b}'\mathbf{p}_0 = \sum p_{it} a_i / \sum p_{j0} a_j$ according to eq. 3.2.7. The same is true for Carli's index.

The formula P_{0t}^{QM} will be referred to in **sec. 5.2** because it is aggregative consistent but not (more restrictively) additive in the sense defined above. P_{0t}^{QM} gives (relative to an arithmetic mean) higher weights to relatives that are far away from unity and lower weights to those that are nearer to one. Hence not surprisingly $P_{0t}^{QM} > P_{0t}^{L}$, unless all price relatives are equal (or equivalently var(a) = 0). In fact, the formula of P_{0t}^{QM} turns out to be simply $\sqrt{\text{var}(a) + \left(P^L\right)^2}$, where var(a) is the variance of price relatives.

Notice further that

- crossing *weights* does not destroy additivity, as can be seen in the case of Walsh's index (the same would be true for the index P^{ME} [Marshall-Edgeworth], or for a harmonic crossing of weights [Geary Khamis] for example), but
- the same is *not* true, however, in general for crossing the Laspeyres and Paasche *formula*, although both indices are clearly additive indices in the sense defined above.

c) Generalization of Bortkiewicz's theorem for additive indices

The theorem of Bortkiewicz, as introduced in **sec. 1.3** part d, turns out to be a special case of a more general law of the ratio of two additive indices (see **fig. 3.2.2**). The interesting result is eq. 3.2.9. It can easily be seen that the special case of **sec. 1.3** was as follows:

$X_0 = \overline{X} = P^L_{0t}, X_t = P^P_{0t}$ and $\overline{Y} = Q^L_{0t}$. Only in this case the coefficients of variation, V_x and V_y, are symmetrically defined, one representing the relative dispersion of price relatives and the other the relative dispersion of quantity relatives.

The theorem (eq. 3.2.9) can of course be applied to *any* pair of indices, where both are of the additive type. Thus, for example applied to the comparison of P^L_{0t} (Laspeyres) with P^W_{0t} (Walsh):

We now have $x_t = p_t, x_0 = p_0, y_0 = q_0$ and $y_t = \sqrt{q_0 q_t}$, hence the relevant variances are

- the variance of the relatives $\frac{x_t}{x_0} = \frac{p_t}{p_0}$ relative to the mean $\overline{X} = X_0 = P^L_{0t}$, and
- the variance of the $\frac{y_t}{y_0} = \sqrt{\frac{q_t}{q_0}}$ measured around $\overline{Y} = \frac{\sum p_0 \sqrt{q_0 q_t}}{\sum p_0 q_0}$.

The extent to which Walsh's index P^W_{0t} turns out to be greater or smaller than Laspeyres' index P^L_{0t} depends on these variances in combination with the covariance between $\frac{p_t}{p_0}$ and $\sqrt{\frac{q_t}{q_0}}$, which is given by $\frac{\sum p_0 \sqrt{q_0 q_t}}{\sum p_0 q_0}(P^W_{0t} \sum p_t - P^L_{0t} \sum p_0)$.

A consequence of this result is for example: if $\frac{p_t}{p_0}$ and $\frac{q_t}{q_0}$ are negatively correlated such that $P^L_{0t} > P^P_{0t}$, the same will be true for $\frac{p_t}{p_0}$ and $\sqrt{\frac{q_t}{q_0}}$ such that $\sum p_0 P^L_{0t} > \sum p_t P^W_{0t}$.

Another example for the comparison between two linear indices is Dutot's index P^D_{0t} as X_t, compared with Laspeyres and Paasche as X_0. In these cases we get

X_t/X_0	The relevant covariance is between
P^D_{0t}/P^L_{0t}	$\frac{p_{it}}{p_{i0}}$ and $\frac{1}{q_{i0}}$
P^D_{0t}/P^P_{0t}	$\frac{p_{it}}{p_{i0}}$ and $\frac{1}{q_{it}}$

Figure 3.2.2: Generalization of Bortkiewicz's theorem
(law of the ratio of two additive indices)

first additive index X_0	second additive index X_t
$X_0 = \frac{\sum x_t y_0}{\sum x_0 y_0}$	$X_t = \frac{\sum x_t y_t}{\sum x_0 y_t}$

weighted (weights $w_0 = x_0 y_0 / \sum x_0 y_0$ throughout) arithmetic mean of the relatives

x_t/x_0 : $\overline{X} = X_0$	y_t/y_0 : $\overline{Y} = \frac{\sum y_t x_0}{\sum y_0 x_0}$ *)

variances (weights w_0) of the relatives

x_t/x_0 : $s_x^2 = \sum \left(\frac{x_t}{x_0} - \overline{X} \right)^2 w_0$	y_t/y_0 : $s_y^2 = \sum \left(\frac{y_t}{y_0} - \overline{Y} \right)^2 w_0$

the covariance is given by

$$s_{xy} = \sum \left(\frac{x_t}{x_0} - \overline{X} \right) \left(\frac{y_t}{y_0} - \overline{Y} \right) w_0 = \frac{\sum x_t y_t}{\sum x_0 y_0} - \overline{X} \cdot \overline{Y}$$

and the ratio of two additive indices

(3.2.9) $$\frac{X_t}{X_0} = 1 + r_{xy} V_x V_y = 1 + \frac{s_{xy}}{\overline{X} \cdot \overline{Y}}$$

where $r_{xy} = \frac{s_{xy}}{s_x s_y}$, $V_x = \frac{s_x}{\overline{X}}$ and $V_y = \frac{s_y}{\overline{Y}}$

*) The formula of $\overline{Y}$ can be derived from $\overline{X} = X_0$ by interchanging x and y.

d) Mean value property (mean value test for price relatives)

By this condition it is required that the index number should take a value between the smallest and the largest individual price relative (a_{0t}^i), that is for P_{0t} should hold

(3.2.10) $\min(a_{0t}^i) \leq P_{0t} \leq \max(a_{0t}^i)$ (*strict* mean value property).

This means the price index should take on values within the boundaries between $\min(a^i_{0t})$ and $\max(a^i_{0t})$, to add "strict" means that the boundaries themselves should be relevant *only* (the equality should hold only) if $a^i_{0t} = a_{0t} = \min(a^i_{0t}) = \max(a^i_{0t})$ for all i = 1, 2, ..., n.

In contrast to the strict mean value property the so called *weak* mean value property does not require that P_{0t} equals $\min(a^i_{0t})$ or $\max(a^i_{0t})$ only in those cases in which all price relatives are identical.[35]

Mean value property expressed more rigourously in mathematical terms means that there exists a real number λ such that the price index is a convex combination of the price relatives

(3.2.11) $$P(\mathbf{p}_0, \mathbf{q}_0, \mathbf{p}_t, \mathbf{q}_t) = \lambda \min(a^i_{0t}) + (1-\lambda)\max(a^i_{0t})$$

where "strict" means $0 < \lambda < 1$, whilst in the "weak" case $0 \leq \lambda \leq 1$ is admitted.

e) Relations between monotonicity, proportionality and mean value property

The following relationship is well known (see also Eichhorn/Voeller 1976, p. 10)

if monotonicity* and proportionality** is met	⟶	then also strict mean value property

* weak monotonicity sufficient

** which is the same as: identity and linear homogeneity

Again the converse relation is not true (if *strict* monotonicity is concerned at least). An example for this is once more Palgrave's index.

This index $P^{PA}_{0t} = \sum \frac{p_t}{p_0} \frac{p_t q_t}{\sum p_t q_t}$ is by construction clearly a mean value of price relatives but nonetheless not strictly monotonous.[36] Another relationship is

given strict mean value property	⟶	then also proportionality

by which proportionality is simply a special case of the *strict* mean value property. An example for the converse *not* being true is again (see part a of this **sec. 3.2**) the index function P_{max}, which meets the mean value property only in the weak sense because P_{max} equals $\max_i(p_{it}/p_{i0})$ even though not all price relatives $a^i_{0t} = p_{it}/p_{i0}$ are identically $\max(a^i_{0t})$ but different despite satisfying proportionality.

[35] According to Pfouts (1966) the mean value property is one of the most essential properties of an index function. The reason is that the mean value property conforms with our intuitive notion of an index to be a measure of a "representative" aggregated change. Furthermore, Pfouts demonstrated that an index conforming with eq. 3.2.10 as well as with linear homogeneity must be a power mean of price relatives.

[36] The following numerical example, **ex. 3.2.4**, will show that P^{PA} may violate strict (but not weak) monotonicity.

Strict mean value property also implies weak monotonicity (but not strict monotonicity, as shown in **ex. 3.2.4** with reference to Palgrave's index)

given strict mean value property	⟶	then also weak monotonicity

The converse is again not true (see Olt 1996, p. 146 for a function[37] satisfying weak monotonicity and weak mean value property, but not strict mean value property). Finally, weak mean value property is to be considered as more demanding than proportionality because

given weak mean value property	⟶	then also proportionality

See again Olt 1996, p. 145 for a function which meets proportionality but not weak mean value property.

Example 3.2.4

This numerical example should show again (like **ex. 3.2.2**) that Palgrave's index may violate strict monotonicity in *current* period prices despite clearly being an arithmetic *mean* of price relatives. Consider the following figures concerning two commodities, A and B

			alternative prices			alternative price relatives		
i	p_0	q_t	p_t^1	p_t^2	p_t^3	p_t^1/p_0	p_t^2/p_0	p_t^3/p_0
A	1	3	2	2	3	2	2	3
B	2	1	1	2	1	0.5	1	0.5

With prices p_t^1 we obtain $P_{0t}^{PA} = 2 \cdot \frac{6}{7} + \frac{1}{2} \cdot \frac{1}{7} = 1.7857$, which exceeds the result obtained when prices are p_t^2, which is $P_{0t}^{PA} = 2 \cdot \frac{6}{8} + \frac{2}{8} = 1.75$ although $\mathbf{p}_t^2 > \mathbf{p}_t^1$. But slight modifications show that strict monotonicity is in most cases fulfilled. Taking prices $\mathbf{p}_t^3$, for example, the result is $P_{0t}^{PA} = 2.75$, which is - as expected - much greater than 1.7857.

In all cases Palgrave's index stays within the interval between the smallest and the greatest price relative. The results for P_{0t}^{PA} with prices p_t^1 through p_t^3 are as follows: p_t^1: 0.5 < 1.7857 < 2, p_t^2: 1 < 1.75 < 2, and p_t^3: 0.5 < 2.75 < 3 because Palgrave's index is a mean of price relatives.

[37] as usual, somewhat farfetched.

This example also shows that Palgrave's index is not able to meet additivity in current prices in the sense of eq. 3.2.3. Taking prices p_t^1 as a reference point, the summation conditions are as follows:

$\mathbf{p}_t^1 = \begin{bmatrix} 2 \\ 1 \end{bmatrix}$, $\mathbf{p}_t^+ = \begin{bmatrix} 0 \\ 1 \end{bmatrix}$ and $\mathbf{p}_t^2 = \mathbf{p}_t^1 + \mathbf{p}_t^+$. The index using $\mathbf{p}_t^+$ yields $P_{0t}^{PA(+)} = 1/2$, but 1.75 differs from 1.7857 by - 0.0357 ≠ + 0.5 . In comparing $\mathbf{p}_t^1$ with $\mathbf{p}_t^3$ we have $\mathbf{p}_t^+ = \begin{bmatrix} 1 \\ 0 \end{bmatrix}$ and get a similar result: using the vector $\mathbf{p}_t^+$, Palgrave's index amounts to 1 and again 1.7857 + 1 ≠2.75. Thus, as stated above, Palgrave's index is *not* an additive (linear) index. ♦

f) The meaning of linear homogeneity

Linear homogeneity is a property not easy to understand. While monotonicity requires an index to show an increase or decrease as individual prices increase or decrease, linear homogeneity consists of comparing two indices measuring different degrees of changes in prices. To make things even more complicated, there is also a relationship between linear homogeneity, proportionality and mean value property, as monotonicity is related to all these properties.

An index function P(...) satisfying

(3.2.12) $P(\mathbf{p}_0, \mathbf{q}_0, \lambda \mathbf{p}_t, \mathbf{q}_t) = \lambda P(\mathbf{p}_0, \mathbf{q}_0, \mathbf{p}_t, \mathbf{q}_t), \quad \lambda \in \mathbb{R}$

is said to be linear homogeneous (homogeneity of degree +1) in the prices of period t. Note that eq. 3.2.12 corresponds to eq. 3.2.1. The axiom corresponding to eq. 3.2.2 is

(3.2.13) $P(\lambda \mathbf{p}_0, \mathbf{q}_0, \mathbf{p}_t, \mathbf{q}_t) = \frac{1}{\lambda} P(\mathbf{p}_0, \mathbf{q}_0, \mathbf{p}_t, \mathbf{q}_t)$,

which is called homogeneity of degree -1 in base period prices.

In the case of monotonicity, both parts of the definition have to be postulated separately, but to express the axiom of linear homogeneity it is in general not necessary to focus in addition on eq. 3.2.13, since

given linear homogeneity and dimensionality	⟶	then also homogeneity of degree -1

Linear homogeneity in combination with dimensionality implies homogeneity of degree -1.

The Laspeyres price index is an example of an index that conforms with linear homogeneity (and also with proportionality) since

$$\frac{\lambda p_{1t}q_{10} + \lambda p_{2t}q_{20} + \ldots + \lambda p_{nt}q_{n0}}{p_{10}q_{10} + p_{20}q_{20} + \ldots + p_{n0}q_{n0}} = \lambda \frac{p_{1t}q_{10} + p_{2t}q_{20} + \ldots + p_{nt}q_{n0}}{p_{10}q_{10} + p_{20}q_{20} + \ldots + p_{n0}q_{n0}} = \lambda P_{0t}^{L}.$$

Linear homogeneity applies also to the following index

(3.2.14) $P_{0t}^{Y} = \sqrt{\frac{\sum p_t^2 q_t^2}{\sum p_0^2 q_0^2}}$,

a formula ascribed to Allyn Young[38]. P^Y does not need to be able to satisfy proportionality except for the weak form: $P(\mathbf{p}_0, \mathbf{q}_0, \lambda\mathbf{p}_0, \mathbf{q}_0) = \lambda$ and (which is the weaker condition) identity and therefore the mean value test neither[39]. There is also no guarantee that P_{0t}^{Y}, as opposed to say $P_{0t}^{Y*} = \sqrt{\frac{\sum p_t^2 q_0^2}{\sum p_0^2 q_0^2}}$ satisfies monotonicity (strict or even weak only), since an isolated rise in the price of the i-th commodity $p_{it} > p_{i0}$, all other prices remaining unchanged, can well be offset by a corresponding fall of the quantity of, say, commodity j (such that $q_{jt} < q_{j0}$) or of several commodities (possibly including the i-th commodity).

g) Linear homogeneity and proportionality

Strict proportionality $P(\mathbf{p}_0, \mathbf{q}_0, \lambda\mathbf{p}_0, \mathbf{q}_t) = \lambda$, implied by linear homogeneity together with identity, is a characteristic that greatly appeals to common sense[40], as shown already in **sec. 3.1**:

Table 3.2.3: Independence of linear homogeneity and proportionality

	linear homogeneity	no linear homogeneity
proportionality	many formulas	1. exponential mean index P_{0t}^{EX} 2. Stuvel's indices[1)] P_{0t}^{ST}, Q_{0t}^{ST} 3. Vartia-I index[2)] 4. $V_{0t}/Q_{0t}^{ME} = P_{0t}^{BA2}$ of Banerjee[41]
no (strict) proportionality	1. P_{0t}^{Y} as opposed to P_{0t}^{Y} 2. value index V_{0t}	many formulas

1) only weak proportionality and identity but not linear homogeneity

2) This holds true for P^{V1} in contrast to the Vartia II index (P^{V2}). Olt 1996, p. 86 erroneously states that the Vartia II index violates linear homogeneity and the Vartia I index violates strict proportionality.

[38] see also **sec. 2.2** for the index function P^{QM2} which is the same formula.

[39] The reason is of course that quantities q_{it} and q_{i0} might differ.

[40] It is widely an expression of common sense but there are chain-index-formulas that violate this fundamental condition. If all prices rise with the same rate of growth, the index does not necessarily change with the same rate of growth.

[41] This factor antithetic index is an example given by Eichhorn and Voeller 1983. It is the Banerjee-2-index P^{BA2} which satisfies identity and proportionality but not linear homogeneity (in fact this index satisfies all four axioms of the system EV-4 of Eichhorn and Voeller but not linear homogeneity of EV-5, cp. **sec. 3.3**).

If all prices rise with the same rate of growth λ, the index should indicate exactly this rate of growth, i.e. it should amount to λ. Hence the following relation is well established:

if linear homogeneity and identity are met	$\longrightarrow$	then also (strict) proportionality

Again the converse is not true (see ex. 3.2.5). This and the opposite situation is summarized in the following tab. 3.2.3.

An example for an index function satisfying linear homogeneity and yet *not* fulfilling proportionality is the value index: due to $\frac{\sum \lambda p_t q_t}{\sum p_0 q_0} = \lambda \frac{\sum p_t q_t}{\sum p_0 q_0}$, the function V_{0t} satisfies linear homogeneity but since $\sum \lambda p_0 q_t / \sum p_0 q_0 \neq \lambda$ may well occur in the case of $q_{it} \neq q_{i0}$, strict proportionality is *not* met.

Figure 3.2.3: Linear homogeneity, monotonicity and proportionality (see also **fig. 3.2.1**)

A statement concerning

one index only, reflecting correctly	difference of two indices
direction of change: **weak monotonicity**	direction: **strict monotonicity**
P_{0t} is increased ($P_{0t} > 1$) whenever any of the prices in t are raised or any of the prices in 0 are lowered	1. $P^*_{0t} > P_{0t}$ if $\mathbf{p}^*_t > \mathbf{p}_t$. or 2. $P^*_{0t} < P_{0t}$ if $\mathbf{p}^*_0 > \mathbf{p}_0$ 2)
amount of change: **proportionality**	amount: **linear homogeneity**
if all prices rise with the same rate1) λ, the index P_{0t} should amount to λ ($\mathbf{p}_t = \lambda \mathbf{p}_0$)	$P^*_{0t} = \lambda P_{0t}$ if prices in t are $\mathbf{p}^*_t = \lambda \mathbf{p}_t$ and in 0 the same, that is, $\mathbf{p}^*_0 = \mathbf{p}_0$ 3)

1) more precisely, λ is the growth *factor* of prices.

2) and the other price vectors (p_0 in 1 and p_t in 2) remain unchanged

3) comparing prices $\mathbf{p}_t$ with $\mathbf{p}_0$ (proportionality) or prices $\mathbf{p}^*_t$ and $\mathbf{p}_t$ (lin. homogeneity)

The (unweighted) exponential mean price index is defined as follows

$$(3.2.15) \quad P^{EX}_{0t} = \ln\left[\frac{1}{n}\sum \exp\left(\frac{p_{it}}{p_{i0}}\right)\right],$$

and will be examined in the following example. This index meets the (weak)[42] proportionality test but not the axiom of linear homogeneity.

Example 3.2.5

Suppose there are only two commodities and

a) both prices rise *equally* by 50%. It is not really necessary to demonstrate the consequence of this for P_{0t}^{EX} by a numerical example, since in this case all price relatives are equally $\lambda_i = p_{it}/p_{i0} = \lambda = 1.5 \ \forall \ i = 1, 2, ..., n$ and therefore P_{0t}^{EX} is simply

$P_{0t}^{EX} = \ln\left(\frac{1}{n}\sum_i e^{\lambda}\right) = \ln(e^{\lambda}) = \lambda = 1.5$ since $\Sigma e^{\lambda} = n\, e^{\lambda}$, such that proportionality is met;

b) both prices rise *unequally*, the first one by 20%, the other by 50%. Thus we get

$$P_{0t}^{EX} = \ln\left(\frac{1}{2}\left(e^{1.5} + e^{1.2}\right)\right) = \ln(3.901) = 1.3612.$$

c) Alternatively we now assume a rise of prices exceeding situation b) in both cases by 50%, that is 80% (since $1.2 \cdot 1.5 = 1.8$) and 125% ($1.5 \cdot 1.5 = 2.25$) which results in

$$P_{0t}^{EX} = \ln\left(\frac{1}{2}\left(e^{2.25} + e^{1.8}\right)\right) = 2.0501 > 1.5 \cdot 1.3612 = 2.0418.$$

Since the result now does not exceed the first index (1.3612) by 50%, but by more (50.6%) than that, the exponential mean is not a linear homogenous index function, although it does fulfill the weak proportionality test. The reason for this is that $e^{1.5}\left(e^{1.5} + e^{1.2}\right) = e^{1.5+1.5} + e^{1.5+1.2} = e^{3} + e^{2.7} \neq e^{1.5 \cdot 1.5} + e^{1.5 \cdot 1.2} = e^{2.25} + e^{1.8}$. ♦

Recall that there were two conditions under which (strict) proportionality was concluded: linear homogeneity *and* identity. Hence it is *not* correct to say linear homogeneity is more demanding, or "stronger" than proportionality, although there exists (as **fig. 3.2.3** shows) a certain resemblance between

- weak and strict monotonicity on the one hand (see **fig. 3.2.1**), and
- proportionality and linear homogeneity on the other.

We may safely conclude that "if strong monotonicity then weak monotonicity" but not "if linear homogeneity then proportionality", unless the index function also meets identity (which however is clearly a special case of proportionality).

In general an index with base 0 makes *several* comparisons, a *series* of comparisons $P_{01}, P_{02}, P_{03}, \ldots$ Thus linear homogeneity and strict monotonicity are both statements having a significant bearing on time series by contrast to proportionality

42 If the index depends on prices only and not on quantities there is no point in distinguishing strict and weak proportionality.

and weak monotonicity, which are requirements referring to a single index taken in isolation.

h) Linear homogeneity, monotonicity and mean value property

Only in combination with strict monotonicity and identity does linear homogeneity entail the mean value property:

if (strict) monotonicity, linear homogeneity and identity are met	⟶	then also (strict) mean value property

This is due to the fact that

1. linear homogeneity *and* identity yield proportionality, and
2. monotonicity and proportionality imply the (strict) mean value property.

As mentioned above it is not sufficient to assume linear homogeneity to conclude an index also meets the mean value property. An example for this is the value index V_{0t}: if all prices p_t^* are λ-fold prices p_t, ($p_{it}^* = \lambda p_{it}$), the value index will also change by λ, irrespective of what the quantities are. Thus V_{0t} is linear homogeneous and also monotonously increasing in prices $p_t^* > p_t$, but[43]

- V_{0t} is *not* a mean value of price relatives, and
- V_{0t} does *not* necessarily indicate a λ-fold change if all prices change equally λ-fold either, because quantities q_{it} and q_{i0} can differ (hence proportionality is not met)

as the following **ex. 3.2.6** will show.

The relation between proportionality and the mean value test is closer than between linear homogeneity and the mean value test:

if (strict) monotonicity and proportionality are met	⟶	then also (strict) mean value property

and

if (strict) mean value property is met	⟶	then also proportionality*

* Propoptionality is clearly an implication of mean value property, but the converse is not true as can be seen by the index function P_{max}.

Strict mean value property does not imply linear homogeneity as it does imply proportionality. An example is the *exponential mean* (eq. 3.2.15) P^{EX}, which undoubtedly is a mean and thus

[43] The example of V_{0t} shows: it is not sufficient to assume (strict) monotonicity and linear homogeneity instead of proportionality to deduce that the mean value test will hold, because identity can be violated and has to be assumed separately.

- P^{EX} is bounded by $\min(p_{it}/p_{i0})$ and $\max(p_{it}/p_{i0})$, and of course
- P^{EX} also fulfills proportionality, but *not* linear homogeneity.

Example 3.2.6

This numerical example will demonstrate that the value index is linear homogeneous but does not satisfy strict proportionality, identity and the mean value test[44] ($\lambda = 1.5$).

i	p_{i0}	λp_{i0}	p_{it}	λp_{it}	q_{i0}	q_{it}	p_{it}/q_{i0}
1	12	18	18	27	10	7	1.5
2	15	22.5	12	18	8	6	0.8
3	20	30	24	36	6	3	1.2

The value index yields the following results with p_{i0} and prices at time t as follows

p_{it} : 270/360 = 0.75 < 0.8, which is the smallest price relative (commodity 2), thus mean value property (mean value of price relatives) of V_{0t} is not met,

λp_{it}: 1.5 · 0.75 = 1.125, thus linear homogeneity holds,

p_{i0} : V_{0t} = 234/360 = 0.65 ≠ 1 despite identical prices $p_{it} = p_{i0}$. Thus identity is violated. If V_{0t} were to serve as a price index, a decline of prices would be the diagnosis although in reality prices remained constant and only quantities diminished substantially.

λp_{i0}: $\Sigma \lambda p_{i0} q_{it} = 126 + 135 + 90 = 351$, hence V_{0t} = 351/360 = 0.975, indicating a decline although each price *rose* by 50%.

In the last situation obviously $P^L_{0t} = P^P_{0t} = 1.5$ such that $Q^L_{0t} = Q^P_{0t} = 0.65$. Note that this is a somewhat exaggerated situation in which consumers react to an increase of prices by reducing dramatically the quantities consumed. Both P^L_{0t} and P^P_{0t} (and therefore also P^F_{0t}) indicate correctly the increase by 50%. This shows by the way that the popular belief that P^L_{0t} will always overestimate the price movement is not generally correct. As P^L_{0t} = P^P_{0t} it is not always better to choose a price index formula which comes closer to the value ratio in that it makes more use of the more recent quantities q_{it} than P^L_{0t} does (P^L_{0t} being determined by "old" quantities q_{i0} only).[45] ♦

[44] Most of the figures are taken from ex. 3.2.3.

[45] This type of reasoning is often taken as one of the main arguments in favor of so called chain - indices.

i) Proportionality with respect to quantity indices, the "value index preserving test"

The value index preserving test[46] requires that a price index should be $P^L = P^P = \mathbf{p}_t'\mathbf{q}_0/\mathbf{p}_0'\mathbf{q}_0$ if no quantity changes and therefore Q = 1. It is an implication of proportionality (and hence also identity) applied to a quantity index if the quantity index in combination with a price index satisfies the product test.[47]

Proportionality in the case of a quantity index Q means $Q(\mathbf{p}_0, \mathbf{q}_0, \mathbf{p}_t, \lambda\mathbf{q}_0) = \lambda$, and when $\lambda = 1$ we should get Q = 1 (identity) and therefore

$$(3.2.16) \quad P(\mathbf{p}_0, \mathbf{q}_0, \mathbf{p}_t, \lambda\mathbf{q}_0) \underbrace{Q(\mathbf{p}_0, \mathbf{q}_0, \mathbf{p}_t, \lambda\mathbf{q}_0)}_{=\lambda=1} = V_{0t} = \frac{\sum p_t q_0}{\sum p_0 q_0} = P_{0t}^L = \frac{\sum p_t q_t}{\sum p_0 q_t} = P_{0t}^P$$

should hold, which means that a price index has to be $P^L = P^P$ (which is the same in the case of all quantities q_{i0} and q_{it} being equal) under such circumstaces.

Table 3.2.4: Summary information on relationships among axioms

assumption(s)	consequence	assumption(s)	consequence
1 circular test + identity	time reversal test	**2** dimensionality + commensurability	quantity dimensionality
3 linear homogeneity + identity	(strict) proportionality	**4** linear homogeneity + dimensionality	homogeneity of degree -1
5 strict mean value property	weak monotonicity, proportionality, dimensionality[1)]	**6** strict monotonicity	weak monotonicity
7 strict monotonicity + proportionality[2)]	strict mean value property	**8** proportionality	identity (simply the special case $\lambda = 1$)

1) and of course also weak mean value property

2) because of 2 also strict monotonicity + linear homogeneity + identity → strict mean value property.

It might be useful to summarize the results obtained so far and quoted above as to which conclusions can be derived from knowing that an index function meets certain "axioms", or "tests". In **tab. 3.2.4** an attempt is made to give such a survey.

46 See Vogt 1978, 1979, who called this test "Wertindextreue-Test", see also Balk 1995.

47 For what follows it is sufficient to assume the product test will hold, the factor reversal test is not necessary.

j) "Value dependence test", another uniqueness theorem for Fisher's ideal index

The idea of the so called "value dependence test" (not to be confused with the "value index preserving test") is not easy to understand, because the requirement expressed in it seems to be rather arbitrary, poorly motivated and not intuitively convincing. By this test should hold:
The function $P_{0t} = P(\mathbf{p}_0, \mathbf{q}_0, \mathbf{p}_t, \mathbf{q}_t)$, $\mathbb{R}^{4n}_{++} \rightarrow \mathbb{R}_{++}$ and the following function
$P_{0t} = f\left(\sum p_0 q_0, \sum p_0 q_t, \sum p_t q_0, \sum p_t q_t\right)$, $\mathbb{R}^{4}_{++} \rightarrow \mathbb{R}_{++}$, or simply f(a, b, c, d) should yield the same result.

By virtue of this test the index formula can be expressed as some kind of combination of four "value" terms. The index functions f( ) are in the case of the index of Laspeyres, $P^L = c/a$, or of Paasche, $P^P = d/b$. Hence P^L and P^P pass this test.

Another index that fulfills the value dependence test is for example the Marshall Edgeworth index in which f turns out to be $P_{0t}^{ME} = \frac{\sum p_t(q_0 + q_t)}{\sum p_0(q_0 + q_t)} = \frac{c+d}{a+b}$ (and $\frac{b+d}{a+c}$ being the quantity index of this type) or the price index of Stuvel (see **sec. 2.6**). This is true also for Fisher's ideal index $P^F = [(cd)/(ab)]^{1/2}$ and $Q^F = [(bd)/(ac)]^{1/2}$, the product of both is d/a, which is the value index $V_{0t} = \Sigma p_t q_t / \Sigma p_0 q_0$ since Fisher's index has the factor reversal property, whilst the Marshall Edgeworth index does not because $\frac{c+d}{a+b}\frac{b+d}{a+c} \neq \frac{d}{a}$.

Thus P^L, P^P, the Marshall Edgeworth formula P^{ME} as well as Stuvel's formula P^{ST} and Fisher's index P^F all satisfy the value dependence test. But an f-function presentation is not possible for example in the case of the logarithmic Laspeyres index, Walsh's index, or the Törnqvist index.

The two Dutch index theoreticians J. van Yzeren and B. M. Balk have shown that Fisher's index is the *only* formula that satisfies simultaneously the following three tests:

- (F) factor reversal
- (H) linear homogeneity and
- (V) value dependence.

This is another uniqueness theorem for P^F, called **UT-2** later in **sec. 3.2**.j. Moreover, the three axioms (or "tests") are components of an axiomatic system which is proved to be independent and consistent.[48] Thus for example P^L satisfies H and V but not F, or P^{ST} satisfies F and V but not H. The example for satisfying F and H but not V is less familiar. Furthermore, we have the following problem not uncommon in the case of uniqueness theorems:

---

[48] This uniqueness theorem concerning Fisher's formula and the fact that the three tests constitute an axiomatic system has been proven by B. M. Balk, A simple characterization of Fisher's price index, Statistische Hefte 26 (1985), p.59 - 63.

The problem of this finding rather seems to be how to find some kind of (intuitively appealing) interpretation that could be given to this result (and to the value dependence test V in particular) and to demonstrate the advantages Fisher's formula P^F has (or allegedly has) owing to the fact that P^F is the only formula that satisfies the three tests, F, H and V. To what extent is P^F better than other formulas because it is the only formula passing just these three tests? Is this advantage sufficient to compensate for disadvantages in other respects (e.g. in deflation and aggregation)?

k) Some new tests in the tradition of Fisher's reversal tests: Partial reversals and additional uniqueness theorems concerning Fisher's ideal index

In the tradition of Fisher's reversal tests and his circular test some new tests have been proposed.[49] In what follows some of them are introduced, in particular

- tests making use of the possibility to interchange only parts of the index function[50] $P_{0t} = P(\mathbf{p}_0, \mathbf{q}_0, \mathbf{p}_t, \mathbf{q}_t)$, like the quantity reversal and the price reversal test, and
- weakened versions of Fisher's tests,

along with some uniqueness-theorems, stating that a certain index function, like P_{0t}^F for example, is the only function for which certain relations will hold, and inconsistency theorems, showing that inconsistencies are raised in the first place by transitivity (circular test) in combination with reversal tests (as e.g. the factor reversal test). We begin with[51] the

1. Quantity reversal test (QRT)

In order to meet the quantity reversal test (Funke/Voeller 1978), a price index P should react on the substitution of $\mathbf{q}_t$ for $\mathbf{q}_0$ and of $\mathbf{q}_0$ for $\mathbf{q}_t$ as follows

(3.2.17) $\quad P_{0t} = P(\mathbf{p}_0, \mathbf{q}_0, \mathbf{p}_t, \mathbf{q}_t) = P_{0t}^* = P(\mathbf{p}_0, \mathbf{q}_t, \mathbf{p}_t, \mathbf{q}_0)$. Thus

> **QRT**: The quantity reversal test requires the price index to be invariant upon interchanging of quantities (if the quantity vectors for the two periods in the price index function are interchanged then the price index remains invariant), or in other words: the index should be reflective of *both* quantity vectors in a symmetric fashion.

---

[49] A weak version of time reversal test has already been introduced in **sec.2.2**. The product test as a (weak) variant of the factor reversal test has been mentioned several times already. Another variant of factor reversibility is the value index perserving test just mentioned in this section.

[50] The tests introduced in what follows appear rather farfetched and difficult to be motivated. The reason is that some tests are used only as elements of so called "uniqueness theorems" (UT). Some of those UT's will be quoted subsequently.

[51] The quantity (QRT) and the price reversal test (PRT) are listed in **tab. 3.1.1** as tests T12 and T13.

It is easy to see that the Laspeyres price index violates this test since $P_{0t}^{*L} = \frac{\sum p_t q_t}{\sum p_0 q_t}$ differs from P_{0t}^{L}. On the other hand, quite obviously, this test will be passed by all index functions derived from averaging of quantity weights, like P^{ME}, P^{W}, P^{GK}. Surprisingly, all "superlative" indices in the economic theory of index numbers belong to this type of quantity-reversible index functions.

It is clear that in the tradition of Fisher, aspects like symmetry and "fairness" (with respect to the two (!) periods 0 and t = 1, 2, ...) are as such valuable characteristics of an index function. However, from another point of view (in particular when the focus is on "pure" price comparison), there is prima facie not much use to be found in this property. The only advantage a price index passing the QRT might enjoy over indices which do not is that such a price index is to a certain extent reflective also of (or "accounting for" somehow) the quantity movement ($q_{i0} \neq q_{it}$), which could be viewed as the task of a quantity index, however.

Moreover, the quantity reversal test again assumes (just like the price reversal test, subsequently introduced) an isolated interchanging of quantites not accompanied by an interchange of prices, hence a certain independence between prices and quantities is assumed in the QRT (as well as the PRT), which is rather contrary to the spirit of the economic index theory.[52]

Again a ***uniqueness theorem*** (**UT**) for Fisher's ideal index can be expressed in terms of this test,[53] the third in our collection (a complete list is given in tab. 3.2.6)

UT - 3	P_{0t}^{F} (Q_{0t}^{F}) is the only index that satisfies the 1. time reversal, 2. factor reversal, and 3. quantity reversal test.

2. Price reversal test (PRT)

If *price* vectors in $P_{0t} = P(\mathbf{p}_0, \mathbf{q}_0, \mathbf{p}_t, \mathbf{q}_t)$ are interchanged such that $\mathbf{p}_0 \to \mathbf{p}_t, \mathbf{p}_t \to \mathbf{p}_0$, and the quantity vectors remain unchanged, then we obtain the index $P_{0t}^{*} = P(\mathbf{p}_t, \mathbf{q}_0, \mathbf{p}_0, \mathbf{q}_t)$, and the two indices P and P* are said to satisfy the price reversal test if

(3.2.18) $\quad P_{0t}P_{0t}^{*} = P(\mathbf{p}_0, \mathbf{q}_0, \mathbf{p}_t, \mathbf{q}_t)P(\mathbf{p}_t, \mathbf{q}_0, \mathbf{p}_0, \mathbf{q}_t) = 1$, or

PRT: An isolated interchanging of prices gives the reciprocal price index.

Note that the time reversal test (TRT) requires $P(\mathbf{p}_0, \mathbf{q}_0, \mathbf{p}_t, \mathbf{q}_t)P(\mathbf{p}_t, \mathbf{q}_t, \mathbf{p}_0, \mathbf{q}_0) = 1$, which differs from the PRT[54]. It is easy to see that the Laspeyres price index satisfies this

[52] It is therefore no surprise that for all tests based on a partial interchanging of one type of vectors only the same type of objections applies.

[53] See Funke/Voeller 1978, Balk 1995 (there theorem 5).

[54] A function that meets the PRT but not the TRT is P^{L}. Functions able to satisfy both tests, TRT and PRT, are among others those that are symmetric in the quantities (meet the QRT) because in this case it does not make a difference whether the order of quantity vectors is $\mathbf{q}_0$, $\mathbf{q}_t$ as in P_{0t} or the other way round, $\mathbf{q}_t$, $\mathbf{q}_0$ (like in P_{t0}).

test (PRT), as in $P_{0t}^{L}=\frac{\sum p_t q_0}{\sum p_0 q_0}$ an isolated interchanging of price vectors leads to $P_{0t}^{*L}=\frac{\sum p_0 q_0}{\sum p_t q_0}=\frac{1}{P_{0t}^{L}}$ such that in fact $P_{0t}^{L}\cdot P_{0t}^{*L}=1$.

Thus P^L is able to pass the PRT, which is not true, however, for the following formulation of the test T13 in **tab. 3.1.1** carrying the same name[55]:

(PRT, alternative definition): If the price vectors for the two periods are interchanged then the quantity index remains invariant.

Interchanging of prices in Q_{0t}^{L} yields Q_{0t}^{P}, not Q_{0t}^{L}. As all price indices which are symmetric in the quantities fulfill the QRT, so do all quantity indices symmetric in the prices fulfil this type of PRT, in particular Q_{0t}^{F}, Q_{0t}^{W} etc.

The meaning and usefulness of the price reversal test (version of eq. 3.2.18, but also in the alternative version) is all but clear and open to debate.

Voeller and other authors described the idea of the test with reference to a "change back of prices" from p_t to p_0. So we might think of a change of prices from p_0 to p_t and back to p_0. The awkward aspect now, however, is the reaction of the quantities in response to prices:

- In the time reversal test it is supposed that the transition $p_0 \rightarrow p_t$ and then back $p_t \rightarrow p_0$ will make quantities to react as follows $q_0 \rightarrow q_t$ and then back $q_t \rightarrow q_0$
- however according to the PRT we assume that a change $p_0 \rightarrow p_t$ in prices (as measured by P_{0t}) and a change $p_t \rightarrow p_0$ (as measured by P_{0t}^{*}) are both accompanied by the same change in quantities from q_0 to q_t, and this can hardly be performed by one single consumer.

As Sato (1980) pointed out, there is no satisfactory interpretation in terms of the economic theory of index numbers: if in a situation in which the prices p_0 (or p_t respectively) prevail it is rational (maximizing-utility-behaviour assumed) to consume the quantities q_0 or q_t respectively, then it clearly cannot be rational to consume q_t (in addition to, or in the place of q_0) or q_0 (just like q_t) at the same time.[56] Thus

From the point of view of the economic index theory it cannot be rational to consume both quantites $\mathbf{q}_0$ and $\mathbf{q}_t$ "symmetrically" under the regime of different price vectors $\mathbf{p}_0$ and $\mathbf{p}_t$ (as required in the QRT) or not to respond to an isolated interchanging of prices, that is, to maintain quantites $\mathbf{q}_0$ in spite of $\mathbf{p}_0 \rightarrow \mathbf{p}_t$ (as required in the RPT).

55 Cp. CPI manual or PPI manual. This version of the PRT in contrast to the definition (3.2.18) is obviously shaped by analogy to the QRT test.

56 "Unlike other index-number tests" Sato says, this test "involves more than one individual" and more rigorous: "the price reversal test is neither required nor desired for an economic index to satisfy".

This makes the following UT more or less into an argument *against* P^F rather than in favor of P^F. Moreover as the required symmetry between quantities $\mathbf{q}_0$ and $\mathbf{q}_t$ in the QRT enables an index function to satisfy both the time reversal and the price reversal test at the same time (despite an underlying behavior which is rather contradictory), the QRT can hardly be a useful criterion. On the basis of this test Voeller 1976 formulated another uniqueness theorem[57] which states that P^F, Fisher's ideal index, is the only index simultaneously satisfying the following three reversal tests:

UT - 4	P^F_{0t} is the only price index that satisfies the 1. time reversal, 2. factor reversal, and 3. price reversal test.

Fisher's ideal index can thus be *characterized* according to the theorem of Voeller as being the *only* index that satisfies simultaneously all three reversal tests. These tests are independent in the sense that any *two* tests can be satisfied by a function which violates the remaining third test (see the following tab. 3.2.5).

Table 3.2.5: Independence of time-, factor- and price reversal test

Index	time reversal	factor reversal	price reversal
Laspeyres P^L	violated	violated	satisfied
Paasche P^P	violated	violated	satisfied
Marshall-Edgeworth P^{ME}	satisfied	violated	satisfied
Walsh P^W	satisfied	violated	satisfied
double reversal index[1)] P^{RR}	satisfied	satisfied	violated
E+V construct [2)] P^{EV}	violated	satisfied	satisfied
antiharmonic mean[3)]	violated	violated	violated
Fisher's ideal index P^F	satisfied	satisfied	satisfied

1) see below eq. 3.2.19. This index also complies with circularity.

2) an index for demonstration purpose only, see below eq. 3.2.20.

3) of price relatives weighted with $p_0q_0/\Sigma p_0q_0$.

The following function, P^{RR}_{0t} is an index formula never considered seriously to be used in practice. However

$$(3.2.19)\qquad P^{RR}_{0t} = \sqrt{\frac{\sum p_t q_t}{\sum p_0 q_0}} = Q^{RR}_{0t}$$

(RR stands for "double reversal"[58]), or in vector notation

$$(3.2.19a)\qquad P^{RR}_{0t} = \left(\frac{\mathbf{p}_t'\mathbf{q}_t}{\mathbf{p}_0'\mathbf{q}_0}\right)^{1/2}$$

[57] See also Eichhorn and Voeller 1983, p. 443f, Balk 1995 (there Theorem 6).

[58] i.e. time and factor reversal.

is of some use as an example for demonstration purposes: P^{RR} violates only the price reversal test but satisfies the remaining two tests, the time- and the factor- reversal test (in addition, P_{0t}^{RR} also satisifes the circular test). But of course it has no use for the practice of index calculation: The index Q_{0t}^{RR} is the same as P_{0t}^{RR}, and neither is P_{0t}^{RR} a measure of pure price change nor is Q_{0t}^{RR} a measure of pure quantity change. It is very difficult to find a formula which satisfies factor reversibility but violates the time reversal test. According to Eichhorn and Voeller (1983), the following construction expressed in vector notation

$$(3.2.20) \quad \left(\frac{\mathbf{p}_t'\mathbf{q}_0}{\mathbf{p}_0'\mathbf{q}_0}\frac{\mathbf{p}_t'\mathbf{q}_t}{\mathbf{p}_0'\mathbf{q}_t}\right)^{1/2}\frac{(\mathbf{p}_t'+\mathbf{q}_t')\mathbf{p}_0}{(\mathbf{q}_0'+\mathbf{q}_t')\mathbf{p}_0}\frac{(\mathbf{p}_0'+\mathbf{q}_t')\mathbf{q}_0}{(\mathbf{p}_t'+\mathbf{q}_t')\mathbf{q}_0}\frac{(\mathbf{p}_0'+\mathbf{q}_0')\mathbf{p}_t}{(\mathbf{p}_0'+\mathbf{q}_t')\mathbf{p}_t}\frac{(\mathbf{p}_t'+\mathbf{q}_0')\mathbf{q}_t}{(\mathbf{p}_0'+\mathbf{q}_0')\mathbf{q}_t},$$

is such an index satisfying factor reversibility but not time reversibility.

3. No change ratios

Let P_{0t}^{*} denote a price index of an unknown type, and $\overline{P}_{0t}^{*}$ the *same* kind of index provided no change in *quantities* takes place, that is $P_{0t}^{*} = f(\mathbf{p}_0,\mathbf{q}_0,\mathbf{p}_t,\mathbf{q}_t)$ and $\overline{P}_{0t}^{*} = f(\mathbf{p}_0,\mathbf{q}_0,\mathbf{p}_t,\mathbf{q}_0)$.

Consider now the corresponding quantity indices Q_{0t}^{*} and $\overline{Q}_{0t}^{*}$ (with no change in *prices*), that is $Q_{0t}^{*} = f(\mathbf{q}_0,\mathbf{p}_0,\mathbf{q}_t,\mathbf{p}_t)$ and $\overline{Q}_{0t}^{*} = f(\mathbf{q}_0,\mathbf{p}_0,\mathbf{q}_t,\mathbf{p}_0)$. This permits us to express another uniqueness theorem (Hacker 1979)[59]:

UT - 5	P_{0t}^{F}, Q_{0t}^{F} is the only pair of indices that satisfies the: 1. factor reversal test $P_{0t}^{*}\,Q_{0t}^{*} = V_{0t}$ and 2. the condition $\frac{P_{0t}^{*}}{\overline{P}_{0t}^{*}} = \frac{Q_{0t}^{*}}{\overline{Q}_{0t}^{*}}$.

Note that according to the definition of "no change"-indices as above we obtain $\overline{P}_{0t}^{L} = P_{0t}^{L} = \overline{P}_{0t}^{P}$ and $\overline{Q}_{0t}^{L} = Q_{0t}^{L} = \overline{Q}_{0t}^{P}$, by which it is easy to verify that the above relation leads to $\sqrt{P_{0t}^{P}/P_{0t}^{L}} = \sqrt{Q_{0t}^{P}/Q_{0t}^{L}}$, a well known equation. Thus this relation which might be called "no change ratio" reduces in the case of the Laspeyres formula simply to $\frac{P_{0t}^{*}}{\overline{P}_{0t}^{*}} = 1$ and $\frac{Q_{0t}^{*}}{\overline{Q}_{0t}^{*}} = 1$, both unity by definition.The equality of $\frac{P_{0t}^{*}}{\overline{P}_{0t}^{*}}$ and $\frac{Q_{0t}^{*}}{\overline{Q}_{0t}^{*}}$ also holds if P_{0t}^{L}, P_{0t}^{P}, P_{0t}^{DR} are substituted for P_{0t}^{*} (and quantity indices for Q_{0t}^{*} correspondingly), but unlike Fisher's index, the indices of Laspeyres, Paasche and Drobisch do *not* pass the factor reversal test.

Finally, to complete this short presentation of some uniqueness theorems for P_{0t}^{F}, the following UT is worth being mentioned. It should be viewed as supplementing **UT-1** by pursuing basically the same idea with the time reversal test as UT-1 did with the product test. Therefore P_{0t}^{F} is the only index which is time reversible and related to

[59] See Balk 1995 (Theorem 7). It is more or less a restatement of UT-1.

P^L_{0t} as follows: $\frac{P^F_{0t}}{P^L_{0t}} = \frac{Q^F_{0t}}{Q^L_{0t}}$. Applied to Paasches indices instead of Laspeyres indices, this leads to the same result since $P^*_{0t}/P^P_{0t} = Q^*_{0t}/Q^P_{0t}$ instead of $P^*_{0t}/P^L_{0t} = Q^*_{0t}/Q^L_{0t}$. Thus the UT equivalent to UT-1 using the time reversal property is to express as follows:

UT - 6	P^F_{0t} is the only price index that satisfies the time reversal test $P^*_{0t}P^*_{t0} = 1$ and $P^L_{0t}P^*_{t0} = P^*_{0t}P^L_{t0}$ (or $P^P_{0t}P^*_{t0} = P^*_{0t}P^P_{t0}$).

Again the problem of this finding might be to find some intuitively appealing interpretation that could be given to this result able to "prove" the specific advantages or even superiority of Fisher's formula. Finally our summary table looks as follows:

Table 3.2.6: Characterization (uniqueness theorems) concerning Fisher's index formulas

	comment
UT - 1[1)]	product test and a relation between P^F, Q^F to P^L, Q^L
UT - 2[2)]	factor reversal test, linear homogeneity, value dependence test
UT - 3	time, factor and quantity reversal test
UT - 4	time, factor and price reversal test
UT - 5	factor reversal test (and a relation between a quantity index and the corresponding quantity index obtained if no price changes in the quantity index[3)])
UT - 6	time reversal test (and the relation between P^F and P^L or P^P, just like between P^F, Q^F to P^L, Q^L as in UT-1)

1) see sec. 3.1

2) see sec. 3.2.i

3) likewise: and a relation between a price index and the corresponding price index obtained if no quantity changes in the price index

The problem with these characterizations of Fisher's ideal index is that there are many of them and the set of possible theorems is perhaps unbounded. Some sort of an integrated interpretation of all axiomatic requirements that appear in such uniqueness theorems would therefore be useful in order to arrive at a deeper understanding of Fisher's index formula. However, this can hardly be provided by the present state of index theory. It is also worth being repeated once more that a function which offers many uniqueness theorems is not necessarily a good index function, because such theorems may comprise also somewhat useless properties and also leave open the possibility that interesting requirements are *not* met.[60]

[60] We don't have an agreed upon list of all valuable properties an index function can possibly have. If there were such a list, the suitability for "deflation" should be considered most important. However, here in particular the performance of P^F is all but impressive (see **sec. 5.4**).

l) Some weak versions of the circular test, and inconsistencies between some of Fisher's reversal tests and the circular test

A final note to Fisher's circular test along with some inconsistency theorems should be added. As already mentioned, Fisher's circular test has always been the object of much controversy. The idea is that in intertemporal comparisons the separate year-to-year index numbers can be joined together by successive multiplications like the links of a chain.[61]

It very soon became apparent that circularity is difficult to attain. Hence attempts to weaken the condition of circularity have a long tradition. Such a test, called **base (period) invariance test**[62], is a weak version of the circular test and requires a ratio (function) f of two indices

(3.2.21) $$\frac{P_{02}}{P_{01}} = \frac{P(p_0, q_0, p_2, q_2)}{P(p_0, q_0, p_1, q_1)} = f(p_1, q_1, p_2, q_2) = f$$

not depending on prices and quantities of period 0, but only on prices and quantities of any two periods, 1 and 2, to be compared. Obviously the conclusion

if the circular test is met	⟶	then also the base invariance test

is valid, but the converse is not true. If circularity is met, we simply have f = P_{12}, which does not depend on prices and quantities of periods other than 1 and 2. The test is weaker than circularity because f( ) does not need to have the same functional form as P( ).

However, by this test the tight restrictions of the circular test are relaxed only *slightly*. It is difficult to find an example of an index function satisfying the base invariance test but not at the same time the circular test. Both indices, the Laspeyres and Paasche index, are not capable of satisfying even this weakened version of the circular test, since

(3.2.21a) $$\frac{P^L_{02}}{P^L_{01}} = \frac{\sum p_2 q_0}{\sum p_1 q_0} = f_L(\ldots) \text{ and}$$

(3.2.21b) $$\frac{P^P_{02}}{P^P_{01}} = \frac{\sum p_2 q_2}{\sum p_0 q_2} \frac{\sum p_0 q_1}{\sum p_1 q_1} = f_P(\ldots)$$

the function f_L does not depend on q_1 and q_2 but on q_0 instead, and f_P depends on p_1, p_2, q_1 and q_2 as required, but also on p_0 and q_0, not in agreement with the test. It also follows that Fisher's index fails this test as $f_F = \sqrt{f_L f_P}$.

[61] This is not to say that chain indices comply with the circular test. On the contrary, chaining leads to path dependency, the very opposite of chainability, which thus should not be confounded with the chain index principle (see chap. 7).

[62] This test was already known to Edgeworth (1896) and Fisher (1911).

The base invariance test is weakened further by the so called **base test**[63], according to which one of the following equations should hold

(3.2.22) *either:* $P(\mathbf{p}_0,\mathbf{q}_0,\mathbf{p}_1,\mathbf{q}_1)\, f(\mathbf{p}_1,\mathbf{q}_1,\mathbf{p}_2,\mathbf{q}_2) = P(\mathbf{p}_0,\mathbf{q}_0,\mathbf{p}_2,\mathbf{q}_2)$[64]

or: $f(\mathbf{p}_0,\mathbf{q}_0,\mathbf{p}_1,\mathbf{q}_1)\, P(\mathbf{p}_1,\mathbf{q}_1,\mathbf{p}_2,\mathbf{q}_2) = P(\mathbf{p}_0,\mathbf{q}_0,\mathbf{p}_2,\mathbf{q}_2)$.

Of course, again the circular test is more restrictive than this "base test", in that the function f(...) has to be the same function as P(...), hence

if the circular test is met ⟶ then also the base test

The converse is not true. In can easily be verified that Fisher's index does *not* even meet this weakened test.

Two other again even more weakened versions of the circular test[65] are known as

(3.2.23) $P_{01}\widetilde{P}_{12} = P(\mathbf{p}_0,\mathbf{q}_0,\mathbf{p}_1,\mathbf{q}_1)P(\mathbf{p}_1,\mathbf{q}_0,\mathbf{p}_2,\mathbf{q}_1) = P(\mathbf{p}_0,\mathbf{q}_0,\mathbf{p}_2,\mathbf{q}_1)$ **(price circularity)**,

where $\widetilde{P}_{12}$ denotes an index measuring the change in prices from 1 to 2 with quantities lagging behind (quantities referring to periods 0 and 1 instead of 1 and 2 respectively), and

(3.2.24) $P_{01}\overline{\widetilde{P}}_{12} = P(\mathbf{p}_0,\mathbf{q}_0,\mathbf{p}_1,\mathbf{q}_0)P(\mathbf{p}_1,\mathbf{q}_0,\mathbf{p}_2,\mathbf{q}_0) = P(\mathbf{p}_0,\mathbf{q}_0,\mathbf{p}_2,\mathbf{q}_0)$ **(weak circularity)**,

a still weaker test where quantities remain unchanged throughout the whole interval under consideration. Note the difference between these equations and the full-blown circular test

$$P_{01}P_{12} = P(\mathbf{p}_0,\mathbf{q}_0,\mathbf{p}_1,\mathbf{q}_1)P(\mathbf{p}_1,\mathbf{q}_1,\mathbf{p}_2,\mathbf{q}_2) = P_{02} = P(\mathbf{p}_0,\mathbf{q}_0,\mathbf{p}_2,\mathbf{q}_2).$$

It is of course clear that indices, like P^L, P^P, P^{ME}, P^F etc. all satisfy the price circularity test, as we get for example in the case of the Laspeyres and Paasche index

(3.2.23a) $\frac{\Sigma p_1q_0}{\Sigma p_0q_0}\frac{\Sigma p_2q_0}{\Sigma p_1q_0} = \frac{\Sigma p_2q_0}{\Sigma p_0q_0}$ for P^L, and $\frac{\Sigma p_1q_1}{\Sigma p_0q_1}\frac{\Sigma p_2q_1}{\Sigma p_1q_1} = \frac{\Sigma p_2q_1}{\Sigma p_0q_1}$ for P^P,

and therefore these indices also pass the even less demanding test of eq. 3.2.24 too.

However, doubts arise as to whether eq. 3.2.23 really is a weakening of the circular test, since for the value ratio we get $\frac{\Sigma p_1q_1}{\Sigma p_0q_0}\frac{\Sigma p_2q_1}{\Sigma p_1q_0} \neq \frac{\Sigma p_2q_1}{\Sigma p_0q_0}$, but the value index satisfies the circular test.

Before discussing inconsistencies as they frequently occur in connection with the circular test[66], a final remark on a variation of the circularity idea should be added. Some more sophisticated theoretical considerations[67] focus on the functions

---

[63] Eichhorn/Voeller 1983, p. 432, Olt 1996, p. 47.

[64] This first equation is of course eq. 3.2.21, by which the base invariance test is defined. The base test in contrast to the base invariance test offers a second posssibility however (the second equation of q. 3.2.22).

[65] See Eichhorn/Voeller 1983, p. 433, Olt 1996, p. 48.

(3.2.25) $f(\ ) = P(p_0, q_0, p_1, q_1)P(p_1, q_1, p_2, q_2)$ and $g(\ ) = P(p_0, q_0, p_2, q_2)$

and examine the k-th (k = 0, 1,...) partial deriviates of f and g with respect to the price vectors in order to show that the reqirements may be satisfied *approximately* if not exactly.

In **sec. 3.1** we already encountered a typical ***inconsistency theorem*** (**IT**). The following second inconsistency theorem

IT - 2	There does not exist a pair of functions P_{0t} and Q_{0t} (Q_{0t} gained from P_{0t} by interchanging prices and quantities) which satisfies simultaneously 1. the identity axiom, 2. the circular test, and 3. the factor reversal test.

is proved in a similar manner. The example of P_{0t}^{RR} does not disprove this inconsistency theorem because P_{0t}^{RR} violates (strict) identity. Another theorem[68] is:

IT - 3	There do not exist index functions P_{0t} and Q_{0t} which satisfy the fundamental requirements* of monotonicity, proportionality, dimensionality and commensurability and also 1. the base test or circular test and the determinateness test, or 2. the base test or circular test and the factor reversal test or product test, or 3. the time reversal, factor and price reversal test**.

* these are the four axioms of the system EV-4 of Eichhorn and Voeller (see **sec. 3.3**)

** and a condition not mentioned above, which is similar to the determinateness test, or the cicular/base test.

To sum up, the circular test is extremely doubtful in the sense that as a rule the circular test and the underlying idea of this test, according to which a regular adjustment of quantity weights[69] should be made, cannot be fulfilled simultaneously by a reasonable index function whatsoever. This is an important result with a significant bearing on chain indices.[70] On the other hand the example of the Cobb Douglas index shows that circularity can be achieved only at the price of weights totally unrelated to quantities.

---

66 In consequence of the intolerable practical implications of the theorems quoted above, Eichhorn and Voeller but also other authors quoted by them (like Funke and Hacker) propose to discard the circular test as well as the base test.

67 Diewert 1978, Lau 1974, Olt 1995, p. 49f.

68 Eichhorn and Voeller 1983, p. 446.

69 A uniqueness theorem relating the Cobb Douglas index P^{CD} to such axioms and the circular test is not in contradiction to the finding of IT-3 since P^{CD} is a "degenerate" price index with weights unrelated to quantites such that the question of factor reversibility does not arise.

70 The implication is simply that the two alleged advantages of chain indices, i.e. to enable price statisticians to make consistent long-range-comparisons by chaining and at the same time to cover all new developments by a continuous updating of weights, are contradictory. The solution is that a consistent aggregation over time does not take place: chain indices are not chainable.

3.3 Systems of axioms

There will always be discussion about which test is more and which less important, and a solution is unlikely to be found. So index theoreticians started to design and examine collections, or so to say "bundles", of axioms, all of which they found to be most fundamental. Such a system of axioms is a tool to better understand certain index formulas and to evaluate their merits. If there exists only one formula able to meet all of the axioms we have a uniqueness theorem[71] by which the index function in question is said to be (completely) "characterized". This section deals with some systems of axioms: after starting with Fisher's system we briefly introduce a system of Marco Martini, discuss at length two systems of Eichhorn & Voeller (EV), and present some additional considerations of B. Olt[72].

a) Irving Fisher's system of axioms (tests)

The complete set of axioms comprises eight "tests"[73] (see **fig. 3.3.1**) constituing a set known to be neither "independent" nor "consistent" (noncontradictory).

We first discuss the justification and the implications of some of the tests. The circular test **F7** is an outsider distinct from the remainder in that it is dealing with the multisituation-case (comparing more than just two situations), whereas all other tests should be regarded as related to binary comparisons (two situations). **F7** can also be viewed as focusing on temporal aggregation (over time)[74], much like **F8** (or all aggregation properties)[75] refers to aggregation over commodities. Tests **F1**, **F2**, **F3** are fundamental prerequisites for all kinds of measurement. In a similar vein, **F4** is most fundamental, making a separate mention of **F2** superfluous, however. In our view, **F5** (time reversal) and **F6** (factor reversal) are the least justified tests[76], being well aware of course that Fisher himself took great pride in precisely these "great reversal tests" as he called them and that some of the most renowned index theoreticians agree with him on this point.

---

[71] The term uniqueness theorems (UT) has been used several times already in **sec. 3.1** and **3.2**. Such a UT allows us to deduce a unique formula (index function) from a set of given conditions (axioms, tests).

[72] Olt's systems were developed on the basis of EV. Interestingly in none of the last mentioned axiomatic systems (Martini, EV, Olt) are the reversal tests (time and factor reversal) or the circular test included, and these were exactly the tests that dominated Fisher's way of assessing index formulas.

[73] It is assumed that the tests are sufficiently explained above already.

[74] In chapter 8 we will deal with interspatial transitivity (i.e. consistency among paired comparisons between any two countries, say AB, BC, AC and so on).

[75] See **sec. 5.4** for more details concerning Fisher's index and deflation/aggregation. We also make mention of this test there.

[76] Besides an analogy between an index and individual relatives, we cannot see much substance in the reasons given for these tests.

Figure 3.3.1: The system of Axioms of Irving Fisher

F1	determinate (determinateness) test[1)]
F2	identity $P_{00} = 1$ ($p_{it} = p_{i0}$ for all i)
F3	commensurability
F4	proportionality (strict version)[2)] $(\mathbf{p}_0, \mathbf{q}_0, \lambda\mathbf{p}_0, \mathbf{q}_t) = \lambda$
F5	Time/country reversal test $P_{t0} = 1/P_{0t}$
F6	factor reversal test[3)] $P_{0t}Q_{0t} = V_{0t}$
F7	circular test[4)] $P_{0t} = P_{0s}P_{st}$, or $P_{0t} = P_{0r}\,P_{rs}\,P_{st}$ etc. for *all* 0, r, s, t
F8	withdrawal-and-entry test

1) or: weak continuity axiom.

2) or: if all period t prices change λ-fold then the value of P is also changed by λ (we get the weak variant if in addition $\mathbf{q}_t = \mathbf{q}_0$ is assumed)

3) Q_{0t} is derived from P_{0t} by interchanging prices and quantities, otherwise product test.

4) also called "transitivity" or "chain test"

He was in particular proud of having found $P^F_{0t} = \sqrt{P^L_{0t}P^P_{0t}}$, being one of the few ever found formulas in accordance with both, the time *and* the factor reversal test. P^F has ever since been seen as a high profile player in the game, enjoying a preferential treatment and much reputation. However, Fisher's "ideal" index does *not* satisfy *the circular test* since

$$\text{(3.3.1)}\quad P^F_{01}P^F_{12} = \sqrt{P^L_{01}P^P_{01}P^L_{12}P^P_{12}} = \sqrt{\left(P^L_{01}P^L_{12}\right)\left(P^P_{01}P^P_{12}\right)} \neq P^F_{02} = \sqrt{P^L_{02}P^P_{02}}$$

because both P^L and P^P are not transitive, that is $P^L_{01}P^L_{12} \neq P^L_{02}$ and $P^P_{01}P^P_{12} \neq P^P_{02}$, and has many disadvantages concerning aggregation and deflation. The fact that his ideal index P^F satisfies all of his tests except for the circular test (and **F8**)

- made his index appear to him superior to P^L and P^P and all other "third generation" indices, and
- led him to the conclusion that his system of tests might be inconsistent, because he was unable to find an index that could pass *all* tests, in particular the circular test he henceforth deprecated (all the more as this was (except for F8) the only test his otherwise ideal index failed), and which he believed to be the source of inconsistency.[77]

Besides these hints to *inconsistency* doubts also arose as to the *independence* of the requirements. Obviously[78]

[77] Fisher already *conjectured* (without being able to give a proof) that adding the circular test to the factor reversal test (or vice versa) engenders an inconsistent system.

[78] This follows from the singularity of a matrix of indices P_{ij} as shown in **sec. 2.2.**

if the circular test and identity is satisfied	→	then also the time reversal test

but the converse is *not* true: Fisher's index is an example for this, since P^F satisfies the identity and time reversal test but not the circular test. Another violation of the independence requirement is

if proportionality is satisfied (**F4**)	→	then also identity (**F2**)

so that there is no need to add identity to the system when proportionality is postulated (F2 is an implication of F4).

Fisher's system, however, failed both requirements, **consistency** in addition to **independence**. To be "consistent" means that there is no inherent contradiction, in which case no index could satisfy all axioms. It was difficult to prove that the tests of Fisher are inconsistent, so that in fact no index exists that will pass all the tests (such that it is impossible to fulfill the system as a whole): however long and intensive the search for such an index will be, it will be unsuccessful. The correct proof of inconsistency has been very difficult and took many years.[79] The situation was that the tests of Fisher apparently seemed to be inconsistent, as could be seen by his ideal index, which violated transitivity.[80] But it was difficult to prove that this defect of Fisher's index was a *necessary* consequence of inconsistency.

In the context of the first inconsistency theorem (**IT-1**, see **sec. 3.1**) we demonstrated how in principle such a proof of, say, requirements R_1, and R_2 being inconsistent can be provided:

- choose admissible vectors $\mathbf{p}_0, \mathbf{q}_0, \mathbf{p}_t, \mathbf{q}_t$, and apply this choice to R_1, and show
- that with requirement R_2 we will then run into difficulties (contradictions) with the consequences of the choices concerning the vectors and R1.

To summarize the defects of Fisher's axiomatic system it should be noticed that his "great reversal tests" **F5** and **F6** are unjustly restrictive in what they rule out (due to **F6**) etc. Drawn on this fact, the modern systems - as mentioned above - refrain from adding **F5** (let alone **F6**) to the list of axioms deemed essential (as a consequence we could go a step further and stop praising or recommending a formula simply because it enables **F5** to be satified). To sum up:

From a modern point of view Fisher's axiomatic system is seriously defective, because

1. it is first and foremost ***inconsistent*** (impossible to be fulfilled as a whole),
2. it lacks ***independence***, or more specifically, there are two redundancies, and finally
3. the "reversal tests" **F5** and **F6** are additions that are ***unnecessarily restrictive***.

---

[79] First attempts were made by Ragnar Frisch (1930) and Abraham Wald (1937) but they could be shown to be erroneous. Convincing ultimate proofs were not performed until the late 60's and early 70's by Swamy and Eichhorn.

[80] Fisher's index also violates the withdrawal-and-entry test and has weak aggregation properties (see **sec. 5.4**).

b) A system of minimum requirements of an index by Marco Martini

According to the Italian statistician M. Martini (1992), a price index should at least meet the following three conditions[81]:

identity, commensurability, linear homogeneity	→	by implication: proportionality*

* as a consequence of identity and linear homogeneity

A definition of these axioms can be given without reference to quantities, except for the case of the value index referred to below. As is common practice, Martini demonstrated the independence of his system of axioms by giving examples of index functions that satisfy any two of the axioms but violate the remaining third axiom.

Examples given by Martini to demonstrate the independence of this system

axioms fulfilled	axiom violated	Example
2 and 3	1: strict identity	value index $V_{0t} = \Sigma p_t q_t / p_0 q_0$
1 and 3	2: commensurability	Dutot's index
1 and 2	3: linear homogeneity	exponential mean index (see **sec. 3.2**)

Since the three axioms are part of other systems of axioms[82], it is easy to show that the system will be satisfied by many well known index functions, such as P^L, P^P, P^{ME}, P^W and the like, as it is simply part of the axiomatic system (EV-5 in figure 3.3.1). A fundamental requirement *not* mentioned in Martini's system is *monotonicity*, which appears to be rather fundamental, however. An example for a function which satisfies all three axioms of Martini but *not* monotonicity is given by Eichhorn and Voeller as follows

(3.3.2) $$P_{0t}^{NM} = \left(\frac{p_{1t}}{p_{10}}\right)^{\alpha_1} \left(\frac{p_{2t}}{p_{20}}\right)^{\alpha_2} \cdots \left(\frac{p_{nt}}{p_{n0}}\right)^{\alpha_n} \text{ where } \sum \alpha_i = 1, \alpha_1 < 0,\ \alpha_2, \ldots, \alpha_n > 0$$

(NM = no monotonicity). **Ex. 3.3.1** will demonstrate that P^{NM} would in fact be an index function admitted in Martini's system.

Example 3.3.1

The following prices are given for three commodities A, B and C:

[81] It is assumed that the meaning and implication of the closely related properties monotonicity, linear homogeneity and proportionality is known (see **sec. 3.1** and **3.2**).

[82] The system of five axioms by Eichhorn and Voeller to be more clear.

	p_{i0}	p_{it}	p^*_{it}	λp_{it}	λp_{i0}	α_i
A	10	12	14	18	15	-0.2
B	15	18	18	27	22.5	0.7
C	20	20	20	30	30	0.5

Show *firstly* that the function P^{NM} violates monotonicity. To this end calculate P^{NM} using prices p_{i0} and various prices at period t, viz. p_{it}, and p^*_{it}.

p_{it}: since the price of the third commodity did not change, the formula simplifies to $P^{NM}_{0t} = 1.2^{-0.2} \cdot 1.2^{0.7} \cdot 1^{0.5} = \sqrt{1.2} = 1.0954$;

p^*_{it}: note that $\mathbf{p}^*_t > \mathbf{p}_t$ but $P^{NM}_{0t} = 1.2^{0,7}/1.4^{0,2}$ = 1.0622 < 1.0954 such that monotonicity is violated;

Secondly, in order to demonstrate that nonetheless linear homogeneity is met, P^{NM} on the basis of prices p_{it} should be compared with P^{NM} on the basis of prices λp_{it}:

(λ = 1.5) $P^{NM}_{0t} = \left(\frac{18}{10}\right)^{-0,2}\left(\frac{27}{15}\right)^{0,7}\left(\frac{30}{20}\right)^{0,5} = 1.5 \cdot \sqrt{1.2} = 1.06432$, thus linear homogeneity is fulfilled, and also proportionality as shown next by substituting λp_{i0} for p_{it}: it is easy to see that P^{NM}_{0t} is now $1.5^{-0,2} \cdot 1.5^{0,7} \cdot 1.5^{0,5} = 1.5 = \lambda$. Finally, note that monotonicity is not *necessarily* violated. It is sufficient that it *can* be violated. With prices p^*_{it} = 12, 18, 24 (instead of 14, 18, 20) we get P_{0t}^{NM} = 1.2 (because *all* prices increased by 20% and proportionality holds), which is greater than 1.0954. ♦

The example shows:

> On the one hand, the axiomatic system of Martini admits index formulas which hardly can be regarded as being useful, like P^{NM}, but on the other hand "linear homogeneity" is a rather restrictive requirement by which some definitely meaningful indices would be ruled out, like the price index of Stuvel P^{ST} for example (see **sec. 2.6** for this index).

c) Two systems of axioms established by Eichhorn and Voeller

The German index theoreticians W. Eichhorn and J. Voeller introduced systems of five, and later also of four axioms (1976), to be referred to as EV-4 and EV-5 for short. They studied also unweighted index numbers (depending on price vectors $\mathbf{p}_0$ and $\mathbf{p}_t$ and thus on price relatives only), that is[83]

[83] Index formulas in which no quantities appear are for example unweighted means of price relatives, like the historical formulas of Dutot or Carli (see **sec. 1.2**). There is, however, an implicit reference to quantities in any index formula (even in the Dutot and Carl index formula), in the sense that the abso-

(3.3.3) $P_{0t} = P(\mathbf{p}_0, \mathbf{p}_t)$, $\mathbb{R}^{2n}_{++} \to \mathbb{R}_{++}$ (index function P)

as opposed to indices depending on quantity vectors as well

(3.3.4) $P_{0t} = P(\mathbf{p}_0, \mathbf{q}_0, \mathbf{p}_t, \mathbf{q}_t)$, $\mathbb{R}^{4n}_{++} \to \mathbb{R}_{++}$ (index function P).

The first four axioms of the five-axioms-system EV-5 as an example can also be established for unweighted indices, see **fig. 3.3.2**.

Figure 3.3.2: Systems of axioms by Eichhorn & Voeller

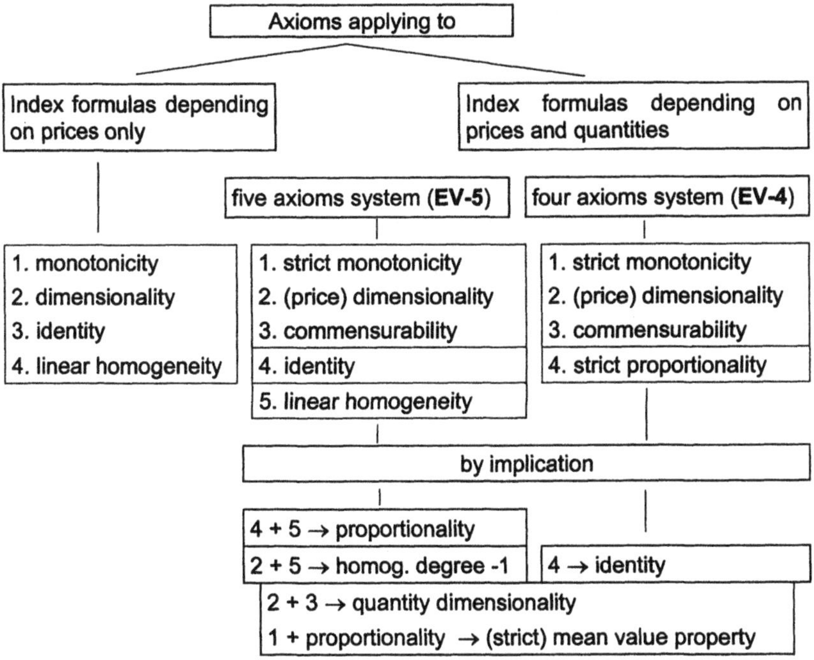

Before going into details of some interesting features of the systems of Eichhorn and Voeller, both EV-4 and EV-5, it should be noted:

1. none of the ***reversal tests*** (time and factor reversal test) nor the circular test of Fisher is mentioned in the EV-systems
2. as in most other modern axiomatic systems, also no mention is given to axioms restricting the type of ***weighting schemes*** wanted for an index, i.e. axioms dealing with quantities (e.g. weights of both periods, 0 and t, should be used,

lute price of a commodity depends on the quantity to which the price quotation refers. It is to such implicit quantities in this meaning to which the "incommensurability" of Dutot's index refers.

and they should enter the formula in a symmetric fashion, more recent variable weights are to be preferred to weights of 0 kept constant and thus becoming increasingly out-dated or so), further

3. though monotonicity is an element of both EV-4 and EV-5 (ruling out for example a function like P_{01}^{NM}) ***no*** attention has been given to ***additivity*** as a special case of monotonicity nor to other useful properties relating to ***aggregation*** and deflation,
4. all index formulas admitted by the five and four axioms system also "automatically" (by implication) satisfy the so called mean value test (in current prices), and finally
5. the five axioms system of Eichhorn and Voeller (EV-5) as well as the four axioms system (EV-4 being a subset of EV-5), are proved ***consistent*** and ***independent***[84].

As indicated (by shadows) in **fig. 3.3.2** the system of five axioms (EV-5)

monotonicity, dimensionality, commensurability	identity and linear homogeneity

and the four axioms system[85] (EV-4)

monotonicity, dimensionality, commensurability	and proportionality

differ by the fact that identity and linear homogeneity in EV-5 have been replaced in EV-4 by proportionality, being weaker and implied in EV-5. Hence

an index function statisfying the five axioms system (EV-5)	⟶	is also able to statisfy the four axioms system (EV-4)

but the converse is not true. Thus indices satisfying the five axioms system constitute a subset of those indices that satisfy the four axioms system.[86]

In **tab. 3.3.1** and **3.3.2** it is demonstrated how independence of the axioms in the sense that any four (EV-5), or any three (EV-4) of the axioms can be satisfied by a function P which does not satisfy the remaining fifth, or fourth axiom respectively is proved by giving appropriate examples.

Indices that do not pass all tests of EV-5 but meet EV-4 are functions to which proportionality applies, but not linear homogeneity, such as Stuvel's indices P^{ST}, the exponential mean index P^{EX}, and some log-change indices.

[84] Consistency is given since many index formulas meet *all* conditions, like P^L, P^P, P^F, and crossed indices other than P^F, and in addition also many indices constructed in a fundamentally different manner (like some log-change indices for example).

[85] The four axioms system of Eichhorn and Voeller was originally proposed by Vogt.

[86] A formula that fits to the four axioms (EV-4) but not to EV-5 is in particular Stuvel's index P^{ST}.

Table 3.3.1: Examples given by Eichhorn and Voeller to demontrate the independence of the five axioms system (**EV-5**)

axioms fulfilled	axiom violated	example
2, 3, 4, 5	1 monotonicity*	$\left(\frac{p_{1t}}{p_{10}}\right)^{\alpha_1}\left(\frac{p_{2t}}{p_{20}}\right)^{\alpha_2}\cdot\ldots\cdot\left(\frac{p_{nt}}{p_{n0}}\right)^{\alpha_n}, \alpha_1 < 0,\ \alpha_2,\ldots,\alpha_n > 0,\ \sum\alpha_i = 1$
1, 3, 4, 5	2 dimensionality	$\frac{\mathbf{p}_0'\mathbf{q}_0}{\mathbf{p}_0'\mathbf{p}_0 + 1}\cdot\frac{1}{n}\sum\frac{p_{it}}{p_{i0}} + \frac{1}{\mathbf{p}_0'\mathbf{q}_0 + 1}\max_i\left(\frac{p_{it}}{p_{i0}}\right)$
1, 2, 4, 5	3 commensurability	$\mathbf{a}'\mathbf{p}_t/\mathbf{a}'\mathbf{p}_0$ where $a_1, \ldots, a_n > 0$ are real constants
1, 2, 3, 5	4 identity	$V_{0t} = \frac{\sum p_t q_t}{\sum p_0 q_0} = \frac{\mathbf{p}_t'\mathbf{q}_t}{\mathbf{p}_0'\mathbf{q}_0}$ if $\mathbf{q}_t \neq \mathbf{q}_0$
1, 2, 3, 4	5 linear homogeneity	$\sqrt{P_{0t}^L}$ since $\sqrt{\frac{\sum \lambda p_t q_0}{\sum p_0 q_0}} = \sqrt{\lambda}\sqrt{\frac{\sum p_t q_0}{\sum p_0 q_0}}$

*see eq. 3.3.2 formula P^{NM}

Table 3.3.2: Examples given by Eichhorn and Voeller to demontrate the independence of the four axioms system (**EV-4**)

axioms fulfilled	axiom violated	example
2, 3, 4	1: monotonicity	same example as in tab. 3.3.1
1, 3, 4	2: dimensionality	same example as in tab. 3.3.1
1, 2, 4	3: commensurability	same example as in tab. 3.3.1
1, 2, 3	4: proportionality	$\sum\frac{p_{it}}{p_{i0}} = nP_{0t}^C$ (P^C = Carli)

d) Combinations of index formulas and transformations of data vectors

The construction of new price index formulas by combining known index formulas P_1, $P_2,\ldots,P_k$ was a method extensively practiced by I. Fisher. An example is the linear combination with weights (factors, constants) $a_1, a_2, \ldots, a_n$, leading to $\widetilde{P} = a_1P_1 + a_2P_2 + \ldots + a_nP_k$.

A more general *combination*, the power mean, is a price index as follows

$$(3.3.5)\qquad (\alpha_1 P_1^\delta + \ldots + \alpha_k P_k^\delta)^{\frac{1}{\delta}} = \overline{P}\quad \begin{cases} \delta \neq 0,\quad \alpha_1 \geq 0,\ldots,\alpha_k \geq 0 \\ \alpha : \text{real constants}, \sum\alpha_i = 1 \end{cases}$$

If P_1, P_2, ..., P_k are price indices, each of them satisfying EV-4 or EV-5 respectively, then $\overline{P}$ will do so as well.

A *transformation* of vectors entering the formula that is permissible in this sense, i.e. that will provide a function in accordance with the axioms, is a transformation of the quantities $\mathbf{q}_0$ with the matrix $\mathbf{A} = \mathrm{diag}[a_1 ... a_n] = \begin{bmatrix} a_1 & ... & 0 \\ ... & ... & ... \\ 0 & ... & a_n \end{bmatrix}$ and $\mathbf{q}_t$ with $\mathbf{B} = \mathrm{diag}[b_1 ... b_n]$ and a transformation of both price vectors $\mathbf{p}_0$ and $\mathbf{p}_t$ with the same matrix **C** (defined correspondingly), i.e. according to Eichhorn and Voeller the indices[87]

(3.3.6) $P(\mathbf{Aq}_0, \mathbf{Cp}_0, \mathbf{Bq}_t, \mathbf{Cp}_t)$ and $P(\mathbf{q}_0, \mathbf{p}_0, \mathbf{q}_t, \mathbf{p}_t)$ have the same properties.

As a consequence of this existence of permissible transformations and combinations, the class of functions obeying the four axioms of EV-4 (or the five axioms of EV-5 resepectively) will in general rapidly grow to an extremely wide set of formulas. It might be desirable therefore to add some additional conditions (tests, axioms) in order to *reduce* the number of functions.

On the other hand, the aim can well be to *widen* the set of axioms in order to include also those index functions that seem to be reasonable despite not being able to pass the EV-4 or EV-5 criteria.

e) Some additional axiomatic systems of B. Olt and final remarks on the choice between systems of axioms

In the tradition of Eichhorn and Voeller, Bernhard Olt (1996) examined several systems which provide less restrictive requirements than EV-4 (and of course also EV-5, see **fig. 3.3.3**.

All axiomatic systems of Olt were proved to be consistent and independent. Olt 1 is derived from EV-4, the point of departure of all of his systems, by replacing strict monotonicity by weak monotonicity, by which Olt arrived at a weakening of the conditions. Since weak mean value property means less than proportionality, Olt 2 is even weaker than Olt 1, and in a similar vein Olt 3 is again weaker than Olt 2 since weak mean value property and proportionality follows from strict mean value property (but not vice versa).

An index function admissible in the defintion "Olt 3" in contrast to "EV-4" is for example Palgrave's index, because of being monotonous only in the weak sense (as [strict] mean value property implies weak monotonicity and proportionality) but not in the strict sense (that is the reason why this index does not comply with EV-4).

To derive a system of axioms is always to some extent a matter of taste and consequently the result may appear somewhat arbitrary. We are at liberty to design new systems and accept them as reasonable, provided we do not fail to show that the system is consistent and independent. There are no clear, generally accepted criteria by which one system of axioms is to prefer to another unless it were possible to

[87] The notation differs slightly from the notation used till now in that we use in eq. 4 Eichhorn's and Voeller's order of occurence of the four vectors.

"prove" that a certain axiom has to be regarded as more important or more indispensable than another axiom.

Figure 3.3.3: Three systems of axioms by B. Olt

Three additional axiomatic systems

Olt 1	Olt 2	Olt 3
1. dimensionality	1. dimensionality	1. dimensionality
2. commensurability	2. commensurability	2. commensurability
3. weak* monotonicity	3. weak monotonicity	3. strict mean value property
4. proportionality	4. weak mean value property	4. symmetry

* Weak instead of strict is the difference to EV-4

The axiomatic approach is not able to make distinctions related to "importance", "relevance" or so, let alone to introduce criteria like being more or less intuitively or (economically) understandable. This is a major shortcoming of the axiomatic approach: formulas are explored with respect to certain axioms, and we may also study certain mathematical relations between sets (systems) of axioms and try to explain the meaning of certain axioms, but it is difficult to say why a certain system as a whole should be preferred to another.

With a smaller number of axioms the class of admissible index numbers will *generally* (not necessarily)[88] be greater, provided the axioms themselves are not too restrictive. An intolerably restrictive condition would be for example: the index should be unity if and *only if* no price changes (it makes sense to require an index to be unity "if[89]" but of course not "*only* if" no price changes, as this would rule out most of the reasonable index formulas[90]). The system of axioms should ideally be so that not

[88] The nature of the axiom in question can of course be such that there are not many functions left to comply with it. In the extreme case, a system may reduce to a single axiom which can be fulfilled by one index function only. An example is the "structural consistency of volumes" gained from the index used as deflator (see **sec. 5.2**): the only index in accordance with such a system now would be the Paasche index P^P. This structural consistency is most restrictive and leaves room only for Paasche's index. It is desirable as an additional property but cannot be an axiom.

[89] which is in fact the meaning of identity.

[90] This applies to formulas like P^L and P^P both providing some sort of compensation, due to which rising prices can be balanced by falling prices and vice versa such that the overall effect is an index amounting to unity.

only *no reasonable* formulas are ruled out but also if possible *all objectionable* formulas are excluded.[91]

3.4 Log-change index numbers I: Cobb-Douglas- and Törnqvist-index

This section firstly introduces some properties of "log-changes", making them a reasonable if not superior alternative to growth rates and price relatives, and then proceeds to certain index formulas based on log - changes and geometric means.

a) Growth rates, log changes, and new index formulas on the basis of log changes

Assuming a time series $y_0, y_1, y_2, \ldots y_t$ we define the growth *rate* r_t and the growth *factor* f_t with time t as a *discrete* variable as follows

(3.4.1) $r_t = \frac{y_t - y_{t-1}}{y_{t-1}} = \frac{\Delta y_t}{y_{t-1}}$, and

(3.4.2) $f_t = \frac{y_t}{y_{t-1}} = 1 + r_t$.

The funcion $f_t = 1 + r_t$ is equivalent to the *links* ℓ_t (t = 1, 2, ...), already introduced in **sec. 1.1**. The function r_t makes use of backward differences. An alternative definition is $r_t^* = \frac{y_{t+1} - y_t}{y_t} = r_{t+1}$, and $f_t^* = \frac{y_{t+1}}{y_t} = f_{t+1}$, using forward differences. *Growth rates* and growth factors, despite being very popular, have the following *two disadvantages*:

- they are *not symmetric*, that is: $\frac{y_t - y_{t-1}}{y_{t-1}} \neq -\frac{y_{t-1} - y_t}{y_t}$ and
- the sum of two (or more) growth rates over time has no meaningful interpretation.

Furthermore, a general notion of growth rate could be as follows:

(3.4.3) $\text{growth rate} = \frac{\text{absolute change}}{\text{level}} = \frac{\Delta y}{A(y)}$,

taking Δy as an indicator of the change (nominator) and y_{t-1} or y_t, or some kind of a mean of both[92] denoted by A(y), as the indicator of the "level" (denominator), for example $\frac{1}{2}(y_{t-1} + y_t)$.

---

[91] As matters stand, this will in general not be achieved. A system of axioms will usually contain also formulas that seem to have no meaningful interpretation, like for example the index of A. Young (P^Y in sec. 3.2), which in fact satisfies all axioms of EV-5.

[92] that is an average of y at the beginning and the end of the relevant interval.

It should be noted that an alternative method to describe relative changes (growth rates) is to use *log-changes* of $y_1, y_2, \dots y_t$. Though much less known than traditional growth rates and growth factors, log changes have two substantial advantages, being able

- to avoid the two principal shortcomings of r_t or r_t^* as mentioned above, and
- in combination with the logarithmic mean (or simply log-mean eq. 1.1.9) as a measure of the "level" they provide an expression, which has an interpretation in the sense of eq. 3.4.3.

In order to see this, recall the definition of log changes and the logarithmic mean in **sec. 1.1**

(1.4.8a) $D\ell_t = \ln\left(\frac{y_t}{y_{t-1}}\right) = \ln(y_t) - \ln(y_{t-1}) = \ln(f_t)$, and of the logarithmic mean of two positive numbers, y_t and y_{t-1}

$$L(y_t, y_{t-1}) = L(y_{t-1}, y_t) = \frac{y_t - y_{t-1}}{\ln(y_t / y_{t-1})} \text{ if } y_t \neq y_{t-1}. \tag{3.4.4}$$

Transformation of eq. 4 shows that log-changes are growth rates in the sense of eq. 3, since

$$D\ell_t = \ln\left(\frac{y_t}{y_{t-1}}\right) = r_t^L = \frac{y_t - y_{t-1}}{L(y_t, y_{t-1})}, \tag{3.4.5}$$

making use of L as a kind of level A(y) instead of for example

$$r_t = \frac{y_t - y_{t-1}}{y_{t-1}}, \text{ if } A(y) = y_{t-1} \text{ or } r_t^A = \frac{y_t - y_{t-1}}{\frac{1}{2}(y_t + y_{t-1})}, \text{ if } A(y) = \tfrac{1}{2}(y_t + y_{t-1}).$$

The only disadvantage of log changes seems to be their lack of intuitive appeal. Log changes seem to have no "level" in the denominator nor an absolute difference in the numerator. But eq. 5 reveals that the impression is erroneous, and things change markedly when the concept of log changes is combined with "logarithmic means". **Tab. 3.4.1** shows that there are quite a few benefits from taking log changes (as compared to traditional growth rates).

Growth factor f and growth rate r of P_{0t}^L, for example, are defined as follows

$$f(P_{0t}^L) = 1 + r(P_{0t}^L) = \frac{P_{0t}^L}{P_{0,t-1}^L} = \sum \frac{p_{i,t}}{p_{i,t-1}} \left(\frac{p_{i,t-1} q_{i0}}{\sum p_{i,t-1} q_{i0}} \right) = \sum \frac{p_{it}}{p_{i,t-1}} \beta_{it} = P_{t-1,t(0)}. \tag{3.4.6}$$

This is a relation worth being examined in more detail, which will be done in the framework of chain indices comparing the rebased index $P_{t-1,t(0)}$ with the "link" of a chain index.

Log-changes[93] $D\ell_y = \ln\left(\frac{y_t}{y_0}\right)$ and growth rates $x := \frac{y_t}{y_0} - 1$ are related to one another in that x provides a Taylor approximation to $\ln\left(\frac{y_t}{y_0}\right) = \ln(1+x)$. The Taylor polynomial (Taylor expansion) about $x_0 = 0$ reads as follows

$$\ln(1+x) = \ln(1+x_0) + \frac{1}{1+x_0}(x-x_0) - \frac{1}{(1+x_0)^2}\frac{(x-x_0)^2}{2} + 2\cdot\frac{1}{(1+x_0)^3}\frac{(x-x_0)^3}{6} \mp \ldots$$

giving $\ln(1+x) = \sum_{k=1}^{\infty}(-1)^k \frac{x^k}{k} = x - \frac{x^2}{2} + \frac{x^3}{3} \mp \ldots \approx x$.

Hence, for small values of x the growth rate x approximates the log-change[94].

Table 3.4.1: Advantages of log changes over traditional growth rates

aspect	log changes	traditional growth rates[1)]
symmetry	$\ln(y_t) - \ln(y_{t-1}) = -\left[\ln(y_{t-1}) - \ln(y_t)\right]$	no symmetry
summation over successive intervals	$D\ell_t + D\ell_{t+1} = \ln\left(\frac{y_{t+1}}{y_{t-1}}\right)$ the sum is a growth related to a time span of *two* periods[2)]	the sum $r_t + r_{t+1}$ is not meaningful
eq. 3.4.3 interpretation	$D\ell_t = \frac{\Delta y}{A(y)} = \frac{y_t - y_{t-1}}{L(y_t, y_{t-1})}$ see eq. 3.4.5 log mean of y_t and y_{t-1} as "level"	lower (or upper) bound, that is y_{t-1} as level A(y) in the denominator

1) r_t in eq. 3.4.1

2) correspondingly the sum of m adjacent log-change-terms measures a change over m periods

Index functions built as geometric means have extremely simple formulas of growth factors. The unweighted geometric mean of price relatives $a^i_{0t} = p_{it}/p_{i0}$ is known as Jevons' index $P^{JV}_{0t} = \prod_i \left(\frac{p_{it}}{p_{i0}}\right)^{1/n} = \sqrt[n]{\prod a^i_{0t}}$. The growth factor of P^{JV}_{0t} is simply f(P^{JV}_{0t}) $= \Pi\left(p_{it}/p_{i,t-1}\right)^{1/n}$, which is a (simple) geometric mean of the link relatives ℓ_t, since

[93] What applies to the preceding period t-1 in equations 1, 2 and 5 is also applicable to the base period 0.

[94] This explains that a log change will not differ much from a growth rate x provided that x is small.

(3.4.7) $$Da_{0t}^{i} = \ln\left(\frac{p_{it}}{p_{i0}}\right) = \ln\left(\frac{p_{i1}}{p_{i0}}\right) + \ldots + \ln\left(\frac{p_{it}}{p_{i,t-1}}\right) = \sum_{\tau=1}^{\tau=t} \ln\left(\frac{p_{i,\tau}}{p_{i,\tau-1}}\right) = \sum_{\tau} \ln(\ell_{\tau}).$$

It turns out that the log-change index corresponding to the unweighted index of Carli P_{0t}^{C}, that is, DP_{0t}^{C}, is Jevons' index P_{0t}^{JV}[95]. Likewise the logarithm of DP_{0t}^{L} resembles the index P_{0t}^{L}, where however the place of "normal" relatives a_{0t}^{i} has been taken up by the $\ln(a_{0t}^{i})$.

In **fig. 3.4.1** an attempt is made to find a system for the various index formulas of this and the following section. The scheme might permit some insights into relationships between the index formulas by indicating the way in which they were derived.

Table 3.4.2: Defintion of a "log-change" (price) index

The logarithm $\ln(DP_{0t}^{*})$ of a log-change (price) index DP_{0t}^{*} is a function of logarithmic price relatives $Da_{0t}^{i} = \ln(p_{it}/p_{i0})$; for example a weighted arithmetic mean of such Da_{0t}^{i} terms:

(3.4.8) $\ln(P_{0t}^{*}) = \sum g_i (Da_{0t}^{i})$ with weights g_i where $a_{0t}^{i} = p_{it}/p_{i0}$.

To give some examples:

* stands for	$\ln(DP_{0t}^{*})$	DP_{0t}^{*}	weights
Carli/ Jevons	$\frac{1}{n}\sum \ln\left(\frac{p_{it}}{p_{i0}}\right)$ Carli-type	$\prod\left(\frac{p_{it}}{p_{i0}}\right)^{1/n}$ Jevons type	$g_i = 1/n$ for all i
logarithmic Laspeyres	$\ln(DP_{0t}^{L}) = \sum_{i=1}^{n} \ln\left(\frac{p_{it}}{p_{i0}}\right) g_i$	$DP_{0t}^{L} = \prod_{i}\left(\frac{p_{it}}{p_{i0}}\right)^{g_i}$	$g_i = p_{i0}q_{i0} / \sum p_{i0}q_{i0}$

The notation DP* is chosen in order to indicate a relationship between a traditional index P* and its "log change" counterpart.

b) Cobb Douglas index P_{0t}^{CD}, constant weights and the circular test

The Cobb-Douglas index[96] was introduced already in eq. 2.2.3 in **sec. 2.2** and is defined as

[95] Or in other words: Jevons' index formula is the "log-change" counterpart to Carli's index

[96] The reason for the name "Cobb-Douglas" index is that the same function (which simply is a weighted geometric mean) is also known in production theory as Cobb-Douglas production function.

(3.4.8) $P_{0t}^{CD} = \prod_{i=1}^{n}\left(\frac{p_{it}}{p_{i0}}\right)^{\alpha_i}$ (and $Q_{0t}^{CD} = \prod_{i=1}^{n}\left(\frac{q_{it}}{q_{i0}}\right)^{\alpha_i}$ correspondingly),

where α_i are any real constants, $\Sigma\alpha_i = 1$ and $0 \leq \alpha_i \leq 1$, not necessarily related to expenditure shares. Note that P_{0t}^{CD} is a weighted geometric mean whilst P_{0t}^{JV} is a simple (unweighted) geometric mean (or $\alpha_i = 1/n$ for all i).

This scheme contains only those index functions that were in fact proposed and recommended for use by somebody. It can easily be seen that the system of weights using arithmetic, geometric, harmonic or logarithmic means in combination with price relatives and logarithmic price relatives (i.e. "log changes"), as well as the possibility to arrive at index formulas by crossing (i.e. by calculating means of other index functions), might generate a great number of additional index functions.

The shortcomings of *both* indices (P_{0t}^{CD} and P_{0t}^{JV}) are as follows

- they are **"degenerate"**[97] price **indices** because they do not depend on quantities
- the product of the price index and the corresponding "quantity index" gained by substituting quantity relatives for price relatives, that is Q_{0t}^{CD} or Q_{0t}^{JV} respectively (which is also degenerated), is not the value ratio V_{0t}, thus the *factor reversal test* is not met, or rather is not applicable,
- the antithetic indices, V_{0t}/P_{0t}^{CD} and V_{0t}/P_{0t}^{JV}, cannot be regarded as quantity indices, such that the less demanding *product* test is not met either.

On the other hand, both indices (or any other log-change index with *fixed* weights), have some advantages

- they fulfill the "circular test[98]", that is, transitivity, and
- the growth factor f_t of P^{CD} (and mutatis mutandis of P_{0t}^{JV}), which is the geometric mean of the growth factors (links) of the individual price relatives, is constant as the coefficients α_i are constants, and in particular
- very much in contrast to chain indices, which are not transitive, the growth factor between any two periods, t and t*, that is P_{0t*}^{CD}/P_{0t}^{CD} or P_{0t*}^{JV}/P_{0t}^{JV}, depends only on prices at times t and t*, and not on prices at time 0 or any other period.[99]

---

[97] According to Balk (1995), P^{CD} is a "degenerate member" of the class of price indices, "because it is *only* a function of prices" and "the derived quantity indicator" V_{0t}/P^{CD} "is in general not an admissible quantity index".

[98] Circularity in combination with identity ensures time reversibility, a property of geometric means that made them eligible in the view of the Boskin Commission for what they called "low level aggregation" (see **sec. 1.2**).

[99] This has often been "held to be one of the superiorities as an index of the geometric average over the arithmetic average" (Rich, 1948, p. 339).

This can easily be verified, since P^{CD} can be expressed as a product of link-indices. Assume for convenience of presentation only two commodities and three periods 0, 1 and 2, then

$$(3.4.9)\quad P_{02}^{CD}=\left[\left(\frac{p_{11}}{p_{10}}\right)^{a}\left(\frac{p_{21}}{p_{20}}\right)^{1-a}\right]\left[\left(\frac{p_{12}}{p_{11}}\right)^{a}\left(\frac{p_{22}}{p_{21}}\right)^{1-a}\right]=P_{01}^{CD}P_{12}^{CD},\ (\alpha_1=a,\ \alpha_2=1\text{-}a)$$

which simply means that P^{CD} satisfies the *circular* test. The growth factor of P^{CD} in the transition from t - 1 = 1 to t = 2 therefore is P_{12}^{CD}, which is the geometric mean of the two link relatives p_{12}/p_{11} and p_{22}/p_{21}. Assuming constant growth factors of the prices such that $\lambda_1=\frac{p_{1t}}{p_{1,t-1}}$ and $\lambda_2=\frac{p_{2t}}{p_{2,t-1}}$, then the growth factor of P_{0t}^{CD} is the geometric mean of the two growth factors $\lambda_1^a\lambda_2^{1-a}$, and *is constant for all periods t*. In contrast, the same conditions prevailing in the case of the Laspeyres price index P_{0t}^{L} will result in the following growth factors:

$$P_{00}^{L}=1\rightarrow P_{01}^{L}:\ \sum\lambda_i\frac{p_{i0}q_{i0}}{\sum p_{i0}q_{i0}}=\sum\lambda_i\beta_{i1},\quad P_{01}^{L}\rightarrow P_{02}^{L}:\ \sum\lambda_i\frac{\lambda_i p_{i0}q_{i0}}{\sum\lambda_i p_{i0}q_{i0}}=\sum\lambda_i\beta_{i2}$$

$$P_{02}^{L}\rightarrow P_{03}^{L}:\quad \sum\lambda_i\frac{\lambda_i^2 p_{i0}q_{i0}}{\sum\lambda_i^2 p_{i0}q_{i0}}=\sum\lambda_i\beta_{i3}\ \text{ and so on,}$$

which is a series of weighted arithmetic means, weights β_{it} being variable as t varies. Or more specifically: assume only two commodities and $\lambda_1 > \lambda_2$ then the growth rate of P_{0t}^{L} will tend to λ_1:

> The growth factor of P^{CD} is a geometric mean with constant weights α_i (for all periods t = 1,2, ...) and therefore constant as well, whereas the growth factor of P^{L} is an arithmetic mean with changing weights and tends to the largest individual growth factor of prices.

In **ex. 3.4.1** this will be demonstrated numerically. This property of the CD-index, or in general of a geometric mean with *constant* weights α_i, of having a geometric mean growth factor $\frac{P_{0t}^{CD}}{P_{0,t-1}^{CD}}=\prod_i\left(\frac{p_{it}}{p_{i,t-1}}\right)^{\alpha_i}$, which is constant when all individual prices also change at a constant factor (rate) $p_{it}/p_{i,t-1}=\lambda_i$ (for all t), is found particularly useful in the case of the so-called **(effective) exchange-rate-indices**, compiled in order to estimate the relative value of a currency vis-a-vis a *group* of other currencies[100] without treating appreciating and depreciating currencies asymmetrically:

[100] This kind of index (geometric mean rather than arithmetic mean) is sometimes also used for indices of stock exchange rates.

Figure 3.4.1: Log-change indices

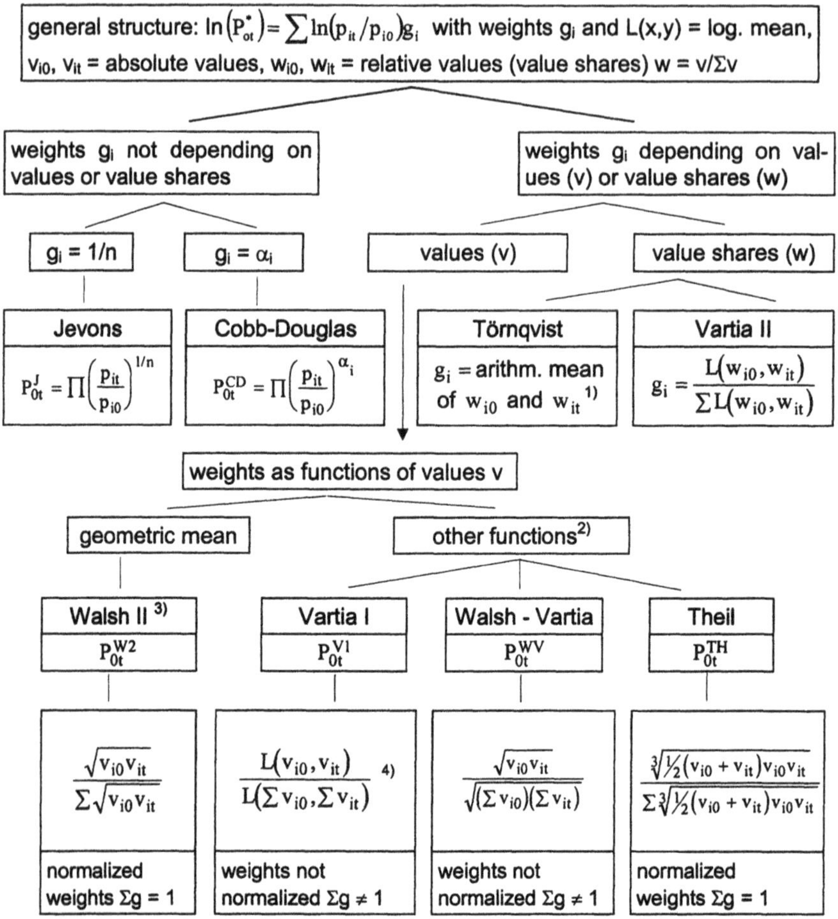

1) $\overline{w}_i = (w_{it} + w_{it})/2$

2) an index function in this context not mentioned here is the index of Rao

3) The name was given because this index has some resemblance to the "normal" index of Walsh

$$P^{W}_{0t} = P^{W1}_{0t} = \frac{\sum p_{it}\sqrt{q_{i0}q_{it}}}{\sum p_{i0}\sqrt{q_{i0}q_{it}}} = \sum \frac{p_{it}}{p_{i0}} \frac{\sqrt{(p_{i0}q_{i0})(p_{i0}q_{it})}}{\sum \sqrt{(p_{i0}q_{i0})(p_{i0}q_{it})}}.$$

4) Note that in general $\sum_i L(v_{i0}, v_{it}) \neq L(\sum v_{i0}, \sum v_{it})$.

In the case of the *arithmetic* mean with constant weights (as e.g. in the case of Laspeyres P^L) the growth factor tends to the greatest of the individual growth factors of prices λ_i, and it can easily be seen that in the case of a *harmonic* mean the growth factor would tend to the lowest growth factor. To avoid such a domination by extremes take the geometric mean.

Example 3.4.1

Consider two commodities with base period expenditure shares $w = w_1 = p_{10}q_{10}/\Sigma p_{i0}q_{i0} = 0.6$ and $w_2 = 1 - w = 0.4$. Prices are increasing at a constant rate of 80% and 20% respectively, such that the constant growth factors are $\lambda_1 = 1.8$ and $\lambda_2 = 1.2$, and the prices are $p_{1t} = \lambda_1^t p_{10}$ and $p_{2t} = \lambda_2^t p_{20}$ respectively. The series of P_{0t}^L now is determined by

$$\frac{P_{0,t+1}^L}{P_{0t}^L} = \lambda_1\left(\frac{A_t}{A_t + B_t}\right) + \lambda_2\left(\frac{B_t}{A_t + B_t}\right) = \lambda_1(\alpha_t) + \lambda_2(\beta_t)$$

where $A_t = (\lambda_1)^t w$, $B_t = (\lambda_2)^t(1-w)$ and $P_{0t}^L = A_t + B_t$. It follows that the growth factor of P_{0t}^L is a weighted ***arithmetic*** mean of λ_1 and λ_2 with variable weights $\alpha_t = \frac{A_t}{A_t + B_t}$ and $\beta_t = \frac{B_t}{A_t + B_t}$. Since $\lambda_1 > \lambda_2$, the weight α_t will grow faster than β_t and the growth factor of a "fixed weighted" direct Laspeyres index will tend to $\lambda_1 = 1.8$:

	t = 0	t = 1	t = 2	t = 3	t = 4
A_t	w = 0.6	1.08	1.944	3.4992	6.2986
B_t	1-w = 0.4	0.48	0.576	0.6912	0.8294
P_{0t}^L	1	1.56	2.52	4.1904	7.128
$P_{0t}^L/P_{0,t-1}^L$		1.56	1.62	1.663	1.701

In contrast to P_{0t}^L the growth rate of a Cobb Douglas index P_{0t}^{CD} will grow at a constant growth factor $(\lambda_1)^w(\lambda_2)^{1-w}$ = 1.5305. Under these conditions, a Laspeyres **chain** index $\overline{P}_{0t}^{LC}$ will also have a variable growth factor, viz. $P_{t+1}^{LC} = \sum_{i=1}^{n} \frac{p_{it}q_{it}}{\sum p_{it}q_{it}}\left(\frac{p_{i,t+1}}{p_{it}}\right) = \sum w_{it}\lambda_i$, because of weights w_{it} varying with time and depending on changing quantities as well as prices.

When we calculate under the same conditions the ***harmonic*** mean with constant weights of the base period (that is index P_{0t}^{HB} in fig. 2.2.1), we get

$P_{01}^{HB} = \left[\lambda_1^{-1}w + \lambda_2^{-1}(1-w)\right]^{-1} = 1.5$

$P_{02}^{HB} = \left[\lambda_1^{-2}w + \lambda_2^{-2}(1-w)\right]^{-1} = 2.16$, and so on, such that

$$\frac{P_{0,t+1}^{HB}}{P_{0t}^{HB}} = \frac{1}{\lambda_1}\left(\frac{A_t^*}{A_t^* + B_t^*}\right) + \frac{1}{\lambda_2}\left(\frac{B_t^*}{A_t^* + B_t^*}\right)$$

where $A_t^* = \left(\frac{1}{\lambda_1}\right)^t w$ and $B_t^* = \left(\frac{1}{\lambda_2}\right)^t (1-w)$. Hence the index P_{0t}^{HB} develops (until t = 5) as follows: 1.5, 2.16, 2.991, 3.999, 5.195, and the growth factors are 1.5, 1.44, 1.385, 1.337, and 1.299, tending to $\lambda_2 = 1.2$. ♦

Some attempts are made to describe a behaviour which will lead to a substitution of the kind assumed in the case of a geometric mean, e.g. a habit to sell securities which have appreciated such that the total value of investment in each security remains equal (Rich op. cit.) or a constant re-allocation of the basket of currencies in a certain manner is stipulated which is, however, difficult to motivate.[101]

Unlike an unweighted mean over commodities (Jevons' index), which may be justified by time reversibility, it appears difficult to justify a *weighted* geometric mean using constant weights. Hence the economic interpretation P_{0t}^{CD} is not satisfactory There is a relation to an indiffference curve of the Cobb-Douglas type $U = x^{\alpha}y^{1-\alpha}$ with "quantities" $x = q_1$ and $y = q_2$ of two commodities, but it is difficult to see why the indifference curve should remain unchanged and possess arbitrary constant coefficients, like α and $1-\alpha$.

Obviously the Cobb-Douglas index P^{CD} (and also its growth factor) is a geometric mean of price relatives (and correspondingly Q^{CD} has a mean of quantity relatives interpretation), but P^{CD} is not a ratio of aggregates. Its economic meaning is far from clear, and the pair P^{CD}, Q^{CD} does not meet the product test, let alone the factor reversal test. Hence P^{CD} is of no use at all for the purpose of deflation.

The attention given to P_{0t}^{CD}, however, is owed in the first place to the fact that this index, though degenerate and worthless from the economic point of view, is the only transitive index being nonetheless also able to meet certain minimum requirements of axiomatic index theory.

c) The circular test and a characterization (uniqueness theorem) of P^{CD}

The index P^{CD} was already known to Irving Fisher, who was interested in this formula (to which Fisher did not give a name) because P^{CD} satisfies the circular test. Fisher conjectured (but he was unable to present a proof) that the only formulas conforming to the circular test are certain index numbers with constant weights, like P^{CD} or

$$(3.4.10) \qquad P_{0s}^{LW} P_{st}^{LW} = \frac{\sum p_s q}{\sum p_0 q} \frac{\sum p_t q}{\sum p_s q} = P_{0t}^{LW} = \frac{\sum p_t q}{\sum p_0 q}$$

---

[101] According to the Swiss authors Baumberger and Keel (1985), the choice of a weighted geometric mean is again characteristic for "a growing tendency to define indices according to criteria of statistical aesthetics instead of constructing them based on an explicit economic question of a possible client".

which is the index function of Lowe, or more general: the fixed (weight) based index, $P_{0t(*)}$, that is, an index with constant (fixed) weights (*) not depending on the periods to be compared (periods 0 and t).[102] Of course, unlike P^{LW}, the index P^L does not satisfy the circular test since

$$P_{0s}^L P_{st}^L = \frac{\sum p_s q_0}{\sum p_0 q_0} \frac{\sum p_t q_s}{\sum p_s q_s} \neq P_{0t}^L = \frac{\sum p_t q_0}{\sum p_0 q_0}.$$

Fisher also already noticed that the main intention of his circularity axiom, that is, the rapid adjustment of budgets or "baskets" (quantity weights), cannot be accomplished. He realized that he could not find any formula that would simultaneously meet the factor reversal test and the circular test. He suspected that the conditions were contradictory and inconsistent, which was a main reason for his eventual abandoning of the circularity criterion. Not only this inconsistency but also a uniqueness theorem has been proven by Eichhorn and Voeller (1976) and Funke et al. (1979). They were able to prove the following uniqueness theorem (**UT-7**):

UT-7: The Cobb Douglas index is the *unique* index function that satisfies
1. the circular test (transitivity) *and*
2. the following five fundamental axioms (EV-5): 1. monotonicity, 2. price-dimensionality, 3. linear homogeneity, 4. identity and 5. commensurability.

The axioms listed under number 1 and 2 constitute a *system* of axioms (independent and consistent) that can be satisfied by one and only one specific index function, the function P^{CD}. This finding is fully in line with another result, called a "second theorem on quantities" (to meet circularity)[103], which might throw some light on the type of index functions required to satisfy circularity, or "chainability" (a consideration important in the context of chain indices). According to this theorem, what is often supposed to be the aim and great advantage of a chain index,

- to establish a consistent comparison between several adjacent periods, 0-1, 1-2, 2-3 and so on by successive multiplication of indices P_{01}, P_{12}, P_{23} and to enable
- a continuous adjustment of quantity weights (adjusted to a new consumption pattern)

cannot be achieved, because of contradictory requirements.[104]

---

[102] In interregional (international) comparison the counterpart of Lowes price index is the ECLA-approach to derive transitive purchasing power parities. Lowe's index, violates for example commensurability and thus P^{LW} can hardly be regarded as a reasonable "price index". There is also no quantity index which can meaningfully be related to Lowe's price index formula P^{LW}.

[103] See Eichhorn/Voeller 1983, p. 434f for a proof; the second theorem was first derived by Hacker and independently by Krtscha. The *first theorem on quantities* (Eichhorn/Voeller 1983, p. 434) runs as follows: every function meeting proportionality (and thus also identity) and the circular test (or base test) only depends on the prices of the base (0) and comparison period (t). This means of course that it does not depend on the quantities.

[104] A similar result is true for multinational comparisons: conforming to transitivity is impossible without sacrificing other desirable criteria (like e.g. proportionality, characteristicity or so).

d) The Törnqvist index P_{0t}^{T}, an "unbiased" index formula in a system of six indices

Definition

The following index function is known as the Törnqvist - index (or Törnqvist - Theil index; the reader may please apologize that we use both spellings Törnqvist and Törnqvist and could not unify the spelling)

(3.4.11) $P_{0t}^{T} = \prod_{i=1}^{n} \left(\frac{p_{it}}{p_{i0}} \right)^{\overline{w}_i}$ where $\overline{w}_i$ is the mean of expenditure shares for period 0

and period t $\overline{w}_i = \frac{1}{2}(w_{i0} + w_{it}) = \frac{1}{2}\left(\frac{p_0 q_0}{\Sigma p_0 q_0} + \frac{p_t q_t}{\Sigma p_t q_t} \right)$, or alternatively

(3.4.12) $\ln(P_{0t}^{T}) = \frac{1}{2}\left[\Sigma w_{i0} \ln(p_{it} / p_{i0}) + \Sigma w_{it} \ln(p_{it} / p_{i0})\right] = \frac{1}{2}\left[\ln(DP_{0t}^{L}) + \ln(DP_{0t}^{P})\right]$, or of course equivalently $\log P_{0t}^{T} = \frac{1}{2}\left[\log(DP_{0t}^{L}) + \log(DP_{0t}^{P})\right]$, where DP_{0t}^{L} denotes the "logarithmic Laspeyres price index" and DP_{0t}^{P} the "logarithmic Paasche price index".[105]

Inequalities, analogy with Fisher's index

As **fig. 3.4.2** shows, there are six index formulas derived in a systematic manner in the style of Irving Fisher. From this two interesting inequalities follow

(3.4.13a)	either $P_{0t}^{L} > P_{0t}^{F} > P_{0t}^{P}$ or $P_{0t}^{L} < P_{0t}^{F} < P_{0t}^{P}$, and
(3.4.13b)	either $DP_{0t}^{L} > P_{0t}^{T} > DP_{0t}^{P}$ or $DP_{0t}^{L} < P_{0t}^{T} < DP_{0t}^{P}$.

As $P_{0t}^{F} = \sqrt{P_{0t}^{L} P_{0t}^{P}}$ arises from the first inequality (13a), so does

(3.4.12a) $P_{0t}^{T} = \sqrt{DP_{0t}^{L} DP_{0t}^{P}}$

from the second (13b). Note that for example $P_{0t}^{L} > P_{0t}^{P}$ does not entail $DP_{0t}^{L} > DP_{0t}^{P}$ but is rather compatible with the opposite inequality, that is $P_{0t}^{L} > P_{0t}^{P} \rightarrow DP_{0t}^{L} < DP_{0t}^{P}$ (as **ex. 3.4.2** shows). Equations 3.4.12 and 12a show the index P^T of Leo Törnqvist.

P^T is the geometric mean of two log-change indices DP^L and DP^P, much in the same way as P^F is the geometric mean of P^L and P^P. Thus the role P^T plays with respect to log-change indices (logarithms of price relatives) is similar to the role P^F plays in the case of normal indices (price relatives). There are, however, limitations to the analogy: P^T is a mean of relatives, P^F has no such mean-of-relatives interpreta-

---

105 Of course DP^L and DP^P can also be calcluated with logarithms [log] as with natural logarithms [ln] without making a difference.

tion.[106] On the other hand, P^F is able to pass the factor reversal test (is "ideal"), while P^T does not even conform to the (weaker) product test.

Notion of "unbiasedness"

Time and again it is pointed out by a great number of authors that both indices, P_{0t}^F and P_{0t}^T, are *unbiased* in the sense of making use of both weights, w_{i0} and w_{it}, in a ***symmetric*** fashion (for example by using the the weights $\overline{w}_i = \frac{1}{2}(w_{i0} + w_{it})$ in the case of P_{0t}^T), or standing in the center of an inequality (eq. 3.4.13a and 13b) whilst the two boundaries, P_{0t}^L and P_{0t}^P, or DP_{0t}^L and DP_{0t}^P respectively, are said to be "biased", upwards or downwards, respectively. It is rather deplorable that the concept of "unbiasedness" is used in this context, because it is common to mistake it as some kind of conceptual purity (biased = impure, imperfect or so), whereas in fact it only implies a sort of non-central "average" as the numerical result.

To strive for "unbiasedness" in this sense (or for symmetry) is again a consequence of the widespread affection for a "compromise" (or "crossing") in index theory. Much of Fisher's work rests on this kind of reasoning which is altogether highly unfortunate and makes it difficult to advocate an "extreme" index, such as P^P or P^L. *Conceptually* (from the point of view of the target concept), we can see no reason why an index habitually rendering particularly high or low results in empirical applications cannot be the "correct" one.

It should also be remembered that in general the concept of "bias" in statistics is different, not oriented at numerical results (values) of different statistical measures but rather at

> the true (population) parameter θ and the expected value of an estimate of this very parameter $E(\hat{\theta})$, where $\hat{\theta}$ is based on sample data (bias means $E(\hat{\theta}) \neq \theta$).

The concept "bias" is used when there is a comparison between different estimates, like $\hat{\theta}_1$, ... for the same well-defined parameter θ, not a comparison between different operational concepts to measure a less clearly defined matter. There is no point in calling one measure "less biased" than another when it is not clear that they both refer to the same target concept.

Properties of Törnqvist's index

The analogy between log-change indices and "normal" indices is also such that DP^P is the time antithesis to DP^L (and vice versa) in the same way as P^P is the time antithesis to P^L (and vice versa).[107] Therefore P^T is able to pass the *time reversal* test

[106] Note also that the average weights used in P^T can be viewed as *crossing of weights* whereas eq. 3.4.12/12a can be regarded as *crossing of formulas*. Thus P^T has two interpretations (in terms of "crossing"), P^F only one.

(3.4.14) $P_{0t}^{T}P_{t0}^{T} = \Pi\left(\frac{p_{it}}{p_{i0}}\right)^{\overline{w}_i} \Pi\left(\frac{p_{i0}}{p_{it}}\right)^{\overline{w}_i} = 1.$

It is, however, *not* true that DP^L and DQ^P or DP^P and DQ^L are factor antithesis to one another

(3.4.14a) $DP_{0t}^{L}DQ_{0t}^{P} = \prod_{i=1}^{n}\left(\frac{p_{it}}{p_{i0}}\right)^{w_{i0}} \prod_{i=1}^{n}\left(\frac{q_{it}}{q_{i0}}\right)^{w_{it}} \neq V_{0t},$

and correspondingly $DP^PDQ^L \neq V_{0t}$ as opposed to $P_{0t}^{L}Q_{0t}^{P} = P_{0t}^{P}Q_{0t}^{L} = P_{0t}^{F}Q_{0t}^{F} = V_{0t}$.

It is therefore no surprise that Törnqvist's index does *not* meet the *factor reversal* test[108], in spite of its construction bearing some resemblance to Fisher's ideal index.

Note that $\log(DP_{0t}^{L}) \neq \log(P_{0t}^{L})$ and $\log(DQ_{0t}^{P}) \neq \log(Q_{0t}^{P})$. Otherwise $\log(P_{0t}^{L}) + \log(Q_{0t}^{P}) = \log(V_{0t})$ would imply the same result for $\log(DP_{0t}^{L}) + \log(DQ_{0t}^{P})$.

e) Quantitative relations between six indices

As Vartia (1978) and Vartia and Vartia (1984) pointed out, there are some interesting relationships between six index formulas listed in **fig. 3.4.2**. The index P_{0t}^{HL} (also denoted P_{0t}^{HB} above) is called the "**h**armonic **L**aspeyres" (**h**armonic/**b**ase-period-weights), and defined as

(3.4.15) $\left(P_{0t}^{HL}\right)^{-1} = \Sigma w_{i0}(p_{i0} / p_{it}).$

The two inequalities quoted in **fig. 3.4.2** (i.e. the two shaded boxes) are *generally* valid because it is always some variance term which is responsible for the gap between any two index formulas in an inequality. Unfortunately they cannot be combined to one single inequality.

There was hardly any attention given to index no. 3. All three indices (1), (2) and (3) are "unbiased" index formulas and their results are generally in close agreement with one another.

The extent to which P^L is greater than DP^L depends on the variance of logarithmic price relatives (log changes in prices $(\dot{p})$). Some differences are related to the covariance between log changes in prices and volumes (that is $\text{cov}(\dot{p},\dot{v})$) or between log changes in prices and quantities (that is $\text{cov}(\dot{p},\dot{q})$). Vartia (1987) has proved the

---

[107] There is no problem with the term "logarithmc Laspeyres", but "logarithmic Paasche" seems to be misleading since an arithmetic mean with weights being *current* expenditure shares would give the price index of the Palgrave index instead of Paasche's index. But the equations above show that nonetheless there is some justification in calling DP^P "logarithmic Paasche".

[108] It does not even meet the product test since the implicit (indirect, antithetic, cofactor) quantity index V/P^T is not proportional in the quantites q_{it} (because they affect w_{it}).

following approximations using covariances and variances, according to which the difference

(3.4.16a) $\log\left(DP^P_{0t}\right)$ - $\log\left(DP^L_{0t}\right)$ is approximately equal to $\mathrm{cov}(\dot{p},\dot{v})$,

and correspondingly

(3.4.16b) $\log\left(P^P_{0t}\right)$ - $\log\left(P^L_{0t}\right) \approx \mathrm{cov}(\dot{p},\dot{q})$.

These are covariances between logarithmic deviations $\dot{p}_i, \dot{q}_i$ and $\dot{v}_i$, i.e. deviations of log-changes from their respective mean, weighted by $\overline{w}_i$, like

$\dot{p}_i = \log(p_{it}/p_{i0})$ - $\log\left(P^T_{0t}\right)$, $\dot{q}_t = \log(q_{it}/q_{i0}) - \log(Q^T_{0t})$, and

$\dot{v}_i = \log(v_{it}/v_{i0}) - \log(V_{0t})$.

Figure 3.4.2: System of six index formulas (Vartia) and the "five tined fork" of I. Fisher

type of mean	type of weight structure	
	w_{i0}	w_{it}
arithmetic	P^L (Laspeyres)	P^{PA} (Palgrave)
geometric	DP^L (log. Lasp.)	DP^P (log. Paasche)
harmonic	P^{HL} (harm.Lasp)	P^P (Paasche)

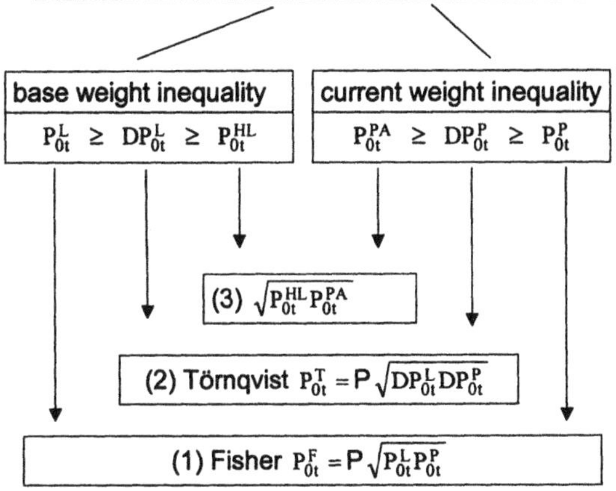

The five tined fork	
tine	formula
uppermost (+2)	P^{PA}
mid-upper (+1)	DP^P
middle (0)	P^L, P^T
unbiased	P^F, P^P
mid lower (-1)	DP^L
lower most (-2)	P^{HL}

Fisher's result was
$P^{PA} > DP^P > P^L > P^T > P^F > P^P > DP^L > P^{HL}$

The *approximation* $\log(P^P_{0t}) - \log(P^L_{0t}) \approx \mathrm{cov}(\dot{p},\dot{q})$ (that is eq. 16a) has some resemblance to the well known *exact* relationship derived by L. v. Bortkiewicz

$$\frac{P^P_{0t} - P^L_{0t}}{P^L_{0t}} = \frac{\mathrm{cov}(p_{it}/p_{i0}, q_{it}/q_{i0})}{P^L_{0t} Q^L_{0t}} = \frac{C}{P^L_{0t} Q^L_{0t}},$$

where C is the covariance between price and quantity relatives (not the *logarithmic* relatives). When C < 0 (which is the more likely case), then $P^P < P^L$ and the corresponding relationship between DP^P and DP^L is true when the covariance $\text{cov}(\dot{p},\dot{q})$ is negative.[109]

Example 3.4.2

The relationships between the six indices (price and quantity indices) should be verified using the following data:

commodity	base period (0) price	quantity	value	current period (t) price	quantity	value
A	10	20	200	12	20	240
B	20	16	320	18	20	360
C	16	30	480	24	25	600

The price relatives are for commodity A: 1.2, for B: 0.9 and for C: 1.5. Hence we get $DP^L = 1.2^{0.2}\ 0.9^{0.32}\ 1.5^{0.48} = 1.2182$, and $DP^P = 1.2^{0.2}\ 0.9^{0.3}\ 1.5^{0.5} = 1.2307$ and the two "traditional" indices $P^L = 1.248$, and $P^P = 1.2$. Interestingly:

though $P^P = 1.2 < P^L = 1.248$ we have DP^P 1.231 > $DP^L = 1.218$.

The two geometric means are $P^T = \sqrt{DP^L DP^P} = 1.224$ and $P^F = 1.224$. For the harmonic Laspeyres index we get $P^{HL} = \left(\frac{0.2}{1.2} + \frac{0.32}{0.9} + \frac{0.48}{1.5}\right)^{-1} = 1.187$ and for Palgrave's index $P^{PA} = 1.2 \cdot 0.2 + 0.9 \cdot 0.3 + 1.5 \cdot 0.5 = 1.26$, so that the third unbiased index turns out to be $\sqrt{P^{PA}P^{HL}}$ =1.223.

As already predicted, the following relationships, $P_{0t}^{PA} > DP_{0t}^{P} > P_{0t}^{P}$ and $P_{0t}^{L} > DP_{0t}^{L} > P_{0t}^{HL}$ hold. Depending on the covariance C between price and quantity relatives either $P_{0t}^P < P_{0t}^L$ or $P_{0t}^P > P_{0t}^L$ might occur. In this example we have $P_{0t}^P < P_{0t}^L$ and the covariance C amounts to - 0,0385.

It may be interesting to have a short look at the quantity indices. On the basis of the quantity relatives (1 for A, 1.25 for B and 0.833 for C) we get $DQ^L = 1.25^{0.32}$ $0.833^{0.48} = 0.98403$ and $DQ^P = 1.25^{0.3}\ 0.833^{0.5} = 0.97607$, hence $Q^T = 0.98004$. Note that $P_{0t}^T Q_{0t}^T$ yields a result which is *almost* equal to $V_{0t} = 1.2$, namely 1.199995. But of course $P^T Q^T = \Pi\left(\frac{p_{it}}{p_{i0}}\frac{q_{it}}{q_{i0}}\right)^{\overline{w}_i}$ is not necessarily equal to

---

[109] Vartia and Vartia 1984 present another result concerning the quantitative relationships between the six indices: $\log\left(\frac{P^L}{DP^L}\right) \approx \log\left(\frac{DP^L}{P^{HL}}\right) \approx \frac{1}{2}s_{0p}^2$ and $\log\frac{P^{PA}}{DP^P} \approx \left(\frac{DP^P}{P^P}\right) \approx \frac{1}{2}s_{tp}^2$, where s_{0p}^2 and s_{tp}^2 denote the variance of log changes in prices $(\log(p_{it}/p_{i0}))$, base-year weighted or current year weighted respectively.

$V_{0t} = \Sigma\left(\frac{p_{it}}{p_{i0}}\frac{q_{it}}{q_{i0}}\right)w_{i0}$. Thus Törnqvist's pair of indices may come close to satisfying the factor reversal test, whilst Fisher's pair of indices always *exactly* satisfies this test. In our example we have $Q^L = 1$, $Q^P = (1.04)^{-1} = 0.9615$, hence $Q^F = 0.98058$ and $P^F Q^F = \sqrt{1.44} = 1.2$. Finally the results concerning the harmonic Laspeyres and the Palgrave quantity index are $Q^{HL} = (1.032)^{-1} = 0.969$ and $Q^{PA} = 0.9917$. Hence $\sqrt{Q^{PA}Q^{HL}} = 0.98026$ and the product of this index with $\sqrt{P^{PA}P^{HL}}$ also comes very close to $V_{0t} = 1.2$.

The inequalities in the case of quantity indices are as follows: $Q_{0t}^{PA} > DQ_{0t}^{P} > Q_{0t}^{P}$ and $Q_{0t}^{L} > DQ_{0t}^{L} > Q_{0t}^{HL}$. The numerical example also conforms with this expected result. It is well known that $P^P/P^L = Q^P/Q^L$ so that $P^P < P^L$ implies $Q^P < Q^L$. The results are:

$P^P = 1.2$	$DP^P = 1.23$	$P^{PA} = 1.26$
$P^{HL} = 1.187$	$D^{PL} = 1.218$	$P^L = 1.248$

$Q^P = 0.962$	$DQ^P = 0.976$	$Q^{PA} = 0.992$
$Q^{HL} = 0.969$	$D^{QL} = 0.984$	$Q^L = 1.0$

♦

f) Three unbiased and superlative indices

As **fig. 3.4.3** shows, it is reasonable to take geometric means of two indices, one belonging to the base weight inequality and the other to the current weight inequality. Hence there are three such "unbiased" indices,

(1) $\sqrt{P^{PA}P^{HL}}$ (2) $P_{0t}^{T} = \sqrt{DP_{0t}^{L}DP_{0t}^{P}}$ and (3) $P_{0t}^{F} = \sqrt{P_{0t}^{L}P_{0t}^{P}}$.

These indices are all members of the family of "*superlative*" indices in the sense that Diewert introduced this term.[110] Applied to data, they are likely to yield similar results, and they are quadratic approximations of each other for small absolute changes in prices and quantities.

Note that both terms, "superlative" and "unbiased" (and also "ideal") are often mistaken for something which is perfect, or at least comes close to *perfection*. It is indeed a great mischief that there is a widespread misuse of such terms in price index theory without questioning the criteria underlying a search of such an "optimum". It turns out that such criteria making a formula "superlative", "unbiased", "ideal" and the like

- focus on purely formal aspects of index functions, or on results to be expected when applied to data,

[110] see Vartia 1978, p. 24 and **sec. 2.4**.

- they are often not really relevant and justified in a convincing manner (like for example the criterion "factor reversal test"), and
- sometimes oriented at aspects of more aesthetic nature, like symmetry, even handedness etc. (for example taking various weighting schemes equally into account),

and there is much fixation, or even obsession, with formulas satisfying such criteria.

Finally, it should be noted that Vartia was able to show that

The difference between P^F and P^T is very small (since both indices are superlative) no matter which data are used - even though both indices are conceptually quite different[111] - and that there is no tendency for P^F to be greater than P^T or vice versa.

Being close to P_{0t}^F also applies to $\sqrt{P^{HL}P^{PA}}$, which is the geometric mean of two indices, both badly biased. On the other hand, the combination of two index formulas, both of which are biased *up*wards, like P^{PA} and P^L, cannot be recommended, and will result in an index biased upwards as well. Vartia and Vartia (1984) have shown that P^{FCI}, the Finish Currency Index (1974 = 100), was an index of such a type

$$P_{0t}^{FCI} = \tfrac{1}{2}\left(P_{0t}^{PA} + P_{0t}^{L}\right) \geq \sqrt{P_{0t}^{PA}P_{0t}^{L}} \geq \sqrt{P_{0t}^{PA}P_{0t}^{HL}} \text{ since } P_{0t}^{L} > P_{0t}^{HL}.$$

In the same way, the combination of two harmonic index functions, like $\tfrac{1}{2}\left(P_{0t}^{P} + P_{0t}^{HL}\right)$ or $\sqrt{P_{0t}^{P}P_{0t}^{HL}}$, will necessarily produce a *down*wards bias.

In **ex. 3.4.2** we observed in fact that the Finish currency index would be $\tfrac{1}{2}\left(P_{0t}^{PA} + P_{0t}^{L}\right) = 1.254$, which is well above the unbiased indices, which range between 1.223 and 1.224. On the other hand, $\tfrac{1}{2}\left(P_{0t}^{P} + P^{HL}\right)$ should be below the range of unbiased indices, which is in fact the case, since this index turns out to be 1.194. Finally, it should be added that the six index formulas P^L, P^{PA}, DP^L, DP^P, P^{HL}, and P^P in **fig. 3.4.2** are special types of the more general *moment mean* (power mean). The general structure of a moment (power) mean is as follows[112]

$$P(\alpha,c) = \left[\prod_i (p_{it}/p_{i0})^{\alpha} c_i\right]^{1/\alpha}$$ which corresponds to the CES[113] function:

α	σ	$c_i = w_{i0}$	$c_i = w_{it}$
-1	2	P^{HL}	P^P
→0	→1	DP^L	DP^P
+1	0	P^L	P^{PA}

111 In our view this in fact speaks volumes of the usefulness of the axiomatic approach: though there are remarkable differences in the conceptual foundation, reflected (?) or not (?) in differences as regards the axiomatic performance of the formulas, two index formulas may nonetheless in practice yield quite similar relsults.

112 see **sec. 2.2**, eq. 2.2.19.

113 Constant elastisticity of substition; there σ = 1 - α denotes the elastisticity of substition.

such that we might think of infinitely many more index functions (for example $\alpha = 1/2$ etc.).

3.5 Log-change index numbers II: Vartia's index formulas

Two well known index formulas have been introduced by the Finish author Yrjö O. Vartia. The first index is known as the "Vartia I index[114]", denoted by P_{0t}^{V1}. The so-called "Vartia II" index, P_{0t}^{V2}, is better known and has some properties that make P_{0t}^{V2} more acceptable than P_{0t}^{V1} for many purposes. Both index formulas have the following two features in common

- they meet the factor reversal test, so that they are both "ideal", and
- they are based on log changes and make use of the so called logarithmic mean of values (v) or value-shares (w).

a) Aggregation of log changes and the logarithmic mean

1. Properties of log means

Recall the logarithmic mean (or simply log-mean) of two positive numbers, x and y

(3.5.1) $$L(x,y) = L(y,x) = \frac{y-x}{\ln(y/x)} \text{ if } x \neq y \text{ and } L(x,y) = x \text{ if } x = y.$$

With L expressed in logarithms at base 10 instead of natural logarithms we would get of course $\log\left(P_{0t}^{V1}\right)$ and $\log\left(P_{0t}^{V2}\right)$ instead of $\ln\left(P_{0t}^{V1}\right)$ and $\ln\left(P_{0t}^{V2}\right)$ respectively.

The function L has the following useful properties[115]:

1	symmetry	$L(x,y) = L(y,x)$
2	linear homogeneous[1)]	$L(\lambda x, \lambda y) = \lambda L(x,y)$
3	monotonically increasing in both arguments	$L(x^*,y) > L(x,y)$ for $x^* > x$ and $L(x,y^*) > L(x,y)$ for $y^* > y$
4	bounded by min(x,y) and max(x,y)[2)]	$\min(x,y) \leq L(x,y) \leq \max(x,y)$
5	bounded by G and A[2)]	$G(x,y) \leq L(x,y) \leq A(x,y)$

1) Or: homogeneous "of the first degree", or "of degree one"

2) This fact establishes the mean value property of L(x,y). Both bounds (under 4 and 5) are not exact in the sense that possibly tighter bounds can be found.

---

114 or "pseudo price index". It was originally "invented" by Montgomery (1937) and was rediscovered by Vartia (1976). The Vartia I index is often quoted as an example of an index not satisfying the equality test though it is consistent in aggregation. Hence P^{V1} is used to demonstrate that both tests are independent (see **sec. 5.2**).

115 See for the theorems stated here Lorenzen 1990 and Lorenzen 1989 where further reference is also given.

where G(x,y) is the geometric mean $\sqrt{xy}$ and A(x,y) is the arithmetic mean $\frac{1}{2}(x+y)$. Moreover, it has also been shown that L(x,y) is a special case of a more general mean. The log-mean is almost completely unknown to standard textbooks of statistics and the same is true for log changes, presumably by reason of their lacking intuitive appeal. However in **sec. 3.4** we already discussed in some detail the interpretation of log-changes in combination with the log-mean L as growth rates with an absolute level L in the denominator and an absolute difference in the numerator

$$\ln\left(\frac{y_t}{y_{t-1}}\right) = \frac{y_t - y_{t-1}}{L(y_t, y_{t-1})} = \frac{\Delta y}{L}.$$

2. Aggregation of log changes (over commodities)

Some advantages of log-changes over conventional growth rates were already mentioned in **sec. 3.4**. It can be argued that the log change should even be preferred to the conventional growth rate because of the denominator $L(y_t, y_{t-1})$ making use of both levels y_t and y_{t-1} and hence being more balanced, whilst r_t is onesided, using information on y_{t-1} only. Another point, however, is aggregation over n commodities (instead of the aggregation over time, examined in **sec. 3.4**). The prejudice is widespread that log changes will not fit to additive models (i.e. to variables $Y = \Sigma y$). Yet the prejudice is not justified.

The weights a_{i0} in **fig. 3.5.1** are necessary to allow for the summation $Y_0 = \Sigma y_{i0}$ and $Y_t = \Sigma y_{it}$. In the case of log changes the weights $\beta_i = b_i/\Sigma b_i$ serve the same purpose. Again the log mean plays an important role in combination with log changes: log means provide the weights needed for aggregation of log changes over n commodities. To prove eq. 3.5.3 proceed as follows: substition of $a_{i0}Y_0$ for y_{i0} and of $a_{it}Y_t$ for y_{it} gives $b_i \ln\left(\frac{a_{it}Y_t}{a_{i0}Y_0}\right) = (a_{it} - a_{i0}) + \ln\left(\frac{Y_t}{Y_0}\right) b_i$, and because of $\sum a_{it} = \sum a_{i0} = 1$ we get

$$\sum b_i \ln\left(\frac{y_{it}}{y_{i0}}\right) = \ln\left(\frac{Y_t}{Y_0}\right) \sum b_i .$$

Then eq. 3.5.3 follows from dividing both sides by Σb_i.

Note that there is again a remarkable difference betweeen the conventional growth rates (using weights a_{i0}) on the one hand and log-changes (using weights $\beta_i = b_i/\Sigma b_i$) on the other in that:

- weights a_{i0} exploit the information given in the data in an asymmetrical manner, while
- weights $b_i/\Sigma b_i$ are "balanced" (using both periods, 0 and t).

Figure 3.5.1: Aggregation growth rates over commodities (additive model)

Notation 1. consider change from 0 to t,
2. variable referring to individual commodity y_i ($i = 1, \ldots, n$),
3. aggregate $Y_\tau = \sum y_{i\tau}$, $\tau = 0, 1$ (additive model)

conventional growth rate	growth rate on the basis of log changes
(3.5.2) $r(Y) = \frac{Y_t - Y_0}{Y_0} = \sum_{i=1}^{n} a_{i0}\left(\frac{y_{it} - y_{i0}}{y_{i0}}\right) = \sum a_{i0} r(y_i)$	(3.5.3) $\ln\left(\frac{Y_t}{Y_0}\right) = \sum_{i=1}^{n} \frac{b_i}{\sum b_i} \ln\left(\frac{y_{it}}{y_{i0}}\right) = \sum \beta_i \ln(y_{it}/y_{i0})$, $\beta_i = b_i/\Sigma b_i$
weights $a_{i0} = \frac{y_{i0}}{\sum y_{i0}} = \frac{y_{i0}}{Y_0}$	weights $b_i = L(a_{i0}, a_{it})$, $a_{it} = \frac{y_{it}}{\sum y_{it}}$

commentary on the type of weights

weights a_{i0} make use of the structural information related to period 0 only	b_i, being an average of a_{i0} and a_{it}, take both periods into account*

* weights are "balanced", i.e. they employ structural data of both periods.

3. Decompositions of total value log changes

Aggregated values V_t and V_0 will play the part of Y_t and Y_0 when the mathematics of log-changes and their aggregation is applied to value changes. It turns out that there are in principle two methods of aggregating terms like $\ln\left(\frac{v_{it}}{v_{i0}}\right)$ to a total value log change $\ln\left(\frac{V_t}{V_0}\right)$:

a) use weights $L(v_{i0},v_{it})$, that is, means of absolute values, which leads to

$\sum_i \ln\left(\frac{v_{it}}{v_{i0}}\right) L(v_{i0}, v_{it}) = V_t - V_0$, and since $L(V_t, V_0) = \frac{V_t - V_0}{\ln(V_t/V_0)}$ we get

(3.5.4) $\ln\left(\frac{V_t}{V_0}\right) = \sum \ln\left(\frac{v_{it}}{v_{i0}}\right) \frac{L(v_{it}, v_{i0})}{L(V_t, V_0)}$ which is the basis for the Vartia-I index;

b) use value shares $w_{it} = v_{it}/\Sigma v_{it} = v_{it}/V_t$ and $w_{i0} = v_{i0}/\Sigma v_{i0} = v_{i0}/V_0$ and weights

$$L(w_{it}, w_{i0}) = \frac{w_{it} - w_{i0}}{\ln(v_{it}/v_{i0}) - \ln(V_t/V_0)}$$ we get

(3.5.5) $L(w_{it}, w_{i0})\ln(v_{it}/v_{i0}) = (w_{it} - w_{i0}) + L(w_{it}, w_{i0})\ln(V_t/V_0)$ because

$$\ln\left(\frac{w_t}{w_0}\right) = \ln\left(\frac{v_{it}/V_t}{v_{i0}/V_0}\right) = \ln\left(\frac{v_{it}}{v_{i0}} : \frac{V_t}{V_0}\right)$$

and since $\sum w_{it} = \sum w_{i0} = 1$

$$\ln\left(\frac{V_t}{V_0}\right)\sum L(w_{it}, w_{i0}) = \sum\left(L(w_{it}, w_{i0})\ln\left(\frac{v_{it}}{v_{i0}}\right)\right)$$, and finally

(3.5.6) $\ln\left(\frac{V_t}{V_0}\right) = \sum\left(\ln\left(\frac{v_{it}}{v_{i0}}\right)\frac{L(w_{it}, w_{i0})}{\sum L(w_{it}, w_{i0})}\right)$, which is the basis for the Vartia-II index.

It should be noticed that the factors $\frac{L(w_{it}, w_{i0})}{\sum L(w_{it}, w_{i0})}$ in eq. 3.5.6 add up to unity. But the same is *not* true for the factors $\frac{L(v_{it}, v_{i0})}{L(V_t, V_0)}$ in eq. 3.5.4 since

(3.5.7) $\sum_i L(v_{it}, v_{i0}) \neq L(V_t, V_0) = L(\sum_i v_{it}, \sum_i v_{i0})$,
which is the reason why the Vartia-I index is unable to pass the proportionality test. Since log means are symmetric we get $L(v_{it}, v_{i0}) = L(v_{i0}, v_{it})$ and $L(V_t, V_0) = L(V_0, V_t)$.

b) The Vartia-I index $\left(P_{0t}^{V1}\right)$

The Vartia-I price index is defined as

(3.5.8) $\ln\left(P_{0t}^{V1}\right) = \frac{\sum_{i=1}^{n} L(v_{it}, v_{i0})\ln\left(\frac{p_{it}}{p_{i0}}\right)}{L(V_t, V_0)}$, and correspondingly the quantity index

(3.5.8a) $\ln\left(Q_{0t}^{V1}\right) = \sum_{i=1}^{n} L(v_{it}, v_{i0})\ln\left(\frac{q_{it}}{q_{i0}}\right) \Big/ L(V_t, V_0)$,

both indices being weighted sums of log-changes, with weights, however, not adding up to unity.[116] There exists an interpretation of P_{0t}^{V1} similar to Stuvel's index and Stu-

---

[116] as mentioned above already. The indices therefore are not weighted *means*.

vel (1989) has systematically studied most of the properties of this formula. The P_{0t}^{V1} index seems to be more or less exotic with respect to practical applications. Some Dutch authors (Al, Balk, de Boer and den Bakker 1986), however, recommended using the chain index version of P_{0t}^{V1} for deflation purposes,[117] because P_{0t}^{V1} satisfies the factor reversal test.

Another aspect would be to require the contribution of the i-th commodity to $\ln(Q_{0t}^{V1})$, that is

(3.5.8b) $$\frac{L(v_{it},v_{i0})}{L(V_t,V_0)}\ln\left(\frac{q_{it}}{q_{i0}}\right)=\frac{L(v_{it},v_{i0})}{L(V_t,V_0)}\frac{q_{it}-q_{i0}}{L(q_{it},q_{i0})}=\Delta\left(\ln\left(Q_{0t}^{V1}\right)\right)$$

to have an interpretation comparable to the contribution of the i-th commodity to Q_{0t}^{L}, which is $p_{i0}q_{it}/\sum p_{i0}q_{it}$. In part d of this section it will be shown that this is not guaranteed, and there seems to be no obvious explanation of what $\Delta\ln(Q^{V1})$ really means.

c) The Vartia-II index $\left(P_{0t}^{V2}\right)$

It has been shown by Vartia in several publications that the log-change in the total value $\ln(V_t/V_0) = \ln(V_{0t})$ has (in addition to eq. 3.5.4) the following exact decomposition

(3.5.9) $$\ln V_{0t}=\Sigma\hat{w}_i\ln(v_{it}/v_{i0}),$$

where $\hat{w}_t=L(w_{it},w_{i0})/\Sigma L(w_{it},w_{i0})$ and $w_{it}=v_{it}/V_t=p_{it}q_{it}/\sum p_{it}q_{it}$, ($w_{i0}$ defined correspondingly). By virtue of eq. 9, the indices P_{0t}^{V2} and Q_{0t}^{V2} are able to pass the factor reversal test. To prove eq. 9 we use the result

(3.5.10) $\ln\frac{w_t}{w_0}=\ln\frac{v_t}{v_0}-\ln\frac{\Sigma v_t}{\Sigma v_0}=\ln\frac{v_t}{v_0}-\ln\frac{V_t}{V_0}$ (the subscript i has been omitted for convenience of presentation) leading to eq. 3.5.5 and summation over all commodities yields

(3.5.6a) $$\Sigma L(w_t,w_0)\ln\left(\frac{v_t}{v_0}\right)=\ln\left(\frac{V_t}{V_0}\right)\Sigma L(w_t,w_0).$$

Taking into account $\sum w_t=\sum w_0=1$ and the first factor representing $\ln V_{0t}$ and the second factor $\Sigma L(w_t,w_0)$ respectively, we get eq. 9, which we intended to prove.

The Vartia II indices are now defined as follows

(3.5.11) $$\ln\left(P_{0t}^{V2}\right)=\Sigma\frac{L(w_{it},w_{i0})\ln\left(\frac{p_{it}}{p_{i0}}\right)}{\Sigma L(w_{it},w_{i0})}=\Sigma\hat{w}_{it}\ln(p_{it}/p_{i0}),$$ and

117 See my book on chain indices (v. d.Lippe, 2001) for more details on chain indices as tools for deflation.

(3.5.11a) $\ln\left(Q_{0t}^{V2}\right) = \sum \hat{w}_{it} \ln\left(q_{it}/q_{i0}\right)$.

The three indices listed in **fig. 3.5.2** are in close relation to one another. The index P_{0t}^{V2} can be regarded as a geometric mean index patterned according to Törnqvist's price index P_{0t}^{T} with weights $\hat{w}_t = L\left(w_{it}, w_{i0}\right) / \sum L\left(w_{it}, w_{i0}\right)$ instead of $\overline{w} / \Sigma \overline{w} = \overline{w}$.

Eqs. 9, 11, and 11a show that $\ln\left(V_t/V_0\right) = \ln\left(V_{0t}\right)$, the log-change of value, is the sum of the log-change of prices and the log-change of quantities because v_{it}/v_{i0} = $(p_{it}/p_{i0})(q_{it}/q_{i0})$. Hence

(3.5.12) $\ln\left(V_{0t}\right) = \Sigma \hat{w}_i \ln\left(p_t / p_0\right) + \Sigma \hat{w}_i \ln\left(q_t / q_0\right) = \ln\left(P_{0t}^{V2}\right) + \ln\left(Q_{0t}^{V2}\right)$, which is tantamount to the Vartia II indices satisfying factor reversibility.

Figure 3.5.2: Törnqvist, Vartia-II and Walsh-II index

general structure: $P(g) = \prod \left(\frac{p_{it}}{p_{i0}}\right)^{g}$

with weights g ($\Sigma g = 1$)

Törnqvist	Vartia II	Walsh II
$g = \overline{w} = \frac{1}{2}\left(w_0 + w_t\right)$	$g = \hat{w} = L(w_0, w_t) / \sum L(w_0, w_t)$	$g = \sqrt{v_0 v_t} / \Sigma \sqrt{v_0 v_t}$

$\sqrt{w_0 w_t} \leq L(w_0, w_t) \leq \frac{1}{2}\left(w_0 + w_t\right) = \overline{w}$, $\sqrt{w_0 w_t} = \sqrt{v_0 v_t} / \sqrt{V_0 V_t}$

Example 3.5.1

Using the data of **ex. 3.4.2** calculate P_{0t}^{V2} and Q_{0t}^{V2}.

i	p_{it}/p_{i0}	v_{i0}	v_{it}	w_{i0}	w_{it}	q_{it}/q_{i0}	$L(w_{it}, w_{i0})$	$L(v_{i0}, v_{it})$
A	1.2	200	240	0.2	0.2	1	0.2	219.393
B	0.9	320	360	0.32	0.3	1.25	0.3099	339.608
C	1.5	480	600	0.48	0.5	0.833	0.4899	537.770
	sum	1000	1200			sum	0.9998	1096.770

The result concerning the Vartia-II indices is $P_{0t}^{V2} = 1.2276$, $Q_{0t}^{V2} = 0.9775$; verify the product of both indices is in fact $V_{0t} = 1.2$. To calculate the Vartia I indices note that[118]

$$\sum L(v_{i0}, v_{it}) = 1096.7705 \neq L(V_0, V_t) = 1096.9630.$$

Thus we get $\ln P_{01}^{V1} = 222.26/1096.96 = 0.2026$, hence $P_{0t}^{V1} = 1.2246$.

Calculation of the Vartia I quantity index results in
$\ln(Q_{0t}^{V1}) = -22.266/1096.96 = -0.0203$ hence $Q_{0t}^{V1} = 0.9799$ and in fact
$P_{0t}^{V1} Q_{0t}^{V1} = V_{0t} = 1.2$. Note that both results P_{0t}^{V1} and P_{0t}^{V2} do not differ much from the unbiased indices P^F, P^T and $\sqrt{P^{PA} P^{HL}}$ in ex. 6.4.2.

The same is true for the quantity indices.

Summary of the results of **ex. 3.5.1** and **3.4.2**

index formula	P_{0t}	Q_{0t}	index formula	P_{0t}	Q_{0t}
Palgrave	1.26	0.9917	logarithmic Paasche	1.2307	0.9761
Paasche	1.2	0.9615	Laspeyres	1.248	1.0
logarithmic Laspeyres	1.2182	0.9840	harmonic Laspeyres	1.1873	0.9690
Törnqvist	1.2244	0.9800	Fisher	1.2238	0.9806
Vartia I	1.2246	0.9799	Vartia II	1.2276	0.9775
Stuvel's index	1.2264	0.9784	no-name*	1.2231	0.9803

* $\sqrt{P^{PA} P^{HL}}, \sqrt{Q^{PA} Q^{HL}}$ ♦

d) Properties of Vartia indices

1. Proportionality[119], linear homogeneity

It can easily be seen that P_{0t}^{V1} (in contrast to P_{0t}^{V2}) is not a weighted arithmetic mean of log changes with weights adding up to unity. Therefore PV1 can (unlike PV2) fail proportionality as well as linear homogeneity (see **ex. 3.5.2**).

Consider alternative price vectors $\mathbf{p}_t$ and $\lambda\mathbf{p}_t$, or $\mathbf{p}_t = \lambda\mathbf{p}_0$ (other things being equal), there is no reason to exclude

(3.5.13) $P_{0t}^{V1}(\mathbf{p}_0, \mathbf{q}_0, \lambda\mathbf{p}_t, \mathbf{q}_t) \neq \lambda P_{0t}^{V1}(\mathbf{p}_0, \mathbf{q}_0, \mathbf{p}_t, \mathbf{q}_t)$ (no linear homogeneity)

---

118 Please pay attention that $w_{i0} = w_{it} = 0.2$ does not mean that $L(w_{i0}, w_{it}) = 0$, because according to eq. 3.51 (see below), equality of the w's gives $L(w_{i0}, w_{it}) = 0.2$ rather than 0. We are grateful to Mr. Dugas from the U.S. National Academy of Science who at first thought our solution of ex. 3.5.1 was erroneous and then realised that eq. 3.5.1 should be observed. We explicitly mention this because it is certainly not unlikely that this equation will be overlooked.

119 We return to this point in **sec. 3.6**.

(3.5.13a) $P_{0t}^{V1}(\mathbf{p}_0,\mathbf{q}_0,\lambda\mathbf{p}_0,\mathbf{q}_t) \neq \lambda$ (strict proportionality violated).

Obviously $\ln\left(P_{0t}^{V1}(\mathbf{p}_0,\mathbf{q}_0,\lambda\mathbf{p}_0,\mathbf{q}_t)\right) = \ln(\lambda)\frac{\sum L(v_{i0},v_{it})}{L(V_0,V_t)}$ does not necessarily amount to $\ln(\lambda)$ because of eq. 3.5.7. Furthermore, in the case of λ-fold prices in period t, the *structure* of weights will change since $L(\lambda v_{it},v_{i0}) = \frac{\lambda v_{it} - v_{i0}}{\ln(\lambda v_{it}/v_{i0})} \neq \frac{v_{it} - v_{i0}}{\ln(v_{it}/v_{i0})}$, because

in P_{0t}^{V1} weights are determinded by *absolute* values rather than value shares. The situation is different in the case of P_{0t}^{V2} where shares of value, like w_{i0}, and

$w_{it} = \frac{v_{it}}{\sum v_{it}} = \frac{\lambda v_{it}}{\sum \lambda v_{it}}$ (for all i = 1, 2, ... , n) determine weights $L(w_{i0},w_{it})/\sum L(w_{i0},w_{it})$

which thus will remain unchanged. Furthermore since
$\ln(\lambda p_{it}/p_{i0}) = \ln(\lambda) + \ln(p_{it}/p_{i0})$, we get

(3.5.14) $\ln\left(P_{0t}^{V2}(\lambda)\right) = \ln(\lambda) + \ln\left(P_{0t}^{V2}\right)$, or $P_{0t}^{V2}(\lambda) = \lambda P_{0t}^{V2}$

as required by linear homogeneity and proportionality.

Example 3.5.2

The purpose of this example is to show that the index P_{0t}^{V1} can violate the requirement of strict proportionality as well as linear homogeneity. Assume two commodities and calculate P_{0t}^{V1} and P_{0t}^{V2} using prices p_{it} and λp_{it} respectively. Since λ = 1.5, in what follows linear homogeneity requires the price index in the second case to result in a value 50% higher.

i	p_{i0}	p_{it}	λp_{it}	q_{i0}	q_{it}
1	10	12	18	20	25
2	20	18	27	16	20

i	w_{i0}	w_{it}	$L(w_{i0},w_{it})$	v_{i0}	v_{it}	λv_{it}	$L(v_{i0},v_{it})$	$L(v_{i0},\lambda v_{it})$
1	0.3846	0.4545	0.4186	200	300	450	246.63	277.46
2	0.6154	0.5454	0.5797	320	360	540	339.61	420.45
Σ	1	1	0.9983	520	660	990	586.24	697.91

$L(V_0, V_t)$ = 587.22 and $L(V_0, \lambda V_t)$ = 729.95 which renders in combination with prices p_0 and p_t: $\ln(P_{0t}^{V1})$ = 0.01564 → P_{0t}^{V1} = 1.01576, and $\ln(P_{0t}^{V2})$ = 0.01527 → P_{0t}^{V2} = 1.015385.

Using prices λp_t we get $\ln(P_{0t}^{V1})$ = 0.39628 → P_{0t}^{V1} = 1,48638, but 50% more than 1.01576 would amount to 1.523646. ♦

2. Aggregation properties

As aforementioned, P_{0t}^{V1} is known for representing a most rare case of an index function which is aggregative consistent and yet fails the equality test. Moreover both indices, P_{0t}^{V1} as well as P_{0t}^{V2}, are unable to yield consistent volumes (structural consistency; see **sec. 5.2**). Those who recommend Vartia indices (or even a chain version of P_{0t}^{V1}) for deflation like Al, Balk, de Boer and den Bakker 1986 frankly admit this shortcoming: "A practical disadvantage of this formula however, is that the indices thus obtained would not yield a table exhibiting additive consistency in real values" (p. 348).

Nonetheless the authors recommend a Vartia I index (in the chain version, complicating things even further) for purposes of deflation. But there is still another problem referring not only to the Vartia I but also the Vartia II index. To give "volume" meaning, it is also necessary to examine the relationship between the amount of an isolated change in the quantity of the i-th commodity on the one hand and the change in aggregated volume as measured by Q_{0t}^{V1} (see eq. 6.5.8b), or Q_{0t}^{V2}, on the other.

Example 3.5.3

Consider n = 2 commodities only and $q_{1t} = \lambda_1 q_{10} = \lambda_1 q_1$, and $q_{2t} = q_{20} = q_2$ and alternatively $q_{1t}^{*} = \lambda_2 q_{10} = \lambda_2 q_1$. We try to demonstrate: in spite of $\lambda_1 > \lambda_2$ the index Q_{0t}^{V2} does not in general display a greater rise in quantity in the first case compared with the second.[120] The reason is that the relative weights of the two commodities $L(w_{1t}, w_{10}) / \sum L(w_{it}, w_{i0})$ and $L(w_{2t}, w_{20}) / \sum L(w_{it}, w_{i0})$ depend also on prices in both periods. Hence, though $\ln(q_{1t}/q_{10}) = \ln(\lambda_k)$ and $\ln(q_{2t}/q_{20})$ = 0, the impact of $\ln(\lambda_k)$ on Q_{0t}^{V2} in the two situations to be compared (k = 1, 2) may differ substantially, depending on the weight w_{1t} given to commodity no. 1 in period t. To see this assume:

	base period		first variant			second variant		
i	p_{i0}	q_{i0}	λ_1	p_{it}	q_{it}	λ_2	p_{it}	q_{it}
1	16	10	1.5	15	15	1.4	30	14
2	10	20	1	11.25	20	1	4	20

expenditure shares base period w_{10} = 4/9, w_{20} = 5/9, first variant w_{1t} = 1/2, w_{2t} = ½, second variant w_{1t} = 3/4, w_{2t} = 1/4.

For the first variant we get $\ln(Q^{V2}) = 0.19144 \rightarrow Q^{V2} = 1.2110$, and for the second $\ln(Q^{V2}) = 0.11503 \rightarrow Q^{V2} = 1.1219$, although $\lambda_1 = 1.5 > \lambda_2 = 1.4$. ♦

120 But to display a greater increase is just what we would expect the index to do according to the principle called "reflection of quantity movement" see **sec. 5.2**.

3. Monotonicity

Log-change indices in general may violate strict monotonicity. To show this in more general terms, consider the following numerical example taken from Olt 1996, p.140.

Example 3.5.4

The example aims to demonstrate that log-change indices are not guaranteed to comply with monotonicity. Consider two commodities, 1 and 2, the price of the second, that is p_{2t}, will be varied (indicated by a shadow in the relevant box).

	p_0	p_t	q_0	q_t
1	30	40	50	20
2	80		4	40

The following values are examined for the price p_{2t} 1, 5, 10, yielding values for various log change index functions as follows. We expect of course that $p_{2t} = 5$ will yield a higher P_{0t} than $p_{2t} = 1$ and again $p_{2t} = 10$ a higher value of P_{0t} than in the case of $p_{2t} = 5$. However, this is not true for the following indices[121]: P_{0t}^{W2} = Walsh II, P_{0t}^{V1} = Vartia I, P_{0t}^{V2} = Vartia II, P_{0t}^{WV} = Walsh-Vartia, P_{0t}^{T} = Törnqvist, and P_{0t}^{TH} = Theil.

p_{2t}	P_{0t}^{L}	P_{0t}^{P}	P_{0t}^{W2}	P_{0t}^{V1}	P_{0t}^{V2}	P_{0t}^{WV}	P_{0t}^{T}	P_{0t}^{TH}
1	1.101	0.221	0.861	0.701	0.837	0.864	0.791	0.813
5	1.110	0.263	0.751	0.454	0.751	0.751	0.750	0.753
10	1.121	0.316	0.744	0.193	0.740	0.748	0.730	0.749

Interestingly, the two indices added for the sake of comparison, that is Laspeyres and Paasche (shaded area), *both* comply with strict monotonicity, but *none* of the log change indices does so. Furthermore, these indices differ tremendously from one another. So virtually *all* log-change index numbers display a decrease rather than an increase in prices when an isolated rise in price p_{2t} takes place from 1 to 5 and to 10. ♦

Example 3.5.5:

To assess the use of Vartia I- indices for deflation (no proportionality of Q_{0t}^{V1}) and to verify some relationships between Vartia I- and Vartia II-index:

Assume n = 2 commodities with prices and quantities at period 0 and period t (the shaded parts of the table will be varied):

---

[121] Some log change indices not discussed above are also included in this example and Olt's original table. To facilitate comparisons the table above also contains price indices of Laspeyres and Paasche.

	p_{i0}	q_{i0}	p_{it}	q_{it}	v_{i0}	w_{i0}	v_{it}	w_{it}	p_{it}/p_{i0}	q_{it}/q_{i0}
1	15	20	18	25	300	0.6	450	05625	1.2	1.25
2	25	8	35	10	200	0.4	350	04375	1.4	1.25

To calculate the logarithmic means necessary for P_{0t}^{V1} we proceed as follows:

$$L(v_{10}, v_{1t}) = \frac{v_{1t} - v_{10}}{\ln(v_{1t}/v_{10})} = \frac{450-300}{\ln(1.5)} = 369.95,\ L(v_{20}, v_{2t}) = 268.04\text{, further}$$

$$L(V_0, V_t) = \frac{V_t - V_0}{\ln(V_t/V_0)} = \frac{800-500}{\ln(1.6)} = 638.29 \neq L(v_{10}, v_{1t}) + L(v_{20}, v_{2t}) = 637.99.$$

Thus the ***Vartia I price index*** amounts to

$$\ln\left(P_{0t}^{V1}\right) = \frac{1}{638.29}\left(369.95 \cdot \ln(1.2) + 268.04 \cdot \ln(1.4)\right) = \frac{157.64}{638.29} = 0.2470 \text{ such that}$$

$P_{0t}^{V1} = \exp(0.2469) = 1.28014$ in contrast to $P_{0t}^{L} = 1.28$, $P_{0t}^{P} = \frac{800}{625} = 1.28$, both indices, P_{0t}^{L} and P_{0t}^{P}, coincide because the covariance C between price relatives and quantity relatives vanishes. This is so because q_{it}/q_{i0} = const.= 1.25. Calculation Q_{0t}^{V1} results in $\ln(Q_{0t}^{V1})$ = 637.99 ln(1.25)/638.29 = 0.2230 hence Q_{0t}^{V1} = 1.24987. Note that $Q_{0t}^{V1} \neq 1.25$, though both quantity relatives are equally 1.25. Hence Q^{V1} does not meet proportionality (in quantites). Deflation should result in 625 (increase of the value V_0 by 25%), which is the result of a traditional deflation using the Paasche index: V_t / P_{0t}^{P}. Deflation with P_{0t}^{V1}, however, results in a volume of 624.93, which is slightly less than 625. Note that the product P_{0t}^{V1} and Q_{0t}^{V1} is in fact 1.6, which is in fact the value ratio V_{0t}.

To now obtain the ***Vartia-II indices***, new weights have to be calculated:

$$L(w_{10}, w_{it}) = \frac{0.5625 - 0.6}{\ln(0.5625/0.6)} = 0.5810, \text{ and } L(w_{20}, w_{2t}) = \frac{0.4375 - 0.4}{\ln(1.09375)} = 0.4185$$

$$\Sigma L(w_{i0}, w_{it}) = 0.9995 \text{ further}$$

$$\ln\left(P_{0t}^{V2}\right) = \frac{1}{0.9995}\left(0.5810\ln(1.2) + 0.4185\ln(1.4)\right) = 0.2469 \rightarrow P_{0t}^{V2} = 1.28$$

Not surprisingly, $P_{0t}^{V2} = P_{0t}^{L} = P_{0t}^{P} = 1.28$ because both quantity relatives are identically 1.25, hence $Q_{0t}^{V2} = 1.25$ (while $Q_{0t}^{V1} < 1.25$) and Vartia II indices satisfy the factor reversal test. ♦

The following table might be interesting in that it gives a summary of the results obtained for various modifications of the original example and it also gives some hints to assess the various index designs with respect to deflation problems. The variants examined are:

Variant	substituting vector	for vector*
1	$\mathbf{p}_t$ = [12 70]	[18 35]
2	$\mathbf{q}_t$ = [30 8]	[25 10]
3	$\mathbf{q}_t$ = [15 20]	[25 10]
4	$\mathbf{p}_t$ = [13 30] $\mathbf{q}_t$ = [30 7]	[18 35] [25 10]
5	$\mathbf{p}_t$ = [12 30] $\mathbf{q}_t$ = [25 15]	[18 35] [25 10]

* $\mathbf{p}_t$ or $\mathbf{q}_t$ vector of the original example

Variant 1 shows that the difference between the Vartia indices and the Laspeyres and Paasche indices is not so great as might have been expected. Again $P_{0t}^{L} = P_{0t}^{P}$ = 1.6 and $Q_{0t}^{L} = Q_{0t}^{P}$ = 1.25 because price relatives and quantity relatives are uncorrelated. In this case as before $Q_{0t}^{V2} = q_{1t}/q_{10} = q_{2t}/q_{20} = Q_{0t}^{L} = Q_{0t}^{P} = 1.25$.

The results are now

	P_{0t}^{V1}	Q_{0t}^{V1}	P_{0t}^{V2}	Q_{0t}^{V2}	P_{0t}^{L}	P_{0t}^{P}	Q_{0t}^{L}
1	1.6110	1.2414	1.6	1.25	1.6	1.6	1.25
2	1.2697	1.2917	1.2705	1.2908	1.28	1.2308	1.3325
3	1.3002	1.4920	1.3093	1.4817	1.28	1.3379	1.45
4	0.8019	1.2471	0.8175	1.2232	1	0.8	1.25
5	0.9827	1.5263	0.9798	1.5309	0.96	1	1.5

In *variant 2* the two quantity relatives are no longer identically 1.25 but 30/20 = 1.5 and 1 respectively, affecting the result conforming with expectations when C is negative: P_{0t}^{L} = 1.28 > $P_{0t}^{P} = 1.2308$. The results of P^{V1}, P^{V2} and P^{L} are again in close agreement.

The *third variant* reflects a considerable variety in quantity relatives, ranging from 0.75 (= 15/20) to 2.5 (= 20/8). There is also a divergence $\sum L(v_0, v_t) = 683.85 \neq L(V_0, V_t) = 709.23$ so that P^{V1} can hardly be regarded as a "mean" of log changes. Yet the results again do not differ much.

The *fourth variant* is constructed in such a way that $P_{0t}^{L} = V_{0t} = 1$. It can be seen that the two Vartia price indices come closer to Paasche's price index than to P_{0t}^{L}.

The *final fifth variant* examines a situation in which $P_{0t}^{P} = 1$. The data are also such that P_{0t}^{V2} and Q_{0t}^{V2} are simple (unweighted) means (because $L(w_{10}, w_{1t}) = L(w_{20}, w_{2t})$). Note that with the exception of variant 1, in all cases both Vartia indices happen to fall into the P^P- P^L-interval.

3.6 Ideal index functions and Theil's Best Linear Index (BLI)

In this section, models of absolute ($\Delta V_t = V_t - V_0$) or relative value change ($V_{0t} = V_t/V_0$) with *three* components, a price- (PC), quantity- (QC) and a residual or structural component (SC) will be introduced and examined. It will be shown how so called "ideal" (i.e. factor reversible) index functions manage to reduce the three components to only two symmetrically defined components, that is to an appropriately transformed PC and QC respectively.[122] Finally, a brief introduction is given into Theil's approach to find a pair of index functions capable of satisfying factor reversibility *approximately* in a least squares sense.

a) Three-component model of value change (the structural component)

1. Additive model (value change)

The *absolute* value change $\Delta V_t = V_t - V_0 = \Sigma p_{it}q_{it} - \Sigma p_{i0}q_{i0}$ can be decomposed as follows (using vector notation):

(3.6.1) $$\mathbf{p}_t'\mathbf{q}_t - \mathbf{p}_0'\mathbf{q}_0 = \mathbf{q}_0'(\mathbf{p}_t - \mathbf{p}_0) + \mathbf{p}_0'(\mathbf{q}_t - \mathbf{q}_0) + (\mathbf{q}_t' - \mathbf{q}_0')(\mathbf{p}_t - \mathbf{p}_0)$$

$$= \mathbf{q}_0'\Delta\mathbf{p}_t + \mathbf{p}_0'\Delta\mathbf{q}_t + \Delta\mathbf{q}_t'\Delta\mathbf{p}_t = PC + QC + SC.$$

The following interpretation of this simple definitional equation is usually given

- the pure price component (PC) is represented by $\mathbf{q}_0'\Delta\mathbf{p}_t = \Sigma p_{it}q_{i0} - \Sigma p_{i0}q_{i0}$
- the pure quantity component (QC) is $\mathbf{p}_0'\Delta\mathbf{q}_t = \Sigma p_{i0}q_{it} - \Sigma p_{i0}q_{i0}$
- the structural component (SC) is $\Delta\mathbf{q}_t'\Delta\mathbf{p}_t = \Delta\mathbf{p}_t'\Delta\mathbf{q}_t = \Sigma(p_{it} - p_{i0})(q_{it} - q_{i0})$.

Dividing both sides of eq. 1 by $\mathbf{p}_0'\mathbf{q}_0$ yields an equation known from Stuvel's approach:

(3.6.2) $$V - 1 = (P^L - 1) + (Q^L - 1) + (V - P^L - Q^L + 1) = PC + QC + SC.$$

Fig. 3.6.1 provides a geometric interpretation of this relationship with three shaded fields representing PC, QC and SQ and shows how Stuvel has (implicitly) allocated a part of the structural component $SC = V - P^L - Q^L + 1$ to the price component (PC), and the remaining part of SC to the quantity component (QC) in order to arrive at a pair of factor reversible indices.

2. Multiplicative model (value ratio)

The *relative* value change $V_{0t} = \Sigma p_{it}q_{it} / \Sigma p_{i0}q_{i0}$ is broken down into three factors as follows

---

[122] As might be assumed, this can be done by assigning (roughly speaking) one "half" of the SC to the PC and the other half of the SC to the QC. The simplest way to do this is the method by which Fisher derived his ideal index formula.

(3.6.3) $$\frac{\mathbf{p}_t'\mathbf{q}_t}{\mathbf{p}_0'\mathbf{q}_0}=\left(\frac{\mathbf{p}_t'\mathbf{q}_0}{\mathbf{p}_0'\mathbf{q}_0}\right)\left(\frac{\mathbf{q}_t'\mathbf{p}_0}{\mathbf{q}_0'\mathbf{p}_0}\right)\left(\frac{\mathbf{p}_t'\mathbf{q}_t}{\mathbf{p}_t'\mathbf{q}_0}\frac{\mathbf{p}_0'\mathbf{q}_0}{\mathbf{p}_0'\mathbf{q}_t}\right)=P_{0t}^{L}Q_{0t}^{L}\left(Q_{0t}^{P}/Q_{0t}^{L}\right)=P_{0t}^{L}Q_{0t}^{L}\left(1-\frac{C}{V_{0t}}\right).$$

Figure 3.6.1: Stuvel's way of deriving an ideal index

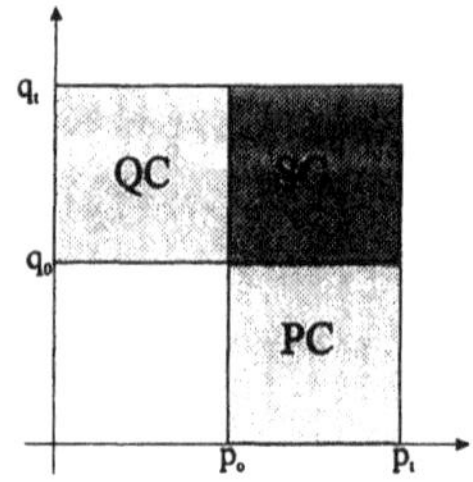

a) Additive approach

(3.6.2) $V - 1 = (P^L - 1) + (Q^L - 1) + (V - P^L - Q^L + 1)$
$= PC + QC + SC$

b) Stuvel's solution

price component PC	structural change component SC	quantity component QC
$P_{0t}^L - 1$	$V_{0t} - P_{0t}^L - Q_{0t}^L + 1$	$Q_{0t}^L - 1$

$\frac{1}{2}\left(V_{0t} - P_{0t}^L - Q_{0t}^L + 1\right)$ $\qquad$ $\frac{1}{2}\left(V_{0t} - P_{0t}^L - Q_{0t}^L + 1\right)$

new price component $P^* = PC + ½SC$	new quantity component $Q^* = QC + ½SC$
$P^* = \frac{1}{2}\left(P_{0t}^L - Q_{0t}^L\right)+\frac{1}{2}\left(V_{0t}-1\right)$	$Q^* = \frac{1}{2}\left(Q_{0t}^L - P_{0t}^L\right)+\frac{1}{2}\left(V_{0t}-1\right)$

The product P*Q* is not meaningful, however, since $P^*Q^* = [(V - 1)^2 - (P^L - Q^L)^2]$. A better solution is

$P^{ST} = \frac{1}{2}\left(P_{0t}^L - Q_{0t}^L\right)+\frac{1}{2}R$	$Q^{ST} = \frac{1}{2}\left(Q_{0t}^L - P_{0t}^L\right)+\frac{1}{2}R$

The product $P^{ST}Q^{ST}$ should equal V, setting $P^{ST}Q^{ST} = [R^2 - (P^L - Q^L)^2] = V$ we get

$$\frac{R}{2}=\sqrt{\left(\frac{P_{0t}^L - Q_{0t}^L}{2}\right)^2+V_{0t}}=\sqrt{\left(\frac{Q_{0t}^L - P_{0t}^L}{2}\right)^2+V_{0t}}$$

R is related to the covariance (C) between price and quantity relatives as follows

$$R=\sqrt{\left(P^L+Q^L\right)^2+4C}$$

The three components are

- the pure price component (PC*) which simply is P_{0t}^{L}
- the pure quantity component (QC*) or Q_{0t}^{L}
- the structural change component (SC*) which is $(V_{0t} - C)/V_{0t}$, or P_{0t}^{P}/P_{0t}^{L}.[123]

Fisher's ideal index can be arrived at by splitting SC* into two equal parts. The two components, PC* and QC*, are simply multiplied by $\sqrt{SC*}$ to yield P^F and Q^F respectively.

b) Distribution of the structural component (additive model): Stuvel's solution

It is well known that the factor reversal test and the (weaker) product test are concerned with only *two* indices, a price and a quantity index, the product of both being V_{0t}:

In order for a pair of index functions to satisfy the factor reversal test the SC has to be distributed evenly to the components PC and QC. Accordingly, an uneven distribution will produce a pair of indices, P and Q, which fails to fulfil the factor reversal test, and satisfies the weaker product test only.

The Laspeyres approach to define a *price* index (P^L) is to allocate (most unevenly) the entire SC to the quantity component QC (generating Q^P), such that the three components, PC, QC and SC in eq. 1, are reduced to two terms by $V - 1 = (P^L - 1) + (QC + SC)$ where $QC + SC = (Q^P-1)P^L$, which indeed is eq. *A1* in Stuvel's analysis (see **sec. 2.6**). In a similar vein Q^L, the Laspeyres *quantity* index, is obtained by assigning the entire SC to the PC (eq. *A2* in Stuvel's analysis)[124]: $V - 1 = (Q^L - 1) + (PC + SC)$ where $PC + SC = (P^P-1)Q^L$.

From this it follows that Stuvel's solution will consist in achieving an *equal* distribution of the structural component (SC) in contrast to the onesided solutions of Laspeyres and Paasche. In **fig. 3.6.1** it is shown that Stuvel's solution is in fact to add ½(SC), suitably transformed[125] however, to the pure price change (PC) to derive a price index and to add the same term to the pure quantity change (QC) to derive a quantity index.

Note that the indices arrived at so far by adding ½(SC) are *not* the indices P^{ME} and Q^{ME} of Marshall-Edgeworth nor the indices P^{DR} and Q^{DR} of Drobisch as might be expected at first sight, but rather P* and Q*:

---

[123] C refers to the theorem of Bortkiewicz and represents the covariance between price and quantity relatives.

[124] Thus eq. *A2* comes to a pair of indices, Q^L and P^P (just like eq. *A1* yields Q^P and P^L). This also shows the "indirect" manner in which the Paasche approach defines a component, e.g. the price component by $PC + SC = V - Q^L = Q^L(P^P-1)$. Price change measured this way is the amount by which the increase of value (e.g. expenditure), that is V, exceeds the increase in quantity (or volume), Q^L. It is pure price change *plus* structural change.

[125] It has to be taken into account that an additional correction is necessary to make the product $P^{ST}Q^{ST}$ equal to the value ratio V_{0t}.

$P^* = \frac{1}{2}(1+Q^L)(P^{ME}-1) = P^{DR} + \frac{1}{2}(Q^L-1)(P^P-1)$ and

$Q^* = \frac{1}{2}(1+P^L)(Q^{ME}-1) = Q^{DR} + \frac{1}{2}(Q^P-1)(P^L-1)$.

c) Distribution of the structural component (multiplicative model): Fisher's solution

Compared with Stuvel's derivation, the way Fisher's ideal index can be arrived at by splitting the SC* into two equal parts is absolutely easy to understand (see **fig. 3.6.2**). Instead of adding ½SC to the pure components, each of the two components, PC* and QC*respectively, is simply multiplied by $\sqrt{SC*}$.

d) Aggregation of a two component multiplicative micro-model: Vartia's solution

There is a third, more complicated method of constructing ideal indices by defining suitable *weights* (again within the framework of a multiplicative decomposition model). It is the method underlying the log-change index functions of Vartia. Part d of this section will deal with them.

Ideal index functions on the basis of log changes, like the two Vartia indices, cannot be derived as easily as Fisher's indices, P_{0t}^F and Q_{0t}^F. The approach is entirely different and does not start with three components, but only two. The problem then is to preserve a *two* component micro-model for a single commodity in aggregating over all n commodities. This is done *by finding appropriate weights* for aggregating individual log changes.

To see how Vartia's indices manage to fulfill the factor reversal test it should be realized that a simple summation of eq. 4 (**fig.3.6.4**)

(3.6.4a) $\Sigma \ln(v_{it}/v_{i0}) = \Sigma \ln(p_{it}/p_{i0}) + \Sigma \ln(q_{it}/q_{i0})$,

is not useful because neither $\Sigma \ln(v_{it}/v_{i0})$ will equal $\ln(V_{0t})$, nor will $\ln p* = \Sigma \ln(p_{it}/p_{i0})$ represent a reasonable price index, or $\ln Q* = \Sigma \ln(q_{it}/q_{i0})$ a reasonable quantity index.

Note that for example $\left(P_{0t}^*\right)^{1/n} = \sqrt[n]{\prod(p_{it}/p_{i0})}$ is simply an unweighted geometric mean of price relatives (Jevons' index). The solution can only consist in introducing weights g_i, ensuring also that is the factor reversal test is satisfied.

The key equations explaining why P_{0t}^{V1} and Q_{0t}^{V1}, or P_{0t}^{V2} and Q_{0t}^{V2} satisfy the factor reversal test and also relate Vartia-I- and Vartia-II weights are

(3.6.5) $\ln(v_{i0}) = \ln(w_{i0}) + \ln(V_0)$ and correspondingly $\ln(v_{it}) = \ln(w_{it}) + \ln(V_t)$

Figure 3.6.2: Fisher's way of deriving an ideal index

a) Multiplicative approach

$$\text{(3.6.3)} \qquad \frac{\mathbf{p}_t'\mathbf{q}_t}{\mathbf{p}_0'\mathbf{q}_0} = \left(\frac{\mathbf{p}_t'\mathbf{q}_0}{\mathbf{p}_0'\mathbf{q}_0}\right)\left(\frac{\mathbf{q}_t'\mathbf{p}_0}{\mathbf{q}_0'\mathbf{p}_0}\right)\left(\frac{\mathbf{p}_t'\mathbf{q}_t}{\mathbf{p}_t'\mathbf{q}_0}\frac{\mathbf{p}_0'\mathbf{q}_0}{\mathbf{p}_0'\mathbf{q}_t}\right) = (PC^*)(QC^*)(SC^*).$$

b) Fisher's solution

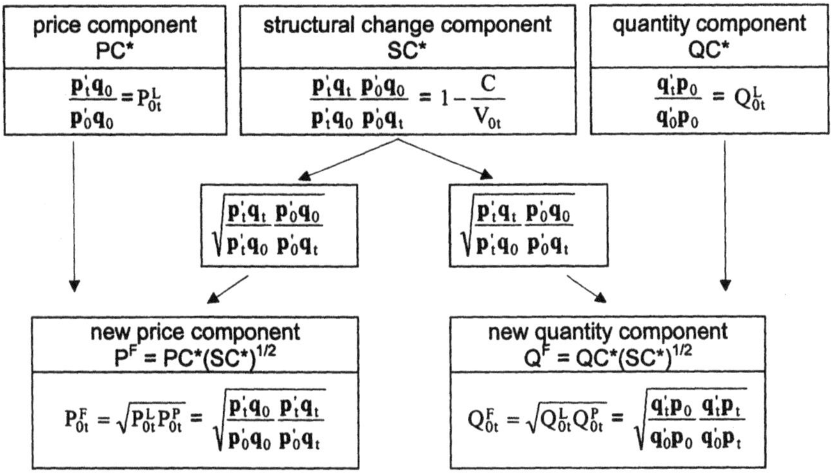

$$\text{(3.6.6)} \qquad \frac{L(v_{it}, v_{i0})}{L(V_t, V_0)} \ln(v_{it}/v_{i0}) = \frac{v_{it} - v_{i0}}{V_t - V_0} \ln(V_{0t})$$

$$= \frac{w_{it} - w_{i0}}{\ln w_{it} - \ln w_{i0}} \ln(v_{it}/v_{i0}) / \sum L(w_{it}, w_{i0})$$

$$= [(w_{it} - w_{i0}) + \ln(V_{0t}) L(w_{it}, w_{i0})] / \sum L(w_{it}, w_{i0}).$$

Summation over all n commodities has to take into account that $\sum(w_{it} - w_{i0}) = 0$ and $\sum(v_{it} - v_{i0}) = V_t - V_0$.

Figure 3.6.3: Vartia's way of deriving ideal log-change-indices

a) Micromodel in log changes

(3.6.4) $$\ln\frac{v_{it}}{v_{i0}} = \ln\frac{p_{it}}{p_{i0}} + \ln\frac{q_{it}}{q_{i0}} \text{ since by definition } \frac{v_{it}}{v_{i0}} = \frac{p_{it}}{p_{i0}}\frac{q_{it}}{q_{i0}} =$$

b) Aggregation problem

$$\Sigma\left(\ln\frac{v_{it}}{v_{i0}}\right) \neq \ln\left(V_{0t}\right) \text{ find weights g such that}$$

$$\Sigma_i\left(g_i \cdot \ln\frac{v_{it}}{v_{i0}}\right) = \Sigma_i\left(g_i \cdot \ln\frac{p_{it}}{p_{i0}}\right) + \Sigma_i\left(g_i \cdot \ln\frac{q_{it}}{q_{i0}}\right) \Rightarrow \text{two solutions (Vartia I and II)}$$

Figure 3.6.4: Weights in aggregating log-changes, Vartia's solution*

Single commodity (i) equation (definition):

(3.6.4) $\ln(v_{it}/v_{i0}) = \ln(p_{it}/p_{i0}) + \ln(q_{it}/q_{i0})$

Aggregation:

$$\Sigma g_i \ln(v_{it}/v_{i0}) = \Sigma g_i \ln(p_{it}/p_{i0}) + \Sigma g_i \ln(q_{it}/q_{i0})$$

Vartia I weights	Vartia II weights
$g_i = \frac{L(v_{i0}, v_{it})}{L(V_0, V_t)}$, $V_0 = \Sigma v_{i0}$, $V_t = \Sigma v_{it}$.	$g_i = \frac{L(w_{it}, w_{i0})}{\Sigma L(w_{it}, w_{i0})}$
$\Sigma g_i \ln(v_{it}/v_{i0}) = \ln(V_{0t})\Sigma\left(\frac{v_{it}-v_{i0}}{V_t - V_0}\right)$ $= \ln(V_{0t})$, since $\Sigma\left(\frac{v_{it}-v_{i0}}{V_t - V_0}\right) = 1$	$\Sigma g_i \ln(v_{it}/v_{i0})$ $= \frac{\Sigma\left(\frac{w_{it}-w_{i0}}{\ln w_{it} - \ln w_{i0}} \ln(v_{it}/v_{i0})\right)}{\Sigma L(w_{it}/w_{i0})} = \ln(V_{0t})$

* The problem is *not* to show that $\Sigma g_i \ln(p_{it}/p_{i0})$ is $\ln\left(P_{0t}^{V1}\right)$, or $\ln\left(P_{0t}^{V2}\right)$, because these indices are *defined* this way. The same is true for quantity indices. It only remains to prove that the weighted aggregation $\Sigma g_i \ln(v_{it}/v_{i0})$ results in $\ln(V_{0t})$, with value index V_{0t}.

e) Brief introduction into Theil's Best Linear Index (BLI)

It may be appropriate to introduce briefly Theils's index-pair in the two situation case (times 0, t). A pair of indices P* and Q* forming a matrix **D*** is defined by minimizing the sum of squared differences of the elements of the two matrices $\mathbf{D}=\begin{bmatrix}1 & Q^L_{0t}\\ P^L_{0t} & V_{0t}\end{bmatrix}$ and **D*** where V_{0t} in **D** is of course $V_{0t}=P^L_{0t}Q^L_{0t}(1+C)$. The following **fig. 3.6.5** (left part) summarizes once more this description of the approach.

Figure 3.6.5: Theils's -pair of "best linear indices"

two situation case (times 0, t)	**multi** situation case (times 0, 1, 2, ..., t)
Find vectors **p** and **q** such that the sum of squared differences between the corresponding elements of matrix **D** and **D*** will be as small as possible	
$\mathbf{D}=\begin{bmatrix}1 & Q^L_{0t}\\ P^L_{0t} & V_{0t}\end{bmatrix}$	$\mathbf{D}=\begin{bmatrix}1 & Q^L_{01} & \cdots & Q^L_{0t}\\ P^L_{01} & V_{01} & \cdots & B_{1t}\\ \cdots & \cdots & \cdots & \cdots\\ P^L_{0t} & A_{1t} & \cdots & V_{0t}\end{bmatrix}$, $A_{1t}=\frac{\sum p_t q_1}{\sum p_0 q_0}, B_{1t}=\frac{\sum p_1 q_t}{\sum p_0 q_0}$
$\mathbf{D}^*=\begin{bmatrix}1 & Q^*_{0t}\\ P^*_{0t} & P^*_{0t}Q^*_{0t}\end{bmatrix}=\mathbf{p}\cdot\mathbf{q}'$	$\mathbf{D}^*=\mathbf{pq}'=\begin{bmatrix}P_0Q_0 & P_0Q_1 & \cdots & P_0Q_t\\ P_1Q_0 & P_1Q_1 & \cdots & P_1Q_t\\ \vdots & \vdots & \ddots & \vdots\\ P_tQ_0 & P_tQ_1 & \cdots & P_tQ_t\end{bmatrix}$, $P_0 = Q_0 = 1$
where $\mathbf{p}=\begin{bmatrix}1\\ P^*_{0t}\end{bmatrix}$, $\mathbf{q}'=\begin{bmatrix}1 & Q^*_{0t}\end{bmatrix}$	$\mathbf{q}'=\begin{bmatrix}1 & Q_1 & \cdots & Q_t\end{bmatrix}$ and **p** correspondingly

The solution for P^* and Q^*, and all the more for $P_1, P_2, \ldots, P_t$ and $Q_1, Q_2, \ldots, Q_t$, is highly complicated

The exact expression of P* and Q* is complicated and of course a function of P^L, Q^L and the covariance C. If $C \approx 0$ (and therefore $P^P \approx P^L$), it can be seen that P* and Q* will in fact lie in between the corresponding interval of Paasche and Laspeyres indices. Theil's approach has also been applied to time series, the problem being the approximation of the matrix **D*** on the right part of fig. 3.6.5 by the product of a column vector **p** and row vector **q'**, or in other words: find the time series of index numbers $\mathbf{p}=\begin{bmatrix}P_0\\ P_1\\ \vdots\\ P_t\end{bmatrix}$ and the row vector **q'** (equivalently defined with quantity indices Q_t = Q_{0t}) such that the matrix **D*** = **pq'** is a least squares approximation to the empirical matrix **D** composed of the index numbers $A_{it}=\sum p_t q_i / \sum p_0 q_0$, and

$B_{it} = \sum p_i q_t / \sum p_0 q_0$ and the conventional Laspeyres indices P^L and Q^L. The vectors (index formulas P_0, P_1, ... Q_0, Q_1, ...) are derived by minimizing the sum of squared differences **D-D***.

Chapter 4

Price collection and index systems in official statistics

Index theory has often been criticized for its focus on the choice of the appropriate index formula whilst in the practice of price statistics many other problems have to be solved, the consequences of which are perhaps much more influential (with respect to the results, published by price statisticians). Therefore it appears appropriate to devote a whole chapter to these types of problems. The following sections will outline

- some general methodological discussions in price statistics, and
- practical problems related to the most important price indices (to be continued in chapter 6 for some indices).

It is just this kind of practical problems we unfortunately would not be aware of if we would almost exclusively look at the problem of price index fomulas.

4.1 The set up of a system of price quotations and price indices in official statistics

The following considerations are devoted to a number of problems that arise in the practical work of official statistics, which has to provide a system of price observations and price indices. The price quotations[1] and compilations of indices have to be timely, regular and systematic. They should consistently cover a wide field of market activities (sales/purchases) and serve a great variety of users and purposes. In some instances it will be more, in other less difficult to arrive at correct prices and weighting schemes for price indices, and quite often there is also a need for compromises, sometimes even for makeshift solutions in the light of conflicting principles in price statistics.[2]

a) The role of official statistics

The results of price statistics are of vital importance to many people, not only when costs of living are concerned, i.e. the general purchasing power of money, but also for example in the case of special indices to which decision making in business and administration will react.[3] Furthermore monetary policy will always be subject to discussions among politicians, economists and the general public in the sense that it

---

[1] They constitute the raw material for price indices.

[2] To give an example, instructions to price reporters have to be "tight" with strict and detailed definitions of the items to be priced in order to guarantee a pure price comparison. On the other hand, instructions should permit a certain degree of flexibility and also vague working definitions in order to facilitate the selection of the most representative item and to handle the disappearance of a product on the market properly.

[3] Due to such an *indexation* in some countries a rise or decline of a price index (in most cases the consumer price index) will have consequences almost automatically for other variables, like wages, rents and the like.

fails or succeeds in fighting inflation or in supporting growth and employment. As there is an ongoing debate on whether inflation is measured correctly or not, where "result-oriented" thinking is quite common, it is generally accepted that price statistics has to be carried out by a *neutral*, competent and *trustworthy* institution which, in the great majority of the countries, is the National Statistical Institute (NSI).

It is not only a tradition, but also a well established conviction that price statistics should be a matter reserved for official statistics because competing private inquiries might not gain that amount of general confidence which is indispensible in this field. On the other hand, the temptation is great to interfere with this business. "Result-orientation" means that - if ever somehow possible - inflation should be lower and real growth should be higher. There are certainly some issues which gain an undue amount of attraction and attention in this country simply because they come from the USA. Examples of such issues (both rather reconsiderations than new inventions) are probably chain indices and hedonic methods to make adjustments for quality change. Official statistics has to avoid the impression of applying questionable methods or of experimenting with concepts and formulas that are difficult to understand and sometimes advanced not without some political interest. All considerations concerning organizational and methodological innovations should therefore take into account factors like

- making methods transparent and understandable for the general public which usually (rightly) reacts extremely sensitively,
- the cost-benefit relation and feasibility aspects, and most important of all,
- the preservation of acceptance and neutrality of official price statistics and thereby also a fending off of result-oriented interventions.

b) The implementation of a system of surveys in price statistics

In general, empirical knowledge of prices is gained by observations (inquiries), as a basis for information on absolute prices, for index compilations and also for imputed prices. In the set up of a system of surveys (inquiries), decisions have to be made on

- what kind of prices should be collected (*scope* of price statistics)
- the *source* of information best to be used (e.g. sellers or buyers)
- the *periodicity* of regular surveys.

The **scope** of price statistics is certainly to observe prices and also to provide information useful for the estimation of unobservable prices. In general, observation of prices is feasible only when a sale takes place. There are also cases, however, like transfers in kind, which apply particularly to redistributive activities of governments where goods or services are provided to other units without receiving money in exchange, or where the money equivalent is less than the actual costs of production.[4]

[4] This applies to subsidies on products or also to the production, services of financial intermediaries paid partly with other revenues of banks and the like.

In all these cases (remuneration in kind, redistributive activities etc.) a transaction takes place although it is not a monetary transaction or "market transaction". Hence a price has to be estimated.[5] Another type of "non market" transaction differs from the type just described in that not only a price has to be estimated but also a transaction itself has to be imputed. An example is the production of goods and services retained *for own use*,[6] which is a typical non - maket production.

Valuation problems also occur with respect to stocks that are composed of goods produced or acquired in previous periods but perhaps no longer available. Because there are no transactions in such goods *now*, a correct pricing of items belonging to the "existing assets" (acquired earlier however) may require finding the price of a *comparable* product, resembling an existing item and (at least theoretically) being capable of replacing it.

Figure 4.1.1: Scope of price statistics and origin of data on prices

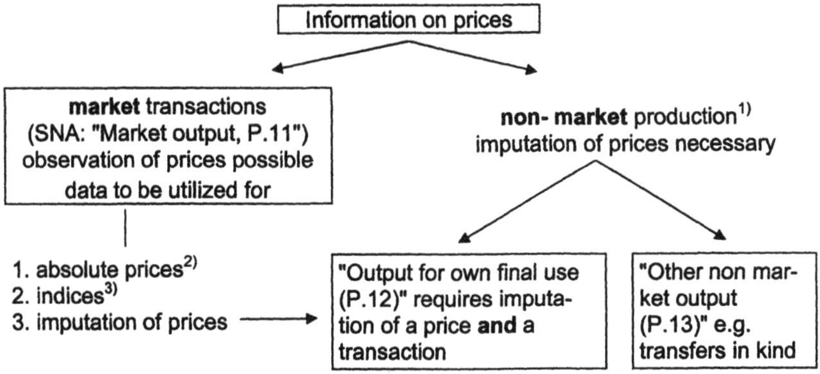

1) Goods and services not actually sold or bartered for other goods or services, but provided free as transfer in kind or produced for own use

2) for information on specific markets

3) among them also indices for purposes of deflation

Closely related to the scope of regular price quotations is also the question of **what type of prices** should be observed and quoted. This leads to problems like

- actual transaction prices (possibly with rebates) as opposed to list- or tender prices

[5] It would not be correct, however, to say that a transaction has to be "imputed", because it is the price, not the transaction, which is imputed.

[6] i.e. allocated by producers for their own consumption or capital formation. In such cases there is no transaction.

- point or interval in time to which the prices refer[7] (prices when a contract is made, the transaction effectively occurs, or when consumption takes place);
- prices excluding or including value added tax (VAT),[8]

and many more questions of this type.

As to the **source** of information on prices (**who should report prices?**) practical experience shows that the type of information collected by price statistics does not only comprise prices as such but also much of the so called price determining characteristics (PDCs as introduced in **sec. 1.3**). This requires a close cooperation between reporting firms and the NSI, which has to rely on expert knowledge of the firms as to the quality of products and the prevailing conditions on the market. It is therefore often better to collect prices from sellers rather than from buyers not only in the case of indices of selling prices, but also in the case of indices of purchase prices (see **tab. 4.1.1** for this distinction).

It is known from opinion polls that consumers tend to criticize the consumer price index (CPI) as having a downward bias, that is, they suspect the CPI to underrate the rise of prices. There are probably only rare occasions known in which buyers said that the official rise of selling prices seems lower to them than what they pretended to "feel". Some empirical investigations revealed what seems to be common sense, namely that rise and decline of prices are experienced emotionally quite differently:

- a rise will be remembered for some months and perhaps also exaggerated in retrospect,
- whilst stability or even decline of prices may not be realized at all.

It is sometimes claimed that in a democratic system consumer prices should be reported by the many buying households and not by the few selling enterprises. The idea is that there are more people involved on the demand side than on the supply side, that consumers' opinions should be given more weight since they are the majority and that it is *their* way to perceive price changes that should be reflected by an index. A CPI, in which consumers not only report prices but also decide on the prices which enter the formula, is expected to give a more realistic picture and to fit better to a democratic society.[9] The realistic point of view is, however, that there are very few cases in which buyers can give adequate and competent information on prices on a regular basis, taking into account also all of the price determining characteristics (PDCs, like quality for example) and the changes in the PDCs.

---

[7] It is common practice to choose a reference day within a month (for example the 10th, 15th or 20th or so). But when on such a selected day there are too few prices to be observed (as happens not infrequently for example in the case of exports, imports or construction work) price quotations might alternatively well refer to the average of a month calculated from observations on a number of days. The issue is known as the problem "point-in-time-" vs."period-prices".

[8] The first is common in the case of a producer price index (PPI) or in foreign trade, while the latter is more often practiced in the case of a consumer price index (CPI).

[9] The argument is also that this would be more consistent with the way weights are established. Normally weights in the CPI are derived from Family Budget Surveys or Family Expenditure Surveys (FES), that is, by asking consumers. So why not ask consumers as well when the selection of goods and the quotation of their prices is concerned?

With respect to rents of dwellings, for example, practical experiences have shown that it is more likely to get reliable information on prices from landlords than from tenants, especially if tenants have been living in their dwellings for a long time. Since rents usually rise when the rental dwellings change hands, collecting rents from tenants would probably create a downward bias. Another example in which it is not appropriate to collect reports from the demand side is a price index for purchases of farm supplies (designed to reflect the level of prices to which a farmer is confronted on the input side).[10]

An exception of the rule that in general most of the prices are better collected by surveys of *sales* rather than by surveying *buyers* are import prices. Very few countries can collate a true price index (instead of a unit value index only) for imported goods at all, and if they have such an index, which is one of the most difficult indexes to be computed, they are normally based on reports from domestic importing firms. The alternative, to organize reports from foreign country exporters, proved to be practically almost impossible.

Regarding the **periodicity** of regular surveys, in official price statistics we mostly have monthly reports. There are cases in which quarterly data will in general be sufficient, however, as for example in the case of rents for dwellings or construction work. On the other hand sometimes more frequent than monthly quotations are required. This applies to share prices on the stock exchange in particular if official price statistics deals with such prices (and indices on their basis) at all.

c) Some general remarks concerning the structure of a system of indices

A price index regularly published should have a well defined **coverage** indicated in general by the **name** of the index in question. The denomination of the index should denote

a) whether goods and their prices refer to the supply side or the demand side,[11]

b) to which of the two sides of the market the weights refer (in general it makes a great difference whether weights are derived from sales or from purchases),

c) the branch or sector to which the index refers and in addition, even more specifically, also the type of business involved.[12]

The third problem refers to the distinction between an institutional as opposed to a functional approach. A wholesaler (in terms of an "institution") for example will sell most of his products to a retailer (and thus act as a wholesaler in a functional sense), but perhaps to a certain degree he may also deliver to final users, which may be a

---

[10] We cannot expect to gain the information required for such an input price index from farmers' reports. There are very few goods and services farmers use to buy every month and if they do so it is highly questionable that their purchases are comparable over time.

[11] As shown above this is not equivalent to which side is asked for reporting prices within the framework of the price statistical survey.

[12] Moreover there are, or at least there used to be, many National Statistical Offices (NSOs) that distinguish between a *retail* price (or consumer price) and a *wholesale* price index, the latter referring to all prices, in trade as well as producer prices (and agricultural as well as industrial producers).

corporation or even a single private household. Hence the question arises: should the wholesale price index include or exclude the second type of turnover?

These are some minimum requirements concerning the name of an index which should at least to some extent indicate the "message" of the index. To avoid discussions it is also wise to publish as many details of the methods applied as possible. An index of which not much more than just its name is known can hardly be used and interpreted correctly. Some other characteristics of the index in question are therefore usually added to its name: base year, perhaps also index formula, periodicity etc.

In contrast to other fields in statistics it is not very common in price statistics to publish a **preliminary** index first and later a more complete and reliable **revised** index or even several revisions and corrections in sequence. This is common practice, however, in the National Accounts field as an example. In price statistics, revisions should be avoided mainly due to the publicity they (esp. the consumer price index CPI) experience. Thus a great effort is usually exerted to present timely and ultimate results and to avoid debates about possible corrections and interpretations.

d) The German system of index calculations in official statistics

As an example of a *system* of indices in which the elements are co-ordinated the German system will be set out in what follows. One of its most prominent features is its structure motivated by needs for providing deflators in National Accounts if at all possible. As **tab. 4.1.1** shows the system is incomplete, however, not covering indices of sales prices and of input (purchases) prices for *all* sectors (e.g. there are only a few indicators for prices in the service industry and for some sectors true input price indices are missing). Although there is no index for services in general, some of the activities of the service sector are covered nonetheless, by

- special indices for hotels and restaurants, sea-water-transport, postal services and telecommunications, or by
- the relevant sub-indices of the CPI, like housing (shelter), health services, renting of cars or "personal" services (like hairdressing) etc.

The main reason for such gaps is a lack of detailed turnover data for most of the service branches, and in default of such figures it is difficult to select items to be priced and to establish weights for indices. It is also true that measurement of pure price change and accounting for changes in quality tends to be more difficult in the case of services than of goods.[13]

Two other special features of the German system are worth being mentioning

1. the existence of a special index of wholesale selling prices, and
2. the compilation of both true price indices as well as unit value indices for export and import.

---

[13] Nonetheless, at any rate in the near future more special indices to measure price movement in service sectors will definitely come into being.

Table 4.1.1: System of price indices in German official statistics

Economic sector[1)]	Purchase prices (**Input**)	Selling (Producer) prices (**Output**)
Agriculture and forestry (NACE-Section A; excluded hunting)	Index of purchase prices of farm supplies	Producer price index of agricultural products, and several Producer price ind. of forestry production
Mining[2)], manufacturing, electricity[3)] (C, D, E)	Input price index for production industries[4)]	Index of producer prices for industrial products
Construction (Section: F)		Price indices for buildings
Wholesale and retail trade (NACE-Section: G)		Index of selling prices in wholesale trade and Retail price index
Hotels and Restaurants (NACE-Section: H)		Price index for hotels and restaurants
Transport, storage and communication (NACE-Section: I)		Indices of sea freight rates (NACE-class: 61.10) Price index for postal services (NACE-class: 64.11) Price index for telecommunication services (NACE-class: 64.20)
Other economic activities of the service sector[5)]		some subindices of the CPI only
Foreign Trade	Import price index[6)]	Export price index[6)]
Private Households	Consumer price index	

1) Kind of economic activity (Sections in the [EU-] NACE Classification).

2) and quarrying

3) plus gas, water supply

4) Price index for goods received (materials and supplies, energy etc.) by production industries.

5) Financial intermediation, real estate; renting and similar business activities etc. NACE-Sections: J+K+M+N, or parts thereof.

6) Unit value indices in addition.

There are only very few countries that have a true "**wholesale price index**" (WPI) defined by sales of wholesalers[14] (as opposed to producers), including commercial mark-ups (not included in producer price indices, PPI). As to their origin, goods covered by the German WPI may be domestic or imported, and as to their purpose, it is assumed that they are sold at the wholesale stage exclusively in Germany. The reason for calculating a special wholesale selling price index in Germany is not only tradition but its usefulness for deflation and forecasting. Experience has shown that

[14] In some countries a "wholesale price index" is compiled which is combining wholesale and producer prices, and not only prices for sales in the domestic market but also in foreign markets (i.e. export prices).

wholesale prices react promptly to changes in demand and supply and therefore they tend to serve well as leading indicators for consumer prices (collected in the main in retail outlets).

Table 4.1.2: Some details of selected German official price indices

Name of index	Price collection	Origin of weights
Price indices in agriculture a) Producer Prices b) Input Prices	a) ≈ 1,600 quotations from 300 reporting units for 240 products (monthly) and b) ≈ 6,100 quotations (mostly quarterly) from 650 reporting units	Structures of farmers' sales and purchases provided by the German Federal Ministry of Agriculture
Producer Price Index[1] (only domestic sales) PPI	≈ 13,600 quotations from 6,000 producers monthly. Prices include excise taxes and duties	Turnover and when broken down to smaller units of commodity classification production values[2]
Input Prices of Production Industries	No separate price collections[3] ≈ 9,000 domestic and import prices	weights derived from National Accounts estimates
Import and Export Price Indices	Monthly reports (mail questionnaire) from 2,500 importing and 3,000 exporting firms[4]. In all ≈ 15,000 price quotations.	Foreign Trade Statistics import/ export values for ≈ 10,000 goods in a breakdown by countries and commodities.
Consumer Price Indices CPIs[5]	≈ 400,000 quotations monthly, price collection mostly decentralized[6], plus ≈ 25,000 rents[7].	Various types of Family Budget (or Family Expenditure) Surveys[8], basket of ≈ 700 items

1) Mining and manufacturing industries

2) Corrected for exports

3) Prices of PPI, Wholesale- and Import Price Index

4) Some prices taken from special publications.

5) Until the index system 2000 = 100 different CPIs for different types of households on the basis of so called "Extended Budget Surveys" (sample for *all* types of households in 5 years intervals) and "Continuous Budget Surveys" (a regular survey for 3 specific types of households each month). Meanwhile the coexistence of different types of budget surveys has been given up in Germany. Nowadays there is only one single CPI in Germany (beginning with the index system 2000 = 100).

6) Regional Statistical Offices responsible.

7) Price quotations collected from landlords and tenants.

8) FBS or FES for short

Computing "genuine" **price indices** in addition to unit value indices for **foreign trade** is done primarily on the following grounds

- unit value indices display more (or too much) volatility and they are said to be[15] less suitable for deflation than true price indices;
- price indices are - in contrast to unit value indices - true Laspeyres indices, reflecting price movement, *not* affected by changes in quantities and qualities of goods, changes in the mix of countries (supplier countries/countries of origin) or changes in the terms of payment and delivery etc., and
- they refer to prices at an early stage (when a contract is made) and to narrowly defined commodity groups, whilst unit value indices as a by-product of trade statistics refer to prices at the time of crossing the border and to aggregates of commodities defined much wider and therefore less homogeneous.

To have price indices in addition to unit value indices is found useful especially in a country like Germany, where foreign trade is not only important but (more important) also concentrated on finished goods differing substantially in their composition from month to month. It is therefore desirable to isolate the price component and to make comparisons over time in a consistent manner.

4.2 Quality adjustment in price statistics

This section deals with quality adjustment, arguably one the most contentious problems of price statistics. From a more conservative doctrine of making pure comparisons (the Laspeyres index philosophy) adjustments have to be made solely for the sake of comparability.[16] In what follows we discuss the need for and scope of adjustments (part a), the *"identification"* problem (part b), and in (part c through e) some more traditional methods to *evaluate the amount* of quality change. Part f covers the "hedonic regression" method and concludes.

a) Need for quality adjustment

The objective of price statistics is to measure *pure* price change for a fixed set of commodities, however, in reality this set will usually not remain unchanged for a longer period in time but inevitably always undergo some changes. *New products* can be seen emerging, sometimes being offered for a certain time simultaneously together with "old" ones they eventually replace. Also some gradual improvement (or worsening) in the quality of existing products is constantly taking place.

---

[15] this is contentious however; see **sec. 6.4** for more details.

[16] This traditional view is now challenged by a doctrine based on the economic index theory (COLI-concept, see **sec. 2.4**), according to which adjustments should not only make goods and services comparable but should also take into account the increased utility of new consumption conditions (such as greater variety of products, more convenient payment etc.), advocating in short a much wider scope and purpose of "quality" adjustments. A *new-goods-bias* is often said to occur when 1. new goods are not brought into the sample with sufficient speed (such that the typical decline in the price of a good at the beginning of its life cycle is not captured), and 2. by not accounting for the welfare gain (additional utility) a new good may provide. The latter aspect will be viewed as "bias" only when one is striving at the COLI-concept as the target concept of a CPI.

The appearance of new products and disappearance of old products shows that the "quality problem" and the "sampling problem" are related: in order to maintain representativity and continue comparable price observations time and again, replacements and re-samplings of items have to be made.

As shown in **fig. 4.2.1** a distinction (though in reality somewhat vague) might be useful between the following three situations:

1. the change is of *minor* importance with little or no relevance for the ability of the good (or a component of it) to perform its function and also for the utility perception of the consumer, i.e. the commodity continues to be the same product, perhaps more or less *improved* or *worsened*, however,
2. the change is *substantial* and we may call the result in a *new variety* of the product, or as the limiting case, a completely *new good* that gains importance at the expense of the old one (in practice it will often be difficult to distinguish "new" products from mere quality "changes").

The appearance of new goods on the marketplace (not on sale in the base period) and the disappearance of other goods may be considered as an extreme case of quality variation and the notion of "change" depends of course on how many aspects and what kind of aspects are called relevant with respect to the quality in question.

Furthermore, there will always be some substitutions *in reaction* to those changes in quality and to the emergence and disappearance of products, and such discontinuities in supply as well as demand will usually occur simultaneously with the observed changes of prices so that it is sometimes very difficult to distinguish between

- "pure price changes" that can and should *not* be attributed to changes in the price determining characteristics (PDCs), such as changes in quality, and
- rises or declines of prices that result from changing PDCs.

The principle of pure price comparison requires that only the first change be displayed, while neutralising the second type of a change of prices. Thus it may not suffice to compare only prices, just as they are observed.[17] It is rather necessary to *estimate* (and deduct or add) that part of the observed change of prices that can rightly be attributed to changes in the PDCs.[18] Hence:

Since the price determining characteristics (PDCs) are not *in fact* the same at two points in time (or at two places) provisions have to be made to adjust prices so that they are comparable, i.e. the task is to estimate that very level of prices we *would have* if the PDCs had in fact remained unchanged.

It is important to note that price statistics cannot simply report the observed prices but has to make adjustments if necessary for the sake of comparability, and such adjustments inevitably are somewhat hypothetical.

[17] The same considerations, of course, also apply to interspatial comparisons of prices.

[18] The emphasis is on PDCs. Quality adjustment is only one example, perhaps the most prominent one, of adjustments that *have* to be made for the sake of pure price comparison.

Figure 4.2.1: Types of quality change (of good A) and adjustment methods

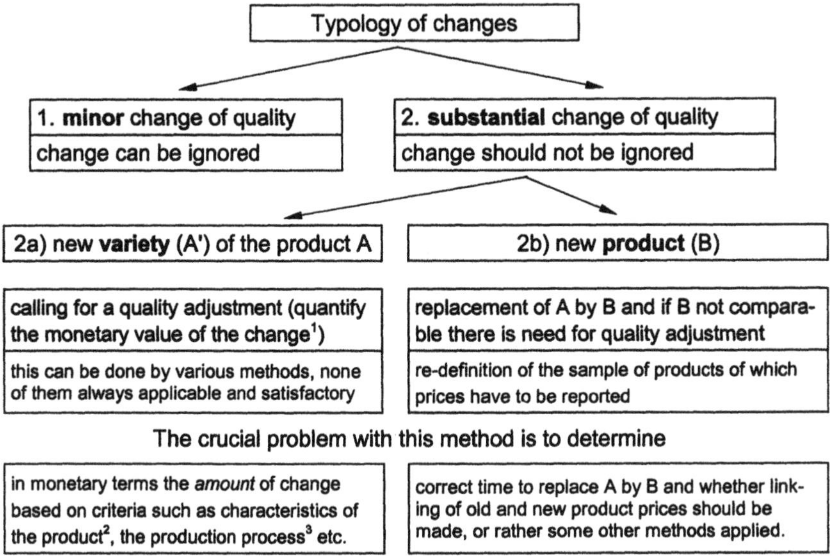

1) quality improved or diminished
2) the hedonic approach rests on physical and other characteristics of the products A' as compared with A
3) an occasionally used method is to take production costs as a proxy of the true change in quality

The task is twofold and requires in detail:

1. **identifying** the physical (or other) changes in the items (commodities) as sufficient to be called a quality improvement or deterioration, that is, to decide which change is to regard as quality changes and which not,
2. **evaluating** changes (or more precisely, to value them) *in monetary terms*,[19] so that an adjustment can be made by lowering (in the case of improvement) or raising (in the case of deterioration) the reported price appropriately.

The *identification* problem is to determine the threshold where a change in (physical) characteristics is no longer negligible but needs to be accounted for. The evaluation-problem consists in estimating the amount of change that has to be "accounted for". The task is to find a margin in accordance with the expected impact on consumer utility (in the case of consumer goods) or producer productivity (for investment goods like capital equipment) by which the price has to be "adjusted".

To sum up: why does an adjustment have to be made? Price indices should measure the pure price change. Consider a price, p_{it} of commodity i differs from p_{i0} such

[19] It is not sufficient to make *ordinal* statements of quality. We rather need an "amount" of quality measurable in terms of currency units so that it can be compared with a price.

that $p_{it} > p_{i0}$, which is solely due to a better quality of i, as appreciated by consumers. In this situation it clearly would not be correct to simply observe the difference in prices $\Delta p_i = p_{it} - p_{i0}$ without taking into account that this difference partly (or maybe even exclusively) reflects an improvement in quality. An increase (decrease) of quality *not* taken into account would make the rise of prices appear higher[20] (lower) than it really is since an increase of quality will usually justify a higher price.

b) For what kind of changes should adjustments be made? (Identification problem)

The problem of "quality adjustment" is also to identify correctly the situation calling for quality adjustment. It should be noted that "quality" change is usually understood in a very wide sense: "adjustments" should not only refer to changes in the (physical) appearance or certain attributes of the commodity (as a rule "quality" should be defined with regard to the use-value the buyer enjoys, and not (or not primarily) with reference to "experts", or producers; it is *incomparability* of *all* kind that should be taken into account.[21] "Quality" therefore should be understood in a very *broad* sense, inclusive also of factors not tied to discrete goods and services such as

- the *variety of supply*, i.e. the availability of many rather than only a few models or makes, among which a choice can be made, the *time and location of sale*,[22] the type of shop in which the sale takes place, the existence or absence of qualified sales personnel, of guarantees or free after-sales services, the convenience as well as the "environment", as for example in the case of a restaurant, and many other aspects of supply and demand of relevance to determine the price (i.e. that might justify different prices);
- adjustments also have to be made for *restrictions of competition*, as this has to be considered as a factor that seriously affects the comparability of prices by *reducing the possibilities of free choice*;[23] prices may be paid that exceed what consumers are *willing* to pay if they had alternatives and a free choice (on the

---

[20] and a production index (quantity index) rise slower.

[21] This is already true for the above mentioned traditional point of view. The economic theory thinking in this field now goes a step further. The problem is not only to ensure comparability but also to estimate a utility-difference to be attributed to completely new products such as text editing on a PC to a traditional type writer with no electronic facilities at all.

[22] Availability at peak times for example, or at a more convenient location (or where there is a greater demand) should clearly be seen as an improved quality which justifies as higher price.

[23] To give an example: it now appears that international comparisons of prices including the former socialist countries and utilising also official price statistics of these countries as often made in the 1970's and 1980's has probably been of very limited use only. It was often said that certain commodities had not been available at the official price (as reported in official statistics) and on the official market in those countries. Quantities supplied had been limited and reduced by rationing and queuing, and a second (unofficial) market came into existence, ironically allocating a great deal of just those products that where urgently needed.

other hand in some countries at least, *skill in bargaining* resulting in a lower price may also be a factor to be accounted for).[24]

In all these situations (limited information, market power, different degrees of competition and limited supply due to rationing) the observed prices will not correctly indicate a pure price movement and reflect the utility for the purchaser and the costs or scarcity on the supply side, and it will therefore be difficult to provide a "pure" price comparison unless some adjustments are made. Obviously it is sometimes extremely difficult, however, to distinguish genuine price discrimination from situations in which different prices do indeed reflect differences in quality.

We now turn to the notion of "quality". A vast amount of business administration literature is devoted to exactly this problem to define "quality". It shows that absolute standards are rare, and we often cannot assess quality by characteristics of the outward appearance,[25] let alone that quite a few aspects among these characteristics are difficult to measure, such as style, reliability, ease-of-use, fashionable look etc.[26] We mostly have to rely on relative standards (whoever is setting them), relating in some way or other the use value of a good to its price.

A concept of quality which certainly provides the theoretical basis for *all* sorts of quality adjustment methods,[27] not only the hedonic method, is based on the (hypothetical) effect on utility of a consumer, evaluated in some sort of a thought experiment. According to this notion what matters is consumer preference. If the consumer feels indifferent between purchasing an item A at $p_A = 100$ and B at $p_B = 120$ we can reasonably conclude that the quality of item B exceeds the quality of A by 20%.

To sum up, in principle *all* differences in prices that are attributable to factors of quality in a *broad* sense should be "adjusted for" somehow for the sake of making pure comparisons. As stated above, this applies even to things not connected with discrete goods, like market conditions, increase or reduction of the possibilities of free choice etc.

It should be noted, however, that there are clearly limits for adjustments to be reasonable and that the correct idea of quality adjustment can easily be misused in a way that eventually all or most of the observed price increase can be denied or "adjusted away". Purism or perfectionism in this respect is *not* advisable on the following grounds:

---

[24] in the sense that prices that skilful people are paying are not comparable with those of less skilful people, and that the potential a purchase is offering for bargaining might be regarded as part of the use value of the product in question.

[25] The existence of measurable characteristics is obviously a necessary condition for the quality concept underlying the hedonic method.

[26] As regards outlets rather than goods, factors such as variety, accessibility, better advice and service etc. come into play. From a statistical viewpoint, a drawback of such aspects is that it is difficult to define them in "operational" terms.

[27] It should be noted, however, that this COLI-concept (**sec. 2.4**) of quality, though possibly appealing from a theoretical point of view, may give little guidance in practice and also give rise to the aforementioned questionable widening of the scope of quality adjustment.

1. Situations are frequent in which not the need for adjustments as such but their exact amount will be an issue of never-ending controversies.[28]
2. With a sufficiently broad understanding of what "change" has to be accounted for in mind, it is easy to make all increases of the consumer price level disappear statistically.[29]

c) Typology of quality changes and adjustment methods

Methods should be distinguished on the following lines: whether a continuation of the price quotation is possible or rather a discontinuation (replacement, re-sampling) called for, in which case we may or may not be able to find an item which is comparable with the replaced one. In the case of incomparability we may find an indication of the amount of quality change in the prices (i.e. "implicitly") or "explicitly", i.e. with reference to other variables (see **fig. 4.2.2.** for an overview).

c1) New or only qualitatively modified goods and services

As demonstrated in fig. 4.2.1 a distinction can be made (at least in theory) between minor (case 1) and significant changes (case 2), the latter in the shape of new products (case 2b) or modified existing ones (case 2a) calling for either a re-selection of the sample of products or for "adjustments" of the reported prices in order to account for quality change (or more fundamentally: for incomparability) in existing commodities. In case 1 it may be legitimate to *ignore* the change and to continue with reporting prices as they are observed (a procedure which is the so called **"direct comparison[30]"** or "unadjusted price comparison"). This is equivalent to deciding that no change has taken place, and the difference in the price is entirely "genuine", not justified by a quality change.[31]

Another possibility is that the commodity selection can remain unchanged but modifications of goods are such that quality change can no longer be ignored (case 2a in fig. 4.2.1) and the market value of the quality change has to be estimated (see part c3 of this section).

---

[28] This applies in particular in the case of medical progress where for example such difficult characteristics as an increased longevity or better *quality* of life is an issue.

[29] It is simple to find some reasons to "justify" any rise of prices by some kind of allegedly better "quality" in the widest sense. This has been common practice in socialist countries, especially in the former German Democratic Republic (GDR), in which, at least officially, hardly any inflation was experienced at all. The reason simply was that statisticians regarded (or were forced to do so) many minor changes as qualitative improvements which were considered to more or less exactly offset the rise of the price.

[30] Note that this term is highly misleading as "direct" is also used in the sense of evaluating the quality difference by some objective method independent of the price difference.

[31] It is sometimes argued that this procedure has to put up with a growing bias due to the fact that "creeping" quality changes "sum up" and "accumulate". With regard to a frequent rebasing of indices and the possibility that quality changes of different sign may cancel themselves out this objection may be not so convincing, however.

If, on the other hand, a completely new good emerges or the good in question simply is no longer available (case 2b in fig. 4.2.1) action is called for concerning the selection of goods. In this case a distinction should be made between addition and replacement. As will be shown in **sec. 4.3** the existence of genuine "new" goods (providing an additional "utility") represents an "*addition*" giving rise to a revision of weights *as opposed to a replacement* where a new good only takes the place (and hence also the weight) of the old one.[32]

c2) Options with respect to "replacements": The matched models(MM) method vs. continually re-sampling and chaining

Some short remarks concerning the "matched models method" (**MM method**), which is the traditional approach of official statistics in dealing with this situation, will be appropriate here. In accordance with the "pure-price-comparison", the strict MM-method requires comparable items *all along* from 0 to t. This should be kept distinct from a method called "[monthly] re-sampling and chaining" (**RC method**), in which only matched models of any two *adjacent* periods are compared (i.e. a monthly updating of the sample is done and the results are linked together).

Figure 4.2.2: Relationship between sampling and quality adjustment

a) the distinction between the matched model method (MM - method) and re-sampling and chaining (RC - method) deals with the appropriate strategy when replacements are made

b) the ideal matched model has been found such that quality can be directly compared

c) if the not perfectly comparable item B is replaces item A, an adjustment has to be made to its price (p_B to be comparable to p_A).

The method of matched models (MM method) consists of

[32] As replacement has to do with re-sampling, the issue of replacement will be dealt with in more detail in **sec. 4.3**, devoted to sampling.

- preferably refraining from a comparison between A (disappearing) and B (or A', the new good or variety) as they are incomparable; thus, the method includes only items that exist in both periods, disregarding all non-matching items; the obvious consequence of this strategy[33] is of course a progressive attenuation (attrition) of the sample as time goes by due to the ever increasing number of disappearing and new goods, such that eventually not much is left of (the only eligible for comparison) ongoing goods, or
- if a replacement is unavoidable for the sake of "pure comparison" the most *similar*[34] (to the replaced one) item should be chosen.

Note that if we carry out a replacement, as done on a regular basis in the RC method of (monthly) re-sampling and chaining, there is again the well-known conflict between "pure comparison" and "representativeness", in that in this case a decision has to be made whether the criterion for a re-selection should primarily be to select

- the most *similar*, most *comparable* item [aiming at matched items for the sake of "pure comparison"],[35] or
- the presently most *representative*, though unmatched one [for the sake of "representativeness"] (the problem with this method is that there remains an unknown price movement of the unmatched items).

The paramount principle is to compare like with like and to select replacement items similar enough to be placed in the same generic group.[36] Only if this is not possible do adjustments have to be made,[37] thus the MM method aims to avoid quality adjustments and to keep to the "basket" constant if ever possible.

The MM method has recently become the object of criticism for a number of reasons:

1. It is most likely[38] that old models and makes are kept for a too long time in the sample which will *render the sample progressively unrepresentative* of the true price development. Hence, the MM method will give rise to a sampling bias which is perhaps its most serious shortcoming.[39] There is no incentive for price collectors to timely introduce new goods because this may trigger a complicated

---

33 MM-method in the strict sense.

34 To make sure of this, a specification of goods should be detailed enough in order to continually identify the same or at least a most similar product in the course of time.

35 An obsolete (therefore disappearing) good may be replaced by a similar and therefore likewise obsolete, unrepresentative good.

36 Note that this is not guaranteed if replacement is governed by utility reasoning.

37 The notion of "quality" as above expounded implies that this adjustment ought to consider all relevant aspects (like e.g. location), and not only the physical characteristics of the goods in question.

38 also in view of the rationale of the MM-method as well as possibly the instructions given to price collectors.

39 It is not impossible that the prices of the matched pairs (representing an old technology) will show hardly any change while at the same time in the universe of items (containing also all unmatched items), to which for example hedonic regression refers, prices are undoubtedly falling. As there are constant changes in this universe, the number of unmatched items will grow unless an update is made by "re-sampling".

quality adjustment problem. Therefore the MM-method is likely to eventually end in a *forced replacement*.[40] It also fails in the case of prices for houses because each house is strictly speaking a unique good.

2. As new products are usually prevented from inclusion in the sample until the next full "index revision" there may be two flaws involved in the procedure:
   - once the prices of new and old models develop in a different manner, the MM method over time is said to result in a bias of the price index,[41] and
   - from the COLI point of view the belated inclusion of new and improved goods with an advanced technology implies that the gain in utility such goods may provide will not be timely accounted for,[42]

   or in short: the two shortcomings are *the price effect and the welfare increase of new goods*, and the MM-method relies on the assumption that the price change of the (included) matching items is representative for the price change of the (excluded) non-matching items. Moreover, the method also decisively relies on
3. the price collectors' ability to select *comparable* (in terms of a given specification) items, and in case they fail to correctly assess the quality difference in a decentralised fashion (i.e. on the spot of the outlet).

In essence the assumption is made that price collectors in their field-work are capable of making sure that like is compared with like.[43] The results then will be incorporated in the index series using a number of more or less subjective "implicit" methods.[44]

c3) Methods of decomposing the observed price difference $\Delta p = p_B - p_A = \Delta k + \Delta r$

When A is substituted (replaced) by the not fully comparable item B (or A by A') and an adjustment for quality change has to be made, we have in principle a choice between the three options of an (implicit) method according to how the observed price difference Δp is to be divided into a part representing quality change (Δk) and a residual part (Δr) reflecting a pure change (see **fig. 4.2.3**), viz.

---

[40] As this happens at times when the supply-demand situation is abnormal or without an overlap period of some length, a belated replacement will also aggravate problems with quality adjustment.

[41] Another suggestion to an early incorporation of new goods is based on the U-shape of the price movement. It is maintained that the CPI should capture the falling first half of the U.

[42] As already mentioned, this timely inclusion of new products is considered requisite within the framework of the COLI concept, it does not, however, apply to a CPI following a non-COLI concept.

[43] The MM-Method thereby introduces an element of subjectivity in assessing the equivalence of quality. The problem also is that in the case that a suitable matching model cannot be found, the comparability of the replacing model has to be assessed in a *decentralized* fashion by price collectors on the spot. The more or less subjective results then will be incorporated in the index using implicit methods such as linking, overlap etc.

[44] Such as linking or overlap etc.(see **fig. 4.2.3**). Dissatisfaction with this situation greatly contributed to the increased attention hedonics were able to attract recently (due also to the famous Boskin Report in 1996 which triggered a new enthusiasm for this method world-wide).

1. the **direct** comparison ignoring a quality change whatsoever ($\Delta k = 0$ and therefore the observed price difference Δp is considered entirely as "pure"), or
2. according to **linking** (or "link to show no [price] change") with or without an overlap,[45] we have the opposite extreme: Δp can be fully accounted for as a quality change (improvement if $\Delta p > 0$, because an improvement commands a higher price, and deterioration if $\Delta p < 0$) such that no true price change has taken place ($\Delta r = 0$); however
3. in practice the most frequent situation is an **intermediate** case (number 3 in fig. 4.2.3) in which the difference in prices is partly attributable to a quality change Δk (and thus justified) and partly residual (Δr) or "true".

Before embarking on discussing such methods it may be pertinent to make a note to the usage of certain terms, as unfortunately, the fact is that terminology differs greatly between statisticians which unduly complicates the surveying of this subject:

- "direct" is not only used to express the possibility (or assumption)[46] of comparing "directly", without adjustment, but rather in the sense of disposing of (objective) factors describing the amount of quality difference (somewhat in the sense of an explicit method) and there are also terms direct/indirect within the scope of the hedonic method,[47]
- a distinction is sometimes made between "linking" and "splicing" depending on whether an overlap exists (p_A and p_B can be observed simultaneously) or not,[48] consequently the term "to show no (price) change"[49] is used in combination with linking or splicing,
- also the term "overlap" is used for a variety of methods, such as "bridged overlap"(which may be viewed as a variant of the imputation methods [**1.B** in **fig. 4.2.4**], or "simple overlap" (i.e. the new good B simply replaces A without linking the price quotations).

In order to decompose the price change (Δp) into a quality change (Δk) and pure change (Δr)

- recourse can be made to external information[50] (**explicit** methods), or

---

45 If there is no overlap some authors prefer the label "splicing" instead of "linking".

46 The assumption is reasonable if consumers can make such a direct comparison, that is, in a situation in which both goods are considered more or less equivalent, they are available simultaneously without limitations and when the difference in the prices is not reflective of differences in utility. The number of goods directly comparable to those of the base period diminishing of course with the passage of time.

47 Thus "direct" and "indirect" is also sometimes distinguished much in the same way as "explicit" and "implicit".

48 splicing = no overlap.

49 An often (in the USA) used term is "link-to-show-no-price-change". The term "no change" is meant in the sense that the price level would be unchanged if a comparison could be made (the whole of the price difference is contributed to the instance of not being comparable).

50 external information = information other than prices.

- the quality effect[51] can be estimated without using additional data, or in other words, the estimate is based on the observed prices only (**implicit** methods).[52]

Figure 4.2.3: Components of the price difference and adjustment methods

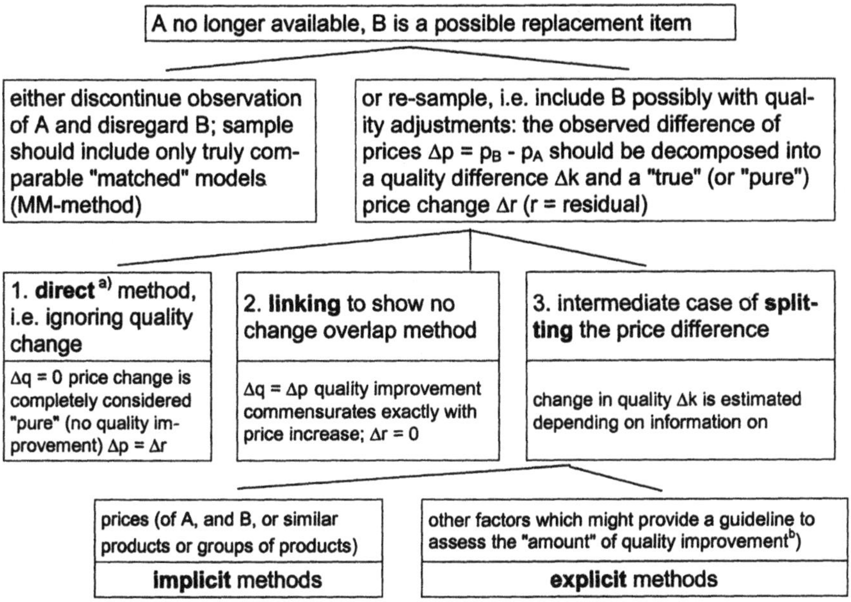

a) better "unadjusted"
b) e.g. hedonic regression oriented at "characteristics" of the good in question

As to the ***implicit methods***[53] (group 1 in **fig. 4.2.4**) it appears plausible to distinguish between methods which compare prices of A and B only, that is, group 1.A, and those methods where the price change is inferred from price change between related goods. This group of methods (1.B) is called imputation methods because a price is imputed assuming a price change between other goods to be similar to the unobservable price change between the goods in question. Depending on how to determine the "other" goods, the following variants of the imputation method are in use:

---

[51] or quality *difference* Δk.

[52] Note that the so called "link method" also ignores any quality change whatsoever (i.e. the "direct method"), that is, the two extreme solutions in **fig. 4.2.3** can also be regarded as special cases of the implicit method.

[53] The list given in **fig. 4.2.4** is only tentative. In the literature we also found methods called "implicit", for example: "monthly chaining and re-sampling" (or "deletion and monthly replenishment"), various types of "overlap" and "link" methods such as "bridged overlap" or "simple overlap" and "link to show no change" or the "carry-forward" method (simply repeat the last quoted price) or the "comparable replacement" (i.e. finding an item which is so similar that a "direct" comparison can be made).

- *overall mean* -, and targeted imputation (comparing goods of the same classification stratum, possibly with [if "targeted"] weights to differentiate between more or less comparable goods) on the one hand and
- *class mean* imputation (based on the price difference of those goods which are judged comparable or are already quality adjusted) on the other.

Figure 4.2.4: System of adjustment methods

A = "old" item, B = (new) replacement item

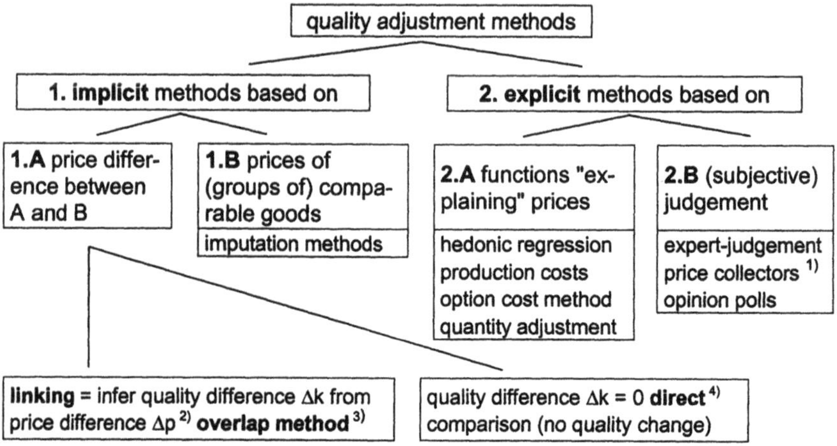

1) decentralized judgmental adjustment of price collectors (assuming that valuation correctly reflects product features as seen by consumers) in contrast to centralised expert judgements.
2) overlap in t-1: $p_{B,t-1}/p_{A,t-1}$ allows to adjust p_{Bt}, $p_{B,t+1}$, ...
3) The usually applied extreme variant of this is *link to show no price difference* ($\Delta p = \Delta k$) or *automatic* linking banned as an unacceptable method (as possibly creating systematic underestimation of inflation) in the framework of the European HICP (see **sec. 6.1**).
4) because prices are considered (or assumed to be) *directly* comparable (without adjustment).

All imputation methods are based on the assumption that the missing item and all the other items of the same group[54] have undergone the same change of prices and that the new (replacement) good B is "improved" to the extent to which the price of B (compared with the old good A) is rising higher than the average of the group of comparable goods. Likewise in all overlap (or linking) methods any difference in price level is assumed to be a measure of the quality difference. However, in practice there are many reasons to render the underlying assumptions (identical price movement unless there is a quality difference, or prices differ only to the extent to which qualities differ) invalid, as for example "strategic" setting of prices (e.g. discounting of old

[54] The definition of the "group" can of course account for different aspects, e.g. a similar origin, or use of the good, the same outlet or a comparable position in the life cycle, and it is obviously this decision on the definition of the group that decisively influences the result.

models to clear inventories or "skimming" of market segments when new goods are first marketed).

The main difference between the linking method and all variants of the imputation method is that the former relies on the comparison between two items only (the old one, A, and the replacement item, B) whereas the latter is based on the comparison with an *average of several* items belonging to the same group.[55]

Hence the basis of the imputed rate of price change is smaller or greater but in any case grounded on observed prices of comparable goods rather than on variables (or characteristics) describing the nature and extent of quality change. And this is in fact an assumption which is not always tenable.

Obviously with the combination of

1a. cheaper and at the same time better products, or those which are

1b. more expensive and worse, or finally where

2. quality has changed but the price remained constant

we will run into difficulties with all those methods in which the amount of quality change is inferred from the price difference.

c4) Explicit methods

These methods (group 2 in **fig. 4.2.4**) may be distinguished in such methods where a mathematical function is established from which the exact quality adjusted price follows (group 2.A) and those methods in which no such explicit function exists (group 2.B) as for example in the case of a more or less subjective expert judgement (or if consumers are asked to give their judgement). In a sense all methods of group 2.A can be regarded as special applications of the hedonic approach.

If, for example, the usual "characteristics" of this approach consist in providing an additional feature or simply more of certain ingredients or so, it is straightforward to make an estimate on the basis of the costs of the feature (production cost,[56] or option cost method) or the quantity in question (quantity adjustment method).

Before going into details of the various methods it might be useful to consider some formulas in a more elaborate manner by recourse to a numerical example in the following digression.

We will add some comments on explicit methods thereafter in part e of this section.

[55] The definition of the "group" can of course account for different aspects, e.g. a similar origin, or use of the good, the same outlet or a comparable position in the life cycle, and it is obviously this decision on the definition of the group that decisively influences the result.

[56] to be kept distinct from "user cost", sometimes explicitly estimated as for example in the case of durable goods or owner occupied housing (see **sec. 6.2**).

d) Digression: numerical example for the implicit methods: quality adjustment by a proportional change of the observed price

1) Assumptions, symbols, *and the general procedure with a numerical example*

	actual price of commodity		fictitious[1)] price of
period in time	A (old)	B (new)	B (comparable with A)
0	$p_0^A = 5$	×	$\hat{p}_0^B = p_0^A / F$
...	...	×	...
t-1	$p_{t-1}^A = 10$	×	$\hat{p}_{t-1}^B = p_{t-1}^A / F$ [2)]
t (overlap period)	$p_t^A = 12$	$p_t^B = 15$	$\hat{p}_t^B = p_t^B \cdot F$
t+1	×	$p_{t+1}^B = 18$	$\hat{p}_{t+1}^B = p_{t+1}^B \cdot F$
...	×	...	
t+k	×	$p_{t+k}^B = 23$	$\hat{p}_{t+k}^B = p_{t+k}^B \cdot F$

1) quality adjusted price (as opposed to the observed price) of B so that B is comparable with A
2) There is no point in calculating this price, however, because in period t-1 no commodity B did exist
× means "not available" or "not applicable"

The general principle goes as follows: multiply (divide) the price by the factor F

(4.2.1) $\hat{p}_{t+k}^B = p_{t+k}^B \cdot F$ (k > 0 after overlap) and $\hat{p}_{t+k}^B = p_{t+k}^A / F$ (k < 0 before overlap)

in order to enable a continuation of the series of item A though no longer available (k > 0) or to construct a quality adjusted price of B in the past (k < 0). F can take different values depending on which of the following methods[57] is used (there two extreme methods, α and β, and one intermediate solution, γ)[58]:

α) overlap method	β) direct method	γ) the "50%-rule"
$F = \frac{p_t^A}{p_t^B} = \frac{12}{15} = 0.8$	$F = \frac{p_t^B}{p_t^B} = 1$	$F = \frac{\frac{1}{2}(p_t^A + p_t^B)}{p_t^B} = 0.9$

In what follows we explain how we can easily arrive at these simple solutions for F and thus can simply treat all three cases as special cases of the same method (correction of the price of B using the factor F, that is to transform $p_{t+k}^B \to \hat{p}_{t+k}^B$.

[57] We will demonstrate only three cases (out of infinitely many).

[58] The 50% rule is only one example for such an intermediate procedure (it is - or was at least - widely in use in Germany).

2) Overlap method *(linking method, price increase equals quality improvement)*[59]

The method amounts to assuming $\hat{p}_t^B = p_t^A$, which means although the observed price of B may well exceed the price of A (is indeed the case as in our example p_t^B = 15 > p_t^A = 12), when the quality change (improvement in the case $p_t^B > p_t^A \Rightarrow$ F < 1 or deterioration if $p_t^B < p_t^A$ and thus F > 1) is taken into account, the price should be unchanged. Hence due to $\hat{p}_t^B = p_t^B \cdot F = p_t^A$ the factor F is given by

(4.2.2) $\hat{p}_t^B = p_t^B \cdot F = p_t^B \dfrac{p_t^A}{p_t^B} = p_t^A \;\Rightarrow\; F = \dfrac{p_t^A}{p_t^B}$, or

(4.2.2a) $\hat{p}_t^B = p_t^A + 0 \cdot (p_t^B - p_t^A)$,

attributing none of the price difference to a real price increase. We may also state the principle as follows

(4.2.3) $\dfrac{p_t^A}{p_0^A} \cdot \dfrac{p_{t+k}^B}{p_t^B} = \dfrac{\hat{p}_{t+k}^B}{p_0^A}$ since $\hat{p}_{t+k}^B = p_{t+k}^B \cdot F = p_{t+k}^B (p_t^A / p_t^B)$.

This equation explains the term "linking" or "chaining" widely used for this method.[60] An alternative way of expressing the underlying principle is

(4.2.4) $\dfrac{\hat{p}_{t+k}^B}{p_0^A} = \dfrac{p_{t+k}^B}{\hat{p}_0^B}$ or $\dfrac{p_{t+k}^B}{p_0^A} \cdot F = \dfrac{\hat{p}_{t+k}^B}{p_0^A} = \dfrac{p_{t+k}^B}{\hat{p}_0^B}$

implying that (in the case of F < 1) the adjusted (fictitious) price of B ($\hat{p}_0^B$) relative to A should have risen less than B relative to A because the higher level of B's price is owed to a better quality. In eq. 4 the (fictitious) base period price of the new (better) good B is imputed such that according to eq. 1 we have $\hat{p}_0^B = p_0^A / F$. It is equally possible to carry the *adjusted* price $\hat{p}_{t+k}^B$

- backward (k < 0 in eq. 1) by dividing p_{t+k}^A by F (base-period adjustment)
- forward (k > t) by multiplying p_{t+k}^B by F (current-period adjustment)

The numerical example gives F = $p_t^A / p_t^B = 12/15 = 0.8$:

	0	t-1	t	t+1	t+k
observed price	A: 5	A: 10	A:12, B: 15	B: 18	B: 23
imputed price	$\hat{p}_0^B = p_0^A / F =$ 6.25	$\hat{p}_{t-1}^B = p_{t-1}^A / F$ = 12.5	$\hat{p}_t^B = p_t^A =$ $p_t^B \cdot F$ = 12	$\hat{p}_{t+1}^B = p_{t+1}^B F$ = 14,4	$\hat{p}_{t+k}^B = p_{t+k}^B F$ = 18.4
price relatives	1	2 = 10/5	2.4	2.88	3.68

59 also known as "link to show no (price) change" method.

60 The change over an interval from 0 to t+k (denoted by [0, t+k]) is derived from linking two subintervals [0, t] and [t, t+k].

The price relatives are $\hat{p}^B_{t+k}/p^A_0 = p^B_{t+k}F/p^A_0$, or equivalently $p^B_{t+k}/\hat{p}^B_0 = p^B_{t+k}/(p^A_0/F)$.

Base-period adjustment is in practice generally more convenient. We only have got to adjust one single base period price $p^A_0 \to \hat{p}^B_0$ and can subsequently[61] compare each *observed* price p^B_{t+1}, p^B_{t+2}, ... with this adjusted (fictitious) price $\hat{p}^B_0$[62] rather than continually adjust each current period price $p^B_{t+1} \to \hat{p}^B_{t+1}$, $p^B_{t+2} \to \hat{p}^B_{t+2}$... and to compare them with an unadjusted base period price p^A_0.

To pull things together we may summarize the method as follows: The idea is that only the adjusted (imputed) prices $\hat{p}^B_t$, $\hat{p}^B_{t+1}$, ... are reflect the true (genuine) increases of the price of B (as compared with A), whereas the observed prices p^B_t, p^B_{t+1}, ... result from an increase of both the price and the quality as follows

period	observed price increase of commodity B	true increase in price (rise of the quality adjusted price of B)	extent to which increase in price of B is owed to an improvement of quality
t + k	p^B_{t+k}	$\hat{p}^B_{t+k} = p^B_{t+k}F$ (F = 0.8)	$p^B_{t+k} - \hat{p}^B_{t+k} = p^B_{t+k}(1-F)$
t	15	12.0	3.0
t + 1	18	14.4	3.6
t + k	23	18.4	4.6

Note that unlike the ratios $p^B_{t+1}/\hat{p}^B_{t+1}$, ..., $p^B_{t+1}/\hat{p}^B_{t+1}$, the differences $p^B_{t+1} - \hat{p}^B_{t+1}$,..., $p^B_{t+k} - \hat{p}^B_{t+k}$ are not constant but 3.6, 4.6,...., increasing as the prices p^B_{t+1}, ... p^B_{t+k} go up.

3) Direct comparison ("No-adjustment-method")[63]

In this situation F is with necessity unity (F = 1) such that there is no quality adjusted price $\hat{p}^B_{t+k}$ which is lower than the observed price p^B_{t+k} for the simple reason that (by assumption) no quality improvement has taken place. The method amounts to

(4.2.5) $$\hat{p}^B_t = p^A_t + 1\cdot(p^B_t - p^A_t) = p^B_t,$$

or regarding all of the difference $p^B_t - p^A_t$ as a true (not quality induced) price increase. Hence $\hat{p}^B_{t+k} = p^B_{t+k}\cdot F = p^B_{t+k}$, and our numerical example gives constantly higher (by 20%) price relatives as in the case of part 2 (where F = 0.8):

---

61 that means to divide each readily available *observed* (with no corrections made) price of B in t+1, t+2 etc. by a fictitious base period price.

62 In the case of a quality improvement, this is equivalent surcharging the value of the quality surplus once and for all to the base period price p_0 of the old product A.

63 price increase is implicitly considered to be totally "genuine".

4) The 50% rule (An intermediate solution):

The adjustment $p_{t+k}^{B} \rightarrow \hat{p}_{t+k}^{B}$ to be made is now explained as follows: given that B is more expensive but also qualitatively superior to A, we may assign half of the difference between the prices, that is $(p_t^B - p_t^A)/2$, to the quality component and half to a real price increase such that

(4.2.6) $\hat{p}_t^B = p_t^A + \frac{p_t^B - p_t^A}{2}$, or equivalently $F = \frac{\frac{1}{2}(p_t^A + p_t^B)}{p_t^B} = 0.9$ in $\hat{p}_t^B = p_t^B \cdot F$,

as to be compared with eqs. 4.2.2a and 4.2.5.

Our numerical example now yields results exactly in-between the two methods above, that is

	overlap (automatic link)	direct comparison no adjustment	50% rule
data (observed prices)	$\hat{p}_t^B = p_t^B \cdot 0.8$	$\hat{p}_t^B = p_t^B$	$\hat{p}_t^B = p_t^B \cdot 0.9$

period	A	B	$\hat{p}_t^B$	relatives	$\hat{p}_t^B$	relatives	$\hat{p}_t^B$	relatives
0	5		6.25		5		5.55	
t	12	15	12	2.4	15	3	13.5	2.7
t+1		18	14.4	2.88	18	3.6	16.2	3.24
t+k		23	18.4	3.68	23	4.6	20.7	4.14*

* example: 4.14 = 20.7/5 or 4.14 = 23/5.555

To derive the price relatives it again proves much more convenient to divide the (sequence of) observed prices of B by a constant fictitious base period price $\hat{p}_0^B = p_0^A/F$ such that the price relatives are given by $\frac{p_{t+1}^B}{\hat{p}_0^B}, \frac{p_{t+2}^B}{\hat{p}_0^B}, \ldots$ instead of relating the adjusted prices of B to the observed base period price of A, that is $\frac{\hat{p}_{t+1}^B}{p_0^A}, \frac{\hat{p}_{t+2}^B}{p_0^A}, \ldots$.

e) More comments on selected methods

e1) Implicit assumptions and justification of linking (overlapping) methods

Whenever price quotations of (qualitatively different) variants of the same product A, that is A' or B overlap in at least one period s (the link period), it may be reasonable to link the series of the price quotations together (to assume proportionality) if

- both goods A and A' (or A and B) are selling in volume simultaneously on a competitive market and they both are in a similar stage of their life cycle[64] and
- they both offer the same price/quality ratio PQR (also called "price-quality-opportunity"),

otherwise the price difference in the overlap period can hardly be taken as an indication of the consumer valuation of the quality difference between the two products in question. The assumptions will not be tenable in situations in which

- B is successfully superseding A precisely because its PQR was more favorable than indicated by $1-F=\left(p_t^B-p_t^A\right)/p_t^B$ which thus understates the quality change, or
- the price of B may be unusually low in order to "penetrate" the market,[65] on the other hand
- it is also possible that A was on sale with discounts *because of* its impending removal from the shelves in which case p_t^A is unduly low (and the new price p_t^B unduly high)[66] so that the quality jump (improvement) is overstated, resulting in a downward bias of the index.[67]

Hence the method runs into difficulties when we have *"strategic" prices* (market penetration, skimming, selling-out etc.) and no stable (or "normal") conditions of pricing in a *competitive* market[68] in the overlapping periods. A problem with implicit methods is not only that there has to be a certain *overlapping period*, but rather that stable conditions of pricing just in precisely those periods may be unlikely (in fact it will be more realistic to assume that just in the case of new or improved products these conditions will *not* prevail).[69]

---

[64] The end or start may give rise to particularly low or high prices. In general the disappearing product A is likely to be priced lower in the overlap period t than B, which will replace A from t+1 onwards.

[65] Underpricing of a new product B is a strategy which is quite familiar if the aim is to "create" a market and to promote the introduction of the new product. This sacrifice of price reductions is sometimes necessary in order to have better conditions to "exploit" the market after having it "created".

[66] The old (discontinued) product A may be sold at discounted prices whilst the new product B may be overcharged, which is likely because creating a new product often gives the chance for price discrimination. If it is possible to divide the market, it would be wise to address the more powerful (with respect to the purchasing power) segments of the market first.

[67] Hence the thrust of the arguments against this approach has to do with the information conveyed by the observed transaction prices (are they indicative of the PQR or not?) It seems to be overlooked that exactly the same kind of arguments also apply to the hedonic method.

[68] In particular it is assumed that the purchasers are well informed, in fact rational and free to choose between goods offered at different prices and there is no monopolistic structure on the supply side of the market, permitting a strategy of price discrimination or segmentation of the market.

[69] Implicit methods based on price differences are clearly not applicable if the new (or replacement) good is entirely new without an earlier counterpart. Furthermore, a weakness of the method is definitely the lack of clear criteria for *the best point* in the overlapping period at which best to make *the switch to the new product*. As product cycles are becoming shorter and shorter it is even more difficult to realise in which period a "normal" price is given.

The link method assumes that differences in prices can be taken as an evidence that qualities differ to the same extent.[70] The argument is that the differences could otherwise not persist, as rational buyers would always buy the lowest priced item. This argument rests, however, on conditions that do not always prevail, and particularly not in the case of new products.

Moreover, a major drawback of all methods relying on an overlap is that it is generally not possible to predict when an overlap will no longer exist. In pursuit of this method and of the ideal of keeping the selection constant, it is not unlikely that price collectors are waiting too long, up to a time in which a parallel price quotation is no longer possible such that there is a *forced replacement* of A by B and B cannot be compared with a product simultaneously at sale, let alone with A.

Finally, we already mentioned that situations in which products become cheaper and at the same time better[71] or more expensive and worse or in which quality changes while the price remains constant will create difficulties. Whenever the difference p_t^B - p_t^A gives no hint to construct $\hat{p}_t^B$, the fictitious qualitatively comparable price of B, *explicit* methods are needed

In establishing the HICP (see **sec. 6.1**), therefore, the *automatic* application of this "*link-to-show-no-change* method" method ($\hat{p}_t^B = p_t^A$), or *carry-forward* procedure ($\hat{p}_t^B = p_{t-1}^A$), where a price of A is lacking, is deemed to impart the false impression of stability and has been banned and found inappropriate.

e2) *Explicit methods*

The most common methods of explicit quality assessment using external indicators (other than prices of the item in question) to be applied when neither linking nor direct comparison is appropriate are oriented at[72]

- α quantity adjustment (package size adjustment)
- β cost and related indicators,
- γ "characteristics" (distinctive properties, mostly related to the physical appearance, technical standards, and the functions of the item); it is in the first place the "hedonic method" which is based on this kind of reasoning,
- δ finally also expert judgement and opinion polls are considered.

Quantity adjustment assumes that 10% more is 10% better and that the good in question is homogeneous (infinitely divisible without loss of quality).

---

[70] We can also regard the options-made-standard method (described as "*option pricing*" below) as a variant of the overlap method because the price difference of models with and without optional features observed simultaneously is considered to commensurate with the difference in utility.

[71] The situation is not uncommon in the case of a *totally improved product generation* (a new production technology) as has been typical for the computer industry.

[72] It should be noted that the imputation method based not on characteristics or elements of goods but on entire but related other goods is sometimes also considered an "explicit method".

Under certain circumstances the change in ***production cost*** may be an estimate of the amount of quality change associated with the modification of the product and does not simply depend on the price differential. To take differences in production costs as an estimate of the difference (increase) in quality appears particularly straightforward whenever the new product has some extra features, the addition of which has caused identifiable extra-costs.[73] The method requires asking manufactures and to add mark-ups plus taxes in order to derive the acceptable price difference in view of the changes in quality. The justification of measuring a quality improvement by production cost is quite similar to the linking method. Again, the argument is that competition will result in cost being roughly in proportion with the true quality differential, as seen by the demand side. The reason is that if the costs were considerably higher (or lower) than the price differential accepted by the market there would be a situation in which either new competitors would be promptly attracted or they would be soon wiped out due to their excessive losses.

The principal objection against this method is always that the producer may be inclined to exaggerate the extra-costs and that the criterion of a CPI (as opposed to the PPI) is the user value rather than the impact on cost. Another drawback is that given technical progress would make the production of an item cheaper, while the "objective" quality remains the same, this would be mistaken for less quality or even quality deterioration compared with the former cost.[74]

A variant of the cost method, where no specific collection of cost information from the producer is needed, is the so called ***option pricing***: when a feature like an airbag for example used to be an option only while it is now standard equipment, it appears logical to take the former option price as an estimate of the amount of quality improvement achieved by providing this feature as a regular equipment[75].

However, to the extent that something is getting standard (offering economies of scale) it may be less expensive. The former option price is then likely to overstate quality improvement and thus understate price increase. Moreover, consumers may now value the additional item less than they did before simply because they no longer can refuse it and they now receive it seemingly free.

The option method therefore is subject to some serious theoretical objections. One can argue that the fact that some buyers did not decide to buy the variant with the optional feature shows that the option price would overadjust the quality improvement the option represents. Thus many statistical agencies take only half of the option price or so for an indicator of the amount of improvement. So the price of an option is not simply a function of its utility irrespective of how many buyers make use of the option. The same is true for the additional cost of providing a certain specification

[73] All cost methods are of limited use whenever "quality" is only vaguely defined and not resulting from a distinguishable new production process.

[74] Moreover, in general cost methods fail when different production technologies are compared. When nowadays chips are produced at a minute fraction of the erstwhile cost, this is of course not a quality deterioration.

[75] Situations in which the cost-method appears suitable, such as the case of some additional features, may also give rise to the option cost method in which a feature formerly (or currently in comparable models) available only as an option has become standard.

because such costs will usually fall as the extra-feature will be made available in more and more goods, or eventually in all goods of a certain kind.

Expert judgement can be made by price collectors or external commodity specialists. The weakness of this type of methods is - all too obviously - that a transparent and objective way of deriving the result is lacking. The same applies to opinion polls, i.e. asking the consumer themselves rather than experts. A major shortcoming of such an utility-oriented approach is that strictly speaking the consumer's utility depends also on a number of social and environmental factors as well as individual and collective circumstances which are not the object of purchases, e.g. health, a reduced crime rate etc. Furthermore it is uncontested that reasoning in terms of utility may well be the only theoretically satisfactory approach but it is in practice of limited use for operational solutions of quality adjustment problems.

The hedonic approach, another explicit method, may well be viewed as another attempt to generate objective quality assessments, thus in a way the very opposite of opinion polls.

f) Comments on the hedonic method

If "quality" can be decomposed into certain "characteristics" that can be described quantitatively as a function of some (k) few measurable characteristics (acting as regressors, i.e. explaining variables) x_1, ... x_k and if sufficient price information on goods representing the different levels (or variants) of these features is available, the regression method of adjustment, also widely known as "hedonic method"[76], is applicable. Under these conditions we may reasonably run the regression of the prices (p_i) on these variables (characteristics)[77] x_1, ... x_k (possibly also together with dummy variables introduced to account for different conditions in different time periods) such that we get

(4.2.7) $\hat{p}_i = b_0 + b_1x_{1i} + b_2x_{2i} + ... + b_kx_{ki}$ or $\hat{p}_i = b_0 + \sum_j b_jx_{ji}$, (i = 1,..., n, j = 1, ..., k)

where $\hat{p}$ represents the estimate (and fictitious) "quality adjusted" price as opposed to the actually observed price p, a function known as the "hedonic price function".

The underlying assumption is that the regression coefficients, b_1, b_2, ... measure the buyers' marginal willingness to pay for a small amount more of the respective characteristics x_1, x_2, ... since they are the partial derivatives (gradients) such that $b_j = \partial\hat{p}_i/\partial x_{ji}$. We may possibly also arrive at a "price" $\hat{p}_g$ of good g which is a hypothetic price of a hitherto non-existent good (such that there is no p_g), and where g is characterized by its specific combination x_{1a}, x_{2a} ... never seen before.

The method theoretically yields good results but in practice the conditions (measurable characteristics as valid descriptions of quality and price data of a sufficient number n of models representing all combinations of characteristics) are rarely met.

---

[76] also known as hedonic price index or "regression-", "characteristic method" etc.

[77] More often labeled z_1, ..., z_k.

The hedonic approach as a regression approach also requires that correlation between the regressors (independent variables) x_{ji} should be small which in turn means that we should have a sufficient number n of goods (models or "varieties") that vary significantly with respect to the characteristics. In actual fact, models of good quality in one aspect (characteristic) will often be good also in other aspects, and correspondingly goods poor in one dimension will be poor in others as well. So regressors are often correlated. This "*multicollinearity*" is notorious in the hedonic regression; it makes it difficult to disentangle the contributions of the various characteristics to the overall quality[78] and yields inefficient though unbiased estimates of the b-coefficients. The rationale of this method leads quite naturally to some fundamental statistical problems involved in this method:

1. The ***selection of*** the relevant ***characteristics*** (variables): Goods for which the hedonic approach is applicable must have distinct characteristics that can easily be observed on a metric scale and that can explain the price to a good deal or ideally almost completely. The best candidates for hedonic analysis are categories of goods for which quality change is frequent but incremental and for which the characteristics' changes are easy to measure. Difficulties may arise with entirely new goods having new and incomparable characteristics. It is sometimes difficult to find a limited number of variables that can be assumed to uniquely determine the price. Moreover, as mentioned above, the variables are often highly correlated with each other (multicollinearity).[79]
2. The ***specification of the*** (regression) ***function***: The way in which the variables "explain" the price should be known and should correspond to the regression model. As a rule, the exact functional form of the relationship between the price and the explanatory variables (regressors) cannot be deduced from theoretical considerations.[80] The choice of the function was often made in an ad hoc manner or for the sake of simplicity and (econometric) convenience. Apart from the linear function according to eq. 4.2.7 we often see the *semi-log model*

   (4.2.8) $\hat{\pi}_i = c_0 + \sum_j c_j x_{ji}$ where $\pi_i = \ln(p_i)$, and the *log-linear model*

   (4.2.8a) $\hat{\pi}_i = \gamma_0 + \sum_j \gamma_j \ln(x_{ji})$,

   and trends and time-dummies are also quite common, that is, models like $\hat{p}_{ti} = b_0 + \sum_j b_j x_{ji} + \sum_t c_t D_t$,

---

[78] On account of multicollinearity we have at times insignificant regression coefficients along with a satisfactory (significant) fit of the overall regression and we have incorrect and unusually wide confidence intervals for the coefficients and more problems with missing variables and errors in the variables. It is never quite clear (unless we had independent regressors) whether or not a variable is reflecting to a certain degree not only its own influence but also the effect of other variables. A variable may also serve as a proxy for a number of other less obvious characteristics with which it is correlated, for example weight in the case of automobiles: the bigger and heavier the car the more we will usually find of the latest gadgets and useful equipment in such a car, more handling ease etc.

[79] It is generally the overall goodness of fit which gets the most attention and multicollinearity is more or less disregarded.

[80] There is, for example, not much known about the functional form (linear, semi-log etc.) that should be chosen in order to correctly reflect the consumers' value of the selected aspects of quality.

where prices of different (j = 1, ...,n) goods on the left hand side may relate to different times t = 1, 2, ... and where $D_\tau = 1$ if $t = \tau$ and $D_\tau = 0$ if $t \neq \tau$.

3. The need for an additional and costly ***collection of meta-data*** on other goods (not included in the regular price statistics) for the sole purpose of better adjusting the included goods for quality; it is not only sometimes quite difficult to define appropriate characteristics, there is also a need for an additional and costly collection of observations of prices and characteristics of goods. This may in practice prevent the hedonic approach from being applied more often and on a regular basis or even across the board in Official Statistics and give rise to more international cooperation since the hedonic function may anyway not be very different across countries.

4. ***Weighted*** objects to account for their different relevance (importance): Should the observations of k characteristics on each of the n objects be given the same weight in the regression? It seems awkward to give a rarely bought Ferrari and a more frequently bought Ford-model the same weight in determining the regression coefficients and thus the price function $\hat{p} = f(x_1.x_2,...)$.

Figure 4.2.5: List of some major problems with hedonics

1	reasoned selection and measurement of (operationally defined) "characteristics"; expensive data collection [a)] of indirect use only
2	functional form [b)], multicollinearity, heteroscedasticity, estimation problems (stepwise regression)
3	weighting (e.g. with sales values; the choice of weights is far from clear [c)])
4	interpretation of the regression coefficients (does theory give guidance?); problems with entirely new goods

a) therefore priority setting (due to limited resources) is an important issue; furthermore there seems to be not much cost benefit analysis carried out in the filed of hedonics
b) in general semi-log or double-log regression functions often provided a better fit than linear functions
c) expenditure weights, for example, automatically give larger weights to expensive models. One of the reasons for preferring the unweighted approach is that the data for the regression are often taken from catalogues, price lists, comparative data collection presented in the internet etc.

It should be recognized on the other hand that hedonics has many advantages, to regress prices or quantities on "characteristics" has the merits of flexibility, objectivity and robustness (applicability under various conditions; there is no need for example to have an overlap or so, and we may even calculate an expected price of a completely new product, at least as long as there are still the "characteristics" of similar products applicable).[81]

[81] Hedonics permits estimating the price of a variant of a commodity for a time, or location (the method has been applied not only in intertemporal but also in international comparisons) at which *this* very variant was not on sale. It allows for "artificial prices" of commodities with *any* mix of the given k variables, whether or not actually observed. As mentioned above the method does not, however, provide

Fig. 4.2.5 provides a list of these problems, and with some of them more comments may be useful.

Quality described with characteristics: The hedonic approach is well suited for example in the case of washing machines (with characteristics like e.g. consumption of electricity, water, or washing powder or the maximum spinning speed), personal computers, printers and similar products, the quality of which depends mainly on capacity, speed and a few other features, refrigerators, vacuum cleaners, cranes (characterized by the lifting capacity, some dimensional characteristics, like weight, height and so on), cars, or prices of houses (location, number of rooms etc.). The hedonic regression was also applied to some non-durables like wines, breakfast cereals etc. Suspiciously in such cases the brand-(name)-dummy in particular is known as (or alluded to) being a catch-all-variable and hence it therefore comes as a no surprise that this variable often accounts for ("explains") a major part of the variation of prices. The brand is not infrequently one of the most influential characteristics, if not the only significant one.

However, for some products it is almost impossible to say which are the elements that constitute the "quality" of the whole product, or the elements themselves are defined only qualitatively. This is true in particular in the case of services, especially those services rendered under specific conditions and with respect to a very specific demand. For example, it might be difficult to establish criteria for good work of a solicitor/barrister or for the advisory service provided by a bank.

Related methods: The concept of the hedonic approach is not so revolutionary as might be thought at first glance, since in principle the same idea is pursued in price statistics of buildings or constructions (more about pricing of "unique" products and the so called "*specification pricing*" in **sec. 6.3**). In this case, prices of the whole (and unique [or fictitious])[82] building are usually estimated by summing up prices for particular works on the building (like painting, electric installations, erecting the walls, roofing etc.). This has been an established method to price unique products for years. The weighted (according to the different "quantities" of the various types of work involved in the building as eventually erected) sum applied here is not radically different from a regression function. It is only the way in which the "weights" or regression coefficients respectively are found empirically that is different. In building price statistics the final invoice paid is the basis to determine the weights and in the case of the hedonic approach it is the scatter diagram of various variants of the product and their corresponding prices that determine the regression coefficients.

Theory of hedonics: Application of hedonic techniques is not in principle linked to a specific kind of "index theory". Regressing prices on product characteristics can be done within a COLI framework as well as within a non-COLI framework. In the former case, however, it may be more important to make sure that the market condition really corresponds to what is theoretically assumed: a competitive market valuation of

---

prices for entirely new goods that cannot adequately be described with characteristics of known goods for a similar purpose.

[82] Both methods are in principle also appropriate to estimate prices of goods, which actually were not sold in that specification at that time or in that place. The result is either a sum of costs for a hypothetic combination (a fancy building) of works or it is a hypothetic price, defined as a weighted sum of prices for certain quality elements.

characteristics should lead to price differences in proportion to utility differences.[83] In a non-COLI framework it may be sufficient to see whether the regression function is stable or not.

Note that prices are necessarily assumed to reflect only differences in utility in the framework of hedonics. They may well also reflect abnormal situations in the market-place, such as the appearance of new and disappearance of old products,[84] or purchasers ill informed or having difficulties in assessing the relative qualities of the goods. A large and statistically significant coefficient for a particular quality attribute in an estimated hedonic price function may well also reflect supply conditions, cost or scarcity, not only consumers' high valuation.

The theory of hedonics (for the most part akin to the "economic" [COLI] approach) is in no small measure also dealing with issues like

- the relationship between the form of the regression function and the type of index function (it is maintained, for example, that a linear hedonic regression function corresponds to an index compiled as an arithmetic mean of price relatives or that a translog hedonic function is the only suitable function for quality adjustment in the case of a Törnqvist index etc.),[85]
- the relationship between the functional form of the hedonic regression and the utility (preference) function governing the consumer's utility maximizing behaviour with respect to quantities of characteristics (not quantities of goods), and
- the role of dummy variables within the framework of hedonic regression.

It should be noted that the theory of hedonics makes somewhat far fetched assumptions such as: economic behaviour (utility maximization for example) relates to characteristics rather than goods as a whole[86] as if they could be separated and combined at will and consumers could make informed choices with respect to charac-

---

[83] If no utility arises from extra cost, the consumer would not be willing to pay this cost. Hence such a "change" should be ignored as irrelevant from a COLI point of view. On the other hand, in a competitive market the existence of different prices can be taken as evidence for different qualities. In a COLI context it is therefore important to consider the market conditions and see whether they are fulfilled. From the above follows that in general hedonics in a COLI (rather than COGI) context requires taking more regressors into account (e.g. convenience, state of competition, availability of alternatives etc.) and thus to measure qualities like variety, convenience, safety, or quality of life (in the case of prices of the health sector) etc.

[84] The new item may be introduced at an unusually high price and the old product may be discounted for example. Interestingly these are exactly the sort of events we have in mind when implicit methods are said to fail due to abnormal prices and therefore hedonics is called for. Why should this criticism of (at times) inadequate interpretation of differences in prices only apply to traditional methods and not also to at least some items and prices of the data base of a hedonic regression?

[85] It is far from obvious why a relationship such as the hedonic function $p_i = f(x_1, x_2 ...)$ should exist as a function in *characteristics* while the price index is a function in price and quantity vectors.

[86] This appears to be root and branch erroneous if not ridiculous. Consumers cannot buy less fuel consumption (more gasoline mileage) as such or more displacement in cubic inch (volume of the cylinder thorough which the piston travels) *in isolation*. They only can buy a complete car and we cannot assess the utility of a single characteristic separated from other characteristics. These characteristics are offered only in constant bundles, i.e. in terms of goods. We cannot buy memory without a computer.

teristics; furthermore, the existence of indifference curves in the characteristics space (as opposed to the goods space), of marginal sub-utilities and marginal acquisition costs of characteristics etc. is assumed. Moreover, some studies make the implicit assumption of all buyers having identical tastes (towards characteristics) which makes it difficult to explain why there are in general not only one but several goods for the same use. More often than not, theory tends to move from the sublime to the ridiculous.

In a number of articles Silver and Heravi advocated a move from the traditional matched models method in official statistics to the so-called "superlative" or "exact" hedonic price indices (SEHI, according to Feenstra 1995). These indices are based on the economic index theory and much like the "usual" COLI approach here a Laspeyres- and Paasche-SEHI-formula again acts as a upper and lower bound. Unlike the superlative indices[87], a SEHI does not use observed prices but rather *adjusted* (for changes in the characteristics) prices in one of the two periods, and unlike the traditional (direct) hedonic approach, in correcting observed prices a weighting system is introduced by which the commodities entering the hedonic equation are given unequal weights.[88]

Functional form of the regression: The choice of the functional form is mostly decided on empirical grounds. It has turned out that semi-log or double-log regression functions often provided a better fit than other functions, such as the linear regression for example. The estimated coefficient of determination R^2 is the most popular indicator of goodness of fit.[89] In practice, however, we not infrequently encounter quite similar results for R^2 although the regressions are markedly different with respect to the functional form and the estimated regression coefficients. This makes it difficult to put faith in any of the regressions.

Direct and indirect method of hedonic regression: as shown in **fig. 4.2.6** a distinction has to be made according to the type of regression function (including or excluding time dummies).

In the indirect method a quality adjusted price is estimated in each period separately. This approach therefore relies a lot on the stability of the function and fits better in backwards looking studies. A choice can then be made to adjust p_{it} to the quality adjusted price using the regression function of period 0, giving $\hat{p}_{it}(f_0(x_{10}, x_{20}, \ldots))$, or using the function of period t, giving $\hat{p}_{it}(f_t(x_{1t}, x_{2t}, \ldots))$.

The direct method on the other hand estimates a regression-coefficient (associated to the time dummy) representing the shift from period t - 1 to period t in terms of the price-dimension. This coefficient reflects that part of price variation that is *not* ac-

87 in the COLI sense, that is Fisher's or Törnqvist's index.

88 It is based on sales data. According to Silver and Heravi (S+H) treating each model in the regression as equally important is perhaps the biggest defect of the hedonic method and the SEHI method will effectively surmount it. As to the data (in particular the sales values) needed for this approach S+H heavily rely on scanner data in retail outlets as they contain not only prices but on account of the many details stored in the bar-code also information on models (makes) and their characteristics.

89 It should be borne in mind that a high value of R^2 is not a proof of having revealed the true nature of the relationship (due to multicollinearity) and not omitted (excluded from the regression) other important variables.

counted for by the explicit regressors (i.e. explained by the quality indicators) and thus directly measures the "true" (quality kept constant) price movement. While the indirect method is a sort of cross-section analysis, the direct method usually requires data of more than one period. With both methods results depend very much on the specific selection of items entering the sample and frequent updates of the estimated regression equations are crucial for this approach.

Figure 4.2.6: Variants of the hedonic method

regression function contains

data from a single reference period in order to calculate the implicit prices of each of the characteristics	data are covering all periods, time dummies are included; regression is re-run each time the index is compiled
indirect -, or individual coefficient -, or simply imputation method	**direct** [a)]-, or time dummy variables method
can be used to estimate a price for an item not on the market or to adjust post-hoc the observed price of a replacement item (method can be combined with other methods of quality adjustment, more appropriate for official statistics)	difference between the time dummies is taken to represent the genuine price change [b)] (quality change excluded); price movements over a number of years are estimated in one regression. Less observations needed than in indirect method

a) because price index is calculated directly from the regression, another term is adjacent year regression (AYR) when only two periods are involved as opposed to the indirect method, where a separate regression is run every period.

b) those price changes that cannot be associated with a change in the characteristics

Final remarks: Some people take the ostensibly advanced theoretical underpinning of hedonics provided in general by the COLI-approach as evidence of superiority of the hedonics over conventional methods. Therefore hedonics has gained an enormous amount of extra attention recently as allegedly having the potential to solve all difficulties with the conventional methods.[90] However "that just because the conventional quality adjustment methods are problematic, this does not by itself make the case for using hedonic methods. Hedonic indexes also have conceptual and empirical problems. Choosing between hedonic and conventional methods requires considering advantages and disadvantages of both methods...".[91]

In an "a-theoretic" approach hedonic methods can also be combined with traditional methods, and in practice various adjustment methods can be utilised in combination and in parallel anyway. However, when the matched model method (MM

[90] In particular the Boskin Commission (1996) in the USA should be given credit for having contributed much to spread the gospel of the upward bias and the superiority of the hedonic method.

[91] Paragraph 135 in an OECD-"Handbook on Quality Adjustment of Price Indexes for Information and Communication Technology Products" 2000, a book written by Jack Triplet dealing primarily with hedonic methods.

method) is not applicable, there is no overlap and therefore an imputation has to be made "hedonically", imputed prices can fill the gap[92] and enter a conventional price index formula. This procedure also corresponds to the practice of official statistics in many countries.

After the impetuous enthusiasm of the Boskin Commission for hedonics we find much more reservation (explicitly on theoretical rather than empirical grounds) in the so-called Schultze Panel[93]: "We recommend a more cautious integration of hedonically adjusted price change estimates into the CPI", and this method is "only justifiably applied to a narrow class of goods", and where new goods are "sufficiently unlike existing ones, there hedonic techniques do not hold much promise."

The Eurostat Handbook of Volume Measurement[94] considers a method for quality adjustments recommended or at least acceptable (that is an A, or B method) if it makes explicit rather than implicit evaluations of the quality change, does so in a verifiable way and takes both user's preference and producer's costs into account. The hedonic approach is an A method, if characteristics can be identified that influence the price and can be quantified.[95]

Finally a remark may be in order to the interpretation of quality adjustment (with whichever method) in politics. To all theoretical considerations of the kind demonstrated above concerning the development of more and more refined methods of quality adjustment, an important caveat should be added. Time and again emphasis should be given to the following consideration:

> Using quality *adjusted* prices leading for example to *statistically displayed* higher growth rates (real GDP), a rise in productivity and lower inflation rates should not induce the misunderstanding that some sort of "surplus" is ready to be distributed. Goods are still sold at the observed market prices not at adjusted prices, and whatever higher or lower adjustments are made there is no corresponding flow of money. Technical progress can be distributed only once in the way of more purchasing power (in the COLI sense) for a given price, but not once more via increased wages. "Profits" resulting from statistical wizardry are always only mere figments.

Quality is also in issue regarding the selection of outlets. Factors such as variety, accessibility, style, reliability and timeliness in supply, quality of service etc. ought to be taken into account, and again representativity and comparability (across time) is considered requisite. This leads us to the sampling problem which for such reasons has some intricacy.

---

[92] The operation of matching may be viewed as a device to minimise imputation. If not applicable recourse has to be made to imputation of one kind or the other. Hedonic regression is basically only one of several ways of imputing a price, viz. on the basis of measurable "characteristics" and an established empirical relationship (function) of prices and characteristics.

[93] Charles L. Schultze & Christopher Mackie (editors), At What Price?, Washington D.C., 2002, p. 6, 7.

[94] see **sec. 5.3** for more details.

[95] The corresponding statement concerning option prices is for example "if separately priced options can be distinguished, and the price of a separately purchased option is not very different when it is included in the product". This amounts somewhat to the fundamental wisdom that methods are good if their assumptions are fulfilled (!).

4.3 Sampling in price statistics

This section deals with problems giving reason for sampling and re-sampling. To date non-random sampling is still common practice in price statistics. Random sampling may be feasible in certain areas (such as the selection of outlets [or more general: reporting units]) while it is less likely ever to be implemented on a large scale in other fields such as the selection of goods. Furthermore, there is a dynamic universe of items (commodities) as well as outlets (reporting establishments) and other price determining aspects so that ideally a frequent re - sampling should take place. We have already addressed problems with appearance and disappearance of products giving rise to re - sampling in the framework of quality adjustment with which they are closely related[96] in **sec. 4.2**. As such practical problems occupy the greater part of the time and effort of official price statisticians, a textbook on price statistics should devote a separate section to these issues.

a) Random sampling: why and what?

The practice of price statistics is still governed by purposive (judgmental, non-random) selection and this has been criticized for long by many index theoreticians.[97] The arguments in favor of random (or probability) samples in principle are:

- Randomness prevents from selecting in a (deliberately) biased manner and thereby making observations which are not representative of the universe
- Price statistics delivers only point estimates of unknown accuracy, as if it were irrelevant whether a price relative were empirically established by 10 or by 1000 observations.
- Selection methods other than random sampling do not allow us to estimate the sampling error[98] of price quotations (and consequently also of the index altogether). Only random selection enables price statisticians to gauge the accuracy of their statistics through the estimates of the sampling error by providing also interval estimates, not only point estimates of their figures.
- Price quotations of unknown accuracy or possibly even biased ones will also yield a price index as based on such data which will inevitably also be biased.

As to the last mentioned consideration it should be borne in mind that we cannot yet dispose of a statistical theory of the sampling properties of certain relative figures

[96] We therefore addressed re-sampling problems already in sec. 4.2. The fact that the universe of goods is not static but rather constantly changing conceptually relates the issues of "sampling" and "quality adjustment" to one another. The problems of sec. 4.2 and 4.3 are also related insofar as in both fields (making adjustments as well as assessing the representativity) the market conditions need to be accounted for (strategic prices such as dumping out of old models or skimming off high entry prices for the new ones etc.), and therefore co-operation with market research organisations, trade associations etc. in the setting up of a meta-information basis seems to be highly advisable.

[97] There emerged some criticism only recently by judges and lawyers, too. This should be viewed against the background of fairness as regards the distribution of response burden.

[98] It has continually been emphasised that true random samples are needed not only for the sake of accuracy but also as they require more definite decisions about procedures, definitions etc. that can easily be ignored in the more flexible and less exact procedures of judgmental sampling.

such as *index* numbers. The situation is different from the estimation of absolute figures such as average wages, turnovers, production values, or the estimation of stocks for example. A price index combines prices and weights, both data are likely to be subject to sampling errors and there is little known about the sampling aspects of the *combination* of price and weights to an index.

Anyway, time and again it has been argued on the part of price-statistics-theory that official price statistics should move to random samples and give up the traditionally prevailing purposive selection. To assess the relevance of this argument, however, it is necessary

- to first pick out the elements (stages) of index calculation in which a selection is involved (see **fig. 4.3.1**), secondly
- to discuss the conditions to be met in order to obtain true random (probability) samples, and finally
- to evaluate the relative weight of sampling and non-sampling errors involved in index compilations respectively (see **fig. 4.3.3** for a typology of errors and biases in index compilations).

Figure 4.3.1: Stages in index calculation where selections are involved

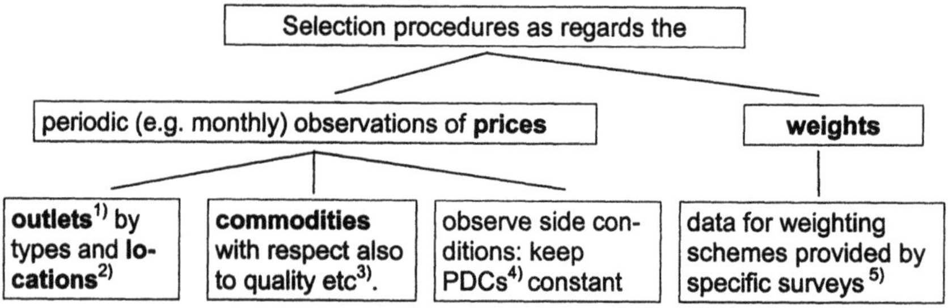

1) reporting firms; they should be representative of the purchases of households (in the case of the CPI) and likely to allow constantly reliable and representative price quotations; to make sure of this is the task of special surveys (if they are carried out), the Point of Sales (POS) Surveys
2) the sample should be representative also of various kinds of geographical areas and municipalities
3) e.g. regular availability (presently but also in future): the problem is that no sampling frame is available and due to the dynamics of the universe a frequent re-sampling is needed
4) in price statistics it is crucial to have all sorts of PDCs (price determining characteristics) constantly correctly represented and kept constant (comparability is called for) as time goes on (changes in the PDCs should be constantly observed and taken into account)
5) in the case of the CPI these are family expenditure surveys (or family budget surveys, FES or FBS) which are mostly quota samples and voluntary surveys only, which therefore suffer in no small measure from non-response problems.

In **fig. 4.3.1** an attempt is made to list stages of index calculations in which a selection has to be made.[99] Before commenting on them it is worth noticing that – as al-

[99] A dimension not mentioned in fig. 4.3.1 is sampling in the time dimension, which is determining a day on which normally observations are made. As a rule, in a non-random manner the 15th of a month

ready mentioned – some conditions have to be met in order to obtain true random samples in price statistics. In particular we have to face the following difficulties:

1. It is crucial to make clear what has to be considered as the *unit of observation*, and what thus constitutes the universe. Are for example two products with a different size, design, colour or style of wrapping to be seen as *two* commodities, or are they only two varieties of *one* commodity? On those things a decision has to be made in order to fulfill a second prerequisite, viz.

2. we should be able to dispose of a *"sampling frame"* (from which a sample can be drawn) in one way or another, that is, a device representing the universe from which a sample should be drawn. However, as a rule a complete and up-to-date list of all units to be sampled is not only lacking in the case of items (goods and services) but as a rule in most countries also in the case of outlets.

3. The *definition of the target population* does not only comprise goods and reporting units, it also requires the observation of all aspects (PDCs) of sales and purchases, in particular a constant monitoring of the (continually changing) representative PDCs. For each transaction we may think of a number of different PDCs (for example different qualities of the goods or of sales services) and there is of course no register out of which sample PDCs could be drawn. Given the need to observe such "*side conditions*", the sampling problem is in practice much more complicated than might be supposed on the face of it. As the selection of commodities (items) from the dynamic universe of goods has got to take place under the restriction of comparable PDCs, sampling in this field seems to remain a domain of purposive (judgmental) sampling in the foreseeable future.

4. A decision has to be made regarding the sample size n. As a rule, the more detail is required (for example concerning the commodity classification or the desirability of making regional comparisons) and the more heterogeneous the universe, the larger the size n has got to be.

5. A number of aspects characterizing the survey in question have to be taken into account: frequency of data collection, method of measurement (personal visits, telephone interview, postal or electronic questionnaires), type of prices to be collected (e.g. effective transaction prices vs. less suitable list or book prices), sample rotation,[100] a voluntary vs. statutory survey etc. to name a few only.

6. Sampling also requires competence in handling problems like missing data and non-response, training of price collectors, data editing, internal checks of accuracy which became more and more mandatory for official price statistics in order to reduce procedural errors. Decisions also have to be made as regards time

(or days near to the 15th) is chosen for this purpose in most countries. To determine a particular day as reference day within a month is known as "point-in-time-prices" (as opposed to "period-prices" on the basis of continual observation). When, however, on such a selected day too few prices are to be observed (as happens for example not infrequently in the case of exports, imports or construction work) price quotations might alternatively well refer to the *average* of a month calculated from observations on a number of days (for example the 10th, 15th or 20th or so).

[100] This may help to cope with problems of sample maintenance such as sample depletion, loss of representativity, establishments becoming resentful of the response burden etc.

and scope (and of course also feasibility) of re-sampling (see below) when the universe has changed significantly.

Sampling is therefore not an easy task in the case of price statistics. Conditions for random sampling are rather unfavorable. Thus true random sampling repeatedly carried out in all stages of index calculation is rather improbable in official statistics.[101] This applies in particular to the case of goods whereas random sampling of communities (municipalities) and outlets therein is a less disputed case for probability sampling in price statistics. In what follows we look at the selection problems in each dimension of fig. 4.3.1 one by one. Part c will be devoted to the selection of goods in particular.

b) Random selection of localities and outlets, derivation of weights from budget surveys of private households

Localities are certainly well amenable to random sampling. In general, first a sample of municipalities is drawn (where the aim is to have each type of community equally well represented, a number of units should be represented at any rate)[102] and then in a second stage, in each selected municipality a sample of outlets is chosen (more or less explicitly taking into account the size and type of the outlet, and aiming at a balanced representation of outlets of all sorts).

Although the selection of outlets may be a least disputed case for random sampling, the sampling of outlets is certainly more troublesome than it might appear at first sight. To illustrate this, in the case of the CPI remember that complete lists of units (to serve a sampling frame)[103] are difficult to obtain, and if they are available it is widely unknown whether they are representative and up-to-date unless point of sales surveys (POS) are made on a regular basis. In general, lists of establishments may be taken from tax filings, an industrial census or a similar enumeration of retail establishments, telephone directories ("yellow pages"), business registers,[104] records of local administrations and organizations or (preferably) specific registers maintained for statistical purposes. Those sources do not, however, provide information about variables suitable to measure the size of the unit[105] and about the structure of the assortment of goods the outlets offer, let alone the degree to which their sales are representative of the purchase of the households.[106] It also should be borne in mind, however, that it does not suffice to find a reporting unit where commodities may be priced, the selected units also should be selling fairly regularly, remain representative and – most importantly - the unit should (continue to be) highly competent

---

[101] It should be noted, however, that experience has shown that non-random sampling does not necessarily perform worse.

[102] It is therefore quite common to select the capital and some of the biggest towns with certainty.

[103] A sampling frame is a representation (enumeration) of the universe from which a sample can be drawn.

[104] They tend, however, to overcoverage of the target population.

[105] Such information is particularly necessary in order to perform probability proportional to size (PPS) sampling.

[106] This is precisely what POS surveys are trying to find out.

and co-operative. It is therefore not unlikely that price statisticians tend to select outlets deliberately because they are known to be good respondents.[107]

Another dimension in which a selection has to take place is the collection of *data for weighting schemes* – in the case of the CPI –as a rule provided in most countries by specific family expenditure surveys (FES) also known as budget surveys. They are generally fairly resource intensive, costly and complicated and will therefore be repeated only in longer intervals in time. The criticism that habitually arises concerning the empirical foundation of the "basket" of consumer price indices in most if not in almost all countries is as follows:

- weights are based on too small a number of budgets reported in surveys like the FES,
- FES often exempt certain households (mostly those who are better off) from responding,
- since participation is in general voluntary[108] the selection will resemble more a quota sample than a random sample.

c) Sampling of goods, new goods and the use of scanner data

The type of goods, i.e. the sort of goods and services the index comprises and the type of prices involved, defines the "type" of index (i.e. its scope and coverage), e.g. a consumer price index (CPI) or producer price index (PPI). Definition and selection[109] of goods and services is therefore crucial. However, as said already, selection suffers here even more from the lack of a sampling-frame than in the selection of outlets. There is no – and probably never will be – a complete and up-to-date list in which *all* (!) commodities are listed and defined with exactly that degree of detail found necessary for price statistics to allow representativity as well as pure price comparisons.[110] And there is of course no such register allowing for differentiating with respect to the PDCs either. As repeatedly pointed out, price statistics has got to take the PDCs into account when a sample of goods is to be drawn.

---

[107] There may also be a tendency for practical reasons to price as many different commodities in the same store as possible. This in turn will ordinarily be introducing a correlation among price quotations.

[108] This is so due to the high response burden and the ease with which respondents could give incorrect answers anyway if the survey were obligatory.

[109] It is obviously absolutely impossible to observe all commodities traded for example in wholesale and much less in retail establishments in order to compile a wholesale price index or retail price index respectively. Even if this were possible, the results would be almost useless because in many instances a price quoted which has no counterpart in most of the other establishments or for which no counterpart can be found in future will only add noise to the data.

[110] As pointed out already, these two principles require different descriptions of commodities in terms of physical features, kind and grade of materials, parts, uses, workmanship etc. Pure price comparison requires fairly detailed descriptions, making sure that the sample does not change while in the interest of representativity a specification should be adopted which allows for some latitude in order to switch to a more representative item if necessary.

Furthermore, whenever a new good (NG) emerges, or only slightly[111] different varieties of "old" products gain importance,[112] and whenever the practice (habits) concerning the PDCs changes *we should draw a new (costly) sample*, which is of course most unlikely to be done in practice. While inclusion of a NG serving another purpose than the old one requires an addition together with a new weight assigned to the good, a new variety or quality improvement of a good no longer available is accounted for by a simple replacement. Another distinction is useful. It concerns

1. the question how (with which aim) a modification of the selection of goods should be performed, and if a re-sampling is made
2. whether the selection should be totally updated or rather individually (that is goods one by one whenever required) and if so whether by carefully choosing an appropriate successor or by simply adding a new good no matter which good is deleted.

The first alternative is addressed in **sec. 4.2** referring to the two extreme strategies, viz. the **MM-method** (matched models),[113] and in a sense the opposite procedure, continuous **re-sampling** of items in order to maintain the representativity of the sample.[114] The underlying conflict behind this alternative is again the conflict between the principles of representativity and pure price comparison.

As to the second point **fig. 4.3.2** tries to explain the alternatives of dealing with the dynamics of products when re-sampling is called for. The advantage of re-sampling is that the sample size is always sufficiently large and new products are introduced earlier.[115] However, the monthly re-sampling that continually has to be performed is unlikely to be random and the method requires a constant and resource intensive monitoring of changes in the universe of goods and services to maintain representativity.[116] Moreover it is resource-intensive and the collected data is only partly directly entering published results. Keeping pace with the dynamics of the changing universe of items requires the collection of a number of *meta-data* relating to the characteristics (not only) prices of the goods.

---

111 This is of course also a question of the amount of detail devoted to the specification of types and qualities of goods and services underlying the collection of price data.

112 As repeatedly pointed out completely new goods (NG), in the sense of genuine innovations are difficult to distinguish from (major) quality changes. There is no clear dividing line. In principle the NG should provide an entirely new service (not only a service similar to an existing one) or satisfy a need hitherto not satisfied (in this sense a NG creates its new market of its own).

113 According to this method the sample should be kept constant if ever possible and therefore include only items that exist in all the periods to be compared. The method is following over time a *fixed sample of items* in order to make sure that *only pure price changes* are recorded, thereby avoiding problems with quality adjustment and disregarding all non-matching items and representativity.

114 For this reason this alternative is also known as Monthly Chaining and Re-sampling method.

115 Due to a constant change strictly speaking only prices of two adjacent months are comparable, and the chained results of the re-sampled prices are also reflecting a structural component.

116 Monthly re-sampling is not the only methodological novelty recommended for use in official statistics that in actual fact amounts to an enormous increase of observations to be collected on a regular basis for methodological purposes, as the same is true basically also for the hedonic method for example.

Unlike the Adding-and-deleting method (no 3 in **fig. 4.3.2**), the Replacements method (no. 2) cannot adequately reflect completely new developments as each newly introduced good should have some relation to a replaced one.

Figure 4.3.2: Dealing with appearances and disappearances of goods

Dynamic changes in the universe of commodities

collectively replacing items	replacing items individually (one-to-one)	
a completely new selection of goods is being carried out	selected new good is replacing old one	enter (delete) goods whenever required by representativity
1. Re-sampling	2. Replacements	3. Adding[3] and deleting
logically the very opposite of MM-method[1] time consuming, expensive, continually maintaining representativity, usually combined with linking[2]	problems with • defining the best individual successor[2] • finding the right time for replacements	disappearing items can just be deleted with no need to find a suitable replacement item; advantage: timely introduction of entirely new goods is easy

1) matched models (see sec. 4.2)
2) "most sold" (most representative) or most like (then the least problems with quality adjustment)?
3) As mentioned in sec. 4.2 in the case of (entirely, truly) "new" goods it is also necessary to make additions to the weighting scheme.

There is a discussion triggered only recently as to the usage of ***scanner data*** (bar codes in retail shops providing information about sales by type of products) in price statistics, possibly as a valuable source of data in order to

- establish the universe of commodities from which a sample has to be drawn (i.e. they may provide a sampling frame for the selection of commodities)[117]
- continually monitor the representativity of the sample and
- detect (identify) and evaluate quality changes, and provide data in order to
- introduce size data needed for probability proportional to size (PPS) sampling and weights of outlets so that in averaging prices the relative importance of the outlets where the purchases take place is adequately accounted for.

The issue is interesting and appears promising. Scanner data are said to improve timeliness, accuracy and representativity of index compilations in official statistics

117 This applies even more so as data gained from electronic scanning by bar-code readers at the cashiers desk are not only readily available but also informative due to the fact that they combine many pieces of information, such as quantities sold and detailed specifications of the goods in question.

and in particular they may be used for a true probability sampling of items to be priced.

It is widely maintained, however, that the data they provide, though not insufficiently detailed, will as a rule not be suitable in order to serve the purposes listed above. Experiments with scanner data related to some selected commodity groups therefore have been carried out by various market research firms and statistical offices. They aimed to examine whether random sampling on the basis of scanner data would perform better than the still prevailing non-probability judgmental sampling (like for example the common method oriented at the most frequently sold items which comes close to a kind of cut-off selection).

It turned out, however, that reality was far from the ideal. In principle, no two different commodities should have the same code number.[118] The problem with the system of scanner codes[119] is rather the opposite situation, in that a different number may be assigned to items which are for all practical purposes identical items, and that these numbers change rapidly (sometimes simply because of a change in price). Furthermore, it is also possible that the same number may represent goods of widely different relevance in terms of their turnover. Thus,

Processing a large scanner database for CPI compilation on a regular basis may be time-consuming. It is possibly also tedious to make sure that like is compared with like by checking this data in which information abounds and which yet *incorrectly* may be regarded as representing the universe.

The use of scanner data is therefore possibly not compatible with keeping price determining characteristics (PDCs) constant over time as required (or at least desirable) in price statistics. Moreover, to find out which are the most frequent transactions in order to ensure representativity in price statistics is only half of the problem. The other half certainly is comparability, and the task is to continually find a fair compromise between representativity and pure price comparison.

d) Sampling designs in general and methods applied in official consumer price statistics

In order to review methods of sampling for their potential use in price statistics it is useful to notice at the outset that index compilations are not only subject to (random) sampling errors but also systematic[120] errors, in particular in the form of procedural errors, which well may even dominate the former. **Fig. 4.3.3** therefore provides an overview of errors (and possibly biases) along with the well known methods of sampling.

The decisive criterion of an estimator $\hat{\theta}$ (sample) of a parameter θ (universe) is the so called mean square error (MSE)

(4.3.1) $$\text{MSE} = E(\hat{\theta}-\theta)^2 = E(\hat{\theta}-E(\hat{\theta}))^2 + (E(\hat{\theta})-\theta)^2$$

---

[118] Nor should of course one code number denote different goods.

[119] In Europe there is the European Article Number (EAN) system and its U.S. counterpart is the UPC code (US Universal Product Code).

[120] In this case a negative error is as a rule not offset by (equally likely) positive errors.

which has two components, the variance $\mathrm{Var}(\hat{\theta}) = \mathrm{E}(\hat{\theta} - \mathrm{E}(\hat{\theta}))^2$ and the squared bias B^2 = $(\mathrm{E}(\hat{\theta}) - \theta)^2$. A small B^2 means that $\hat{\theta}$ is on average close to θ and a small $\mathrm{Var}(\hat{\theta})$ means that $\hat{\theta}$ does not vary much. Both qualities are desirable. The focus in sampling therefore is on minimizing the MSE, the sum of B^2 and $\mathrm{Var}(\hat{\theta})$.

Figure 4.3.3: Errors and sampling methods in price statistics

- Errors
  - sampling errors
    - arising from the exclusion of elements of the universe (i.e. from not performing a complete enumeration)[1)]
  - non-sampling errors
    - **observation errors**: over-coverage, conceptual defects, response- and processing errors
    - **non-observation errors**: under-coverage, item-non-response[2)] or (complete) unit-non-response

1) the specific advantage of random sampling in contrast to non-random sampling is that probability theory applies to this kind of error which thereby can be numerically assessed

2) the unit is not informing about a particular item (element of the questionnaire) but willing to cooperate in general (otherwise unit-non-response)

- Sampling
  - random
    - equal inclusion probability (n/N): simple[1)] sampling (SIS): each unit has the same inclusion probability
    - unequal inclusion probabilities: sampling proportional to size (PPS), stratified sample[2)]
  - non - random
    - judgmental (purposive) selection (currently the most frequent method), cut-off sampling[3)], quota sampling

1) or equal probability sample

2) Cluster (area) samples are rarely mentioned two stage sampling designs; their main advantage is less fixed cost (e.g. travelling time) due to concentration of observations on a selection in the first stage of the procedure and no need for a complete sampling frame. By the same token price quotation is preferably taking place in larger units and it is also advisable to collect a number of prices at a time in the same outlet rather than only few prices in a great number of outlets.

3) Units with a size above a certain threshold are selected with certainty whereas those below are not included at all.

The variance $\text{Var}(\hat{\theta})$ can normally be reduced by increasing the sample size,[121] while this does not apply to the squared bias B^2. Moreover, experience shows that not infrequently B^2 is quantitatively more influential than the variance. This is the case for well managed smaller samples rather than costly large samples.

Although in theory random sampling is often regarded as overdue in official price statistics, the difficulties in implementing true random samples for commodities at least are still insurmountable. Thus the majority of European National Statistical Offices (NSOs) continue to rely on purposive (non-random) sampling in this field. This is known from an inquiry carried out by Statistics Netherlands in 1997 among NSOs concerning the sampling procedures used for CPI compilations.[122] The issues explored in detail were methods of sampling relating to

1. geographical localities (towns, municipalities),
2. outlets (a task which is anything but simple given the side-conditions that have to be observed regarding the outlet managers, e.g. the ability and willingness to give competent price information on a regular basis),
3. *items*[123] (i.e. commodity or service, the specification of which is usually provided by the central NSO in both cases, if prices are observed de-centrally by price collectors, or if they are observed centrally; note that to establish and update this list of "specifications" requires meta-data which may also be used for item substitutions in the course of re-sampling and for hedonic studies),
4. *varieties* (concrete product within the item description selected in an outlet by the price collector using a random, or - more often than not - a non-random selection method).

In addition the questionnaire also asked for a description of methods applied in order to deal with missing prices ("non-response" due to non-availability of a price quotation) or to decide on the size of the sample of outlets, items etc. (i.e. criteria used in determining the optimal sample size).

In the majority of EU countries the sample of locations and of outlets is non-random.

Only a minority of countries made use of probability techniques (*random* samples) for the selection of outlets such as[124]

---

[121] It is also in general difficult to estimate and a complicated function of the variance within an outlet (between products) and between outlets (for the same product) with weights determined by the sampling procedure.

[122] A report on this survey of sampling designs prepared by M. Boon had been presented at a joint ECE/ILO Meeting on CPIs in Geneva 24-27 November 1997 and can be found in the Internet .

[123] The term "item" may cover what is meant by "basic heading" or "elementary index" (the smallest level in a classification for which weights will be introduced when an aggregation is made), but it may also denote more specificly a particular commodity representing (in lieu of other commodities) this so defined "item".

[124] The USA may in fact be the country in which the implementation of probability sampling in this field is most advanced. Many countries still rely on the judgement of experts (at the central office or locally in the selected outlets) for selecting the items representing a commodity group.

- simple (equal probability) sampling (SIS) or
- sampling with probability proportional to size (PPS)[125]

in the case of a stratified or not stratified universe. Typical *non-random* selections were on the other hand

- judgmental[126] (based on the judgement of experts) selection
- cut-off sampling (limiting observations to the largest units)
- quota sampling (aiming at a sample which is representative of the universe with respect to the structure, i.e. to "quotas" and in which certain types of units ought to be represented to a certain extent or excluded a priori).[127]

Reasons given for non-random sampling are many, such as non-availability of an appropriate sampling frame, or a randomly selected item may well soon after being selected disappear from the market and there is need for frequent revisions whenever the universe has undergone a change. Moreover, judgmental samples may benefit more from the expertise of the local price collector in finding a reasonable compromise if necessary than centrally monitored random samples with very limited latitude left for the staff in their fieldwork.

The EU countries also differ from one another as regards the incidences giving rise to and the principles governing the carrying out of a re-selection (revision of the selection) of localities, outlets and items. In some countries new outlets (or perhaps also new items) are only introduced into the sample at base year revisions[128] while in other countries such changes take place on an ongoing basis. Items were mostly chosen at the central office not only in the case of central but also de-central price collection and they were prescribed to the field representatives in a more or less tight specification. The selection at the central stage was in general based on criteria such as representativeness (much like, in principle, the case of quota sampling) and at the local stage (the price collector) it was mainly judgmental (not infrequently with the support of the outlet manager).

---

[125] In this sampling design the inclusion probability is proportional to a variable describing the relevant "size" of the unit of observation such as number of employees or retail floor space in the case of establishments, the turnover of (or shelf space occupied by) the commodity in question. Some countries, e.g. the USA, Sweden Finland or the UK have already gained much experience with PPS-sampling. It is important that sales value data in order to arrive at a size variable for PPS-selection or weighting are collected over a not too short interval in time because otherwise special campaigns with unusual low prices would unduly affect the results.

[126] also called "representative item method" or "purposive sampling". For this method (no doubt the still predominant one) it is crucial how tight or loose (flexible) the product specification is the price collector has to follow. There are obvious problems with extreme solutions.

[127] In this sense for example villages or other small municipalities often were excluded from the sample of localities or the intention is to have localities of all size classes (equally well) represented. A quota sample differs from a stratified sample in that the selection within the a priori (centrally) fixed number of units to be selected was not made at random but rather delegated to the local price collector (i.e. de-centrally and judgementally).

[128] It should be noted that the inquiry took place before binding European regulations of the HICP (see **sec. 6.1**) came into force requiring a timely inclusion of "new" goods and services.

e) The Generalized Fisher index

The universe of goods and services to be priced and from which a sample may be drawn is continually undergoing changes in that new goods are emerging and old ones disappearing. With the passage of time the composition of this dynamic universe will constantly vary and the sample of goods on which a price index will be based will therefore cover different sections of this universe. The Dutch statistician de Haan introduced the Generalised Ideal Index in the tradition of Fisher (P^{GF} for short)[129] in order to cope with the problem of changes in the sample of goods. His index formula explicitly accounts for new and disappearing goods in the following way:

Denote $\mathbf{q}_0^D$ and $\mathbf{p}_0^D$ the vectors of quantities and prices respectively of goods available in 0 but not in t, hence, comprising disappearing goods (and services). Likewise $\mathbf{q}_0^C$, and $\mathbf{p}_0^C$ are vectors related to "ongoing", or continually observable in both periods 0 and t, such that we also have prices and quantities of such goods in t and thus, observed vectors $\mathbf{p}_t^C$ and $\mathbf{q}_t^C$. New goods (in t but not yet available earlier) are labelled N so that $\mathbf{p}_t^N$ and $\mathbf{q}_t^N$ denote vectors of prices and quantities of such goods. As certain quantities do not exist (are zero by definition $\mathbf{q}_t^D = \mathbf{q}_0^N = \mathbf{0}$) the vectors $\mathbf{p}_t^D$ and $\mathbf{p}_0^N$ have to be estimated (the Laspeyres formula requires a forward imputation of prices of disappearing goods $\hat{\mathbf{p}}_t^D$ and the Paasche formula a backward imputation of new goods' prices $\hat{\mathbf{p}}_0^N$ respectively. So that de Haan's GF formula is in a sense (need for imputation) "symmetric". For a better understanding of the aggregates involved and their inherent relationship it may be useful to introduce the notation shown in **fig. 4.3.4**.

Figure 4.3.4: Notation for de Haan's Generalized Fisher Index

$A_1 = \mathbf{p}_t^{C'}\mathbf{q}_t^C$	$B_1 = \mathbf{p}_0^{C'}\mathbf{q}_t^C$	$C_1 = \mathbf{p}_t^{C'}\mathbf{q}_0^C$	$D_1 = \mathbf{p}_0^{D'}\mathbf{q}_0^D$
$A_2 = \mathbf{p}_t^{N'}\mathbf{q}_t^N$	$B_2 = \mathbf{p}_0^{N'}\mathbf{q}_t^N$	$C_2 = \mathbf{p}_t^{D'}\mathbf{q}_0^D$	$D_2 = \mathbf{p}_0^{D'}\mathbf{q}_0^D$

	prices 0		prices t		Laspeyres	Paasche
goods	q_t	q_0	q_t	q_0	P^L	P^P
new N	$B_2^{(*)}$		A_2			A_2/B_2
ongoing C	B_1	D_1	A_1	C_1	C_1/D_1	A_1/B_1
discont. D		D_2		$C_2^{(*)}$	C_2/D_2	

(*) the asterisk means "requires imputation"; as quantities of N-goods in 0 are 0 (and so are quantities of D-goods in t) the shaded fields must necessarily (by definition) represent 0 or 0/0.

[129] Some other authors (for example Silver and Heravi) have repeated this exercise using the formula Walsh and Törnqvist.

The conventional Paasche index excludes goods which disappeared in the time between 0 and t as P^P is based on quantities relating to period t, and in a similar vein the Laspeyres index ignores new goods. Hence,

$$(4.3.2)\quad P_{0t}^{P} = (\mathbf{p}_t^{C'}\mathbf{q}_t^{C} + \mathbf{p}_t^{N'}\mathbf{q}_t^{N})/(\mathbf{p}_0^{C'}\mathbf{q}_t^{C} + \mathbf{p}_0^{N'}\mathbf{q}_t^{N}) = (A_1 + A_2)/(B_1 + B_2)$$

a formula which requires the vectors of prices of new goods in 0 to be estimated (imputed), that is the vector $\mathbf{p}_0^N$ in B_2, and

$$(4.3.2a)\quad P_{0t}^{L} = (\mathbf{p}_t^{C'}\mathbf{q}_0^{C} + \mathbf{p}_t^{D'}\mathbf{q}_0^{D})/(\mathbf{p}_0^{C'}\mathbf{q}_0^{C} + \mathbf{p}_0^{D'}\mathbf{q}_0^{D}) = (C_1 + C_2)/(D_1 + D_2)$$

where the vector $\mathbf{p}_t^D$ in C_2 calls for imputation.[130] Obviously P_{0t}^P does not account for disappearing goods, much like P_{0t}^L does not reflect prices of new goods. Fisher's index now is gained as usual by taking the geometric mean of both indices. This index

$$(4.3.3)\quad P_{0t}^{GF} = \sqrt{P_{0t}^{L} P_{0t}^{P}},$$

called the "generalised Fisher index" is related to the Fisher index using ongoing (C-goods) commodities only (i.e. the "normal" or "matched" Fisher index confined to the "intersection universe", thus labelled $P_{0t}^{F,int}$)

$$(4.3.4)\quad P_{0t}^{F,int} = \left[(\mathbf{p}_t^{C'}\mathbf{q}_t^{C})(\mathbf{p}_t^{C'}\mathbf{q}_0^{C})/(\mathbf{p}_0^{C'}\mathbf{q}_t^{C})(\mathbf{p}_0^{C'}\mathbf{q}_0^{C})\right]^{1/2} = \sqrt{A_1 C_1 / B_1 D_1}$$

in such a way that the generalised index P_{0t}^{GF} may be factored into three factors as follows

$$(4.3.5)\quad P_{0t}^{GF} = P_{0t}^{F,int} \cdot \sqrt{\frac{(A_1 + A_2)/A_1}{(D_1 + D_2)/D_1}} \cdot \sqrt{\frac{(C_1 + C_2)/C_1}{(B_1 + B_2)/B_1}} = P_{0t}^{F,int} \cdot \alpha \cdot \beta .$$

De Haan tries to give the following interpretation: The second factor (α) "re-scales" $P_{0t}^{F.int}$ "for the fact that the expenditures of new and disappearing goods, have not been taken into account".[131] Note that the second factor is composed of observed values only, unlike the third factor, β, which "contains imputed prices" (backward imputed for N - goods in the B_2 aggregate, and forward imputed prices in C_2) and "is needed to handle new and disappearing goods in the correct way". Obviously this (third) factor β can be written as $\beta = \sqrt{\frac{1 + C_2/C_1}{1 + B_2/B_1}}$, where ratios such as C_2/C_1 compare prices of disappearing goods with ongoing ones on the basis of base period quantities much like B_2/B_1 compares prices of new goods with ongoing ones on the basis of current period quantities.

---

130 In P^P certain base period prices of N-goods are fictitious, and so are current period prices of D-goods in PL. Note that aggregates B2 as well as C2 contain imputations (indicated by shadows) while all 1-aggregates A1, ..., D1 as well as A2 and D2 refer to observed values.

131 That is A2 for new and D2 for disappeared goods respectively, which means that values such as A2 (new goods in t) and D2 (disappeared goods in 0) are needed to correct a current period term (A - term) in the numerator and a base period term (D - term) in the denominator.

In a similar manner, the second factor α can be expressed as $\sqrt{\frac{1+A_2/A_1}{1+D_2/D_1}}$ and the factors can easily be rearranged in order to get an alternative decomposition (so that $\alpha\beta = \mu\lambda$)

(4.3.5a) $$P_{0t}^{GF} = P_{0t}^{F,int} \sqrt{\frac{1+A_2/A_1}{1+B_2/B_1}} \sqrt{\frac{1+C_2/C_1}{1+D_2/D_1}} = P_{0t}^{F,int} \cdot \mu_t \cdot \lambda_t .$$

"The factors μ and λ can be seen as the effects of new and disappearing goods, respectively, on the generalised Fisher price index". Moreover, the factors μ and λ are related to P_{0t}^L and P_{0t}^P as follows

(4.3.6) $$P_{0t}^L = \frac{C_1+C_2}{D_1+D_2} = \frac{C_1}{D_1} \cdot \lambda_t^2$$ and (4.3.6a) $$P_{0t}^P = \frac{A_1+A_2}{B_1+B_2} = \frac{A_1}{B_1} \cdot \mu_t^2 .$$

In other words:

If there were no difference in a Laspeyres-index of ongoing goods on the one hand and discontinued goods on the other, that is if $C_2/D_2 = C_1/D_1$, we also have $C_2/C_1 = D_2/D_1$, or $\lambda_t = 1$. Likewise, if the Paasche index of the new goods equals the Paasche index of the ongoing ones, we have $\mu_t = 1$.

μ and λ therefore are related to the extent to which the Laspeyres index (Paasche index) is likely to overstate (understate) the price movement as compared to the COLI, or a proxy for it such as Fisher's ideal index. From

$\mu_t < 1$ follows $A_2/A_1 < B_2/B_1$ or comparing two Paasche type indices $A_2/B_2 < A_1/B_1$ of new and ongoing goods respectively,[132] or according to eq. 6a

$$\mu_t^2 = \frac{P_{0t}^P}{A_1/B_1} = \frac{(A_1+A_2)/(B_1+B_2)}{A_1/B_1} = \frac{\text{Paasche (new and ongoing)}}{\text{Paasche (only ongoing)}}$$ and from

$\lambda_t > 1$ follows the relation $C_2/D_2 > C_1/D_1$ of two Laspeyres type indices comparing old to ongoing goods or $\lambda_t^2 = \frac{P_{0t}^L}{C_1/D_1} = \frac{\text{Lasp.(old and ongoing)}}{\text{Lasp.(only ongoing)}}$.

It may not be necessary to explain in detail the reasons given by de Haan for his expectation to hold $\mu < 1$ as well as $\lambda > 1$, given a competitive demand orientated economy. The phenomena of new and disappearing goods are thus similar to the effect of substitution (among ongoing goods only) and as prices evaluated using current quantities $\mathbf{q}_t^C$ tend to be lower ($P_{0t}^P < P_{0t}^L$) so prices of new goods tend to lower a relation compared to prices of old goods.

132 on the basis of qt, hence, for Paasche type indices. The corresponding relation with $\mu < 1$ rather than $\mu > 1$ easily follows.

Hence, we may say that what matters and might be responsible for the difference between the usual matched Fisher index $P_{0t}^{F,int}$ and de Haan's generalised index P_{0t}^{GF} is the

- imputed behaviour of N-goods (in 0) as compared with C-goods evaluated in a Paasche manner, i.e. referring to quantities q_t, and the
- imputed development of D-goods (in t) as compared with C-goods evaluated in a Paasche manner (with reference to q_0).

De Haan contends that his P^{GF}- index, though in need of imputing prices ($\mathbf{p}_t^D$ and $\mathbf{p}_0^N$), is better suited for handling dynamic aspects of price statistics in the universe of commodities than the traditional (matched) Fisher index. The GF-index is also related to a chained version of Fisher's index in that it constantly captures new developments and deletes obsolete commodities. In de Haan's view the most prominent advantages of chaining (see chapter 7 for an alternative view) are

1. to cause the set of ongoing goods not to shrink unduly over time and
2. to incorporate new goods as quickly as possible without allowing the set of new and disappearing goods to grow too large.

f) Accuracy of a price index

Situations in which an estimate of the accuracy of an index calculation is called for are not uncommon. The present author was asked for example by the judges of the German Constitutional Court to give an estimate of the accuracy of the regional CPI comparisons (Munich compared with other municipalities of Bavaria) he has worked out on request of the court in Dec. 2006 using of the monthly Bavarian CPI-statistics. Such questions usually are liable to place us in an uncomfortable situation because official statistics does not provide us with quantitative estimates of how accurate the data are that enter their index computations let alone with some rough idea about the confidence interval of the index as a whole.[133] An index is a point estimate and does not make a difference between 10 or 1000 prices collected for the same item, although it stands to reason that a price based on 1000 observations should be more reliable than if based on 10 only. This situation is clearly unsatisfactory and many authors have found it deplorable.

As Balk and Kersten (1985) rightly pointed out there is not much thought given to a theory of the accuracy of index compilations, or more precisely, there is astoundingly little attention paid to this issue compared with others like for example the "substitution bias" or the impact of quality adjustment. Worse even, as they also point out, it should be borne in mind that "much empirical work ... becomes futile when the statistical accuracy of the data used is taken into account" and when "the difference between the Laspeyres and Paasche price indices ... is statistically insignificant" (p. 20). It also is problematic to request a "more frequent updating of the weights" (in

133 Not only has the practice of official price statistics but also index theory left us in an uncomfortable position. Time and again an indication of accuracy attached to each index has been requested by theorists but they themselves have not been successful in delivering the appropriate instruments.

order to produce a chain index) or "the propagation of a different type of indices" for example for different socio-economic groups (types of households) in the case of a CPI without knowing much about the impact of the weighting scheme on the overall accuracy of an index, or a CPI in particular.[134]

The task to assess the degree of accuracy quite naturally leads to considerations in terms of random variables and sampling theory although very different sources of variation give rise to inaccuracy, not all of them (if not the smaller part of them) can be viewed as random variation. Indices are basically deterministic (non-stochastic) models such that there is no room for a random term reflecting an "error in equation" (due to "other" influences not explicitly accounted for in a regression equation)[135] as assumed in the "stochastic index theory" (sec. 2. 3). What the judges had in mind, to come back to this issue of inter-regional comparisons of the CPI, was not this kind of error resulting in a poor goodness of fit of the *model*, but rather an error due to an incomplete and not "representative" *data* basis. In this context we again think in terms of random variables, however, on the following grounds: in order to estimate a confidence interval the price quotations should be conceived as sample of all prices (possessing a probability distribution).

With little or no knowledge about the sampling error or possibly also systematic (biases!) errors in price quotation and weights the user of official statistics has almost no clue to estimate the range of the likely error. In the case of Munich's price level as compared to "comparable" larger towns and the average of Bavaria there were fortunately some aspects that could be considered as reducing the likely error. So a number of prices were uniform all over Bavaria or even nationwide (for example telecommunication, cigarettes etc.) and for all regional indices the same "basket" (structure of weights) was used. Also the plausibility of the rents for housing entering the index[136] could be checked from other sources. However, nothing was known about the variances and covariances of the other prices of goods (branded and non-branded) and (probably more divergent) prices of services. This is particularly regrettable as the Statistical Office – not the user of its data – could *in principle* at least estimate variances and covariances as it collects in general *a great number* of prices referring to the same good, or assess the impact of the latitude in the item description implying that items (and outlets) are possibly not strictly comparable across municipalities.

[134] Balk and Kersten, op. cit., p. 33. We also agree with their demand (addressed to official statistics) that every index number should be published with a measure of its accuracy. Apparently not much has changed ever since. Index compilations require a number of estimations in various steps of the compilations. We are not even equipped with reliable information about the confidence intervals involved in certain steps taken in isolation and much less we know about the over-all accuracy.

[135] Note that this sort of error is also effective in the case of data gained by a full enumeration (i.e. it also exists in the universe from which a sample may be drawn). It is a fundamental vagueness of the stochastic index theory that this approach does not clearly distinguish between this (*model*) error – when we wish to "explain" the observed price p_{it} as a function of certain regressors - and the sampling error (*data* error) where the price p_{it} is again a "random variable", however, for another reason.

[136] Interestingly – and good for our attempts to measure accuracy – they were both, perhaps the main source of regional disparity of consumer prices and representing a dominant share in the CPI-basket

Chapter 5

Deflation and aggregation

Price indices are compiled for two different purposes: firstly, to provide measures of the year to year movement of specific price *levels* (e.g. the rise or decline of consumer prices), and secondly to "deflate" aggregates. Chapter 6 is devoted to the first problem, and chapter 5 to the second. It is not surprising that in general an index formula is not able to serve both purposes equally well and we should therefore regard the two major uses of index numbers (measurement of price levels and deflation of aggregates respectively) as two distinct topics in index theory, each of which has a "logic" of its own.[1] There are also some mathematical properties of index functions, like "aggregative consistency" and (to be kept distinct) "structural consistency" that turn out to be of particular importance in the context of deflation and give rise to some additional "axioms".[2] This chapter also deals with international recommendations concerning the deflation of specific aggregates in National Accounts,[3] and once more with the so called "ideal index" of Irving Fisher.

5.1 Introduction to deflation methods

The focus of this section is to show what deflation aims at, and the methods that can be used to this end. There are several approaches to deflation and their suitability depends on the type of aggregate to be deflated.

a) Objectives and types of deflation

The operation called "deflating" or "deflation" is usually defined as isolating the volume component (that is quantity and quality) in an aggregate.[4] In this sense we introduced the concept "deflation" already in **sec. 1.1**: the task is to derive an aggregate at constant prices (quantities q_{it} valued at prices p_{i0})[5] from an aggregate at current prices symbolized as follows:

[1] In our view it is therefore (contrary to a widespread belief) not really an advantage of an index formula if it complies with factor reversibility and thus allows the construction of an index which may be used for price level measurement as well as deflation. See part c of sec. 5.1 for more details.

[2] In this respect chap. 5 also is an extension of the axiomatic approach introduced in chap. 2 and 3.

[3] For more about the recommendation of chain indices for the purpose of deflation see sec. 5.3.

[4] It should be noticed that the model of only two components in values and value change, that is a volume and a price component, may be regarded as too simplified. We will come back to this point.

[5] In the light of new approaches to deflation it is no longer generally accepted that "volumes" should refer to constant prices of a certain base period.

given: $\sum p_t q_t$	→	result: $\sum p_0 q_t$
value, an aggregate at current prices		volume, an aggregate at constant prices

The summation should take place over the same i = 1, ... , n commodities in both cases, a condition not necessarily given whenever chained indices are used as deflators because in this index design the collection of commodities may well vary from one link to another.

We are dealing with essentially the same problem if quantities referring to country A (denoted by q_{iA}) are to be valued at prices p_{iB} (i.e. another country B) instead of prices p_{iA} (the same country, A). Again, it is required that prices p_{iB} and p_{iA} should refer to the same n commodities. In practice it is, however, much more difficult to ascertain that the commodities (i) in the price-quotations p_{iB} and p_{iA} are identical than to make sure that prices p_{it} and p_{i0} really refer to the same commodities.

It may be useful therefore to regard deflation as the intertemporal variant of a much wider concept, which might be called **"revaluation"**. Generally speaking, the aim of revaluation is to compare values under different regimes of prices.[6] The valuation at constant prices of aggregates originally valued at current prices is only an example of revaluation.

A distinction should be added between two fundamentally different types of deflation, the

- **"volume approach"**, i.e. the calculation of "volumes" (a proxy for "total quantity"), and
- the **"real income approach"**, i.e. the measurement of "real" aggregates (or "real income"), in the sense of the amount of money a value would represent, if the purchasing power of money had been kept constant, or in other words, if inflation had not taken place.

In contrast to the concept of "isolating the quantity component", deflation in "real" terms (at constant purchasing power of money) applies also to so called **non commodity flows (NCFs)**[7], not only to aggregates representing a sale or purchase of (a certain quantity of) goods or services, that is, to **commodity flows (CFs)**.[8] Thus deflation in the sense of making adjustments (or neutralising) for the effect of inflation (that is for the diminishing purchasing power of money)[9] may be regarded as a more general approach. In the case of NCFs there is no observable quantity by definition. Therefore deflation cannot aim at "volumes" but has to focus on "real values". A typi-

---

[6] We may speak of "different regimes of prices", 1. in the case of prices that refer to different points in time and that differ due to inflation, and 2. if we are dealing with different countries.

[7] The term "flow" in this context is not meant as the opposite of "stock". The term "flow" is widely used in this context, although the considerations apply to stocks as well.

[8] This once more shows that in order to choose the correct method of deflation it is important to take into account the economic nature of the aggregate (see sec. 5.3).

[9] Note that strictly speaking the intention to "neutralise" inflation (as if it had not taken place) is – taken seriously – far more ambitious than just to divide an aggregate by an appropriate price index.

cal NCF is simply an amount of money where no well defined "quantity" component exists, like income, tax payments, or bank deposits.[10]

Figure 5.1.1: Notion of "deflation"

a) ***Types of "revaluation"***

Revaluation of aggregates for

intertemporal comparisons (= **deflation**)	interspatial (international) comparisons
calculation of $\Sigma q_{it} p_{i0}$ with current (t) quantities q_{it} and base period (0) prices p_{i0}	calculation of $\Sigma q_{iA} p_{iB}$ with reference to country quantities q_{iA} and base country (B) prices p_{iB}

b) Types of deflation

Deflation of aggregates

in volume terms	**in real terms**
"at constant prices"	at constant purchasing power of money
Volume-approach (quantity interpretation of the result intended); isolation of the quantity component	Real-income-approach (adjusting income for inflation): the aim is to estimate (adjust for) the effect of inflation
Applicable to value changes only, that can be decomposed into price- and quantity-changes, i.e. to commodity flows (CFs)	Applicable also to values that do not have price and quantitiy dimensions on their own, i.e. to non commoditiy flows (NCFs)
Deflate aggregates $A_1, A_2, ..., A_m$ (representing CFs) by using m price indices P_i (i = 1,...,m) for the m aggregates (P_1 refers to the commodities in A_1 and P_2 to A_2 and so on)	Deflate aggregates $A_1, A_2, ..., A_m$ (all or many of them NCFs) using *one single* deflator that accounts for the general development of the purchasing power

the typical problem in applying the method is

double deflation (i.e. deflating C = A - B by deflating A and B separately)	**choice of *the*** general (universal) **deflator**, that has to be applied to *all* aggregates

Abbreviations: CF = commodity flow, NCF = non commodity flow

[10] This is an example of a non commodity stock. Deposits are by definition not yet spent for specific goods. It therefore would not make much sense to construct some physical "quantity" for this NCF.

Note that the distinction between two types of deflation refers not only to questions of applicability (CFs or NCFs) of one approach or the other, but also to fundamentally different interpretations of the objectives and results of "deflation" and to different methods, data and indices that are required to pursue this approach and the other. Each approach has a typical problem to solve: in the volume approach it is the problem of double deflation, and when the real income approach is applied, the problem is which price index can be taken as *the* (universal, general) deflator.[11]

In some instances the same aggregate may well be deflated using either or both deflation methods. Thus, for example, the external balance (export [X] less import [M]) can be deflated in real terms yielding E_R, and in volume terms by deflating X and M separately with a price (or unit value) index of exports and imports respectively, giving a volume E_V. The difference E_R - E_V is known to be the "terms of trade effect" (T). The gross domestic product (GDP) at constant prices (volume) plus T is the "real value" of the GDP or sometimes also called "income" (GDI), from which we arrive at the (gross) national income (GNI) by adding the real value of the balance of primary incomes (net income received from the rest of the world deflated by using a suitable measure of purchasing power).

Note also that deflation in real terms does not give rise to problems with "additivity" (or more precisely "structural consistency") as discussed in **sec. 5.2** because in this case the structure of prices (relating to period t) is left unchanged and all subaggregates are adjusted in unison by exactly the same rate. The situation is similar to the case of all price relatives being equal. Just as there is no index problem[12] once all prices change at the same rate, so aggregation will be easy once all subaggregates are deflated using the same deflator. Hence we may also say that deflation in volume terms as opposed to deflation in real terms accounts for changes in the level as well as the structure of prices.

b) Direct and indirect volume measurement, new products

The following considerations are restricted to the volume approach and to CFs so that there is no problem in defining "quantities" of goods "belonging" to the aggregate in question. In principle there are three ways of arriving at an absolute volume, or at an index reflecting the development of the volume, i.e. at a quantity index[13] as shown in **fig. 5.1.2**. Theoretically it should be possible to calculate for each individual commodity i the product $p_{i0}q_{it}$ instead of $p_{it}q_{it}$ and to sum up the volumes to $\Sigma p_{i0}q_{it}$. This is called method 1 (repricing individual expenditures). Of course this is a "method" of theoretical interest only, having no relevance for the practice of price statistics, where deflation is usually done with the help of a comprehensive (covering a whole group of commodities) "deflator" price index (i.e. by method 2).

[11] This is also indicated in fig. 5.1.1.

[12] In the sense of the choice of an index formula simply because all resonable formulas will yield the same result under such conditions.

[13] Deflating is also the task of converting a value ratio (or index) V_{0t} (instead of an absolute value) into a quantity index Q_{0t} (instead of an absolute volume).

Figure 5.1.2: Alternative methods to estimate volumes or volume (quantity) indices

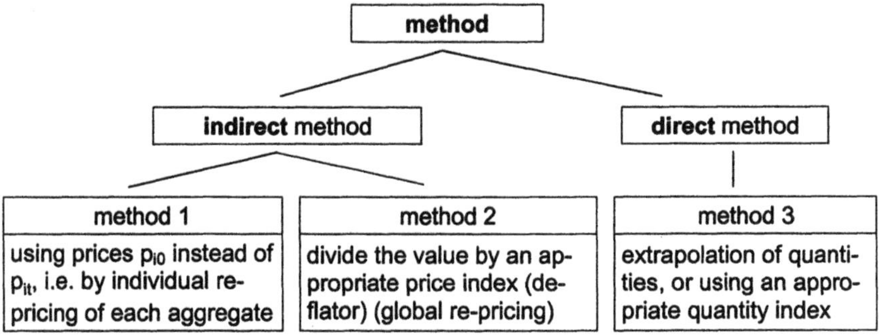

It is, of course, more convenient to provide a revaluation of all items (i = 1,...,n) in a more global manner by dividing $\Sigma p_{it}q_{it}$ by the appropriate (comprising the same n commodites) deflator-index P_{0t}. The deflator should be a price index according to Paasche (P_{0t}^{P}), but many other formulas are recommended as well.[14]

If a choice has to be made between a direct measurement of volumes (by observing quantities) and the indirect method of deflating values (by using price indices) the indirect method (that is method 2 in **fig. 5.1.2**) may be preferred on the following grounds:

1. Statistics of values and prices may be more readily available than those of quantities.
2. Both statistics of prices and of quantities will usually be collected to be representative, that, is a selection or (random) sample will provide the necessary data. By the very nature of production as an aggregate in contrast to the "price level" (a sort of an average), however, the direct approach ought to aim at a complete coverage whilst for the indirect approach it suffices to have observations representative of the general trend. Thus less data have to be collected. Moreover, since price relatives tend to vary less than quantity relatives, the indirect method may be associated with a smaller sampling error.
3. It is in general much easier to deal with new products and changes in the quality of existing products in the framework of price indices, than to account for such phenomena correctly in quantity indices.

A short remark on **new products** is in order. Assume two prices and quantities are observed but commodity 2 is new in t and had not been on sale in period 0 such that $p_{20} = q_{20} = 0$. The quantity relative of commodity 2, that is, q_{2t}/q_{20}, would tend to infinity because $q_{20} = 0$. To deflate the value ratio $(p_{1t}q_{1t} + p_{2t}q_{2t})/(p_{10}q_{10})$ with a price index of the remaining commodities (in this case: the price relative p_{1t}/p_{10}, a deflator ignoring p_{2t}) would result in the following "volume" (quantity) index:

[14] In fact, international recommendations (U.N. and European Union) prefer the direct or even the chain version of Fisher's index.

$$(5.1.1)\quad \frac{p_{10}\left(p_{1t}q_{1t}+p_{2t}q_{2t}\right)}{p_{1t}p_{10}q_{10}}=\frac{p_{10}q_{1t}+\left(p_{10}\frac{p_{2t}}{p_{1t}}\right)q_{2t}}{p_{10}q_{10}}\quad \text{(given } p_{20}=q_{20}=0\text{)}.$$

which is equivalent to making the assumption $\hat{p}_{20}=p_{10}\frac{p_{2t}}{p_{1t}}$, or that both prices vary in proportion $\frac{p_{1t}}{p_{10}}=\frac{p_{2t}}{\hat{p}_{20}}$, such that $\frac{p_{10}q_{1t}+\hat{p}_{20}q_{2t}}{p_{10}q_{10}}$.

c) Inflation measurement and deflation (volume measurement)

It is open to discussion whether price level measurement and deflation should be solved in one step only with the same price index, or rather be treated separately, as two different problems requiring different price indices. The SNA takes the first position and regards the implicit deflator gained as a byproduct of the (direct or indirect) measurement of volume aggregates (like real GDP) also as a valid measurement of inflation (increase of "general price level").

Figure 5.1.3: Alternative views concerning uses of price indices

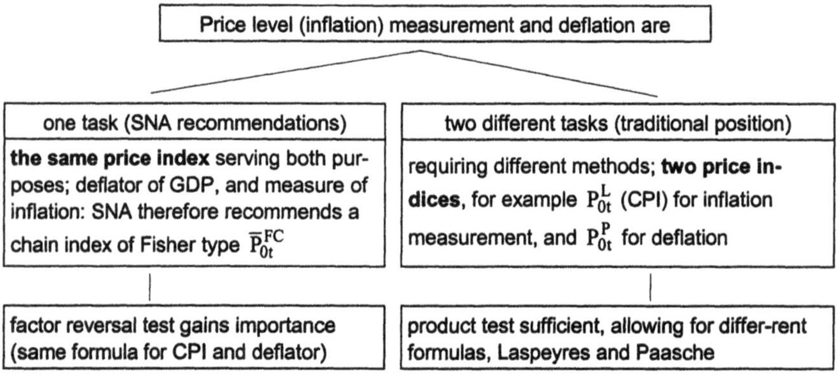

For such a position, the **factor reversal test** seems indispensable. The inspiration for this test is the desire to use the same index for both problems, deflation and price level measurement. As the SNA sets great store by factor reversibility, it takes pride – precisely for this reason – in its recommendation of (chained) Fisher indices for *direct* (using $\overline{Q}_{0t}^{FC}$), and *indirect* (using $\overline{P}_{0t}^{FC}$ as deflator) volume measurement respectively.

The traditional view, which we advocate with emphasis, is, however, that factor reversibility is of limited use and that the two tasks should rather be separated. There is a number of reasons why both problems (deflation and inflation measurement) cannot or should not be solved with the help of the same price index, and generally speaking there is no satisfactory multi-purpose (for different [or all] types of economic analysis) solution of index problems anyway. To name only a few of the aspects in which deflation and inflation measurement differ: in National Accounts (NA) aggregates we find both, CFs and NCFs, requiring two different approaches, a volume and a (real) income approach. However, only CFs appear in inflation measurement with a consumer price index (CPI). Furthermore we have imputations, or to account, for inner-sectoral and non-market transactions (where no price exists by definition) in NA, in contrast to the CPI situation. On the other hand, the GDP deflator is generally not considered suitable for monetary policy, and unlike NA aggregates a CPI usually refers to a certain type of household. Thus there are several reasons for *two different* price indices and there is no point in striving for factor reversibility; the product test suffices.

d) Deflating National Accounts aggregates

Both types of deflation, that is, the isolation of a "quantity component" ("in volume terms") as well as "deflating" in the sense of adjusting for inflation ("in real terms"), are particularly important in the context of National Accounts (NA).

As shown in **fig. 5.1.4** there are mainly two different situations:

1. In the so called production approach and the expenditure approach (see **fig. 5.1.4**) we encounter mainly commodity flows (CFs) and balancing items,[15] thus aggregates clearly calling for deflation in volume terms and double deflation[16] respectively, which are widely or almost generally accepted procedures for this part of the NA.
2. In the income approach to the NA, however, a breakdown of national income is made into *types* of primary incomes[17] and sectors receiving these incomes from domestic production activities as well as from the rest of the world. Hence[18] this part of NA clearly deals with NCFs only and therefore a volume approach to deflation seems to be inappropriate altogether. Most statistical offices apply here deflation in terms of real income, that is, the application of one single deflator to all aggregates, or they refrain from calculating a "deflated" income.

---

[15] They have to be treated as non commodity flows (NCFs).

[16] The so called **double deflation** method (see **sec. 5.2.b**) consists of applying the volume oriented method to aggregates that are defined as differences between two aggregates (e.g. value added). The justification of this method is highly controversial, since double deflation may yield somewhat awkward results (e.g. real value added being negative).

[17] These are wages and salaries on the one hand (or more precisely "compensation of employees"), and property incomes, i.e. interest and rents, profits and the like on the other. These are inherently non-commodity flows (NCFs), which applies even more pronouncedly to balancing items, such as "mixed incomes" (comprising both operating surplus accruing to the owner and compensation for the owner's labor) and the remaining parts of the operating surplus that have not been distributed as primary incomes to other units.

[18] The distinction made between CFs and NCFs widely corresponds to the distinction between the two methods of deflation.

Figure 5.1.4: Production, expenditure and income approach in National accounts (NA)

NA breakdown into aggregates, predominantly

- commodity flows (CFs) and balancing items
- non commodity flows (NCFs)

Production approach: supply side	**Expenditure approach: uses side →GDP = F - M**	**Income approach: distribution of incomes**
Total supply (S)	Final expenditure (F), GDP	National income
S = Y + M + T;	F = S –Z = C + G + K + X	NDP + PI

Definitions

C	Final consumption of households (Private consumption)
G	Final consumption of government (Collective consumption)
FC	Final comsumption = FC = C + G
F	Final expenditure F = FD + X
FD	Final domestic expenditure FD = FC + K
PI	net Primary incomes from the ROW[b)]
T	Taxes less subsidies on products[c)]
VA	value added (prominent example of an NCF)
Z	Intermediate consumption (P.2)

Definitions, Identities

VA	value added (= Y – Z), a balancing item)
GDP NDP	Gross domestic product GDP = F – M, Net domestic product NDP = GDP – D
K	(gross) Capital formation (P.5)[a)]
D	Consumption of fixed capital (K.1)
M	Imports (P.7)
X	Export
S	Total supply
Y	Output (P.1)

a) of which gross fixed capital formation (= investment, P.51), Changes in inventories (P.52) and Acquisition less disposal of valuables (P.53)

b) ROW = rest of the world

c) Aggregates D.21-D.31

It follows from above that for deflation of aggregates in National Accounts (NA) holds:

NA are perhaps the most important field in which deflation becomes relevant. Furthermore, all types of aggregates (CFs and NCFs) and deflation methods (volume-approach and real-income-approach) introduced so far coexist in a single comprehensive system of accounts. This imposes the discipline of both conceptual as well as numerical consistency.

In practice the results of both methods of deflation are not always consistent with one another. Data are drawn from different sources and the methods differ with respect to their rationale. Discrepancies are therefore common, but in some cases they give rise to interesting interpretations, as for example in the case of the so called "terms of trade effect".[19]

e) Quantities and qualities, price level and price structure

According to some authors, it is insufficient to distinguish in the case of a value change $V_{0t} = V_t/V_0$ (or $V_t - V_0$) between two components only, a price component and a volume component. In order for the movement of values to be analyzed correctly Schimmler (1973), for example, maintains that the change of two other "elements" has to be considered:

1. "structural changes in physical *quantities*" or "changes in *quality* structure" (which is in general accounted for in the volume [quantity] component), meaning that the mix of goods assembled in the "basket" may undergo a change by quantity and by type,[20]
2. "change in *price structure*", an aspect which might be described by the variance of price relatives (if all prices would rise or fall at the same rate, the variance would be zero).

However Schimmler's view induces a number of problems:

- The extent to which a structural component can be made visible is a question of the *level of disaggregation*. An index A with aggregates broken down into n subaggregates A_i ($i = 1,2,\ldots,n$) will be more influenced by "structural" changes than an index B in which only m subaggregates B_j ($j = 1, 2, \ldots, m$ and $m < n$) are distinguished.
- Furthermore, it will be difficult to find a solution for the problem of how to measure a pure *structural* component, as opposed to the "*level*" component,[21] and to distinguish "more" or "less" change, not only "change" and "no change" of structure.

To solve the second problem, the difference between a unit value index[22] and the corresponding price index might be seen to reflect to a certain extent a "structural change" in the sense that consumers bought more (or less) of the high priced (low

[19] discussed in part g of this section.

[20] Note that these are in fact two components, since change in the variety of goods is not dependent on a change in the quantities.

[21] In general the phenomenon a price index reflects will always be a combined effect of changes in the level as well as in the structure of prices, if not some other changes too (such as changes in the qualities, the composition of the "basket" and the quantities in the basket). All these change occur simultaneously and they can hardly be separated conceptually, let alone in terms of their "amount", or "extent".

[22] See **sec. 6.4** for a discussion of unit value indices.

priced) products. But does this also reveal a change in quality? Finally, it is well known that if the price structure remains constant, all price indices would be equal (for example $P^L_{0t} = P^P_{0t}$), but if on the other hand the structure changes, is then the degree of change a simple function of the difference $P^L_{0t} - P^P_{0t}$, or is the "degree of change" something more complex?

f) Deflation in terms of "real income"

Income can not always be factored into a price and a quantity component. The underlying "quantity", if there is any, can be conceived as

- either an amount of labor calling for an index of wages, or an "Employment Cost Index" (ECI) as a deflator, or
- an amount ("basket") of consumer goods and services, making a consumer price index (CPI) the more appropriate deflator.

The latter approach leads to a definiton of real income as follows:

(5.1.2) $$\text{real income} = y_r = \frac{\text{nominal income}}{\text{consumer price index}} = \frac{y_n}{\text{CPI}}.$$

It should be noted that y_r cannot be interpreted as a "quantity" because neither nominal nor real income has a "quantity structure" (like the "basket" in the CPI). Even if y_n were a sum of price-quantity-products, $\sum p_{kt}q_{kt}$, and the CPI as usual a P^L_{0t} index, we would get

(5.1.2a) $$\frac{\sum p_{kt}q_{kt}}{P^L_{0t}} = \frac{\sum p_{kt}q_{kt}}{\sum p_{it}q_{io}} \sum p_{io}q_{io} = S_1 \neq S_2 = \sum p_{ko}q_{kt}$$

which differs from a real income $\sum p_{k0}q_{kt}$ (as a fictitious commodity flow) for a number of reasons, such as the fact that summation in y_n and y_r would take place over the products k = 1, ..., N, whereas P^L_{0t} refers to products i = 1, ..., n. Thus, even if income were simply a sum of purchases, real income S_2 will in general differ from S_1. Less doubtful, as compared with eq. 2a is, however, the *relative* expression Q^P_{0t} in $Q^P_{0t} = \frac{V_{0t}}{P^L_{0t}}$.

g) Terms of trade effect

The terms of trade effect (TE) is an interesting example of how economic analysis makes use of the different methods of deflation. The TE measures the influence of the terms of trade (T) on gross national income (GNI) or gross national product (GNP). The TE consists essentially of the GNP and GNI being different. The extent to which the real value (with respect to the purchasing power of money) of the GNP, which is called GNI, differs from the GNP at constant prices (volume) is called TE.

Let

(5.1.3) $N = X - M$

be the external trade balance at current prices, with exports (X) including primary incomes received and imports (M) including primary incomes paid. If net primary incomes from abroad were excluded we would study the relationship between aggregates on a domestic base (GDP and GDI) instead of "national" concepts.

To measure N in volume terms, X and M have to be deflated using the appropriate price indices P_X and P_M respectively. Thus we get

(5.1.4) $$\frac{N}{P_N} = \frac{X}{P_X} - \frac{M}{P_M} = X_V - M_V = N_V \text{ (V = volume).}$$

The implicit deflator of the trade balance P_N is of course again an index obtained by double deflation.[23] To calculate N in real terms we have to divide both X and M by the same deflator P giving

(5.1.5) $$N_R = \frac{N}{P} = \frac{X - M}{P}.$$

The excess of real expenditure (income) N_R over real product (in volume terms), that is, N_V, is called terms of trade effect (= G)

(5.1.6) $$G = N_R - N_V$$

where $G > 0$ represents a gain, and $G < 0$ a loss from external trade. The points of interest are mainly

- the relationship between G and the so called terms of trade (T), and
- the type of indices to be used to calculate N_V and N_R.

There are of course various definitions of terms of trade commonly in use. For our purposes it will be sufficient to focus on only two of them, the so called net (barter) terms of trade

(5.1.7) $$T_N = \frac{P_X}{P_M} \text{ (net barter terms of trade), and}$$

(5.1.8) $$T_G = \frac{X_V}{M_V} \text{, the gross (barter) terms of trade.}$$

It is often said that an improvement in the terms of trade ($T_N > 1$) affects the trade balance favorably, by which is meant an increasing surplus ($N > 0$) or a decreasing deficit ($N < 0$).

It turns out that more complication is involved in saying "favorably" than may appear at first sight. The difficulty lies in that it depends on whether a surplus or a deficit has been assumed as starting point. The following holds true:

[23] A similar result will be obtained in double deflation. Note that P_X may be regarded as a harmonic mean of P_N and P_M. However, since X and M also contain income flows (NCFs), the deflators P_X and P_M are themselves already derived by some sort of double deflation.

In principle *T will increase* if
$P_X > 1$ and $P_M < 1$ that is the prices of X rise, those of M fall
P_X and P_M both rise but P_X more than P_M, such that $P_X > P_M > 1$
P_X and P_M both fall but P_M more than P_X, hence $P_M < P_X < 1$.
There are situations, however, in which the reaction of T is not uniquely determined.

To see how the balance at current prices will react, net export N has to be expressed in terms of T_N and T_G. From eq. 3, 7 and 8 we obtain

(5.1.9) $$N = M_V P_M (T_N T_G - 1) = M(T_N T_G - 1).$$

We refrain from giving a numerical example and consider the three cases outlined above only by giving a summary table. There are two situations in which a change might take place in either direction:

situation	surplus	deficit
1: $P_X > 1$, $P_M < 1$	nominal surplus becomes greater	deficit becomes smaller or changes into a surplus
2: $P_X > P_M > 1$	greater surplus	change either way
3: $P_M < P_X < 1$	change either way possible, surplus greater or smaller	improvement, i.e. diminishing deficit or surplus

h) The choice of "the" deflator (the notion of the "general price level")

Deflation in terms of "real income" (to adjust for a general decline in the purchasing power of money) is implicitly based on the assumption of a "general price level", the existence of which is not self evident. There is, for example, nobody facing "all" prices in his purchases or sales. Hence it is difficult, if not impossible, to conceive and establish empirically a "basket" and "weights" comprising all (!) sorts of goods and services in exactly the proportions as required by "the" general price level. In what follows, we present the pros and cons of various price indices from the point of view of their aptitude to represent "the" general purchasing power of money. Price indices that can be considered such a comprehensive measure are primarily:

- the GDP deflator,
- the deflator of domestic final demand, and
- the overall (all-items) CPI.

By convention, the CPI is considered a broad measure of inflation and therefore has a long tradition in the focus of central banks in particular if inflation targeting monetary policy is pursued. On the other hand, the coverage of the CPI is limited, not comprising for example intermediate consumption, capital formation, collective consumption etc. In contrast, the Producer Price Index (PPI) for total output (see **fig. 5.1.4**), or YPI[24] for short, is more comprehensive and may thus better serve as a de-

[24] The YPI, however, is broader in scope than a "total" PPI which in general covers market-production only.

flator in the real income approach. On the other hand, PPIs (and hence also the YPI) only cover domestically produced goods, whereas the CPI is also affected by imports.[25]

In a price index for final uses (FPI for short) intermediate consumption is excluded. On the other hand, detailed product information for a breakdown of the FPI by products is more readily available than for a YPI, for example. In contrast to the CPI, or the Private Consumption Deflator (PCI in tab. 5.1.1) of National Accounts, the FPI is also influenced by Collective Consumption as well as capital formation (including inventories).[26]

Table 5.1.1: Comments on possible general deflators

price index (PI)	coverage	interpretation as "the" deflator
CPI, Consumer Price Index	Only individual consumption (definition differs from to PCI).	Unlike PCI a Laspeyres index
SPI, for Total Supply	Unlike YPI[1)], covers imports in addition to domestic products.	Fot Input-Output-Tables rather than real income deflation.
FPI, for Final Uses (or: Final Expenditure), F	Comprises final comsumption, capital formation (= final domestic expenditure) *and*[2)] exports.	In the framework of the so-called "asset inflation" it is desirable to include capital formation as in FPI.
FDPI, Final Domestic Demand	Final Domestic Demand is FD = F – X, hence also comprises investment	This index shows to what extent inflation is "homemade".
GDP deflator	Gross domestic product differs from FPI by the exclusion of imports.	Results from double deflation, and therefore subject to estimation errors
PCI, Private consumption[3)]	Differs from CPI as regards some taxes, payments for services etc.	Like value added gained by double deflation.
ECI, Employment Cost Index	Compensation of employees[4)] by occupations.	Measures the price of labor services only, no other factors.

1) YPI is the producer price index of total output (see above)

2) unlike final *domestic* expenditure,exclusive of exports

3) more precisely "Final consumption of households"

4) comprising wages and salaries and employer's social contributions (Note that the U.S. ECI is different from the European Labor Cost Index LCI)

[25] The price index of total supply, SPI (as opposed to output), covers output and imports valued at basic prices (as opposed to purchasers' prices in the CPI) and thus can be defined (and derived) as a weighted mean of an output index (YPI) and an import index (MPI). An index of the SPI type is useful for analytical purposes, supply- and use-, or input-output-tables and double deflation. As broad measures of inflation, however, price indices for final uses are more appropriate.

[26] Such items are relevant for the so called "asset inflation" (see sec. 6.2). It should be added that an FPI (as well as its components such as a PCI) does not include consumption from own production but does include for example owner occupied dwelling.

Indicators of labor costs such as the ECI in tab. 5.1.1 are suitable to deflate wages and salaries and therefore relevant for the perspective of most of the general public when "inflation" is meant. The ratio of the ECI (or LCI) with (divided by) the CPI indicates the purchasing power of wages in terms of consumption of goods and services.

5.2 Deflation in volume terms, aggregation and double deflation

We first (in part a) consider what might be understood by "volume" in the case of commodity flows (CFs), where an isolated quantity component of the aggregate exists. Another problem discussed in this section (part b) is the problem of double deflation. Finally (parts c through e) some notions like "aggregation" and "additivity", mentioned repeatedly already, will be explored in more detail.

a) Interpretation of "volumes", axiomatic considerations

To examine under which conditions the result of deflation could be interpreted as a measure of (or substitute for) "quantity"[27], we assume it is possible to observe quantities directly and to aggregate them to a meaningful total quantity Σq_{i0} or Σq_{it}, such that indicators of the rise or decline of quantity ($M_{0t} = \Sigma q_{it} / \Sigma q_{i0}$) and of the structure of quantities ($q_{i0}/\Sigma q_{i0}$ and $q_{it}/\Sigma q_{it}$) can be provided. Under such conditions it may be interesting to compare

- the *volume* index (not necessarily Q_{0t}^{L}) resulting from deflation with M_{0t}, and
- the *structure* of volumes with the structure of quantities.

Concerning prices and quantities in period t compared with period 0, we assume

1. prices (quantities) change by the same rate ($p_{it} = \lambda p_{i0}$ and $q_{it} = \omega q_{i0}$); note that constancy ($p_{it} = p_{i0}$ and $q_{it} = q_{i0}$) is simply a special case ($\lambda = \omega = 1$), or
2. a change with different rates will take place.

Hence we have the four situations of **fig. 5.2.1** along with the properties of volumes derived from deflations that appear reasonable to require in these respective situations. In general volume (as opposed to value) should be affected by changes in quantities only (or at least "isolate" in some way the quantity component). Hence in cases 11 and 21 it appears reasonable to require a volume index to be $Q = M = \omega$ whatever the change of prices may be. It can easily be shown[28] that notably a deflation with a *chain* (unlike a *direct*) Fisher price index $\overline{P}_{0t}^{FC}$, as recommended in the SNA, can well violate this requirement of proportionality.[29]

[27] As already mentioned, it is sometimes said that "volume" reflects both quantity and quality. Note that the following assumptions (existence of sums Σq) are generally not met in practice.

[28] In **sec. 5.4** this will be demonstrated with a numerical example, and we come back to this point in **sec. 7.2**.

[29] For more about this axiom see **sec. 3.1**.

Applied to the inherent *quantity index*[30] (given that there are at least three links multiplied to a chain). In our view, this is an aspect the SNA unfortunately has not given the attention it deserves.[31]

Figure 5.2.1: Requirements reasonably interpretable "volumes" should fulfill

Prices	Quantities	
	(1) same* rate ω	(2) different rates
(1) same* rate λ	case 11	case 12
(2) different rates	case 21	case 22

* the case of constant prices/quantities is the special case of $\lambda = 1$, or $\omega = 1$ respectively

How should deflation reflect the situations distinguished above? The following criteria are increasingly restrictive **PQC** $\subseteq$ **RQM** $\subseteq$ **PQ**.

Case(s)	Requirements	Comment
11 and 21	Volumes should be proportional in the quantities (**PQ**, proportionality applied to the **q**uantity index)	If all quantities are changing uniformly at the same rate ω, deflation should result in a volume index reflecting the *same* rate of change ω. Not even this "common sense" criterion is met in the case of deflating with a Fisher chain index
12	Reflection of differential quantity movement [**RQM**]	Volumes (of subaggregates) should rather rise above (below) average as quantities rise above (below) average.
22	Pure quantity comparison [**PQC**]	Volumes are linear (additive) in the quantities[a)]. Equivalent to structural consistency of volumes (SCV), often called "additivity"[b)] too.

a) The quantity index reflecting the volumes and resulting from deflation should be

b) or **A3** in **fig. 5.2.3**.

RQM in case 12 the structure of volumes should not remain constant whenever quantities change with *different* rates and prices may change uniformly. Dynamics and structure of volumes should be equal to dynamics and structure of quantities, because there is no differential price movement to disturb this relation between quantities and volumes. It should be noted that PQC is more demanding than RQM: by RQM it is only required that a *differential* (with respect to subaggregates) change in quantities should lead to a *differential* rise or decline in volumes. It is *not* said that

[30] It can easily be seen that this is equivalent to the so-called value preserving test (Vogt) and that it is satisfied in the case of deflating with a direct Paasche or direct Fisher index because the situation all quantities $q_{it} = \omega q_{i0}$ results in $P^L = P^P$ and therefore also $P^P = P^F$.

[31] In contrast to the non-additivity problem for example, arising from a deflation with a chain as well as a direct Fisher price index.

quantities and volumes should vary in proportion as required in PQC. Obviously from PQC follows RQM (such that PQC → RQM) but the converse is not true. When prices change differently it might be desirable to have volumes that change in parallel with quantities by eliminating the effect of a changing *structure* (not only level) of prices. PQC is a rather restrictive requirement; it can be met only by using the *same* single set of prices (say p_{i0} as in the case of the traditional deflation with P_{0t}^{P})[32] when volumes at different points in time are calculated and consistently compared with one another. In this case, volumes are linear in the quantities (q_{it}), with prices p_{i0} as constant weights.

The following **ex. 5.2.1** demonstrates that under the conditions of case 12 (prices change uniformly by $\lambda = 1.5$), deflation using the appropriate (direct) Paasche index P_{0t}^{P} and deflation using the (direct) Fisher index P_{0t}^{F} will yield the same result, completely in line with RQM (because under such conditions $P_{0t}^{F} = P_{0t}^{P}$). Note that a *chain* Fisher index fails this test.

Example 5.2.1

The following prices and quantities of four commodities, A_1, A_2, B_1 and B_2 classified into two groups, A and B are given:

	p_{i0}	q_{i0}
A_1	30	70
A_2	50	30
B_1	90	20
B_2	120	30
	Σ	150

The point of interest is not only how indices will reflect the movement of values, volumes, prices and quantities, but also the structure of these variables as measured by the shares of group A and B respectively at time 0 (shaded):

group	values		quantities	
	$\Sigma p_0 q_0$	share	Σq_0	share
A	3600	2/5 = 0.4	100	2/3 = 0.667
B	5400	3/5 = 0.6	50	1/3 = 0.333
sum	9000	1	150	1

32 Deflation using P^P results in a quantity index Q^L which is linear in the quantities q_{it}. To require deflation to meet PQC, is also consistent with the so-called direct deflation by observing quanties or "updating of volumes" (see part f of this section). To enable a deflator to provide volumes in accordance with PQC volumes ought to be constantly weighted sums of quantities (weights need not be p_{i0} if only they are constant), such that volume movement will not be affected by price movement.

To study the cases 12 and 22 we make the following assumptions concerning the prices and quantities at time t:

	prices/quantities in case 12		prices/quantities in case 22		price-/quantity relatives in case 22	
A1	45	84	30	84	1.0	1.2
A2	75	48	40	48	0.8	1.6
B1	135	36	81	36	0.9	1.8
B2	180	24	168	24	1.4	0.8
prices	uniform change		non-uniform change			
quant.	non-uniform change		non-uniform change			

In the case of a uniform change in prices by $\lambda = 1.5$ (case 12) and quantities changing at different rates (from 0.8 to 1.8), price indices P^L and P^P and consequently also P^F are necessarily equal ($P^L = P^P = P^F = 1.5$). Hence in case 12 deflations by P^P and P^F will yield the same result:

	value	V_{0t}	$\Sigma p_t q_0$	**volume** $\Sigma p_0 q_t$	volume structure
A	7380	2.05	5400	4920	0.446
B	9180	1.70	8100	6120	0.554
sum	16560	1.84	13500	11040	1

Comparison of volumes and quantities reveals that the rise (from base period 0 to t) in volume by 22.67% is no longer equal to the rise in quantity, which was 28%.

	volumes			quantities		
	structure (time 0)	structure (time t)	Q_{0t}^L	structure (time 0)	structure (time t)	M_{0t}
A	0.4	0.446	1.3667 = 2.05/1.5	0.667	0.6875	1.32
B	0.6	0.554	1.1333 = 1.7/1.5	0.333	0.3125	1.20
			1.2267 = 1.84/1.5			1.28

The group of commodities for which this rise of quantities was above average (that is group A) also experienced an increase in volume above average and therefore a higher share with respect to volumes (0.446 > 0.4) as well as quantities. This is in line with the requirement RQM and with what we expect deflation should produce. ♦

The structure of volumes will generally not coincide with the structure of quantities.[33] This can easily be seen by expressing Q_{0t}^L and M_{0t} as arithmetic means of quantity relatives x_i and weights w_i or v_i respectively in the following way:

(5.2.1) $Q_{0t}^L = \Sigma x_i w_i$ where $x_i = q_{it}/q_{i0}$ and $w_i = p_{i0}q_{i0}/\Sigma p_{i0}q_{i0}$

[33] A (trivial) case of both structures being identical and also changing identically is given once all base period prices are the same, that is $p_{i0} = p_0$ for all commodities i.

(5.2.2) $M_{0t} = \Sigma x_i v_i$ where $v_i = q_{i0}/\Sigma q_{i0}$.

Hence: The difference between volumes and quantities is a result of the structure of base period prices. The same increase in quantity will be weighted higher if a high priced good is involved and lower in the case of a low priced good (prices of the base period). Prices p_{i0} (acting as constant weights) will also play a part when we try to justify that a "volume" can reasonably be taken as a proxy for "total quantity".

Note that according to eq. 5.2.1, the result of deflation by using Paasche's index, that is, Q^L_{0t}, depends only on quantities q_{i0} and q_{it} and on *constant* prices p_{i0}; it therefore complies with PQC. Such volumes as linear combinations of quantities according to PQC appear reasonable regarding the standard interpretation of volumes as substitutes for (aggregate) quantities, otherwise unobservable:

It appears justified to regard volume as a measure of the aggregated "quantity" just *because* the sums $\Sigma q_{it}p_{it}$ and $\Sigma q_{it}p_{i0}$ are different only due to different prices, p_{it} and p_{i0} respectively, and each element in the sequence $\Sigma q_{i0}p_{i0}$, $\Sigma q_{i1}p_{i0}$, $\Sigma q_{i2}p_{i0}$, ... differs from the other ones only by the set of quantities.

It is interesting to note (see **ex. 5.2.2**) that deflation using Fisher's (direct) price index ("Fisher deflation"), in contrast to traditional "Paasche deflation", is unable to meet PQC, since volumes are not structurally consistent.

Example 5.2.2 (= ex. 5.2.1 ctd)

We now examine case 22 to see whether the conclusions drawn in case 12 will still be valid. Now not only quantities but also prices vary at different rates and we get:

	value	V_{0t}	P^P	volume P^P	volume P^{F*}
A	4400	1.233	0.9024	4920	4881.7
B	6948	1.287	1.1353	6120	5871.7
sum	11388	1.265	1.0315	11040	10658.6**

* (deflation with Fisher index) for purpose of comparison only (see ex. 5.51 for more details)

** obtained by deflating 11388 with the overall Fisher price index, but 4881.7 + 5871.7 = 10753.4 ≠ 10658.6.

It should be noticed that with respect to the *quantity* movement, this case (22) does not differ from the preceding case 12. Therefore, the results of deflation using a Paasche price indices remain the same (indicated by shadows).[34] But this is not true for deflation with a Fisher price index in spite of identical quantity movement in both cases, 12 and 22. Moreover, the example shows that Fisher deflation

[34] Apparently it is the *quantity* change (not the price change) that matters in deflation with P^P and the influence of different price structures on 0 and t is eliminated. This is an effect of PQC (pure quantity comparison) which appears meaningful and strongly appeals to our intuitive understanding of deflation in volume terms.

violates structural consistency of volumes (SCV), requiring the sum of (separately) deflated sub-aggregates to equal deflated total-aggregate. ♦

In part e of this section it will be shown that the rather restrictive requirement SCV can *only* be met by using P_{0t}^{P} as deflator, resulting in Q_{0t}^{L} as the quantity index. In the case of a non-uniform change of prices (non-zero variance of price relatives), using P_{0t}^{F} as deflator results in

(5.2.3) $$Q_{0t}^{F} = Q_{0t}^{L}\sqrt{1 + r_{ab}V_aV_b}$$

And may well differ from Q_{0t}^{L}, depending on the correlation r_{ab} between price- (a) and quantity relatives (b) and their respective coefficients of variation.[35] Tab 5.2.1 tries to sum up the results.

Table 5.2.1: Three deflators in the light of criteria for good deflation (in volume terms)

Criterion	direct Paasche P_{0t}^{P}	direct Fisher P_{0t}^{F}	Fisher chain $\overline{P}_{0t}^{FC}$
Proportionality in the quantities PQ and volumes reflect quantity movement [RQM]	yes	yes	no
quantities and prices change differently → volumes linear in the quantities q_{it} (PQC)	yes	no	no

b) Double (indirect) deflation method

It is useful to describe first the nature of non commodity flows (NCFs) and thereafter to distinguish between two variants of deflation in volume terms

1. Deflation of C, where C is a *difference* $C_1 = A_1 - B_1$, which is called *double* deflation or indirect[36] deflation, the most prominent examples of which are value added (see below) and final domestic demand (expenditure) FD = F – X (F = final expenditure, X = export), and
2. deflation of C defined as a *sum* $C_2 = A_2 + B_2$, which is called a "*direct*" deflation.

We first consider double deflation as such and then the situation in which double deflation may be compared with the application of a direct deflation.[37]

[35] In the rather trivial case 21 where (due to $q_{it} = \omega q_{i0}$ and therefore $V_b = 0$) the Fisher quantity index (as the result of "Fisher deflation") is $Q^F = Q^L = \omega$ (According to eq. 5.2.3 another case of $Q^L = Q^F$ would be zero-correlation between price and quantity relatives). PQC thus is more demanding than RQM.

[36] The terms "direct" and "indirect" in this context are misleading because they are also used with a different meaning (see **fig. 5.1.2**).

[37] When data necessary for direct deflation are available, which in general is not realistic.

1. Value added as a non commodity flow (NCF)

The characteristic feature of commodity flows (CFs) is that they represent (at least in some abstract sense) a sale or purchase of a well defined quantity of commodities. It can be broken down into a quantity- and a price-component. A "quantity – component" does not apply, or is at least highly questionable, in the case of NCFs, e.g. income in general, profits, rents and interest payments in particular, but also all kinds of tax payments, subsidies and other transfers. NCFs are mere amounts of money with no possibility to specify for which commodities it will eventually be spent.

This definition may also apply to a flow C_1 obtained by subtracting B_1 from A_1, where both, A_1 and B_1 are CFs. An example of this is value added, defined as Y = O - I. Note that value added, Y, is defined as the difference between output O and input I, both being CFs.[38] Whilst O and I are CFs, the difference Y is not. Y is a balancing item, a kind of "surplus", an income category, and therefore a NCF. CFs and NCFs require different approaches to deflation. In general, deflation of NCFs, like VA, in the same manner as applied in the case of CFs, that is, by using the appropriate price indices, covering the same types and quantities of goods, is neither possible nor meaningful.[39] Alternatively we may think of various procedures (**fig.5.2.2**):

- According to the SNA an option is the *direct approach* by observing *quantities* to arrive at "deflated" VA, but it is questionable if this is really an alternative, since VA as NCF has no quantity dimension by definition.[40]
- VA may be regarded as the sum of factor inputs (labor, capital), each of which may have a quantity dimension (at least in some abstract understanding). Deflation by deflating the additive components is (also) called *direct deflation*.
- NCFs like VA, being the difference 0 – I, may be deflated using the price index of output P^0 and the price index of input P^I, since both output O and input I are CFs (this is essentially the idea of *double deflation*).
- Finally, as VA has an income interpretation, it may also be *deflated in real terms* with an underlying quantity component (which is, however, not easy to define).[41]

---

[38] They mainly consist of shipments and sales on the output side and of raw materials, energy consumption and the like on the input side.

[39] Moreover, a typical NCF- aggregate like value added (VA) offers an alternative to deflate it: wages (as part of VA) can be treated as production costs and could be deflated at constant input prices (using a *wage index*) to some kind of labor "quantity" (volume), but wages are also receivable income, requiring a real-income-approach to deflation, that is a *consumer price index* (oriented at expenditures and measuring the purchasing power).

[40] Another interesting element of the SNA recommendations for deflation of value added is to make use of *chain indices*, especially of the Fisher form (both price- and volume chain indices). This will be discussed in great detail in chapter 7.

[41] The division by a consumer price index (which generally is of the Laspeyres type) will be suitable in the case of wages or other personal incomes (perhaps even with government consumption), but it will definitely not be appropriate with other income aggregates. The central problem of this (real - income) method therefore is the search for an appropriate index to deflate *all* kinds of income flows.

Figure 5.2.2: Alternative estimations of value added (VA) or GDP at constant prices as recommended in the rev. SNA 1993

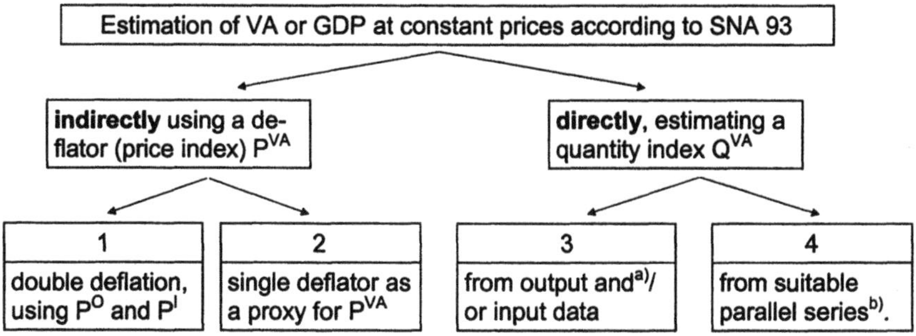

a) SNA calls this the double indicator method
b) like wages, consumption of raw material and energy etc

2. Double deflation considered in isolation

Let the symbol O(s,k) denote output with quantities relating to time s and prices related to time k. Input I and value added Y will receive equivalent symbols. Then we have by definition

(5.2.4) $Y(t,t) = O(t,t) - I(t,t)$

and we define value added at constant prices Y(t,0) as follows

(5.2.4a) $Y(t,0) = O(t,0) - I(t,0)$.

Value added at constant prices gained this way is said to be derived by double (or "indirect") deflation. Thus the central idea of this method is to

1. require volume (deflated, constant price) aggregates to satisfy the same equations as nominal (current price) aggregates, and
2. to deflate output and input separately, getting Y(t,0) using eq. 5.2.4 and finally
3. the implicit VA-deflator by

   (5.2.5) $$P_{0t}^{impl}(Y) = \frac{Y(t,t)}{Y(t,0)} \ .$$

Double deflation is essentially deflating by making use of defintional relations (consistencies) that hold for nominal as well as deflated aggregates.

Let $P_{0t}^{P}(O)$ be the Paasche price indices of output (or output deflator YPI in sec. 5.1) and correspondingly $P_{0t}^{P}(I)$, the price indices of input. Deflating Y goes as follows

(5.2.6) $$Y(t,0) = \frac{O(t,t)}{P_{0t}^{P}(O)} - \frac{I(t,t)}{P_{0t}^{P}(I)} = O(t,0) - I(t,0) .$$

Let i denote the input quota i = I(t,t)/O(t,t), that is, the percentage (share) of output received as input from other firms at time t, with both prices and quantities referring to period t. The implicit deflator (price index) of value added is then given by

(5.2.7) $$P_{0t}^{imp}(Y) = \frac{Y(t,t)}{Y(t,0)} = \frac{P_{0t}^{P}(O) \cdot P_{0t}^{P}(I) \cdot (1-i)}{P_{0t}^{P}(I) - i \cdot P_{0t}^{P}(O)}, \text{ where } i = \frac{I(t,t)}{O(t,t)}.$$

A more convenient interpretation of the implicit deflator reads as follows[42]:

(5.2.8) $$\frac{1}{P_{0t}^{P}(O)} = i\frac{1}{P_{0t}^{P}(I)} + (1-i)\frac{1}{P_{0t}^{imp}(Y)}.$$ Hence

The output-deflator $P^P(O)$ can be regarded as a weighted harmonic mean of the input deflator $P^P(I)$ [both indices P(O) and P(I) of Paasche type], and the implicit value added deflator $P^{imp}(Y)$, the weights being the quotas i and (1-i) respectively.

However the interpretation of this implicit double deflation value added deflator shows that the method may well yield some awkward results

- the denominator in eq. 7, $P_{0t}^{P}(I) - i \cdot P_{0t}^{P}(O)$, can possibly be negative, resulting in a negative "real" value added
- $P_{0t}^{imp}(Y)$ can be less than unity, indicating a decline of prices although both output and input show increase of prices, for example i = 0.7, $P_{0t}^{P}(I)$ = 1.4 and $P_{0t}^{P}(O)$ = 1. 2 results in $P_{0t}^{imp}(Y)$ = 0.9 < 1.
- It is also possible to obtain $P_{0t}^{imp}(Y)$ > 1 with declining prices of both output *and* input (let $P_{0t}^{P}(O)$ = 0.8, $P_{0t}^{P}(I)$ = 0.9 and i = 0.7 as before; then $P_{0t}^{imp}(Y)$ = 1.56).

As a result of deflating by using *Fisher* price indices we get the implicit VA-deflator, $P_{0t}^{imp(F)}(Y)$ and $\frac{1}{P_{0t}^{P}(O)} = i\frac{1}{P_{0t}^{P}(I)}R_1 + (1-i)\frac{1}{P_{0t}^{imp(F)}(Y)}R_2$ where $R_2^2 = P_{0t}^{P}(O) / P_{0t}^{L}(O)$ and $R_1^2 = R_2^2\left[P_{0t}^{P}(I) / P_{0t}^{L}(I)\right]$, an equation not too easy to understand and that can no longer be regarded as a weighted harmonic mean because the sum iR_1 + $(1-i)R_2$ does not necessarily amount to unity.

Another issue is to use *chain indices*. As is well known, the method is recommended in the SNA in order to better deal with a situation in which the two points in time (0 and t) are quite distant. In this case (long term comparisons of VA) it is assumed the inherent defects of any method to deflate a non commodity flow will aggravate considerably, and – unfortunately (and unjustifiedly) – chain indices are considered superior in this situation.[43]

[42] Another interpretation is that we get a price index P^{imp}, where the output deflator has a positive weight and the input deflator a negative weight.

[43] Conspicuously, the reason often given for this runs as follows: because they limit the estimation to two periods, t-1 and t, which are only one interval apart (this, however, applies only to a link, but not to the chain).

3. Consistency requirements

Consider an aggregate C, which is given as a difference $C = A_1 - B_1$ as well as in the shape of the sum $C = A_2 + B_2$. It follows from above that in order to be consistent, a deflator price index $P(A_1)$ has to be the harmonic mean of P(C) and $P(B_1)$, or $P(A_1) = H\{P(C), P(B_1)\}$. In the same manner, $P(C) = H\{P(A_2), P(B_2)\}$ should hold. Hence the three indices are unequivocally related to one another: $P(A_1) = H\{[H\{P(A_2), P(B_2)\}], P(B_1)\}$. In practice it is not guaranteed that the three independently compiled price indices $P(A_2)$, $P(B_2)$ and $P(A_1)$ will satisfy this constraint.

c) Aggregative properties of index functions and deflation methods

For a long time, axiomatic theory did not pay much attention to aggregative properties of index functions. The reason presumably is that price level measurement used to be the only or at least the dominant purpose for which price indices were needed. As deflation became a more and more important field of price statistics, so more emphasis was put on aggregation problems. But it is also desirable for analytical purposes that price indices can be decomposed or aggregated without difficulties. Much analysis is done by compiling indices in various experimental versions, by excluding or including certain series of prices which usually are, or are not, included in the official index. Comparing such calculations will give some insight into the structure of prices as well as the causes of an increase or decline in the overall price level.

The term "additivity" is used in widely differing meanings and the relationship between these types of "additivity" is by no means self explaining. We may distinguish between three types of additivity, and **fig. 5.2.3** may help to systematize these distinctions. Notably "aggregative consistency" (ACF, **A1**) and "structural consistency" (SCV, **A3**) are often mixed up and indiscriminatingly called "additivity". It should also be borne in mind that SCV is most restrictive, leaving room only for the direct Paasche index as deflator.

The analytical use of aggregative consistency of an index function (**A1** or **ACF**) as a measure of a price "level", for example, consists of facilitating to examine the behaviour of indices differently composed of and broken down into sector indices. It is for example interesting to study an index in which some price series are excluded and to compare it with another index which possibly might include some additional price series. Much of the work done by analysts is of precisely this kind. **Fig. 5.2.4** aims to clarify graphically the idea of "aggregating" subindices, or sector indices $(P_1,...,P_K)$ to calculate the global (overall) index P.[44]

The concept **A3** or **SCV**[45] deals with deflation and means that it is certain that if certain definitional equations are satisfied by values, so they will also hold for volumes (deflated values) derived from values by using the appropriate sectoral and global price indices. It can easily be shown (though the result is very little known in its significance and implication) that this condition is met by Paasche price indices *only*.

---

[44] It is sufficient for the function $f(P_1,...,P_K)$ in **A1** to be the *same* (not necessarily linear) function by which price relatives are combined to sector indices and the overall (or global) index.

[45] The SNA also calls this "additivity".

d) Consistency in aggregation (aggregative consistency of the index function, A1)

Consistency in aggregation, also frequently called simply "additivity", is related to the aggregation of sector-indices to global indices. In official statistics, the *global* (total) aggregate often is subdivided into *sectors* (subaggregates), such as "Consumption of private households" (total, global) into subaggregates like 1. food, beverages, tobacco, 2. clothing and footwear, 3. rent and energy consumption etc., and each subaggregate may again be subdivided into sub-subaggregates etc. The total structure of an aggregate in practice contains several levels, four or five levels as an example.

Figure 5.2.3: The usage of the terms "additivity" in (price)[1)] index theory

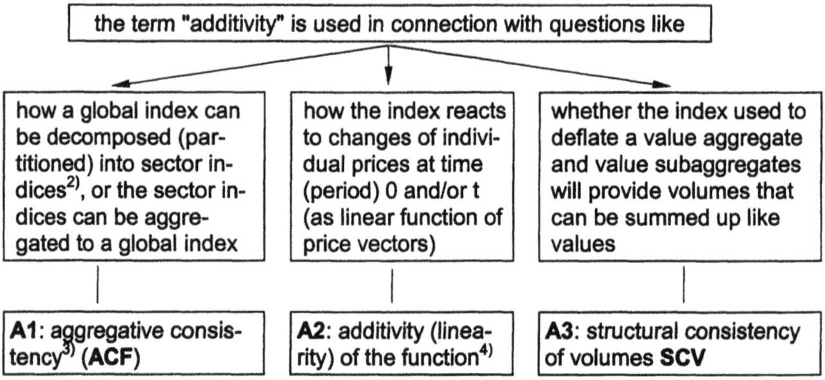

1) the same applies to quantity indices
2) or sub-indices
3) of the index function
4) see **sec. 3.2** for additivity (or equivalently linearity) as a variant of monotonicity

To introduce the idea of consistency in aggregation we consider n commodities that can be classified into K nonoverlapping (distinct) and exhaustive (all n goods are included, such that $n = n_1 + n_2 + ... + n_K$) groups or "sectors" ($k = 1,...,K$). Such a breakdown of an aggregate into K "sectors" or subaggregates is usually performed by a classification or "nomenclature" of commodities. Both properties A1 and A2 refer to the index function, not to properties of aggregates resulting from deflation, as A3 does. The three types of additivity are related as follows:

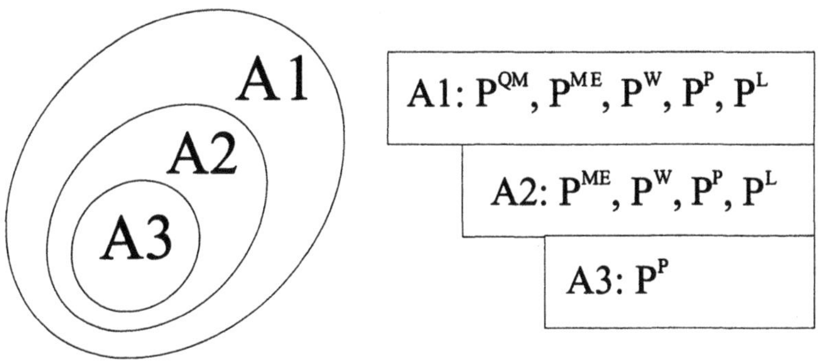

Price indices that are aggregative consistent (A1) but not linear in the current prices (A2) include the quadratic mean index P^{QM} but also the logarithmic Las-peyres index (not listed above). A2, but not A3, applies in the case of Laspeyres (P^L), Marshall-Edgeworth (P^{ME}) and Walsh (P^W). Index functions that do not comply with any of these additivity concepts are for example Fisher's ideal index or the formula of Drobisch $(P^L + P^P)/2$.

In such cases consistency of an index function defined as follows is deemed useful:

The funtion P_{0t} is called **aggregative consistent** (or consistent in aggregation) if the same formula (function f) can be used to derive

1. the overall (global) price index P_{0t} as a function of n price relatives $a_{0t}^i = p_{it}/p_{i0}$ (i = 1, ..., n), that is, $P_{0t} = f(a_{0t}^i)$ for all n commodities
2. the K sector indices $P_{0t}^1,\dots,P_{0t}^K$ for the K subaggregates (sectors), calculated separately as functions f of $n_1,\dots,n_K$ price relatives where $n = n_1 + \dots + n_K$
3. the global price index P_{0t} as a function f of the K sector indices $P_{0t}^1,\dots,P_{0t}^K$, such that $P_{0t} = f(P_{0t}^k)$, k = 1,...,K.

Thus aggregative consistency means that the global index P is a function *of the same type* of the K sector-indices P_{0t}^k as P is a function of the n price relatives or the K sector indices are functions of the individual price relatives. The idea is that the operations 1, 2 and 3, as visualized in **fig. 5.2.4**, are essentially the same operation.

Fisher's ideal index P^F is not additive in the sense of A1, let alone A2.[46] We are unable to calculate the global index P^F from K sector indices, P^{Fk}, and from n price relatives in *the same way*. It can easily be seen (and the numerical example **ex. 5.2.3** will demonstrate this), that the global index P^L (Laspeyres) is a weighted arithmetic mean of price relatives (with weights $g_i = p_{i0}q_{i0}/\Sigma p_{i0}q_{i0}$), but P^L can also be calculated as a weighted arithmetic mean of the sector indices P_{0t}^{Lk} (k = 1,...,K) with weights de-

[46] None of the three aggregation properties A1 to A3 will be satisfied by Fisher's "ideal index". In **sec. 5.4** below we will try to summarize some deficiencies of the so called "ideal" index.

fined analogously. The same is also true for the Paasche index P^P, with weights, $w_i = p_{i0}q_{it}/\Sigma p_{i0}q_{it}$, defined differently, however. But in the case of Fisher's index this is *not* possible. With n individual price relatives we get

$$(5.2.9) \qquad P^F_{0t} = \sqrt{\left(g_1 a^1_{0t} + g_2 a^2_{0t} + ... + g_n a^n_{0t}\right)\left(w_1 a^1_{0t} + w_2 a^2_{0t} + ... + w_n a^n_{0t}\right)},$$

Figure 5.2.4: The concept of aggregative consistency (K = 3 sectors)

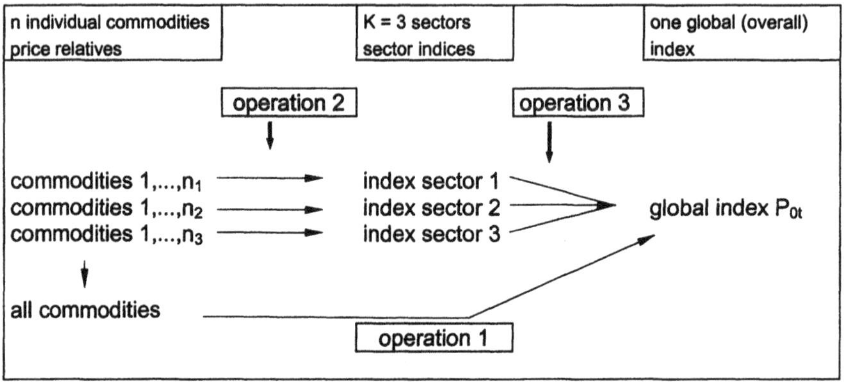

where the weights are $g_i = p_{i0}q_{i0}/\Sigma p_{i0}q_{i0}$ and $w_i = p_{i0}q_{it}/\Sigma p_{i0}q_{it}$. The result of the first expression (in brackets) is P^L_{0t} and the result of the second expression is P^P_{0t}. But using K Fisher type sector indices the formula for P^F_{0t} is, in the case of aggregation of sector indices instead of individual price relatives, not given by

$$(5.2.9a) \qquad \sqrt{\left(g_1 P^{F1}_{0t} + g_2 P^{F2}_{0t} + ... + g_K P^{FK}_{0t}\right)\left(w_1 P^{F1}_{0t} + w_2 P^{F2}_{0t} + ... + w_K P^{FK}_{0t}\right)}$$

with "sector Fisher indices" defined by $P^{Fk}_{0t} = \sqrt{P^{Lk}_{0t} P^{Pk}_{0t}}$, but rather by

$$(5.2.10) \qquad P^F_{0t} = \sqrt{\left(g_1 P^{L1}_{0t} + g_2 P^{L2}_{0t} + ... + g_K P^{LK}_{0t}\right)\left(w_1 P^{P1}_{0t} + w_2 P^{P2}_{0t} + ... + w_K P^{PK}_{0t}\right)}$$

where $P^{Fk}_{0t} = \sqrt{P^{Lk}_{0t} P^{Pk}_{0t}}$. Thus Fisher's index is *not* aggregative consistent.[47] It might be conjectured that the fact that there is no weighted mean of price relatives representation of P^F is responsible for P^F's failing A1, and that all index functions able of such a representation will satisfy A1. This is *not* true, however, as the following example of Drobisch's index (P^{DR}) shows. $P^{DR}_{0t} = \frac{1}{2}\left(P^L_{0t} + P^P_{0t}\right)$ is in terms of individual price relatives (i = 1,...,n) given by

$$(5.2.11) \qquad P^{DR}_{0t} = \frac{1}{2}\left[\left(\sum_{i=1}^{n} \frac{p_{it}}{p_{i0}} g_i\right) + \left(\sum_{i=1}^{n} \frac{p_{it}}{p_{i0}} w_i\right)\right] = \frac{1}{2}\left(P^L_{0t} + P^P_{0t}\right) = \sum_{i=1}^{n} \frac{p_{it}}{p_{i0}} \frac{g_i + w_i}{2}$$

[47] P^F also fails the even weaker "equality test" (see part e of this section). This will be demonstrated in **ex. 5.2.3**.

showing clearly that P^{DR} is an arithmetic mean with weights $(g_i + w_i)/2$. With n commodities grouped into two subindices, such that j = 1, ..., m belongs to group A, and k = m+1, ..., n to group B respectively we therefore get

$$P_{0t}^{DR} = \frac{1}{2}\left[\left(\sum_{j}^{m} \frac{p_{jt}}{p_{j0}} g_j^*\right)\sum_{j}^{m} g_j + \left(\sum_{k}^{n} \frac{p_{kt}}{p_{k0}} g_k^*\right)\sum_{k}^{n} g_k + \left(\sum_{j}^{m} \frac{p_{jt}}{p_{j0}} w_j^*\right)\sum_{j}^{m} w_j + \left(\sum_{k}^{n} \frac{p_{kt}}{p_{k0}} w_k^*\right)\sum_{kl}^{n} w_k\right],$$

where $g_j^* = g_j / \sum_j g_j$, and g_k^*, w_j^*, w_k^* are defined correspondingly. We may also write

$$(5.2.12) \qquad P_{0t}^{DR} = \tfrac{1}{2}\left[P_{0t}^{LA} g_A + P_{0t}^{LB} g_B + P_{0t}^{PA} w_A + P_{0t}^{PB} w_B\right],$$

and with two subindices only we have $g_A + g_B = 1$ and $w_A + w_B = 1$, and as

$P_{0t}^{DRA} = \tfrac{1}{2}\left(P_{0t}^{LA} + P_{0t}^{PA}\right)$, and with P_{0t}^{DRB} defined analogously, this yields

$$(5.2.13) \qquad P_{0t}^{DR} = P_{0t}^{DRA} + P_{0t}^{DRB} - \tfrac{1}{2}\left[P_{0t}^{LA} g_B + P_{0t}^{LB} g_A + P_{0t}^{PA} w_B + P_{0t}^{PB} w_A\right]$$

which is in general *not* to be equated to $P_{0t}^{DRA}\left(\frac{g_A + w_A}{2}\right) + P_{0t}^{DRB}\left(\frac{g_B + w_B}{2}\right)$, unless the weights g_A and w_A equal ½. In conclusion:

Aggregative consistency (ACF) is *not* automatically satisfied when an index function can be expressed as a weighted mean of relatives. As will be shown in more detail in part g of this section, it is necessary but not sufficient for an index function to satify A1 that it can be expressed in terms of relatives.[48]

Example 5.2.3:

For the sake of simplicity we assume K = 2 and call the two sectors A and B. Prices and quantities of four commodities (goods) 1, 2, 3 and 4, classified into the two groups (or sectors) A and B, are given as follows:

		base period (0)			current period (t)			
	good	p_0	q_0	p_0q_0	p_t	q_t	p_tq_t	p_0q_t
A	1	10	16	160	15	12	180	120
A	2	16	15	240	18	14	252	224
B	3	20	10	200	24	12	288	240
B	4	15	20	300	21	15	315	225
sum (Σ)				900			1035	809

The following index formulas should be used to calculate both sector indices and the global index: the formula of Laspeyres P^L, Paasche P^P, Fisher P^F, the logarithmic Laspeyres index DP^L, and the quadratic mean price index P^{QM}. It should

[48] Not all indices that can be expressed as means of price relatives can also be aggregated consistently. On the other hand: we don't know of any index function not being a mean of relatives (like Fisher's index) which is nonetheless aggregative consistent.

also be shown how the global index can be derived from the sector indices. Solution: The price relatives a_i and the weights $g_i = p_0q_0/\Sigma p_0q_0$ and $w_i = p_0q_t/\Sigma p_0q_t$ are

good	a_i	g_i global	g_i within sector	w_i global	w_i within sector
1 (A)	1.5	0.1777	0.4 = 160/400	0.1483	0.3488
2 (A)	1.125	0.2667	0.6 = 240/400	0.2768	0.6511
3 (B)	1.2	0.2222	0.4 =200/500	0.2967	0.5161
4 (B)	1.4	0.3333	0.6 =300/500	0.2781	0.4839

We get the following sector indices

Sector price index	sector A	sector B
Laspeyres	P_{0t}^{LA} = 510/400 = 1.275	P_{0t}^{LB} = 660/500 = 1.32
Paasche	P_{0t}^{PA} = 432/344 = 1.2558	P_{0t}^{PB} = 603/465 = 1.2968
Fisher	$P_{0t}^{FA} = 1.2654 = \sqrt{P_{01}^{LA}}\sqrt{P_{01}^{PA}}$	$P_{0t}^{FB} = 1.3083 = \sqrt{P_{01}^{LB}}\sqrt{P_{01}^{PB}}$
Logarithmic Laspeyres	DP_{0t}^{LA} = 1.2622	DP_{0t}^{LB} = 1.3163
Quadratic mean	P_{0t}^{QA} = 1.28817	P_{0t}^{QB} = 1.32361

With weights g_i and w_i given by

weights*	sector A	sector B
Weights g	g_A = 400/900 = 0,4444	g_B = 500/900 = 0,5556
Weights w	w_A =344/809 = 0.4252	w_B =465/809 = 0.5748

* expenditure shares

Laspeyres' (Paasche's) global index can be calculated as weighted arithmetic means of sector indices. The resulting global indices are therefore:

Laspeyres $P_{0t}^{L} = 0{,}4444 \cdot 1.275 + 0{,}5556 \cdot 1.32 = 1.3 = 1170/900$ and

Paasche $P_{0t}^{P} = 0{,}4252 \cdot 1.2558 + 0{,}5748 \cdot 1.2968 = 1.2794 = 1035/809$.

The logarithmic Laspeyres sector indices are

$DP_{0t}^{LA} = 1.5^{0.4}\, 1.125^{0.6} = 1.2622$ and $DP_{0t}^{LB} = 1.2^{0.4}\, 1.4^{0.6} = 1.3163$.

Also, in the case of this index, the global index can be expressed as a simple function of the sector indices: $1.2622^{0.4444}\, 1.31628^{0.5555} = 1.2919$. This aggregation of the two sector indices yields the same result as a calculation directly using all four price relatives does, viz. $DP_{0t}^{L} = 1.5^{0.1777}\, 1.125^{0.2667}\, 1.2^{0.2222}\, 1.4^{0.3333}. = 1.2919$.

This demonstrates that DP^L is aggregative consistent. Getting the same result in both cases rests on the fact that the weights are related to one another as follows:

$0.1777 = 0.4 \cdot 0.4444$, $0.2667 = 0.6 \cdot 0.4444$ and

$0.2222 = 0.4 \cdot 0.5555$, $0.3333 = 0.6 \cdot 0.5555$.

A similar argument applies to the quadratic mean index (QM-index), P_{0t}^{QM}, where using individual price relatives we get

$$P_{0t}^{QM} = \sqrt{\frac{400}{900}\cdot\left(1.5^2\cdot\frac{160}{400}+1.125^2\cdot\frac{240}{400}\right)+\frac{500}{900}\cdot\left(1.2^2\cdot\frac{200}{500}+1.4^2\cdot\frac{300}{500}\right)} = 1.307988,$$

and the same result is achieved in aggregating the sectoral QM-indices (i.e. the terms in brackets) to the global QM-index $P_{0t}^{QM} = \sqrt{\frac{4}{9}\left(P_{0t}^{QMA^2}\right)+\frac{5}{9}\left(P_{0t}^{QMB^2}\right)}$, which clearly shows that the QM-index is aggregative consistent, though this index is not an additive index.

Things are much less convenient in the case of Fisher's so called ideal index. The geometric mean of Laspeyres and Paasche index gives us the global Fisher index $P_{0t}^{F} = \sqrt{1.3\cdot 1.2794} = 1.2896$. But no simple formula (like a weighted geometric mean) exists, by which the K sector indices of Fisher P_{0t}^{Fk} (k = 1,...,K) can be "combined" to a global index of Fisher. An aggregation can only be performed, according to eq. 5.2.10, with the K sector indices of the Laspeyres-type (P_{0t}^{Lk}) and of the Paasche-type (P_{0t}^{Pk}). The same is true for Drobisch's index, and notably for chain indices, but all of them are obviously not aggregative consistent. ♦

The example shows, as stated above, that aggregative consistency (**A1**) applies to DP^L as well as to P^{QM}. This shows that the property is not confined to index formulas that can be expressed as *linear* functions of the price relatives, nor to those functions that have a dual interpretation: the weighted mean of price relatives and the ratio of expenditures interpretation of an index. The common feature of the indices is that they have a weighted mean interpretation. Neither DP^L nor P^{QM} possesses such a "ratio of expenditures" interpretation. Both indices are non-linear index functions and yet both functions are aggregative consistent.

Fisher's ideal index, however, does not meet *any of these* requirements (as opposed to P^L, P^P etc. which comply with *all* criteria): additivity, dual interpretation and aggregative-consistency (ACF).

P^F is not even capable of satisfying a weak variant of ACF, i.e. the "equality test". Hence the difference in "test-performance" is remarkable.

e) Equality test (ET)

Whilst linearity (**A2**) can be regarded as a more restrictive condition than aggregative consistency (**A1, ACF**), the so called equality test is a weak version of aggregative consistency of the index function P. The function P satisfies the equality test if

(5.2.14) $P_{0t} = f(P_{0t}^1, P_{0t}^2,...,P_{0t}^K) = f(\lambda,\lambda,...,\lambda) = \lambda$ (equality test)

Given all sector indices P_{0t}^k (k = 1,2, ... ,K) equal λ, so the global index should yield $P_{0t} = \lambda$. This seems to be a rather weak condition, such that

if aggregative consistency is given	⟶	then the equality test is also met

and in fact there is no index that passes the equality test and also is not aggregative consistent.[49] The opposite situation appeared likewise impossible, and for a long time there was no index function known, that satisfied aggregative consistency and yet was unable to meet the ET, such that ET and ACF seemed to be independent. The existence of the Vartia - I index[50] has shown, however, that it is possible to satisfy ET and *not* ACF. There are index formulas that

- comply with both criteria (ET *and* ACF), such as the indices of Laspeyres and Paasche, and those
- to which neither ET nor ACF applies, like Fisher's index, where thus the aggregation performance is the poorest we can think of.

A simple example may be adequate in order to demonstrate that Fisher's "ideal" index indeed fails the equality test, a fact not easy to understand at first glance.

Example 5.2.4

Consider two commodities and weights $g_1 = 0.6$, and thus $g_2 = 0.4$, to calculate P^L_{0t} using sectoral indices, and correspondingly weights $w_1 = 0.4$, $w_2 = 0.6$ to calculate P^P_{0t}. Then assume $P^{L1}_{0t} = 1.25$ and $P^{P1}_{0t} = 1.2$, and likewise $P^{L2}_{0t} = 2$ and $P^{P2}_{0t} = 0.75$, such that $P^{F1}_{0t} = P^{F2}_{0t} = \sqrt{1.5}$. But the global index P^F_{0t} is not $\sqrt{1.5}$, as required by the equality test, but rather $\sqrt{1.55 \cdot 0.93} = \sqrt{1.4415}$, because the global indices are $P^L_{0t} = 1.55$ and $P^P_{0t} = 0.93$.

Note that we obtain the global indices P^L_{0t}, P^P_{0t}, and thus also P^F_{0t}, from the sectoral ones using a *weighted arithmetic* mean, whilst the sectoral indices P^F_{0t} are gained from the sectoral Laspeyres and Paasche indices by taking an *unweighted geometric* mean. This is the reason why P^F_{0t} is unable to meet the equality test. ♦

f) Structural consistency of volumes (A2), a case for using Paasche indices in deflation

The notion of structural consistency of volumes (**A3** in **fig. 5.2.3**, or **SCV**) was already introduced in part a of this section. It is now our aim to find the conditions under which SCV is met. They turn out to be highly restrictive and a strong case for (direct) Paasche indices. The demonstration is extremely easy, yet apparently not too well known, and even more importantly, its implications for the choice of a deflator-formula are substantial.

---

[49] We found in literature sometimes the remark that the Edgeworth Marshall index is an example for this situation. But this is definitely not true.

[50] also known as index of Montgomery index (Stuvel 1989, Balk 1995).

Let V_1, V_2, ... ,V_K denote *values* (aggregates at current prices) referring to *sub*-aggregate 1 to K, and V_T to the *total* (T) aggregate respectively, such that by definition

(5.2.15) $V_1 + V_2 + ... + V_K = \sum V_k = V_T$ (k = 1, 2, ... , K).

Each volume is defined by dividing a value by its corresponding price index (deflator), P_1, P_2, ... , P_K. To satisfy SCV, the following equation

(5.2.16) $$\frac{V_1}{P_1} + ... + \frac{V_K}{P_K} = \frac{V_T}{P_T}$$

has to hold for P_T, the "total deflator". Next, consider value shares (or "weights") w_k to describe the fact that total value V_T is broken down into K subaggregate' values

(5.2.17) $$\frac{w_1 V_T}{P_1} + ... + \frac{w_K V_T}{P_K} = \frac{V_T}{P_T}, \text{ where } w_k = \frac{V_k}{V_T}$$

and after division of both sides of the equation by V_T we get

(5.2.18) $$w_1 \frac{1}{P_1} + ... + w_K \frac{1}{P_K} = \frac{1}{P_T}$$

which simply means that P_T has to be a *weighted harmonic* mean of sectoral indices (deflators) with weights being value shares $p_t q_t / \sum p_t q_t$, that is, $w_k = V_k / V_T$.

This reduces the class of admissible indices to a unique index formula: the deflators have got to be Paasche indices. Starting with the lowest level of aggregation, that is, individual price relatives up to the highest level of the overall index, a Paasche index is always a harmonic mean of sectoral indices (or relatives) of the preceding level of aggregation. Hence it can easily be verified that:

The only deflator price index capable of producing structurally consistent volumes at all levels of aggregation is the Paasche index, due to the fact that this index is based on a harmonic mean (of price relatives or sub-indices [sectoral deflators] respectively): P^P is therefore *characterised* by SCV.

Hence SCV is most demanding: If and *only* if the corresponding direct Paasche price indices are introduced for P_1, P_2, ..., P_K will eq. 5.2.16 hold.

If, on the other hand, the price indices P_1, P_2, P_T are replaced by sectoral *Fisher* deflator-indices $P_k^F = \sqrt{P_k^L P_k^P}$ (k = 1,...,K), we get

(5.2.19) $$\frac{V_1}{P_1^P}\sqrt{\frac{P_1^P}{P_1^L}} + ... + \frac{V_K}{P_K^P}\sqrt{\frac{P_K^P}{P_K^L}} \neq \frac{V_T}{P_T^P}\sqrt{\frac{P_T^P}{P_T^L}},$$

with the result that the left hand side (LHS) will usually differ from the right hand side (RHS). To see this, consider K = 2. With Paasche price indices we get the following result

(5.2.18a) $$\frac{V_1}{P_1} + \frac{V_2}{P_2} = \frac{V_T}{P_T}, \text{ or solved for } P_T \text{ we get } P_T = \frac{V_1 + V_2}{\frac{V_1}{P_1} + \frac{V_2}{P_2}} = \frac{1}{\frac{w_1}{P_1} + \frac{w_2}{P_2}}.$$

(5.2.18b) $P_T = \dfrac{V_1 + V_2}{\dfrac{V_1}{P_1} + \dfrac{V_2}{P_2}} = \dfrac{1}{\dfrac{w_1}{P_1} + \dfrac{w_2}{P_2}}$. The LHS of eq. 19 yields

(5.2.19a) $\dfrac{w_1 P_2^F + w_2 P_1^F}{P_1^F P_2^F} V_T$, in contrast to the RHS: (5.2.19b) $V_T \dfrac{1}{P_T^F}$.

On the other hand

(5.2.19.c) $\dfrac{1}{P_T^F} = \sqrt{\dfrac{P_1^P P_2^P}{w_1 P_2^P + w_2 P_1^P} \dfrac{1}{g_1 P_1^L + g_2 P_2^L}}$ where $g_k = \dfrac{\sum_j p_{j0} q_{j0}}{\sum_k \sum_j p_{j0} q_{j0}} = \dfrac{V_k^B}{\sum_k V_k^B}$

(g_1 and g_2 are shares of the subaggregates referring to *base period* [B] values), which will differ from the corresponding term in eq. 19a. The total Fisher index P_{0t}^F is not a harmonic mean of sectoral Fisher indices, as required in eq. 18 and 18a, and therefore deflation using P_{0t}^F as a deflator does not result in structurally consistent volumes. In addition, the P_{0t}^F index function is not additively consistent either. Note that

Structural consistency of volumes is equivalent with a quantity index Q^L being linear in the quantites. This property is valuable in that it enables total volumes to be derived consistently from the updating of sectoral volumes.

Interestingly another property described in **sec. 5.2a** as "reflection of quantity movement" (**RQM**, in the case of uniform price movement) and "pure quantity comparison" (**PQC**, in the case that prices change non-uniformly) is most closely related to structural consistency. Given some base period values (= volumes) $V_k^B = \sum p_{k0} q_{k0}$ for any k = 1,...,K, such as private consumption (V_1^B) or investment (V_2^B) at period 0 (quantities and prices relating to 0) as a starting point, it might be desirable to "update" these aggregates using suitable quantity indices Q_k, such that

(5.2.20) $V_1^B Q_1 + ... + V_K^B Q_K = (V_1^B + V_K^B) Q_T$.

The only total-aggregate (Q_T) quantity index permitting this type of consistent "updating" of base period (subaggregate) volumes $\Sigma p_0 q_0$ to current period volumes $\Sigma p_0 q_t$ needs to be an *arithmetic* mean of Q_1, Q_2, ..., with weights g_k as defined in eq. 5.4.8c, hence a *Laspeyres quantity index* Q_{0t}^L. Thus the harmonic mean in P corresponds to an arithmetic mean in Q, such that we get the pair P_{0t}^P, Q_{0t}^L. In our view, it seems much more reasonable to seek a pair of indices P and Q, such that P is aggregative consistent (SCV), and Q is linear in the quantities (PQC), than to require factor reversibility.

Any pair of indices other than P_{0t}^P, Q_{0t}^L (for example P_{0t}^F and Q_{0t}^F, or the corresponding chain indices, $\overline{P}_{0t}^{FC}$ and $\overline{Q}_{0t}^{FC}$) will violate either or both requirements. They pass the factor reversal test but fail SCV (and thus also PQC). The resulting (from deflation by P_{0t}^{F}) index Q_{0t}^{F} is at least *proportional* in the quantities, while $\overline{Q}_{0t}^{FC}$ is unable to meet even this criterion. Nevertheless, according to the SNA "the preferred measure of year to year movements of real GDP is a Fisher volume index" chained together (thus $\overline{Q}_{0t}^{FC}$) and "the preferred measure of year to year inflation for GDP is therefore a Fisher price index", i.e. $\overline{P}_{0t}^{FC}$.

Before having a closer look at the type of deflation provided by (direct) Fisher indices (see **ch. 7** for the chain version) in **sec. 5.4**, it will be useful to find a relationship between Paasche and Laspeyres indices on the one hand and Fisher's indices on the other. Again, the theorem of Bortkiewicz (**sec. 1.3**) will be useful as a starting point. According to this theorem, the value index V_{0t}, the price indices and the covariance C are related as follows $R = \frac{P_{0t}^{P}}{P_{0t}^{L}} = \frac{V_{0t}}{V_{0t} - C}$. The following relationship between P^F deflated aggregates (Fisher volumes V^F) and P^P- deflated aggregates (Paasche volumes V^P) is given (on all levels of aggregation)

$$(5.2.21)\qquad V^{F} = V^{P}\sqrt{\frac{V_{0t}}{V_{0t} - C}}.$$

If the covariance is negative [positive], a volume gained by deflation with a Fisher index tends to fall short of [exceed] a volume derived from deflation with a Paasche index. Conversely $P_{0t}^{F} > P_{0t}^{P}$ as $C < 0$ and $P_{0t}^{F} < P_{0t}^{P}$ as $C > 0$.

g) Some new results concerning aggregative consistency[51]

The advantage of aggregative consistent index functions (notion **A1** in **fig. 5.2.3**) is that the results of index compilations are independent of which aggregation levels are being distinguished and in which order aggregation is being carried out. Invariance with respect to the kind and sequence of aggregation is most attractive for analytical purposes in particular.[52] To demonstrate this and (once more) the well known shortcomings of Fisher's formula, we start with a numerical example taken from L. von Auer):

Example 5.2.5

i		p_{i0}	q_{i0}	p_{it}	q_{it}
1	Non durable goods	100	130	165	40
2	Durable goods	100	95	85	160
3	Services	100	40	60	120

[51] The following part is drawn on an article of Ludwig von Auer (2004) on "consistency in aggregation".

[52] An analysis of the impact and relative importance of certain influences requires the consistent aggregation of sub-indices to total (all item) indices.

Commodities 1 and 2 are combined in the sub-aggregate "goods", while commodity 3 constitutes a sub-aggregate of its own. It can easily be verified that the price index of Fisher amounts to $P_{0t}^{F(1)} = 1.016$, calculated in one single stage using the three individual price relatives, whereas the same formula applied in two stages (price relatives → sub indices, and thereafter sub indices→ all item index) yields $P_{0t}^{F(2)} = 0.995$.[53] ♦

Aggregative consistency requires the one-stage (or direct) compilation to equal the two-stage (or indirect) compilation of an index number. This property is clearly violated in the case of P_{0t}^{F}, while both the Paasche and the Laspeyres formula are able to satisfy this test. At the outset of his considerations, von Auer noticed that

> In order to satisfy certain conditions of additivity (in the sense of aggregative consistency), it is necessary for an index function to be expressible in terms of price relatives. For this in turn to be ensured, it is indispensable that the function satisfies the commensurability axiom.[54]

Furthermore, the index functions $P(\mathbf{p}_0, \mathbf{q}_0, \mathbf{p}_t, \mathbf{q}_t)$ (demonstrated with P_{0t}^{L} as an example) are *able to be expressed in at least three different ways* in terms of price relatives $r_i = p_{it}/p_{io}$ and value (expenditure) shares $v_i^{st}/V^{st} = p_i^s q_i^t / \Sigma p_i^s q_i^t$

(5.2.22) $P_{0t} = \widetilde{P}(r, v^{00}, v^{0t})$

for example $P_{0t}^{L} = \Sigma \frac{p_t}{p_0} \frac{p_0 q_0}{\Sigma p_0 q_0} = \sum r_i (v^{00}/V^{00}) = f(r, v^{00})$

(5.2.23) $P_{0t} = \hat{P}(r, v^{tt}, v^{t0})$ or $(P_{0t}^{L})^{-1} = \sum \frac{p_0}{p_t} \frac{p_t q_0}{\sum p_t q_0} = \sum \frac{1}{r_i} \frac{v^{t0}}{V^{t0}}$, and finally

(5.2.24) $P_{0t} = \overline{P}(r, v^{00}, v^{tt})$ in the case of P_{0t}^{L} identical to eq. 22,

or in a more general notation, P(a, b), with a and b representing v-definitions $\widetilde{P}(00,0t)$, $\hat{P}(tt,t0)$ and $\overline{P}(00,tt)$. The interesting, novel feature is now this:

> Though evidently all three representations are perfectly equivalent and one can easily be transformed into the other, this - most surprisingly - does not mean that an index function able to satisfy aggregative consistency in one specific r-v-representation will do so in another r-v-representation as well.

As eq. 5.2.24 (i.e. the $\overline{P}$ form) was the first to be examined in detail in the context of aggregative consistency (by B. Balk, to whom v. Auer explicitly refers), von Auer called a function *specific consistent*[55] if its one-stage and multi-stage formulas of the $\overline{P}$-type are equal. This applies, for example, in the case of the Laspeyres formula, where A denotes the set of all (n) price relatives and B denotes the set of sub-

---

[53] The example is particularly striking as the results differ not only in amount but also in the sign of the growth rate, i.e. + 1.6% as opposed to - 0.5%.

[54] which is a precondition fulfilled in the case of all the index functions to be dealt with in what follows.

[55] which is in actual fact a specific variant of weak consistency.

indices, then clearly $\sum_{i\in A} r_i v_i^{00} = \sum_{k\in B} P_{0t}^{L(k)} v_k^{00} = P_{0t}^{L}$, that is, the single stage index equals the two stage index where the subindices $P_{0t}^{L(k)}$ replace the price relatives and the base-period values v^{00} are adjusted (summated) accordingly.

Table 5.2.2a: Laspeyres, Paasche and M-E index in r-v-representations*

	Laspeyres	Paasche	Marshall-Edgeworth
(22) $\tilde{P}(r,v^{00},v^{0t})$	$P_{0t}^{L} = \sum \frac{p_{it}}{p_{io}} \frac{p_{io}q_{io}}{\sum p_{io}q_{io}}$ $= \sum r_i v_i^{00} / \sum v_i^{00}$	$P_{0t}^{P} = \sum \frac{p_{it}}{p_{io}} \frac{p_{io}q_{it}}{\sum p_{io}q_{it}}$ $= \sum r_i v_i^{0t} / \sum v_i^{0t}$	$P_{0t}^{ME} = \frac{\sum r_i (v_i^{00} + v_i^{0t})}{\sum (v_i^{00} + v_i^{0t})}$
(23) $\hat{P}(r,v^{tt},v^{t0})$	$\left(\sum \frac{1}{r_i} v_i^{t0} / \sum v_i^{t0}\right)^{-1}$	$\left(\sum \frac{1}{r_i} v_i^{tt} / \sum v_i^{tt}\right)^{-1}$	$\frac{\sum (v_i^{tt} + v_i^{t0})}{\sum \frac{1}{r_i} (v_i^{tt} + v_i^{t0})}$
(24) $\bar{P}(r,v^{00},v^{tt})$	$\sum r_i \frac{v_i^{00}}{\sum v_i^{00}}$ (= eq. 22)	$\sum \frac{1}{r_i} \frac{v_i^{tt}}{\sum v_i^{tt}}$ (= eq. 23)	$\frac{\sum (r_i v_i^{00} + v_i^{tt})}{\sum (v_i^{00} + v_i^{tt} / r_i)}$

* Note that in the case of P^L row 1 and 3 (and in the case of P^P row 2 and 3) are identical. This, however, need not be the case, as the somewhat more complicated formula P^{ME} (or P^F as geometric mean of P^L and P^P) already shows.

Table 5.2.2b: Index formulas and their aggregative properties

Criterion (type of consistency)	equality of one- and multistage compilation of the ... form	Formulas satisfying the criterion
Strict consistency	all three forms	L, P (= L Laspeyres, P Paasche)
Specific consistency	only 5.2.24 ($\bar{P}$ form)	S,V1,WV (= S Stuvel, V1 Vartia I, WV Walsh Vartia)
Weak consistency	one or two forms* but not all three forms	ME (Marshall-Edgeworth, here only eq. 22 [$\tilde{P}$] applies)
No consistency whatsoever	none of the formulas 5.2.22 through 5.2.24	Banerjee (BA2), Drobisch, Fisher , Törnqvist, Theil, Vartia II, Walsh (Walsh 1), Walsh 2

* if $\bar{P}$ in particular, then "specific" (so that "specific" is a special case of "weak")

If an index function is able to satisfy consistency (equality of single-and two-stage aggregation) in one or two but not all representations, the function is said to comply with *weak consistency*. It is of course much more restrictive to satisfy consistency in all three r-v-representations, a property called *strict consistency*. As usual, strict consistency implies weak consistency but the converse is not true.

Finally, it is interesting to asses a number of well known index formulas as regards their aggregation properties (**tab. 5.2.2b**). Conspicuously, precisely those formulas which are generally worshipped as "superlative" or so perform remarkably bad. This applies in particular to the formulas of Fisher, Törnqvist and Walsh.

5.3 Harmonization of deflation methodology in Europe

The European System of National Accounts (ESA) attempts to harmonize and unify methods of National Accounts and therefore also the ways to calculate aggregates at current prices and at constant prices. In what follows the recommendations, standards and provisions concerning deflation in general and of specific aggregates set up so far for the EU (European Union), and European Monetary Union (EMU) member countries are presented in some detail.[56] Harmonization of deflation methods gained importance considerably due to the progress in European integration. In particular, the so called "Excessive Deficit Procedure", as part of the "Stability and Growth Pact" (SGP) calls for truly comparable volume figures.

a) General approach to harmonization of deflation methods in the EU

The EU Commission *Decision 98/715*, which is the centrepiece of all harmonization-rules set in force so far, and the "*Handbook Volume Measures* in National Accounts" (HVM, 2001) specified *three* main *principles*:

1. (*level of disaggregation*) the measurement of prices and volumes should refer to a level of aggregation with as much detail as possible (for example, at least some 60 items in the case of output) in order to capture satisfactorily the impact of structural changes;
2. (*Laspeyres/Paasche*) In contrast to the vigorous support Fisher`s formula received in the SNA, the Europeans were less inclined to abandon traditional methods. The Laspeyres approach continues to be relevant in the case of a quantity (or volume) index and the Paasche formula is considered eligible for a deflator price index. The next principle, however, implies turning away from tradition in that
3. (*updating of weights, chaining*) weights should be derived from the previous year, such that annual links are calculated and multiplied to a chain index of prices or volumes respectively.

Valuation concepts, the scope of deflation

The supply side (production approach) is valued at basic prices (BP) while the uses (expenditure approach) are valued at purchasers' prices (PP). To derive PPs from BPs, non-deductible VAT as well as transport and trade margins have to be added

[56] We are not dealing with special problems of quarterly national accounts. It should be noted that the rules laid down in the ESA, clarified by commission or council decisions/regulations and described in a handbook (HVM) depicted in what follows are binding for member countries of the EU and EMU.

(and to be deflated).[57] The scope of deflation is the entirety of aggregates and transactions recorded in the System of National Accounts (SNA), primarily, however, the GDP and its components according to the "production approach" and the "expenditure approach", that is output, intermediate and final consumption, capital formation, but also taxes, subsidies and other distributive transactions.

Unfortunately, the move to chain indices aggravated problems with accounting constraints by adding another (third) difficulty to the problem of making consistent aggregates based on (1.) different concepts, and (2.) different data sources etc. statisticians in the field of National Accounts have to cope with.

As to the methods, in general *"indirect" methods using "deflator" price* indices are preferred to "direct" methods of extrapolating quantities, as price indices are more readily available and subject to less variance than (direct) quantity indicators. However, in quite a few instances, the "direct" method of (directly) observing "quantities" was preferred. This applies even to those aggregates in which it might be difficult to discover "quantities", such as (non-market) services in education, administration and health for example.

Ranking of methods: Another feature of the procedures laid down to harmonize volume measurement in Europe is the distinction made between three types of methods

A methods:	most appropriate methods
B methods:	those methods which can be used when A methods cannot be applied, and
C methods:	those methods which shall not be used.

For a method to be eligible as an A method the following conditions should be met:

- coverage should be complete with regard to the aggregate in question,
- quality change should be taken into account,
- consistency with definitions and equations of national accounts has to be ensured, and finally, especially in the case of output-deflation,
- producers' prices (so-called basic prices) rather than purchasers' prices or input cost should be used.

Decisions on some of the more complicated problems of deflation methodology, such as deflation of non-market output, were initially postponed until results of a special *research programme* are available.[58]

[57] In principle taxes are deflated by applying the base period rates to the flows that constitute the basis for taxation. Changes in the tax rates as well as in exemptions ought to be treated as belonging to the volume component.

[58] However, it was not considered premature to settle another most influential issue, viz. the *chain index* formula problem. Germany was outvoted in its attempt to defer a decision on using the chain index approach for deflation purposes, and to wait and see the results of a test phase in which experiments could have been made with chain indices.

b) Aggregates related to the "production approach"

The relevant aggregates that need deflation in this framework are output (P.1),[59] intermediate consumption (P.2), and also taxes, subsidies and some other distributive transactions as well as "value added" as a balancing item. However, aggregates such as final consumption (P.3/P.4) or capital formation (P.5) are components of the so-called "expenditure approach" (see part e).

1. Output

Market production (as opposed to non-market output, of which output for own final use is a part) of goods and services seems less difficult to deflate. The paramount principle is: Each product should be deflated separately by a Producer Price Index (PPI), broken down into sufficient detail and making adjustments for quality changes.[60] Hence PPIs need to be available in a sufficient breakdown. In contrast, using a Consumer Price Index (CPI) is generally considered to be only a B-method, on the following grounds

- The CPI normally refers to purchasers' prices rather than basic prices, and
- the CPI's weighting structure as well as its coverage differ from what is required in order to deflate output.[61]

Not only CPIs are more or less inadequate to serve as deflators instead of PPIs - the same is true for *unit value indices* and price indices related to input prices (e.g. labor or raw material related to the production of the output in question).

The most important and generally encountered problems in deflation of outputs indirectly with price indices are

- the existence of unique products,[62]
- difficulties with the adjustment for quality and productivity changes and
- the valuation of services, where in general a clear separation of "price" and "quantity" cannot be found.

In general, *volume extrapolation* (*direct approach* to deflation) is inferior to indirect methods, unless we are dealing with very homogeneous products with little or no quality changes and readily available quantity information, or with products provided for "free" (see below). The same is true for *input methods*, where the costs of produc-

[59] The labels P.1, P.2, etc. refer to the classification of types of transactions used in the SNA and the CPC, and their meaning is as follows: P = production-, K = capital-, D = distributive transactions.

[60] By the same token the preferred method of deflating input (intermediate consumption) is product by product, using prices collected from purchases and taking differences in prices for different purchasers into account. The problem with this, however, is that satisfactory price indices of this kind rarely exist.

[61] As will be shown below, the concept of "private consumption" within the CPI-framework is not totally in line with the concept of final consumption that applies in National Accounts.

[62] This applies in particular to investment goods. The HVM prefers a method called "model pricing" as described in **sec. 6.3** in the context of price indices (PPIs) for the construction sector.

tion are used to approximate basic prices. An exception of this rule is, however, the case of (other) *non-market-output*.[63]

In those cases (i.e. non-market services) indicators of inputs (e.g. number of employees), activity (e.g. number of operations performed in hospitals) and output (e.g. amount of teaching) may be acceptable, provided a reasoned adjustment is made for quality changes, progress made as regards productivity, and a sufficiently large number of indicators is used in order to improve reliability and to avoid erratic and extreme estimates. Hence, while output indicators are the preferred approach of deflation in the case of those non-market services that are attributable to individuals (e.g. health, education), indicators of costs (deflated inputs) are considered acceptable in the cases of truly collective final consumption by the government (e.g. defence, administration etc.). A key problem in using the aforementioned indicators then is to verify that quality changes are in fact correctly taken into account and to find methods of valuation of this change. In the field of health, education and so on, quality assessment can easily be highly subjective and inconsistent over time. The dimensions in question, like longevity and "quality of life", "standards" of education, frequency and severity of crimes etc., are by their very nature difficult to measure, let alone to quantify in monetary terms.

Problems in deflating output and net value added are still left undecided, particularly in the case of complicated aggregates such as *financial intermediation* services, dwelling, real estate and other business services, health and education as well as social and personal services.[64]

2. *Intermediate consumption*

According to the HVM, intermediate consumption should be deflated product-by-product, preferably by making use of a genuine price index compiled from purchasers´ data and distinguishing between domestic and imported inputs. Price indices for goods and services received as inputs are rare, however, compared with PPIs, and if available such indices often lack a sufficiently detailed breakdown by types of inputs to serve as deflators.

3. *Value added*

Value added as a non-commodity flow gives rise to the so called double deflation method. Double deflation is regarded (in the SNA/ESA and consequently in the HVM handbook) as theoretically the correct method without reservation, and hence all "double indicator methods" based on independently observing both output and input indicators were called "A methods". The advantage of such methods was, among other things, seen in the possibility to derive a reliable estimate of productivity

[63] The SNA makes a distinction between two types of non-market production, i.e. "output for own final use" and "other non-market output", which mainly comprises certain services provided by the government and non-profit institutions serving households (NPISH).

[64] Note that just in these areas, in lieu of appropriate price indicators, the use of *quantity indicators* (or even indicators of production *costs* in some cases) is expressly recommended.

changes (output and input *independently* assessed). Consequently, the "real income approaches" to deflate value added using a single deflator or a single indicator of volume were ruled out, or regarded as B or C methods.

c) Aggregates related to the "expenditure approach"

1. Final consumption: households

To deflate household final consumption, in many cases the CPI is found appropriate. However, differences between a CPI (or the HICP in Europe) and the required deflator for National Accounts are substantial. They refer to

- the *coverage* (own produced goods, income in kind, national [SNA] versus domestic [HICP] concept, exclusion of net premiums of life insurances) as well as
- differences in the *concept*[65] and the *formula* (Laspeyres vs. Paasche),

and therefore the CPI is, though recommended, not considered an A method unless some adjustments are made and account is taken of the quality of products. Note that the HVM does not recommend deflation of private consumption[66] in terms of real income, by analogy to output, the volume approach is preferred

2. Final consumption: government

Final consumption by government and non-profit institutions[67] (sectors S. 13 and S. 15 in the SNA) is conventionally valued as the sum of costs less any true "sales" with fees involved (i.e. the *input approach* is applied). Hence it is acceptable in the case of these aggregates to use *inputs* in order to estimate the *value* ("at current prices"). However, this does not mean that *volume* ("at constant prices") is established by considering inputs as well.[68] The HVM does not accept simply summating deflated inputs to get a constant price "volume", for the simple reason that this approach would fail to account for quality and productivity improvements. It recommends rather the use of output indicators[69] (adjusted for quality changes).[70]

---

[65] For example, the treatment of owner occupied dwelling, where the SNA makes use of the rental equivalence method or the estimation of the value und volume of non-life insurance services.

[66] or more precisely: final consumption expenditure by households.

[67] Non profit institutions serving households (NPISHs).

[68] Though the value (at current prices) of the aggregate may be estimated on the basis of inputs, a proper deflation is not made by simply deflating the inputs.

[69] If ever possible one should strive at establishing independent measures of input and output in volume terms.

[70] Thus for final consumption of government and NPISH's, in principle the same considerations apply as for the deflation of non-market production.

3. Gross fixed capital formation (GFCF, P.51)

GFCF is subdivided into

1. tangible assets (P. 511), comprising construction, machinery and equipment (among which also cultivated assets in agriculture)
2. intangible assets (P. 512), including mineral exploration, computer software, entertainment plus literary and artistic originals,
3. major improvements to tangible non-produced assets, e.g. land, and costs associated with the transfer of such assets.

Obviously the predominant type of aggregate is a commodity flow, calling for PPIs broken down at a detailed product level. However, to be suitable deflators, PPIs (based on basic prices) have to be transformed to purchasers' prices. The problem with deflation of capital formation is that genuine producer price indices of investment goods are sometimes difficult to compile because of the small number of truly comparable goods.[71]

4. Other components of capital formation (P.52, P.53)

4.1. Changes in inventories (P.52, CI) are the second type of investment and require a deflation of inventories in materials and supplies (on the input side), and finished goods (output side) as well as *goods for resale* and *work-in-progress* (including livestock). The HVM clearly states that CI should not be estimated as a residual item but rather by independently observing entries and withdrawals. Ideally entries and withdrawals should be deflated separately using the appropriate (PPI) price indices.

4.2 Acquisition less disposals of valuables (P.53), a new investment category posing some difficult deflation problems, many are related to the unique goods problem. The HVM recommends a suitable PPI for industry producing valuables[72] as an A method.

5. Net export

Net export does not introduce specific problems of deflation. It should be mentioned, however, that in many countries unit value indices (UVIs) instead of true price indices of export and import are used (see **sec. 6.4**), and there is some discussion about which type of index is more appropriate for deflation purposes. The HVM recommends PPIs if adjusted to record border values at the exporter's customs frontier (i.e. the so called f.o.b. valuation)[73], PPIs in general reflect prices on the domestic market, but they may be adjusted to reflect prices charged for exports too, and in contrast to UVIs they usually better account for quality changes and are also more likely to be

---

[71] As will be shown in **sec. 6.3**, most of the construction work, but also some equipment, of which investment consists, is unique, or at least produced in limited numbers.

[72] However, precious metals and stones as well as antiques and other art objects are by definition scarce and thus not items typically covered by price indices.

[73] Imports on the other hand should be valued on the basis of c.i.f. values (including transport and insurance services to the importer's border).

capable of isolating the price component. The same applies generally to the deflation of *imports* where, however, the direct use of a PPI could be better justified, as prices on the domestic and import market trend to be quite similar.

6. Taxes and subsidies

The HVM discusses at length the deflation of taxes and subsidies *on products* being part of the difference between the basic price of a product and its purchasers' price. They exist mainly in two forms, based on the value ("ad valorem", V-type) and on the quantity (Q-type) of products. In the latter, less complicated case it is sufficient to apply a constant (base period) tax rate to the actual quantity in order to get the volume, i.e. the tax revenue at constant tax rates. Changes in the rate or the scope of a tax ought to be recorded as price effects and therefore be eliminated. The V-type of taxes on products, however, requires an additional adjustment of the price component in the tax determining value.

Other taxes and subsidies *on production* (as opposed to "on products") as e.g. payroll taxes, tax on ownership or use of assets, or licenses (if not combined with services provided) are components of *value added*, which should preferably be deflated in volume terms by double deflation.[74] By definition these taxes (and subsidies) are not linked to the value or quantity of production. Subsidies on "production" (as opposed to "products") may be granted for reducing pollution or other characteristics of the production process. The problem with them is that no simple rules are known to isolate a volume component.

7. Compensation of employees

Compensation of employees comprises employers' social contribution in addition to wages and salaries. The HVM discards the real income method and instead recommends an extrapolation of the "quantity" of hours worked of a given type and level of skill, that is, a kind of output-indicator-method. The problem is then of course to account for quality (skill, grade, occupation or so). The second best solution is considered indirect deflation, making use of a suitable wage index reflecting the pure "price" movement in cash wages and salaries.[75]

---

[74] The HVM discards the income approach even in the case of non-market production where the value ("at current prices") of production may be estimated on the basis of inputs. In spite of known difficulties with this approach, the HVM prefers (physical) output indicators for an estimation of volumes.

[75] As regards wage indices, we may distinguish an approach of the unit value index (UVI) type (average wages method, AW) on the one hand and a deflator coming closer to a true price index (wage rate method, WR) on the other. Not surprisingly, in view of the UVI-theory (see **sec. 6.4**), the problem with AW is the degree of detail, or homogeneity of the categories of employees, while the main WR-problem is whether the wage rate movement refers to a representative selection of categories or not.

8. Consumption of fixed capital

Finally, consumption of fixed capital is a component of value added that has to be deflated in terms of volume. Given that the perpetual inventory method to estimate the stock of fixed assets is widely used, all the requisite data on fixed capital formation as well as the respective price indices and estimates of the probable economic life of assets should be available.

d) Summary of the HVM and some remarks concerning direct volume measurement by extrapolation of output indicators

As to recommendations concerning output by product classification (CPA groups), an attempt is made to summarize the key aspects and the recommended methods in **tab. 5.3.1**. It should be kept in mind that the HVM makes a strong case for adopting output methods where suitable PPIs are lacking (as necessarily in the case of non-market production). It also does so because the observation of output is a method treating non-market production more similarly to market-production, where it is again output rather than input that matters. To get a feel of the type of indicators under discussion, in what follows we present some examples of indicators in two fields:

Education: pupil- or student-hours[76] (number of hours spent being taught), the number of examinations, and the amount of those which passed or resulted in a "move up" (passing through to a higher year). To the extent that a sort of market production takes place by levelling fees or so such cost data should be taken into account too (in such cases the respective prices may even be part of a PPI or the CPI and making use of them would be an A-method).

Health: patient treatments by type, or simply number of visits or consultations applicable only in the case of the general practitioner, number of occupant days (spent in hospital) or number of discharges (if not further detailed by type of disease, treatment, level or care etc. such indicators would definitely be too crude).

e) Move from "fixed base" indices to chain indices

As to the chain-index and the non-additivity issue, it should be mentioned that the HVM is well aware of the problems resulting from the SNA recommendation:

> "some countries have adopted the practice .... (to publish) ..: non-additive chained series, that give the headline growth figures, as well as chained series in which the mathematical discrepancies are removed in some way. The last series can be useful to some users, for

[76] A distinction is made between output (amount of teaching) and outcome (knowledge and skills resulting from successful teaching). Indicators should preferably cover the outcome dimension. In the case of health output would be the number of treatments, but outcome would ideally be gains in "quality adjusted life years" (QALY). Though QALY attributable to a specific treatment would be difficult to define, this number is considered preferable to the more readily available deflated costs of health service.

example econometric model-builders, but the growth figures of those series do not correspond to the headline figures".[77]

The European Commission clearly already voted at an early stage of the harmonization process for a rather substantial change in methodology in adopting the chain index approach.[78]

Without anticipating the discussion of the chain index issue (and thus the implications of this choice of the index formula) in chapter 7, it may be useful to summarize in short the two rival positions. The main arguments in favor of deflating by using chain price indices were

- volumes based on previous year prices give the "best", or most "accurate" estimates of changes in volume from one period to the next, a kind of comparison which frequently is considered the most[79] desired (this argument in the case of deflation corresponds to the idea that chain indices in the context of inflation measurement [price level measurement] are to be preferred because of providing *better growth rates* due to making use of the most "representative" or "relevant" weighting structure);
- the use of chain indices would considerably *ease the treatment of quality changes* and new or disappearing products.

Interestingly, these arguments were found compelling enough, although very little seems to have been known about the degree of increase in "accuracy", "relevance" or "representativity"[80]. By which exact criterion, for example, is $\sum p_{t-2}q_t$ more "relevant" a volume than $\sum p_{t-3}q_t$ or $\sum p_0q_t$? And why is the value (not volume) $\sum p_tq_t$ not even more "accurate" and "relevant" than $\sum p_{t-2}q_t$ as this logic would imply?

In the same way, the term "treatment" in the second argument is not particularly clear. The argument owes much of its appeal to the vagueness of "treatment". The price of greater ease in treating new or disappearing products is that care is no longer taken when comparing like with like. The majority of member states were well aware of some disadvantages of the chain index approach (see **sec. 7.2**).[81] The present author devoted a whole book to the analysis of the properties of this index ap-

---

[77] Note that this procedure is in line with the SNA, which in paragraph 16.75 explicitly recommends the publication of alternative series (see digression of chapter 5 page 369).

[78] As the USA did shortly beforehand. The relevant decision is the EU Commission Decision (as concerns the principles for measuring prices and volumes) Nr. 2223/96 approved by the Statistical Programme Committee in Stockholm, 27th May 1998. As mentioned above, Germany was the only Member State opposed to the idea of chaining on some fundamental conceptional grounds. But Germany was outvoted.

[79] Note that the "most" desired type of analysis does not mean: the only legitimate. It is noteworthy that it is by no means less important to make consistent comparisons of growth rates between *a series of successive years* (for example over a whole business cycle), and not only between two adjacent years at a time. In such a type of analysis chain methods lose much of the attraction they might have when the sole aim of statistics is supposed to compare one period to its directly preceding period.

[80] Evidently these are notions which do not even appear sufficiently defined in operational terms as to make them satisfactorily measurable.

[81] The minority position of Germany in this issue not only emphasised the costs and additivity problem but mainly stressed lack of comparability of volumes over time.

proach in which the appropriateness of chain indices for deflation is also explored in detail.

Table 5.3.1: Selected problems of output deflation by product

Type of output	Major problems to be solved for an appropriate deflation	Recommended method of deflation in the HVM
Agriculture fishing, forestry	Production process spread over more than one accounting period; seasonality, many subsidies	Suitable price indices (PPIs) are often available, quality has to be taken into account
Mining and quarrying	Less problems with insufficient availability of data and heterogeneity	Pure quantity data (direct approach), but also unit value indices acceptable
Manufacturing	Definition of quality for some products such as fashion goods or high-tech goods. STS-Regulation[1] requires detailed monthly statistics on production and prices. Only few data on major capital goods[2]	Quantity indicators if available or price indices (sometimes difficulties with large equipment goods; model or specification pricing)[3]. Problems also with rapid productivity growth (if input indicators are used)
Computers and related equipment	Quality adjustment is no small task, due to the rapid progress in technology in this field	Hedonic method recommended, judgemental approach possible but considered much less appropriate
Energy, gas, water etc.	Price discrimination (e.g. bulk users pay less than private customers), division between producer and distributor	Deflation should treat production and distribution (service) differently; there are many vertically integrated firms
Construction	Long duration and uniqueness of projects; more input than output data available	Input price indices as deflators should be abandoned in favor of true output price indices
Wholesale and retail trade services	Trade margin is considered the equivalent of the value of trade services. Value and volume of services often assumed to correlate with sales[4]. Price differences between outlets cannot be taken as indication of differences in the quality of service provided.	Double deflation of sales and purchases in constant prices if two separate indices are available; changes in the quality of services should be taken into account (explicit adjustments still are rare to date); problems with new trade channels (e-commerce)
Hotels and restaurants	Wide coverage of extremely different standards: discounts for block booking and many small producers	Price indicators as part of the CPI available yet efforts should be made to directly measure output[5]

Table 5.3.1 (ctd.): Selected problems of output deflation by product

Transport, communication	Complicated system of prices ("one-off", season, group tickets) giving different rights[6] under varying conditions. Discounts and subsidies are common[7].	Output as number of tickets of a specified type multiplied by price of each ticket in the base period. Appropriate PPIs are A-method; volume indicators (more widely available than price indices) B-method.
Financial intermediation	Management of accounts, loans, savings, investment etc. directly charged, or indirectly measured[8]	These services are not well covered by PPIs. Suitability of various volume indicators currently explored.
Postal and telecommunication services	Genuine PPIs available in many countries only recently. Different prices according to quality differences and different packages of services	CPI adjusted to basic prices or PPIs A-method (provided that quality adjustment is made). Volume indicators comprising the full range of services are B-methods
Insurances and pension funds	Direct volume measurement and deflator of gross premiums inappropriate. Bundling of products (various insurances) very common	No A-method available, output indicators such as the number of acquisitions, policies etc. is considered a B-method
Real estate, renting and business service	If separate prices exist for charged services, deflation by a PPI is A-method[9], often fees are related to property prices. Mostly knowledge-based services. For letting of residential buildings CPI information available. For owner occupied dwelling[10] rental equivalent approach acceptable	While output price indices for real estate services are rarely available, prices of the corresponding assets will be eligible as B-methods. Acquisition price of new dwellings is explicitly downgraded to be C-method, though often regarded as a solution of the owner occupation problem (!), see sec. 6.2.
Computer and related services	Bundling computer software with hardware and consultancy common. PPI data exist for standard (non-customised) software. Other software and consultancy service is mostly unique[11].	Hedonic methods are widely used in this field. Model pricing of well defined services could be considered A-method. Simply making use of foreign (US) indices relating to comparable products is to be avoided.
Research and development	Output is by nature unique. Model pricing not applicable[12], sometimes charge-out rates and hourly fees available (B-method)	Output data such as number of patents, publications, diplomas, citations and the like are (surprisingly) deemed acceptable in the HVM

Table 5.3.1 (ctd.): Selected problems of output deflation by product

Other business services[13]	"Representative pricing" applied to a list of products (contracts, tax returns) would be A-method. Difficult to ensure constant quality (standards). A most heterogeneous group of services with different indicators of output.	Crude volume measures such as the number of tax returns, or projects (in the case of technical consultancy), or the amount of advertising space as well as rates, fees etc. are unsatisfactory, due to the rapid productivity changes in this field
Public administration, education, health and social work	Weighted (with costs) sum of output indicators (relatives). Due to ever increasing interest in the efficiency of government more indicators of output, quality and productivity will be available	Output indicators (if quality is accounted for and if closely related to the kind of administrative work) are considered A-method. Input methods are B only (even if broken down into sufficient details)
Recreational, cultural and sporting services	Comprising widely different activities such as services provided to an audience, libraries, gambling, films, radio and TV, news agencies and the like	Often number of tickets sold by type of event is a good output indicator. Some activities are encompassed in CPI, such as personal services (e.g. hairdressing)

1) Short Term Statistics in the European Union, Commission Regulation 1165/98.

2) like airplanes, ships etc. Explicitly excluded in the STS Regulation.

3) See **sec. 6.3** for problems with unique goods where making pure price comparisons calls for these above-mentioned methods of "pricing".

4) A most widely used method implies that the volume of margins follows the volume of sales and assumes that no quality change in the services takes place.

5) Crude volume data (e.g. number of clients) and input data for deflation are considered C methods.

6) e.g. to travel in a defined area or within a limited time period only.

7) In air traffic e.g. depending on flexibility, or time before the date of travel, time of travel (peak, peak-off), destinations etc. Such factors are regarded as indicating different quality.

8) This is the case called FISIM= Financial Intermediation Service Indirectly Measured, an item in the SNA which causes considerable problems as regards volume measurement. Value is assumed to correspond to the difference between loans' and deposits' interest rate.

9) This applies also to services auxiliary to financial intermediation.

10) See **sec. 6.2** for more details concerning OOD. Classification of dwellings should be sufficiently detailed (floor area, number of rooms, location etc.).

11) See **sec. 6.3** for a discussion of unique products.

12) There is of course no price for achieving an already known research result again. On the other hand futile R&D activity without a useful result should still be regarded as a kind of "output".

13) Such as legal services, (advice etc.), accounting, tax consultancy, management, architectural, engineering and related technical consultancy, labor recruitment, advertising, cleaning, security services etc.. Some services are basically of a routine nature (for example, certain book-keeping work) other services are tailor-made and therefore by nature unique. In the case of management services there is no "market" for this kind of work and thus a deflation of inputs acceptable.

5.4 Fisher's ideal index far from being ideal

In **sec. 5.2**, and in particular in **tab. 5.2.1**, it has been shown that deflation using direct[82] Fisher price indices P_{0t}^{F} (or "Fisher deflation" in contrast to "Paasche deflation"), let alone chain Fisher indices $\overline{P}_{0t}^{FC}$, is unable to meet certain requirements,[83] like PQC[84] and SCV[85], or even RQM[86]. This will be demonstrated in more detail in parts a through c in this section. Furthermore, it is interesting to note that Fisher's formula has many more shortcomings in addition to the aforementioned problems with aggregation and deflation (they will be listed in part d of this section).

a) Fisher deflation (direct Fisher price indices) and quantity movement

It has been shown in **ex. 5.2.1/2** that there are situations in which volumes resulting from Fisher deflation will not follow the pattern of volumes gained by Paasche deflation and differ substantially from quantities. A distinction has been made between case 12 with a uniform increase of prices by 50% ($P^P = P^F = 1.5$)[87], and case 22 with a non-uniform increase.

	base period (0)		period t: case 12		period t: case 22	
	prices	quantities	prices	quantities	prices	quantities
A1	30	70	45	84	30	84
A2	50	30	75	48	40	48
B1	90	20	135	36	81	36
B2	120	30	180	24	168	24

According to **ex. 5.2.2** volumes and quantity indices in case 22 were as follows:

	volumes			
	Paasche deflation		Fisher deflation	
	abs.	Q_{0t}^{L}	abs.	Q_{0t}^{F}
A	4920	1.3667	4881.7	1.3560
B	6120	1.1333	5871.7	1.0872
sum	11040	1.2267	*)	

*) as shown in **ex. 5.2.2** two different solutions possible (i.e. the structural consistency issue see also below **ex. 5.4.2**): using the "total" (overall) P^F index (amounting to 1.0684) yields 10658.6 while summation of 4881.7 and 5871.7 results in 10753.4.

---

[82] See chapter 7 for the distinction between "direct" and "chain" indices.

[83] The following criteria appear particularly reasonable if volumes are to be interpreted as substitutes for (aggregate) quantities, otherwise unobservable.

[84] Volumes represent pure quantity comparison or linearity (additivity) in quantities.

[85] Structural consistency of volumes.

[86] Volumes reflect quantity movement (when all prices change *uniformly* [at the same rate], the change of volumes will equal the change of quantities).

[87] In this case we have identical results of Paasche and Fisher deflation respectively.

Note that the result of Paasche deflation now (case 22) does not differ from case 12 in spite of quite different assumptions relating to the price movement. Hence volumes in Paasche deflation (as opposed to Fisher deflation) are not affected by differential price movement. It is only the quantity change that matters. Volumes can be interpreted as substitutes of quantities in a quite straightforward manner. "Pure quantity comparison" (PQC)[88] is met.

In the case of Fisher deflation the situation is fundamentally different, however. It is not only the well known fact that Fisher volumes are liable to be structurally inconsistent (to be shown in more detail in part b and **ex. 5.4.2**) which makes them doubtful, it is also no longer true that the same quantity movement is reflected in the same manner (resulting in the same volumes) irrespective of how prices changed. It should be possible to construct a situation in which deflation by P^P and P^F will result in changes of volumes being contradictory. This is done in the following example.

Example 5.4.1

Ex. 5.2.2, discussed above, will now be slightly modified in order to demonstrate the fact (and its causes) that Fisher volumes will not necessarily yield results in line with pure quantity comparison (PQC). Assume prices at time t as follows:

	p_{it}	q_{it}	p_{it} / p_{i0}	q_{it} / q_{i0}
A1	60	84	2.0	1.2
A2	10	48	0.8	1.6
B1	180	36	2.0	1.8
B2	72	24	0.6	0.8

Shadowed boxes are introduced to indicate which part of the **ex. 5.2.2** has been changed. The following table shows that - due to the correlation between price and quantity relatives - PQC is not met in the case of Fisher deflation[89] whereas Paasche deflation yields exactly the same results of ex. 5.2.2.

	$\Sigma q_t/\Sigma q_0$ (1)	$V = \Sigma p_t q_t$ (2)	P^P (3)	P^F (4)	V/P^P (5)	V/P^F (6)	(7)
A	1.32	6000	1.2195	1.2752	4920	4645.0	1.046
B	1.20	8208	1.3412	1.1961	6120	6862.5	1.271
Σ	1.28	14208	1.2870	1.2288	11040	*)	

*) again two different solutions possible (i.e. structural consistency of volumes [SCV] is not met)

[88] PQC in the sense of the influence of price movement being eliminated strongly appeals to our intuitive understanding of deflation in volume terms. Conversely, the results of Fisher deflation do *not* remain the same. The impact of (differential) prices is influential. Volumes depend on both movements in quantities as well as in prices.

[89] In contrast, the results of Paasche deflation are of course not affected by this modification.

The point is that while A has risen in quantity more (+ 32%) than B (+ 20%), which is also reflected by Paasche deflation (V/P^P), Fisher deflation (V/P^F) suggests the opposite to be true: Fisher volume of A has risen less (4.6%) than B (27.1%). ◆

Not measuring a pure quantity movement (PQC) in the case of the Fisher quantity index Q_{0t}^F resulting from Fisher deflation is equivalent to not linear (additive) in quantities, in the sense of

(5.4.1) $$Q\left(\mathbf{q}_0, \mathbf{q}_t^*\right) = Q\left(\mathbf{q}_0, \mathbf{q}_t\right) + Q\left(\mathbf{q}_0, \mathbf{q}_t^{\Delta}\right) \text{ where } \mathbf{q}_t^* = \mathbf{q}_t + \mathbf{q}_t^{\Delta}.$$

Obviously the Laspeyres quantity index resulting from Paasche deflation is linear in quantities, since with elements q_{it}^{Δ} of the vector $\mathbf{q}_t^{\Delta}$ we get in the case of Q_{0t}^L

(5.4.2) $$Q_{0t}^L(\mathbf{q}_t^*) = \frac{\sum q_{it}^* p_{i0}}{\sum q_{i0} p_{i0}} + \frac{\sum q_{it}^{\Delta} p_{i0}}{\sum q_{i0} p_{i0}} = Q_{0t}^L(\mathbf{q}_t) + Q_{0t}^L(\mathbf{q}_t^{\Delta}).$$

A sequence of Paasche deflation volumes is given by $\sum p_0 q_2$, $\sum p_0 q_3$,...., in contrast to the corresponding sequence of Fisher deflation volumes, which is

(5.4.3) $$\sum p_0 q_2 \left[\left(\frac{\sum p_2 q_2 \sum p_0 q_0}{\sum p_2 q_0 \sum p_0 q_2}\right)^{1/2}\right], \ \sum p_0 q_3 \left[\left(\frac{\sum p_3 q_3 \sum p_0 q_0}{\sum p_3 q_0 \sum p_0 q_3}\right)^{1/2}\right], \quad \ldots \quad ,$$

or as compared to the Paasche sequence ($\sum p_0 q_2 \sqrt{Q_{02}^P / Q_{02}^L}$, $\sum p_0 q_3 \sqrt{Q_{03}^P / Q_{03}^L}$.

which does not appear to be particularly easy to understand. It is also by no means evident why this should be called "at constant prices *of period 0*". In the case of a deflation with a Fisher chain price index $\overline{P}_{0t}^{FC}$ as recommended in SNA 93, things will become even more complicated (see part c of this section, as well as **sec. 7.2**).

b) Structural inconsistency of Fisher - deflation

Structural consistency means that the sum of deflated subaggregates should equal the deflated total aggregate.

Example 5.4.2

The calculations for **ex. 5.2.2** as quoted above show that Fisher deflation is not structurally consistent. They yield:

	value $\Sigma p_t q_t$	P^F	volume
	(1)	(2)	(1)/(2)
A	4400	1.233	4881.7
B	6948	1.287	5871.7
sum	11388	1.265	10658.6

Summing up the deflated subaggregates gives 4881.7 + 5871.7 = 10753.4 whilst deflating the total aggregate, dividing by the total P^F, gives 11388/1.0684 = 10658.6 which falls short of 10753.4 by almost 1%.

To find the reason for this, it is useful to study Bortkiewicz's theorem. The covariance[90] $C = V(P^P/P^L - 1)$ depends on the relations R = P^P/P^L, and the results for this example are as follows:

	relation R	$R^{1/2}$	covariance C
A	0.9845	0.9922 (- 0.78%)	-0.01914
B	0.9205	0.9594 (- 4.05%)	- 0.08368
total	0.9321	0.9654 (- 3.45%)	- 0.077637

The column $\sqrt{R}$ explains why the result of Fisher deflation is about 1%, or 4% smaller than the Paasche volumes $\Sigma p_0 q_t$. The same exercise made for **ex. 5.4.1** reveals[91]:

	value $\Sigma p_t q_t$	P^F	volume V/P^F	$\sqrt{R}$	C
A	6000	1.2752	4645.0	0.944	-0.243
B	8208	1.1961	6862.5	1.121	+0.362
Σ	14208	1.2288	*	1.047	+0.153

*The sum of 4645 and 6862.5 is 11507.5 whereas 14208 divided by 1.2288 gives 11562.2. ♦

To sum up: In view of all criteria established for a good deflation, traditional deflation with Paasche price indices P^P seems to be much more justified and intuitively appealing than deflation with Fisher indices P^F as recommended in the SNA. Even more obscure are results derived from deflation using the *chain* index of Fisher $\overline{P}_{0t}^{FC}$.

c) The chained Fisher's ideal index as deflator

According to the recommendations of SNA[92] *chained* Fisher price indices should be used for both price level measurement and deflation.[93] In contrast, we take the position that traditional solutions for deflation are more reasonable and more easy to implement and to understand. The SNA recommendations should be viewed with strong suspicion for several reasons. Among them: results of deflation may well be eccentric in the economic interpretation, and $\overline{P}_{0t}^{FC}$ as a deflator is not even able to ensure volumes that are proportional in the quantities (according to **tab. 5.2.1** the merits and demerits of deflators may be summarized as follows: P_{0t}^{P} best with respect to all

[90] omitting the subscripts 0 and t for convenience.

[91] The motivation of the modification of ex. 5.4.1 (in ex. 5.4.2) was to arrive at a high negative (positive) covariance in group A (B) so that deflation with P^F will well differ from deflation with P^P.

[92] See **sec. 7.1.f** and **7.2.a** for more detail.

[93] It is also maintained in the SNA that as an alternative to Fisher price indices, in the framework of indirect deflation Fisher quantity indices should be used whenever direct deflation appears possible. These indices are also recommended to avoid the well known awkward results of double deflation of value added (VA).

criteria, P^F_{0t} much less acceptable though better than the chain version $\overline{P}^{FC}_{0t}$,which is definitely the worst of all deflators).

To demonstrate the violation of proportionality (or identity, as a special case of proportionality in the following example), it may be useful to anticipate **ex. 7.2.7** and **tab. 7.2.4** in which we show that despite

- the same quantities in 0 and t ($q_{A0} = q_{At} = 5$ and $q_{B0} = q_{Bt} = 15$), and
- a uniform rise in both prices by 50% (that is $p_{At} = 1.5p_{A0}$ and $p_{Bt} = 1.5p_{B0}$)

deflation using $\overline{P}^{FC}_{0t} = 1.564 > 1.5$ results in a decline of the volume by 4.1%, whereas both deflators, P^P_{0t} as well as P^F_{0t}, amount identically to 1.5 such that in the example the volume is $\sum p_t q_t / P^P_{0t} = \sum p_t q_t / P^F_{0t} = \sum p_t q_t / 1.5 = \sum p_0 q_t = \sum p_0 q_0$, thus showing *no increase because all quantities were unchanged* and only prices went up by a uniform rate of 50%. The deflator $\overline{P}^{FC}_{0t}$>1.5 on the other hand, indicating an increase of prices by more than 50%, produced a decrease in "volume" by 4% (since 0.96 - 1= -0.04) because 1.5/1.564 ≈ 0.96. Moreover, **tab. 7.2.4** shows that the formulas of the sequence of "volumes" resulting from $\overline{P}^{FC}_{0t}$ deflation appear somewhat difficult to interpret (as compared to the traditional deflation using P^P_{0t}, which gives simply $\sum p_0 q_1$, $\sum p_0 q_2$, $\sum p_0 q_3$,... The Fisher-chain-"volume" for t = 3 of this sequence is for example $\left(\Sigma p_3 q_3 \Sigma p_2 q_3 \frac{\Sigma p_0 q_0}{\Sigma p_1 q_0} \frac{\Sigma p_0 q_1}{\Sigma p_2 q_1} \frac{\Sigma p_1 q_2}{\Sigma p_3 q_2} \right)^{1/2}$ and will be even more complicated for t = 4 etc.

d) Fisher's ideal index all but "ideal"

The SNA recommends deflation by $\overline{P}^{FC}_{0t}$ because (among other reasons) the Fisher formula P^F_{0t} or $\overline{P}^{FC}_{0t}$ meets factor reversibility. The only reasonable (though rarely, if ever, mentioned) legitimating of the factor reversal test seems to be that then both problems, deflation and price level measurement, can be solved using the same index formula. This, however, is a poor argument for a formula otherwise suffering from many shortcomings.[94] In particular, both formulas P^F_{0t}, and even more $\overline{P}^{FC}_{0t}$, are unable to provide a pure price comparison, and the same applies to Q^F_{0t} and $\overline{Q}^{FC}_{0t}$ with respect to quantities. It is not only a change of price (quantities) to which P^F_{0t} (or Q^F_{0t}) will react. In the case of *chain* Fisher indices situations will be even more arcane. It is difficult to disentangle the complex of influences that make a Fisher index rise or decline.[95]

---

[94] There are many reasons lying in practical problems of official statistics as well as in interpretation for taking two differently constructed indices for these purposes.

[95] On the other hand, monetary policy as well as economic analysis in general, however, aims at an interpretation in terms of "goals" and "controls". There is a need for empirical figures that can be viewed as effects of some instruments or can be used as instruments themselves (such as the volume

According to Pfouts 1966, an index should answer a specific question, but "the purpose of the ideal index is at best opaque". To interpret P^F there is not much left apart from saying, it is the geometric mean of P^L and P^P. No weighted mean of relatives interpretation (price or quantity relatives) can be provided.[96] Nor is it possible to interpret P^F as expenditure for a certain (not necessarily constant) given basket of commodities either (that is, no ratio-of-aggregates expression exists).

Moreover, the formula P^F is not additive (in all connotations distinguished in **sec. 3.2** and **sec 5.2**) in the index function nor are volumes gained by Fisher - deflation structurally consistent.

According to Pfouts, Fisher's admiration of P^F was mainly inspired by his compliance with his reversal tests and the belief that "if one index measures above the 'true' value and another measures below it, then an averaging of the two should yield a result closer to the 'true' value than either component." The foundation of both ideas is weak, the idea of reversal tests as well as the "rectifying" of formulas by "crossing" as Fisher himself called his procedure. As to the reversal tests see **sec. 2.2** and as far as the "true" index is concerned no generally accepted view exists. "If we abandon the notion of a 'true' index, as has been done by many writers, much of the theoretical appeal of the ideal index is lost" (Pfouts). To sum up, Pfouts reached the conclusion that "this index should be abandoned".

It is true that P^F simultaneously satisfies many axioms, perhaps more of them than most of the other well known formulas. It is also true that most of the known "uniqueness theorems" refer to P^F. But both arguments are not sufficient to look at P^F as being "ideal", since many axioms are controversial, not related to the specific task of deflation and most of the functional equations postulated to arrive at uniqueness theorems have little or no interpretation.

Fisher's index has poor aggregative properties: desirable properties like aggregative consistency (**A1** in **fig. 5.2.3**)[97] and structural consistency of volumes (**A3**) are not fulfilled. It has also been demonstrated that other indices unable to be aggregative consistent can at least be interpreted in terms of a "weighted mean of price relatives", e.g. Drobisch's index $P_{0t}^{DR} = \frac{1}{2}\left(P_{0t}^{L} + P_{0t}^{P}\right)$ (see eq. 5.2.11).

P^F does not even pass the weaker (compared to **A1**) equality test (**ET**). The function P satisfies the *equality test* if $P_{0t} = f(P_{0t}^1, P_{0t}^2, \ldots, P_{0t}^K) = f(\lambda, \lambda, \ldots, \lambda) = \lambda$, or in words: given all sector indices P_{0t}^k equal λ, so the global index should yield $P_{0t} = \lambda$.

---

of money, as measured by M_1, M_2, M_3 and so on) It is also desirable that an index is understandable and will be accepted by the general public. This again requires a clear and comprehensive mechanism connecting empirical influences and the effects they produce in terms of some statistical measures. It is unlikely that Fisher indices will serve such purposes better than some more traditional methods.

[96] An attempt to express P^F (or Q^F) in terms of price (or quantity) relatives shows that weights would then be influenced by changes in prices in case of P^F (or of quantity change in the case of Q^F.

[97] We are unable to calculate the global index P^F from K sector indices, P^{Fk} (k = 1, ..., K) and from n price relatives in *the same way* (using a formula of the same structure).

Figure 5.4.1: Fisher's ideal index - far from being ideal

Some shortcomings of Fisher's ideal index

interpretation	use for deflation	other properties
No weighted mean of relatives , nor a ratio of expenditures 1)	no structural consistency (equations for values will not hold for volumes) 2)	not consistent in aggregation, not even weak equality test 4)
only "opaque" interpretation of being $\sqrt{P^L P^P}$	resulting quantity index Q^F is not linear in quantities 3)	more difficulties in compiling data 5)

1) changing cost of a certain "budget"

2) aggregation of *volumes* follows the same rules as aggregation of *values*.

3) In **sec. 5.2** we introduced various interpretations of volumes in terms of "quantities", and it could be shown that Paasche deflation allows *all* of these interpretations whereas Fisher deflation satisfies *none* of them

4) no simple function exists by which sectoral indices of P^F-type can be aggregated to a total P^F-index.

5) applies to situations only where hitherto the use of Paasche- or Fisher-indices was not a common practice (for example Consumer Price Indices)

According to Stuvel[98], P^F is also unable to pass the (compared to ET) even weaker "Withdrawal-and-entry test" (**WT**). A test becomes "weaker" as additional conditions are introduced in its description. We arrive from ET to WT assuming a subaggregate consists of only one single commodity. Once *all* subaggregates consist of one commodity only we get proportionality (**PR**), making the assumption of identical price relatives λ. As $\lambda = 1$ an even more special condition is set: identity (**ID**). Hence from left to right tests will become weaker:

$$\mathbf{ET} \rightarrow \mathbf{WT} \rightarrow \mathbf{PR} \rightarrow \mathbf{ID}$$

It is possible to conclude in this order: an index to which WT applies will also meet PR and I. The converse order of conclusion is not permitted. Thus the Fisher index satisfies only the last (rightmost) two criteria, **PR** and **ID**, Paasche and Laspeyres indices satisfy *all* of them, and chain indices (of all kinds) *none*. Hence the difference in "test-performance" is remarkable.

In a more recent survey of the axiomatic approach, also containing a more detailed inquiry into the problems of consistency in aggregation, a criterion hitherto somewhat neglected in index theory, Balk 1995 reached the conclusion that

- "the evidence for choosing the Fisher price index as the ultimate index is not conclusive" and that
- "the use of Laspeyres and Paasche indices in official statistical systems can be legitimized".

98 Stuvel 1989, p. 49 ff.

The comments on Fisher's ideal index set out so far only relate to the "direct" index, not yet to the chain version $\overline{P}^{FC}_{0t}$. It is very likely that the "message" of $\overline{P}^{FC}_{0t}$ will be even more obscure than of P^{F}_{0t}. This applies for example to deflation: Q^{F}_{0t} resulting from deflation by P^{F}_{0t} is at least *proportional in the quantities*, while $\overline{Q}^{FC}_{0t}$ (resulting from deflation by $\overline{P}^{FC}_{0t}$) is unable to meet even this fairly weak criterion. Notwithstanding, according to the SNA "the preferred measure of year to year movements of real GDP is a Fisher volume index chained together" (thus $\overline{Q}^{FC}_{0t}$) and "the preferred measure of year to year inflation for GDP is therefore a Fisher price index", i.e. $\overline{P}^{FC}_{0t}$. However:

The philosophy underlying this SNA-recommendation is fundamentally different to our approach: The emphasis of SNA was *not* put on deflators yielding structural consistency of volumes or on permitting a consistent aggregation and disaggregation *but rather* on criteria like factor reversibility, or on the desire to avoid constant prices p_{i0} of a past base period in the definition of "volumes". In our view, however, exactly in opposition to the SNA criteria like

- comparability, consistency requirements, interpretation of volumes in terms of "quantities", and isolation of a pure "quantity" component and a pure "price" component should come first, whereas
- criteria like reversibility, symmetry, up-to-date weights etc. should come second.

Digression: The SNA and ESA on how to deal with structural inconsistency

Extracted from the book: Peter von der Lippe, Chain Indices, 2001, p. 214:

The SNA though most enthusiastic about $\overline{P}^{FC}_{0t}$- deflating was well aware of this defect [i.e. structural inconsistency of volumes] (in SNA called "non-additivity"). It is interesting to see what kind of remedy is recommended ... The SNA firstly noticed that non-additivity "can produce counter intuitive and unacceptable results in the long run" (p. 16.65) but

- a Fisher chain index still "may provide the best volume measure of value added from a theoretical point of view" (p. 16.67), and
- "Although desirable from an accounting viewpoint, additivity is actually a very restrictive property" (p. 16.55).

Interestingly there *firstly* was no doubt emerging as to the appropriateness of the theoretical viewpoint despite its counter-intuitive results and there was little attention only given to the accounting viewpoint. ... Instead of calling the "theoretical" point of view in question ... the SNA emphasized the idea that the traditional Q^{L}_{0t}, P^{P}_{0t}- approach also has to put up with lack of additivity when the base year is changed. It is called an "advantage ... that chaining avoids introducing apparent changes in growth or inflation as a result of changing the base year" (p. 16.74). The "advantage" consists in ... *always* having non-additivity, not only when direct indices are linked together, but in *every* period. ...

Secondly the SNA discussed several ways out of the non-additivity problem, but rejected all of them. It is most interesting to have a look at the sort of remedies discussed (see no. 16.58):

1. publish structurally inconsistent (non-additive) volumes "as they stand without adjustment";
2. to "distribute the discrepancies over the components at each level of aggregation";
3. to "eliminate the discrepancies by building up the values of the aggregates as the sum of the values of the components at each level of aggregation" a possibility the SNA itself rejected because this would not only contradit the results obtained by deflating the aggregate as a whole but would also make the results "depend quite arbitrarily" on the level of disaggregation".

The SNA also discussed the possibility to "publish data only in the form of index numbers" and to avoid absolute figures, but SNA rejected this idea "because it may merely conceal the problem from users" (p. 16.56). *Thirdly* and finally the SNA ended up with recommending:

> "Although the preferred measure of real growth and inflation for GDP is a chain Fisher index, ... it must be recognised that the lack of additive consistency can be a serious disadvantage for many types of analysis ... It is therefore recommended that disaggregated constant price data should be compiled and published in addition to the chain indices for the main aggregates. The need to publish two sets of data ... should be readily appreciated by analysts engaged in macro-econometric modelling and forecasting" (p. 16.75).

... As the European System of National and Regional Accounts (ESA) consists in most parts in identical recommendations as laid down in the SNA it is no surprise that in principle the same preference for a Fisher chain index deflation is expressed in the relevant documents. By a Commission Decision clarifying a Council Regulation (No. 2223/96) certain rules of deflation are prescribed to all Member States. ... Chain indices of Laspeyres and Paasche are only qualified as "acceptable alternatives" [to $\overline{P}_{0t}^{FC}$], and to the direct version of those indices is not given mention at all (p. 10.65). In view of "the well-known 'non-additivity' problem" (text of the a.m. Decision) it is recommended in ESA

- "...that disaggregated constant price data, i.e. direct valuation of current quantities at base-year prices, are compiled in addition to the chain indices for the main aggregates" (p. 10.66.), and
- "...it will have to be explained to users why there is no additivity in the tables. The non additive 'constant price' data is published without any adjustment. This method is transparent and indicates to users the extent of the problem" (p. 10.67).

The solution does not appear being the ultimate wisdom. Results are to be published showing discrepancies. These discrepancies have to be explained and there are also additional series without these discrepancies.

Chapter 6

Price indices and unit value indices in official statistics

Arguably the best known price indices of official statistics are the following three indices: Consumer Price Index (CPI), Producer Price Index (PPI) of industrial and agricultural production, as well as construction for example, and the price indices in foreign trade (XPI for export and MPI for import). In the case of foreign trade but also regarding salaries and wages, we always have an alternative between a true-price-index and a unit-value-index methodology. It appears adequate to devote a whole chapter to the CPI and general problems of inflation measurement, to the PPI (**sec. 6.3**), and unit-value in contrast to price-indices (**sec. 6.4**).

6.1 The Consumer Price Index (CPI) and the Harmonized Index of Consumer Prices (HICP) in Europe

The CPI is a well known instrument for measuring inflation and for deriving "deflators" for National Accounts.[1] Not surprisingly, attempts to make national CPIs comparable have started early. Moreover, the (expanding) European Union (EU, the former European Community) and the implementation[2] and possible enlargement of the European Monetary Union (EMU) gave rise to

- a harmonization of the member countries' CPIs, and
- the desire to establish a Harmonized Index of Consumer Prices (HICP) aggregated over a group of countries such as the EU, the EMU Member Countries, or the EEA which serves also as an integrated measure of *the* EMU inflation rate (if there is such a phenomenon).[3]

Thus integration of monetary policy makes it necessary to compile harmonized indices and even more so to integrate them over the group of EMU- countries. Such attempts to harmonize national CPIs have revealed that there was initially a considerable amount of divergence in the national practices and national methodological traditions to compile a CPI as well as in the institutional and organizational framework that price statistics have to deal with. In what follows, we describe some fundamental issues concerning CPI compilation at the national level, along with strategies and performances in the harmonization process in Europe.

---

[1] Note that for this purpose many other price indices will be of great interest as well.

[2] A low national inflation rate (assessed using the HICP) is one of the criteria ("Maastricht criteria") to be fulfilled in order to be admitted to the Monetary Union (EMU).

[3] Different levels of aggregations of national HICPs across countries are distinguished (in this order comprising more and more Member Countries): EMU, EU (comprising also e.g. Denmark, Sweden and the United Kingdom as well as the recently acceded countries of Eastern Europe) and the European Economic Area (EEA) to which Iceland and Norway also belong. The corresponding HICPs are called MUICP, EICP and EEAICP respectively.

a) Political need for harmonization of CPI compilations in Europe

As European integration progresses, the need for harmonized statistical procedures and indices, especially a harmonized CPI (that is the Harmonized Index of Consumer Prices, **HICP** for short), increases. Such indices are necessary in order to provide a constant monitoring of social and economic performance across the member states, to provide information on living conditions, real wages and social concerns, guarantee a balanced growth, and to decide upon enlargement (accession of new member states) of the EU and EMU.[4] Harmonization of CPIs plays an important part in the integration process because European integration first and foremost implies a growing harmonization of national economic policies. Comparisons among member countries have to be based on harmonized data even though differences in economic institutions and practices still exist and will continue to exist. This applies mainly to:

- supply and "purchase" of financial services (insurance in particular),
- the organisation of health and educational services (in some countries dominated by private institutions, in others a domain of governmental activities),
- paying for governmental services by fees, taxes and the like and provision of local and infrastructure facilities, such as transport, energy, communication, and finally
- the organisation of private or semi-governmental social and welfare institutions.

As integration progressed from the "single market" (in force as of 1. 1. 1993) to a currency unit (EMU), the more and more harmonized national CPIs became an important instrument to assess the convergence process and the effectiveness of national policies.[5]

b) Main aspects of index construction that need to be harmonized

However, the process of harmonization had been surprisingly slow in the past and not before the end of the 1980s attempts were made to examine the "state of the art" in CPI compilation for the EU Member Countries[6] in order to accelerate harmonization. Questionnaires filled by the NSIs revealed many dissimilarities in national CPI practices in the following seven fields:

---

[4] The HICP is to be monitored carefully in order to assess convergence progress (with respect to present and possible future EMU member states), check fulfilment of the Stability and Growth Pact and to effectively combat inflation by the European Central Bank (ECB). Of course, the degree of comparability of national statistics required to successfully operate a supranational entity is much higher than what seems necessary to make international comparisons among independent countries (see **ch. 8**).

[5] This applies firstly to the monetary policies, but also to (plans of more harmonization in) taxation, social security policies etc. As free movement of persons becomes more important, the so called "emergence of a European social dimension" and the growing economic and social "coherence" will become more and more important too. There is therefore a growing need for harmonization of statistical indicators, among which the CPI plays a key role.

[6] In what follows our presentation relies heavily on a report prepared by Ruud Teekens in 1989 for Eurostat in which methods applied by NSIs in their CPI compilation were reviewed.

1 Index formula,
2. Coverage (field of observation)
3. Index population
4. Family expenditure surveys (FES) to establish weighting structure (or "basket")
5. Organisation of price surveys
6. Calculation of regional indices in addition to the national index, and finally
7. Revision and validation.

These points should now be considered one by one.

ad 1: Index formula

The Teekens Report did not put much emphasis on this point but remarked that countries using the chain principle would have less problems in providing (fixed based, i.e. direct) Laspeyres indices, "since they have all the necessary material for this purpose".

ad 2: Coverage

The selection of the commodities and services to be included in the calculation of an index is of paramount influence. "Coverage" means, among other things, the treatment of

- accommodation, esp. "imputed rents"[7] on owner occupied dwellings,
- educational services, health, social security and expenditures on insurances[8]
- interests and direct and indirect taxes[9]
- remuneration and other benefits in kind as well as own consumption[10], and
- the treatment of existing (second hand) commodities

as parts of "consumption". It is widely accepted that "consumption" in the case of family budget and consumer price statistics should be defined in a way that fits to the

[7] In 1987, in only six of the then twelve EU members were imputed rents included. Moreover, the practice of estimating imputed prices differed considerably (from the rental equivalent method as applied e.g. in Germany to the housing cost method in the U.K.). See **sec. 6.2** for more details.

[8] Only few countries made an attempt to break down the insurance expenditures (gross premium) into its components, the net premium (compensation for financial services) and reimbursements (consumption). As to the type of insurances included there was also a great variety, ranging from nil to eight (such as liability-, building-, home content,- private- and public health insurances). Public health insurance should be treated like taxes, while life insurance on the other hand is part of capital formation rather than consumption.

[9] *Interests* should be excluded unless used as an indicator for cost of housing (imputed rents). Likewise, *taxes* are expenditures but in general they cannot be regarded as consumption. Views differed as regards vehicle taxes and expenditures which are "tax-like" (cost of refuse, sewerage etc.).

[10] *Own consumption and benefits in kind* are to be included according to SNA concepts but most countries seem to find it difficult to collect appropriate data. The same is true for *existing commodities (second hand goods)*.

National Accounts (NA). Since the performance regarding price stability was one of the so called "convergence criteria" requisite for being admitted to EMU membership according to the "Maastricht Treaty", the need for making the coverage of the national CPIs comparable had been reinforced significantly. On the other hand the differences were quite fundamental, such that they could not be eliminated in *one single* step. Thus a *two-stage-strategy* was adopted to establish the aggregated HICP for the Monetary Union.

ad 3: Index population

Most of the countries refer to[11] "all private households", some to urban or middle class employee households only, and only a minority provide additional indices for special types of households, specified by size, income and social position.

ad 4: Information as regards the basket and FES

The definition of the index population is of course closely related to the type of budget surveys carried out more or less regularly. There are (or were at least in 1987), however, also some countries in which the CPI is based more or less exclusively on NA data (like in France for example). In this case it appears difficult to refer to the representative index household. On the other hand most countries make use of both budget survey data and NA data, the one used to check the other, as the source to determine weights. In 1987 only a minority of the member countries carried out FES annually (UK, Italy, Netherlands).[12] The pure number of "items" (commodities) in the basket is not a valid indicator of quality because it depends on the underlying definitions and details of classifications.[13]

ad 5: Price surveys[14]

Another aspect responsible for an insufficient comparability of national CPIs is the organisation of price surveys. It is generally agreed that non-random selection as regards items to be priced is tenable[15], although random sampling is considered theoretically preferable. An attempt is also made to cover a wide range of reporting units as far as the type of outlets (including all kinds of supermarkets, mail order houses etc.), their geographical distribution and size is concerned. Prices are mostly surveyed directly by entering the respective shops, to a certain degree also indirectly by collecting price lists, which is acceptable in the case of standardised products distributed nation-wide. Special surveys for housing to collect data on rents are also carried out in many countries, mostly less frequently than reports on other prices.

---

[11] The text of course refers to the situation in 1987 prior to developing the HICP.

[12] Without doubt there is a tendency to use the chain principle in those countries having annual FES and/or relying heavily on NA estimates to define the CPI-basket.

[13] In view of a wide definition a good may be regarded as a single item which could otherwise be counted as two or more goods. Furthermore, positions distinguished in FES are sometimes such that they can be represented by various goods. Hence there is always some vagueness involved in what has to be counted when the number of "items" or "goods and services" of a basket has to be identified.

[14] This point comprises sampling method, treatment of quality changes, new and seasonal products, specific housing surveys to collect data on rents, paid and imputed and other details.

[15] The same applies to the selection of households for the FES (see above).

ad 6: Regional indices

Apart from Italy and Spain there is mostly no tradition of calculating regional indices on a regular basis. In most countries index calculation is confined to a national index in which regional data enter are weighted with the size of the population of the regions.

ad 7: Revision and validation

Not surprisingly, those countries who apply the chain index principle feel no need to revise the methodology and to compare index calculations using different base periods (and baskets). In the framework of the direct Laspeyres index - then (1987) still the predominate method - it was mostly only the introduction of a new base which gave an opportunity to revise other methodological aspects too and to calculate the "new" index back for some periods in the past. None of the member countries was capable of quantifying satisfactorily the overall accuracy of their own index calculation, which probably still applies today.

c) The harmonization strategy

In view of this point of departure in 1987 it is not surprising that the establishment of the European HICP took time. In order to better understand the harmonization project, it is useful to shortly describe the decision making procedure implemented for this purpose. As by the end of 1991 the so called "Maastricht Treaty" laid down certain "convergence criteria", among them also the inflation rate, which ought to be met in order to be admitted to the EMU, some concrete steps were necessary to make national CPIs more comparable. A Working Party (WP) was established to promote work on CPI harmonization. It held its first meeting in June 1993, attended by delegates of NSIs of all member countries and observers of various European institutions. The WP had to make proposals that were to be approved by the Statistical Programme Committee (SPC) of the Directors-General of EU member countries' NSIs. Eventually the rules[16] laid down by the WP and approved by the SPC should become Council Regulations. To speed up the rather lengthy legislative process needed for such a Council Regulation, in May 1994 it was decided to be content with a Commission Regulation when a majority can be found and to leave only truly controversial issues to be decided by Council Regulations. The WP established several Task Forces to deal with details of CPI methodology, such as coverage and items difficult to be priced such as insurance services, monopoly prices, owner occupied dwelling, clothing (problems with fashion change, outselling etc.), discounts granted, standards for dealing with quality change, deriving "elementary aggregates",[17] sampling, updating weights and timely incorporation of new goods in the index, and rules for data delivery to Eurostat (degree of detail, submission time etc.).

Before commenting on the contents of these regulations some general remarks on the objectives and criteria of harmonization appear appropriate.

[16] "Rules" mostly taking the form of Commission Regulations are mandatory. "Guidelines" for the construction of Harmonized Indices of Consumer Prices (HICPs) are less obligatory.

[17] i.e. the lowest level of detail for which expenditure weights are established,

The objective of the HICP clearly was not (or at least not immediately) to replace the national CPI (which may be continued if found useful or necessary) but to provide a comparable measure in order to monitor progress in price stability of different countries. No existing national CPI was considered to be the model that should be followed by all other member states. Therefore the strategy chosen consisted of

- making *all countries move* to a common measurement concept approved by all, or at least most, of them, and
- *gradually* and stepwise (in *two stages*) introducing more and more elements of an eventually fully comparable HICP.

Table 6.1.1: Principles for the Harmonization Process

Principle	Description, comment
Necessity and subsidiarity*	Rules should be made only where there is evidence that non - comparability may arise and they should only specify the kind of output-quality found desirable, leaving it to the member countries to find ways how to best achieve the requirements. As a rule, harmonization should stipulate "*minimum* standards" and may also consist in banning acknowledged "bad practices" but should refrain from making too detailed rules. Once a need for regulations is established, the choice among various rules should be made, paying due regards to the subsequent three principles.
Best practice	Regulations should follow the best of the existing practices and ban all the bad practices. This should contribute to a raising of standards in general but doubts may arise as to what is "good" or even "best". In order to make a majority decision tolerable for countries being outvoted the following criteria were also considered necessary.
Practicability	Rules should be made only when NSIs can accept and follow them, they should not conflict with or contradict other rules (*coherence*), and they should be specific and restrictive enough to ensure comparability but also sufficiently general as to allow variations as long as comparability is not endangered (*specificity*).
Efficiency	Rules should be capable of implementing without undue costs, and whenever costs are likely to exceed existing costs appropriate arrangements have to be made before adopting the rule.

* The two principles are in a sense conflicting. The idea of "necessity" (for comparability) may give rise to detailed regulations whilst subsidiarity is meant as avoiding too much intervention.

In a *first stage* starting March 1996 member states had to provide a so called "Interim Index" compiled for just one year and based wholly on existing national CPIs, from which in particular owner-occupied housing, health and educational services and some other items[18] were excluded. By *stage two* (January 1997) the coverage was

[18] Mainly items hitherto not yet included in *all* national CPIs e.g. alcoholic drinks, tobacco etc.

enlarged[19] – a process which is ongoing, so we have in a sense stage 2a, 2b etc. - and comparability was enhanced due to the fact that an agreement could be reached in some additional methodological areas. The HICP then already covered about 80-90% of the expenditures for Private Consumption.[20]

The harmonization procedure leading to mandatory rules set up as regulations should follow a number of principles or criteria, which sound wise and which are listed in tab. 6.1.1.

d) Rules for the Harmonized Index of Consumer Prices (HICP)

1. Coverage

Consumption Concept

The SNA and the European System of National Accounts (ESA) make a distinction between final consumption *expenditure* (aggregate P.3 in the SNA/ESA) and *actual* final consumption (P.4). The latter, P.4, includes (in addition to P.3) the (value of the) expenditures incurred by government units and non-profit institutions serving households (NPISH) on goods and services provided to households as social transfers *in kind* (or expenses initially incurred by households but subsequently reimbursed by social security, such as some medical expenses) as these goods and services might be viewed as consumed eventually by individual households. The HICP applies the expenditure concept (P.3).[21] The HICP is designed to cover all sorts of final consumption (as opposed to investment) *expenditures* for individual (private) consumption using a new classification known as COICOP[22].

Domestic Concept

By virtue of Regulation No 1688/98 (row 6 in **tab. 6.1.2**) the domestic concept (DC), as opposed to the resident concept (RC), in National Accounts has to be applied (starting from March 1997) in compiling the HICP[23]. The difference can best be illustrated by the following scheme

---

[19] Additional items introduced were insurances for cars and dwellings, package holidays, banking services, some educational goods and services such as evening classes and health goods obtainable without prescription.

[20] In stage two some new rules were introduced, for some countries at least possibly in contradiction with rules hitherto followed to compile the national CPI, which was no problem because it is not the intention of the HICP to replace national CPIs.

[21] Some still existing gaps in coverage, such as items in the area of health and education, will also eventually be filled since at least the privately financed part of such goods and services is clearly an element of consumption.

[22] Classification of individual consumption by purpose.

[23] Simulation studies in Germany have shown that the effect of a move from RC to DC is rather small.

	consumption expenditure	
Expenditure by	in the member state*	abroad
residents**	DC, RC	RC
non-residents	DC	

* in the economic territory of the member state

** the resident status has to be kept distinguished from nationality.

As the "domestic concept" is pursued, all final consumption in the territory, irrespective of nationality or residence status, is included. Expenditures the resident citizens spend abroad have to be excluded whilst those of non-residents within the territory have to be included in the CPI of the country in question.

Notion of Household

The HICP should also cover *institutional* households. They are, however, often not included in budget surveys and there are doubts as to whether sufficiently reliable price indicators can be obtained, which are different from those used for private households. Yet the extra-effort of including such households in the HICP appears justified.

To sum up: The Consumption concept of the HICP is known as the household final monetary consumption expenditure (HFMCE). It covers consumption expenditure

- by private and institutional households irrespective of their nationality
- in monetary transaction on the economic territory of the member state
- on goods and services used for the direct satisfaction of individual needs
- in one or both of the time periods being compared, and
- unlike the National Accounts deflator, the HFMCE covers only that part of the consumption expenditure that is considered relevant to inflation measurement (imputed cost of owner occupied housing is therefore excluded).[24]

2. *Index formula and weights*

The major part of the WP was evidently impressed in no small measure by the chain index methodology and the ideas of the Boskin Commission (see **sec. 1.3**). By regulations quoted in **tab. 6.1.2** (rows 2 and 4) it therefore became mandatory for member states to

- systematically identify *"newly significant goods"*[25] to check the significance of new goods reported by any other member state and if necessary to include

[24] Differences between the HICP and National Accounts (NA) are also given as regards the domestic concept (HICP) vs. national concept (NA) and among other things the treatment of life insurance and pension funds, gambling, prostitution and narcotics.

[25] These are goods and services not yet explicitly included and the estimated consumers' expenditure, which has become at least one part per thousand of the expenditure covered by the HICP.

those goods in the national HICP (obviously a concession to the famous "new good bias" of the Boskin Commission),[26]

- avoid weights which are seven years old or more and to check the so called *"critical" weights* annually.[27]

It is most noticeable that the regulation in question is called "minimum standards for the *quality* of HICP weight", that is "quality" means up-to-dateness.[28] An item or sub-index is said to possess a "critical" expenditure weight when its impact on the over all result of the index amounts to 0.1 percentage point or more on the average change in the HICP over one year (the repeatedly used "one part per thousand threshold").[29]

However in practice there are only few items which simultaneously have a sufficiently high weight and display a substantial amount of change in the prices in order to be "critical".[30] Thus the Regulations are not too restrictive in practice and are perhaps the result of a compromise. Eurostat wanted to

- aim at "ensuring that HICPs keep broadly in step with each other and up-to-date in terms of market developments" (therefore the inclusion of newly significant goods) or at promptly and constantly "maintaining the 'reliability and relevance' of the [weights of the national] HICPs", but at the same time Eurostat admitted
- that CPIs "are, in general, fairly insensitive to errors in weights" and that imposing "the cost of high precision for all weights or frequent updating of weights is not therefore justified".[31]

It is clear that differences in the speed of taking new products into account[32] might well produce an element of incomparability of inflation rates across countries. On the other hand, the two regulations discussed here, i.e. annual revision of weights and inclusion of new products, clearly do impair comparability over time. But as Eurostat

[26] Eurostat is organising an exchange of information on goods identified as "newly significant" in one country with all other countries, which then have to check their relevance for inclusion in their HICP.

[27] The relevant Commission Regulation concerning weights (row 4 in tab. 6.1.2) requires member states to carry out not only regular observations but also revisions of weights if necessary.

[28] To require a complete annual review of weightings is of course a first step to set up the chain index principle for all European countries.

[29] As stated above, the rule also applies to the sales volume of "newly significant" goods and services (Regulation No 1749/96 row 2 in tab. 6.1.2).

[30] German experience has shown that we have items (or sub-indices) which either exhibit a price change or a weight above average but rarely a high level of both factors determining the critical weight. The price of flat screen monitors went down remarkably for example. Nonetheless, purchases did not increase to an extent that would justify a higher weighting of this item.

[31] All quotations refer to the Progress Report on the HICP of Eurostat.

[32] In compliance with the Regulation concerning newly significant goods and services the German CPI included some new television and computer equipment such as pay-TV, subscriptions to the internet, computer games, mobile phones and the like. On the other hand some ecological products such as milk, butter "from happy cows" or "light wine" were not included or were deemed to be represented adequately in other items although such products proved significant in other EU member countries (Eurostat therefore asked Germany to check a possible inclusion of such items).

is a vehement advocate of chain indices comparability over time is dwarfed by flexibility and "relevance".

3. Some comments on selected Regulations and on technical issues

A regulation (see row 3 in **tab. 9.1.2** below) requires the transmission of indices to Eurostat in a detailed breakdown into (about 160) sub-indices. The detailed statistics impose a considerable burden to the NSIs. The combination of flexibility and at the same time a lot of detail in information seems to be characteristic for the European approach to inflation measurement. It might turn out to be difficult to handle and to interpret, especially in view of the poor aggregation properties of chain indices.

Some details of regulations are worth being mentioned (see also **tab. 6.1.2**)[33]

- At the level of "*elementary aggregates*", where relatives are derived from various price quotations referring to the same index item, it is acceptable to use Dutot's or Jevons' formula but not Carli's formula[34].
- The common practice of simply carrying forward the last observed price in dealing with seasonal non-availability has been banned[35].
- In a regulation regarding "minimum *standards for sampling*", random sampling is called desirable though it is difficult to carry out in practice.[36]
- By the stipulation of "minimum standards for procedures of *quality change*", in particular the practice of "automatic linking" was banned, since this method tacitly ascribes the complete difference in prices to the quality component.
- Another problem which required a solution in the course of harmonization is the treatment of the so called "*income related prices*" frequently encountered in the field of social and welfare activities.[37]

---

[33] Evidently many important aspects of index methodology are harmonized already by regulations and the relevant decisions have to be viewed against the background of a preference for chain indices and the US discussion of the Boskin report (which in turn was in many methodological considerations expressly motivated by the desire to lower the inflation rate).

[34] This means that arithmetic or geometric mean prices are allowed whilst the arithmetic mean of price relatives (Carli's formula) has been banned. The decision was again obviously influenced by the Boskin Commission's discussion of this issue. The motivation there apparently was the "time reversal test", but we don't hear much about why just this criterion of Fisher's index theory should indeed be so indispensable.

[35] The regulation of "minimum *standards for prices*" – row 2 in tab. 6.1.2 - says that where prices are not observed they have to be estimated and cannot be "carried forward".

[36] However, the representativeness of outlets has to be checked on a regular basis in order to avoid attrition of the sample.

[37] Such prices are linked to one or more income thresholds, as in the following case: persons with an income exceeding 3000 € pay for example 50 €, while those having a lower income only pay 20 € or so. To ensure a pure price comparison now would not only require keeping the consumption pattern of households (i.e. the type of goods and their weights) constant but also to keep the income level and other socio-economic characteristics of the households constantly under consideration because such variables affect the prices to be paid by these households.

- As to the inclusion of *insurance services*, it is an open question to define (and estimate) the part of the gross insurance premiums that can be regarded as the price for the service taken in isolation.[38]
- The problem with *financial services* is not only the definition of a boundary (which types of services ought to be included or excluded respectively) but also the fact that many prices (for example fees for bank services) depend on the transaction value or the frequency of transactions.[39] And finally, a still highly controversial and a notoriously difficult issue is
- the treatment of *owner occupied housing.*[40]

Table 6.1.2: Harmonization of Consumer Price Indices (Council Regulations [A], Commission Regulations [B], Guidelines)

All legal acts can be found at http://europa.eu.int/celex/

Type	Number, date	Title, comments
1) A	No 2494/95 of 23 Oct. 1995	defines aim, comparability; timetable, procedure etc. of harmonization but no details of compilation of indices
2) B	No 1749/96 of 9 Sept. 1996	on initial implementing measures for Council Regulation No 2494/95; this regulation defines • initial coverage of goods and services, • practices for updating the coverage and inclusion of newly significant goods and services (NSGS), • minimum standards for the procedures of quality adjustment (for example automatic linking was banned), • minimum standards for sampling and for the prices used, • formula for compiling price indices for elementary aggregates • missing observations: where price observation fails because of non-availability prices must be estimated appropriately (for the first and second month and replacement prices shall be used from the third month), amended by the Council Regulation [A] No 1687/98 of 20 July 1998; see row 5 below. The NSGS standard relates to additions, not to replacements, and refers to one part per thousand (1‰) of total consumers' expenditure

38 This also applies to the weights. Another problem is the treatment of changes in the specified amount insured (should an increase be regarded as a "quality improvement" of the insurance?).

39 Some prices are defined only as proportions of the transaction value and not in absolute terms. Hence there is a clear relationship between quantities consumed and prices paid, while in general in index theory these variables are supposed to be independent.

40 See **sec. 6.2** for more details. A solution still has to be found. The rental equivalence approach used in Germany for example is unlikely to gain the support of the majority of the EU member states.

3) B	No 2214/96 of 20 Nov. 1996	concerning transmission and dissemination of subindices of the HICP; defines breakdown of HICP into sub-indices. Amended by: Commission Regulation [B] No 1749/99 of 23. July 1999 (see row 8 below)
4) B	No 2454/97 of 10 Dec. 1997	minimum standards for the quality of HICP weights • defines a maximum age of weights (7 years) and, • requires an annual checking of "critical" weights
5) A	No 1687/98 of 20 July 1998	concerning the coverage of the HICP; defines • additional coverage of HICP for 1999, • use of expenditure concept
6) A	No 1688/98 of 20 July 1998	concerning the geographic and population coverage of the HICP • defines use of domestic concept, • inclusion of expenditures of people in institutional households
7) B	No 2646/98 of 9 Dec. 1998	concerning the treatment of tariffs (tariff prices) in the HICP; defines obligation of respondents to provide structural information of tariffs and data related to consumption pattern to the NSOs. Rules are needed lest changes in tariffs do not falsely indicate a change in the consumption pattern of consumers.
8) B	No 1749/99 of 23. July 1999	• adopts the final version of COICOP, • takes regard of extended coverage (Council Regulation [A] No 1687/98 see row 5 above)
9) B	No 1617/99 of 23. July 1999	concerning treatment of insurances in the HICP; defines • use of service charges for calculation of weights and • use of gross premiums for compilation of prices supplemented by Guidelines for the implementation of the insurance regulation (problem of index-linked contracts)
10) A	No 2166/99 of 8. Oct. 1999	laying down detailed rules for the implementation of Council Regulation No 2494/95 (see row 1 above) as regards standards for the treatment of goods and services in the education, health and social protection services in the HICP, decides • which reimbursements should be deducted from the "price", • how to handle "income related" prices (or reimbursements)
11) B	No 2601/2000 17. Nov. 2000	time of entering purchaser prices into the HICP (prices ought to be included when they are observed [goods] or when the service has been rendered, (i.e. the month in which the consumption of the service commenced) in the case of goods the acquisition principle is followed, as for services, the standards differ from National Accounts (where the *completed* service counts).

12) B	No 2602/2000 17. Nov. 2000	treatment of price reductions or discounts (they should be taken into account when they are attributable to the purchase of an individual good/service, non-discriminatory [available to all potential consumers], known to the buyer, claimable and their impact on quantities purchased is substantial)
13) B	No 1921/2001	policy of revisions of the HICP (revisions have to be approved and there is no quantitative assessment of the impact of revisions unless a revision affects the results by more than 1 per thousand). Mistakes should be corrected, revisions due to new or improved information should be implemented, changes in harmonization rules should in general not require revisions of published HICPs, unless the impact is significant
14) B	No 1920/2001 28. Sept. 2001	treatment of Financial Services in the HICP and of prices for services depending on the volume of transactions (service charges proportional to transaction values): prices should be defined as proportions, multiplied by the value of a representative unit transaction; flat rates should be included, inertest-like charges and interest payments excluded, however.

* referring to Council Regulation No 2494/95 concerning minimum standards

A – Council Regulation B – Commission Regulation

Some Guidelines not mentioned above

	Guidelines for the treatment of
1	extreme (rejected) price observations in HICPs: in general, the prices reported by the price collector should be accepted (automatic corrections of unusually high or low prices are not permitted), it should be proved that corrections should not differ systematically from a correct HICP by more than 0.1 percentage points on average.
2	data processing equipment and especially PCs in HICPs (such items have to be included and the relevant expenditure weight must be checked annually)
3	clothing in HICPs; this calls for rules for • inclusion of seasonal and fashion clothing (carry forward the last observed price of winter/summer until next winter/summer), • replacements of items and • problems of comparing (special) sales prices with regular prices.

Some legislative projects

minimum standards for sampling and treatment of seasonal items in the HICP
treatment of owner occupied housing in the HICP
minimum standards for sub-indices of the HICP
(new rules concerning) treatment of newly significant goods and services
agreement on best practices for quality adjustment of specific goods (cars, books, computers, clothing etc.) and pilot calculations of acquisition prices of housing
better systems to assess EU member states' compliance with existing regulations

It should be noted that a most positive side-effect of the harmonization procedure in the EU is this: the process entailed much discussion on methodological issues in price statistics in all member states (as well as applicants for EU enlargement) and thereby contributed a lot to a clarification and better understanding of methods. There is a remarkable inconsistency however between the attention which is rightly paid to comparability across countries (and the progress achieved in this area) on the one hand and the defects of the adopted methods of aggregation (over countries) and the chain index methodology with respect to comparability over time on the other. These problems seem to be less well solved.

e) Aggregation over time (the chain index approach)

The HICP makes use of the chain approach in creating a summary measure of inflation. The national HICPs as well as the weighted mean (M) of these indices (see part f of this section), ought to be designed as chain indices with December (m = 12) being the month when two baskets will be linked together.[41] Since the basis for linking is December (12) and not an average over the year, in a comparison of a month $m \neq 12$ in year t with the value of the month m in the previous year (t - 1) the baskets of at least two years[42] are involved.

Moreover, in practice the "age" of the basket will be two years rather than one year only. Hence we get the following formula of the national HICP (H for short) for year t and month m

$$H_{0t,m} = \frac{\sum p_{t,m} q_{t-2}}{\sum p_{t-1,12} q_{t-2}} \frac{\sum p_{t-1,12} q_{t-3}}{\sum p_{t-2,12} q_{t-3}} \left(\frac{\sum p_{t-2,12} q_{t-4}}{\sum p_{t-3,12} q_{t-4}} \cdots \right), \text{ and}$$

$$H_{0,t-1,m} = \frac{\sum p_{t-1,m} q_{t-3}}{\sum p_{t-2,12} q_{t-3}} \left(\frac{\sum p_{t-2,12} q_{t-4}}{\sum p_{t-3,12} q_{t-4}} \cdots \right).$$

[41] Note that we are not dealing with methods of linking deflated values of Quarterly National Accounts here. To explain the relevant methods like "Annual Overlap" or "Over the Year" etc. in this field would not be worthwhile in this context.

[42] This is a point which will be made more transparent in the chain-index chapter. Since a Laspeyres chain index is chosen there will be two baskets, q_{t-2} and q_{t-3} in eq. 6.1.1. In the case of a Fisher chain index we would have three baskets. See also v. d. Lippe, 2001, p. 140.

The expressions in brackets will cancel out when a ratio of two price indices, both for a month m, is formed. Such a ratio of two indices is given by

$$(6.1.1) \qquad H_{0t,m}/H_{0,t-1,m} = \frac{\sum p_{t,m}q_{t-2}}{\sum p_{t-1,12}q_{t-2}} \Bigg/ \frac{\sum p_{t-1,m}q_{t-3}}{\sum p_{t-1,12}q_{t-3}}.$$

Table 6.1.3: Some quotations taken from regulations concerning the HICP

Extract (epitome) of the regulation text
Comparability (RegulationNo 2494/95 **row 1 in tab. 6.1.2**) "HICPs shall be considered to be comparable if they reflect only differences in price changes or consumption patterns between countries. HICPs which differ on account of differences in the concepts, methods or practices used in their definition and compilation shall not be considered comparable. The Commission (Eurostat) shall adopt rules to be followed to ensure the comparability of HICPs under the procedure laid down in Article 14. Differences in concepts, methods or practices, which would affect the change in the HICPs by more than 0.1 percentage point on average over one year against the previous year cannot be accepted."
Inclusion of newly significant goods (NSG) and services Reguation No. 1749/96 **row 2 in tab. 6.1.2**) ***Definition, sales volume of 1 per 1000 of total consumers' expenditure*** "Newly significant goods and services are defined as those goods and services the price changes of which are not explicitly included in a Member State's HICP and the estimated consumers' expenditure on which has become at least one part per thousand of the expenditure covered by that HICP." ***Compulsory checks (once a Member State reports NSG) and adjustments*** "Member States shall: (a) systematically seek to identify newly significant goods and services and (b) check the significance of goods and services reported to be newly significant in other Member States. The HICP shall be compiled to include the price changes of a newly significant good or service ... This shall be accomplished within 12 months of their identification either by adjusting the weights... or by assigning part of the weight specifically to the newly significant good or service." **Price indices for elementary aggregates** accepts formulas of Dutot (ratio of average prices) or Jevons (geometric mean of price relatives) but bans formula of Carli (arithmetic mean of price relatives).

Minimum Standards for the Quality of HICP Weights (No. 2494/97 **row 4**)

1) maximum age of weights

"Each month Member States shall produce HICPs using weightings which reflect consumers' expenditure patterns in a weighting reference period ending no more than seven years before."

2) frequency of revision

"Each year, Member States shall carry out a review of weightings in order to ensure that they are sufficiently reliable and relevant to meet the comparability requirement."

3) object of revision

"In the review, Member States shall check whether or not there have been any important changes since the weighting reference period in current use regarding price developments of each major component index relative to the HICP, or sustained market developments in each major component group."

4) obligatory adjustment of weights

"Where reliable evidence shows ... [that a weighting change] ... would affect the change in the HICP by more than 0.1 percentage point on average over one year against the previous year, Member States shall adjust the weightings of the HICP appropriately."

Hence the annual growth rate $H_{0t,m}/H_{0,t-1,m}-1$ describing the change from month m in t - 1 to month m in year t is a function of quantities (weights) relating to *two* years, not just one, that is, to the years t - 2 and t - 3. There is no "'pure" price comparison and the weights are not really most up-to-date.[43] Moreover, in our view the fact that the HICPs have to be chain indices is the most serious shortcoming. Disadvantages of this index design abound (see chapter 7), such as

- path dependency (the index depends on its "history")
- poor aggregation properties, and
- ambiguity in interpretation

to mention only some of the drawbacks. The last point should be commented on: An increase of H by say + 10% may result from a great variety of influences, such as changes in the selection of goods and their quantities, the "history" of prices and quantities and finally also the prices (p_t compared with p_0).

There is always a *number* of factors that may influence the result of a chain index. Not all of them (if not the least of them) are clearly related to inflation and theoretically justified. Chain indices violate the principle of pure price comparison.

The usefulness of a chain approach may be questioned even more once a third kind of aggregation (again using *variable* weights) comes into play, viz. aggregation over countries in addition to aggregation over items (sub-indices) and time-intervals (by chaining).

[43] In commenting on this result the German Federal Bank pointed out: "The advantage of 'pure' price comparison is lost even in the case of short term inflation measurement; ... But weighting structures having an age of two years are not necessarily more suitable than those of three or four years" (Bundesbank 1998, p. 65 [translation of the present author]).

f) The Monetary Union (EMU) Index of Consumer Prices (MUICP)

The purpose of this index compiled by Eurostat is to combine all national HICPs denoted by H_m (m = 1, 2, ..., μ for EMU member state)[44] in order to provide a measure of the European ("Euro") inflation rate and serve as a guideline for monetary policy of the European Central Bank (ECB). The MUICP is calculated as a weighted average of the HICPs of the EMU member states. As all national HICPs are chain indices there is an additional element in the MUICP changing each year, the changing country weights.[45] The weight of a EMU member state in the MUICP is the proportion of final consumption expenditure of households[46] in the EMU total.

It should be noticed that the MUICP differs substantially from the methods traditionally used for international comparisons discussed in chapter 8 as the task now is to aggregate over countries rather than to compare countries. There is no effort made for example to ensure comparability on the level of individual (groups of) commodities and to price comparable items by using "community prices". The same item, for example traffic expenses, clothing or shelter, may have an entirely different character and significance in one country than in another. There is no EMU-price[47] or EMU-basket in the MUICP. There are only many national prices, baskets, and indices. Furthermore, following the agreed upon methodology, an additional country in the Euro-area can be incorporated easily by "linking in", i.e. by entering its HICP in the weighted mean of HICPs and by changing (downscaling) weights of the existing μ countries appropriately so that the "new" index will now be a weighted average of μ+1 rather than μ HICPs (and the new μ+1 weights c_{mt} again add up to unity).

The MUICP was originally based on 1996 = 100 (t = 0),[48] and country weights c_{mt} of the member countries m therefore started with data related to 1996. In December 1996 the first update was made to arrive at weights used in 1997 (t = 1). Hence to compare 1996 (t = 0) with 2001 (t = 5) for example the MUICP (denoted M) is

$$(6.1.2) \qquad M_{05} = (\Sigma c_{m0} H_{m01})(\Sigma c_{m1} H_{m12})(\Sigma c_{m2} H_{m23})(\Sigma c_{m3} H_{m34})(\Sigma c_{m4} H_{m45})$$

where the summation in each bracket takes place over a (varying) number of countries m = 1, 2, ...μ and t = 1 is the time span January 1997 through December 1997 (using the weight of t = 0 = 1996), t = 2 refers to the interval from Dec. 1997 through Dec. 1998 (using the weight related to t = 1 (Dec. 1996)) and so on such that t = 5 means 2001. Note that M_{05}, which is supposed to compare the price level in 2001 (t =

---

[44] The MUICP refers to member countries of the European Monetary Union (EMU) having adopted the Euro, a subset of countries belonging to the European Union to which the "European Index of Consumer Prices" (EICP) refers. As pointed out at the outset we have a variety of indices differing only with respect to the countries included.

[45] in addition to changes in weights for commodity groups and in the composition of goods and services in each national HICP.

[46] Private households' domestic consumption expenditure.

[47] In contrast to a national HICP no attempt is made to establish *the* prices p_{it} and p_{i0} for the whole Euro zone, and to combine those prices to an index using *one* single basket. Thus the situation of the MUICP as to the interpretation is completely different to a national CPI (HICP).

[48] But since almost nothing related to period 0 is held constant, the choice of the base period is more or less irrelevant in the chain index design.

5) to 1996 (t = 0), makes use of no less than 5μ (that is 55 with μ = 11 Euro-countries as it originally was the case) different national baskets and 55 country weights as well as prices in 1997, ..., 2000 in addition to the prices in 1996 and 2001. Hence a result like for example $M_{05} = 1.2$ may well be produced by a number of very different combinations of weights, time paths etc.,[49] not only by prices referring to the same items being 20% higher in 2001 than in 1996. According to eq. 6.1.2 the MUICP is the product of links, each of them being a weighted sum of national HICPs itself. Hence there are three aggregations made, that is aggregations over

- commodities
- countries, and
- time (sub-) intervals to arrive at the total interval.

In the first two cases variable weights are used and in the third case integration over time is done by chaining (as if chain indices were transitive). Thus the inflation rate of the Euro-area will depend on

- the *prices* in each member country in each period, and
- changing weights of the *commodities* in each country and each period,[50]
- the change of weights of the *countries*, and
- the *path* of the index, since a chain index always depends on its "history".

Furthermore, all sorts of changes in the "domain of definition", that is, in the selection of goods (due to emergence of new goods etc.) and outlets, have an impact.[51]

Another criticism is lack of theoretical foundation. A "changing cost of a particular basket" – interpretation does not apply to the MUICP, as there is no common basket, let alone a *constant* basket, related to the consumption of a specified "All-European index family" facing the type of "price level" the index is measuring.[52] Therefore theories developed to explain a traditional CPI in terms of utility maximizing households,

[49] As usual in a chain approach the accuracy of the product also depends on the accuracy of the existing factors when a new factor is multiplied with the chain to get a longer chain. A direct approach on the other hand would consist of making an independent estimate for each t.

[50] As weights are updated in national HICPs on an *ongoing* basis, the weighting structure of the MUICP is in principle not related to a particular year (weights are of different age).

[51] Such changes should be accounted for in an index in which great store is set by having weights adapted *continuously*. The impression clearly is that the frequency of updating weights in the MUICP as well as in the HICP was considered far more important than questions of "pure" comparability and avoiding ambiguity in interpretation. Another criticism is the lack of additivity of chain indices which makes it most difficult, if not impossible, for external users of statistics to compile sub-indices for parts of the commodities covered by the HICP. Comparisons among various differently constructed (ad hoc) sub-indices are frequently made in order to discover "causes" of inflationary processes.

[52] In a sense this may also exist in the case of a single country where a household living in one part of the country will likewise not face exactly the "average" prices entering the formula of the national CPI. But there are reasons to expect greater differences when various *countries* with still widely differing living conditions and standards are involved as in the case of different *regions* of one single country.

as the COLI-concept cannot simply be transferred to the case of aggregating over different countries with different baskets.[53]

Moreover, where is the economic process which corresponds to the averaging over countries with different index-baskets? From the point of view of consumers in France, for example, high prices in their country will not be "averaged out" by low prices in another country, e.g. Ireland, as implicitly assumed in the MUICP.

In the development of the MUICP the choice and empirical determination as well as the up-dating of country weights has played an important part. Questions to be settled were for example

- Should National Account aggregates be used or alternatively population data?
- How can we avoid a growing weight given implicitly to such countries where inflation is above average?[54]

To sum up: The interpretation of the index is - to say the least - not straightforward or related to a clear theoretical concept of the index construction and the empirical results are affected by a number of influences where it is not quite clear how they are related to a European "inflation".[55] This problem should not be underrated: the less clear the "message" of an index, the greater, as a rule, the temptation to interpret the result in view of what is *politically* desired and the greater the temptation to interfere *politically* in a bluntly result-orientated manner.[56] On the other hand a moderately aggregated statistical figure with a clear though limited interpretation might be preferable. It is perhaps better to analyse national indices separately with respect to the causes, nature and effects of the price movement rather than to lump those indices together to get an overall aggregate which at best reflects a *mix* of influences difficult to unravel.

[53] It should be noted in passing that they should also refer to the consumption concept "actual consumption" rather than to the expenditure concept underlying the HICP (see above).

[54] In a sense the choice of a (chain) index formula was made in order to avoid such consequences. Assuming constant rates of all individual price changes then it is well known – and shown in **sec. 3.4** - that the growth rate of a Laspeyres-index as an *arithmetic* mean (in contrast to a *geometric* mean) tends to be dominated by the highest individual growth rate. This certainly contributed to not using the Laspeyres formula but a chain index instead. The conclusion that a (direct) geometric mean index would also avoid such a gradually growing implicit weight of a high inflation country was discussed but could not impair the triumph of the chain index ideology in the development of the MUICP. It probably was not realized that another option would have been *constant* country weights.

[55] Interestingly, a comparison of the European HICP with an experimental U.S. index constructed in a HICP-manner has shown that contrary to expectation the U.S. inflation measured in the European manner was not higher but lower than assessed according to the official U.S. methodology (Monthly Labor Review [of the BLS] May 2006, p. 20).

[56] We have not yet experienced really high inflation rates in the EMU. The real test of the ECB is still to come. I venture the bet that the scapegoat will then once again be the official statistics (as it was in the case of the Boskin Commission controversy) and there will be many (above all professors of statistics and economics) who will again discover an upward bias in the CPI, that is, in the MUICP this time.

6.2 Some controversial issues in inflation measurement

In this section we will present three conceptual and technical problems with regard to inflation measurement along with proposals to deal with them. In the majority of the issues there still is no consensus on methods. The selected topics are the measurement of the so-called "core inflation", the inclusion or exclusion of asset prices and the treatment of owner occupied dwelling in a consumer price index (CPI).

a) Core inflation and "net price indices"

The notion of core inflation became popular in the 1970's when a number of sector specific disturbances, the most prominent of which was the oil-price-crisis, engendered turbulences and volatilities. Since then it has appeared desirable to identify and eliminate the more volatile and only temporary influences on "inflation" because they are not of primary concern to central bankers and might be viewed as only clouding or disturbing the statistical picture of inflation. A distinction therefore is common to all concepts of core-inflation, i.e. the distinction between a "core" - inflation[57] component Π_{0t} and a distortion ε_{0t} such that the observed inflation rate P_{0t} is viewed as being contaminated somehow by this second component ε_{0t},

(6.2.1) $$P_{0t} = \Pi_{0t} + \varepsilon_{0t},$$

and the task is to purge the observed inflation from those disturbances.

An alternative concept to identify a "core" component goes as follows: the individual log-change of the i-th price (commodities i = 1, 2, ..., n) is given by

(6.2.2) $$\ln(p_{it}/p_{i,t-1}) = \ln(\Pi_t/\Pi_{t-1}) + \ln(\varepsilon_{it}/\varepsilon_{i,t-1}),$$

and the task is to identify the link Π_t/Π_{t-1} which is independent of the specific commodity i, and which is known as core inflation.[58] A conceptualisation of a price index Π_{0t} (direct index), or $\Pi_{0t} = \Pi_1 \cdot \Pi_2 \cdot ... \cdot \Pi_t$ (chain index) in this manner by making use of a residual component ε_{0t}, or $\varepsilon_{it}/\varepsilon_{i,t-1}$ respectively is well known from the so called "stochastic theory of index numbers". The common feature of core-inflation, asset inflation and the stochastic approach is the lack of theoretic underpinning.[59]

The methods proposed to identify and eliminate the idiosyncratic component of the "headline" CPI (or in the case of the ECB, the European HICP) and thereby to isolate the core inflation can be broadly classified into two categories (see **fig. 6.2.1**)

---

[57] It is also said that this component is the result of stable expectations built on the basis of experience with monetary policy in the past. The idea is that it will be difficult to reduce this particular part of inflation as the expectations might prove quite resistant and insoluble.

[58] Note that in this model reference is made to the previous period t-1 such that we have logarithms of "links", that is, log changes throughout. Note further that this equation and the older stochastic theory of index numbers (**sec. 2.3**) amount to very much the same.

[59] The idea is abandoned that the CPI should reflect the rising cost of achieving a given level of satisfaction or of buying the same basket under two price-regimes.

Figure 6.2.1: Notion of core inflation and methods of measurement

observed ("headline") inflation P_{0t} as sum of

"core" component Π_{0t}	"idiosyncratic" component ε_{0t}
or: persistent, common, systematic, "underlying" inflation, "monetary" or truly inflationary component	or transitory, sector-specific, erratic, incidental, random or "non-monetary" component 1)

method	T methods	C methods
I	Dynamic Factor (DF-) Index	Exclusion Method (EM) Trimmed Means (TM) Variability adjusted Means (VA)
G	Smoothing techniques (ST) 2) VAR-methods 3)	

1) caused by non-monetary factors
2) Moving averages, filtered series
3) vector autoregressive methods

1. time series-based methods(**T-methods**), and
2. cross-section-based methods (**C-methods**).

The distinctions made in the first group are defined along the line persistent vs. temporary on the basis of the wavelength of fluctuations while the second group of methods differentiates in terms of common vs. sector-specific according to the type of goods and services. The ECB introduced a second distinction for a classification of methods, that is, between **I-methods** (using individual price relatives), and **G-methods** (global methods focusing on the all-item-index)[60] respectively.

Apart from the so called DF-Index (see below) the variables of the two classification are highly correlated (combinations I-C and G-T are the most common ones).

The *ST approach* exists in the shape of moving averages- and other filtered series-methods (like a Hodrick-Prescott filter or so), thus conceptualizing "core" inflation as a sort of trend in the observed headline inflation. The aim then is to find an estimate of the trend by eliminating short-term fluctuations superimposed on (and adding "noise" to) the trend. The objection against this approach is primarily that it is backward-looking.[61] In the *DF approach* it is assumed that both the idiosyncratic components and the common component follow a certain time series model.

The same is true for the *VAR approach*, in which the core inflation is estimated on the basis of its non existent cumulative impact in the long run on certain real vari-

[60] mostly additionally using some explanatory exogenous variables as regressors.

[61] Moreover, on account of inherent problems with defining a "trend", results of estimations are available only with a certain delay. The ST-method is also said to be slow in capturing new developments.

ables such as output. The interesting feature of this method – making it perhaps theoretically more attractive - is that it also uses information other than the inflation rate itself in order to conceptualize core inflation.

In the variability adjustment approach (*VA approach*) weights are assigned to individual prices amounting to the inverse of the variance σ_i^2 of the prices of the series in question (that is of the item i) over a certain period in time (in general from 0 to t). Thus the price relative p_{it}/p_{i0} is given the weight $1/\sigma_i^2$ relative to the sum of all such inverse weights $\sum_i \frac{1}{\sigma_i^2}$ (i = 1, ..., n). However, it is far from clear to which time horizon the σ_i^2 (which results from summating squared price relatives of i over a certain period in time) should refer and if so how such variability weights (or V-weights for short)[62] should be related to expenditure weights (E-weights). The latter in general account for the "relevance" of a commodity and they define the "basket". We have learnt that detailed, empirical and most up-to-date E-weights have to be established with the utmost care, however, now a possibly completely different set of weights, the V-weights, are said to be equally or even more important. Should they totally replace E-weights, or work as additional weights in some sort of a double weighting system? Should the influence of certain goods with high E-weights be played down only because they are volatile?[63]

In the trimmed means - or the *TM- approach* a, say, 10% symmetric trimmed mean excludes the upper and lower 10% of observations from each side of the distribution, making it more robust and less sensitive to outlayers. The method is tantamount to applying zero-weights to extremes.[64] In contrast to the EM approach, in the case of the TM method the list of excluded series (items) may well vary over time. The limiting case of a trimmed mean is the median in which on both sides (almost) 50% is excluded, such that there is a single item remaining. This means that the core inflation rate defined in such a way as the *median price relative* is possibly determined by a different commodity each month. In February inflation is measured by the price of a bus ticket, in March it is a shirt, in April the only relevant price is the price of bread and so on. Worse even, we will provide such measures without any theoretical justification whatsoever. Furthermore, if there is a price trend of a commodity i such that $p_{it}/p_{i,t-1}$ is consistently higher or lower than the average (for all t) we will easily fail to capture this particular development. The so-defined core inflation will fall short of representing this trend altogether, and it will become more and more dissimilar to the actually observed (headline) inflation rate.

---

[62] An index using V-weights is sometimes called a "(Neo-) Edgeworth(ian) index", not to be confounded with the (Marshall-) Edgeworth index.

[63] In a similar vein (determination of a system of weights for the sole purpose of ensuring certain empirical results) there are some other approaches to measure core inflation (not mentioned above) which make use of the lead-lag-structure of series for example. They stipulate weights such that the weighted average of price relatives will represent the best sub-collection of leading indicators and thereby permit good forecasts of future inflation rates.

[64] The percentage of series to be trimmed (excluded) on both sides of the distribution is of course open to debate.

Finally, the *EM approach* consists of excluding certain series from the overall - (or "all items") index due to theoretical reasons (or perhaps simply because the selected series are deemed to be highly volatile). The most common candidates for exclusion are energy products, unprocessed food, alcoholic beverages, and tobacco. As a variant of the EM- method therefore, the *index net of* (excluding) *the effects of changes in taxation* (taxes on products) can be viewed. Such an index is compiled in some countries on a regular basis, for example in the Netherlands, where a "CPI exclusive of tax changes" is used for wage negotiations. The purpose of this "net price index" is to prevent an increase of (indirect) taxes from automatically escalating wages. The problem of the EM approach (probably the most popular of the methods to measure core inflation) is the justification of an exclusion from the headline index on theoretical grounds. Generally, exclusion based methods are attractive as no complicated estimations are required and are therefore are easy to operate, much like the VA and TM approach.

To assess the aforesaid methods the following aspects should be borne in mind

- timely availability of data and estimates (in this respect for example the C-methods should be preferred to the T-methods),
- whether or not a re-estimation is necessary once new data are available,
- the extent to which the results the method in question is yielding will differ from those of other methods, and if so how this may be explained plausibly,
- the robustness to changes in assumptions made by the method, such as the degree of aggregation (over commodities), the length of the time span over which computation takes place etc., and finally
- the extent to which the method is based on theoretical considerations and permits an identification of the true sources of inflationary pressures.

In the light of these criteria we may draw the conclusion that the concept of core inflation, though possibly illuminating, does not yet lead by itself to a fully acceptable and satisfactory measure suitable for *official* statistics. Such measures should therefore better be in the domain of *non-official* experimental estimations.

b) Asset inflation and "total inflation"

Another interesting topic which surfaced in price statistics literature only recently is whether or not asset prices should be included in a consumer price index in addition to the usual *consumer* goods and services which traditionally make up a measure of "inflation".

The general idea of "asset inflation"

The starting point of those who argue in favor of asset inflation (AI) is that inflation is defined as a fall in the value of money irrespective of how (for which purposes) the money is spent. Hence inflation should not only mean a rise in prices relevant to consumers when they buy non-durable goods and services, but also a rise of prices of non-financial and financial assets.

There is already one prominent exception in confining oneself to "consumption" in that most CPIs in one way or the other take owner-occupation of dwellings into account. Housing should be viewed as a sort of investment rather than consumption and the received wisdom is that such prices should be included in an inflation gauge; so why not allow for private investment other than owner occupation and also for investment expenditures of enterprises and government as well?

According to this view the correct measure of inflation should reflect *all* sorts of prices, "genuine" consumer goods and services and investment in financial and non-financial assets alike, because money can be spent for all those things without difference and a measure therefore should be affected by rising prices of non-durable consumer goods and also of new (produced) assets as well as of existing ones.[65] Thus, there is also of late much talk of "*total* inflation" as opposed to "consumer inflation". As a consequence of this reasoning there is no longer a need for a distinction between various sorts of indices such as the CPI, PPI, or price-indices for construction and the like. Furthermore, in *"total" inflation* there is no recourse made to a specific basket of a consumer[66] and the focus is rather on a more general level of prices such as the coverage of the "domestic product deflator" (GDP deflator) or the national income deflator rather than the traditional CPI.

In our view the idea that money can be spent for *anything* and thus an indicator of purchasing power of money should cover *all* purchases of goods and services of whatever kind or purpose is not convincing though clearly economics of the unpretentious sort. Those who advocate the alleged need to measure "asset inflation"(AI) make, however, some additional points:

1. According to this view utility is not only a function of present consumption but also of claims to *future consumption*, the price of which is said to be embodied in asset prices.[67] AI is therefore assumed to be a necessary extension of inflation measurement in that it accounts for *expectations* and future prices.[68] And

[65] Thus if prices of future housing services or of a unit of future dividend return on equity for example go up this should indicate inflation in just the same manner as rising prices of non-durable consumer goods.

[66] Also the inflation measurement can no longer be related to considerations in terms of utility and compensation for losses in real income of private households.

[67] Furthermore it is emphasised that inflation not only affects households but also companies, the government and virtually all sectors of the economy. The same is true for making expectations and taking actions in anticipating future inflation.

[68] The argument against this contention is that asset prices not only reflect a change in the price of future consumption or in the expected future inflation but they are also affected by *a number of* other factors such as shifts in tastes, shifts in the preferences for different types of assets and changes in the asset quality which cannot reasonably be related to inflation.

it is maintained that an indicator of the only relevant "inflation" should have a broad coverage (e.g. the GDP deflator) instead of a narrow focus (such as the CPI).

Figure 6.2.2: Classification of non financial assets in the SNA 1993

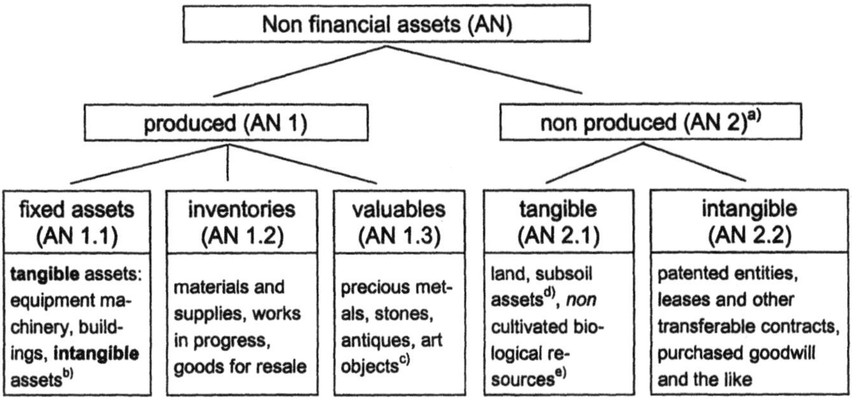

a) The SNA makes a distinction between assets according to whether they are "produced" or "non-produced" and "tangible" or "intangible". The code numbers (AN [= assets non-financial] + number) of the relevant aggregates refer to the Central Product Classification (CPC). In general non produced assets are more difficult to be priced correctly because of their limited availability, uniqueness (all circumstances properly accounted for) and irreproducibility.

b) computer software, literary and artistic originals etc.

c) and other rare and valuable objects giving rise to considerable valuation problems (it also may be questioned if they are really "produced" or rather due to their scarcity almost "non produced").

d) the quantity of which often is unknown.

e) like water resources etc.if cultivated they belong to AN 1.1.1 (tangible fixed assets).

2. The notion of AI is also made popular on the following grounds: households explicitly invest in valuables and other (tangible) assets in order to evade inflation by making use of possible holding gains from such assets. This behaviour is at the same time both cause (of future inflationary processes) and effect of (past) inflation. It explains phenomena like "volatility", "infection" and "chain reaction" etc. in an inflationary process and it is therefore wise, so it is argued, to observe asset prices in addition to the usual consumer prices.

The weaknesses of this AI idea become more and more apparent when an attempt is made to determine the asset boundary and the weights to be assigned to the different types of assets. Those who advocate the AI concept prefer in general a wide notion of "assets" as a target concept of AI, including even non marketable, non produced and intangible assets of all sorts, even more metaphorical, figurative and highly dubious ones such as for example "human capital". Hence it is by no means clear how to establish a defensible measure of AI.

The "asset boundary" as a problem of asset inflation

To explain the sort of difficulties of measurement and regular price observations entailed in AI based on "assets" in a broad sense, we should start by looking at the SNA classification of non-financial assets. **Fig. 6.2.2** demonstrates the relationships between various non-financial assets the SNA 1993 distinguishes between.

The notion of intangible non produced assets (AN 2.2) leads to the well known discussion of "welfare" measurement. Since the very beginning of National Accounts it has been argued that quite a number of dubious items should be included in (or explicitly excluded from) the GDP in order to arrive at an (allegedly) "true" measure of "wealth" and "well-being". Such items claimed to be positively or negatively accounted for were for example: production of households for their own use, environmental deterioration, increase of leisure time, "human capital" as "invested" in health (longevity more precisely) and education, pensions to be expected by elderly people, the contribution of peace, intact political organizations and social security to wealth and many other items.

The reason why proposals like these were not accepted was in most instances the insurmountable valuation problem involved when they were included in the relevant aggregates like consumption or investment. But there are also quite a few theoretical deliberations which convincingly lead to the conclusion that such vague "assets" or "benefits" should rather not be contemplated in the case of National Accounts or inflation measurement.

Discontinuities and unique goods

There is a problem with discontinuities in the purchase of consumer durables like cars, household equipment, furniture and the like. Households are in general considerably affected by outlays of this kind in the month in which the purchase takes place. However, they usually undertake high priced acquisitions only discontinuously and in long intervals. In contrast the usual CPI will in general implicitly treat discontinuous high priced acquisitions as if they were occurring constantly with a rather moderate weight. Thus the way in which the increase of prices is experienced by the households will differ from the picture drawn by the official price index.[69]

The phenomenon of a discontinuous relevance of prices (and purchases) in the eyes of households, temporary non-availability of goods and "unique" goods where no price comparison can be made by definition encountered to some extent in PPI indices[70] will automatically enter the CPI methodology and severely aggravate the difficulties of index compilation once we move to an AI measurement instead of the traditional consumer-price-inflation.

Theoretical considerations to justify AI measurement

In order to decide which weight should be assigned to which type of assets the traditional "basket" (or COLI) approach is often completely abandoned in favor of the so

[69] The index therefore may underestimate the rise of prices, at least compared with how the price movement is experienced by those households who actually presently undertake such acquisitions.

[70] See **sec. 6.3** for some solutions in the case of price indices for the construction sector (where at least the composition or arrangement of the parts is unique whereas the parts or components themselves may be standardised).

called "stochastic approach" and weights are made dependent on the ability to forecast future output and/or future inflation rather than on the actual consumption pattern.[71] Thus weights of items (goods, services and assets) should be related to their relative contribution to an inflation forecast. Here and in this stochastic-index-theory-approach, how to measure inflation becomes to a huge extent an empirical rather than a conceptual issue. This also makes the argument somewhat "circular":

- defining inflation is reduced to the statement "inflation is the ability to forecast inflation", and
- interest rates strongly influence inflation as measured in the AI approach but on the other hand the inflation rate is used for planning and monitoring actions of the monetary authorities which clearly aim at a desired level of interests.

To summarize: AI is likely to create more problems than it is able to solve.

c) The treatment of owner occupied dwelling (OOD)

Households living in dwellings owned by themselves are numerous and millions of households have invested their savings in this kind of property which in general ranks top of desirable objects. Therefore the phenomenon of owner occupied dwelling (OOD)[72] is by no means negligible[73] and the correct treatment of it in the framework of a (consumer) price index is of considerable importance. There is little doubt that the costs of housing not only in the case of *actually paid* rents, but also in the case of owner occupation, where rent is *imputed*, ought to be included in a CPI. There is, however, some debate on how to estimate OOD costs. At the start of the European harmonization project (construction of the HICP) some countries, e.g. France, Austria, Italy, Greece, Luxembourg and Switzerland, used to exclude owner occupation from their national CPI altogether due to the difficulties of estimating these (imputed or "fictitious") costs of shelter. In principle a choice has to be made among the four methods displayed in **fig. 6.2.3** and on which now comments will be given one by one.

1. Net acquisition approach (NA)

In the case of non-durables a distinction between acquisition and use has no relevance because the consumption takes place simultaneously or shortly after the pur-

---

[71] Interestingly in both cases, core inflation and asset inflation, there is a tendency to establish weights *not* related to the (most up-to-date) expenditure structure of consumers, which is in other cases (for example in the cases of chain indices) deemed to be of utmost importance. We admit of course that it might be difficult to define "quantities" and expenditure weights in the case non-financial assets like housing and financial assets such as equity, loans, shares an the like.

[72] or owner occupied housing (OOH) or simply "owner occupation".

[73] Owner occupation makes up already about one half or even more than that of the whole costs of dwelling in many developed countries. In Europe the share of owners (as compared with tenants) varies between 80% in Ireland and about 40% in Germany.

chase. However, when households buy goods with an extremely long lifetime and at very high costs,[74] such as shelter, things will differ substantially.

Figure 6.2.3: Methods to deal with Owner Occupied Dwelling (OOD)

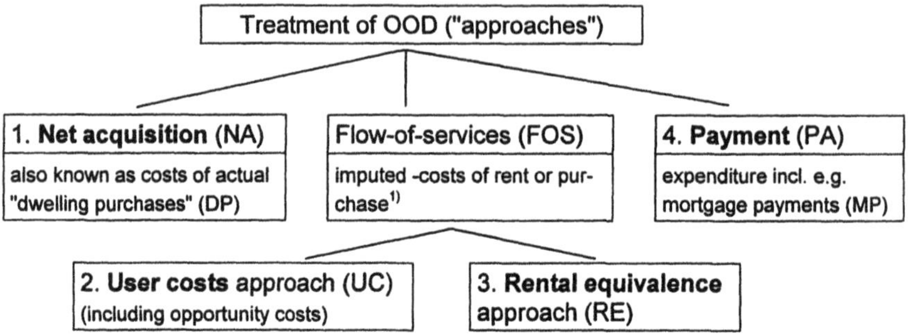

Method	General idea	Comments
Net acquisition approach (NA)	Value of OOD estimated on the basis of observed prices of (new) houses. Weights derived from construction statistics.	Current costs (e.g. for repairs), inheritance instead of purchase, quality changes and holding gains should also be taken into account.
User costs approach (UC)	Net operating and opportunity costs of owner occupier plus net holding gains (imputed rather than actually observed).	Comprises actually observed costs (e.g. mortgage costs) "costs" which need a more or less complicated estimation or imputation.
Rental equivalence approach (RE)	Observed rents spent by tenants are used to impute what owner-occupiers would pay if they were tenants.	Existence of well established and intact rental housing markets facilitates estimation of imputed rents. In line with the practice of NA[2)].
Payment approach (PA)	Actual payments by owners due to having incurred liabilities (e.g. mortgage payments). Poor estimates of OOD.	Makes the price of a service flow dependant on factors more or less unrelated to this service[3)]. Ignores home-value appreciation.

1) A problem is inclusion of holding gains (appreciation of the housing)
2) National Accounts (There RE approach is common practice to estimate output of housing sector)
3) as e.g. portfolio changes (reducing or increasing the level of equity)

Using dwelling purchases as a proxy for shelter costs is to treat houses like cars or furniture or other consumer durables[75] or just like non-durables. The job of the statis-

[74] These costs are usually spread over time by taking out mortgages for example.

ticians then consists simply[76] in observing *prices* for new as well as existing houses and flats, and it is also necessary (and perhaps even more difficult) to establish (expenditure) *weights* on the basis of such purchases (of households).

To mention some shortcomings of this method, it should be seen in the first place that most of the European countries are not very experienced in applying this method.[77] Furthermore it does not take into account that many households are lucky to inherit property from dead relatives or to obtain funds (reflecting a rise in the market valuation of their property) from selling their previous homes. A deduction of holding (capital) gains from the price of housing is also in general not made. Moreover, the purchase[78] of houses and flats should be viewed as an *investment* rather than *consumption* (as required in the framework of a *consumer* price index, CPI).[79]

In addition to the price at the start of an investment in OOD the current costs for maintenance and repair should be taken into account. An owner-occupier clearly continues to spend money on the *operating* of a house while he is living in it. He does so not only once and for all at the time of its acquisition. Hence the NA approach based on prices paid for new houses (purchases less sales of existing ones) does not fully reflect what owner-occupiers really have to pay for their housing.

We observe some volatility of prices for new houses which certainly does not reflect inflation, at least not as experienced by those households not presently acquiring property but actually residing in a property they have acquired (or inherited) some time ago already.[80]

Housing purchases should be kept distinct from housing "services" and the investment portion in owning a house must be separated from the consumption portion. This is the starting point of the two flow-of-services approaches, that is the RE and UC approach.

---

[76] The NA method would be "simple" only if there were no problem of comparability over time (as regards quality for example).

[77] This method has long been in use for example in Australia and New Zealand.

[78] A distinction is often made between "purchase" and "acquisition", in which the former is restricted to *new* accommodations. Sales of *existing* houses and flats by the household sector will in general not equal purchases of such items because they can also take place between different sectors, not only within the household sector. It is therefore not sufficient to simply observe the construction of new buildings in order to establish a weighting scheme and it is far from easy to provide comparable prices of various types of houses on a regular basis. What matters is the difference between purchases and sales (by private households from/to *other* sectors such as the corporate or government sector) in the prices and weights. This is meant when we talk of "*net*" acquisition.

[79] In general the CPI should not contain prices paid for other purposes than consumption of goods and services. This criticism applies to all methods other than the flow-of-services methods, that is to the net acquisition as well as the payment approach.

[80] Once we try to properly account for such increases in prices of existing houses and thus for capital (holding) gains, the NA-approach runs into difficulties very similar to those of the user-cost-approach where such capital gains also are responsible for the observed extreme volatility of the results. Holding gains can even be so high that user costs can become negative.

2. Flow-of-services approach I: User costs approach (UC)

The paramount idea of this method is that shelter is to be seen as consumption of *services* rather than goods and that capital theory may provide an estimate of user costs in the time period[81] t (denoted by C_t) which is given as follows

(6.2.3) $$C_t = r_t^m M_t + r_t^e E_t - A_t + Z_t,$$

where M_t and E_t are the (remaining) mortgage amount and the equity (or unmortgaged price) respectively (such that the actual price P_t used as an indicator of OOD in the NA method is $P_t = M_t + E_t$), while r_t^m and r_t^e are the mortgage interest rate and the opportunity costs of equity (r_t^e is the forgone return of other investments, that is, the amount of interest we would have got if we had invested in deposits, bonds or other financial assets instead of housing). It is also maintained that the interest rates r_t should be *real* interest rates rather than simply *nominal* interest rates. In contrast to other valuation methods, user costs (UC) also deduct capital gains (A_t, the "appreciation rate" due to the rise in the average price of a comparable house over the period) accruing to property from costs, and they allow for depreciation (as a part of Z_t) and to account for all other current costs expressed in Z_t.

Obviously the most difficult variables in this approach are A_t and r_t^e, and they therefore always found a lot of attention by those criticising the UC-approach. Given an active rental and resale market and no uncertainty or frictions then user costs can be defined as follows

(6.2.3a) $$C_t^* = r_t^* P_t - A_t + Z_t,$$

where r_t^* is the single rate of interest in period t. Hence the opportunity costs of holding and using the house are given by the capital costs $(r_t^* P_t)$ plus operational costs (Z_t) less the increase in its value and under the conditions of perfect competition and equilibrium the rent paid by a tenant R_t^T should equal the rent received by a landlord R_t^L and C_t^*.

When the assumptions of no uncertainty and no friction are relaxed we can allow for different asset yields (no single r_t^*) and the possibility of $R_t^L < R_t^T$. We arrive at equation 6.2.2 for C_t in this manner. There the troublemaker variable is however r_t^E, that is, the opportunity costs (the alternative returns forgone) of equity capital which has to be estimated.[82]

---

[81] Period t is begins at point t-1 and ends at point t.

[82] Hence evidently UCs depend on financial and operating costs as well as on possible holding gains and *expected* inflation rates, and it is not clear that these factors will correctly reflect "inflation" and nothing but inflation, at least in the sense in which the term is usually understood. Note also that the focus is generally on the *actual* total user costs (UC) and not on some *expected* marginal opportunity costs taking into account the likely returns on all sorts of investment other than home ownership. The concept last mentioned should suit even more the microeconomic approach to costs-of-living than the concept of UC. Yet it seems to be difficult to apply in practice.

UC are more in line with the "true cost of living index" (COLI) of the "economic theory of index numbers" because utility is more likely to be provided by the stream of "uses" (consumption services) rather than by the purchase as such.[83] Yet such services are far from being actually observable transactions and therefore UCs have to be approximated adequately using related data.[84]

The problem with this method is mainly the repeatedly observed volatility of its results for which capital gains (A_t) and certain interest rates (r_t) are primarily responsible. As mentioned already the A_t component can be such that the user cost C_t can even become negative. Moreover it should be noted that the estimated sum C_t is clearly not an amount of money actually *spent* by somebody in the real world in t and thereby influencing inflation in t or later. Another objection is that methods in which interest rates enter should be avoided[85] because to the extent that government and monetary authorities try to combat inflation by raising interest rates an indicator comprising interest rates r might mislead into thinking that the trend of inflation has risen. This is considered to be a perverse result. This was precisely the argument why user costs were rejected in the HICP debate,[86] because to adopt this method would introduce interest rates into the HICP (via approximating the true opportunity costs of equity).

3. Flow-of-services approach II: Rental equivalence approach (RE)

The method to take actually paid rents of comparable objects to estimate imputed rents of owner-occupiers is a simple and straightforward solution which has already been used by a number of European countries for many years.[87] The idea is simply that - given an intact and active competitive rental market[88] - an owner would not be willing to pay significantly *more* than a tenant because he otherwise would rent rather than purchase the dwelling. Conversely in the case of an imputed rent which is significantly *less* than the rents of tenants there would be a massive incentive for tenants to buy housing rather than to rent it. Therefore, given competition and rational

---

[83] It is sometimes said that the UC approach has its merits as regards the *theoretical* purity while *in practice* UCs may be most difficult to implement.

[84] The argument is not only difficulty in getting the necessary data on a regular (monthly) basis but also that we have to make a recourse on data like rents actually paid or rents imputed in order to estimate the opportunity costs of equity. Hence we may ask: why not simply move to the RE-approach when we need such data anyway?

[85] This applies in particular also to the payment approach, PA [see below], but also to asset inflation.

[86] Central banks dislike inflation measures that depend on interest rates. They treat interest rates as an instrument in their monetary policy and they therefore disapprove an indicator which is explicitly influenced by what the policy is aiming at. The above quoted view was surprisingly undisputed in the HICP development. However, there is of course an interdependence between interest rates and prices at least of certain services such as rents. As r rises (for whatever reason) prices such as rents will rise as well. So why should we prevent a CPI from reflecting this undoubtedly existing phenomenon?

[87] This approach was used in the U.K. before 1975 and it was (or still is) in use in Germany, Denmark, or (until recently) in the Netherlands for example. RE are also estimated in the USA.

[88] In countries in which the most part of the population consists of owner-occupiers we may well run into difficulties with this assumption, as this method gives an unduly large weight to OOD and thus imputations will influence the overall CPI to an intolerable extent.

behaviour in the long run, rents are (theoretically at least) unlikely to differ much from user costs. Hence (theoretically) the rent cannot be less than R_t^L (what the landlord will receive) because the owner-occupier forgoes this amount by not leasing his house but residing in it instead, and on the other hand it cannot be greater than R_t^T (what the tenant is paying) because he could alternatively rent an equivalent object at this price. Thus a satisfactory measure M_t of OOD should be bounded by R_t^L and R_t^T such that

(6.2.4) $$R_t^L \leq M_t \leq R_t^T,$$

and taking R_t^T as a proxy of OOD costs as done in the RE-approach amounts to making an estimate tending to the upper rather than the lower bound given that the rental market is of course profitable for landlords.[89]

It has been claimed in favor of this method that the RE approach provides a good and easily observable approximation to user costs (UC) and that observations of the rental market are needed anyway in all reasonable methods to deal with OOD, if not directly (RE-approach) then at least indirectly (for opportunity costs in the framework of the UC-approach). Given that in many countries data on rents are available on a regular basis for a great variety of dwellings it is also straightforward to estimate the rent that is (or could be) charged as a function of characteristics of the unit, the neighbourhood and location and those of the tenants and the type of contract.

In other words, the RE-method gives rise to the application of the *hedonic method* (see **sec 4.2**) and thereby also to the construction of imputed prices of fictitious units representing combinations of characteristics normally not offered on the rental market. This may also help to overcome an argument often advanced against the RE approach, such as:

- The difficulty of the RE method is that owners generally do not live in the same type of houses than those people who have to rent dwellings, simply because they (the tenants) cannot afford to buy them, or
- the rental market may not offer that much variety and such high quality standard of housing that is needed for valid imputations of "owner rents".

It should in principle be possible to estimate by hedonic regression "rents" for all sorts of high-quality dwellings as a function of characteristics such as size, location, age, equipment etc.

Furthermore, it has been argued that the estimates achieved by imputing rents do not correspond to actual expenditures (which only are responsible for inflation), and that they are less volatile than results derived by other estimation methods for the simple reason that rents (and thus also rental equivalents) generally fail to reflect promptly and to the correct extent all changes in the actual costs of housing.

[89] It has been shown, however, that the increase of the imputed rents (RE-approach) was lower than the estimates derived from a "hedonic rent index".

The main objection[90] against this method is however that an imputation has to be made and the received wisdom is that imputations in a price index ought to be avoided if at all possible. Moreover, in quite a few countries owner-occupation is a highly significant phenomenon, and precisely in those countries the rental market may be too small and too dissimilar to establish reasonable imputed rents.[91]

4. Payment Approach (PA)

This method (also known as the cash flow method) consists of measuring the actual cash-outflows resulting from OOD. It is perhaps the method which for many appears to be the easiest, most comprehensible, and most straightforward method as it reflects the costs that housing owners are actually facing. Yet it is also a method against which conspiciously many objections can be raised. The payment consists primarily of mortgage payments[92] for which data are readily available, however

1. It is difficult to ensure consistency across purchases: other goods, like cars, furniture etc. possibly also sometimes bought on credit are in general not treated as more/less expensive as credit costs rise/fall.[93]
2. In a similar vein it appears difficult to treat interest payments consistently across persons (The method focuses on the borrower's situation. He is indeed worse off when interest rises while the lender is better off).[94]
3. The method also ignores capital gain/losses on account of the price movement of similar houses in the meantime. On the other hand, doubts may be raised as to whether such phenomena really reflect consumption instead of investment.
4. It should be noticed that Family Expenditure Surveys (FES) and National Accounts generally make the same distinction between consumption and investment, that is, they exclude mortgage costs and related payments from consumption.[95] Thus not only prices but also weights are difficult to establish in the case of estimating OOD on the basis of the PA.

---

[90] At least the main objection in the course of the European harmonization project.

[91] It has been maintained that it is unwise to compile up to (depending on the proportion of tenants as opposed to owner-occupiers) a fifth or so of an index on the basis of "imputations".

[92] The main difference between mortgage payments (MP) and mortgage cost (MC) is that MC are calculated under the assumption that the value of the house is mortgaged completely and that repayment of the principal is excluded in MC. The first assumption is necessary in order to arrive at an estimate that is comparable to rents tenants actually pay their landlords. In reality, however, only a part of the average value of houses (initially approximately two thirds) is mortgaged, depending of course on the length of the period in which repayment of the debt (which is capital formation and therefore not included in MC) has already taken place.

[93] In general the price of a purchase and how the purchase has been financed should be kept distinct.

[94] which the method ignores, however. This well known argument is certainly doubtful since it is not clear why a CPI should not only reflect changes in the situation of the borrowing household (consumer) but also of the bank or whoever acting as the money lender.

[95] Moreover the FES in most countries also exclude families with a particularly high disposable income, exceeding a certain threshold and these are just the type of households for which the problem of owner occupation becomes most relevant.

Table 6.2.1: Aspects to be considered in the choice of a method to deal with OOD

Problem	Comment
1. The relevant data are readily available	Applies to mortgage payments, maybe also to rents, prices of new houses etc., but comparability often questionable.
2. Comparability of shelter quality[1)]	Goods are rare in which so fundamental differences as regards quality, location etc. can be observed as in the case of housing.
3. Definition and statistical source of "quantities" (i.e. "weights")	As usual a separation of changes in prices (the observed variable, to be displayed by the index) from changes in volumes (reflected in the weights) is required. Weights are in general less well established in FES[2)] as sources of data.
4. Actual expenditures vs. imputations	In general, if possible *actual* expenditure should be preferred to *imputed* costs.
5. Volatility[3)] should not unduly affect CPI	Otherwise the inclusion of owner-occupation would simply add "noise" (unrelated to true inflation)
6. Avoid treating comparable situations too differently[4)]	The method should be comparable across tenants and owners and over different types of the latter (as regards the kind of acquisition and financing)
7. Method should be defensible from an *economic-theory* point of view[5)]	A lot of factors such as the ever increasing complexity of financing methods, the accrual of capital gains and its taxation, and the depreciation of the housing stock will make it most difficult, if not futile to achieve theoretical purism.

1) to which the observed expenditures are related.
2) Family Expenditure Surveys. Methods making use of purchases and payments related to debts will in general pose more problems. Since house or dwelling acquisition are not consumption but rather investment, such expenditures serving capital formation are not included in FES in general. It will therefore not be easy to establish weights for this part of an index.
3) Nor should the method underrate volatility of course. This has been argued for example in the case of the RE approach. Rents of tenants may be less volatile than actual costs owners have to face.
4) This is for example not the case when simply dwelling purchases are observed, because unlike buyers of houses, those who inherited houses are treated as if their housing were completely free. The method should be consistent across persons as well as across purchases.
5) As to this criterion, the UC approach clearly is an admirable concept but unfortunately none of the UC-components is directly observable (without certain corrections) and the sum of these components does not correspond to actual expenditures. Nothing reflects any actual transaction.

5. A measure of inflation should not depend on the age structure of mortgages and houses, nor should it reflect structural changes referring to aspects such as acquisitions vs. inheritance, or the speed of down-payments[96], second hand purchase as opposed to first-time buying of houses, and inheriting vs. buying property and other distinctions of this kind.

[96] Repayments should be deducted from actual expenditure as they do not represent consumption but rather investment (capital formation).

6. There are also mortgages for home improvement and it is also possible that people move from low price to high price houses and therefore have to incur more liabilities. Inflation measures should not reflect influences other than rising costs of the *same* quality of housing.
7. Methods of financing houses are becoming increasingly more complex and sophisticated such that it will be more and more doubtful and suspect to measure OOD indirectly by the costs entailed in financing this kind of service.
8. It turned out that the method introduces a lot of volatility, and its results may differ from the development of house prices in the course of the business cycle.

Conclusion

To sum up, there is no method which is satisfactory with respect to the all aspects worth being considered. Therefore, not surprisingly, an undisputed internationally agreed upon methodology is still lacking. To make a choice from among the great variety of methods, each of which with its respective merits and demerits, it is pertinent to stipulate some criteria though it may, and in fact will be, impossible to meet them all.

The criteria (or at least most of them) brought into play in the discussion of the aforesaid methods are listed in **tab. 6.2.1**. There is quite naturally a conflict (see row 4 in tab. 6.2.1) between estimates that properly reflect the *true expenditures* incurred by owners and those that take account of "costs" and benefits of a more hidden and theoretical nature, not necessarily leading to payments in the period in question.

In a similar vein a distinction can be made between "observable" (or "operational") versus "purely theoretic". From a theoretical point of view the UC method is perhaps the preferred concept[97] but the implementation in operational terms is clearly unsatisfactory, and estimation of user costs will perhaps remain troublingly difficult in practice forever.

6.3 Producer Price Indices (PPI)

The purpose of this section is to introduce some methodological issues in three selected fields of the producer price indices (PPIs) viz. in agriculture, manufacturing and other production industries, and in particular some problems specific for measuring prices of (new) buildings and other constructions (roads, bridges etc). Some other PPIs will be discussed along with problems arising from so-called "unit value indices"[98] in sec. 6.4.

[97] It can be argued, however, that the RE approach fits the COLI concept better, which attempts to price the current consumption. Interest rates and asset appreciation, both entering UC formulas, are out of the scope of a COLI. A correct individual UC estimation also has to account for tax benefits of home owners, the future selling date, which in turn will be influenced by a lot of idiosyncratic factors such as the portfolio composition of the household.

[98] Unit values are already defined in sec. 1.1. Price and quantity indices on the basis of unit values are quite common in the case of foreign trade and wages and salaries.

a) Price indices in agriculture

In the majority of countries both price indices, an index of output prices (producer prices) as well as an index of input prices (purchases of farm supplies), are available and the income situation of farmers is analyzed by comparing the two types of indices.[99] In general, the more interesting methodological problems of agricultural price indices are encountered in the case of output price indices (PPIs) rather than in input price indices. Thus in what follows the focus is primarily on the agricultural PPI.[100] The problems are primarily

- the definition of the coverage of an index and the empirical derivation of its weights,
- seasonal fluctuations (or even absence) of sales and purchases of agricultural products, and the temporal aggregation of monthly indices to annual indices.[101]

Coverage, weights, and procedures of price collection

There are two concepts in use guiding the selection of activities and sales/purchases to be covered by Producer Price Indices (PPIs) in agriculture, viz. the so-called

- "*average farm concept*" (*AFC*), according to which all sales of farmers are supposed to be covered irrespective of the type of buyer (in particular this "gross sector approach" also covers all intrasectoral purchases of farmers[102]), or
- the alternative concept known as the "*federal farm concept*" (*FFC*), which may be viewed as a sort of "net sector approach".[103]

Problems of coverage of a different type (by kind of activity, or type of expenditure) are also involved in the case of the input price index in agriculture:

- Should the index also comprise inputs for ancillary (or "secondary", i.e. non-agricultural activities in the strict sense, such as trade with households in so-called "farm-outlets", accommodation and entertainment services etc.)?
- Which of the following inputs should be included in addition to the cost of material and other produced supplies and intermediate goods and services: expenditures for investment goods such as transport equipment, new buildings, wages, taxes, and insurance premiums?[104]

---

[99] As to such analyses, there are quite a few reservations to be made. See below for more details.

[100] Nonetheless, some problems of input price indices will be discussed hereafter, when appropriate.

[101] This applies especially to the agricultural PPI. In the case of an input index the most prominent problem is to ensure constancy (comparability) of quality and other price determining characteristics.

[102] Intrasectoral transactions in agriculture are for example seeds, animals and animal feed.

[103] Germany used to apply (until the index system 1991 = 100) both concepts, the AFC for national purposes as well as the FFC, which in those days was requested by Eurostat. As soon as the system based on 1995 came into force, the office began to apply the AFC only (since then Eurostat standard).

[104] As a rule, investment goods are included while the other aforesaid expenditures should be excluded. Prices of all sorts of investment goods in this Agriculture *Input* Price Index are usually taken from the Industry-(Output)-PPI (i.e. the *output* price index of the Production Industries).

To establish a weighting structure is likely to be particularly difficult in agricultural price indices. This is because in many countries farmers are exempted from book-keeping, or the relevant obligations are relaxed compared to other sectors. Hence the empirical basis for weights is likely to be small and of limited reliability. Another problem is clearly that the structure of sales and purchases varies widely over the year. This problem of "seasonality" (see below) has to be considered as regards price collection, weights and the compilation of annual indices from monthly indices. Given a substantial seasonal variability of expenditures and revenues the question arises: what is the relevant "basket"? Are in fact weighting structures obtained by averaging monthly structures meaningful?

Finally, for practical reasons[105] prices are mostly recorded at selling units like agricultural co-operatives, wholesalers, cattle markets and the like, rather than by asking farmers themselves.[106]

Seasonal absence of sales and/ or purchases

Especially when compiling a producer's price index of agricultural products, the varying representativity of the index and (as the limiting case) the non-availability of products at certain times is a severe problem for which a generally accepted solution has not yet been found. Weather conditions but also consumer habits prevent certain goods from being offered on the market at certain periods of the year. Thus it is not always possible to provide information on *all* prices *each* month as in general tacitly assumed in the construction of a price index.

The problem not only refers to food and other agricultural products but also to clothing, footwear, travel expenditure, vacation resorts and other goods on the following grounds: Strictly speaking the problem is not only an absence of transactions in commodities of a similar kind but more generally a changing representativity[107] of the transactions incorporated in the index.

Ideally the prices recorded in a regular (monthly) index should refer to strictly comparable transactions with respect to quality, typicality and representativity every month. From a purist standpoint it may otherwise be questioned whether the observed (unadjusted) price movement will represent a "true" change of prices or rather also reflect (to a certain degree) a change of conditions of sale and purchase. The fact that sales are more or less frequent over the time span of a year (i.e. depending on seasonal factors affecting supply and demand) will have an impact on prices and their comparability which should be taken into account appropriately. Thus, in principle the phenomenon of reduced representativity or non-existence of a price quotation, e.g. for prices for agricultural products during the winter season, should call for adjustment actions of price statisticians in much the same way as quality change is taken into account (or "adjusted").

---

[105] Already mentioned in **sec. 4.1**.

[106] In general there are also less problems with quality change or new products as there are in the case of consumer prices or producer prices of industrial goods and services. This applies, however, to the output rather than the input price index in agriculture.

[107] We use the terms representativity and representativeness as synonyms.

The effect of non-availability, or more general of a changing degree of representativity is to impair comparability. In this respect the situation is similar to quality change, but unlike quality change reduced representativity is usually not taken into account, at least not by making adjustments on observed prices.

The varying degree of representativity of a commodity and its price will generally not call for adjustments or corrections of one type or another unless the commodity is no longer for sale. An exception is the case of the seasonal absence of certain prices where price statisticians proposed several solutions to deal with the problem[108]. Some comments on these methods (see **fig. 6.3.2**) are appropriate:

Figure 6.3.1: Various types of reduced representativity

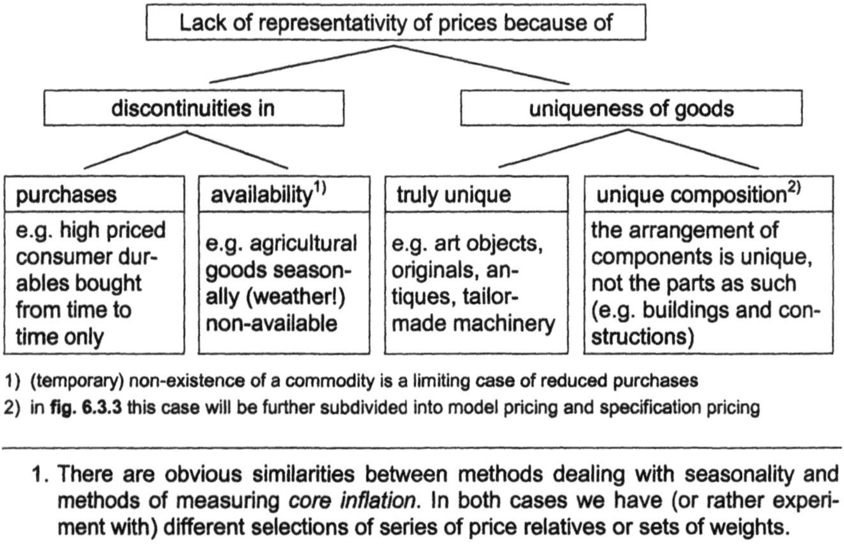

1) (temporary) non-existence of a commodity is a limiting case of reduced purchases
2) in **fig. 6.3.3** this case will be further subdivided into model pricing and specification pricing

1. There are obvious similarities between methods dealing with seasonality and methods of measuring *core inflation*. In both cases we have (or rather experiment with) different selections of series of price relatives or sets of weights.
2. The weakness of the **C method** is that it may provoke some discontinuities (whenever the set of *new* prices enters the index formula we may have an abrupt movement upwards or downwards after having displayed a delusive stability due to keeping prices constant though no prices were observed in reality) and interruptions, perhaps with some kind of seasonal pattern, whereas the true development might be a lot smoother. The opposite (i.e. giving the false impression of more stability than actually exists) is supposed to be more likely in the case of the M-method[109] and the E-method rather than the C-method.

[108] The problems of seasonal fluctuations arise firstly in consumption groups such as fruits and vegetables, but also to other goods. In principle a choice can be made among the three (or four if the P-method is included) possibly best known methods of **fig. 6.3.2**.

[109] However we also find the opposite assertion that the M-method is likely to *intensify* rather than *smoothen* the cyclical movement in the course of a year, provided the price relatives of those products

Figure 6.3.2: Methods to deal with (seasonal) non-availability of (agricultural) products

Accounting for discontinuities (in the availability)

- a **single** set of **weights** for the whole year (allocation of constant weights) and making corrections for the prices only
  - carry forward (hold constant *) the last (before disappearance) observed price
  - omit (i.e. ignore) the goods in question (to be referred as E-method later), or impute the unobserved price
- changing (on a monthly basis) weights over the year, **multiple** (or moving) **weights** (M-method)

* C – method; if there is a trend of rising or declining prices on the market, the "price freezing" clearly creates a downward or upward bias respectively. It is therefore in general preferable to impute the average movement in prices for those goods with prices that are temporarily not available.

Method	Description
exclusion **E-method**	Calculate the index without the prices of non available (though in other months available) commodities after having adjusted the weights of the remaining commodities such that they add up to unity[1)]. This is equivalent to the assumption that the price movement of the non available products equals the average price movement, i.e. the price index as calculated on the basis of the remaining products.
Continuation[2)] **C-method**	To continue the calculation of the index with the last reported price of the commodity in question during the period of its not being traded until a new price is able to replace the old one.
Multiple weights **M- method**	Provide different weighting schemes each month (hence a variable "basket") according to variations in supply and demand over a year.
P- method	"Parallel series" simply a combination of the C and E method[3)]

1) Otherwise (without redefining the weights) the calculation would implicitly treat the price relatives of missing goods as if they were zero.
2) Because the method requires us to continue with (carry forward) the last available price.
3) Two series of index numbers are compiled, one including products that are not available throughout the year but which may be seasonally absent (in this case the C-method is applied) and another series exempt from this kind of products (thus in this respect we follow the E- method).

3. The **E-method** on the other hand is perhaps likely to create an unduly moderate (or underestimated) seasonal pattern and tend to reveal a less distinct cycle by

differ from the average in the months in which they are relevant (i.e. when the product to which they belong is available).

implicitly treating absent price changes (non observable changes) as *average* price changes (see below).[110]

4. The **M-method** (also called the Rothwell index) is said to be the most demanding method (as regards data requirements). In most countries, however, price statistics will hardly dispose of so many survey data as to establish not only one but rather no less than twelve "baskets" empirically. Weights w_{it} are in general given to products i and months t = 1, ..., 12 in such a way that a relatively high (low) weight is given to product i in a month t in which i is more (less) available.
5. The M-method allows us to account for different *degrees* of absence (or availability)[111]. On the other hand, and most importantly, it should be noted that the method does not comply with the principle of pure price comparison: Index results referring to successive months will not reflect a pure price change but also to an unknown extent differences in the weights which on their part are not designed to reflect variations in prices but rather variations in supply and demand, which is clearly is an element of the quantity (volume) component instead of the price component. In other words: the extent to which results differ from month to month will not only reflect differences in the level of prices but also in the structure of weights.[112]

In sum, none of the methods outlined above can be recommended without hesitation.

Digression: Omission of (non-available) prices in an index

It can easily be seen that the **E-method** is tantamount to assuming a price movement of the omitted commodity, say no. 1 (or a group of omitted commodities) which precisely equals the average price movement of the remaining commodities (i.e. the goods no. 2 and 3 for example):

Suppose commodity no. 1 is seasonally not available and the basket also contains goods 2 and 3 with price relatives (or if 2 and 3 denote *groups* of commodities, sub-indices rather than price relatives) P_{0t}^2, P_{0t}^3 and weights w_2 and w_3 respectively. The E-method is equivalent to assuming the implicit unknown price relative (sub-index) P_{0t}^1 equals the (average) price index P_{0t} derived from the commodities no. 2 and 3, that is

[110] We may also say that the E method – in contrast to the M-method – amounts to giving a more moderate (and temporarily vanishing) weight constantly (throughout the year) to a certain commodity.

[111] i.e. not only the case of *complete* but also of partial seasonal absence by attaching a zero- or non-zero-weight to the commodities in question.

[112] Israel moved from the M-method to the allocation of *constant* weights in 1988 because 1. most goods hitherto seasonally unavailable were in the mean time on sale throughout the year, and 2. the allocation of monthly varying weights (according to the shares of the value or the quantities) introduces an inconsistency with the idea of pure price comparison because under such conditions the price index will be influenced by both prices *and* weights and could well indicate a change although no price changed, simply because the baskets of successive months are different. While Israel abandoned the M-method it is still in use for example in Germany.

(*) $$P_{0t}^2 \frac{w_2}{w_2 + w_3} + P_{0t}^3 \frac{w_2}{w_2 + w_3} = P_{0t},$$

the index with commodity no. 1 excluded and with weights adjusted. On the other hand P_{0t} is by definition

(**) $$w_1 P_{0t}^1 + w_2 P_{0t}^2 + w_3 P_{0t}^3 = P_{0t}.$$

It can easily be seen now that the E-method implicitly relates the unknown (missing) price relative (or sub-index) P_{0t}^1 to P_{0t} such that $P_{0t}^1 = P_{0t}$.

Upon substitution of P_{0t} for P_{0t}^1 in equation (**) and using $w_2 + w_3 = 1 - w_1$ it follows $w_2 P_{0t}^2 + w_3 P_{0t}^3 = P_{0t}(1 - w_1) = P_{0t}(w_2 + w_3)$ giving equation (*). ♦

Ratio of input and output price index

It is most popular to consider the ratio R of the producer- and purchaser price index in agriculture (often [incorrectly] named "terms of trade in agriculture"[113]). The common interpretation given to this figure is that an improvement (R > 1) brought on by producer prices rising more than prices of supplies is tantamount to an improved profit situation for farmers. On the other hand R < 1 is often taken as proof for the need of subsidies and viewed as indicating a worsening of the situation for farmers. Of course R can only reflect an isolated *price* movement of selected goods. To assess the profit situation it is also necessary to look at quantities of sales and procurements in addition to prices. It should also be taken into account that some inputs - as mentioned above - are not included in the input-price index and of course that productivity does not remain constant. The true cost-profit-relation of a farmer introducing product and process *innovations* may well differ from what is indicated by R.

b) Producer Price Index (PPI) for the Production Industries

As to the sector "Production Industries", in most countries a Producer Price Index (PPI) is available. The term PPI often refers exclusively to the industrial sector, comprising not only manufacturing industries but also the energy and water supply industries, mining, and in some countries also the construction sector. The products of the latter are in a way so specific that an approach different from other parts of the Production Industries appears justified (see part c for the PPI of Construction Industries). In principle four different price indices for the Production Industries may be distinguished as regards the geographical coverage:

input prices	output prices (PPI)
1. domestic purchases	3. domestic sales
2. imports	4. exports

[113] The term is a misnomer because the activity in question is "production" and not "trade".

The role of index no. 4 may be taken by an index of export prices (designed to describe the function of "trade" rather than "production"). In general it is an advantage to monitor prices of sales on the domestic market and sales abroad (exports) separately because some interesting insights are gained from analyzing differences in the short-term development of prices for the two types of sales.

A short remark about input indices may be useful. A combined index covering field 1 and 2 is known as "Price index of commodities received in production industries" in Germany. This index measures the price development of materials and supplies purchased within the domestic market or abroad by all producing industries (including construction), except for investment goods and services. Weights are derived from National Account estimates.

c) Price indices for buildings, improvement of dwellings and other construction work

This section deals with prices of new constructions (buildings, road construction) and reconstruction work. The methods are for analytical reasons quite interesting. Separate PPIs for these industries are commonly in use. In each case the price of land as a non-produced tangible asset (AN. 2.1) is excluded. This part of price statistics can be defined *functionally*[114]. But there are also systems of price statistics in which an *institutional* approach is chosen such that prices observed (along with production) refer to the output of all those branches that belong to the construction industry. In what follows the discussion is restricted to *output*-price indices (or more specifically PPI for construction work or construction industries), leaving open problems of a specific *input*-price index different from an output index, if such a duality of indices exists at all.[115]

What are the fundamental problems in pricing of construction?[116] In general an index should comprise many products, and for each product there should be a sufficient number of price quotations to make reliable estimates. However

Structures (roads, buildings, dwellings, bridges etc.) do not come in mass-produced goods, sold at the same location and at the same time, hence the standard approach to collect prices of (more or less narrowly specified in order to make sure that prices are comparable) finished products fails here.

It is hardly possible to find the same object in period t and in t+1, let alone to solve the problem of comparing prices of a number of goods over *several* periods. This

[114] that is, by the kind of activity (in which case construction work counts, performed by whatever productive unit) to be priced. In this sense (functional vs. institutional) we also distinguish so-called *commodity-based* and *industry-based* indices, a distinction which also refers to different ways to define subindices.

[115] In fact in most countries the only existing type of price index of construction work is an output index. However, difficulties in pricing output often require prices of inputs in order to value outputs suitably, such that a PPI (as an output price index) may well be made up of input-prices.

[116] As stated repeatedly each class of price indices has its own specific type of problems. The particular problem with price indices in construction is that comparable prices are difficult to establish in this field.

means that prices of buildings and (even more so) other constructions (such as roads, bridges and the like) are particularly difficult to compare, and therefore

- we have to provide a *representative* selection,[117] that is, a sufficiently large number of commodities bearing a certain resemblance to justify an index calculation altogether, as well as to
- comply with the principle of *"pure" price comparison* in collecting price quotations on a fairly regular basis for essentially the same product.

The same is true for investment goods[118] other than buildings, such as equipment and inventories, but also to an even greater extent to prices of *land* for agricultural or construction use.[119] Another problem is that the market is subdivided into many submarkets, for example the market for dwellings may be subdivided into the social rent-, private rent-, and owner occupied submarket. They differ *widely* with respect to all kinds of price determining characteristics, such as the size and quality of dwellings as well as the price (cost) of constructing them.

As **fig. 6.3.3** shows, a distinction can be made between three methods of dealing with the difficulties mentioned above. To comment on them it should be pointed out:

- Method 1 treats buildings as *finished goods*, and makes corrections for all sorts of significant differences thus considered as "quality" aspects. Lack of comparability is regarded as a problem of (change in) quality. Whether the method makes sense or not depends largely on the possibility of making reasonable adjustments for quality differences, for example with *hedonic regression*;
- *model pricing* (method 2) consists of asking producers what they would charge for a certain well defined "model" building (defined in terms of output); and
- in the case of *specification pricing* (method 3) the price of a *real* (in contrast to method 2) project is broken down into a number of key elements or "components",[120] the prices of which are recombined to a "price" of the whole building using constant weights[121], where these weights are derived from data on payments in actual contracts between builders and building owners.

---

[117] This applies in particular to the case of road construction and similar construction work (ports, bridges etc.). The main problem in this case seems to be the definition (commodity classification) of a "representative" set of commodities, and of course the selection of goods to be priced regularly (is random sampling possible at all?).

[118] It should be borne in mind that all sorts of constructions are "investments" (tangible fixed assets).

[119] Land is always unique by definition. This is also the reason why many countries – among them Germany – deliberately refrain from compiling indices in this field but content themselves with publishing only certain selected price relatives for land and certain constructions.

[120] These are representative operations such as earthwork (excavating a m^3 of earth under average ground conditions), laying a m^2 of floor tiles, putting in a window, masonry, casting a m^2 of inner wall, plumbing in kitchen and bath room, roofing, installation of electricity, gas, water etc.

[121] While a "model" (method 2) is described in terms of output characteristics, a "specification" (method 3) is rather a list of works to be done and therefore elements of cost (input) elements.

Figure 6.3.3: Methods to compile output-price indices in construction

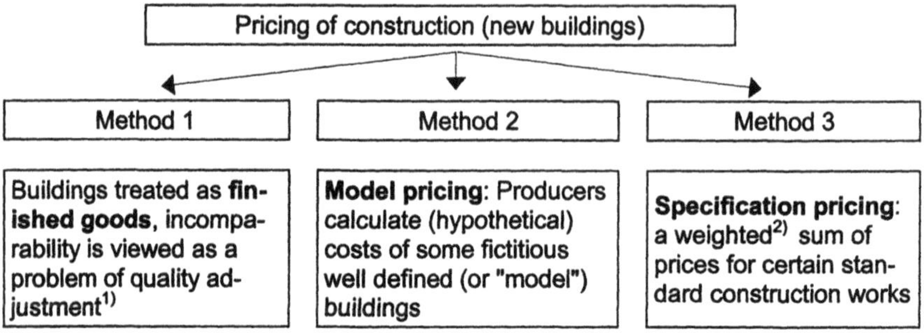

1) different houses are seen as different qualities (calling for adjustments of prices); the method is applicable in particular to pre-manufactured buildings.

2) using weights from actual contracts.

There are obvious similarities between the basic ideas underlying methods 2 and 3 on the one hand and the hedonic approach on the other, in that it is requisite for all these methods that a product can be (exhaustively) described in terms of certain quantitative "characteristics" (or works contributing to the total project) z_1, z_2, ... and the price of the whole product is some sort of a (weighted) sum of partial prices for z_1, z_2, ... etc. An essential difference, however, is the way the "weights" are established empirically.

We now comment on each of the three methods in turn:

1. ***Method 1*** is of course fully applicable in the case of pre-manufactured buildings[122] (as opposed to conventional constructions). Even if quality differences may ideally be decomposed into a set of (output) characteristics (as assumed also in method 2), it may be difficult to estimate the monetary equivalent of quality differences of the following sort: size of windows, type of frames (wooden, aluminium, plastic etc.), types of floors, heating, aspects of design, functionality, beauty, originality and the like.[123] Moreover, very detailed information should be available for making corrections, and as prices usually vary widely there is perhaps a wide range of estimates possible for "corrected" prices of the finished good. Again we see that a method apparently easy to understand quite unexpectedly turns out to be difficult in operation. In practice it is anything but easy to guarantee comparability of prices (this is the problem specific to this method).
2. The problem with ***method 2*** is that on the one hand the selection of models should be representative of all buildings *and* regularly updated, but on the other hand they should remain the same for some successive periods in order to maintain comparability. Furthermore, construction firms are asked to make cal-

[122] It is in fact a situation in which construction comes close to the usual industrial "finished goods", however, we even see in this case a rapid change of "models" and increasing individuality of houses.

[123] This shows that while it may appear not too difficult to make corrections (of prices) for location, number of rooms or so, there are other dimensions not readily amenable to a separate valuation.

culations for buildings for statistical purposes only (for "virtual buildings"), and statisticians will have some difficulties checking whether the calculations of the firms are serious, complete, and based on similar assumptions concerning input of materials, manpower and quality. This also shows that it may be difficult to make a clear separation between an output and input price index, since all these aspects are of course "costs". An index based on costs may be viewed as coming closer to an input rather than an output price index. The same is true for the third method.

3. According to ***method 3***, while the building in its entirety is unique, it is made up of a number of generic sub-activities and the uniqueness arises from the combination of these elements rather than from the elements themselves. Looking at buildings as being decomposable in such a manner makes this method similar to the hedonic approach. To be applicable, a complete list of well defined standard elements ("works" or operations) of any construction is needed and the price finally charged for a building as a whole should explicitly refer to these elements. However such conditions are not a matter of course.[124] Another necessary condition is that weights can be derived from a representative set of actual contracts in the base period[125] in order to meaningfully aggregate such partial-prices of distinctive "works". Moreover, a price index of the Laspeyres type requires keeping these weights constant for some periods.

To sum up: The major problems of the three methods are as follows. In 1 it is difficult to observe *comparable* prices and adjust them for differences in quality. Method 2 is mainly considered an input approach with limited possibility to assess the validity of the data. Method 3 requires a certain legislation and practice of pricing (based on standardized operations) in the construction sector and contracts of real buildings to provide weights for the aggregation of prices. Again, there is no single method which is fully satisfactory.

6.4 Price indices and unit value indices, foreign trade and wages indices

This section introduces the elements of a theory of unit value indices (as opposed to true price or true quantity indices).[126] We then proceed to official indices of the unit value index type which are for a number of reasons, not least the costs and convenience of index compilation, often found in two fields: indices measuring prices of export and imports, and indices of wages and salaries.

---

[124] They are in fact given in some countries, for instance in Germany, where the final invoice has to be detailed in terms of these elements, such as earthwork, roofing etc. as mentioned above.

[125] Experience in Germany where this method has been applied for many years already shows that this condition is not easy to fulfill. It turns out to be strenuous and time consuming to collect a sufficiently great number of contracts in order to arrive at empirical "weights". To dispose of a sufficient number of real and representative contracts is in practice possibly even more restrictive than the detailed classification of key elements of a "specification".

[126] The term "unit value index" (PU) is in general used for an alternative to a price index (P). If unit values are used instead of prices *as weights for quantities* we have a unit value quantity index (QU) instead of a true quantity index (Q).

a) Unit value vs. price indices

In some cases, in which the collection of n goods is confined to goods of a similar kind, having a common unit of measurement of quantities data on sums $Q_0 = \sum_{i=1}^{n} q_{i0}$ or $Q_t = \sum q_{it}$ may have a meaningful interpretation and may be readily available. This is the case in some parts of foreign trade statistics for example. However, normally neither the sum of prices Σp nor the sum of quantities Σq can be attached any significance. The summation of prices suffers from being affected by the units in quantity to which the price quotation refers (this is known as the commensurability problem) and the sum of quantities is often not even defined due to the fact that a common unit of measurement is lacking.

The so-called unit values (a kind of average prices), $\tilde{p}_{k0}$ and $\tilde{p}_{kt}$, for the k-th sub collection of related (homogeneous) commodities[127] are defined by[128]

$$(6.4.1)\quad \tilde{p}_{k0} = \frac{\sum p_{kj0} q_{kj0}}{\sum q_{kj0}} = \frac{\sum p_{kj0} q_{kj0}}{Q_{k0}} \text{ and } \tilde{p}_{kt} = \frac{\sum p_{kjt} q_{kjt}}{\sum q_{kjt}} = \frac{\sum p_{kjt} q_{kjt}}{Q_{kt}},$$

where the summation takes place over $j = 1, 2, \ldots, m_k < n$ goods being a sub-collection of all n goods. Consider now K groups (k = 1, ..., K), each containing m_k commodities such that there are $n = \sum_{k=1}^{K} m_k$ commodities altogether. The groups are usually delimited according to a given classification (e.g. a foreign trade classification). The value index formula on the basis of unit values (instead of prices) is then

$$(6.4.2)\quad V_{0t} = \frac{\sum_k \tilde{p}_{kt} Q_{kt}}{\sum_k \tilde{p}_{k0} Q_{k0}} = \frac{\sum_k^K \sum_j^{m_k} p_{kjt} q_{kjt}}{\sum_k^K \sum_j^{m_k} p_{kj0} q_{kj0}} = \frac{V_t}{V_0}$$

(j = 1, 2, ..., m_k commodities within the k-th group). A Laspeyres type price index based on unit values of groups of commodities, to be called PU_{0t}^L is given by

$$(6.4.3)\quad PU_{0t}^L = \frac{\sum_k \tilde{p}_{kt} Q_{k0}}{\sum_k \tilde{p}_{k0} Q_{k0}} = \frac{\sum_k^K \left(\sum_j^{m_k} \frac{p_{kjt} q_{kjt}}{Q_{kt}} \right) Q_{k0}}{\sum_k^K \sum_j^{m_k} p_{kj0} q_{kj0}} = \frac{\sum_k V_{kt} \frac{Q_{k0}}{Q_{kt}}}{V_0},$$

and the corresponding Paasche index (which is in practice more important than the Laspeyres formula in eq. 3) by

[127] Or "group of goods" henceforth.

[128] As it is well known and introduced already in **sec. 1.1**.

(6.4.4) $$PU_{0t}^{P}=\frac{\sum_k \tilde{p}_{kt}Q_{kt}}{\sum_k \tilde{p}_{k0}Q_{kt}}=\frac{\sum_k^K\sum_j^{m_k} p_{kjt}q_{kjt}}{\sum_k^K\left(\sum_j^{m_k}\frac{p_{kj0}q_{kj0}}{Q_{k0}}\right)Q_{kt}}=\frac{V_t}{\sum_k V_{k0}\frac{Q_{kt}}{Q_{k0}}}.$$

Table 6.4.1: Comparison of true price and unit value (price) indices

	Price index	Unit value index (UV [price[1)]] index)
What is measured?	how prices of a given (fixed) collection of products develop over time	how unit values of all products of a certain type (e.g. all goods exported) in t differ from 0
Merits	pure price comparison by keeping selection constant and making adjustments for quality changes	satisfies "representativity" by inclusion of *all* products (complete coverage [3)] instead of a selection)
Demerits	reliability is said to depend on representativity of the selection [2)]; more demanding regarding price collection, empirical derivation of weights and quality adjustment	influenced by changes in the composition of the products such that a structural change is reflected in the UV (price) index rather than in the quantity (volume) dimension [4)]

1) Note that UVs instead of prices can be used in the compilation of a sort of price index as well as a sort of quantity index, such that there is UV-price index (PU_{0t}) as well as a UV-quantity index (QU_{0t}).
2) and (allegedly) impaired by a lack of representativity (an argument advanced in particular advocates of the chain-index approach).
3) In some countries such exhaustive data files are available, for example in the case of wages, and it will be difficult *not* to make use of them by compiling a UV-wage-index and to prefer a wage-index of the *true* price index type instead (which is also more complicated to compile).
4) A mere switch from cheaper to more expensive products within a group of commodities for which a unit value (UV) is established is reflected in a rise in the UV (and thus in the price dimension which thereby is overstated since prices remained unchanged); using PU_{0t} (instead of P_{0t}) as deflator may therefore overstate price and understate volume change.

Note that unit values appear as building-blocks[129] in numerator and denominator of PU_{0t}^{L} (and PU_{0t}^{P}). They differ from the unit value index of Drobisch[130] P_{0t}^{UD} which is the ratio of universal (all-items) unit values $P_{0t}^{UD}=\frac{\tilde{p}_t}{\tilde{p}_0}$ where $\tilde{p}_t=\frac{\sum_k\sum_j p_{kjt}q_{kjt}}{\sum_k\sum_j q_{kjt}}$ and $\tilde{p}_0$ is defined correspondingly. It is now easy to compare the three formulas

[129] Each of them referring to a *sub-collection* of goods, *not* to *all* n goods.

[130] As introduced in sec. 1.1. The indices PU^L (PU^P) should be kept distinct from P^{UD}. In some recent papers of Balk and Silver we find the unit value index erroneously defined as P^{UD} (rather than PU^P [or less frequently PU^L] which is actually compiled as an index in official statistics).

$$PU_{0t}^{P}=\frac{\sum_k \tilde{p}_{kt}Q_{kt}}{\sum_k \tilde{p}_{k0}Q_{kt}},\quad PU_{0t}^{L}=\frac{\sum_k \tilde{p}_{kt}Q_{k0}}{\sum_k \tilde{p}_{k0}Q_{k0}},\text{ and } P_{0t}^{UD}=\frac{\sum_k \tilde{p}_{kt}Q_{kt}}{\sum_k \tilde{p}_{k0}Q_{k0}}\cdot\frac{\sum_k Q_{k0}}{\sum_k Q_{kt}}\text{ which gives}$$

(6.4.4a) $$P_{0t}^{UD}=PU_{0t}^{P}\frac{QU_{0t}^{L}}{\sum Q_{kt}/\sum Q_{k0}}.$$

So it is not admissible to simply conclude: what applies to P^{UD} also applies to PU^{P} and vice versa. This would require that two measures of the quantities viz. the ratio of summated quantities $\Sigma Q_{kt}/\Sigma Q_{k0}$ and the Laspeyres type unit value quantity index QU^{L} (see below), would be the same, which is unlikely because

$$QU_{0t}^{L}=\sum_k \frac{Q_{kt}}{Q_{k0}}\cdot\frac{\tilde{p}_{k0}Q_{k0}}{\sum_k \tilde{p}_{k0}Q_{k0}}\quad\text{whereas}\quad\frac{\sum_k Q_{kt}}{\sum_k Q_{k0}}=\sum_k \frac{Q_{kt}}{Q_{k0}}\cdot\frac{Q_{k0}}{\sum_k Q_{k0}}.$$

We may conclude, however, that both the two unit value indices PU_{0t}^{P} and PU_{0t}^{L} and Drobisch's index PU_{0t}^{UD} will *not* necessarily meet the mean value condition. To see this[131] we express PU_{0t}^{L} as the weighted mean of price relatives as follows[132]

(6.4.3a) $$PU_{0t}^{L}=\sum_k^K\left(\sum_j^{m_k}\frac{p_{kjt}}{p_{kj0}}p_{kj0}q_{kjt}\right)\frac{Q_{k0}}{Q_{kt}}\Bigg/\sum_k^K\sum_j^{m_k}p_{kj0}q_{kj0},$$

where the denominator is $\sum_k^K\sum_j^{m_k}p_{kj0}q_{kj0}=V_0$. The sum of the weights in eq. 3a is

(6.4.5) $$S=\sum_k^K\left(\sum_j^{m_k}p_{kj0}q_{kjt}\right)\frac{Q_{k0}}{Q_{kt}}=\sum_k^K\left(\sum_j^{m_k}p_{kj0}q_{kjt}\frac{\sum_j^{m_k}q_{kj0}}{\sum_j^{m_k}q_{kjt}}\right)$$

and there is no reason why this sum should necessarily be equal to V_0, the sum of all weights in eq. 3a and given by $V_0=\sum_k^K\left(\sum_j^{m_k}p_{kj0}q_{kjt}\frac{q_{kj0}}{q_{kjt}}\right)$. Hence the value of PU^{L} can well be less than the smallest, or greater than the greatest, individual price relative (and the same is true for PU^{P}). Moreover, unit value indices PU are affected by changes in the composition of quantities within the K sub-collections (groups) and they can indicate a rise (decline) of prices although all prices remained constant[133] (i.e. they can *violate identity*), simply due to changes in quantities. There is no certainty that a change in unit values (and thus a unit value index) really represents a change in prices, let alone in prices only. In sum

---

131 as far as Drobisch's formula is concerned see sec. 1.1.

132 to show this for PU^{P} can be done analogously.

133 In this case all price relatives are unity, and yet unit value indices (such as PU^{L}) do not need to satisfy identity (unless S = 1 in eq. 5).

- unit value indices PU may violate the mean value property, and they therefore do not satisfy proportionality (and by definition they do not satisfy identity either). A change in unit values can well result from structural changes in the quantities, such that PU generally does not reflect a pure price movement.[134]
- The reason for this is aggregation. Unit value indices combine average prices while true price indices aggregate over individual prices. Therefore: the more detailed the product groups are defined (the classification is broken down), that is the more homogeneous the groups, the closer unit value indices PU will come to true price indices P (and QU to true quantity indices Q). The bias of PU measured against P due to an uncontrolled change in the mix of commodities within a group will of course disappear, as groups become more distinguished.[135]

The first observation gave rise to the rather negative view of the SNA:

Unit value indices are "affected by changes in the mix of items as well as by changes in their prices. Unit value indices cannot therefore be expected to provide good measures of average price change over time" (SNA 1993, paragraph 16.13).[136]

Unit value indices PU are sensitive to changes in trade classification. As pointed out already, they can indicate a change even though all prices remained constant, $p_{kjt} = p_{kj0}$ $(\forall k, j)$ because relative quantities m_{kjt} changed (compared to the base period m_{kj0}) since with these assumptions we get

$$(6.4.6) \qquad \tilde{p}_{kt} - \tilde{p}_{k0} = \sum_j p_{kj0}\left(\frac{q_{kjt}}{Q_{kt}} - \frac{q_{kj0}}{Q_{k0}}\right) = \sum_j p_{kj0}\left(m_{kjt} - m_{kj0}\right).$$

Thus an aggregate $\tilde{p}_{kt}$ combining two prices $p_{k1t} = p_{k10}$ and $p_{k2t} = p_{k20}$ respectively may differ from $\tilde{p}_{k0}$ only due to the relative quantities m.

In particular $\tilde{p}_{kt} = p_{k1t}m_{k1t} + p_{k2t}m_{k2t} \neq \tilde{p}_{k0} = p_{k1t}m_{k10} + p_{k2t}m_{k20}$ if for example q_{k1t} rises at the expense of q_{k2t} (such that $m_{k1t} > m_{k10}$ while $m_{k2t} < m_{k20}$).[137] Thus unit values do not reflect price movements only and we may conclude:

As a rule ***PU will understate*** the rise of prices (as compared to a true price index P) if there is a tendency to buy (import) or sell (export) more and more of the cheaper goods (at the expense of the more expensive ones within the same group of goods), that is if the structure (m-coefficients) has changed. Analogously volume (quantity) indices weighted with unit values, to be called ***QU, will overstate*** a rise in quantities compared to a true quantity index Q.

[134] Nor does QU represent a pure quantity movement.

[135] In the limiting case of each group with only one commodity ($m_K = 1$ $\forall k$, $K = n$) and therefore perfectly homogeneous groups, the PU and QU indices and the P and Q indices will be identical of course.

[136] However the SNA apparently (and amazingly) did not realise that the same type of objections can also be raised against chain indices, advocated with great vigor by the SNA.

[137] That is, if there is a shift from one good, or quality, or variant of an item to another (both goods or variants respectively falling into the same k-th subgroup).

Conversely, PU will overstate price movement (and QU understate the change in quantities of export or import respectively) whenever the structure changes in favor of the more expensive (in the base period) goods. This can easily be seen along with the violation of identity ($\tilde{p}_{kt} \neq \tilde{p}_{k0}$ although $p_{ikt} = p_{ik0}, \forall i = 1,\ldots,n_k$) in what follows when the effect of a changing structure of quantities on unit values is examined:

Demonstration

Assume only two commodities comprising a commodity group with prices $p_{10} = p_{1t} = p$ and $p_{20} = p_{2t} = \lambda p$, such that *prices of both goods in fact remain constant.* Furthermore consider shares $\alpha_{10} = \alpha_0$ and $\alpha_{20} = 1 - \alpha_0$ of the total quantity $Q_0 = q_{10} + q_{20}$ (such that $\alpha_0 = q_{10}/Q_0$) in the base period 0, and the shares $\alpha_{1t} = \alpha_t$ and $\alpha_{2t} = 1 - \alpha_t$ of the total quantity Q_t in period t respectively. Or to be more specific we may also assume α_0= ½ and λ = 1.5. The situation is then as follows

good	price in 0	quantity share in 0	price in t	quantity share in 0
1	p	$\alpha_0 = q_{10}/(q_{10} + q_{20})$= ½	p	$\alpha_t = q_{1t}/(q_{1t} + q_{2t})$
2	λp = 1.5p	1 - α_0 = ½	λp = 1.5p	1 - α_t
unit value	$p[\alpha_0 + \lambda(1-\alpha_0)]$ = 1.25p		$p[\alpha_t + \lambda(1-\alpha_t)]$	

The ratio of unit values of this k-th group is then given by

$R = \tilde{p}_t/\tilde{p}_0 = (\alpha_t + \lambda(1-\alpha_t))/(\alpha_0 + \lambda(1-\alpha_0)) = 0.8 \cdot (\alpha_t + 1.5 \cdot (1-\alpha_t)) = 1.2 - 0.4\alpha_t$.

R is by the way identical to the Laspeyres quantity index Q_{0t}^L in this case. This gives the following results:

α_t	0	0.2	0.4	0.5	0.6	0.8	1
R	1.2	1.12	1.04	1	0.96	0.88	0.8

$\alpha_t < 0.5 \rightarrow R > 1$ *	$\alpha_t > 0.5 \rightarrow R < 1$ *

* though all prices in t are the same as in 0!

As α_t > 0.5 (move to the cheaper commodity 1) R declines from 1 to 0.8 (if α_t = 1), indicating thereby a decline in the price level, although all prices remained unchanged. In the same manner: as α_t goes down from 0.5 (R = 1) to 0 then R rises from 1 to 1.2, and thus indicates a rise in the price level change solely due to a change in the structure in favor of the more expensive commodity 2 (the share of which at period t is 1- α_t) .

Figure 6.4.1: The structure of indices on the basis of unit values*

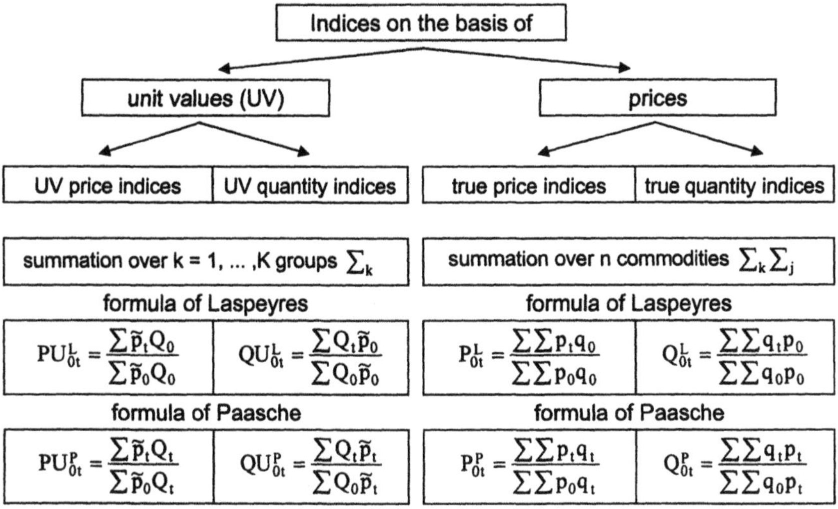

* The universe of n commodities is partitioned into K groups (sub-collections) of related commodities; the subscript k = 1, 2, ..., K denotes the number of the group and the subscript j the j-th commodity within the k-th group.

What has been shown in this demonstration in the case of a single unit value will be generalized to an index (combining a number of unit values) in **ex. 6.4.1**. As indicated in **fig. 6.4.1** in both types of indices, price index as well as quantity index, unit values can play the part of prices, so that a distinction between four indices has to be made. In PU unit values are weighted with quantities just as in QU quantities are weighted with unit values (instead of prices). Therefore the following identities hold

(6.4.7) $$V_{0t} = PU_{0t}^L QU_{0t}^P = PU_{0t}^P QU_{0t}^L = \frac{\sum p_t q_t}{\sum p_0 q_0}$$, in the same manner as

(6.4.7a) $$V_{0t} = P_{0t}^L Q_{0t}^P = P_{0t}^P Q_{0t}^L$$

holds by definition. Hence the same value decomposition as known for price- and quantity-indices holds also for unit value indices. Eq. 7 also explains why PU^P is also used as a deflator (esp. in the case of external trade). The result of deflation using PU_{0t}^P (instead of P_{0t}^P) is QU_{0t}^L, (instead of Q_{0t}^L). The expression QU^L is also known as "*volume index*"[138] and quite common for example in foreign trade statistics.

The different behaviour of PU in contrast to P and QU as opposed to Q is demonstrated also in the following numerical example.

---

[138] The term "volume index" is highly ambiguous, however. In practice the word is used for Q *and* for QU as well. QU^P (instead of QU^L) is uncommon.

Example 6.4.1

Consider a group of commodities denoted by A composed of two commodities 1 and 2 respectively, and a second group B with only one commodity, called 3.

	p_0	p_t	q_0	q_t	p_t/p_0
1 (A)	8	10	5	q	1.25
2 (A)	4	7	5	10-q	1.75
3 (B)	6	9	5	5	1.5

The parameter q enables us to check various changes in the composition of group A and their effects on the UV price index and the UV quantity index respectively. Obviously the following results do not depend on the choice of q:

- the denominator of P^L_{0t}, PU^L_{0t}, QU^L_{0t}, and V_{0t} is $\sum \tilde{p}_0 Q_0 = \sum\sum p_0 q_0 = 90$
- the denominator of PU^P_{0t} is $\sum \tilde{p}_0 Q_t = 6 \cdot 10 + 6 \cdot 5 = 90$
- the numerator of P^L_{0t}, $\Sigma p_t q_0$ (or $\Sigma\Sigma p_t q_0$) amounts to (10+7+9)5 = 130, such that P^L_{0t} = 130/90 = 1.444 irrespective of what is assumed for q.

Furthermore we get $\tilde{p}_{A0} = 6$, $\tilde{p}_{At} = 7 + 0.3q$ and $PU^L_{0t} = (10\tilde{p}_{At} + 45)/90$. A structural change in favor of the more expensive commodity 1, that is, q > 5, will yield $\tilde{p}_{At} >$ 8.5 and therefore $10\tilde{p}_{At} + 45 > 130 \rightarrow PU^L_{0t} > P^L_{0t}$, while q < 5 (change in favor of the cheaper commodity 2)[139] will yield $\tilde{p}_{At} < 8.5$ and therefore $PU^L_{0t} < P^L_{0t}$. The results at a glance are

normal (true) index type		unit value index type	
$P^L_{0t} = \frac{130}{90} = 1.444$	$Q^L_{0t} = \frac{70+4q}{90}$	$PU^L_{0t} = \frac{115+3q}{90}$	$QU^L_{0t} = 1$
$P^P_{0t} = \frac{115+3q}{70+4q}$	$Q^P_{0t} = \frac{115+3q}{130}$	$PU^P_{0t} = PU^L_{0t}$	$QU^P_{0t} = 1$

The reader should verify equations 7 and 7a as giving $V_{0t} = (115+3q)/90$. The identity of PU^L_{0t} and PU^P_{0t} (as well as of QU^L_{0t} and QU^P_{0t}) is only accidental because it was assumed[140] $Q_0 = Q_t = 15$ and $q_{B0} = q_{Bt} = 5$ such that $QU^L_{0t} = QU^P_{0t} = 1$.

Having developed the general formulas it is now easy to try out some values of q to show the influence of various assumptions:

1) q = 2, that is, consumption moving to the less expensive commodity 2[141] yields

[139] Commodity 1(or 2 respectively) is more expensive (less expensive) in 0 as well as in t.

[140] In this example we only have a change in *structure*, but not in the total *amount* consumed. Our aim was to work out the influence of the structure on the results taken in isolation.

prices	$PU^L_{0t} = 121/90 = 1.3444 < P^L_{0t} = 1.4444 < P^P_{0t} = 121/78 = 1.5513$
quantities	$QU^L_{0t} = 1 > Q^P_{0t} = 0.9308 > Q^L_{0t} = 0.8667$

2) q = 8, that is, a change in favor of (a rise in consumption of) the more expensive commodity 1 yields[142]

prices	$PU^L_{0t} = 1.5444 > P^L_{0t} = 1.4444 > P^P_{0t} = 1.3627$ *
quantities	$QU^L_{0t} = 1 < Q^P_{0t} = 1.0692 < Q^L_{0t} = 1.1333$

* Note that P^L is not necessarily an upper bound

As a consequence of $QU^L_{0t} = QU^P_{0t} = 1$ we get $PU^L_{0t} = PU^P_{0t} = V_{0t}$. Note that P^L_{0t} is the same in the case of q = 2 and 8 because this index is not affected by changes in the quantities ($q_{it} \neq q_{i0}$) and substitutions within the groups of goods. The result is fully in line with the considerations set out above already:

> Whenever the *structure* of quantities within a group of commodities *changes in favor* of relatively[143] *less expensive* commodities we get PU < P and QU > Q. Conversely, a change to more expensive goods leads to PU > P and QU < Q.

Hence a structural change *within* the groups of commodities results in an understating of unit value *price* indices PU (as compared with true price indices P) and overstating of unit value *quantity* indices QU (as compared with a true quantity index Q) and vice versa. ♦

It should be noted that the general result q < 5 → PU < P and QU > Q (and q > 5 → PU > P and QU < Q respectively) will also hold when all prices rise by the same rate, say 50% (such that all price relatives are identically 1.5). We then get

normal (true) index type		unit value index type	
$P^L_{0t} = P^P_{0t} = 1.5$	$Q^L_{0t} = Q^P_{0t} = \frac{70+4q}{90}$	$PU^L_{0t} = PU^P_{0t} = \frac{105+6q}{90}$	$QU^L_{0t} = QU^P_{0t} = 1$

and for the situations q = 2 and q = 8 respectively

q = 2	$PU^L_{0t} = 117/90 = 1.3 < P^L_{0t} = 1.5$	$QU^L_{0t} = 1 > Q^L_{0t} = 78/90 = 0.867$
q = 8	$PU^L_{0t} = 153/90 = 1.7 > P^L_{0t} = 1.5$	$QU^L_{0t} = 1 < Q^L_{0t} = 102/90 = 1.1333$

Things will become more complicated, however, once there are several commodity groups (not only one, such as A) with varying types of structural changes. See part b

[141] Commodity 1 is more expensive in 0 and t, but it experienced the *lower rise of price* (price relative 1.25 instead of 1.75) than commodity 2. q < 5 e.g. q = 2 means less of the more expensive commodity (reduced from 5 to 2) and more (8 instead of 5) of the less expensive than commodity 2.

[142] More consumption of the commodity with a lower rise of price implies a negative covariance between price and quantity relatives resulting in P^P being less than P^L instead of $P^P > P^L$ as above.

[143] compared with other commodities in the same group in the base period.

of this section for an attempt to describe these situations in terms of a "Laspeyres-" and a "structural component".

Example 6.4.2

The following modification of ex. 6.4.1 will provide an illustration of the possibility that PU^L does not necessarily satisfy the mean value condition. Again the first two commodities are grouped together to the group A whilst the third commodity forms a single-commodity group, B, on its own:

	p_0	p_t	q_0	q_t	p_t/p_0	q_t/q_0
1 (A)	55	60	3	6	1.0909	2
2 (A)	4	4.8	10	5	1.2	0.5
3 (B)	6	5	9	5	5/6 = 0.833	5/9 = 0.555

A price index should now take a value between 0.833 and 1.2, and a quantity index between 0.5 and 2. The unit values for commodity group A are $\tilde{p}_{A0}$ = (55·3 + 4·10)/13 = 15.769 in period 0 and $\tilde{p}_{At}$ = (60·6 + 4.8·5)/11 = 34.909 in period t respectively, indicating an enormous rise by 121.3%[144], much bigger than 20%, the largest individual price relative. The reason is because the quantity redoubled ($q_t = 2q_o$) in the case of the much higher priced commodity no. 1, whereas the quantity of the lower priced commodity (no. 2) is now only half of what it was in the base period (q_{Bt} = 5 whilst q_{B0} = 10).[145]

The relation between the rise of unit values $\tilde{p}_{kt}/\tilde{p}_{k0}$ as a function of the rise of the two[146] price relatives of goods belonging to k-th group is given by

$$(6.4.8)\qquad \frac{\tilde{p}_{kt}}{\tilde{p}_{k0}} = \frac{p_{k1t}}{p_{k10}}\left(\frac{p_{k10}}{\tilde{p}_{k0}}\cdot\frac{q_{k1t}}{Q_{kt}}\right) + \frac{p_{k2t}}{p_{k20}}\left(\frac{p_{k20}}{\tilde{p}_{k0}}\cdot\frac{q_{k2t}}{Q_{kt}}\right)$$

Note that the weights (terms in brackets) do not add up to unity.[147]

In the present example, $\tilde{p}_{k0}$ = 15.769 is a function of two very different prices, p_{k10} = 55 and p_{k20} = 4, which gives the first price relative (1.0909) an extremely large weight amounting to[148] $\frac{p_{10}}{\tilde{p}_0}\cdot\frac{q_{1t}}{Q_k} = \frac{55}{15.77}\cdot\frac{6}{11}$ = 1.902 such that the ratio $\frac{\tilde{p}_t}{\tilde{p}_0} = \frac{34.91}{15.77}$ $= 2.21$ is well beyond the upper boundary of the span between the individual price relatives (which is p_{2t}/p_{20} = 1.2). The same is true for the Laspeyres-type unit value

[144] Note that this is *much* more than the highest individual price relative which is 1.2 (thus +20% instead of the above mentioned + 121.3%) for commodity 2.

[145] Assume the opposite reaction- the quantity of 1 is halved, and of 2 is redoubled. The result would be 8.651/17.769 for the ratio of unit values, indicating a decline of 45% although both prices went up.

[146] Obviously eq. 8 holds for any n_k, not just n_k = 2.

[147] Their sum is in the example 1.902 + 0.115 = 2.0177 ≠ 1.

[148] The subscript k (or A) to denote the commodity group is deleted for convenience of the presentation.

price index $PU^L_{0t} = \frac{34.909 \cdot 13 + 5 \cdot 9}{15.769 \cdot 13 + 6 \cdot 9} = \frac{498.82}{259} = 1.926$, displaying a rise by 92.6% (1.926 > 1.2). To summarize the results:

unit value price index*	true price index
$PU^L_{0t} = 498.82/259 = 1.926$	$P^L_{0t} = 273/259 = 1.054$
$PU^P_{0t} = 409/203.46 = 2.010$	$P^P_{0t} = 409/380 = 1.076$

unit value quantity index	true quantity index
$QU^L_{0t} = 203.46/259 = 0.7856$	$Q^L_{0t} = 380/259 = 1.467$
$QU^P_{0t} = 409/498.82 = 0.8199$	$Q^P_{0t} = 409/273 = 1.498$

* Both unit value price indices (PU^L and PU^P) violate the mean value condition. However, QU^L and QU^P do not violate the mean value condition applied to quantity relatives. Moreover, it is easy to verify that in fact $PU^L\,QU^P = PU^P\,QU^L = P^L\,Q^P = P^P\,Q^L = V_{0t} =$ 409/259 = 1.579. ♦

Hence price indices on the basis of unit values dramatically overrate the rise of prices as compared with true price indices P^L and P^P (moreover, the rise of both PU^L and PU^P exceeds the highest individual price relative, 1.2) and the opposite is true for quantity indices on the basis of unit values (both QU^L and QU^P show a decline in quantities though Q^L and Q^P indicate a rise of 47 and 50% respectively).[149]

b) Difference between unit value indices and price indices in foreign trade statistics

In many countries unit value indices are used as a proxy of price indices, especially in foreign trade statistics. It is this field of official statistics where the relevant data for compiling unit values are mostly readily available (due to customs statistics). Thus unit value indices are often by-products of external trade statistics. The German foreign trade statistics use indices of the following types: V_{0t}, PU^P_{0t}, and as a by-product QU^L_{0t}, in addition to true (direct, not chained) price indices of the Laspeyres type, both for exports as well as for imports. It should be noted, however, that PU^P_{0t} and P^L_{0t} do not only differ with respect to the formula but also as far as the prices and some other aspects are concerned. Prices entering the unit value index formulas PU^P_{0t} and QU^L_{0t} refer to the time of passing the border whereas prices in the true price index are recorded at the time of contracting.

[149] All these results are obviously a consequence of the fact that within the aggregate (commodity group A) the structure of quantities changed in favor of the expensive good 1 such that $q_{1t} = 6 > q_{10} = 3$ whereas $q_{2t} = 5 < q_{20} = 10$.

Furthermore in the establishment based (gained by surveying establishments) true price index P^L_{0t} a quality adjustment of the prices is made, whereas foreign trade unit value indices abstains from it, such that prices in PU^P_{0t} are not quality adjusted.[150] This explains to a great deal the greater volatility of (customs based) unit value indices compared to true price indices.

In sum, it may be stated and has been shown that

True price indices are more in line with pure comparison (like with like), whereas unit value (UV) indices are greatly contaminated by a number of non-price influences. They are more volatile than true indices, P and Q, there is no long-rung (cointegrating) relation between UV-indices and true indices, and the former are in general not helpful in predicting the latter.

Hence UV indices can seriously misrepresent price and quantity changes, and discrepancies between them and true indices are substantial. The main reason for not replacing UV indices by true indices in many countries is resource restraint. True indices are more costly to compile.

In order to establish a relationship explaining the discrepancy between PU^P_{0t} and P^L_{0t} it is useful to make a recourse to eq. 6.4.7 as well as to the famous equation of L. v. Bortkiewicz, according to which the covariance C between price and quantity relatives is given by $C = Q^L_{0t}\left(P^P_{0t} - P^L_{0t}\right)$. This leads to

$$\text{(6.4.9)} \qquad D = \frac{PU^P_{0t}}{P^L_{0t}} - 1 = \left(\frac{C}{QU^L_{0t}P^L_{0t}}\right) + \left(\frac{Q^L_{0t}}{QU^L_{0t}} - 1\right) = L + S.$$

The term L is referred to as the Laspeyres- or simply ***L-effect***. A negative covariance may arise from rational substitution among goods in response to price changes on a given (negatively sloped) demand curve. A second component of the discrepancy comes into play, which may well reinforce but also counteract the L-effect. This factor is called the structural component, or ***S-effect*** for short, and refers to changing quantities within a group of goods k = 1,...,K (for which unit values are established).

In order to understand the meaning of S, recall that $\tilde{p}_{ko} = \sum_j \left(\frac{q_{jko}}{Q_{ko}}\right) \cdot p_{kjo} =$ $\sum m_{kjo} p_{kjo}$ such that the difference between QU^L_{0t} and Q^L_{0t}, which determines S, is simply

$QU^L_{0t} = \sum_k Q_{kt} \sum_j m_{jko} p_{jko} \big/ \sum_k \sum_j q_{jko} p_{jko}$ as opposed to

$Q^L_{0t} = \sum_k Q_{kt} \sum_j m_{jkt} p_{jko} \big/ \sum_k \sum_j q_{jko} p_{jko}$ [151].

---

150 They are therefore subject to changes in the mix of products as regards qualities, makes, emerging and disappearing goods and countries involved. There is no adjustment for these factors impairing comparability. Extreme outlayers are discarded, however.

151 Note $Q^P = Q^L$ implies $P^P = P^L$ but equality $QU^L = Q^L$ is compatible with $PU^P \neq P^L$. In this case D is simply determined by L only.

Hence whenever $m_{jkt} = m_{jk0}$ (for all j and k), that is, the structure of quantities within the groups of commodities did not change[152] from 0 to t, we would get $QU_{0t}^{L} = Q_{0t}^{L}$ and consequently also S = 0. So S is rightly called the "structural" component because it reflects a structural change. This change, S, is responsible for possibly different signs of D and L, that is

although $C < 0 \rightarrow L < 0$ and $P_{0t}^{P} < P_{0t}^{L}$, it is possible that $PU_{0t}^{P} > P_{0t}^{L}$

and correspondingly $P_{0t}^{P} > P_{0t}^{L}$ is compatible with $PU_{0t}^{P} < P_{0t}^{L}$.[153]

A negative L-effect ($P_{0t}^{P} < P_{0t}^{L}$) may well be offset or outstripped by a positive S and vice versa. The problem of eq. 9 is, however, that we are lacking data as there is no Q_{0t}^{L} in addition to a QU_{0t}^{L} index compiled in practice and we have no knowledge about the covariance C. It is therefore not possible to estimate L and S empirically.

c) Indices of wages and salaries as price indices or unit value indices

In general the objective of price statistics is the regular reporting of prices of goods, that is, commodities and services, but not of prices of an hour worked, i.e. of wages paid.[154] Hence index numbers for wages and salaries are not price indices in the strict sense. On the other hand, the advantages of representing data in terms of index numbers rather than of simply calculating average wages can be made clear particularly easily.[155] It rests primarily on the fact that an index, in contrast to an (absolute) average, is independent of changes in the structure, or in other words, an index provides a *standardized* comparison.

It is a very common impression that a wage index does not offer any peculiarities compared with an "ordinary" price index.[156] The truth, however, is that indices for wages and salaries are closer related to unit value indices than to genuine price indices. Wages and salaries are not only paid for a certain quantum of homogeneous work performed but also include payments for preparatory and evaluatory work, for standby times, or for social purposes.[157] Thus the problem with true price indices in the case of wages is:

---

152 or equivalently: if each group of goods would consist of $m_k = 1$ good only.

153 Here we have L > 0, but D < 0 because S < 0.

154 The price paid for a service offered by an entrepreneur and the wage paid to an employee for a certain amount of hours worked have of course much in common, however, especially if the enterprise is small and if it comprises only very few employees or even only one single own account worker. Furthermore index theory has a long tradition, especially in the field of wage statistics.

155 The advantage of a wage-index instead of ratios of average wages has been shown in sec. 1.1.

156 The reason is that in compiling an index for wages the same principles of selection and "pure" comparison that govern price statistics are relevant in wage statistics as well. It is therefore suggested to conceive a wage index simply as a specialized version of a price index.

157 In general they also comprise remunerations depending on individual conditions such as family status, qualification, duration of employment, distance between home and business and the like.

It will be extremely difficult (more difficult than in the case of goods) to define "quantities" to which the so-defined "wage" applies, to compile data for a "basket" of selected and strictly comparable types of work, and to keep this basket constant over time. The methodology of wage indices therefore cannot be viewed as a perfect analogy to price indices.

In practice, the basket is defined by qualification groups and age of the employees. The task of the statistician is then to select a sample of wage groups that may be regarded as representative. In doing so the statistician will in general make use of so-called "wage groups" to which also the wage agreements between employers and trade unions refer. Such groups are, however, not sufficiently (for a genuine price index) homogeneous with respect to qualification, working conditions etc.

Thus there are a number of aspects which make wage indices more similar to unit value indices rather than to true price indices:

- the problem of defining weights and heterogeneity of wage groups,
- the raw material of data often consisting in wage sums $L_{kt} = \sum p_{kjt} q_{kjt}$ and numbers of employees (as weights), $Q_{kt} = \sum q_{kjt}$, allowing for the calculation of $\tilde{p}_{kt} = L_{kt}/Q_{kt}$, and $\tilde{p}_{k0}$ correspondingly, that is, of unit values.[158]

The wage sum has a price and a quantity element and is affected by shifts between tariff groups.[159] A "true" index of wages or labor costs of the price index type, however, should be influenced by the price component only.[160] As stated above, a unit value index may approach a true index to the extent that the groups of goods become more and more homogenous (and small as regards m_k). This holds true again for the relation between UV-wage indices and true wage indices. To the extent to which wage groups in wage indices are more broadly or narrowly defined (and therefore more or less heterogeneous), and variable or constant in the course of time, a wage index will resemble more a unit value index or more a true price index respectively.[161] This again confirms the result that there is no clear cut distinction between a *price* (or true wage) index on the one hand and a *unit value* index on the other. It all depends on the degree of disaggregation of the "population" in question through which prices, or wages and weights are defined.

---

158 In situations in which wages (for example for services of lawyers or consultants) are calculated on the basis of well specified hourly fees we have a price index approach, while in situations in which only the turnover is known that can be divided by the number of hours worked, we rather follow the UV-index methodology.

159 If a set of employees is shifted from one tariff group to another the process is only a structural change, but it has an impact on the wage sums L_{k0} and L_{kt}.

160 To avoid ambiguity due to structural changes a true wage index should refer to wages, paid to *individuals* (rather than to averages referring to wage groups), taking into account individual conditions and characteristics (such as qualification), and a weighting scheme broken down to individuals, that is, the most "homogeneous" units are needed, which of course is beyond the scope of wage statistics.

161 Along these lines, the difference between U.S. Employment Cost Index (ECI) and the European (European Union) Labor Cost Index (LCI) should be seen. The ECI comes closer to a true "price" index, whereas the LCI (to be delivered by the EU member states as of January 2007) is less ambiguous and resembles more a unit value index. In several respects the LCI method lags behind the US-ECI, an example it originally wanted to follow if they had not shied at the higher cost.

Chapter 7

Chain indices

In this chapter an abbreviated account of some parts of my book on chain indices will be given. As I became more and more a non-chainer after having examined the arguments advanced by chainers,[1] it should be useful to consider first the reasons given for the superiority chain indices purportedly have over direct indices (**sec. 7.1**) along with a possible rebuttal of such arguments. It then remains for us to give a survey of the most important properties of chain indices (**sec. 7.2**), on which agreement or disagreement of this index approach is grounded. It is assumed that the presentation in **sec. 2.5** and the definitions, notations etc. introduced there are familiar to the reader.

7.1 Arguments in favor of the chain index approach

We first comment on some general observations concerning the rivalry of arguments followed by a survey of (alleged) advantages of chain indices compared to traditional (direct) index formulas, discussing each argument one by one.

a) Overview of arguments for chain indices and some general observations concerning the arguments of chainers

It is difficult to find a system for the arguments brought into play in favor of chaining. Nonetheless, an attempt is made in what follows. In principle there are five groups of arguments to support the idea of chaining (see **fig. 7.1.1**):

A links These reasons given for the pretended superiority of chain indices focus on links rather than on the chain and claim an advantage of chain indices arising from the simple fact that the interval (0, t) is subdivided into a number of sub-intervals and the index is derived from multiplication of links,

B base arguments stating that the "base" to which a time series of indices or of year-to-year growth rates refers is more relevant and realisitic in the case of chain indices than in the case of traditional direct indices,

C weights here (alleged) advantages are derived from a superior flexiblity and adaptability as regards the structure of weights and the appearance of new and disappearance of old goods (these are considerations endorsed by lots of people even more if, as usual, combined with the assertion that chain indices better fit to our modern times),

[1] On the face of it these arguments appear plausible. However, it turns out – in our view at least - that the praise of chain indices is largely built on vagueness and inconsistencies.

D results the arguments under this heading concern the likelihood of certain desirable results of chain indices as opposed to direct indices, the Laspeyres-index in particular, when applied to empirical data,

E deflation arguments are stressed, above all in the SNA recommendations, relating to alleged superiorities of chain indices when used for deflation.[2]

This system allows us to give some hints to the critical notions, referred to in the arguments in question, already:

A careful examination[3] shows that most, if not all, of the arguments known to us are simply built on some confusion with notions like "chainability", "base" etc., and on some mysticism built on the simple fact that links are multiplied to form a chain.

Before going into details of the arguments and their rebuttals it is worthwhile making note of some general observations. In reviewing the arguments of chainers it is eye-catching that

1. the justification of chain indices is not theory-driven, not only is there no "new" conceptual foundation designed esp. for chain indices (as for example the COLI-approach[4] is "new" compared with the "stochastic approach"), but we also encounter in no small measure inconsistency and onesidedness, and
2. "advantages" of chain indices are mainly derived from a critique of the *fixed basket* (direct *Laspeyres*) approach P_{0t}^{L}, they do not apply to certain "superlative indices" like P_{0t}^{F} which unlike P_{0t}^{L} do make use of new weights (q_{it}) and not only of old ones (q_{i0}).

   Some arguments[5] even seem to have lost of sight what primarily has been the concern, viz. to get certain (alleged) defects of the fixed basket Laspeyres approach remedied, and finally
3. it sometimes appears to have been overlooked that problems purportedly "solved" by chain indices are not really solved but rather "dissolved": This applies for example to the argument of less problems with quality adjustments and changes in the universe of goods. Once the aim of making "pure" price com-

---

[2] Ironically it is precisely in this context of deflation where disadvantages of chaining abound (see **sec. 5.4**).

[3] As mentioned already, it was primarily this examination of the chain index rhetoric which gave rise to my conversion to a "non-chainer".

[4] That is, the so called "economic theory approach" consisting of comparing expenditures necessary for maintaining the same *utility* level (rather than buying the same basket-*quantities*). Chain indices do not add a new index theory: they are based on the same basket-concept as P^L, the main target of its attacks, modifying it only by the simple idea that it is primarily the updating (instead of the constancy) of the basket that matters. It is sometimes maintained that the Divisia index is capable of providing such a theoretic foundation for chaining. However, this seems to be rather delusive, see **sec. 2.5**. Moreover, I devoted a whole chapter (ch. 4) in my chain index book to the search for a "theory" – if there is any – underlying the chain index approach.

[5] In particular those advocating chain indices for short run comparisons (like **A2**).

parisons has been given up there is not much left of the problem of quality adjustment.

Figure 7.1.1: Twelve arguments in favor of chain indices, an overview

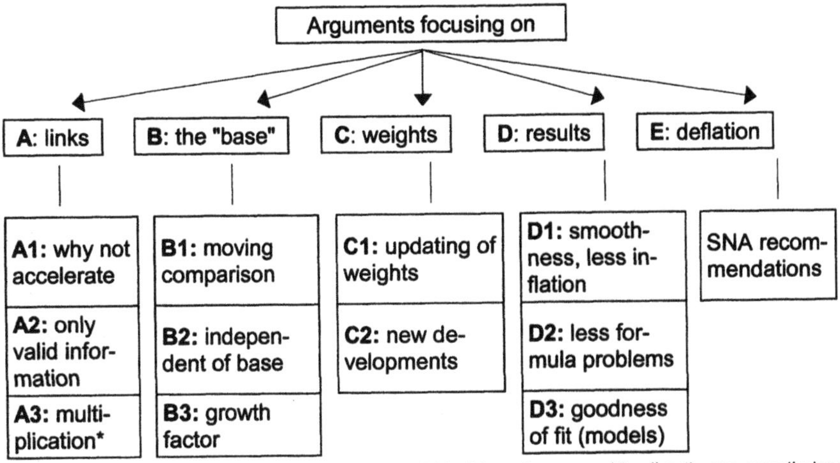

* i.e. the "multiplication mystery", according to which things incomparable directly are nonetheless comparable indirectly and (related to this argument and to B1) some valuable additional information is gained by making such indirect comparisons rather than direct ones.

The most striking example for *inconsistent* and *inconclusive* statements is the standpoint the SNA (rev. 1993) took with respect to unit value indices as opposed to chain indices (as quoted already in sec. 2.5): Unit value indices were rightly rejected by the SNA as being

> "affected by changes in the mix of items as well as by changes in their prices. Unit value indices therefore cannot be expected to provide good measures of average price change over time" (paragraph 16.13).

Most amazingly nobody seemed to have realized that exactly the same would also apply to chain indices, advocated, however, with great vigor by the very same SNA.

A most fundamental inconsistency is addressed in the uniqueness-theorem of Funke et al. concerning the Cobb-Douglas index, and according to which the ideas of chainability and the constant adjustment (up-dating) of weights are incompatible, in the sense that there is no index formula known with variable weights which simultaneously passes the circular test.[6]

[6] See **sec. 7.2.j** for this theorem. Even so, the combination of both ideas certainly has some intuitive appeal. This inconsistency also sheds some light on the multiplication mystery (what is directly incomparable is nonetheless comparable indirectly), which is essentially built on the same combination of two ideas (chaining and variable baskets), which are clearly not reconcilable. Thus this most popular argument cannot possibly hold water either.

Onesidedness and playing down of counter-arguments can be seen by the fact that as good as none of the known disadvantages is mentioned in the parade of arguments above (**fig. 7.1.1**):

- no mention is given, for example, to problems with consistency in aggregation (over commodities), that is, to the decomposition of the index formula in order to distinguish different sources of variation, to path dependance, or to additivity in volumes, and
- not much attention, if any, is given to problems of data collection, let alone the increased cost official statistics has to incur[7] when P_{0t}^{L} is to be replaced by $\overline{P}_{0t}^{LC}$.

In view of the undoubted increase in computation difficulties and costs there should be some *significant* improvements to justify a move from fixed base direct indices to chain indices, and the latter should be advantageous especially in those cases in which direct indices P_{0t}^{L} fail, notably

- when comparisons over *long* intervals in time rather than short ones are wanted, and
- consumption patterns change *rapidly* and *fundamentally* rather than smoothly in response to changes in prices.[8]

Thus, to be worth the increased burden and expense of compilation, chain indices should guarantee a substantial progress to direct indices *in exactly these fields*, where direct indices face their specific weaknesses and (allegedly) intolerable disadvantages. But surprisingly, many proponents of chain indices seem to have forgotten what originally was under discussion (i.e. the keeping constant of a basket over a *long* interval in time that matters such that it is the continuous updating of the basket that is needed).[9]

In view of this it is not fair to claim superiority of the chain approach over the direct approach when the interval between 0 and t is *short*, and weights have not changed dramatically,[10] let alone that the idea that updating weights does not necessarily require a chain index; that a "superlative index" (like Fisher P^F or Törnqvist P^T) in which

---

[7] Interestingly and most regrettable: there is little or no "cost-benefit-analysis" made in order to show whether a move from a direct Laspeyres index to a chain index will be worth the indisputedly higher cost of updating the weights in much shorter intervals in time, a price for the alleged "overwhelming advantages" (as Forsyth and Fowler 1981, p. 230 found appropriate to put it) of chain indices.

[8] In the case of slow changes of minor importance we can hardly blame the direct index approach for making a comparison on the basis of an *irrelevant* historical basket.

[9] The most prominent representative of this logic is the "only-valid" argument **A2**, stating that chaining is primarily good for the short run. By the same token Stuvel made objections to chain indices for *long* term comparisons, but in the case of *short* time series they "might be preferable, since the direct comparisons in such series are between adjacent years" (Stuvel 1989, p. 73).

[10] Note for example that also arguments like **D1** and **D2** actually require conditions under which the specific defects for which the direct Laspeyres index is blamed are less likely.

the current year weights q_{it} also enter (in addition to q_{i0}) will serve the same purpose.[11]

The example of indices P^F and P^T (taking weights of 0 and t symmetrically into account) shows that from arguments focusing mainly on the timely and constant renewal of weights it does not follow that chain indices should take precedence over *all* kinds of direct indices. Stressing this point is at best capable of establishing superiority over P^L.

Thus: demonstrating the importance of the up-to-dateness of weights is not tantmount to "proving" a *general* desirability of chaining.

As stated above (**sec. 2.5**) most of the chain index rhetoric ensues from a critique of P_{0t}^L and is not applicable to these "superlative" direct indices. Other arguments for $\overline{P}_{0t}^{LC}$ owe their appeal directly to the comparison with P_{0t}^L, in that they explicitly pretend to have solved problems we usually have to face primarily in the P_{0t}^L approach only (such that the problems are rather "dissolved"). To this group of "solution vs. dissolution" arguments belong in particular

- the pretension a chain index has solved the problem of *choosing a base*, and
- the alleged solution of problems with making *quality adjustments.*[12] However

In the chain index approach "base" has another meaning, such that there is no longer anything which makes the choice of the base period a problem. Furthermore, once the fixed basket (constant over time) is abandoned, there is no longer any need for comparability over more than just two adjacent periods, t-1 and t, and hence there is no problem in making adjustments for changes in quality or other price determining characteristics.

We now turn to the five groups of arguments listed in **fig. 7.1.1** and discuss them one by one in detail.

b) Arguments of group A: Focus on links rather than on the chain

The common trait of the following arguments (and some related arguments as well, e.g. **B3**) consists of disregarding the existence of *two* elements in defining a chain index, the link and the chain. Many statements pretending advantages of chain indices over direct indices in fact focus on the links rather than the chain, and they compare P_t^C instead of $\overline{P}_{0t}^C$ with the direct index P_{0t}, or they consist of giving too wide an interpretation of the fact that $\overline{P}_{0t}^C$ is derived by multiplying links, if not in a mystification of this multiplication.

---

[11] The Boskin Commission for example (see **sec. 1.3**) which stressed the importance of a rapid and constant updating of weights no less, did *not* decidedly recommend chain indices but rather (direct) "superlative" indices, like P^F or P^T, with weights from both periods, 0 and t respectively.

[12] This remark refers to arguments **B2** and **C2** respectively. They are indebted to a great deal to some vagueness of the term "solution".

A1: The "why not" or "limiting case" argument

1. The argument itself

One of the most widespread and intuitively appealing arguments goes as follows: The chain approach is simply a "fixed base" index rebased more frequently. Instead of rebasing at five year intervals or so, updating is simply done at shorter intervals, viz. every year. Thus the chain index is the "limiting value" or "borderline case" to which a direct index tends: "Chaining is merely the limiting case where the base is changed each period" (PPI/CPI Manual). Or as Allen 1975, p. 177 put it: "why not accelerate and go for annual chaining? There is no reason why not." We call this most insinuating and enticing argument simply for short the *"why not"* – argument.[13]

On this basis it is more often than not argued that a chain index is (compared with the direct index) simply the more general approach. In this sense the SNA (para 16.77) states for example:

> "In effect, the underlying issue is not whether to chain or not but how often to rebase. Sooner or later the base year for fixed weight Laspeyres ... indices ... has to be updated because the prices of the base year become increasingly irrelevant ... Long runs of data therefore almost inevitably involve some form of chain indices. Annual chaining is simply the limiting case in which rebasing is carried out each year instead of every five or ten years."
>
> So why not accelerate?

2. Rebuff of argument A1

There are basically two drawbacks of this argument, firstly a generalization of index designs is made such that the direct index approach is placed in a bad light, and secondly a direct index linked together still differs from a chain index by degrees, if not fundamentally.

> Argument **A1**, esp. as expressed in the SNA quotation above, is a misinterpretation of the guiding principle of the fixed-basket-approach: what is primarily wanted in this approach is comparability *within* an interval, not so much the long term comparison across a number of such intervals. In **A1** however it is tacitly assumed that the chaining of links was the general rule in index compilation, while keeping a basket constant was only the exception.

1. As pointed out repeatedly the guiding principle of the fixed-basket-approach is comparability within an interval rather than across intervals. The primary concern is keeping constant what might disturb comparability, not chaining to arrive at a comparison of two periods very far away. Constancy is the rule, chaining the exception. The latter applies to all direct indices, not only to the fixed-basket-approach. Direct indices referring to different base periods will generally

[13] It is our intention to show that there are indeed *several* good reasons "why not", and to work them out in detail.

not be multiplied. In contrast, in the chain approach links are necessarily chained together in order to derive an index.

What was an exception is now the rule, and what used to be the rule (keeping things constant for making "pure" comparisons) is now abandoned completely. Thus, bluntly speaking, argument A1 is a misinterpretation of the direct index approach and its intentions.

2. Even if chaining of direct indices were done, for example a new base defined at t = 5 and the direct indices at base 0 were chained to those at base 5, there would still be a remarkable difference: $\overline{P}_{09}^{LC}$ is the product of nine factors, using information given in no less than ten sets of prices and nine sets of quantities, such that altogether 19 vectors affect the result $\overline{P}_{09}^{LC} = \frac{\sum p_1 q_0}{\sum p_0 q_0} \frac{\sum p_2 q_1}{\sum p_1 q_1} \cdots \frac{\sum p_8 q_7}{\sum p_7 q_7} \frac{\sum p_9 q_8}{\sum p_8 q_8}$, while $P_{09}^{L*} = P_{05}^{L} P_{59}^{L}$ is the result of chaining two direct indices, and hence the product of *two* factors only. P_{09}^{L*} is influenced by five vectors $\mathbf{p}_0$, $\mathbf{p}_5$, $\mathbf{p}_9$, $\mathbf{q}_0$ and $\mathbf{q}_5$ only. The result is worth being summarized as follows:

The index	is affected by	
$P_{09}^{L*} = P_{05}^{L} P_{59}^{L}$ 1)	prices $\mathbf{p}_0$, $\mathbf{p}_5$, $\mathbf{p}_9$ (3 vectors)	quantites $\mathbf{q}_0$, $\mathbf{q}_5$ (2 vectors)
$\overline{P}_{09}^{LC}$ 2)	$\mathbf{p}_0$, ..., $\mathbf{p}_9$ (10 price vectors)	$\mathbf{q}_0$, ..., $\mathbf{q}_8$ (9 quant. vectors)

1) two direct indices with base periods 0 and 5 linked together
2) chain index annual links

There are 14 more vectors entering the formula of $\overline{P}_{09}^{LC}$. Different data for $\mathbf{p}_1$, ... , $\mathbf{p}_4$, $\mathbf{p}_6$, ... , $\mathbf{p}_8$ as well as for certain quantity vectors will yield different results for $\overline{P}_{09}^{LC}$. But they will not affect P_{09}^{L*}. Finally, there is an interesting exception to the why-not-logic:

For adherents of chaining who find the idea convincing that annual chaining is better than chaining at five year intervals, it should be surprising that monthly chaining is *not* better than annual chaining, and weekly chaining not better than monthly chaining.

By the way, it is again the base in the *link* P_t^{LC} and not of the chain that is updated every year, and where an updating might be accelerated. The base of the *chain* $\overline{P}_{0t}^{LC}$ is always 0, as it is in P_{0t}^{L}; it is not updated more frequently.

A2: The "only valid information" argument (accuracy issue)

1. The argument itself

According to this argument (put forward by Mudgett in particular) the chain index gives the only valid obtainable information, the direction of *change* from year to year, whereas the fixed based index concentrates on more doubtful information, like the actual *level* attained in terms of the base period level (doubtful especially when the base period is years ago).[14]

2. Rebuff of argument A2

The prolem of a chain is that a weak link is able to weaken the whole chain. Thus, for example, errors in the compilation of one particular link will have consequences for all successive values of the chain. Each value of a direct index on the other hand is an independent estimate on its own. Moreover, the logic of this argument does not seem to be compelling.

If this "only valid information" argument were correct we should rather refrain from multiplying links to a chain (an operation according to **A1** characteristic for index calculations of all sorts).[15] Given that links were in fact the only meaningful measures, whereas P_{0t} is not valid simply because of 0 being too distant from t, why then should $\overline{P}_{0t}^{LC}$ (unlike the links P_1^{LC}, P_2^{LC},..., the product of which is $\overline{P}_{0t}^{LC}$) be able to provide a valid information (and not P_{0t}^{L})?

In some sense or other with this argument the same problem arises as with the multiplication mystery and Martini's theory of indirect comparison[16] on the strength of which we are allegedly able to compare validly *indirectly* (by chaining) only something that is on the other hand totally incomparable *directly*. There is also some resemblence with argument **A3**, suggesting that the mere subdivision of the interval is able to do wonders.

A3: The multiplication mystery

1. The argument itself

In some arguments the trivial operation of multiplying links is regarded as an advantage as such. Banerjee 1975, p. 48 for example wrote: "The chain is thus the resultant of a series of comparisons, each between two consecutive periods. The chain method thus eliminates the limitations involved in the comparison between two distant periods".

---

[15] Hence in a sense from **A2** ensues that long term comparisons should better *not* be made while **A1** states that exactly this is the essence of all sorts of index calculations.

[16] see **sec. 4.1** of my book on chain indices where I dealt with the argument of Marco Martini.

It is immediately apparent that much of the persuasive power of this argument is owed to the vagueness of the term "resultant". On a level with this idea is also the (unproved) assumption that chaining results in a smaller total *error* in computations using data each liable to errors, or in a smaller deviation in terms of the factor reversal test or the time reversal test.

The last mentioned criterion is sometimes referred to as "formula error". To quote once more Banerjee 1975, p. 59: "if the comparison is made between two consecutive periods, the formula error should be the least. This incidentially, is a justification for the use of the chain index"[17]. Interestingly this argument is again (like **A2**) very much stressed by Bruce Mudgett, one of the most convinced proponents of chaining ever. In his view, "the rule that comparisons of price and quantity change should be made between consecutive periods or areas" (Mudgett 1951, p. 57)[18] also helps to improve what he calls "formula consistency", i.e. to have a small difference between P_{0t}^{L} and P_{0t}^{P} (see also **D2**), and "homogeneity", i.e. more common commodities when two *adjacent* periods are compared (Mudgett, 1951, p. 43, 48, see also under **C2**). According to Mudgett there are two sources of "advantages" of chain indices, both in terms of "accuracy" and "validity": chain indices are superior as they

- make comparisons exclusively between two *consecutive* periods (concentrating on links), that is by focusing on *short* period comparisons, or on the "only valid information", and *at the same time* as they
- provide "a valid procedure" for making comparisons over *long* series or distant areas by multiplication of those links (Mudgett 1951, p. 58).[19]

To sum up:

It appears inconsistent to take the link for the whole chain, and to ignore the aspect of multiplying links, as done in argument **A2** (and in a way also in **A1**), and at the same time (as argued especially under **A3**) to derive anew some "advantages" of the chain approach from the simple fact that links are multiplied to form a chain.

Therefore

We call argument **A3**, the "multiplication *mystery*" argument, because it consists of deducing some far-reaching and almost marvelous consequences from the simple fact that the interval (0, t) is subdivided, and links considered for the subintervals are multiplied.

---

[17] Note that Banerjee, like Mudgett 1951, defines the extent to which an index deviates from the time reversal, factor reversal, or circular test as various types of "formula error". It is clear that in general the "error" will be smaller the shorter the interval over which a comparison is made. See also the discussion of argument **B1**.

[18] Mudgett not only expressly treats intertemporal and interspatial comparisons equally (in contrast to "consecutive periods" it is not clear what could be meant by "consecutive areas" if this term would be used in a perfect [in our view inappropriate though] analogy between intertemporal and interspatial considerations), it is also the emphasis he placed on Fisher's "reversal" ideas with which we most decidedly disagree.

[19] More than that, Mudgett even claims that it is sometimes the only possible procedure (as also Martini argues), or at least a more accurate procedure.

The alleged advantages consist of: more accuracy (**A3**), and – closely related to this argument – in making long distance comparisons possible which are otherwise (i.e. directly) impossible, in utilizing the information represented in the time series data in a more efficient way, and thereby also in providing valuable "additional information" (argument **B1**), the contents of which remains to be made clear, however.

2. Rebuff of argument A3

The idea of **A3** (and implicitly in related arguments, like **B1**, **C1** and **C2** as well) is often found highly convincing and also the key to Martini's thinking: chain indices permit long distance comparisons by comparing any two consecutive periods only, and aggregating over such short distance comparisons. Thus in some mysterious manner, what is incomparable directly becomes comparable indirectly. The solution of this obvious contradiction is of course that different notions of "comparability" are used. We will come back to this thinking under **B1**.

The idea[20] that chain indices are capable of making otherwise incomparable things comparable, by linking partial comparisons with a certain overlap (made intuitively clear with the help of the following picture[21] (**fig. 7.1.2**)) is flatly wrong.

Figure 7.1.2: Successive comparisons of partially overlapping circles

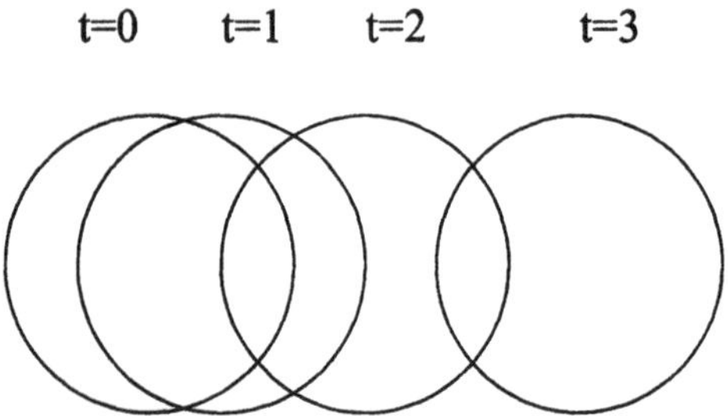

Note that by making indirect comparisons we don't remove a single impediment which might prevent us from making a direct comparison (such as what Martini calls logical requirements).[22]

[20] Expressed, among others, most vigorously by Marco Martini.

[21] The message of the picture is simply this: The first circle on the left hand side and the last on the right hand side have virtually no overlap at all (so they may be called "incomparable"). But as linking proceeds from left to right they are – in the view of "chainers" at least - made comparable because each link (pair of circles) has a certain overlap. Chaining is thus supposed to be an elegant method to overcome problems with direct comparisons.

[22] This means in the first place identity of the domain of definition (to make situations *conceptually* comparable). Beside these prerequisites there are also some "empirical conditions" of comparability in Martinis thinking, such as the existence of an *overlap* of identical varieties of goods and services. It is far from evident that multiplying links achieves the miracle of removing such logical and empirical obstacles of comparability.

That situations, curiously enough, incomparable *directly* turn out to be nonetheless comparable *indirectly* (gained by linking successive direct comparisons) rests on

- overlooking assumptions[23] necessary to accept an indirect comparison as valid, and
- blurring the different nature of the result of direct and indirect comparison respectively (indeed *the result is substantially different* and obviously has a different logical status: path dependence of the indirect method, unknown to the direct method, clearly is an indication of this fact).

Moreover, there is an alternative possibility to ensure comparability over quite a long period in time, which Martini obviously completely ignored, viz. making adjustments (such as the well known adjustments in the case of changes in quality). The only reason why an indirect comparison may be preferred to a direct one is that we may feel entitled to take less care for comparability.[24]

Finally it should be noted that "comparability" is not something constant. For two things to be comparable it is necessary that there are aspects which they have in common and aspects in which they differ. All depends on how strict the criteria we apply to define the *common* aspects and the *differences* are. Martini is very scrupulous with respect to the limitations of direct comparisons (focusing on differences and thus making almost everything *incomparable directly*) but at the same time most lenient as to indirect comparisons (ignoring differences between successive links). Conspicously there is not even a limit for the length of a chain,[25] almost everything is *comparable indirectly*. This shows that there must be something wrong with this (Martini's) kind of reasoning:

Applying "logical conditions" for comparison suitably permits indirect "comparability" of virtually everything with everything. However, comparability of numerator and denominator of a **link** does not make a **chain** a legitimate comparison over a number of links.

c) Group B arguments: Ambiguities concerning the notion of "base"

While the common feature of arguments of group **A** is to emphasize the existence of "links" and their multiplication, a second group of arguments in favor of chain indices (group **B**) rests on some confusion related to the concept of a "base".

The impression that there is no need to refer to a base in the chain type of comparisons, or that there is some automatic choice of the base period, is again a result of taking the link for the chain, and of some vagueness and embroilment created by

---

[23] For example, we in essence tacitly assume that commodities *not* represented in the overlap or violating conditions of comparability will undergo the same change in prices as commodities included in the overlap.

[24] There is, for example, no care taken for *conceptual* comparability, i.e. the abovementioned constant domain of definition of an index function. Hence the result of indirect comparison can hardly have the same quality a direct comparison would have.

[25] Martini gave no hint to how long a period over which a comparison can be made *indirectly* is allowed to be.

notions like "base", "chainability" etc. The fundamental idea of this group of arguments is that

> Chain indices provide a different type of comparison by making use of a "moving" base, and that they thus have "solved", or rather made irrelevant, a problem of sizeable difficulty to all sorts of direct indices, that is, the choice of an appropriate (reference) base period 0, which should be a "normal year".

In our view this is bluntly a misunderstanding of the term "base". Chain indices in fact represent another way of analyzing time series data on prices and quantities, yet by no means a more elegant way. What matters is not the base but the consistency or inconsistency of the aggregation over time. The interesting feature is not the "moving" base of the links but rather the path dependence of the resulting chain, that is, the fact that not only the chain in t depends on its values in the past, (its "history" in t-1, t-2, ...), but it is also a specfic mathematical function of its past values subject to certain cumulative effects.

B1: Moving comparison, additional information

1. Substance of the argument

Chain indices are said to reflect time series or "*runs*" of index numbers instead of simple (direct) binary comparisons, and to provide some kind of "*moving*" or "rolling" comparison (Allen 1975. 177), apparently independent of some arbitrary point in time, called the "base".[26]

It is noteworthy here already, without starting the discussion of this argument, that there is some relationship between **B1** and **A1** and **A2**, in that the idea of a "moving" or "rolling" base again applies to the links, not to the chain:

> It is only the "base" of the link, that is t-1 in P_t^{LC}, which "moves", but of course *not* the base 0 of the chain $\overline{P}_{0t}^{LC}$. Moreover, why is the sequence $\overline{P}_{01}^{LC}$, $\overline{P}_{02}^{LC}$, ... called a "run", while P_{01}^{L}, P_{02}^{L}, ... is not?

By argument **B1** is possibly also meant that an index should account for not only periods 0 and t, but somehow (how exactly?) also for all intermediate periods 1, 2, ..., t-1. But note that this does not necessarily mean that *links* have to be *multiplied* to form a chain. There are some, curious though, index formulas considered in literature[27] which also may allow for taking the whole time series of prices and quantities (weights) into account, and which are very different from chain indices.

---

[26] It is even pretended by some authors that users of index calculations are freed from choosing a base year whatsoever (see also argument **B2**).

[27] This remark makes reference to some more or less singular proposals in the history of index numbers like the so-called "blend-system" or "broadened base system" (more details in my book on chain indices).

Argument **B1** also comprises the somewhat arcane idea that chain indices provide *valuable additional information* because they make better use of all time series data. However, to our knowledge none of the authors making use of this quite popular and prevalent argument gave any indication of the type of information he or she has in mind, let alone wherein its value lies.

2. Rebuff of argument B1

Chain indices do not represent a superior type of analysis of a time series of index numbers. What makes the difference between $\overline{P}_{0t}^{LC}$ and P_{0t}^{L} is not a "moving" or "rolling" pattern in the comparison but rather the path dependence. We have a binary comparison (of two states, 0 and t) in both cases, in $\overline{P}_{0t}^{LC}$ no less than in P_{0t}^{L}. What counts is only that P_{0t}^{L} is *not* affected by prices and quantities of periods other than 0 and t, while $\overline{P}_{0t}^{LC}$ is, because it is path dependent. There is not much use in accounting for intermediate periods 1, ..., t-1 in addition to 0 and t when 0 and t are compared. The problem with the strange "valuable additional information" is not only that there seems to be nobody able to specify the very nature and value of this "information" allegedly gained by chaining, there also seems to be a false inference underlying this argument[28]:

if there are more empirical data (reflecting more phenomena) entering a formula, i.e. we have **more *data input***	⟶	**then** the resulting statistic is more "informative", that is, we then also get **more *information output***

This is blatantly erroneous: statistics affected by *many* (possibly uncorrelated) influences are in general *inferior* rather than superior to statistics determined by a few known factors only, and in general much is done to purge a statistic from irrelevant, contaminating influences.

B2: Independence of a base

1. Substance of the argument

It is argued that the chain index is *independent of the base*,[29] in the sense of being shiftable without error because

$$(7.1.1) \qquad P_{34} = \frac{P_{04}}{P_{03}} = \frac{P_{14}}{P_{13}} = \frac{P_{24}}{P_{23}}$$

holds for chains $\overline{P}_{0t}$. The interpretation given to eq. 7.1.1 or equivalently to[30]

---

[28] The following misunderstanding seems to be quite widespread, so we also find the additional-information and better utilizing all time series data argument in an article of four Dutch authors Al, Balk, de Boer and den Bakker (1986), showing that argument **B1** seems to be subliminally convincing.

[29] We also found in literature the most revealing misunderstanding that a chain index has *no* base (!), or that its base is automatically always t-1.

(7.1.1a) $\overline{P}_{04}^{C}/\overline{P}_{03}^{C}=\overline{P}_{14}^{C}/\overline{P}_{13}^{C}=\overline{P}_{24}^{C}/\overline{P}_{23}^{C}=P_{4}^{C}$

is usually this: A chain index is no more tied to one base than to another, and the result P_{34} is the same irrespective of whether 0, 1, or 2 is the base[31]. Hence there is no need to bother with the problem of choosing an appropriate base period.

This alleged "solution" or "irrelevance" of the problem of choosing a base period is often viewed as a great advantage of the chain principle. We sometimes also find the idea of a non-existent base enabling to argue away definite shortcomings of chain indices. Craig, 1969, for example, claimed that the chain index' violation of identity is no problem on the following grounds: to get $\overline{P}_{0t}^{LC}=1$ if prices in 0 and t are identical is not necessarily desirable because 0 can well be badly chosen as a base period, and it is better *not* to return to such a base. This "argument" for a chain index completely disregards the fact that $\overline{P}_{0t}^{LC}$ has of course the same base as P_{0t}^{L}.

Argument B2 is related to the not infrequently heard objection against traditional deflation using a direct Paasche price index, that once the base is changed all volumes will change and "history has to be re-written". Whenever the choice of the base year matters it should not be surprising that volumes "at constant prices of 2005" and those "at constant prices of 2000" will not be the same. If this inequality is found inconvenient and therefore chain indices are preferred, the underlying idea seems to be: the base period should not matter; a "new" base should not change my results.

2. Rebuff of argument B2

From such confusions with the notion of "base", some other inconsistencies ensue:

In chain indices the *reference* base (RB) is deemed irrelevant, and chain indices are praised not least for being independent of the RB (argument **B2**). On the other hand it is the increased attention given to the *weight* base (WB), and the utmost importance of its up-to-dateness[32] which is above all the justification for chain indices.

Hence the pretended "solution" of the "problem of choosing the base period" RB (or rather irrelevance of the RB) seems to contradict the greatly increased emphasis put on the *weight* base (WB). Such inconclusiveness gives rise to the question of whether, in the case of chain indices, the term "base" still has the same meaning as it used to have in the direct index framework. It is widely accepted that there is no point in comparing indices having different bases in the sense of different RBs, but surprisingly there is not much scruple in studying time series of indices like ... $\overline{P}_{0,t-1}^{LC}$, $\overline{P}_{0,t}^{LC}$, $\overline{P}_{0,t+1}^{LC}$ ..., in which each element has in fact a different WB. Note that

[30] The symbol C is used to denote that this equation applies to a chain index of any type, not only the Laspeyres chain index (LC).

[31] Note that this is simply implied by the *definition* of a "chain" (or the operation of "*chaining*" as opposed to the property of "*chainability*"). Moreover, there is no reason given in **B2** as to why P_{04}/P_{03} should equal P_{14}/P_{13}, and P_{24}/P_{23}. Taken in isolation, the relationship given by eq. 7.1.1 is not desirable, though in a sense it guarantees irrelevance of the reference base.

[32] As in particular stressed in argument **B3**, but also in **C1** where the chain index is praised for making use of the most recent, and hence most "relevant" WB.

1. the irrelevance of the RB is *not* inherent in the data but rather results from the construction of the index, i.e. is derived from the fact that the chain is *defined* by the multiplication of links. It ensues from restrictive assumptions of proportionality imposed on the time series P_{01}, P_{02}, ...[33]. Moreover, different subdivisions of a given interval between 0 and t will yield different results of $\overline{P}^{C}_{01}$.

2. the difference regarding the relevance or irrelevance of RB and WB respectively gives rise to the question of the meaning of "base" in the chain index context. The point becomes apparent when we ask questions like: What makes the choice of the base period difficult in the direct index framework? Is it possible to choose a "wrong" (inadequate) base in the chain index framework?

> The numerical result of a direct index, like P^L, is said to reflect the price level in period t measured *in terms of* the level in 0. The value of P^L in t is expressed "in percent (in units) of the base period value". Hence the price level in period 0 acts as a sort of yardstick, such that it matters which period is taken as the "base".

If "base" were understood in this way we may require the weights to be taken from 0, as this is the period chosen as a *standard* (100%); irrelevance of the RB then does not appear desirable and it should not go unquestioned but should rather be surprising that growth rates between two periods (3 and 4 for example) referring to different base periods (e.g. periods 0, 1, or 2) are equal, as is the case in eqs. 7.1.1 and 1a.

In a sense the notion of a "base" is fundamental for all "comparative", or "relative" figures as indices with necessity are, and this also implies some kind of *constancy*. In other words:

> "A change through time can be recognized only by means of some element which persists unchanged ... by means of a measuring rod which is transported through time without deformation" (Afriat, 1977, S. 3).

In the chain index approach, however, the "base" RB is simply the beginning of the chain, and WB is no longer a fixed measuring rod. Once a decision is made on the starting point (RB) the sequence of weights (WB) is also *uniquely* determined. We cannot legitimately say that $\overline{P}^{LC}_{0t}$ represents a level in percent (in terms of the structure of some "basket") of the *base* period 0, since there now are structurally changing baskets of various periods affecting the result, and we cannot choose a "wrong" base (in the sense of "inadequate" weights) either. To sum up:

> In a sense the choice of a base is not solved but rather dissolved by the simple reason that the "base" no longer has the significance it used to have in the direct approach. In the case of chain indices there is simply nothing left (like disproportionality of time series with different RBs, or a non-automatic determination of the WB, once the RB is fixed) which might make the choice of a base period important, and at the same time difficult.

[33] See **sec. 2.2** and the singularity of matrix P. This result incidentially throws cold water on the argument (**B1**) of better making use of the information given in the (time series) data.

B3: More relevant growth factor

1. Substance of the argument

A chain index (in fact the link, however) is said to directly represent the growth factor[34] of an index (in relation to the preceding year) and this growth factor should be expressed in terms of the most recent (most "relevant", most "representative") quantities in the case of price indices[35] or prices in the case of the growth rates of a volume (real value) respectively.[36]

What (allegedly) makes the growth factor

(7.1.2) $\frac{\overline{P}^{LC}_{0t}}{\overline{P}^{LC}_{0,t-1}} = P^{LC}_t = \frac{\sum p_t q_{t-1}}{\sum p_{t-1} q_{t-1}}$ "better" than (7.1.3) $\frac{P^L_{0t}}{P^L_{0,t-1}} = \frac{\sum p_t q_0}{\sum p_{t-1} q_0}$

are of course the weights $q_{i,t-1}$ in contrast to q_{i0}. The same applies to weight $p_{i,t-1}$ in contrast to p_{i0} in the case of quantity indices

(7.1.2a) $Q^{LC}_t = \frac{\sum q_t p_{t-1}}{\sum q_{t-1} p_{t-1}}$ as opposed to (7.1.3a) $\frac{Q^L_{0t}}{Q^L_{0,t-1}} = \frac{\sum q_t p_0}{\sum q_{t-1} p_0}$.

The argument seems to be attractive. Eurostat voted in favor of deflation by using chain price indices mainly on the basis of the following two arguments[37]:

- volumes based on previous year prices give the "best", or most "accurate" estimates of changes in volume from one period to the next (which is argument **B3**),[38] and
- the use of chain indices would considerably ease the treatment of quality changes and new or disappearing products (which is argument **C2**).

2. Rebuff of argument B3

Interestingly, there is much scruple in calculating actual growth rates of e.g. volume, which ought to be compiled on the basis of the *most recent (relevant)* prices, but there is no scruple in comparing successive growth rates in which this structure of prices is constantly changing, or to compare successive price levels in which this structure of quantities (as weights) varies constantly. As to the quantities in eqs. 7.1.2

[34] This is of course trivial because according to eq. 7.1.1 the chain is *defined* by multiplication of such growth factors (links).

[35] That is, in the measurement of the year-to-year increase of a price level.

[36] The argument with the "more relevant growth" factor f (or growth rate r = f - 1) is more often applied to the latter case (volumes rather than prices). In my book on chain indices I tried to show that there are some assumptions necessary to conclude "most recent = most relevant" with respect to the consumption pattern.

[37] EU Commission Decision (as concerns the principles for measuring prices and volumes) Nr. 2223/96 approved by the Statistical Programme Committee in Stockholm, 27th May 1998.

[38] This kind of comparison is considered to be more important than, for example, to make consistent comparisons of growth rates between successive years.

and 7.1.3, $\overline{P}_{0t}^{LC}$ may in fact be preferable to P_{0t}^{L} because of "obsolete" weights q_0 in P_{0t}^{L}. However, the whole consideration would not apply to the direct Paasche index[39] (instead of P_{0t}^{L}) because

(7.1.3b) $\frac{P_{0t}^{P}}{P_{0,t-1}^{P}} = \frac{\sum p_t q_t \sum p_0 q_{t-1}}{\sum p_0 q_t \sum p_{t-1} q_{t-1}}$ while (7.1.2b) $\frac{\overline{P}_{0t}^{PC}}{\overline{P}_{0,t-1}^{PC}} = P_t^{PC} = \frac{\sum p_t q_t}{\sum p_{t-1} q_t}$.

Hence if the sole objective were to calculate growth rates (of volumes [quantities] or prices) on the basis of non-obsolete weights, a direct Paasche index could do the job equally well. There is no need for a chain index approach for this reason. The more fundamental deliberation concerning this argument should focus however on the

- limitations of the year-to-year comparisons, and correspondingly on the
- need for consistent comparisons of annual growth rates over a **number** of sucessive years (over a business cycle of four to six years or so for example.

As to the last point it may be interesting to have a look at annual GDP growth rates of Norway reported some years ago. This growth rate of the "real GDP" of Norway alternatively calculated using fixed prices of 1984 (row A) on the one hand, and calculated using previous year prices (row B) on the other is given by[40]:

		1987	1988	1989
A	constant base period prices*	4.9	3.0	5.2
B	previous year prices	3.9	1.8	0.9

* fixed prices of 1984

Note that there is no specific chain index problem unless the previous year growth rates are multiplied to form a chain. In this case there is however a substantial difference. Multiplying growth factors on the basis of B gives: $1.039 \cdot 1.018 \cdot 1.009 = 1.067$, in contrast to 1.136 obtained by multiplying the factors of row A. This gives an average annual growth rate of 2,2% and 4,2% respectively for the period 1986 to 1989. Which is the correct average rate? According to A growth was higher in 1989 (5.2%) than in 1988 (3%), whereas judged by B the opposite is true. What was really going on?[41] There is certainly a point in preferring B to A.[42] The substantially smaller rate of B, amounting to 0.9% by 1989 may be found more realistic and more reliable. It has of-

---

39 The same is also true for certain "superlative" direct indices in which weights are taken from both periods, 0 *and t*. Hence it should be remembered that **B3** again does not give sufficient reason to prefer chain indices over *all* sorts of direct indices.

40 Data taken from an internal paper distributed in a task-force meeting of Eurostat.

41 According to B the Norwegian economy seems to be unreletingly running into difficulties, whilst growth measured by A displays a much more comfortable or even promising picture.

42 In the specific case of Norway we also have to take into account that in the base year (1984) prices of crude oil (a most important commodity for Norway's export) was high (with the consequence of a high weight) and dramatically fell thereafter. Hence there is a strong plausibility that A is overstating growth in 1989 substantially due to the high weight given to oil exports in 1984.

ten been observed that growth of more recent years measured by chain methods falls short of growth measured for the same period using fixed prices of an earlier year. The reason is that commodities for which output grows more rapidly tend to be those for which prices increase slowly or even decrease. Hence taking more recent prices as weights will give commodities with strong growth less weight and thereby result in a smaller overall growth rate.[43] Notice, however:

> Arguments referring to the allegedly superior "relevance" of a growth rate are not really convincing, as the problem is not (or not only) to take a single growth rate in isolation, but to make comparisons over a number of periods in a consistent manner.

Lack of comparability of different periods (i.e. in aggregating over time) should clearly be viewed as a serious drawback of the chain method. The question is not only which is the best measure of the *actual* growth rate for the current year. A question of no less importance is: are successive growth rates of say five years or so (the length of a business cycle) *comparable* in the sense of truly reflecting the movement of "volume" (quantity and quality), or are they also affected by changes in prices, the elimination of which is the prime concern of deflation?

Growth rates on the basis of constant prices are comparable and reflective of pure quantity movement only, even if the basis of prices is several years old and kept constant. The argument with the most relevant growth rate (of volume) is simply the deflation (or *price* weighted quantity indices) counterpart to the idea that in inflation measurement (with *quantity* weights) use should be made of the most "representative" or "relevant" quantity weights. Again the aspect of comparability (by keeping weights constant) is dwarfed by "relevance" of weights.

Digression on annual growth rates of figures compiled monthly

In the most popular comparison of the value a (chain) price index takes in a certain month m in year t to the value of the corresponding month of the previous year, t-1 may also be a function of weights of *two* years instead of only one. In a comparison of sub-annual intervals (months for example)[44] we have two quantity structures q_{97} and q_{98} influencing the result.

Assume: 1. compare a month (say May) with the same month of the preceding year, 2. an updating of weights (= average [symbol Ø] of the preceding year) takes place in December.

Thus we have to compare:

Prices	Jan. 99	...	May 99	...	Dez. 99	Jan. 00
Weights	Ø 98	...	Ø 98	...	Ø 98	Ø 99

[43] The problem, however, is that correlation between prices and output varies. Furthermore the effect described is likely to be different in the case of consumer goods compared with investment. And in order to make comparisons between growth rates of different aggregates, additivity would be useful.

[44] This applies to the European Consumer Price Index.

with

Prices	Jan. 98	...	May 98	...	Dez. 98	Jan. 99
Weights	Ø 97	...	Ø 97	...	Ø 97	Ø 98

Hence there are **two** baskets involved: Ø 97 and Ø 98.

d) Group C arguments: Adjustment of weights, new products and changes in quality

The common feature of this group of arguments is the contention that chain indices are able to solve almost insurmountable problems involved in fixed basket indices and at the same time these arguments give chain indices an image of flexibility, adaptability and thus also of a much greater suitability to modern needs.[45] It is claimed that chaining is an elegant device to elude the trouble of keeping a basket constant, despite changes in types and qualities of goods, in the relevance of certain outlets and the like.

As mentioned above, a move to a new method of deflation in Europe in 1998 was primarily made on the basis of exactly this twofold argument[46] of type C.

C1: Most frequent update of weights

1. Substance of the argument

Especially the rev. SNA brought this argument into play, praising emphatically chain indices as (see above) "indices whose weighting structures are as up-to-date and relevant as possible".

2. Rebuff of argument C1 (i. e. more frequently updated weights = better weights)

Note that this argument again compares a direct index with a chain index (or rather a link), as if they both had a single weighting scheme only. To say that we always make use of the most up-to-date previous year weights is only correct for P_t^{LC}, not for $\overline{P}_{0t}^{LC}$, where the weights are cumulative. Hence we once more have a reasoning in which the link is mistaken for the chain.

---

[45] Aspects like these are often regarded as most convincing in support of the idea of chaining. At first sight one is sorely tempted to advocate for chain indices precisely on those grounds: chain indices seem to far better match a rapidly changing world like ours and *at the same time* they also avoid problems inherent in a fixed basket approach which is (in contrast to the chain approach) unfit for keeping pace with the times.

[46] That is: a more modern and flexible index design with weights constantly updated (**C1**) *and* at the same time an elegant solution of a burdensome problem we always had to cope with in the case of the old design of a fixed basket (**C2**).

The problem with all arguments focusing on the most recent (e.g. previous year) and *thus* most "representative" and most "relevant" weights, or simply *the better* basket of a chain index, is

- The implicit assumption, $\overline{P}_{0t}^{LC}$ has a *single* (and better) WB, just like a direct index P_{0t}^{L} or P_{t}^{LC} is simply false ($\overline{P}_{0t}^{LC}$ rather has a *cumulative* WB)[47].
- There is no clear concept to define or measure the *degree* of "representativity" or "relevance", hence the argument in itself cannot give any hint concerning the best frequency of a renewal of weights. Moreover, the length of the time-interval to which a link should refer is also determined by other considerations (e.g. cyclical price movement).
- Given that a price index ought to reflect new *quantity* weights (as required by argument **C1**), what then is the task of a *quantity* index? This raises in a way the question of the "division of labor" between a price index and a quantity index.[48]

Finally there are two objections to be made from the point of view of official statistics:

- Argument **C1** is clearly double-edged; the other side being the increased demand for empirical studies in consumption. Experience shows that compiling representative *weights* repeatedly in short intervals is more burdensome, expensive, and also more liable to errors than reporting *prices* on a regular (e.g. monthly) basis.
- An index design allowing weights, types of goods, outlets and many other features of an index to change constantly is in danger of giving rise to (the impression of) manipulation of various sorts.[49]

C2: Less problems with new developments, quality adjustment less difficult

1. Substance of the argument

The SNA also found worth mentioning that chain indices make it "possible to obtain a much better match between products in consecutive time periods ..., given that products are continually disappearing from markets to be replaced by new products, or

[47] See **sec. 2.5** this is so since each link has its own WB, the chain as the result of multiplying many links has consequently a number of WBs.

[48] Forsyth and Fowler 1981, p. 234 had been criticizing the direct Laspeyres price index as "unacceptable because it utterly fails to represent the changes taking place over the time span" (by which they meant the changes in *quantities* [in the consumption pattern]). The question should be whether or not the reflection of this is really the task of a price index (as it is bound to reflect a change in *prices*) or rather of a quantity index.

[49] In principle, any method which lays itself open to manipulation should be viewed with the greatest suspicion, in particular when applied in official statistics and in the case of inflation measurement. It is one of the main advantages of P^L that the rather rigid "constant basket" design leaves little room for more or less ojectionable manipulations.

new qualities". The chain principle is said to facilitate (or accelerate) the adaptation to new developments by making use of the most recent weights, and to handle better the withdrawal of old and the entry of new commodities. In contrast, the fixed basket approach of the Laspeyres direct index inevitably runs into difficulties as new products emerge, old ones are no longer available, and the characteristics of goods change substantially with an impact on consumers' utility level and on the classification of commodities used to select the specific basket. Hence every index design that appears able to prevent such type of problems from existing will be welcomed.

On the face of it the chain principle thus not only allows for a constant adjustment of quantity weights and for the comparability of non-overlapping selections of items (commodities) in 0 and t by comparing 0 with 1, 1 with 2, ..., t-1 with t, it also pretends to "solve" one of the most intractable problems in practical price statistics, that is, to make adjustments for changes in the selection of goods (and specification of their quality), outlets, etc. It is said that

> "A chain index naturally incorporates changes in the qualities and types of goods as it follows changed consumption patterns through the years" (Forsyth 1978, p. 349), or it "automatically incorporates" changes in consumption (Forsyth and Fowler 1981, p. 229).

2. Rebuff of argument C2

A second thought reveals that the problem with such statements just quoted is that it is not clear but left vague as to what is meant by "incorporates" (or "automatically", "naturally" or so).[50] It is also left open in what sense the practical problems in question (i.e. quality adjustments and the like) of the fixed basket approach of P_{0t}^{L} have been "solved". Instead of "quality adjustments less difficult", the argument **C2** should rather read as: "quality adjustments less necessary" as this is a typical "solution vs. dissolution" argument and it is particularly unfair, because

- it turns out that new products and new qualities are "incorporated" without any adjustments, since as the basket is allowed to (or even bound to) change constantly there is no point in making sure of comparability of the basket in t with the basket in 0; worse even, it is even praised as one of the main advantages of $\overline{P}_{0t}^{LC}$ as opposed to P_{0t}^{L} *not* to be in need of ensuring comparability for more than just two adjacent periods,[51]
- thus the problem, the "solution" of which is allegedly achieved, is in fact simply no longer existent once the fixed-basket approach is abandoned (or comparability is discarded).

---

[50] Vagueness is not only typical for this argument in advocating chain indices, we also find it in considerations like chain indices use more "relevant" weights (how to measure relevance?), they utilize time series data "better" (in which sense?), or they provide some additional information (which information?).

[51] When it is an advantage *not* to aim at long term comparability we cannot construct another additional advantage consisting of the alleged "solution" of a problem which exists only because of aiming at long term comparability.

Making adjustments for changes in the quality of goods is necessary in the fixed basket approach in order to ensure comparability over time. When the fixed basket is abandoned, and comparability is no longer at issue, there is no need for more than just two adjacent periods, t-1 and t, to be comparable. In short: the problem has simply disappeared. To sum up:

The "solution" of a tricky problem (to account for changes in the qualities and types of goods, outlets etc.) involved in keeping a basket comparable over time consists of replacing the method which gave rise to the problem in question, by a method in which comparability is no longer relevant.

It is true that the above mentioned changes make the Laspeyres type direct index P_{0t}^{L} increasingly more difficult to compile with the passage of time. It is deceptive, however, to speak of a "solution" of such difficulties if comparability is simply taken less seriously. It is true that the compilation of a chain index in practice is less troublesome as less care has to be taken for comparability of commodities over time. The increase in convenience has to be contrasted, however, with the fact that

- chain indices require more frequent and more expensive empirical studies of consumption (for the updating of weights), and that
- comparability, which the direct index aims at, is relaxed if not abandoned.

In conclusion: $\overline{P}_{0t}^{LC}$ has not solved a problem of P_{0t}^{L}, it has only replaced one problem (difficulties with maintaining comparability) with a new type of problem (less emphasis on "pure" comparison), and it is far from clear wherein the advantage of this should lie.

e) Group D arguments referring to expected and desired results of index calculations

The focus of the following group of arguments is on the numerical results to be expected when chain indices are used. The arguments do not refer to conceptual aspects whatsoever. Nor do they apply to *all* sorts of empirical data, like axiomatic considerations do for example. Thus arguments of group **D** are not independent of the kind of data entering an index formula, and they are in a way ambiguous, in that they imply advantages as well as disadvantages.

D1: Smoother development, smaller inflation rates

It is sometimes maintained that chain indices are likely to yield this type of result, unless some cyclical movement or regular fluctuation occurs. Chain indices are found unfavorable in the case of fluctuations[52] but useful (in terms of desirable numerical results, like low inflation), especially "if individual prices and quantities tend to increase or decrease monotonically over time" (SNA 93, para. 16.44). Interestingly,

[52] As is well known this is why the SNA does not recommend linking more frequently than annually. If cycles exist chain indices are *not* useful. On the other hand, it is implicitly assumed that when no cycles occur the converse is true, such that chain indices do have advantages.

these are precisely those conditions under which the heavily criticized P_{0t}^{L}- approach is not that defective.

D2: Less formula problems, smaller Laspeyres-Paasche-gap (LPG)

It is also often conjectured that the chain index version of various index formulas will yield less divergent results than the corresponding direct index version. For example, the Laspeyres and Paasche chain indices, $\overline{P}_{0t}^{LC}$ and $\overline{P}_{0t}^{PC}$, will have less different results than P_{0t}^{L} and P_{0t}^{P} (the differences are called "gap") if applied to the same data. Under certain conditions the spread of index formulas (or the "gap" between them) will be smaller for chain indices as opposed to direct indices, with the result that "the choice of index number formula assumes less significance" (SNA 93, no. 16.51). The PPI/CPI manual explicitly makes this point too: "The main advantage of the chain system is that under normal conditions, chaining will reduce the spread between the Paasche und the Laspeyres indices".[53] For this argument (which may be called "less formula problems" or ironically "reduction of conceptual expense"[54]) it is of course important that P^L and P^P are viewed as "equally plausible" or "equally justifiable", as (unfortunately) done repeatedly in the manuals. In our judgement this is not true since the period t and the period 0 are not of the same nature. The conceptual background of P^L and P^P is different (see **sec. 1.3**) and it therefore cannot be an advantage in itself to have a result exactly in between.

D3: Goodness of fit in an econometric model

According to (proponents of the stochastic approach) Selvanathan and Prasada Rao 1994, p.125, satisfactorily high values of the coefficient of determination r^2 found in econometric models "reinforce the general claim of superiority of the chain base index numbers over fixed base indices, especially when the current period is further away from the base period". Note that a higher goodness of fit in a model is taken as a proof of *conceptual* superiority, and there is no consideration of concepts and axioms. As a justification of methods this is clearly insufficient.[55]

General remarks concerning arguments of type D

Arguments of type **D** are unsatisfactory for a number of reasons. There is simply not sufficient evidence that chain indices will produce smooth time series under realistic

---

[53] In this manual the argument has also been advanced that circularity is only exactly met in the case of the Cobb-Douglas index (we quoted this uniqueness theorem (UT) in **sec. 3.4** as UT-7), and that a chain Törnqvist index should be used on the ground that this index *approximately* satisfies circularity.

[54] This argument was also decisive for the Australian Bureau of Statistics (see Mc. Lennan 1998).

[55] Nobody would claim that A is a better "domestic product" than a differently defined aggregate B, only because the correlation of A with, for example, money stock M3 is higher than correlation between B and M3.

conditions.[56] If there were "advantages" with respect to the numerical results it is still not clear whether they can offset the many conceptual deficiencies of chain indices. Moreover, the arguments are not only weak in referring to the numerical "behaviour" (instead of the conceptual foundation) of chain indices only. This behaviour also greatly depends on the type of data to which chain indices are applied. Moreover, they require conditions[57] under which the (alleged) defects the direct Laspeyres index is blamed for would be less likely.

f) Chain indices recommended for deflation (SNA, argument of type E)

The SNA recommendation adds another argument to the alleged advantages of chain indices: superiority of chain indices for purposes of deflation. The main reason for preferring chain indices is again the desire to guarantee a continuous update of weights as soon as possible. The reasons given by the SNA 93 have been demonstrated and commented upon at length in chapter 5.

To sum up we may say: In the opinion of the SNA for index formulas there is little if anything of equal importance to the up-to-dateness of weights. To use the most recent quantity weights in order to measure inflation and the most recent prices in the framework of deflation is capable of outweighing all possible shortcomings involved in adopting the chain index principle.

The idea is to avoid on some pseudo-theoretic grounds terms like $\Sigma p_0 q_t$, in which prices and quantities refer to different periods,[58] thus making the products somewhat artificial, as opposed to $\Sigma p_0 q_0$ or to $\Sigma p_t q_t$, where prices and quantities refer to the *same* period. It is intriguing to contemplate this a bit further: To make this idea (prices and quantities should not belong to too different periods) a desideratum or even a guiding principle eventually amounts to discarding all attempts of calculating "volumes" (being *necessarily* different from "values" $\Sigma p_t q_t$, as they combine prices and quantities of *different* periods), and hence of deflation, if not of indices altogether.

g) Concluding remarks concerning the nature of the arguments

In what follows we try to set out and criticize five strategies to argue for/against an index approach.

---

[56] Experiments with chain indices reveal rather that "chain indices can be more biased than their direct counterparts, which is both contrary to popular belief and to the purpose of linking" (Szulc 1983, p. 555).

[57] Such as smooth and non-cyclical changes, a short rather than long time series, a current year's basket not too different from the base year's basket etc.

[58] As pointed out in chapter 5 the SNA said: a process of production which is efficient at one set of prices may not be very efficient at another set of relative prices; an artificial aggregate like $\Sigma p_0 q_t$ may not only be improbable, it may even be (technically/economically) *impossible* (if prices p_0 really would prevail, the production process that created q_t would not be used). As shown above this kind of reasoning, exemplified in the SNA, would eventually result in an alleged "impossibility" of deflation. All sorts of "volumes" are with necessity some *artificial* combinations of prices and quantities other than the *observed* combination $\Sigma p_t q_t$.

1. Some of the arguments put forward to support chain indices *appeal to emotions*, and make use of vagueness (if not obscurity) e.g. the "multiplication mystery"[59] or the notion of "valuable additional information" or "making more use" of existing data. An appeal to emotions is also implied in concepts like "balanced", "symmetric", "compromise" and so on, or applied to weights "flexible", "up-to-date", "more relevant" and "more representative" and the like.[60] Present time gains dominance over past time to an extent that it even fell into oblivion that an index by definition always has to make some kind of comparison with the *past*, and that the need for some kind of "base" almost disappeared altogether.[61]

   With this some kind of rhetoric choice is predetermined. Furthermore, the alternative (to a move to chain incides) is set out in a conspicuously incomplete and onesided manner. Moreover, as with all emotional notions the underlying idea is usually difficult to describe in operational terms. To our knowledge there is no agreed upon measure of "representativity" or "relevance" of weights, let alone any attempt to quantify the degree of harm done by using "less" (to which degree?) relevant weights.[62] Nonetheless, the obsession with the updating of weights is endemic, and seldom if ever questioned.

2. Another observation is that people tend to knowingly mistake an analytical device for a statement describing the real world, and they have difficulties with accepting that *all indices rest on considerations in terms of "models"*. Strictly speaking the only variable exclusively representing "raw material" of observations is the value index V_{0t}, while all kinds of decomposing it into a price and quantity component have to rely on models of one kind or the other.

   To set out an example: It is a well known fact that "life expectancy" cannot be defined and measured by asking people how long they expect to live. The only possible method to make use of the model of a "life table"-population (or stationary population) in which death risk depending on age is kept constant.[63]

   Without such a model, though clearly in contradiction to observation and real world conditions (such as increased longevity as a result of progress in medicine), measurement of life expectancy would be impossible. It is nonsense to say that life expectancy was incorrectly or "inaccurately" measured because the assumptions of the underlying model are unrealistic. But exactly this is done in

---

[59] This is the (indirect) comparability curiously emerging from (direct) incomparability.

[60] When a choice has to be made among two approaches, one flexible and referring to the latest weights and the other making use of "old" and unchanged weights, we can easily predict that the number of those voting for "inflexible" and "out-dated" will tend to zero. Furthermore, it is most unlikely to find somebody who prefers to make *less* use of data, or to do without certain valuable extra information delivered free as a kind of by-product. Such ideas clearly *sound* convincing, though we hardly hear anything about the kind of information in question.

[61] This alludes to a particularly absurd offspring of the obsession with speedy adjustment of weights, viz. the so-called "axiom of simultaneity", invented by U. P. Reich (see my book on chain indices, there sec. 7.3a).

[62] It is also conspicuous that none of the axioms introduced in index theory expressly refers to abovementioned concepts, like "representativity", "relevance" etc.

[63] the same for each cohort, such that same age = same risk, irrespective of the birth cohort to which one belongs.

the case of the Laspeyres-index, where the "assumption" of a constant basket has much the same function in a model as the constancy of death risks.[64]

Once the instrumental nature of the "fixed basket" as a tool is recognized it is clear that the Laspeyres formula cannot, or should not, be attacked with such trivialities like "the American economy is flexible and dynamic".[65] There is no need to deny this if we decide to defend the Laspeyres index P^L. It is bluntly nonsense to insinuate that P^L requires ignoring facts and that other index formulas can do without some type of model. It is notably true that

> To assume a fixed basket for analytical purpose does not mean that consumption is not responding to changing prices, that an isolated change of one variable is likely to occur in reality, or that the economy as a whole is static.[66]

We should not confound "model" and "reality". Apart from values, value ratios and unit values there are no other statistics in which prices and quantities correlate precisely as they correlate empirically. If all isolated variation of one variable were to be banned, we would have no "volumes", "life expectancies" etc., because some hypothetical element is involved in all such calculations. The idea underlying the so called "true cost of living index" (COLI) is for example that there exists a minimum-cost combination of goods to attain a given utility level, and that households suffer from inflation to the exact extent that these minimum cost are risen. This concept may be found useful,[67] but it should be noticed that

- the assumptions made here are by no means more realistic than the fixed basket, and
- the COLI were *not* proved irrelevant or antiquated if we succeeded in showing that more and more households are unable to maximize utility.[68]

3. There is also a tendency to *mix up concept- and result-oriented aspects* in justifying the choice of a formula, just like there is a tendency not to keep clearly distinct "tool for measurement" on the one hand and "description of reality" on the other. As a rule a conceptual foundation for a formula is desirable and it is given

---

[64] To derive life expectancy at time of birth we account for that amount of mortality which persons, presently at an age of 0, 1, 2, ..., face. We assume that such figures will also prevail when the person just born in question will reach the respective age, that is 1, 2, ..., 100 or so years later. Thus conditions observed at the moment are assumed to remain constant for up to about 100 years. And there is no criticism. But the analogous constancy of a "basket" for only some five years or so, just as much assumed for the sake of simplifying analysis, is almost universally condemned.

[65] Advisory Commission (or simply "Boskin Commission").

[66] The present author has often heard an argument of a very similar kind: if Laspeyres would live in our days he would not propose his static index formula once more but rather a chain index.

[67] Interestingly, there is a constant request for an underlying economic background of *index* formulas. We rarely meet people complaining about lack of microeconomic foundation of the covariance, for example.

[68] This behavior is "assumed" in the framework of the COLI model, but not meant as a description of reality. Things change, however, once the empirical result of a traditional index function, like Laspeyres or Paasche, is identified as the value the COLI is taking. This can be done only if "utility maximizing" behavior is assumed.

by showing how the formula in question fits to what it is intended to measure. Playing off empirical results against the conceptual foundation of a formula cannot be convincing. It is for example always possible to find a formula able to yield a constantly low inflation rate, to display a "smooth" development, or to remain consistently within the span of the Laspeyres and Paasche formula, whatever the data may be. Thus such kind of criteria can never be convincing, nor can they make up for a lack of conceptual basis.[69] Interestingly, much of the justification of chain indices is of exactly this kind, however.[70]

4. A "re-discovery" of index formulas is not infrequent. The reason for this probably is that a great number of conflicting "axioms", "tests" or so on that can be "discovered" and found plausible or reasonable. Not surprisingly, axiomatic theory cannot come up with one single formula passing all tests and being the "best" without any doubt.[71] As there are always criteria met by an index and at the same time those violated, it is popular of course to *selectively exaggerate advantages and play down possible disadvantages of index formulas.* This game can be played infinitely because no rule can be established according to which a certain axiom or property should be regarded as more important (for example the up-to-dateness of weights) at the expense of another (for example aggregational properties).

5. The final observation is related to (deliberately bringing into play) *inappropriate parallels and generalizations*[72]. An example is the SNA stating that "sooner or later" each type of index has to be linked together to form a chain. The common feature of considerations like these is that they place the direct index in the bad position of a rather strange and unfounded *exception* from the rule, the much better reasoned, more general, and also more "natural" concept of a chain index. There is for example no obvious motivation or any good reason given why P_{0t}^{L} should *not*

   - constantly make adjustments for changes in quantities as well (not only of prices),[73] and
   - update weights more frequently (the "why not" – argument), as this is not only possible, but also common practice in the case of chain indices.

   In both cases the chain index appears to be doing what is more logical and more natural, whereas the direct index absolutely unintelligibly refrains from doing so. The same trick is applied when for some reason or another the "eco-

[69] It is conspicuous that not much conceptual consideration can be found to justify the typical feature of the chain approach, that is, the operation of "chaining" (multiplying links).

[70] Thinking in terms of numerical results is possibly a still widespread after-effect of Fisher's ideas, as is the tendency to prefer "compromise"- or crossed formulas, which – interestingly – is done exactly when sufficient reasons cannot be found for either of the two formulas.

[71] It would be naive to expect another result. Thus, contrary to some often expressed opinion, this is absolutely irrelevant in deciding about the usefulness of the axiomatic approach.

[72] We already quoted in **sec. 2.5** an example of this strategy, according to which chain indices may be regarded as the more general concept (Makaronidis, 1999).

[73] This refers to the "generalization" quoted in the preceding footnote.

nomic theory" or COLI index is deemed to be the most general and fundamental concept. Mr. Laspeyres and Mr. Paasche did *not* start with the notion of the "true cost of living index", and it was not their intention to find an upper bound (P_{0t}^{L}) or a lower bound (P_{0t}^{P}) respectively of this index. The rationale of their formulas is different and we cannot do justice to them if we value them from the "economic theory" point of view only, where they necessarily end up as inferior (extreme) indices. In conclusion, P_{0t}^{L} is not correctly valued if primarily seen as an extreme approximation of the COLI, or as an index which is by some oddity only imperfectly and too infrequently updated.

7.2 Properties of chain indices

In Europe chain indices are made mandatory for official statistics, so a critique of chain indices is a bit like swimming against the tide. On the other hand, it is never too late for an analysis of methods of statistics, in particular when the shortcomings of a method are so obvious, as in fact is the case with chain indices. This section is devoted to the demonstration of such drawbacks (as seen from the perspective of a "non-chainer"). An attempt is also made here to find more of what might be called a "theory" underlying chain indices[74] (without much success however).[75]

a) Introduction

Chain indices are recommended by the revised SNA of 1993. Some National Statistical Offices (NSOs) had been making use of them for quite a long time already[76] with much enthusiasm, whereas a minority of other NSOs (esp. Germany) have a tradition of disapproving of this type of index such that they followed the move to chain indices only reluctantly. The SNA recommendations (paragraph 16.73) read as follows

> "(1) the preferred measure of year to year movement of real GDP is a Fisher volume index, changes over longer periods being obtained by chaining: that is, by cumulating the year to year movements;
>
> (2) the preferred measure of year to year inflation for GDP is therefore a Fisher price index, price changes over long periods being obtained by chaining the year to year price movements: the measurement of inflation is accorded equal priority with the volume measurements;
>
> (3) chain indices that use Laspeyres volume indices to measure movements in real GDP and Paasche price indices for year to year inflation provide acceptable alternatives to Fisher indices..."

[74] Practical aspects important for official statistics such as costs, confidence of users, understandability of the meaning of the index function and so on are not covered here.

[75] There is perhaps not much going beyond the fixation on, not to say obsession with, up-to-dateness of weights, a dogma for chainers that largely goes unquestioned.

[76] For example France, the United Kingdom and Sweden.

Table 7.2.1: Summary of shortcomings of chain indices

Problem, property	part[1)]
theoretical justification and interpretation	
Interpretation in terms of mean value (of relatives) and ratio of expenditures	b
Determinants of the drift (temporal correlation, growth factors)	i
Justification of chaining, inconsistency with constant adjustment of weights	j
Acceptance, understandability, room for manipulation[2)]	k
Axioms, aggregation over time	
No identity and monotonicity, axioms apply to links only (not to the chain)	c
Cyclical movement of prices/quantities	d
No transitivity, path dependence versus circularity	e
Nonlinearity (the determinants of an increase/decrease of the price level)	f
Performance of chain indices in deflation and aggregation[2)]	
Inconsistency in aggregation	g
No structural consistency (of volumes) and proportionality in quantities	h

1) of **sec. 7.2**

2) see also **sec. 5.4**

In short the recommendations express a general belief of

- the chain principle being superior to the direct binary comparison and
- being universally applicable for both measurement of price levels as well as deflating values to volumes, and finally a firm belief that
- Fisher's "ideal" index is better than traditional formulas (Laspeyres, Paasche), which are qualified as second best solutions only.

In our view all three statements are false or they need at least some further clarification.[77] In detail the statements require us to examine the following properties of chain indices in general and the Fisher chain indices in particular:

[77] There are other authors who share our view in this respect (arguments to support the alleged superiority of chain indices). The "arguments usually advanced to substantiate this superiority are not ... very convincing" (Szulc 1983, p. 537).

- the theoretical justification and interpretation (e.g. economic plausibility), and the performance of chain indices
- in terms of "axioms" and in comparing over long periods in time, and finally
- in deflation and aggregation (e. g. consistent aggregation of sub-indices to the overall index) and with respect to some practical considerations.

The three points can be broken down into more points as shown in **tab. 7.2.1**, where we also indicate in which part of this section the property will be discussed in detail.

We start by showing that chain indices are not amenable to standard interpretations of indices and then discuss some other formal aspects. Finally, comments are made on some *practical* aspects, in particular in view of the role of official statistics.

b) Chain price indices and the two traditional interpretations of a price index

In what follows an attempt is made to describe a chain index in terms of a ratio of expenditures ("changing cost-of-a-budget") and of a mean of (price-) relatives.

While P^L (as well as P^P for example) has the advantage of allowing both interpretations (ratio of expenditures and mean of relatives), neither of the interpretations applies to the direct Fisher index P^F nor to chain indices[78] of whichever sort.

Ratio of expenditures (a rather farfetched interpretation)

In $P_{0t}^{L} = \frac{\sum p_t q_0}{\sum p_0 q_0}$ it is common practice to call $\sum p_t q_0$ the cost of buying the base period basket q_0 at time t. In contrast to the value index $V_{0t} = \sum p_t q_t / \sum p_0 q_0$, the quantities q_{it} in the numerator of P_{0t}^{L} may be interpreted as "reduced" to q_{i0} *individually* (for each i separately) to the *actual* level of period 0 (such that $q_{it} \rightarrow q_{i0}$).

The equivalent expression in the case of $\overline{P}_{0t}^{LC}$ would be

$$(7.2.1) \qquad \overline{P}_{0t}^{LC} = \frac{\sum p_t q_0^{LC}}{\sum p_0 q_0} \text{ where } q_{i,0}^{LC} = \frac{q_{i,t-1}}{\overline{Q}_{0,t-1}^{PC}}$$

where quantities $q_{i,0}^{LC}$ in the numerator could be interpreted as a set of quantities q_{t-1} "reduced" *globally* (or *collectively*) to a *fictitious* level of period 0. We may say that the level is traced back from t to 0 but not the structure. Moreover, it should be noted that the sum of ratios $p_0 q_0^{LC} / \sum p_0 q_0$ with which the price relatives p_{it}/p_{i0} are weighted does not add up to unity but to $Q_{0,t-1}^{L} / \overline{Q}_{0,t-1}^{PC}$, which already shows that $\overline{P}_{0t}^{LC}$ is unable to permit a "mean of price relatives" interpretation.

---

[78] There is an interpretation in terms of "cost of a budget", somewhat farfetched, however, but it is definitely *not* possible to describe a chain (price) index in terms of price relatives.

It is worthwhile to spell out in detail what the fictitious "quantities" q_{i0}^{LC} mean, viz.

$$(7.2.2)\qquad q_{i,0}^{LC} = q_{i,t-1}\frac{\sum q_0 p_1 \cdot \ldots \cdot \sum q_{t-2} p_{t-1}}{\sum q_1 p_1 \cdot \ldots \cdot \sum q_{t-1} p_{t-1}},$$

which seems to be neither particularly meaningful nor significantly more up to date than q_{i0}, since in addition to the most recent quantities $q_{i,t-1}$ there are also prices and/or quantities related to past periods 0, ... , t-1 entering the formula. In the case of a chain Fisher's index $\overline{P}_{0t}^{FC}$, the underlying quantities q_{i0}^{FC} in the "ratio-of-expenditures" interpretation are even more arcane (and complex even in the case of t = 2, that is, in $\overline{P}_{02}^{FC}$):

$$(7.2.2a)\qquad q_{i0}^{FC} = \sqrt{q_{i2}^2 \frac{\sum p_0 q_0 \sum p_1 q_0 \sum p_2 q_1}{\sum p_0 q_1 \sum p_1 q_2 \sum p_2 q_2}}.$$

No "mean of relatives" interpretation

To see this, consider the most simple chain of two links only and two commodities with weights (expenditure shares) as follows: a and 1-a for good 1 and 2 respectively at period 0, and correspondingly b and 1-b at period 1. The direct Laspeyres index obviously is given by

$$(7.2.3)\qquad P_{02}^{L} = \frac{p_{12}}{p_{10}}a + \frac{p_{22}}{p_{20}}(1-a) = m_1 a + m_2(1-a),$$

where m_1 and m_2 denote price relatives. This should be compared with $\overline{P}_{02}^{LC}$, which is by definition $\overline{P}_{02}^{LC} = \left[\frac{p_{11}}{p_{10}}a + \frac{p_{21}}{p_{20}}(1-a)\right]\cdot\left[\frac{p_{12}}{p_{11}}b + \frac{p_{22}}{p_{21}}(1-b)\right]$, or in terms of price relatives[79]

$$(7.2.4)\qquad \overline{P}_{02}^{LC} = m_1 a[b + g(1-b)] + m_2(1-a)[(1-b) + b/g] = m_1 a\, f_1 + m_2(1-a)\, f_2$$

$$\text{where } g = \frac{p_{11}p_{22}}{p_{12}p_{21}} = \frac{p_{22}/p_{21}}{p_{12}/p_{11}}.$$

Given g > 1 (price 2 rising more than price 1) we get $f_1 > 1$ (hence $f_2 < 1$), which associates a greater weight to commodity 1 (to price relative m_1) in the chain index as compared to the direct index. Conversely, as price 1 rises less than price 2 (g < 1, $f_1 < 1$) $\overline{P}_{02}^{LC}$ will give more weight to commodity 2. Thus we expect $\overline{P}_{02}^{LC} < P_{02}^{L}$ due to giving a relatively lower weight to the commodity whose price is rising fastest in $\overline{P}_{02}^{LC}$ compared to P_{02}^{L}. Note that g = 1 is tantamount to $\overline{P}_{02}^{LC} = P_{02}^{L}$ irrespective of weights b (and thus also 1-b).

---

[79] The term g denotes a ratio of price relatives. In our book on chain indices we gave a more general representation of the relationship in matrix notation (of which eq. 7.2.4 is only a special case).

It follows that the chain index is not an average of price relatives since the sum af_1 + $(1-a)f_2$ does not necessarily equal unity but instead amounts to

(7.2.4a) $af_1 + (1-a)f_2 = 1 + ab(2-g) - a(1-g) - b(g-1+a)/g \neq 1.$

The sum of the weights is in general $Q^L_{0,t-1}/\overline{Q}^P_{0,t-1}$, and therefore in this case $Q^L_{01}/\overline{Q}^P_{01}$. Hence surprisingly yet easily demonstrable

> A chain index may well exceed the greatest individual price relative or turn out to be smaller than the smallest price relative.

This is demonstrated in the following numerical example and in particular in ex. 7.2.2.

Example 7.2.1

Assume the following prices and quantities of n = 2 commodities (i = 1, 2)

	period 0		period 1		period 2	
i	prices	quantities	prices	quantities	prices	quantities
1	2	10	12	3	12	
2	5	4	7	10.29	14	

Quantities in period 2 are irrelevant for the calculation of $\overline{P}^{LC}_{02}$ as well as for P^L_{02}. The direct index obviously is given by eq. 7.2.3

$$P^L_{02} = \frac{p_{12}}{p_{10}}a + \frac{p_{22}}{p_{20}}(1-a) = m_1 a + m_2(1-a) = \frac{12}{2}0.5 + \frac{14}{5}0.5 = 4.4 \text{ such that}$$

$$14/5 = 2.8 < P^L_{02} = 4.4 < 12/2 = 6$$

This should be compared with the chain index $\overline{P}^{LC}_{02}$, given by

$$\overline{P}^{LC}_{02} = \left(6\cdot\frac{1}{2} + \frac{7}{5}\cdot\frac{1}{2}\right)\left(1\cdot\frac{1}{3} + 2\cdot\frac{2}{3}\right) = 6.167 > 6.$$

The terms of eq. 7.2.4 are $af_1 = 0.833$ and $(1-a)f_2 = 0.4167$ such that $af_1 + (1-a)f_2 = 1.25 \neq 1$, and $6\cdot 0.833 + 2.8\cdot 0.4167 = 6.167$, and g =2. ♦

Two points (responsible for the awkward result of $\overline{P}^{LC}_{02} = 6.167 > 6$) should be considered

- the "cumulative" (see **sec. 2.5**) nature of the inherent weights (in the case of two periods P_{02} they are affected by *both* sets of weights a and b), and
- in addition to these weights a and b, it is also the differential price movement of the commodities (over the two periods) as measured by g which determines the index.

And note also: No system of weights using *single* weights (a or b respectively in isolation) *or* multiple ones in the form of a *mean of weights* for example $(a+b)/2$ or $\sqrt{ab}$ would be capable of producing such a counterintuitive result as in ex. 7.2.1.

It is not only a theoretical possibility that chain indices may fail the mean value test, this has been shown empirically already at least once (Szulc 1983, p. 560). The (often quoted) results of the Canadian Consumer Price Index (a chain index) March 1978 were as follows

Goods	171.1
Services	171.4
Goods and Services	170.8

In view of such findings Hill 1988, p. 14 gave the following comments:

> "There is little doubt that most users find these kinds of results perverse. More seriously, it may be inconvenient or even impossible to work with chain indices within the framework of an accounting system or economic model in which different aggregates are defined, or constrained, by accounting identities when the indices for these aggregates do not respect the same identities."

It should be remembered that Peter Hill was one of the chief authors of the SNA recommendations, which in essence amounts to wiping out completely this sort of objections. The interesting conclusions which might be drawn from the violation of the mean value property are twofold:

Aggregation *over time*

the idea of chain indices is to use, if possible, the most recent and thus (!) most *"representative" weights in each link* (period), however

the product, that is, *the chain, is by no means* necessarily *representative* (in the sense of typical or averaging) *of the price relatives* (over the whole time interval)

Aggregation over commodities (or subindices)

Due to the *cumulative* nature of weights in the chain approach, the resulting implicit weights of the chain (as opposed to the links) will in general not sum up to unity, which is the reason for inconsistency in aggregation.

The question of course arises, what is more important, to have representative weights in each period ("new" instead of "old" ones), or to have a representative mean of the relatives p_{1t}/p_{10}, p_{2t}/p_{20}, ... , p_{nt}/p_{n0}, that is, to have a representative[80] result over the whole course of time?

[80] In the sense of "typical" or "average" with respect to the commodities over the full time span.

c) No identity and monotonicity: axioms apply to links only (not to the chain)

Many arguments advanced to justify chain indices suffer from a lack of axiomatic tools to evaluate properties of formulas, which should be seen as a serious disadvantage.[81] There are two problems that need to be considered:

- Axioms as discussed in chapter 2 and 3 apply to the direct approach only.
- As already pointed out in **sec. 2.5**

Only the link is an index in the sense of satisfying or violating certain "axioms" of the type discussed in chap. 2 and 3. But a chain is *not* an index and can violate axioms, despite being "made" of links that satisfy them all.

It is therefore remarkable and should be well borne in mind, that it is

- *of little or no use* to discuss the relative merits of various formulas for links in terms of axioms,[82] although deliberations whether for example a Vartia II *chain* index rather than a Törnqvist *chain* index or so should be preferred are not unusual exercises, however,
- conspicuously we frequently see fictitious numerical examples, simulations etc. in the case of chain indices (in contrast to direct indices) because the *general* approach of axiomatic reasoning is widely inapplicable.[83]

The idea of ***identity*** applied to situations 0 and 2 (i.e. two links)[84] with a situation (t = 1) in between would require $\overline{P}_{02} = 1$ to hold whenever prices in 0 and 2 are equal. In a similar manner, ***monotonicity*** requires $\overline{P}_{02}$ to differ from unity when a single price p_{i2} differs from p_{i0}, all other prices being equal. The following example demonstrates that a chain Laspeyres price index violates both axioms together with the mean value property, although the direct index P^L_{02} of course satisfies all these criteria.

Example 7.2.2

Given the following prices and quantities of n = 2 commodities (i = 1, 2)

---

[81] Properties of direct indices are in general easily inferable from the formula used, while those of chain indices are often "rather obscure" (Szulc 1983, p. 539f.). It is therefore useful to utilize of some tools in order to assess the behaviour of chain indices.

[82] or even to "prove" mathematically the alleged superiority of chain indices over direct indices with the help of certain axioms, as for example the Italian statistician Marco Martini tried to do in an expertise delivered to Eurostat (Martini 1996). Given the fundamental difficulties with axioms in the case of chain as opposed to direct indices, undertakings of this sort are certainly highly dubious.

[83] It should be noted that this is a considerable disadvantage of chain indices because one of the uses of some fundamental axioms is to allow a clear prediction of results of index calculations given some simple idealized or "stylized" conditions. When on the other hand axioms are not applicable we are reduced to the above mentioned wide spread practice to calculate numerical examples and study results of simulations, which is clearly less convincing.

[84] It is not unusual to define axioms for the chain index case by analogy to the corresponding axioms for the direct index case.

	period 0		period 1		period 2	
i	prices	quantities	prices	quantities	prices	quantities
1	8	6	6	10	8	
2	12	4	15	5	12	

The fields involved in variations are set off by shadows. Quantities in period 2 are irrelevant for the calculation of $\overline{P}_{02}$. However we will try different variants concerning prices p_{i2} and quantities q_{i1} (see **ex. 7.2.3**). The results are as follows:

Direct index: P_{02}^{L} = 1 due to identity of prices in 0 and 2, the weights (expenditure shares) in notation of **ex. 7.2.1** are in a = 0.5 and b = 4/9 for commodity 1 in period 0 and 1.

Chain index: $P_1^{LC} = P_{01}^{L} = 1$ and $P_2^{LC} = 1.037$ and therefore $\overline{P}_{02}^{LC} = P_1^{LC}\, P_2^{LC} = 1.037$. $\overline{P}_{02}^{LC}$ indicates a rise in prices, although both price relatives $p_{i2}/p_{i0} = 1$ (i = 1, 2) show no change and therefore the direct Laspeyres index *correctly* shows that prices remained constant[85] (by resulting in $P_{02}^{L} = 1$).

With factors f_1 and f_2 as defined above we get $g = \frac{12/15}{8/6} = \frac{3}{5} = 0.6$ giving $f_1 = \frac{7}{9}$

and $f_2 = \frac{35}{27}$ and eq. 7.1.12 reads as follows in this example

$$\overline{P}_{02}^{LC} = 1 \cdot \frac{1}{2} \cdot \frac{7}{9} + 1 \cdot \frac{1}{2} \cdot \frac{35}{27} = \frac{28}{27} = 1.037 \neq 1$$

which is simply the sum of the "new weights", a f_1 and (1-a) f_2.

Consider two modifications of the prices of the example as follows[86]

Variant a: $p_{12} = 8$, $p_{22} = 11$, variant b: $p_{12} = 7$, $p_{22} = 12$.

Variant a is designed to show that a chain index can well violate monotonicity. The chain index yields $\overline{P}_{02}^{LC} = 1$ although the price of commodity 2 is clearly declining, the price of commodity 1 remaining constant (P_{02}^{L} = 92/96 = 0.9583).

The interesting point in ***Variant b*** is that a chain index, though in line with monotonicity, may well be smaller than the smallest individual price relative $m_1 = p_{12}/p_{10} = 7/8 = 0.875$ (whilst m_2 remains unity). We then get $P_1^{LC} = 1$ and P_2^{LC} = 155/180 = 0.86111 while P_{02}^{L} = 90/96 = 0.9375. Hence a chain index can violate the mean value property not only in the case of *identical* price relatives $m_1 = m_2 = 1$

[85] It is by the way easy to modify the example such that we find a *decline* in the prices though prices in 0 and 3 are identically 8 and 12 respectively in both periods. Simply assume quantities 5 and 10 instead of 10 and 5 in period 1 and we get 16/180 = 0.888 for the second link (and thus for the chain). Note that this variant differs only with respect to quantities, not to prices and therefore f_1 = 2/3 and f_2 = 10/9. Moreover, the "new weights" are af_1 = 1/3 and $(1-a)f_2$ = 5/9 as (compared with a = (1-a) = ½) summing up to 0.888.

[86] All other prices and quantities remain unchanged.

by exceeding this result (1.037 in **ex. 7.2.1**) or falling short of it. This (violation of the mean value property) can also happen in the case of $m_1 = 0.875 < m_2 = 1$. ♦

This example (ex. 7.2.1) shows that a chain Laspeyres index (as all other chain indices) violates identity as well as monotonicity and even the mean value property. Axioms are defined here for the interval (0,2) with two subintervals (0,1) and (1,2) in just the same manner as in the case of direct indices for the interval (0,2).

Another way of expressing the fact that $\overline{P}_{02}^{LC} \neq 1$ (given identical prices in 0 and 2!) results from the Laspeyres formula not being time reversible[87]. It is precisely in view of this shortcoming of the Laspeyres and Paasche chain index that the SNA 93 recommends the use of the chain version of the Fisher- $\overline{P}_{0t}^{FC}$, or Törnqvist index $\overline{P}_{0t}^{TC}$. However, there are other difficulties involved in using index formulas of this kind.[88] From

$$(7.2.5)\quad \overline{P}_{02}^{LC} = \frac{\sum p_1 q_0}{\sum p_0 q_0} \frac{\sum p_0 q_1}{\sum p_1 q_1} = \frac{P_{01}^{L}}{P_{01}^{P}} = \frac{Q_{01}^{L}}{Q_{01}^{P}} = 1$$

it follows that in order to meet identity $\overline{P}_{02}^{LC} = 1$ we should have $P_{01}^{L} = P_{01}^{P}$.

Note that what matters is not the direction of change (rise or decline), but the correlation between price and quantity relatives, because depending on this correlation we have $P_{01}^{L} > P_{01}^{P}$ and thus $\overline{P}_{02}^{LC} > 1$ or $P_{01}^{L} < P_{01}^{P} \Rightarrow \overline{P}_{02}^{LC} < 1$.

It is interesting to note that not only identity and monotonicity but also ***proportionality*** is not met in the case of chain indices, that is $P(\mathbf{p}_0, \lambda\mathbf{p}_0) = \lambda$ will not be met because $P_1^{LC} P_2^{LC} = \frac{\sum p_1 q_0}{\sum p_0 q_0} \frac{\sum \lambda p_0 q_1}{\sum p_1 q_1}$ will not in general result in λ if all prices change λ-fold.[89]

Linear homogeneity requires $P(\mathbf{p}_0, \lambda\mathbf{p}_2) = \lambda P(\mathbf{p}_0, \mathbf{p}_2)$ and is fulfilled because

$$\frac{\sum p_1 q_0}{\sum p_0 q_0} \frac{\sum \lambda p_2 q_1}{\sum p_1 q_1} = \lambda \frac{\sum p_1 q_0}{\sum p_0 q_0} \frac{\sum p_2 q_1}{\sum p_1 q_1}$$ in the case of $\overline{P}_{02}^{LC} = P_1^{LC} P_2^{LC}$.

d) Cyclical movement of prices

The purpose of the following example (**ex. 7.2.3**) is to show that

---

[87] SNA 93, remark no.16.46.

[88] One of them is to find a good interpretation for values deflated with the help of a chain Törnqvist index (see also Forsyth und Fowler 1981, two prominent proponents of the chain index approach).

[89] $p_{i2} = \lambda p_{i0}$ for all i. The chain index will in general not amount to λ unless the change of quantities also meets certain very special conditions.

1. the above demonstrated violation of axioms may also occur in the case of an unweighted chain index, such as the chain index formula of Carli[90], such that it is in general the principle of chaining itself, not the existence of weights, which causes the violation of the axiom (for example identity if time reversibility is not met),[91] and
2. to demonstrate the well known fact that a chain index can "explode" in the case of *cyclical* price movement, such that we cannot always assume that a chain index will show less inflation or a smoother development.[92]

Example 7.2.3

Consider again two commodities. The price of the first will be redoubled in 1, such that $p_{11}/p_{10} = 2$. In a similar vein, in 2 the price p_{12} is half of p_{11}, or in other words, it returned to the original level $p_{12}/p_{11} = 0.5$ or $p_{12} = p_{10}$. The opposite process will be assumed for the other (second) commodity such that $p_{21}/p_{20} = 0.5$ and $p_{22}/p_{21} = 2$.

The Carli chain index now is simply $\overline{P}_{02}^{CC} = \frac{1}{2}(2+0.5)\cdot\frac{1}{2}(0.5+2) = (1.25)^2 = 1.56 > 1$.

Assume further the process carries on endlessly such that we get the following prices $\overline{P}_{02}^{CC}$

commodity	0	1	2 (= 0)	3 (= 1)	4 (= 0)
no. 1	6	12	6	12	6
no. 2	8	4	8	4	8

It should be noted in particular that we get (due to identical prices) for P_{0t}^{L} (and all other reasonable direct indices of course) $P_{02}^{L} = P_{04}^{L} = P_{06}^{L} = \ldots = 1$. Likewise $P_{01}^{L} = P_{03}^{L} = \ldots = 1.25$. However, obviously the Carli chain index (or all other chain indices) will become greater and greater (in the case of $\overline{P}_{02}^{CC} > 1$) as such a cycle of two periods will be repeated again and again, and it will become smaller and smaller if $\overline{P}_{02}^{CC} < 1$.

In our example we get $\overline{P}_{04}^{CC} = (1.56)^2 = 2.44$, $\overline{P}_{06}^{CC} = (1.56)^3 = 3.81$ and so on.

90 also known as the Sauerbeck (chain) index. A similar exercise is to be found in Szulc 1983, p.535f. In the practice of price statistics, however, there is no point in considering an unweighted chain index because chain indices are motivated by being capable of adjusting weights rapidly. In the absence of weights there can be no such advantage of chaining.

91 Note that the following effects are also due to Carli's unweighted *arithmetic* mean and it would not occur in the case of an unweighted geometric mean of price relatives (Jevons' index as chain index).

92 There are many popular prejudices concerning the behaviour of chain indices, such as chain indices moving generally much smoother than direct indices. Szulc 1983, p. 555 for example found out that this is not true, and that "chain indices can be more biased than their direct counterparts, which is both contrary to popular belief and to the purpose of linking".

If weights are introduced the situation is basically the same (which will be demonstrated again in **ex. 7.2.4**). Assume for example weights as follows: a = (1-a) = 0.5 in period 0, 2, 4,... and b = 0.25 , (1-b) = 0.75 in period 1, 3, 5,... respectively. This yields a base weighted chain index $\overline{P}_{02}^{LC} = \left(2\cdot\frac{1}{2}+\frac{1}{2}\cdot\frac{1}{2}\right)\left(\frac{1}{2}\cdot\frac{1}{4}+2\cdot\frac{3}{4}\right)=2.03$ incidentially exceeding both individual price relatives (amounting identically to unity).[93] The index continues as follows: $\overline{P}_{04}^{LC}=2.03^2=4.126$, $\overline{P}_{06}^{LC}=2.03^3=8.381$ etc. ♦

This example is a most simple demonstration of some well known results of chaining, also taken into account by SNA 93 when it says[94] that a chain index should

- ***not*** be used when prices are ***cyclically*** moving (rising and declining, and thereafter returning to a certain level in some regular manner) in contrast to
- a (moderate) *monotone* rise or decline of prices, in which case a chain index is recommended, or in summary SNA 93 arrived at the following rule

when the relative prices in the first and last periods (0,t)	a chain index
1) are very different from each other and chaining involves linking periods in which prices and quantities are intermediate between those of 0 and t	should be used
2) are similar to each other (and very different to an intermediate period t*); example: seasonal variation	should not be used

This amounts to requiring a more or less smooth development in which direct indices are not so objectionable (in contrast to abrupt and substantial changes). It reminds us of the remark made at the beginning of **sec. 7.1**, according to which a chain index is praised to be a good medicament, effective in those cases where no illness exists. In summary the SNA concludes:

"On balance, situations favorable to the use of chain Laspeyres and Paasche indices over time seem more likely than those that are unfavorable".[95]

The question is, however, why such most likely situations should be exactly those which are unfavorable for direct indices of the type Laspeyres and Paasche.

e) No transitivity but path dependence

Many writers erroneously conclude from a chain index being defined as a product that the chain index is transitive (in the sense of Fisher's chain test, transitivity means

93 It is of course easy to construct an example in which the index will be less than unity, the result we get for the relatives. Assume for example weights of 1/4 and 3/4 in period 0 and 3/4 and 1/4 in period 1 respectively. This gives $\overline{P}_{02}=(7/8)^2=0{,}766$.

94 no. 16.47 – 49.

95 SNA paragraph 16.49. According to Banerjee 1975 "the circular test is ... satisfied by the chain index" (p. 55).

path *in*-dependence[96], or in other words *consistency in aggregation over time* by chaining), or "chainable" *by definition* since

(7.2.6) $\overline{P}_{0t} = \overline{P}_{0k}\overline{P}_{kt}$

holds. However the idea of the chain test is that the result ($\overline{P}_{0t}$) should be the same for *any* k, irrespective of how the interval (0, t) is partitioned. Conspicuously, precisely *this* is generally *not* true in the case of chain indices, because

- It can easily be shown that, for example, $P_{01}P_{12}...P_{56}$ and $P_{02}P_{24}P_{46}$ do *not* yield the same result $\overline{P}_{06}$ (see **ex. 7.2.4**).
- Moreover, it can be shown (see part j of this section) that the idea of chainability as a requirement is necessarily in conflict with other important aspects of index construction, like additivity and adaptability of weights.

The name "chain index" is misleading: multiplication ("chaining") may give rise to the impression chainability were met. But this is not true. Different subdivisions of an interval will yield different results and these results will also differ from a direct comparison.

A correct understanding of transitivity (chainability, circularity) requires that *each* chaining over the *same* interval in time will yield the same result, that is, indirect comparisons of *all* sorts (for whatever partition of the interval) should be consistent with direct comparison and also consistent among themselves. And precisely this property is not found in the case of chain indices.

"Circularity" in aggregating over time intervals means irrelevance of the path. The original idea of chaining was to make a long period comparison between 0 and t consistent with a series of short term comparisons. This is what might be called a **temporal aggregation** (over time). In the case of simple price relatives this can be done without any restriction simply by multiplication, since

(7.2.7) $$\frac{p_{it}}{p_{io}} = \frac{p_{is}}{p_{io}}\frac{p_{it}}{p_{is}} = \frac{p_{ir}}{p_{io}}\frac{p_{is}}{p_{ir}}\frac{p_{it}}{p_{is}}$$

will necessarily yield the same result, irrespective of how the interval between 0 and t is partitioned into two (0-s, s-t), or three (0-r, r-s, s-t), or whatever number of subintervals. The result is *unique*, not depending on the type and number of subintervals.[97] Obviously chain indices (instead of simple relatives) do *not* comply with precisely this idea. Hence non-chainability (non-circularity) has two meanings

1. it is often defined in the sense of divergence of *chain* indices from *direct* indices, but
2. another meaning is that different chain indices (using different partitions of the same interval) will also diverge, as demonstrated in what follows.

---

[96] Path dependence therefore is obviously the opposite of transitivity.

[97] The underlying idea of chaining links is to make sure that a result referring to a year for example should not be affected by whether it is produced by data referring to months, or to quarters of a year, or any other subdivision.

Example 7.2.4:

Assume the following "data"

t=0		t=1		t=2		t=3		t=4	
p	q	p	q	p	q	p	q	p	q
2	10	4	12	3	20	1	16	2	10
5	20	3	15	4	10	4	12	5	20

The direct index is of course $P_{04}^{L} = 1$ because all prices (and also quantities) in 4 equal those in 0 (indicated by shadows). However, the chain index not only violates identity but also yields different results depending on how the interval is subdivided:

(a) $\bar{P}_{04}^{LC}(a) = P_{02}^{LC} P_{24}^{LC} = \frac{110}{120}\frac{90}{100} = 0.825$, with only *two* intervals (0, 2) and (2, 4),

but

(b) $\bar{P}_{04}^{LC}(b) = P_{1}^{LC} P_{2}^{LC} P_{3}^{LC} P_{4}^{LC} = \left(\frac{100}{120}\frac{96}{93}\right)\left(\frac{60}{100}\frac{92}{64}\right) = \frac{88.89}{120}\frac{86.25}{100} = 0.7419$

upon dividing the *same* interval into *four* subintervals (0, 1), ... , (3, 4). ♦

Hence the situations 0 and 4 are not uniquely compared. The result depends on the way the interval is subdivided. This is clearly not in line with chaining of relatives and what was originally intended by "chainability" and made chain indices appear desirable. To summarize what we have learned from the example:

> Due to **path-dependence**[98], chain indices depend on how an interval is subdivided; they provide a summary description of a process rather than a comparison of two situations. A chain index in action not only depends on which subintervals are combined, it is also anything but a "pure" price comparison.

Being "path-dependent" is the reason for many shortcomings of chain indices, and it should also be noted that demand for chain indices ensues primarily from a desire to make comparisons over *any* interval of any length, particularly in cases where direct comparisons fail. Given these purposes of chaining it is clearly a disadvantage if the chaining of indices far from provides unambiguous temporal aggregation.

This leads to the more fundamental problem of what kind of measurement an index should aim at. Should this be the (relative) ***level*** of prices in period t compared with period 0, or is an index a summary expression of the shape of a ***time series*** (see **fig. 7.2.1**)?[99]

---

[98] The term describes a more general feature of chain indices, of which several consequences (such as violation of identity) had been introduced already. An interesting question is of course how "chainers" deal with non-circularity (path dependence) of chain indices. In my book on chain indices I quoted some most revealing remarks (p. 177).

[99] In this sense it is sometimes maintained (for example Diewert and Nakamura, 1993) that chain indices serve another purpose: they focus on *multi*-lateral situations rather than *bi*-lateral situations. If this

A sequence of chain indices $\overline{P}_{01}, \overline{P}_{02}, ..., \overline{P}_{0t}$, in contrast to the corresponding sequence of direct indices $P_{01}, P_{02}, ..., P_{0t}$, does not represent a consistent *temporal* aggregation. Moreover, it reflects a *mixture* of various influences, prices and influences other than prices. This can hardly be an advantage or even a gain in information compared to a separation of such influences and an attempt to measure each of them as "purely" as possible.

Figure 7.2.1: Objectives and principles of index construction

Objective of a price index	
direct index to compare prices in two periods*	**chain index** a summary measure of a **time series**
$P_{0t} = P_{0\tau}$ should hold if prices in t equal prices in τ, otherwise we should have $P_{0t} \neq P_{0\tau}$.	It is not clear under which conditions the two sequences of observations $\overline{P}^C_{0t}$ and $\overline{P}^C_{0\tau}$ should yield the same (different) result.
This idea is known as the principle of "pure price comparison" (chain indices violate it *in a number of* ways; see **fig. 7.2** and **7.2.3)**	In the case of identical individual prices in t and τ, $\overline{P}^C_{0t}$ need not be equal to $\overline{P}^C_{0\tau}$, nor is $\overline{P}^C_{0t}$ different from $\overline{P}^C_{0\tau}$ if prices are different.

* the symbols t and τ denote different periods, or different data relating to the same period.

f) Nonlinearity (the determinants of an increase/decrease of the price level)

Due to the linearity (in the prices of period t) of the direct Laspeyres index, an equal amount of change in prices denoted by $\Delta p_t = p_t - p_{t-1}$ in period t and $t^* \neq t$ has the same effect as long as t or t* are periods in the range of an index of the same base 0 (0 < t* < t). We get

$$\text{(7.2.8)} \qquad P^L_{02} = 1 + \frac{\sum q_0 \Delta p_1}{\sum q_0 p_0} + \frac{\sum q_0 \Delta p_2}{\sum q_0 p_0} = P^L_{01} + \frac{\sum q_0 \Delta p_2}{\sum q_0 p_0}$$

and by the same token in the case of three periods

$$\text{(7.2.9)} \qquad P^L_{03} = P^L_{02} + \frac{\sum q_0 \Delta p_3}{\sum q_0 p_0} \text{ etc.}$$

really were the case there would be no logic, however, in saying that one approach (chain) is superior to the other (direct). It is nonsense to say, for example, that the variance is better than the correlation coefficient, two measures designed for different purposes.

For example, a change in the prices of two commodities of, say, Δp_{1t} = -2 and Δp_{2t} = +4 weighted with constant quantities (6 and 4) has the same effect irrespective of the period in which the change takes place. With weights $q_0 / \sum p_0 q_0$ for the commodity 1 and 2 being[100] 6/96 and 4/96, we get

$$P_{01}^L - P_{00}^L = P_{01}^L - 1 = \frac{6}{96}\cdot(-2) + \frac{4}{96}\cdot(+4) = \frac{1}{24} = 0.04167$$

when the change takes place in period 1 (denoted by Δp_1 in eq. 8). And when the same change again occurs in period 2 (such that $\Delta p_2 = \Delta p_1 = \Delta p$) we get a cumulative change as follows $P_{02}^L - P_{00}^L = 0.04167 + 0.04167 = 0.08333$ or

$$P_{02}^L = 1 + 2\left(\Delta p \frac{q_0}{\sum q_0 p_0}\right).$$

Table 7.2.2: Relation between changes of individual prices and the price index

t	direct Laspeyres	chain Laspeyres
1	$P_{01}^L = 1 + \sum \Delta p_1 \frac{q_0}{\sum q_0 p_0} = 1 + K_1^0$	$\bar{P}_{01}^{LC} = 1 + \sum \Delta p_1 \frac{q_0}{\sum q_0 p_0} = P_{01}^L = 1 + K_1^0$
2	$P_{02}^L = P_{01}^L + \sum \Delta p_2 \frac{q_0}{\sum q_0 p_0}$ $= 1 + K_1^0 + K_2^0 = P_{01}^L + K_2^0$	$\bar{P}_{02}^{LC} = \bar{P}_{01}^{LC}\left(1 + \sum \Delta p_2 \frac{q_1}{\sum q_1 p_1}\right)$ $= \left(1 + K_1^0\right)\left(1 + K_2^1\right) = \bar{P}_{01}^{LC}\left(1 + K_2^1\right)$
3	$P_{03}^L = P_{02}^L + \sum \Delta p_3 \frac{q_0}{\sum q_0 p_0}$ $= 1 + K_1^0 + K_2^0 + K_3^0 = P_{02}^L + K_3^0$	$\bar{P}_{03}^{LC} = \bar{P}_{02}^{LC}\left(1 + \sum \Delta p_3 \frac{q_2}{\sum q_2 p_2}\right)$ $= \left(1 + K_1^0\right)\left(1 + K_2^1\right)\left(1 + K_3^2\right) = \bar{P}_{02}^{LC}\left(1 + K_3^2\right)$

In the direct index case the same changes in prices at different times have the same effect on the price index. The corresponding relationship in the case of a Laspeyres *chain* index is much more complicated and it is interesting to examine what happens in the case of a Laspeyres *chain* index when prices change to the same extent in two consecutive periods (see **ex. 7.2.5** and **ex. 7.2.6** for a numerical illustration). Obviously a constant change by, for example, -2 and +4 will *not* have a constant effect. The effect is rather given by

$$(7.2.10) \quad \bar{P}_{02}^{LC} = \left(1 + \frac{\sum q_0 \Delta p_1}{\sum q_0 p_0}\right)\left(1 + \frac{\sum q_1 \Delta p_2}{\sum q_1 p_1}\right) = \bar{P}_{01}^{LC} + \left(\sum \Delta p_2 \frac{q_1}{\sum q_1 p_1} \bar{P}_{01}^{LC}\right).$$

As the chain becomes longer, the relationship becomes more complicated

[100] the figures refer to **ex. 7.2.2**.

(7.2.11) $\overline{P}_{03}^{LC} = \overline{P}_{02}^{LC}\left(1+\frac{\sum q_2 \Delta p_3}{\sum q_2 p_2}\right) = \overline{P}_{02}^{LC} + \left(\sum \Delta p_3 \frac{q_2}{\sum q_2 p_2} \overline{P}_{02}^{LC}\right).$

As can easily be seen *in general* we obviously get in the case of a chain index *the product*

(7.2.12) $\overline{P}_{0t}^{LC} = \left(1+\frac{\sum q_0 \Delta p_1}{\sum q_0 p_0}\right) \cdots \left(1+\frac{\sum q_{t-1} \Delta p_t}{\sum q_{t-1} p_t}\right) = \left(1+K_1^0\right) \cdots \left(1+K_t^{t-1}\right),$

in contrast to the sum

(7.2.13) $P_{0t}^{L} = 1+\frac{\sum q_0 \Delta p_1}{\sum q_0 p_0} + \ldots + \frac{\sum q_0 \Delta p_t}{\sum q_0 p_0} = 1 + K_1^0 + \ldots + K_t^0$

in the case of the corresponding direct index. This shows that:

- P_{0t}^{L} is a linear combination of changes $\Delta p_1, \Delta p_2, \ldots$ with constant "weights" $q_0 / \sum p_0 q_0$ such that in the case of a constant change $\left(\Delta p_1 = \Delta p_2 = \ldots = \Delta p_t\right)$ in all periods a constant term is added, and the result is independent of how the interval (0, t) is subdivided into sub-intervals.[101] In contrast
- in $\overline{P}_{0t}^{LC}$ changes of individual prices $\Delta p_{i1}, \Delta p_{i2}, \ldots$ are *multiplied* by weights which are in general *not* constant. Even the effect of a constant change $\left(\Delta p_1 = \Delta p_2 = \ldots = \Delta p_t\right)$ will not be independent when the change takes place unless all quantities also remain constant such that $q_0 = q_1 = \ldots = q_{t-1}$. Hence:

In a direct index, such as P^L for example, the change of the index from t to t+1 and so on is dependent on (a function of) the changes of individual prices Δp_{it} (i = 1, ..., n, t = 1, 2,...). In particular, a constant change has the same constant effect on the overall index, thereby making its result independent of how the interval (0, t) is subdivided into sub-intervals. This does not apply, however, to chain indices. There the same change will not have the same effect, and the result is not invariant with respect to how aggregation takes place over periods in time (temporal aggregation) nor is the resulting index functionally related to the individual price changes Δp_{it} in a constant and obvious manner.

For example the term to be added to $\overline{P}_{0t}^{LC}$ in order to get $\overline{P}_{0,t+1}^{LC}$ (in eqs. 10 and 11)

(7.2.14) $\sum_i \Delta p_{i,t+1} \frac{q_{it}}{\sum q_{it} p_{it}} \overline{P}_{0t}^{LC} = \overline{P}_{0,t+1}^{LC} - \overline{P}_{0t}^{LC}$

[101] In other words it is irrelevant whether a change of +10$ from 1995 to 2000 is considered or a constant change of +2$ in each of the five one-year-intervals.

is much more difficult to understand (and reflective not only of quantities $q_{i,t-1}$ and prices $p_{i,t-1}$ in the past but also even of the level the chain index has already reached so far)[102] than the term added to P^L_{0t} to get $P^L_{0,t+1}$, which is simply

$$(7.2.15) \quad P^L_{0,t-1} - P^L_{0,t} = \sum_i \Delta p_{it}\left(\frac{q_{i0}}{\sum q_{i0}p_{i0}}\right) = \sum_i \Delta p_{it} w_i$$

with constant weights w_i as a consequence of P^L_{0t} being linear or "additive".

There is no theory known to explain why the *same* amount of change of prices should be treated quite *differently* as shown in eq. 14, and (more importantly) why it should be treated in the way a chain index does, that is, depending on

- current (or most recent) quantities, and
- on the "history" of the chain index and on how the interval (0, t) is subdivided.[103]

The question is not to justify why a *constant* increase in prices should be treated differently depending on *when* the increase occurs. The question is rather by which economic theory it appears justified to do temporal aggregation (summation over time intervals) precisely in the way indicated in eq. 7.2.14.

Example 7.2.5

The following numerical example is again a modification of **ex. 7.2.2**, and the salient feature is that the price of commodity 1 uniformly (in both periods 1 compared with 0 and 2 with 1) declines by - 2 and the price of commodity 2 constantly rises by + 4.

	period 0		period 1		period 2	
i	prices	quantities	prices	quantities	prices	quantities
1	8	6	6	10	4	
2	12	4	16	3	20	

The direct Laspeyres index yields, as shown above, 1.0417 and 1.0833 rising constantly by 4.17 percentage points because the weights of the price are given by (note that $p_{10}q_{10} = p_{20}q_{20} = 48$)

$\frac{q_{10}}{\sum q_{i0}p_{i0}}$ = 6/96 and $\frac{q_{20}}{\sum q_{i0}p_{i0}}$ = 4/96 = 0.0417 such that

[102] The weights attached to price differences Δp are not only changing, they are also influenced by the history of the chain.

[103] In the following part j of this section we try to show that this is a property of chain indices which definitely was *not* intended in the 19th century discussion of transitivity in index theory. The problem discussed here is not lack of transitivity (or inconsistency in "temporal aggregation" over (sub-) intervals) but rather the theoretical justification of precisely that type of temporal aggregation which underlies the procedure of chaining.

$$\Delta p_1 \frac{q_{10}}{\sum q_{i0}p_{i0}} + \Delta p_2 \frac{q_{20}}{\sum q_{i0}p_{i0}} = (-2)\frac{6}{96} + (+4)\frac{4}{96} = 0.0417 .$$

Hence the direct index treats the overall effect of an absolute change of +4 and –2 as a rise of prices by 4.17 percentage points. The chain index, however, develops in the following way

t	$\overline{P}_{0t}^{LC}$	weights*) given in $\overline{P}_{0t}^{LC} - \overline{P}_{0,t-1}^{LC}$ to	
		commodity 1	commodity 2
1	$\frac{100}{96} = 1.0417$	6/96 = 0.0625	4/96 = 0.04167
2	$\frac{100}{96} \cdot \frac{100}{108} = 0.9645$	$\frac{10}{108} \cdot \frac{100}{96} = 0.09645$	$\frac{3}{108} \cdot \frac{100}{96} = 0.02894$

*) This is the term $\Delta p_{it}\left(q_{i,t-1} / \sum p_{i,t-1}q_{i,t-1}\right)\overline{P}_{0,t-1}^{LC}$ for $i = 1,2$.

Note that $\overline{P}_{02}^{LC}$ = 0.9645 indicates a decline in prices as opposed to a rise in P_{02}^{L} = 1.0833. This is due to the fact that the weight of commodity 1 (getting cheaper) rose by 54.4% (from 0.0625 to 0.09645) whereas the weight of commodity 2 (getting more expensive) declined by 30.55%. $\overline{P}_{02}^{LC}$ differs from $\overline{P}_{01}^{LC}$ by - 0.07716 (while P_{02}^{L} exceeds P_{02}^{L} by the constant amount of 0.0417) due to $-0.07716 = (-2)\cdot 0.09645 + (+4)\cdot 0.02894$.

The following example shows that we may well get a sequence $\overline{P}_{0t}^{LC}$, t = 1, 2,... indicating a decline in prices although the direct index is not changing. ♦

Example 7.2.6

Consider the following variant of ex. 7.2.5: prices change time and again by –2 and +4 such that $\Delta p_1 = \Delta p_2 = ... = \Delta p_t$ and the quantity of commodity 1 (the price of which is decreases by –2 in each period) rises by 1 in each period:

i	p_0	q_0	p_1	q_1	p_2	q_2	p_3	q_3
1	8	6	6	7	4	8	2	9
2	12	3	16	3	20	3	24	3

For the direct Laspeyres index we get $P_{01}^{L} = P_{02}^{L} = P_{03}^{L} = 1$ which means a *constant* price level.[104] This is so because the example is constructed such that $\sum \Delta p_i q_{i0} = (-2)\cdot 6 + 4\cdot 3 = 0$. The chain index $\overline{P}_{0t}^{LC}$, on the other hand, takes into account that the quantity of commodity 1 is rising, contributing to a growing extent to

[104] The constantly declining price of commodity 1 is exactly offset by a corresponding rise of the price of commodity 2.

an overall *decline* in the price level. It seems plausible that more consumption of an even cheaper product may be seen as a lowering of the general price level. A direct index such as P_{0t}^P or P_{0t}^F will also show a decline in prices:

index	t = 1	t = 2	t = 3
P_{0t}^L	1	1	1
$\overline{P}_{0t}^{LC}$	1	88/90 = 0.977	0.9353
P_{0t}^P	0.9783	0.9583	0.9375
P_{0t}^F *	0.9891	0.9789	0.9682

* in this case $P_{0t}^F = \sqrt{P_{0t}^P}$.

The example shows that there is more than one way to introduce a variable effect of a constant change in prices, and it is not clear which should be the correct one.

♦

g) Inconsistency in aggregation

Whenever more than one item (commodity) is involved in chaining, or we compare two periods indirectly rather than directly, the approach seems to be much more difficult. Problems with consistency in aggregation (over types of commodities) will inevitably emerge. This can be shown alluding a bit to the original conceptualization of chaining by Westergaard (1890) - a Danish statistician known for having first used the principle of chaining - which was developed in terms of stocks and flows. Denote the initial stock by A and net increases by Z_1 and Z_2 referring to the first and second half of the year respectively. The stock develops as follows:

period	stock	relative
0	A	$m_{00} = A/A = 1$
1	$A + Z_1$	$m_{01} = (A + Z_1)/A$
2	$A + Z_1 + Z_2$	$m_{02} = (A+Z_1+Z_2)/A$

and for the relatives we find $m_{02} = \frac{A+Z_1}{A}\frac{A+Z_1+Z_2}{A+Z_1} = m_{01}m_{12}$. This simple situation[105] becomes much more difficult, however, when stocks and flows are broken down into two sectors with stocks $A = A_1 + A_2$, and two flows with $Z_1 = Z_{11} + Z_{12}$ and $Z_2 = Z_{21} + Z_{22}$ respectively.

[105] Note that the result above only shows that the stock may be calculated directly *or* indirectly. Neither of the two procedures deserves to be preferred to the other. It is also irrelevant in which and into how many subintervals the interval is divided.

period	sector 1	sector 2
1	$m_{(1)01} = (A_1 + Z_{11})/A_1$	$m_{(2)01}$ analogous $m_{(1)01}$
2	$m_{(1)12} = (A_1 + Z_{11} + Z_{21})/(A_1 + Z_{11})$	$m_{(2)12} = (A_2 + Z_{12} + Z_{22})/(A_2 + Z_{12})$

Though relatives for each sector (i = 1, 2) obviously remain transitive $m_{(i)02} = m_{(i)01} m_{(i)12}$, it is difficult to see why multiplication of relatives referring to different structures with weights $\frac{A_i}{\sum A_i}$ for m_{01}, and $\frac{A_i + Z_{1i}}{\sum A_i + Z_{1i}}$ for m_{12} leading to

(7.2.16) $$m_{02} = m_{01} m_{12} = \frac{A_1 m_{(1)01} + A_2 m_{(2)01}}{A_1 + A_2} \frac{(A_1 + Z_{11}) m_{(1)12} + (A_2 + Z_{12}) m_{(2)12}}{(A_1 + Z_{11}) + (A_2 + Z_{12})}$$

should be any better than going the direct way

(7.2.17) $$m_{02} = \frac{A_1 m_{(1)02} + A_2 m_{(2)02}}{A_1 + A_2}.$$

Once the variable in question is made up of *sub*-aggregates doing things with chaining becomes much more complicated than in the case of direct computations. There is no doubt that chain indices are more difficult to handle than direct indices whenever aggregation or disaggregation comes into play, such that structural aspects have to be accounted for. It does not suffice to know the sectoral (partial) *links* (instead of relatives) such as $m_{(1)12}$, $m_{(2)12}$, ...for the items (commodities, sectors) i = 1, 2, ..., m (in our simple example m = 2), you also need to know the changing weights of items as shown in **tab. 7.2.3.**

Table 7.2.3: Development of weights in aggregating over sectoral links in order to calculate a chain index $\overline{P}_{0t}^{LC} = \prod_{\tau=1}^{t} \sum_{i=1}^{n} m_{(i)\tau-1,\tau} w_{i\tau}$

sectoral link*	weight of the i-th sectoral link (i) $w_{i\tau}$
$m_{(i)01}$	$\frac{A_i}{\sum A_i}$ $(i = 1, 2, ..., n)$
$m_{(i)12}$	$\frac{A_i + Z_{1i}}{\sum A_i + Z_{1i}}$
$m_{(i)23}$	$\frac{A_i + Z_{1i} + Z_{2i}}{\sum A_i + Z_{1i} + Z_{2i}}$
$m_{(i)34}$	$\frac{A_i + Z_{1i} + Z_{2i} + Z_{3i}}{\sum A_i + Z_{1i} + Z_{2i} + Z_{3i}}$
General $m_{(i)t-1,t}$	$A_i + \sum_{\tau=1}^{\tau=t-1} Z_{\tau i} \Big/ \sum_i A_i + \sum_{\tau=1}^{\tau=t-1} \sum_i Z_{\tau i}$

* or subindex i, i-th component of the index

Furthermore, we have to combine sectoral links and their changing weights and multiply the links over time such that we get the following equation (7.2.18), which is in the simple case of only n = 2 components already sufficiently complicated

$$(7.2.18)\ m_{0t} = \frac{A_1 m_{(1)01} + A_2 m_{(2)01}}{A_1 + A_2} \cdot \ldots \cdot \frac{(A_1 + \sum_{\tau=1}^{\tau=t-1} Z_{\tau 1}) m_{(1)t-1,t} + (A_2 + \sum_{\tau=1}^{\tau=t-1} Z_{\tau 2}) m_{(2)t-1,t}}{(A_1 + \sum Z_{\tau 1}) + (A_2 + \sum Z_{\tau 2})}$$

as opposed to the simple equation

$$(7.2.18a)\qquad m_{0t} = \frac{A_1 m_{(1)0t} + A_2 m_{(2)0t}}{A_1 + A_2}$$

in the case of a direct index. To sum up:

To arrive at a relative $m_{0t} = y_t/y_0$ relating to an *aggregate* y composed of n components $y_1, y_2, \ldots, y_n$ (i = 1, 2, ..., n), given *links* of components (subaggregates, sectors) $m_{(i)t-1,t}$ (sectoral links), data on the changing weights w_{it} of sectors are needed and in principle we have *to work out again the complete calculation*, that is, to

- aggregate (summate) over the sectoral links i = 1, ..., n, and to
- aggregate (multiply) over time using $(\Sigma w_{i0}\, m_{(i)01}) \ldots (\Sigma w_{i,t-1}\, m_{t-1,t})$.

Obviously aggregation over commodities (sub-indices) and therefore trying ad-hoc indices in various compositions of sub-indices is much easier to handle in the case of a sequence of sectoral *relatives* (each based on 0) $m_{(i)01}$, $m_{(i)02}$, $m_{(i)03}, \ldots$ to be combined using a *constant* set of weights $A_i/\sum A_i$ *for all t* than in the case of having to aggregate over sectoral *links* (each based on t-1) with constantly *changing weights*.

h) No structural consistency (of volumes) and proportionality in quantities

In **sec. 5.2** a distinction was made between aggregation *of the index function* and (structurally) consistent aggregation *of volumes*. As the first aspect, i.e. aggregative consistency[106] (how to aggregate over sub-indices) has just been dealt with in the preceeding part of this section, we now turn to the structural consistency, which is of particular interest in the case of deflation. Recalling the main results of **sec. 5.2** it is not surprising that chain indices are most unfavorable as far as deflation (for example in National Accounts) is concerned. In **sec. 5.4** a critique was given for the *direct* Fisher index deflation and we now try to extend our criticism to the chained Fisher index as recommended by the SNA as well. The result is as follows

[106] It is a well known fact that chain indices are not "consistent in aggregation", let alone "additive". Even the weaker version of this test, known as the "equality test" (see sec. 5.2.e), is not met, and as shown above chain indices may well also fail the mean value property. Also, the inability of chain indices to provide structurally consistent results in deflation is well known.

	aggregative properties of Fisher index deflation	
deflator	Consistent aggregation *of volumes*	Proportionality* in volumes
direct P_{0t}^{F}	no	yes
chain $\overline{P}_{0t}^{FC}$	no	no

* and consequently also identity

As the only index function able to fulfill the criterion of structural consistency in volumes is the *direct* Paasche price index P_{0t}^{P}, it is not surprising that

> No chain index construction is structurally consistent, not even the *chain* version of Paasche (as opposed to the *direct* Paasche index) is structurally consistent, let alone the *chain* Fisher price index as a deflator. Hence it is not ensured that the components of real values (deflated by use of chain indices) will add up to the total of that aggregate, *whatever* type of chain index is used.

It should be clear that Fisher type *chain* indices $\overline{P}_{0t}^{FC}$ as deflators are even worse than direct Fisher deflators P_{0t}^{F}. *In addition* to the structural inconsistency (of the direct as well as chain Fisher deflator) there is at least one more shortcoming when a chained Fisher index is used to deflate values to volumes i.e. the *value index preserving test* (Vogt 1978, Balk 1995) or *proportionality in the quantities*. Since a value index is by definition $V_{0t} = \mathbf{p}_t'\mathbf{q}_t/\mathbf{p}_0'\mathbf{q}_0$ and simply observation in the sense that there cannot be doubt which formula to take for calculating V_{0t}, the following axioms seem to be plain logic: If all quantities change identically by λ such that $\mathbf{q}_t = \lambda\mathbf{q}_0$, then a quantity index should yield $Q(\mathbf{q}_0,\mathbf{p}_0,\lambda\mathbf{q}_0,\mathbf{p}_t) = \lambda$, which also includes the case of no change $\lambda = 1$ (identity), in which Q should amount to $Q(\mathbf{q}_0,\mathbf{p}_0,\mathbf{q}_0,\mathbf{p}_t) = 1$ and[107] $P(\mathbf{p}_0,\mathbf{q}_0,\mathbf{p}_t,\mathbf{q}_0) = V_{0t} = \mathbf{p}_t'\mathbf{q}_0/\mathbf{p}_0'\mathbf{q}_0 = P_{0t}^{L}$.

> Given $\lambda = 1$ (or $\mathbf{q}_t = \mathbf{q}_0$) we get $Q_{0t}^{L} = Q_{0t}^{P} = Q_{0t}^{F} = 1$, thus also $P_{0t}^{F} = V_{0t} = \mathbf{p}_t'\mathbf{q}_0/\mathbf{p}_0'\mathbf{q}_0$[108], however $\overline{Q}_{0t}^{FC} \neq 1$ and $\overline{P}_{0t}^{FC} \neq V_{0t}$ is certainly possible ($t \geq 3$).

The following example will demonstrate this.

Example 7.2.7

Assume prices of both goods A and B are rising by 50% from 0 to 3, and quantities are remaining constant such that $V_{03} = P_{03}^{L} = P_{03}^{P} = P_{03}^{F} = 1.5$. Fisher- and Paasche-deflation (*direct* indices) will therefore produce the same result. In contrast, however, deflating using a chain Fisher index, $\overline{P}_{03}^{F}$ results in *volumes* indicating a *decline*, despite

[107] It is assumed that the pair of index functions P, Q will pass the product test.

[108] The direct Fisher index will always pass the value index preserving test, however, the chain index will not, as shown above.

- zero change in quantities, and
- prices rising uniformly by 50%.

The "logic" or "economic common sense" of this result is difficult to discover.

	period 0		period 1		period 2		period 3	
good	p	q	p	q	p	q	p	q
A	30	5	40	3	50	2	45	5
B	10	15	5	20	10	13	15	15

As $\sum p_3q_3 = \sum p_3q_0 = 1.5\sum p_0q_0$, $\overline{P}_{03}^{FC}$ simplifies to

$\overline{P}_{03}^{FC} = \sqrt{1.5\frac{\sum p_1q_0 \sum p_2q_1 \sum p_3q_2}{\sum p_0q_1 \sum p_1q_2 \sum p_2q_0}}$, not necessarily amounting to 1.5. The results for links and chain indices are

	P_1^C	P_2^C	P_3^C	chain ($\overline{P}_{03}^C$)
Laspeyres	275/300	350/220	285/230	665/368 = 1.807
Paasche	220/290	230/145	450/400	2277/1682 = 1.354
Fisher	0.834	1.589	1.181	$\overline{P}_{03}^{FC} = \sqrt{2.4463} = 1.564$

Dividing of $\sum p_3q_3 = 450$ by $\overline{P}_{03}^{FC}$ gives a chain-index Fisher volume of 287.71, hence a reduction compared with $\sum p_0q_0 = 300$ by 4.1% in spite of quantities remaining unchanged, and prices as well as values rising by 50%. Note that since $450/1.354 = 332.35 \neq 300$ the Paasche *chain* index $\overline{P}_{03}^{PC}$ does not fare much better than $\overline{P}_{03}^{FC}$ as a deflator. It is the *chain* principle, not the type of formula (Fisher vs. Paasche) which is responsible for this shortcoming (violation of identity).

In order to better understand this result of the foregoing example it appears useful to express the sequence of "volumes in constant prices of the base period 0" and the corresponding quantity indices. Deflation using a direct Paasche index (DP-method) yields

	t = 0	t = 1	t = 2	t = 3
volume	$\sum p_0q_0$	$\sum p_0q_1$	$\sum p_0q_2$	$\sum p_0q_3$
quantity index	1	$\frac{\sum p_0q_1}{\sum p_0q_0}$	$\frac{\sum p_0q_2}{\sum p_0q_0}$	$\frac{\sum p_0q_3}{\sum p_0q_0}$

The sequence of volumes as well as the quantity indices clearly exclusively reflect the change of quantities. **Tab. 7.2.4** demonstrates, however, that this is no longer the case whenever a chain index is used in order to deflate aggregates, that is, in order to derive a sequence of volumes from the values $\sum p_3q_3$, $\sum p_4q_4$ etc. In **tab. 7.2.4** two chain methods are compared with two direct methods of deflation. ♦

There is no such table in the SNA in which volumes or quantity indices, like $\overline{Q}_{03}^{FC}$ etc, are presented in terms of all aggregates involved. The table shows that a change of CF-volumes may well be a result of a variety of influences. Therefore it is difficult to state what exactly has changed and to what extent when a volume of this type indicates a change of, say, 5%. It is also clear from this table that there is a *cumulative* pattern in all chain index volumes, such as

(7.2.19) $$\overline{Q}_{03}^{FC} = \sqrt{\frac{\sum p_3 q_3}{\sum p_0 q_0}} \sqrt{\frac{\sum p_0 q_1 \sum p_1 q_2 \sum p_2 q_3}{\sum p_1 q_0 \sum p_2 q_1 \sum p_3 q_2}}$$

(7.2.20) $$\overline{Q}_{04}^{FC} = \sqrt{\frac{\sum p_4 q_4}{\sum p_0 q_0}} \sqrt{\frac{\sum p_0 q_1 \sum p_1 q_2 \sum p_2 q_3 \sum p_3 q_4}{\sum p_1 q_0 \sum p_2 q_1 \sum p_3 q_2 \sum p_3 q_2}}$$ and in general

(7.2.21) $$\overline{Q}_{0t}^{FC} = \prod_{k=1}^{k=t} \left(\frac{\sum p_{k-1} q_k}{\sum p_k q_{k-1}} V_{0t} \right)^{1/2}.$$

Table 7.2.4: Comparison of different deflation methods (deflators P_{0t})

DP = direct Paasche; DF = direct Fisher, CP = chain Paasche, CF = chain Fisher

P_{0t}	t = 3 $(\Sigma p_3q_3/P_{03})$	t = 4 $(\Sigma p_4q_4/P_{04})$
DP	$\Sigma p_0 q_3$	$\Sigma p_0 q_4$
DF	$\Sigma p_0 q_3 \left(\frac{\Sigma p_3 q_3}{\Sigma p_3 q_0} \frac{\Sigma p_0 q_0}{\Sigma p_0 q_3} \right)^{1/2}$	$\Sigma p_0 q_4 \left(\frac{\Sigma p_4 q_4}{\Sigma p_4 q_0} \frac{\Sigma p_0 q_0}{\Sigma p_0 q_4} \right)^{1/2}$
CP	$\Sigma p_0 q_1 \left(\frac{\Sigma p_1 q_2}{\Sigma p_1 q_1} \frac{\Sigma p_2 q_3}{\Sigma p_2 q_2} \right)$	$\Sigma p_0 q_1 \left(\frac{\Sigma p_1 q_2}{\Sigma p_1 q_1} \frac{\Sigma p_2 q_3}{\Sigma p_2 q_2} \frac{\Sigma p_3 q_4}{\Sigma p_3 q_3} \right)$
CF	$\left(\Sigma p_3 q_3 \Sigma p_2 q_3 \frac{\Sigma p_0 q_0}{\Sigma p_1 q_0} \frac{\Sigma p_0 q_1}{\Sigma p_2 q_1} \frac{\Sigma p_1 q_2}{\Sigma p_3 q_2} \right)^{1/2}$	$\left(\Sigma p_4 q_4 \Sigma p_3 q_4 \frac{\Sigma p_0 q_0}{\Sigma p_1 q_0} \frac{\Sigma p_0 q_1}{\Sigma p_2 q_1} \frac{\Sigma p_1 q_2}{\Sigma p_3 q_2} \frac{\Sigma p_2 q_3}{\Sigma p_4 q_3} \right)^{1/2}$

Note also that sometimes (e.g. in compiling production indices) for some series at least it was common practice to estimate the physical quantity as a substitute for net production. Hence it is sometimes desirable to relate a production index to quantity change. It is not easy to discern a "quantity" in an expression like $\overline{Q}_{04}^{FC}$. Nor is it easy to understand why such an expression is said to be a measure of an aggregate at constant prices "of t = 0".

Another interesting observation is that volume changes in DP-deflation from period 2 to period 3 by a factor

(7.2.22) $$\frac{Q^L_{03}}{Q^L_{02}} = \frac{\Sigma q_3 p_0}{\Sigma q_2 p_0} = \Sigma \frac{q_3}{q_2} \frac{p_0 q_2}{\Sigma p_0 q_2} = Q^L_{23(0)},$$

(a Laspeyres-quantity index Q^L_{03} rebased on base 2) depend on *base* period prices only. The corresponding factor in the case of chain Fisher index deflation (CF deflation) is

(7.2.23) $\sqrt{Q^L_{23} Q^P_{23}} = Q^{FC}_3$, depending on *two* price structures[109], both *other* than p_0.

This is of course entirely in line with the chain index logic, according to which (price) weights of quantities have got to change and to adapt to the present situation: the weight assigned to a quantity change at time t > 0 should not depend on prices of the past.

We now turn to some issues which are more or less interesting only from a theoretical point of view.

i) Determinants of the drift (drift-function, temporal correlation, growth factors)

A drift[110] function is a most useful instrument to compare the chain, for example $\overline{P}^{LC}_{0t}$, with the direct index P^L_{0t} (instead of comparing the link P^{LC}_t with the direct index as is often falsely done). The ratio

(7.2.24) $$D^{PL}_{0t} = \overline{P}^{LC}_{0t} / P^L_{0t}$$

is known as a drift-function for a Laspeyres price index. Correspondingly, the drift of a Laspeyres quantity index is given by $D^{QL}_{0t} = \overline{Q}^{LC}_{0t} / Q^L_{0t}$, and the drift-function of a Paasche price index by $D^{PP}_{0t} = \overline{P}^{PC}_{0t} / P^P_{0t}$, and accordingly $D^{PF}_{0t} = \sqrt{D^{PL}_{0t} D^{PP}_{0t}}$ is the drift for the Fisher price index. Consider g^k_t as being the growth factor of prices from t-1 to t on the basis of weights (quantities) belonging to period k, then we get

(7.2.24a) $$D^{PL}_{02} = \frac{\overline{P}^{LC}_{02}}{P^L_{02}} = \frac{\Sigma p_1 q_0}{\Sigma p_2 q_0} \frac{\Sigma p_2 q_1}{\Sigma p_1 q_1} = \frac{P^L_{01} P^L_{12}}{P^L_{02}} = \frac{P^L_{12}}{P^L_{02} / P^L_{01}} = \frac{g^1_2}{g^0_2}, \text{ and}$$

(7.2.24b) $$D^{PL}_{03} = \frac{\overline{P}^{LC}_{03}}{P^L_{03}} = \frac{\Sigma p_1 q_0}{\Sigma p_1 q_1} \frac{\Sigma p_2 q_1}{\Sigma p_2 q_2} \frac{\Sigma p_3 q_2}{\Sigma p_3 q_0} = D^{PL}_{02} \frac{P^L_{02} P^L_{23}}{P^L_{03}} = D^{PL}_{02} \frac{g^2_3}{g^0_3},$$

[109] The SNA said in the light of this result: "Only in the special case in which time series of fixed base Laspeyres volume indices are used ... it is legitimate to equate ... real GDP with ... GDP 'at constant prices'. When chain indices are used, it is not appropriate to describe real GDP as GDP at constant prices" (p. 16.71).

[110] The term "drift" does not imply that one of the two indices in question is regarded as being correct and the other incorrect.

which clearly proves the *recursive* and *cumulative* nature of a Laspeyres drift starting with $D_{01}^{PL} = 1$; we get successively $D_{01}^{PL} = 1$, $D_{02}^{PL} = D_{01}^{PL} \frac{g_2^1}{g_2^0}$, ..., $D_{0t}^{PL} = D_{0t-1}^{PL} \frac{g_t^{t-1}}{g_t^0}$, as both parts of D_{0t}^{PL}, i.e. $\overline{P}_{0t}^{LC}$ (numerator), and P_{0t}^{L} (denominator) can be represented as products (see **sec. 2.5**) such that

$\overline{P}_{0t}^{LC} = g_1^0 g_2^1 g_3^2 \cdot ... \cdot g_t^{t-1}$	$P_{0t}^{PL} = g_1^0 g_2^0 g_3^0 \cdot ... \cdot g_t^0$

What matters is simply the difference between two types of growth factors, both of which can be regarded as arithmetic means of link-relatives (of prices)

$g_2^1 = \sum \frac{p_2}{p_1} \frac{p_1 q_1}{\sum p_1 q_1}$, and $g_2^0 = P_{02}^L / P_{01}^L = \sum \frac{p_2}{p_1} \frac{p_1 q_0}{\sum p_1 q_0}$.

The situation is less easy to understand in the case of the Paasche price index drift. We get

$D_{02}^{PP} = \frac{\overline{P}_{02}^{PC}}{P_{02}^{P}} = \frac{P_2^{PC}}{P_{02}^{P} / P_{01}^{P}}$, the nominator being $P_2^{PC} = \frac{\sum p_2 q_2}{\sum p_1 q_2}$, but it is more difficult to give an interpretation to the denominator $\frac{P_{02}^{P}}{P_{01}^{P}} = \frac{\sum p_2 q_2 \sum p_0 q_1}{\sum p_0 q_2 \sum p_1 q_1}$.

An analysis of the structure of a drift function reveals that drifts (D) and chain price indices ($\overline{P}$) are related as follows

(7.2.25) $D_{0t}\overline{P}_{0t} = (D_{0s}\overline{P}_{0s})(D_{st}\overline{P}_{st})$ hence $D_{0t} = D_{0s}D_{st}$

as by definition $\overline{P}_{0t} = \overline{P}_{0s}\overline{P}_{st}$ holds for any three periods 0, s, t. This may help to assess the behavior of the drift over long intervals. Moreover, the price and quantity index drifts are inversely related to one another. Since the value index V_{0t} is always transitive (unlike most of the price and quantity indices) and the value index can be decomposed into a P^L and a Q^P index we get[111]

(7.2.26) $D_{03}^{PL} = \frac{P_{01}^L P_{12}^L P_{23}^L}{P_{03}^L} = \frac{Q_{03}^P}{Q_{10}^P Q_{12}^P Q_{23}^P} = \frac{1}{D_{03}^{QP}}$.

The drift functions (like for example D^{PL}, or D^{QP} etc.) depend (much like the chain index function to which they refer) on the length of the interval (0, t) in question, on how it is subdivided into subintervals, and on the path (pattern of the time series) of p's and q's as well as the formula of the chain index used.

[111] Hence there is again a reciprocal or "antithetic" relation between Laspeyres and Paasche. As a *Laspeyres* chain *price* index drifts away upwards biased from the corresponding direct price index we might reasonably expect that a chain *Paasche quantity* index will drift away downwards biased, and vice versa, both possibly with a cumulative bias as time goes on and more and more links are linked together.

In general we expect a chain index to draw progressively away from the direct index, such that the drift grows as t becomes more and more distant from 0, but it is difficult to draw a valid conclusion under *general* conditions.

As the drift function is defined as a ratio it might also be reasonable to examine a difference, for example $\overline{P}_{03}^{LC}$ - P_{03}^{L}. However, $\overline{P}_{03}^{LC} - P_{03}^{L} = P_{03}^{L}\left(D_{03}^{PL} - 1\right)$ shows that the difference does not convey any additional information compared to the drift function.

The drift D_{02}^{PL} for example can also be expressed in terms of the famous relationship found by L. v. Bortkiewicz. Denote

growth ***factors*** of individual prices	quantity ***relatives*** (i.e. ***cumulative*** changes)
$x_{i,12} = \frac{p_{i2}}{p_{i1}}$ likewise $x_{i,23} = \frac{p_{i3}}{p_{i2}}, \ldots$	$y_{i,01} = \frac{q_{i1}}{q_{i0}}$ correspondingly $y_{i,02} = \frac{q_{i2}}{q_{i0}}, \ldots$

and consider weighted means, weights being (hybrid) expenditure shares $p_{i1}q_{i0} / \sum p_{i1}q_{i0}$ like

(7.2.27) $\bar{x}_{12} = \sum \frac{p_2}{p_1} \frac{p_1 q_0}{\sum p_1 q_0} = \frac{\sum p_2 q_0}{\sum p_1 q_0} = P_{02}^{L} / P_{01}^{L} = P_{12(0)}$, and

(7.2.27a) $\bar{y}_{01} = \sum \frac{q_1}{q_0} \frac{p_1 q_0}{\sum p_1 q_0} = \frac{\sum p_1 q_1}{\sum p_1 q_0} = Q_{01}^{P}$

it can then be easily shown that the covariance between growth factors of prices and the *cumulated* change in quantities is given as follows (omitting i for convenience)

(7.2.28) $\mathrm{Cov}(x_{12}, y_{01}) = \sum (x_{i,12} - \bar{x}_{12})(y_{i,01} - \bar{y}_{01}) \frac{p_1 q_0}{\sum p_1 q_0} = \frac{\sum p_2 q_1}{\sum p_1 q_0} - \bar{x}_{12}\bar{y}_{01}$

$= \frac{Q_{01}^{P}}{P_{01}^{L}}\left(P_{01}^{L} P_{12}^{L} - P_{02}^{L}\right) = Q_{01}^{P}\left(P_{12}^{L} - P_{12(0)}^{L}\right)$, and

(7.2.29) $D_{02}^{PL} = \frac{\mathrm{Cov}(x_{12}, y_{01})}{\bar{x}_{12} \cdot \bar{y}_{01}} + 1$ (that is row 1 in tab. 7.2.5).

In a similar manner we get x_{23} and y_{02} defined accordingly using weights $p_2 q_0 / \sum p_2 q_0$

(7.2.29a) $\mathrm{Cov}(x_{23}, y_{02}) = \sum (x_{i,23} - \bar{x}_{23})(y_{i,02} - \bar{y}_{02}) \frac{p_2 q_0}{\sum p_2 q_0} = Q_{02}^{P}\left(P_{23}^{L} - P_{23(0)}^{L}\right)$,

and D_{03}^{PL} as shown in tab. 7.2.5 (row 2).

Tab. 7.2.5: Cumulative structure of the drift as a function of the (temporal) covariance

t	drift function D^{PL} and covariance cov(x, y)
$t=2$, D_{02}^{PL}	$\frac{Cov(x_{12}, y_{01})}{\bar{x}_{12} \cdot \bar{y}_{01}} + 1$ (since $D_{01}^{PL} = 1$)
$t=3$, D_{03}^{PL}	$D_{02}^{PL}\left(\frac{Cov(x_{23}, y_{02})}{\bar{x}_{23} \cdot \bar{y}_{02}} + 1\right) = \left(\frac{Cov(x_{12}, y_{01})}{\bar{x}_{12} \cdot \bar{y}_{01}} + 1\right)\left(\frac{Cov(x_{23}, y_{02})}{\bar{x}_{23} \cdot \bar{y}_{02}} + 1\right)$
$t=4$, D_{04}^{PL}	$D_{03}^{PL}\left(\frac{Cov(x_{34}, y_{03})}{\bar{x}_{34} \cdot \bar{y}_{03}} + 1\right) = \left(\frac{Cov(x_{12}, y_{01})}{\bar{x}_{12} \cdot \bar{y}_{01}} + 1\right)\left(\frac{Cov(x_{23}, y_{02})}{\bar{x}_{23} \cdot \bar{y}_{02}} + 1\right)\left(\frac{Cov(x_{34}, y_{03})}{\bar{x}_{34} \cdot \bar{y}_{03}} + 1\right)$

Though the system is becoming apparent, the problem is that it is difficult to draw general conclusions concerning sign and amount of drift.[112] Recall that

1. the variables x and y refer to different periods - x refers to prices and periods t-1 and t, but y is a measure of *cumulative* change over an interval (0, t-1), and therefore
2. the relevant covariance may be calculated by chaining and may well change sign, such that for example $\bar{P}_{02}^{LC} > P_{02}^{L}$ (or equivalently $D_{02}^{PL} > 1$) because of $Cov(x_{12}, y_{01}) > 0$ and at the same time $\bar{P}_{03}^{LC} < P_{03}^{L}$, because $Cov(x_{23}, y_{02}) < 0$;
3. the covariance does not give a straightforward explanation for a chain index drifting further and further away when a cyclical movement is repeated time and again.

As a rule, the amount of drift depends on the cumulated change of quantities associated with the change of prices and on the type of time series given for prices and quantities[113] (either monotonically increasing or decreasing, or with a cyclical movement respectively).

Example 7.2.8

Calculate covariance and means for **ex. 7.2.2** and its variant a (see **ex. 7.2.3**).

1) Calculation for **ex. 7.2.2**

[112] The same is true for the "spread" (or "gap") between the direct Laspeyres and Paasche index formula (more details cp. my book on chain indices).

[113] The problem with such a statement is, however, that empirical findings concerning the shape of time series of prices and quantities are based on *indices* of prices and quantities: given a cyclical movement a chain index might not reflect cycles but rather a perhaps growing upward or downward bias.

Since $P^L_{01} = P^L_{02} = 1$ we also have $\bar{x}_{12} = P^L_{02}/P^L_{01} = 1$ and $\bar{y}_{01} = Q^P_{01} = 135/96 = 1.4063$, $P^L_{12} = 140/135 = 1.037$. Hence Cov(x,y) = $Q^P_{01}(P^L_{12} - 1) = 5/96 = +0.0521$, and $D^{PL}_{02} = P^L_{12} = 1.037 > 1$ because $\bar{P}^{LC}_{02} = 1.037 > P^L_{02} = 1$.

2) Calculation for variant a, **ex. 7.2.3**

$\bar{x}_{12} = 1$, $\bar{y}_{01} = Q^P_{01} = 180/96 = 1.875$, $P^L_{12} = 160/180 = 0.888$, and

Cov(x,y) = $-20/96 = -0.208$, $\bar{P}^{LC}_{02} = 0.8888 < P^L_{02} = 1$, $D^{PL}_{02} = -20/180 + 1 = 0.888$.

♦

Digression on deflation and the "correct" and most "relevant" growth rate of real values (volumes); once more argument B3 in favor of chain indices.

An argument described and criticized already as argument **B3** above goes as follows: to measure the growth of deflated values (of "real GDP" for example) correctly, the growth rate should be calculated using previous year prices,[114] resulting in a *chain* quantity index such that the ratio of two successive growth rates amounts to

$$\frac{Q^{LC}_t}{Q^{LC}_{t-1}} = \frac{P^{PC}_{t-1} + \Delta_t}{P^{LC}_{t-1} + \Delta_{t-1}} \text{ where } \Delta_t = \frac{\sum(q_t - q_{t-1})\cdot p_{t-1}}{\sum q_{t-1}p_{t-2}} \text{ and } \Delta_{t-1} = \frac{\sum(q_{t-1} - q_{t-2})\cdot p_{t-1}}{\sum q_{t-2}p_{t-2}},$$

where the links P^{PC}_{t-1} and P^{LC}_{t-1} measure the change of prices from t - 2 to t – 1. Thus, if this chain approach in calculating growth rates is preferred it should be recognized that price movement seems to be incompletely eliminated.[115] Yet the argument B3 was found convincing and for example the US changed their official measure of real GDP from a constant price to a chain weighted measure.[116] Since this change coincided with a revision of the national income accounts, the exact impact of the change in growth rate calculation was not easy to assess.[117]

The US change in methodology was mainly motivated by the observation that some important commodities experienced a remarkable change in prices, in particular oil prices increased and prices of computers and their peripherals decreased. Under such conditions growth rates on the basis of a more recent price structure will of

[114] In the example of Norway quoted above the two methods (fixed prices of a "base" period, possibly a long time ago vs. previous year prices) were called A and B respectively. Unfortunately, the choice between these two methods is often described as a choice between "simplicity" (fixed prices approach) and "accuracy" (the case of the chain approach) cp. A. Young 1992, p. 36. However, "accuracy" may well be called into question when the problem of "non-additivity" is considered.

[115] In a comparison of any two successive growth rates *at least two* price structures are involved, p_{t-1} and p_{t-2}. Once the whole series of growth rates (of real growth) is contemplated there are of course many more price structures p_{t-3}, ... influencing the result.

[116] In January 1996 the US Bureau of Economic Analysis (BEA) adopted a chain Fisher quantity index as measure of real growth (as recommended by SNA).

[117] In A. Barr 1996 a comparison is made showing that growth rates of the then more recent periods (1993 - 1995) had to be revised downwards considerably which "made the current recovery look particularly weak".

course differ from growth rates based on "old" prices, and it appears reasonable to avoid an overestimation of growth by taking more recent prices.[118] Or in other words: when prices change rapidly, growth rates $g_1^0, g_2^1, g_3^2, \ldots$ tend to be preferred to $g_1^0, g_2^0, g_3^0, \ldots$. Furthermore, it is often maintained that deflation with a direct Paasche price index of prices p_{i0} of a fixed base period 0 has the following shortcomings

1. choice of the base period affects results of deflation,
2. each rebasing entails revising the entire "history" of GDP and ensues new growth rates when calculation is carried back than those hitherto published, and
3. the traditional fixed-base year method of calculating real GDP does not account for substitution and therefore tends to be biased.

The first point is of course trivial. It is only because of continuously *changing* the price structure that chain based measures of real growth *seem to be* independent of prices. The second and third point refer to what is intended to be measured in the case of "deflation". If the objective of deflation is to be understood as a fixed set of prices operating as weights, such that price movement is eliminated, there is clearly no reason for accounting for substitution and the "regular rewriting of economic history" (Barr), though inconvenient, is simply a consequence of choosing another constant set of prices.[119]

j) Justification of chainlinking, uniqueness theorem of Funke et al. concerning the Cobb-Douglas index

The purpose of this section is to discuss some problems of what might be called a "theory of chaining" (if there is any). It turns out that ideas to justify the operation of "chaining" don't seem thoroughly thought out, and in addition they are (if based on the notion of chainability) inconsistent with constant adjustment of quantity weights.

To consistently link two subintervals

(7.2.30) $P_{0t} = P_{0s}P_{st}$ it is implicitly assumed that $\frac{P_{0t}}{P_{0s}} = \frac{P_{st}}{P_{ss}}$, $P_{ss} = 1$ holds,

which is tantamount assuming that indices at *two* different bases (and mostly with different weights) will change *in proportion*. Likewise, in the case of three subintervals we have

---

[118] It has often been observed that growth of more recent years measured by chain methods falls short of growth measures for the same period using fixed prices of an earlier year (see preceeding footnote). Given that commodities for which output grows more rapidly tend to be those for which prices increase slowly or even decrease, taking more recent prices as weights will give commodities with strong growth less weight and thereby result in a smaller overall grow rate. The problem, however, is that correlation between prices and output varies. Young 1992 ended up after detailed comparisons with some general remarks like: "A difference between two measures of real GDP is not evidence that one is wrong", or the worth of either method "lies in whether or not it proves useful in analysis. Viewed in this way, there can be more than one useful measure" (p. 35).

[119] Once volume change is measured referring to "constant" prices, a "rewriting" cannot be avoided. We cannot have both volumes based on a specified set of *constant* prices and volumes *independent* of which set of prices has been chosen.

(7.2.31) $P_{0t} = P_{0r}P_{rs}P_{st}$ which implies $\frac{P_{0t}}{P_{0r}} = \frac{P_{rt}}{P_{rr}}$ ***and*** in addition $\frac{P_{rt}}{P_{rs}} = \frac{P_{st}}{P_{ss}}$,

implicitly assuming that indices of *three* different bases (with different sets of weights) will change in proportion. However, ironically chain indices are motivated on the assumption that indices having a different base (and weight structure) will in general *not* change in proportion (otherwise there would be no point in claiming a constant updating of weights). They will *not* behave like unweighted relatives. Switching to a new base will make a difference and according to "chainers" it will definitely matter and will be better than keeping the old weights.

Thus a contradiction impossible to overlook exists

- between the *operation of chaining* and the assumption that weights matter in the case of indices (as opposed to the case of relatives, where weights are non-existent); there is however also a conflict
- between the *property of chainability* (violated of course in the case of path dependent chain indices) and the continual adjusting of quantity weights.

It was not until the proof of Funke et al. 1979, according to which

> the only index satisfying the minimal requirements monotonicity, linear homogeneity, identity and commensurability and being able to meet chainability is the so called **"Cobb-Douglas index"** given by $P_{0t}^{CD} = \prod_{i=1}^{n} \left(\frac{p_{it}}{p_{i0}} \right)^{\alpha_i}$, where "weights" α_1, α_2 ,... α_n are *arbitrary* real constants *not* dependent on period 0 nor t or any other period, and $\Sigma\alpha_i = 1$.

It is viewed as one of the pretended main advantages of chain indices that they can solve both problems, simultaneously

- making consistent long term comparisons (chainability), and
- allowing for a constant adjustment of weights to new situations, an aspect of dynamics allegedly best served by chain indices.

The latter aspect has been one of the main reasons for Alfred Marshall to advocate chain indices. However, apparently chainability is necessarily in conflict with adaptability of weights. They are contradictory properties not able to be reconciled, you never get both "advantages" simultaneously. As I. Fisher already conjectured: There are chainable indices where the weights must be constant and on the other hand there are indices with variable weights necessarily violating chainability, or in the words of Funke et al. there is no index allowing consistent inter-temporal aggregations over multiplying subintervals and accounting for changes in weights[120]. Hence the history of chain indices started with a dilemma and misunderstanding as regards possible properties of index functions.

---

[120] such "...that the main intention of the circular test, that is, the adjustment of the quantity weights to the new situation in each new dual comparison around a circle of periods or places cannot be accomplished. There simply does not exist such a formula..." (p. 685).

k) Aspects of official statistics: cost-benefit-considerations, understandability, and a system of chain indices

This section is devoted to *practical* aspects relevant mainly for assessing the suitability of chain indices for *official* (price) statistics.[121] Three points are particularly noteworthy:

1. Chain indices will require *more statistical surveys* and thereby cost more, both directly for statistical offices and indirectly for respondents facing an increased response burden;
2. for public acceptance it is important that concepts are *understandable* and the office is seen as *neutral* and impartial; this applies in particular to inflation measurement;
3. a change of methods has far-reaching implications as official statistics has to provide a whole *system* of indices (not only price indices, let alone only consumer price indices, but also indices of production, new orders etc.) which should fit together.

As to *point 3*, apparently there is not much thought given to the question as to what a *universal* application of the chain index approach in official statistics would really mean to us. Furthermore, it is conspicuous and no coincidence that for example Bruce Mudgett, one of the most fervent admirers of chain indices, plays down precisely *the first two points*, viz. *costs* and lack of "*understandability*".[122] In contrast, we are strongly convinced that just this interpretation and understandability issue is fundamental for the acceptance of statistics by the general public and therefore essential, especially for official statistics.

k1) Data requirements due to more frequent updates of weights and point of sales surveys

It is widely accepted that more frequent surveys of consumption patterns (so-called family [or household] expenditure surveys [FES]) are not only costly but also time consuming and laborious for both the offices and the respondents[123]. Hence chain

[121] It is remarkable that such aspects, which are most important in view of the conditions (national) statistical offices face, are habitually neglected and ignored in index *theory*.

[122] The quite often quoted words of Mudgett 1951, p. 37 read as follows: "The present writer is not too greatly impressed with either the 'understanding' or the 'calculation' reasons for choosing index number formulae ... As for the 'understanding' issue, we all use automobiles, telephones, radios, and many other gadgets without knowing too much about how they are made." The parallel with some technical devices, like automobiles etc. is most revealing. Mudgett completely disregards the importance of trust and confidence (in impartiality of statisticians) of the general public in the "statistics" business (as a *public* good).

[123] Mainly on such grounds in some countries, especially in those which until recently favored the direct (Laspeyres) index approach FES were carried out in five years intervals or so only. Note also that the (more frequent) establishment of weights is known to be precisely that part of index compilation which is (compared with the regular reporting of prices) possibly the most costly.

indices are difficult to reconcile with principles adopted only recently, such as the reduction of statistical surveys and the response burden. Moreover, it is highly unlikely that FES, even if carried out annually, will be capable of *timely* providing a renewed basket and that for a more frequent update of CPI "baskets" recourse can reasonably be made to other statistics than FES, for example National Accounts, production and turnover statistics. In practice therefore it is more realistic that detailed weighting schemes will have a *two* year lag rather than lagging behind by *one* year only, and possibly only parts of the weights will be renewed annually,[124] or baskets may be provisional only, and subject to subsequent revisions after some months or so.[125]

Furthermore it is also maintained that a "flexible" chain index should take into account all sorts of changes in the conditions of purchases, such as types of outlets, terms of payment etc. These changes should be observed in "point of sales" statistics (POS) and – unlike in the case of quality change - influence the result of the index calculation.[126] Only very few countries are able to conduct such POS–surveys on a regular basis. To blame the Laspeyres index for ignoring important aspects of changes in a dynamic world is clearly only one side of the coin. The other is the indisputably increased costs and practical difficulties in an index compilation which accounts for such dynamic changes.[127] On the other hand: For "chainers" increasing costs and difficulties with an appropriate data source may be offset by the fact that less care can be taken for comparability over time in the case of chain indices. Once the idea of "pure" price comparison is discarded (or dismissed as old fashioned) we may well hail flexibility, especially in times of an increasing variety in commodities, and more rapid changes of the above mentioned conditions of purchases.

k2) Acceptance, understandability, and leaving no room for manipulation

Understandability and a clear meaning of an index are aspects which should be regarded, especially in official statistics. It is unwise to play down their importance. Interestingly both groups, non-chainers as well as chainers, make use of this "understandability argument". A quite common argument of chainers is that marketing for

[124] This will result in *weights of different ages* as a full update may perhaps be possible only in longer time intervals. As pointed out in sec. 6.1, in the Harmonized Index of Consumer Prices (HICP) in Europe, and thus the European inflation rate calculated by combining national HICPs, we have to put up with such a situation.

[125] In price statistics (in contrast to National Accounts) it is for a number of reasons most uncommon to have "revisions" of indices after their first being published.

[126] The effect of quality change should be neutralized (by "adjustment") at least when two adjacent periods are compared. It is not quite clear whether the same applies in practice to a change in the POS: gains resulting from moving to a less expensive outlet will in general not be adjusted for a possible decrease in quality (of service). While in the Laspeyres concept a substitution of outlets is avoided precisely because of difficulties with quality adjustment in this situation, the chain methodology encourages or even demands such substitutions.

[127] Another point is the increasing complexity and refinement of measurement concepts motivated by international harmonization. The constant extension of the HICP coverage, for example, raised questions related to the treatment of health services, the inclusion of some more complex types of services rendered such as subscriptions to the Internet and services of financial intermediaries, and the inclusion of certain imputed transactions.

P_{0t}^{L} will be difficult in our dynamic economy, although it is often acknowledged that this formula has the advantage of simplicity (alluding to the rising-cost-of-a-fixed-basket interpretation)[128]:

> It is sometimes said that in our days the general public will not understand and accept the fact that a Laspeyres index is based on an "old" basket with obsolete weights. However, it does not seem much easier to explain to non-statisticians
> - that the value of a chain may well reflect quite *a number of* different influences and
> - such features of the chain index like inapplicability of axioms or the standard interpretations of index formulas, path dependence in general and violation of identity in particular, poor aggregation properties like non-additivity, counter-intuitive result when used as deflators etc.

It is not sure that such peculiarities are less difficult to understand and accept than the constancy of weights for the duration of a couple years in a direct index[129], such as P_{0t}^{L} in particular. In fact, if we set great store by understandability of a formula, a theoretical background and the idea of a "pure price comparison", the formula $\overline{P}_{0t}^{LC}$ has little to recommend it.

Chain indices do not measure the increase in an overall price (or expenditure) of the same basket in two situations, they rather combine *several* baskets. They neither represent a mean (or linear combination) of price relatives, nor do they reflect minimum costs to attain the same level of utility in exactly (and solely) two situations, 0 and t (or: under two price regimes only). Although to refer to the latest basket is certainly a point,[130] as there are other no less relevant questions like the following: Instead of asking for the most recent and therefore (!) most relevant[131] basket we should ask more fundamentally whether there is a "basket" *at all* to which the index refers! Is the index reflecting the rising costs of something like a "basket" (whatever its age may be)?

There are problems of "understanding" in abundance with chain indices. To name only some of them: Numerical results of $\overline{P}_{0t}^{LC}$ (for successive periods t) may increase or decrease beyond limits in the case of cyclical price movements, they can violate identity and monotonicity, and there is no economic theory to support the idea of

[128] It should be noted that in our days (unfortunately) "simple" is often found disadvantageous.

[129] Remember that chain indices are regarded as superior to all other indices, i.e. to *all* direct indices, although the arguments concocted to justify this chain index approach are first and foremost aimed at ridiculing the Laspeyres formula, which is only one particular type of a direct index.

[130] Interestingly chainers' fixation on the latest possible weights goes largely unquestioned such that it is rather uncommon to ask under which conditions the latest basket can validly be regarded as the most "relevant" one. See sec. 4.4a of my book on chain indices for such conditions.

[131] There is not even a clear concept of how to measure "relevance" or "representativity" of a basket. However degrees of "relevance" should be defined if we ask for empirical evidence of the widespread idea that fixed-base weights become progressively more "irrelevant" as time goes on.

chaining,[132] as the index cannot be said to reflect a behaviour of a continuously utility maximizing consumer.

Understandability has two additional dimensions worth being mentioned:

- In the framework of the rather rigid fixed basket model it is clearly more difficult to manipulate and wield political influence, *if so wanted*, and
- the fixed basket approach represents in a most straightforward manner the ideas of "pure" comparison (they are described in **fig. 7.2.2** and **7.2.3**).

Surprisingly, rigidity is a great advantage for official statistics that should not be underestimated. In the context of official statistics it is important to realize that:

The principle of keeping a basket constant is first and foremost a kind of safeguard against all sorts of manipulation. In view of the distrust statistical offices more often than not have to face, offices should welcome both aspects of the Laspeyres formula, viz. its straightforward interpretation and its built-in barrier against manipulation by changing the composition of the basket.

Official statistics should always reckon with users that tend to be sceptical and distrustful as far as the neutrality and impartiality of statisticians is concerned.[133] On the other hand, politicians and governmental bodies are often tempted to present lower "empirical" inflation rates.[134] A more flexible index design may offer a greater chance to successfully lower inflation (or what is statistically measured as such).

To better understand why it is most important that official statistics is to be *seen* to be undoubtedly free from such influences of "result-orientation", possibly politically motivated, a personal remark may be in order. The present writer has learnt an awful lot about politics and statistics from a study concerning official statistics in the former GDR. It turned out that it was by no means uncommon in the GDR that inflation was prevented from existing by two rather simple devices:

- a rising price was consistently offset by an equal rise in "quality", and/or
- the commodity becoming more expensive was replaced by another, the price of which "had dropped or at least remained the same".[135]

By all accounts the suspect of this sort of manipulation (and some even more spectacular tricks) not only invalidated statistical figures, but – what is worse even -

[132] As to the *theory underlying chain indices*: if there is any deliberate concept going beyond the urge to (not to say fixation with) using the most up-to-date weights, the arguments seem contradictory (like e.g. the relation between chaining and chainability) and doubtful to say the least (as for example in the case of the arguments put forward to plead for chain indices, as shown in **sec. 7.1**).

[133] In this respect the situation seems completely different to Mudgett's automobiles, telephones or radios etc.

[134] It should also be remembered that many ideas of the Boskin Commission for example were decidedly motivated by the desire to lower the inflation rate. To introduce new goods and new outlets most rapidly, as well as to account for substitution by more frequently changing the basket, was discussed expressly under the headline "correction for overrating inflation". The title "Ways to Lower Inflation" would have described the intentions perhaps more directly and correctly.

[135] *For the selection of goods it was important that the price declined.* By the same token in output figures only those goods were reported where production had *risen*.

gradually ruined the trust in statistics altogether. Wherever continuing strongly held and diverging views are met, like in inflation measurement,[136] rigidity and simplicity in methods is nothing inconvenient, but rather an invaluable advantage.

k3) The idea of "pure" price comparison

In **fig. 7.2.2** it is shown that chain index comparisons are "impure" in *two* respects as they are influenced by other variables than prices, and by more than just the two points in time, 0 and t, which are to be compared.

Figure 7.2.2: Comparability and chain indices (part 1. Dimensions of comparability)

Chain indices violate "pure" price comparison in the sense of not only being affected by*

prices	periods 0 and t
but also by changes in (the structure of) quantities (weights), qualities, types of products, outlets etc.**	but also by referring to other periods and depending on the path connecting 0 and t (not only on the endpoints 0 and t)**
no elimination of structural change	path dependence (no chainability)

* as the first aspect (i.e. prices) refers to the aggregation over commodities, this (second) notion of "pure" refers to the temporal aggregation (over intervals in time).

** this applies to unit value indices as well for example

Chain indices are "impure" in the sense of reflecting a number of other influences than prices and being "contaminated" by structural change. This is particularly unpleasant as it is inconsistent with

- the (rightly made) objections against unit value indices and the correct rejection of the (erroneous) idea of simply using average prices rather than indices (see sec. 1.1, part e),
- and also inconsistent with the treatment of the *single* (n = 1) commodity case where comparability over time (as regards types and qualities of goods) is taken most seriously, very much distinct from the case of a *basket* (case of n ≥ 2 commodities) where it is popular to ridicule the constancy of the basket.

In the case of a single commodity or the critique of the unit value index (as opposed to a true price index), the need is in general strongly emphasized that reference should be made to possibly exactly the *same* goods in both periods 0 and t in

[136] Of course nobody is arguing for chain indices *because of* their offering more opportunities to manipulate but on the other hand – for some users at least - to fall back on more flexible methods of price statistics, if possible with the intention to lower inflation rates, is not unattractive.

order to compare like with like.[137] But curiously the same writers who emphasize such elementary ideas often seem negligent if not oblivious of them once a *basket*, a "*composite* commodity" is concerned (see **fig. 7.2.3** for this fundamental inconsistency in the chain index rhetoric).

Figure 7.2.3: Comparability and chain indices (part 2: Inconsistencies with respect to the number of commodities)

Comparability on the

Single commodity level	Composite commodity (basket-) level
In reporting individual prices much heed is paid to ensure comparability by carefully making • adjustments for quality change, or • detailed specifications of goods*	A fixed basket approach is criticized for its keeping constant of the • selection of goods, and other price determining characteristics (PDC) • weights attached to the goods

* this applies in particular to goods deemed acceptable and suitable for international comparisons.

"Representative" new weights are deemed much more important than aiming at pure price comparison. The fact that in our days everything changes quickly seems to justify to jettisoning principles of fair comparisons in statistics. Why is this reasoning inconclusive? A closer look at the second notion of "pure" in **fig. 7.2.2** (concerning the comparison of two *points in time*, or "temporal aggregation") may help to make this clear. In essence, "pure" comparison here is the idea of making a comparison uniquely determined. However, the result of comparing 0 and t using chain indices is not uniquely determined (in the sense of being affected by these two situations, 0 and t, *only*). Unlike direct indices, chain indices are "path dependent", i.e. their result is dependent on the shape of the time series of prices and quantities connecting the points 0 and t.

"Pure" comparison is to avoid vagueness and ambiguities by eliminating the influence of all factors not in the focus of the comparison in question, that is, in the case of a price index P_{0t} other variables than prices and other periods than 0 and t respectively. The fact that things change more rapidly nowadays does not make principles of meaningful comparisons invalid.

What are those principles? In order to be meaningfully comparable, two things must have certain common aspects (CA) and at the same time aspects with respect to which they are different (DA). The latter (DA) refer to factors which should affect the result while the former (CA) are related to all those aspects which should not influen-

137 This seems to go without saying for example in case of the quality adjustment or the making of detailed specifications of commodities in order to ensure international comparability, but interestingly not when the issue of weights is concerned.

ce the result. The DAs should be reflected while the CAs are necessary in order to compare like with like. There is no point in comparing things which don't have much in common. Comparability requires a minimum number of CAs and not too many of DAs. Ideally we should have one DA only. By "pure" is meant that situations to be compared should differ in only *one* aspect in order to avoid difficulties (ambiguities) of interpretation and to *make sure that like is compared with like.*

Note that this is in line with the general philosophy of statistics. Much if not most of statistics is devoted to precisely this task of making things comparable by estimating a "systematic" component as opposed to "disturbances". The idea to isolate a DA and "control" (keep constant) CAs (by elimination of structural changes) gave rise for the creation of "indices" as a method in statistics instead of simply comparing averages $\bar{x}$ or unit values. A result influenced by an unknown *mix* of a *number* of influences (DAs) is much more difficult to explain, to analyze and to work with.

k4) Applying the chain principle to all kinds of indices and to double deflation

Once chain indices are introduced for price level measurement and deflation purposes there is a strong temptation to make use of this principle in *all* kinds of indices, not only price indices but also production indices, indices of new orders and the like.

Imagine *all* types of indices were compiled as chain indices. This would have significant consequences for

1. results of statistics defined as relations between (e.g. ratios of) two indices or certain methods combining two or more indices, like for example double deflation, or for
2. defining growth rates, endpoints of intervals, turning points, leads/lags, phases of the business cycles etc. which usually is made on the basis of certain indices.

Concepts based on two or more indices

A number of concepts are measured in terms of indices, such as the *"terms of trade"*, *"productivity"*, *real* wages and salaries, or the *real* exchange rate of a currency, which are taken to measure competitiveness. Such ratios of indices rely on the fact that each index is reflecting a well defined phenomenon of its own such that the combination of two such indices has a clear meaning. If, however, both indices, the numerator and denominator index, are path dependent and represent a complex mix of influences the interpretation will run into difficulties, in all probability more than if a single index is taken in isolation. While *productivity* compares output and input in volume (quantity) terms, preferably on the basis of the production structure of the same period, a comparison of output and input *prices* is involved in the *double deflation* method as described in **sec. 5.2** and commented upon in what follows with special regard to the possibility that both price indices are *chain* indices.[138]

[138] Conspicuously, there is not much known about the consequences of a combination of chain index and double deflation, both methods renowned for possibly yielding counter-intuitive results such as negative value added at constant prices. This applies even if all deflators were direct price indices of

Double deflation of value added is regarded in both systems, the SNA and the European System of National and Regional Accounts (ESA), as the only "theoretically correct method".[139] Thus, on the output side each aggregate should be deflated separately by a *chained* Producer Price Index (PPI) of output-prices , $p_{it}(0)$, broken down into sufficient detail. By the same token, input (intermediate consumption) should be deflated product by product, using a *chained* PPI on the basis of input-prices, $p_{jt}(I)$[140], collected from purchases.[141] It seems widely unknown how results will differ if double deflation is made with *chain* rather than *direct* indices of input and output prices. Therefore it might be interesting to spell out in detail what substitution of chain Fisher indices of output and input, that is $\overline{P}_{0t}^{FC}(O)$ and $\overline{P}_{0t}^{FC}(I)$ respectively, for direct Paasche price indices, $P_{0t}^{P}(O)$ and $P_{0t}^{P}(I)$, really means.

Note that in what follows emphasis is not put on "the well-known 'non-additivity' problem"[142] but rather on the *interpretation* of the underlying formula of "real value added" if derived from chain index deflators by showing which influences are affecting the result. To make formulas simple we consider a chain of two links only. It is easy to derive the corresponding more general formula. The deflated output (DO) is given by dividing $\sum p_2(0)q_2(0)$ by

$$\overline{P}_{02}^{FC}(O)=\left(\frac{\sum p_1(0)q_0(0)}{\sum p_0(0)q_0(0)}\frac{\sum p_2(0)q_1(0)}{\sum p_0(0)q_1(0)}\frac{\sum p_2(0)q_2(0)}{\sum p_1(0)q_2(0)}\right)^{1/2},$$

and in exactly the same manner dividing $\sum p_2(I)q_2(I)$ by the corresponding index $\overline{P}_{02}^{FC}(I)$ yields the deflated input (DI). Finally, the implicit deflator price index of value added (Y) is given by

$$P_{02}^{imp}(Y)=\frac{\sum p_2(0)q_2(0)-\sum p_2(I)q_2(I)}{DO-DI}=\frac{\text{nominal value added}}{\text{real value added}}$$ or in terms of Fisher chain price indices and i_t denoting the input share of output (at current prices) at time t,

(7.2.32) $$P_{02}^{imp}(Y)=\frac{(1-i_2)\overline{P}_{02}^{FC}(O)\overline{P}_{02}^{FC}(I)}{\overline{P}_{02}^{FC}(I)-i_2\overline{P}_{02}^{FC}(O)}$$ or in detail

---

Paasche and if all sorts of measurement problems were absent. Such problems arise in great number however, for example the valuation of output for own use, government consumption, services of financial intermediaries, definition of a price and volume component in the case of taxes, subsidies etc.

139 ESA, para 10.28.

140 The subscripts i and j for output and input goods respectively will be omitted for convenience of presentation.

141 It was also found desirable to take into account that different purchasers will pay different prices. The problem with this, however, is that the necessary variety of price indices usually does not exist.

142 As a result of "non additivity", it is laid down in ESA and the Commission Decision of the European Union that "disaggregated constant price data" (gained from using direct deflators) "are compiled in addition to the chain indices for the main aggregates", and "...*it will have to be explained to users why there is no additivity*" (see digression of chapter 5 page 369).

$$(7.2.33)\quad P_{02}^{imp} = \frac{(1-i_2)\sqrt{\frac{\sum p_2(0)q_2(0)}{\sum p_0(0)q_0(0)}}}{\sqrt{\frac{\sum p_0(0)q_1(0)\sum p_1(0)q_2(0)}{\sum p_1(0)q_0(0)\sum p_2(0)q_1(0)}} - \sqrt{i_0 i_2}\sqrt{\frac{\sum p_0(I)q_1(I)\sum p_1(I)q_2(I)}{\sum p_1(I)q_0(I)\sum p_2(I)q_1(I)}}}.$$

Of course with more than two links the result will be all the more complicated. The formula shows that the implicit deflator P^{imp} is reflecting a number of influences in addition to prices of a given selection of inputs and outputs at two periods in time, which does not seem to be thoroughly investigated and well understood yet. Furthermore, this should be viewed against the background that the result of a chain Fisher deflation, even in the case of a single aggregate, is in itself not easy to understand. Double deflation of the "non-commodity flow" Y amounts to the combination of two methods, both doubtful and in a way contradictory among themselves, because

- double deflation is based on the existence of accounting constraints (such that a method which ensures additivity seems to de desirable),[143] while on the other hand
- to apply chain indices as deflators brings with it the "well-known non additivity-problem", i.e. the violation of accounting constraints.

To sum up: the results of such a procedure (use of chain indices for output as well as input prices)[144] are influenced by no less than three factors

a) differences in the development of output- and input-prices,[145]

b) the "path" of prices and volumes in the intermediate intervals, 1, 2, ..., t-1 as they habitually affect the value of chain indices (in contrast to direct indices), and finally by

c) the degree of violation of balancing constraints as a result of using the Fisher formula and a chain index approach instead of direct Paasche index deflators.

Hence there is a structural influence (ensuing from problems with aggregation over time as well as over commodities) in addition to some "usual" problems of interpreting sometimes implausible results of double deflation. However, interestingly there is not much discussion (and thus also knowledge) about the implications of this. The same is true for other results, such as different conclusions concerning lead-lag-structures drawn on the basis of chain index methods as opposed to direct indices.

Phenomena measured by making use of indices

When a chain production index is used for defining (endpoints and thus the length of) *phases in business cycles* we may get different "turning points" depending on

---

[143] according to which the difference between output and input should equal value added, both in terms of current prices (nominal aggregates) as well as constant-prices (real aggregates).

[144] This is as aforesaid made mandatory for all member countries of the European Union.

[145] This factor is already identified as being responsible for some curious results of double deflation even if direct indices are used.

whether a direct or a chain index of production is used. The same is true for the quite often discussed situation of two sets of *growth rates* of GDP, using a direct approach and a chain approach. As already mentioned, the European Commission voted for the adoption of chain weighted measures of real GDP as two years previously the US statistics also moved from a constant price to a chain index measure of volume.[146] Experience has shown that growth rates resulting from new as opposed to "traditional" methods may well differ substantially and there was no clear superiority of one method over the other. Things are likely to become even more difficult to interpret when not only one single set of growth rates (for example of GDP) is analyzed but the relation between two or more sets of chain-based growth rates, as for example in the case of establishing the lead-lag-structure[147] between two time series. Again, little is known about how the lead-lag-structure detected by chain indices will differ from the corresponding result of the hitherto generally applied direct index approach.

It is in fact surprising that much is left vague in this field, since the practice of indices always deals with a *number* of indices and with relating them to one another. Thus, when superiority of a new formula is asserted one would expect that at least some thought has been given to such aspects. But this does not seem to be the case.

---

[146] EU Commission Decision (as concerns the principles for measuring prices and volumes) Nr. 2223/96 approved by the Statistical Programme Committee in Stockholm, 27th May 1998. In January 1996 the US Bureau of Economic Analysis (BEA) adopted a chain Fisher quantity index as measure of real growth (as recommended by SNA).

[147] To draw conclusions in terms of "leads" or "lags" is for example one of the main reasons why it is found useful to compare time series of new orders (considered as *leading* indicator) with time series of turnover (they are supposed to reflect the end of a sequence beginning with demand [new orders] via production and ending with sales, and thus turnover will be *lagging* behind new orders). Assume both indicators, the leading and the lagging, take the form of *chain* indices (instead of the hitherto more commonly used direct indices) known to reflect not only the conditions prevailing in periods 0 and t respectively, but also in the intermediate periods, which may be different in the two series compared and which may have in a sense a cumulative effect. What then will be the meaning of the difference?

Chapter 8

Interspatial comparisons of prices and volumes

The presentation of index formulas set out in the preceding chapters was focused on intertemporal comparisons. In index theory the measurement of differences in price levels or production levels (quantity indices) of different countries and making consistent (transitive) *multi*-lateral comparisons between *a number of* different *countries* has been neglected for quite a long time, but the issue experienced a revival in the 1970s and 80s, due to growing activities in international comparisons and to European integration.[1]

Interestingly, the methods proposed to solve problems of multinational comparisons on the level of individual commodities and national CPI-baskets (in order to ensure "pure" comparability) as well as the aggregation over countries seem to have fallen into oblivion when the euro- and thus the HICP-project emerged and chain indices became fashionable.

The chapter starts with a general introduction (**sec. 8.1**). In **sec. 8.2** we try to give an overview of methods as well as desirable properties of index functions used without going into details and will then give more information concerning selected methods in the following sections.

8.1 Introduction into interspatial comparisons

In some aspects intertemporal (comparing points or intervals in time) and interspatial (interregional or international) comparisons face equivalent problems such that methods developed for one case can easily be translated into the other case. But there are also dissimilarities we are going to work out first in this section (part a). The next point will be to compare *bi*-lateral and *multi*-lateral international problems (part c). This will also shed some light on the relevance of certain axioms and the purposes we aim at in international comparisons.

a) Differences between intertemporal and interspatial comparisons

The most significant differences between intertemporal and interspatial comparisons are as follows:

[1] To make consistent (transitive) comparisons between countries is for example of vital interest in the case of member countries of a community like the EEC (or parts thereof), sometimes also between countries in worldwide comparisons. For a "block" of countries (even in the case of a common currency unit, as for example in the case of the €) there is also the desire to compile and aggregate "volumes", in the sense of expenditures of different countries where the different purchasing power of money is accounted for. A revaluation of aggregates using common prices in order to make them comparable is sometimes also called "regional deflation".

1. Time is ordered, a sequence is uniquely defined, t precedes t+1 whereas, for example, any of the 6 permutations (orderings) of three countries A, B, C is equivalent in the sense that for example the sequence B-C-A is neither more nor less reasonable than the sequence A-B-C. The number of meaningful comparisons that can be made between n countries is significantly greater than the number of meaningful comparisons between n periods. Furthermore, making indirect comparisons, and therefore transitivity, gains importance. There is also no counterpart to interspatial *multi*-lateral instead of *bi*-lateral comparisons in the case of time.

   This is not to say, however, that there is no equivalent to the *chain index* approach. We can well define a set of chained (linked) comparisons between any two countries (now called a *spanning tree*) and make a choice among different spanning trees. In a sense, therefore, one might regard the *method of minimum spanning trees* (MST) as a kind of inter-regional chain index.

   Another consequence of the non-existence of a natural order or sequence is that certain axioms will be of greater importance in interspatial comparisons as compared with the intertemporal case. *Time reversibility* (redefined as *country reversal test*) and *circularity* (transitivity), for example are important aspects in international comparisons. Both axioms reflect a similar underlying *invariance* idea, in the sense that an index function should be independent in some way of the base and the sequence of comparisons:

   - ***Country reversibility*** means that it does not matter whether country A or B should be the base country in a comparison between A and B.[2] It is important *not to confound* this property with base land invariance, a rather ambituous property we prefer to call *strict* transitivity (as opposed to the usual [or "weak"] transitivity).
   - In international comparisons, ***transitivity*** means that *any* indirect comparison of country A with country D (via B, or C, or B *and* C etc.) should produce the same parity as the direct comparison.[3]

2. Space (like countries A, B, C) is necessarily discrete only. Time on the other hand can be conceived as a discrete *or* as a continuous variable. It is the latter case with which the Divisia approach deals, but there is no counterpart to this part of index theory in international comparisons.

3. If an additional (m+1) country has to be taken into account, the results concerning the first m countries will be affected in general[4], whilst in intertemporal comparisons a new period t+1 will not give rise to the need to perform recalculations concerning the first t periods. As to this aspect there is also a sub-

---

[2] This is often found a desirable property relieving us from the problem of choosing a base country since in comparing countries we usually don't have any reason to prefer country A to B as a base.

[3] From the point of view of consistency among parities it is of course highly desirable to obtain the same parity between any two countries, no matter whether we compare them directly or indirectly.

[4] If this is the case and all countries to be compared ought to enter the index calculation, the method is called *full scale* and *closed*.

stantial difference between bilateral (m is fixed to m = 2, hence there is no m+1) and multilateral international comparisons.

Fig. 8.1.1: Some differences between intertemporal and interspatial comparisons

1	order (defined sequence), different axioms gain importance
2	discrete/continuous variable
3	problems with additional countries (full scale[1]) closed comparisons)
4	different and modifiable size of countries, aggregation over countries
5	stable relationships less likely, "economic theory approach" doubtful
6	price indices are no longer pure numbers
7	notion of "regional deflation" and "real aggregates", purchasing power parities (PPP) versus exchange rates (ER)
8	an additional dimension of "consistency": aggregation over periods in time *and* over countries (*both* types ought to be consistent)[2])

1) taking *all* countries to be compared into account

2) also called "super-" or "two-dimensional-transitivity"

4. Time intervals are usually of the same length. Months are compared with months, years with years. To compare a country like e.g. France with Luxembourg is however equivalent to comparing a year with a week or so. It is not desirable that a country exerts more or less influence on the results of a comparison than appears justified by its economic potential. Thus, to avoid this, *weights* should be attached to countries in order to adequately take into account the different size of the units.

   Moreover, countries are *modifiable*: they can be disaggregated (partitioned)[5] into smaller regions or aggregated into whole blocks, and there is a desire to compare *aggregated* countries, not only individual countries with one another.

5. Stable relationships between prices and quantities are more likely to exist over time than across various countries. In addition, it seems to be less troublesome to keep quality, selling conditions and the like constant in intertemporal comparisons than in comparing countries. This applies in particular when those countries are geographically located wide apart, or differ greatly with respect to the standard of living. Not only differences in prices but also in the structure of prices and quantities of goods and services (and their quality) tend to be greater between countries than in intertemporal comparisons.

[5] An interesting property of an index function is for example how it "behaves" in the case of a country in which the quantity vector $\mathbf{q}_t$ may be partitioned into various sub-vectors $\mathbf{q}_{kt}$ (k = 1, ..., K) while the price vector $\mathbf{p}_t$ is assumed to be the same for all K parts of the country (according to Diewert's country partitioning test CPT in the case of K = 2 and equal structure of quantities in both parts such that $\mathbf{q}_{1t} = \mu\mathbf{q}_t$ and $\mathbf{q}_{2t} = (1-\mu)\mathbf{q}_t$, then the quantity index should also reflect the relation $Q_{1,0t} = \mu Q_{0t}$ and $Q_{2,0t} = (1-\mu)Q_{0t}$ where μ and $1-\mu$ are scalars. See criterion D6 in part f of this section (Diewert's axioms).

It also appears more difficult to explain such differences with the help of the microeconomic theory of consumer behaviour. It is doubtful that a true "economic theory approach" really exists in international comparisons.[6]

6. A problem in inter-country comparisons not to be faced in intertemporal comparisons is dependence on the currency unit. Price indices (in contrast to quantity indices) refer to a currency unit, they are no longer pure numbers as they are in the intertemporal case.[7]

7. Deflation is to be defined differently to the intertemporal case where it means removing statistically the effect of inflation in aggregates compared over time, and where we think of aggregates "at *constant prices*" in the sense of prices related to a *fixed* base year, whereas in the inter-country comparison deflation refers to the application of *common prices* for all countries (if the price levels differ) and all periods in question. Hence we should distinguish

   - constant (for all points in time) prices, and
   - common (for all countries) prices,

   and note further that common prices are not the same as a common currency unit or currencies converted by a given exchange rate:

Mostly there is a choice between purchasing power parities (PPP) and official exchange rates (ER) to make aggregates comparable (see below for more details). Only in the case of PPPs are the aggregates called "real", in the case of ERs, however, a "nominal" aggregate, revalued in terms of another currency unit remains "nominal".

8. We encounter another, in fact an *additional*, dimension making "consistency" (or "coherence") in the case of inter-country comparison more difficult. According to this criterion (sometimes called "super-" or "two-dimensional-transitivity")[8]

   - to apply national price indices P_{01}, P_{02}, P_{03}, ... consistently over time in order to get a series of "volumes" (real values) should be consistent with
   - the sequence of real values for that country gained from the time series of PPPs,

   or comparisons across countries and for all periods in time should fit together (successive international comparisons are not in conflict with data gained from extrapolating or updating of national data). Hence it should not make a difference whether we first derive a sequence over time and then aggregate volumes

---

[6] It does not make much sense to assume a household living simultaneously in different countries or to assume households in different countries as having the same preferences. On the other hand, it appears resonable to say what really matters in the case of the purchasing power parity is not to enable a household to buy the same *goods* but rather *to be as well off in terms of utility* (or welfare) with the same amount of money *in both countries* in question.

[7] For example formerly, before the €, the situation was such that if Germany (with DM as currency in those days) was the base country (= 100) and the country to be compared (reference country) was France, the unit of measurement of a price index was "Francs per DM".

[8] i.e. in the time and space dimensions.

or underlying PPPs consistently over countries or if we proceed the other way round.[9]

It comes as no surprise that consistency in all these dimensions, that is over sub-indices (consistency in aggregation for each country) and volumes (structural consistency [in volumes] for each country) and in the sense of the factor reversal test (again for each country in each period) will be difficult if not impossible to achieve.

b) Uses and limitations of international price comparisons; PPPs and exchange rates

Price indices measuring a "price level" of a country compared to other countries (currencies) or to a block of countries are used for the following purposes:

1. the measurement of the *purchasing power parity* (PPP) to be compared with the official exchange rate (ER) of the currency,[10]
2. to serve as a means to *deflate* values[11] and thereby to perform volume comparisons of
   - absolute levels and growth rates of volumes for specific aggregates between countries
   - stuctures of expenditures at constant (comparable) prices of a community between member countries of a community.

Volumes as well as deflators (for example PPPs) also serve as instruments to measure and assess international competitiveness, mostly in combination with price indices for external trade or other indicators of economic performance (e.g. labor costs etc.).

Note that the two objectives of international price comparisons, i.e. measuring of purchasing power and deflation of aggregates, clearly correspond to the two objectives in the case of intertemporal comparison, that is, price level measurement and deflation.

As mentioned in **fig. 8.1.2**, practical problems frequently give rise to calculations of PPPs, as for example the remuneration of employees,[12] or imigration and tourism

---

[9] One of the puzzles of the 1996 OECD comparisons between Canada and the USA was that Canada moved relative to the US, from 77.2 to 79.5% (USA = 100%) between 1993 and 1996 benchmarks, both based on EKS methods (see below). So Canada's average growth rate should have been higher than the US growth rate. On the other hand, according to the Statistic of National Accounts, it was in fact 2.7% less than that of the USA.

[10] Price indices (PPPs) will usually be compared with exchange rates (ER) to derive "gains" or "losses" in purchasing power.

[11] Volumes on the basis of ERs may widely differ from those based on PPPs or other price indices. Bilateral, and even more multilateral volume comparisons, are usually based on PPPs (as measures of "real" values).

(these cases will in general call for different methods, because it is reasonable to assume that an immigrant will change his consumption habits whereas the tourist will not).

> It is generally accepted that "regional deflation" (i.e. volumes comparable with respect to the price level) is more difficult than estimation of PPPs as it is more important to meet certain requirements of consistency and to take the size of countries into account.[13]

Moreover, consistent *multi*-lateral comparisons are much more demanding than just to estimate the PPP of one single country relative to another country.[14]

Figure 8.1.2: Uses of international comparisons of prices

International comparisons of prices

purchasing power parities (PPP)	(regional) deflation of aggregates
the purpose is to estimate • an equivalent remuneration of employees when working abroad; • gains or losses of purchasing power (compared with the exchange rate); • PPP as indicators of performance in fighting inflation and preserving international competitiveness.	deflation (mostly within the framework of National Accounts) • to make comparisons of level and to structure aggregates* • to aggregate volumes across countries (taking into account the different size of countries)**, as e.g. in the case of the EEC

* hence consistency in aggregation should be obeyed more strictly than in the case of PPPs

** in the case of PPPs taking the size of countries or transactions involved into account is in general not necessary.

Why not simply take exchange rates?

Exchange rates (ERs), as opposed to PPPs, are generally considered less suitable for international comparisons as they are influenced by disturbing factors like capital movement, foreign trade, revaluations and devaluations, speculation etc. In the case of a re- or devaluation, the respective national "volumes" of GDP, consumption etc. expressed in a common currency unit and converted by using the official ERs will in

[12] The so called "Ford wage problem" in the late twenties was one of the first practical applications of place to place comparisons in index theory. The wage policy (of Ford) wanted to provide its employees with the same standard of living, whether they lived in Detroit, Cologne or London.

[13] It is also often argued that deflator indices should satisfy the factor reversal test. Unfortunately, there is mostly no reason given why the weaker product test will not suffice to achieve a decomposability of the value change into a price and a quantity change.

[14] Contrary to widespread belief, however, it is *not* the case that a method successfully dealing with multilateral comparison problems will also, as a by-product, provide a satisfactory binary comparison.

general rise or decline although in actual fact no national aggregate has really changed in terms of quantitites. Furthermore, ERs are in general much more sensitive and volatile than PPPs.

It is therefore desirable to have an estimate of the PPP independent of official ERs and the use of PPPs is preferred to ERs in deriving volumes.

c) Bilateral and multilateral price comparisons

It is common pratice to make a distinction between bilateral and multilateral (or multinational) comparisons, because the latter require some additional criteria to be fulfilled. In contrast to intertemporal comparisons, in interregional comparisons between any two countries (or places) the following two types of comparisons,

- *direct* comparison between A and B, denoted by A-B, and
- *all* possible *indirect* comparisons via C, via C and D and so on, that is comparisons A-C-B, or A-C-D-B and the like,

are both legitimate, and in use; they should be consistent with one another.

With four (m = 4) points in time involved, 0, 1, 2, and 3, it makes sense to compare the following periods with 0 as a fixed period: 0-1, 0-2 and 0-3, but not, as an example, 3-0. In general, there are m-1 such comparisons with a fixed base which will be of interest (the shaded parts in fig. 8.1.3, upper left part). There are another m-1 comparisons between m consecutive time periods which also will usually be made, that is, comparisons of the type 0-1, 1-2 and 2-3 (comparable to "links" in the framework of chain indices).[15]

The situation is totally different in the case of m countries, however. With four (m=4) countries involved, A, B, C and D, all 6 direct comparisons between any two countries (or in general all $\binom{m}{2} = \frac{m(m-1)}{2}$ possible pairs) should be treated as equivalent (shaded parts in **fig. 8.1.3** right hand side) where identity and country reversiblity is assumed such that $P_{AA} = 1$ and P_{BA} can be derived from P_{AB}. Hence in the case of 4 countries we have (as shown in **fig. 8.1.3**) 6 direct comparisons (shaded) plus

- m - 2 = 2 indirect comparisons via *one* third country for each of the six pairs, like A-C-B and A-D-B in the case of the pair A-B
- (m - 2) (m - 3) = 2 indirect comparisons via *two* third countries for each of the six pairs, like A-C-D-B and A-D-C-B,

hence altogether 6+12+12 = 30 which have to be consistent with one another.[16]

---

[15] Thus we have altogether at best only 2(m - 1) - 1 comparisons being of any interest, provided that 0-1 is not counted twice.

[16] Correspondingly, in the case of m=5 countries the number of direct and indirect comparisons between two countries that have to be consistent with one another grows up to 160, and with m=6 no less than 975 reasonable comparisons.

Figure 8.1.3: Number of comparisons in intertemporal and interspatial case

a) Direct comparisons

	four points in time			
	0	1	2	3
0	1	P_{01}	P_{02}	P_{03}
1	P_{10}	1	P_{12}	P_{13}
2	P_{20}	P_{10}	1	P_{23}
3	P_{30}	P_{31}	P_{32}	1

	four countries			
	A	B	C	D
A	1	P_{AB}	P_{AC}	P_{AD}
B	P_{BA}	1	P_{BC}	P_{BD}
C	P_{CA}	P_{CB}	1	P_{CD}
D	P_{DA}	P_{DB}	P_{DC}	1

b) Indirect comparisons (via one or two "third" countries)

pair	0 (direct)	1	2
A-B	A-B	A-C-B, A-D-B	A-C-D-B, A-D-C-B
A-C	A-C	A-B-C, A-D-C	A-B-D-C, A-D-B-C
A-D	A-D	A-B-D, A-C-D	A-B-C-D, A-C-B-D
B-C	B-C	B-A-C, B-D-C	B-A-D-C, B-D-A-C
B-D	B-D	B-A-D, B-C-D	B-A-C-D, B-C-A-D
C-D	C-D	C-A-D, C-B-D	C-A-B-D, C-B-A-D
sum	6	12	12

For any two fixed countries, say A and B, there exist m-2 comparisons via one third country, (m-2) (m-3) comparisons via two "third countries", and (m-2) (m-3) (m-4) comparisons via three third countries and so on.

In summary, in the case of four countries we end up with 30 reasonable comparisons as compared with only five in the case of intertemporal comparisons between four points in time, viz. P_{01}, P_{02}, P_{03}, P_{12}, and P_{23}, or $2(m - 1) - 1 = 6 - 1 = 5$ in the case of m = 4.

d) Some methods of bilateral international comparisons

In what follows we only deal with binary comparisons (two countries at a time), not yet with multilateral comparisons. In comparing A (as the "base") with B, the "other", or "reference", country it is useful to distinguish[17] between

[17] This distinction refers also to what is meant by the "Laspeyres" and "Paasche approach". In this context the terms refer to the country from which the list of commodities is taken (in the first case) or to the country from which (this list and) the weights are taken (in the second case). Hence in both cases, number 1 and 2 respectively, we will again meet the terms Laspeyres and Paasche, with a different meaning, however.

1. using information on prices only (without quantities), or of observing parities for specific *commodities* (single commodity case) or *unweighted* averages of them, and
2. estimating a set of transitive price *indices with weights* reflecting the importance of countries or of estimating an average price level for a community.

Case 1 is primarily only the first step in a method to calculate transitive indices (case 2). It is mostly not of interest as such.

1. Unweighted indices, basic parities for EC member countries

The purpose of this subsection is to describe the methods by which Eurostat arrives at binary parities of m member states. Experience shows that it is not sufficient simply to collect information on prices provided by the National Statistical Offices (NSOs) because it is decisive to make sure that the products to which prices refer

- are specified in detail, and with a minimum precision in order to guarantee *comparability* across countries, and they also should be
- *representative and characteristic*, preferably to the same extent for both countries to be compared.

The first step in calculating purchasing power parities (PPPs), if possible for every pair of countries, is therefore the establishment of product lists and specifications of products for which prices are collected as well as procedures for price collections to be followed in the respective member states. The development of a product list is a difficult step and has implications for the representativeness of the price comparisons. Attempts to derive transitive parities cannot be made before this task has been performed. Thus the procedure starts with lists provided by NSOs containing "their" national products.[18] These "preferred" products are in general those commodities which are also included in the national CPI. However, the product specifications necessary for national intertemporal comparisons of consumer prices are usually less detailed than those for products proposed for international comparisons. It is always somewhat astonishing to see how much attention is given to this step in order to ensure comparability *across countries* given the little emphasis placed on comparability *in the course of time* by those who advocate chain indices or scanner data.

Next, the NSOs are asked to compare "their" list with lists of other countries' preferred products and to establish *two* lists, one of "*other*" products which can be priced, and one of products not available. On the basis of these lists a list of so called *basic headings* is established.

Needless to say that such lists and the products chosen on this basis will in general exhibit some bias in favor of those products which are internationally marketed by firms operating in various countries. "Homespun" products on the other hand are likely to be under-represented.

---

[18] The idea of the "preferred" products a NSO has to name was to come close at least to what might be considered representative and characteristic for the country in question.

Basic headings constitute the most detailed categories of aggregation. They are defined in such a way that

- a breakdown of basic headings into smaller categories is not intended although each basic heading may be represented by *a number* of products,[19]
- there is (also due to lack of statistical data)[20] no "weighting" within the basic headings to account for the relative importance of products.

Figure 8.1.4: Usage of notions, like Laspeyres and Paasche and illustration of the notion "basic heading"

	1. Unweighted indices (parities)	2. (weighted) price indices
Principle	the country from which the list of commodities is taken	the country from which this list and the weights (expenditure shares) are taken
Laspeyres	(8.1.1) $\;{}_{A}L_{B}=\left(\prod_{i=1}^{n_A}\frac{p_{Bi}}{p_{Ai}}\right)^{1/n_A}$	(8.1.2) $\;P_{AB}^{L}=\frac{\Sigma p_B q_A}{\Sigma p_A q_A}$
Paasche	(8.1.1a) $\;{}_{A}P_{B}=\left(\prod_{k=1}^{n_B}\frac{p_{Bk}}{p_{Ak}}\right)^{1/n_B}$	(8.1.2a) $\;P_{AB}^{P}=\frac{\Sigma p_B q_B}{\Sigma p_A q_B}$

A = base country, B = reference country, p_A and p_B are absolute prices in A and B; commodities $i = 1,\dots, n_A$ preferred by country A; $k = 1,\dots, n_B$ = preferred by country B

Products and "basic headings"

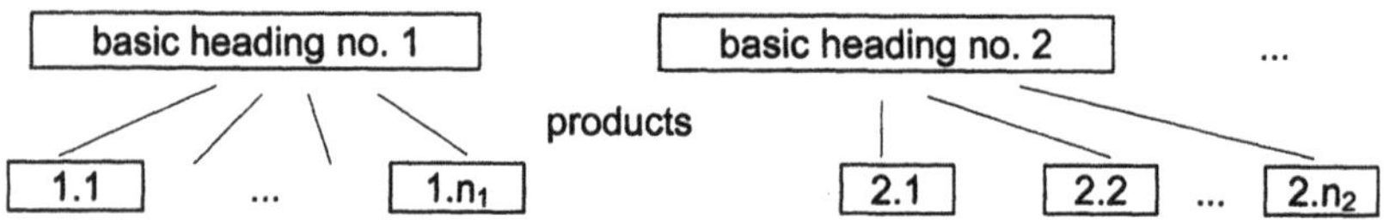

Furthermore, the collection of basic headings may be viewed as a sort of "universe" from which the products that eventually enter into the price comparison are taken as a (non-random) sample. It is neither feasible nor pertinent to strive for random sampling of basic headings or even of products, for the simple reason that there is no EEC-wide common list of products from which a random sample might be drawn.

Hence the approach to price comparisons adopted in 1975 (and maintained thereafter) was to establish a list of basic headings which should be represented by pro-

[19] Basic headings do *not* represent "products" at their most detailed level of specification and there is some freedom to choose the specific products to represent a certain basic heading (see **fig.8.1.4**).

[20] In fact, no comparable statistical information is available on all product specifications in order to assign weights to them; hence weighting might not be necessary, even more, it is not possible.

ducts, both as comparable and representative *as possible* in the countries to be compared.[21]

In comparisons between EEC countries (1975, 1980) the intention was originally to define (relative) prices for each basic heading and each country to be compared, that is, to establish a *complete price tableau*. Due to practical difficulties, this idea was later abandoned in favor of an *incomplete tableau*, i.e. for a given list of basic headings (each containing a number of products), each country will provide prices for only a number of the products. It is now sufficient to have at best the same preferred and other products in a comparison *of any two* countries.

The problem with this method is, however, that it is difficult to say to what extent the collection of products is representative of the consumption of a given country,[22] as the products actually chosen to represent the basic headings are of course not representative to the same extent in each specific country. The intention was to aim not only at a common list of basic headings but to have it also represented (if possible) by the products which are "on average" representative in all countries.

The definition of a parity between A (base country) and B (reference country) given in **fig. 8.1.4** refers to such lists when a *geometric* mean is taken over a set of n_A or n_B goods respectively.[23] Note that the prices p in eq. 8.1.1 etc. are in *absolute* terms, expressed in currency units of the respective countries. From the two estimates of a parity between A and B, ${}_AL_B$ and ${}_AP_B$, we quite straightforwardly attain the Fisher parity by

(8.1.3) $${}_AF_B = \sqrt{{}_AL_B \cdot {}_AP_B},$$

which is regarded as "equi-characteristic".

If for all possible pairs of participating countries a Fisher parity is to be calculated, the price tableau must be complete although the products introduced to represent a given basic heading need not always to be the same. In practice, however, even in the framework of the ECE for the most basic headings a complete matrix of binary Fisher-parities is not available.[24]

---

[21] Prior to 1975 the method was to define a unique list of products to be priced by all countries. As more and more countries joined the EEC and the EEC thereby became more heterogeneous, this procedure had to be abandoned because of difficulties in defining such a universe of products. A similar change of methods took place at the level of worldwide comparisons, that is, in the ICP project. In its phases I to III the ICP project aimed at comparing all countries on the basis of a unique list, mainly inspired by US products. In the subsequent phases work has been decentralized and different product lists were established for different groups of countries.

[22] We not even have agreed on a definition for the "population" to which representativity refers. The notion of (and any measure to assess) "*representativity*" and "*characteristicity*" depends on how detailed and how comprehensive such a description of the population would be, and finally, there is no weighting (to reflect the degree of representativity) between and within basic headings, let alone a weighting of specific products to represent a basic heading.

[23] As mentioned already, the Laspeyres (Paasche) approach takes into account the preferred products of the base (reference) country.

[24] Making parities transitive also in fact solves the problem of a tableau being incomplete. Furthermore, the EKS-method described later will be able to handle simultaneously both problems, the tableau being incomplete and parities being intransitive.

2. Solutions for weighted price indices

In **fig. 8.1.4** we find index formulas traditionally known as the Laspeyres and Paasche approaches to interspatial price comparisons. Note that both, numerator and denominator differ with respect to prices (expressed in different currency units), not with respect to quantities. So the indices P^L_{AB} and P^P_{AB} are no longer pure numbers. If A is Germany and B is the USA, the unit of measurement of P^L_{AB} and P^P_{AB} will be \$ per €. However, the corresponding quantity indices $Q^L_{AB} = \frac{\Sigma p_A q_B}{\Sigma p_A q_A}$ and $Q^P_{AB} = \frac{\Sigma p_B q_B}{\Sigma p_B q_A}$, will refer to the same currency unit, € in Q^L_{AB} and \$ in Q^P_{AB} in both numerator and denominator, and thus represent pure numbers.

With a pair of P^L and P^P indices it is of course not possible to meet the time (country) reversal condition, $P_{BA} = 1/P_{AB}$, a fact which is (erroneously) often viewed as a "shortcoming" of this approach and therefore there is a strong tendency to average P^L and P^P somehow, like in

(8.1.4) $\quad P^F_{AB} = \sqrt{P^L_{AB} P^P_{AB}}$ (Fisher parities) or

(8.1.5) $\quad P^{DRO}_{AB} = \frac{1}{2}\left(P^L_{AB} + P^P_{AB}\right)$ (Drobisch [or Sidgwick] parities),

or some other compromise-formulas. The rationale of this "solution" is to use both quantities q_A and q_B in a "balanced" or "symmetric" fashion because it is difficult to decide between one of them at the expense of the other.

It may be doubtful, however, whether this is a wise decision. Take m > 2 instead of m = 2 for example: obviously in *multi*-lateral comparisons the problem to get a single parity between A and B can hardly be solved by including *all* possible *third* countries we might think of, like C, D etc. in the considerations. Furthermore, though it is no doubt troublesome to have *two* parities for each pair of countries, it is also questionable why this duality should be eliminated artificially. Country reversibility requires that no matter which of the two budgets q_A or q_B we refer to, we always get the *same* parity (i.e. one is the reciprocal of the other). However, in real situations it is unlikely that q_A (and q_B respectively) is equally *characteristic* (or "*representative*") of both countries, A *and* B, alike. Thus *country reversibility* and "*characteristicity*" are clearly in conflict with one another.[25]

The underlying ideas of the country reversal test as well as "equicharacteristicity" is to treat countries symmetrically. However, as already mentioned, two countries may have economies of very different size and performance, and it is therefore less obvious that they should be treated symmetrically like two time-intervals of equal length.

[25] The country reversal test is met if a comparison of prices between country A and B is made in a way that it is irrelevant whether A or B is the base country. The result, however, will in general not be "characteristic", "specific" or "representative" for either of the two countries, A or B. Using *different* quantities (as in P^L and P^P) will in general violate the country reversal test but can be defended for the reason of characteristicity.

On the other hand, for asymmetric solutions, it may be difficult to say why A should be singled out as the base country and why q_A in P^L_{AB} or q_B in $P^P_{AB} = 1/P^L_{BA}$ should be the better choice. Moreover, for all practical purposes it is widely accepted that a unique parity between any two countries is highly desirable, even at the expense of the structure of quantities underlying the index compilation not really being representative for any of the two countries compared. Thus "symmetric" Fisher parities appear preferable to "asymmetric" Laspeyres and Paasche parities.

e) Consistency in multinational comparisons (the meaning of transitivity)

It can easily be seen that "symmetric" and "country reversible" (see **fig. 8.1.6**) Fisher parities are *not transitive*, since (taking only four countries in consideration)

$$\hat{P}^F_{AB(C)} = \frac{P^F_{CB}}{P^F_{CA}} = P^F_{AC}P^F_{CB} = \sqrt{\frac{(p_C{'}q_A)\cdot(p_B{'}q_C)\cdot(p_B{'}q_B)}{(p_A{'}q_A)\cdot(p_A{'}q_C)\cdot(p_C{'}q_B)}} = \sqrt{V_{AB}}\sqrt{\frac{(p_C{'}q_A)\cdot(p_B{'}q_C)}{(p_A{'}q_C)\cdot(p_C{'}q_B)}}$$

$$\hat{P}^F_{AB(D)} = \frac{P^F_{DB}}{P^F_{DA}} = P^F_{AD}P^F_{DB} = \sqrt{\frac{(p_D{'}q_A)\cdot(p_B{'}q_D)\cdot(p_B{'}q_B)}{(p_A{'}q_A)\cdot(p_A{'}q_D)\cdot(p_D{'}q_B)}} = \sqrt{V_{AB}}\sqrt{\frac{(p_D{'}q_A)\cdot(p_B{'}q_D)}{(p_A{'}q_D)\cdot(p_D{'}q_B)}}$$

will in general differ and they also will differ from the direct Fisher index, which is

$$P^F_{AB} = \sqrt{V_{AB}}\sqrt{p_B{'}q_A/p_A{'}q_B}\,.$$

Transitivity is thus a more demanding property than time (country) reversibility. As shown in **sec. 2.2** and **3.2**, a function that confirms with identity and transitivity also fulfills time reversibility but the converse is not true (again obvious in the case of Fisher's ideal index). As is well known, an index meets the circular test if for example $P_{04} = P_{01}P_{12}P_{23}P_{34}$ holds and by virtue of the time reversal test, that is, $P_{40} = 1/P_{04}$, also $1 = \underbrace{P_{01}P_{12}P_{23}P_{34}}_{P_{04}}P_{40}$, which explains the name "circular test" or circularity.[26]

Transitivity is not only a special requirement in *multi*-lateral comparisons (not yet relevant in the bilateral case), it also has different meanings (such that we might speak of strict and weak transitivity, see **fig. 8.1.6**).[27] In **fig. 8.1.5** and **8.1.6** a distinction is made as to whether comparisons are symmetrical (balanced) or not and whether country reversiblility is satisfied or not (both criteria apply to the bilateral case).

---

[26] The equation indicates the source of the name "circular" test: the round circle trip should bring the index back to its starting point.

[27] Strict transitivity is sometimes called *base-land-invariance*. It is of great importance in providing a consistent one-dimensional ranking of countries with respect to the values (results) of indices. When transitivity is given, the product of indices (indirect comparisons) can also fill gaps in binary comparisons.

Figure 8.1.5: Bilateral and multilateral comparisons Laspeyres- and Paasche-approach in bilateral international comparisons[1)]

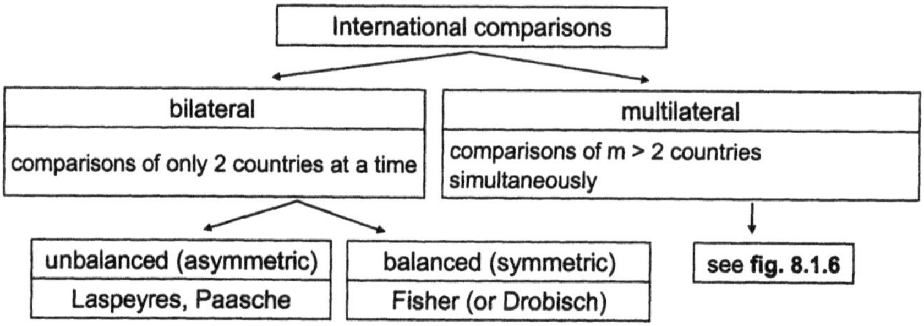

	Laspeyres approach	Paasche approach
Interpretation	Base country budget q_A. Tourist model: a person P of country A visits B as a tourist[2)] and maintains his consumption habits	Reference country (B) budget q_B. Migrator model: assimilation to the new conditions; Migrator P will change his consumption habits
Commodities	Selected commodities are representative of country A	Selected commodities are representative of country B
Country reversibility	(8.1.2) $P_{BA}^{L} = \frac{1}{P_{AB}^{P}} = \frac{\Sigma p_A q_B}{\Sigma p_B q_B}$	(8.1.2a) $P_{BA}^{P} = \frac{1}{P_{AB}^{L}} = \frac{\Sigma p_A q_A}{\Sigma p_B q_A}$
Deviation from reciprocal index[3)]	$P_{BA}^{L} / \left(P_{AB}^{L}\right)^{-1} = P_{BA}^{L} P_{AB}^{L} = \frac{\Sigma p_A q_B}{\Sigma p_B q_B} \frac{\Sigma p_B q_A}{\Sigma p_A q_A} = D$	$P_{BA}^{P} / \left(P_{AB}^{P}\right)^{-1} = P_{BA}^{P} P_{AB}^{P} = \frac{\Sigma p_A q_A}{\Sigma p_A q_B} \frac{\Sigma p_B q_B}{\Sigma p_B q_A} = 1/D$

1) Formulas 8.1.2 and 2a see also **fig. 8.1.4**

2) Or a person P who used to live in A and gets a job in B for a limited period of time.

3) given (summation over an) identical commoditiy lists

According to this scheme, satisfying *strict transitivity* (concerning multinational comparisons only) is only possible if *all* indirect comparisons of purchasing power parites, PPPs for example, or volume indices between any two countries A and B, that is, P_{AB} (or Q_{AB} respectively) obtained by using other countries, like C, as a link (i.e. by multiplying P_{AC} by P_{CB}, or Q_{AC} by Q_{CB} respectively) should be equal to the direct index. **Fig. 8.1.6** also indicates that none of the standard indices in common use (like Laspeyres, Paasche and Fisher) is able to ensure weak (let alone strict) transitivity.

Figure 8.1.6: From properties of bilateral comparisons (between place A and place B) to desired properties of multilateral comparisons

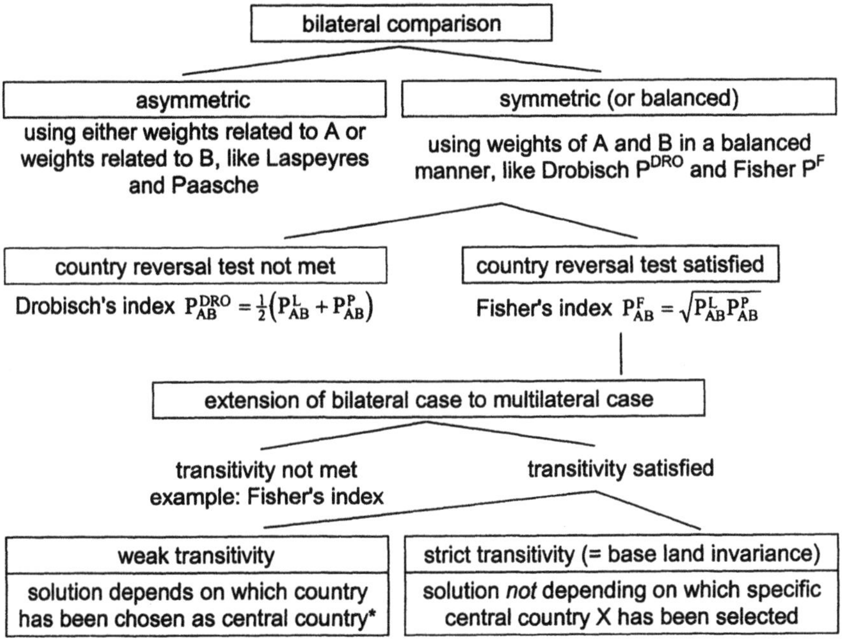

*example for this method: Central Country Method (CCM) making indirect comparisons between any two countries via a (the same in all comparisons) third "common" country X.

The "central country method" (CCM, see **sec. 8.3.a**) ensures only *weak* transitivity in that the result of the method dependends on the choice of the "central" country X. Another country, say Y, chosen as the "central country" will produce different results. *Strict* (or strong) transitivity, however, means that transitivity holds irrespective of which central country has been chosen, and is thus a highly desirable property for multilateral comparisons.

According to **fig. 8.1.6** a necessary but not sufficient condition for the

- **country reversal test** (in the bilateral case) is a symmetric treatment of the two countries involved by using formulas reflective of weights of both countries in a symmetric fashion like Walsh, Törnqvist etc, or by averaging the Laspeyres and Paasche formula[28]
- **transitivity** (in the multilateral case) furthermore requires symmetrically accounting for all country reversible comparisons (e.g. by averaging over the di-

[28] interestingly this is true only for Fisher's (geometric mean) but not Drobisch's (arithmetic mean) approach.

rect and all possible indirect binary country reversible comparisons between any two countries).

Transitivity is typical of multilateral comparisons. If there were only two countries to be compared and thus no need for transitivity, the methods known from intertemporal comparisons could easily be adopted to the multilateral case.

f) Conditions, axioms and required properties in multinational comparisons

As a consequence of the different purposes a multinational comparison has to serve many different and conflicting criteria as proposed in this field (see **fig. 8.1.7**). The requirements (axioms) depend on the aims of the comparison and they do *not* constitute a consistent system of axioms.

Before introducing some methods suitable for international price and volume comparisons it might be useful to enumerate some axiomatic requirements often mentioned in the multinational comparison framework (listed in **fig. 8.1.7**).

Characteristicity requires that commodities and quantity weights are used so that

- not only all countries to be compared are represented, but also that
- commodities and weights provide an adequate coverage and representation of the consumption in the different countries under consideration.

Whenever absolute figures of volumes are to be published additivity (test **5a** in the sense of structural consistency of volumes)[29] is indispensible. This is a sufficient condition also for the volume ratios to meet the average (mean value) test **4b**:

when the deflator index P meets additivity (= struct. consistency, **5a**)	⟶	then also the resulting quantity index meets the mean value property (**4b**)

The converse is not true. The mean value condition (tests 4a, 4b as opposed to additivity, i.e. tests 5b for price or quantity indices) is an interesting and sufficient (weaker criterion) property if only structures matter, not absolute figures.

Another criterion, sometimes mentioned, is for instance the applicability of a method in the case of negative balances (not uncommon in National Accounts).

A set of tests formulated by Diewert for multinational comparisons (focusing on volume comparisons) comprises the following axioms (called D as they are credited to Diewert)[30]:

D1 positivity and continuity (in all arguments) of volume shares $Q_i(i=1, \ldots, m)$,

D2 symmetric treatment of countries test (i.e. invariance of volume shares to permutation of countries),

D3 symmetric treatment of commodities test (the same with respect to [w.r.t.] permutation of commodities),

---

[29] The numbers of tests refer to **fig. 8.1.7**.

[30] The first five axioms refer to some fundamental formal aspects, the following axioms need to be inspected in more detail.

D4 monetary unit test (= invariance to changes in scale test) corresponds to price dimensionality in the intertemporal case: replacing the vector $\mathbf{p}_i$ by $\alpha_i\mathbf{p}_i$ and $\mathbf{q}_i$ by $\beta\mathbf{q}_i$ should not affect the volume share of country i.[31]

D5 invariance to changes in units of measurement tests (plays the part of the commensurability axiom),

D6 country partitioning test[32] (splitting a country into two parts with the same structure w. r. t. quantities): let country h be partitioned into part (province) 1 with a quantity vector $\lambda\mathbf{q}_h$ and a part 2 with $(1-\lambda)\mathbf{q}_h$, and the same price vector for both parts, then the quantity shares should be λQ_h and $(1-\lambda)Q_h$ (or: small countries should not influence the volume shares of large countries unduly),[33]

D7 irrelevance of tiny countries test: denote $\lambda\mathbf{q}_k$ the quantity vector of a (tiny) country k; if $\lambda \to 0$ the quantity shares of all countries should tend to the quantity shares we get if calculations are done exclusive of country k,

D8 (*weak*) *proportionality* (and hence also identity) test w.r.t. prices (or equivalently w. r. t. quantities): upon substitution of $\mathbf{p}_i$ by $\alpha_i\mathbf{p}$ (i = 1, ..., m denotes the country) - and quantities in the same manner, i.e. substitution of $\mathbf{q}_i$ by $\beta_i\mathbf{q}$ - that is, equality of the *structure* (*not* the *level*) of prices or quantities across all m countries - should result in quantity shares dependent on β_i only (= D8 "w.r.t. quantities")[34] and not on α_i (= D8 "w.r.t. prices"). Finally, the criterion most likely to be violated[35] is

D9 *proportionality* test: replace for country h the vector $\mathbf{q}_h$ by $\lambda\mathbf{q}_h$ and the scalar (non-normalized) country weight g_h by λg_h, then the quantity share of this country h should change from Q_h to $\lambda Q_h/1+(\lambda-1)Q_h$ and the share of any other country $i \neq h$ should change accordingly.

Axiom D9 is the most critical and makes sure that changes in the mere size of a country h (expressed in a uniform λ-fold change in all its quantities) should exhibit the effect as above.[36] It appears that there is no method which is able to pass all of the tests. In particular, tests D9 on the one hand and D6/D7 on the other seem to be inconclusive. This does not mean, however, that the inconsistency of the whole system has been formally proved.

[31] or: different inflation rates α_i but equal quantity growth rates β leave the quantity shares unchanged.

[32] For example, using average prices according to Geradi (eq. 8.2.9 and 8.3.28a) would violate this test. The same is true for the Generalized EKS (or GEKS for short) indices (see below **sec. 8.4**).

[33] See also footnote 5 in part a of sec. 8.1.

[34] The assumptions make sure that the quantities of the countries differ only with respect to (w.r.t.) their level, not to their structure. Hence the quantity shares should depend only on the (country specific) quantity determining proportionality factors β_i.

[35] The list neglects the monotonicity test (violated for example in the case m = 2 of the GK method as shown in **sec. 8.3** part b2).

[36] It appears difficult, however, to understand the reason for this requirement.

Figure 8.1.7: Criteria and requirements for international comparisons

This list should not be misinterpreted as a set of non-contradictory (consistent) and independent axioms like, for example, the axiomatic system of Eichhorn and Voeller (see **sec. 3.3**).

Axiom	Meaning	Remark
1 Time (country) reversal test	Unique parity among two countries A and B, such that PAB = 1/PBA (choice of base country irrelevant)[1].	Not to be reconciled with no. 2, at least not in the case of widely differing countries[2].
2 Characteristicity (typicality or equidistance)	Indices should be representative for both (all) countries to be compared with respect to kind and quantities of commodities[3].	Condition referring to type of weights (unusual in axiomatic theory)
3 Unbiasedness **4** Mean value (or: average) test (**4a** for prices **4b** for quantities)	A parity (price index) should lie within the interval between P^P and P^L, it should also meet the mean value condition {= test 4a} (the same is desired for quantity indices[4] test 4b {independent of 4a}).	Relevance and meaning of no. 4 is difficult to distinguish from no. 3; not clear why no. 4 should apply to PPPs too.
5a Additivity (I) (or: structural consistency [of volumes])	Structural consistency in the sense of sec. 5.2, useful to make consistent National Accounts for a block of countries like the EEC.	Required when deflators or volume indices are used in an accounting framework.
5b Additivity (II) (or: aggregative consistency [of the index formula] see sec. 5.2)	Required in particular for quantity indices Q, such that the overall Q can easily be decomposed into sub-indices measuring the quantity movement of sub-aggregates	Useful if comparisons are made at varying levels of aggregation[5] (as for example in National Accounts)
6a Weak Factor Reversal Test (WFR)	Price (P) and quantity index (Q) are related to the value index (V) by PQ = V; also known as the product test	Allows for a breakdown of volumes into a P and Q component
6b Strict (Strong) Factor Rev. Test (SFR)	Like WFR, in addition: P can be obtained from Q by interchanging prices and quantities (likewise Q from P)	Implies that the method to derive P (or Q respectively) is symmetric (balanced)
7 Transitivity **7a** weak, **7b** strict transitivity	The direct comparison and indirect comparisons of all[6] (strict transitivity) sorts should yield the same (identical) result.	More important than in intertemporal comparisons, but difficult to achieve.

Footnotes see next page

1) Not only the country but also time reversal test has been criticized: History cannot be made undone, there is no "run backward" and more often than not no meaningful result to expect (it is absurd to ask for the price of a flight-ticket in 1890).

2) It is argued that the result cannot be trustworthy exactly *because* the country reversal test is satisfied. This is particularly convincing in the case of very different countries (e.g. Germany and India).

3) If any two countries are treated symmetrically with respect to quantities q_A, q_B, as in the case of Fisher (or Drobisch) indices, we say that the approach is *equidistant*. Equidistant indices usually satisfy the product, or even the factor, reversal-test.

4) This is supposed to be a necessary condition to perform real-value comparisons between various countries on different levels of aggregation.

5) Requires a method using a single vector of prices (like the vector of average prices of the community in the Geary Khamis method)

6) Known as strict transitivity, or *base-land-invariance* (a criterion that should be kept distinct from country reversibility). It is possible that transitivity is given only by indirectly comparing with a *specified unique* "third" country (for example the "central" or "star" country). This is known as *weak transitivity.*

In any case, a choice has to be made with an implicit emphasis on one requirement at the expense of others. So those authors who set great store by additivity for example[37] are in favor of the GK-method, while others stressing the Gerschenkron-effect (the analogon to the famous substitution effect in the intertemporal framework) are more likely to favor Fisher's formula and methods based on it.[38]

g) Engel-Gerschenkron effect[39], Paasche - Laspeyres - Spread

In the framework of national inter-*temporal* comparisons a negative correlation between price and quantity relatives is more likely because consumers tend to purchase relatively more (less) of goods that are relatively cheap (expensive). The same negative correlation between price relatives of commodity i, that is p_{iB}/p_{iA} (A = base country, B = reference country), and quantity relatives q_{iB}/q_{iA} has been observed in the case of inter-*national* comparisons as it has been established for many years already in the intertemporal case, and the relationship is known as "Engel-Gerschenkron effect". The effect is in complete correspondence to the relation found by L.v. Bortkiewicz (see **sec. 1.3**), hence given the correlation is negative, the result $P^L > P^P$, and $Q^L > Q^P$ is to be expected.

It is also commonly agreed that the spread (difference) between a Laspeyres and a Paasche index tends to be greater in interspatial comparison as compared to intertemporal comparison.

[37] As for example Khamis, whose criticizm as regards Fisher's formula (as quoted in Balk 1996, p. 213) also deserves attention (and approval).

[38] See Balk 1996, ibid. For his preference for van Ijzeren (van Yzeren) and the simultaneous usage of the GK and the EKS method.

[39] See **sec. 8.3** for more about the Gerschenkron effect.

Consequences drawn from these observations are usually

- to seek for balanced or "equi-characteristic" comparisons on the basis of some compromise between the Laspeyres and Paasche approach,
- to refrain from comparisons between countries too far apart with respect to the standard of living, or to the consumption pattern (culture, climate) and the like, and
- to recognize that volume comparisons not only reflect true quantitative differences, but also differences in the level and structure of prices.

Since the levels of wealth of different countries and their respective price levels are often positively correlated; a tendency also has been observed that volume comparisons (of per capita GDP for example), based on exchange rates instead of purchasing power parities will be biased.[40]

The Gerschenkron effect arises because consumption patterns change in response to changes[41] (regional differences) in relative prices and consumers tend to switch towards relatively cheaper goods. Finally, it is also regarded as resulting from the Gerschenkron effect that international per capita income differentials tend to be larger if

- the average *price* vector is similar to the price vector of the *poorest* country and conversely if
- the average basket (*quantity* vector) is relatively more similar to the *richest* country.

8.2 Overview of methods proposed for multinational comparisons

This section tries to systematize the best known and most used methods and to expound their rationale. The sections **8.3** and **8.4** will deal with some of these methods in greater detail.

a) Introduction into methods to solve the transitivity problem

Strict transitivity can be achieved, or at least approximated, by

1. taking as a central country X not just one of the countries to be compared but an artificial country (or "block") constructed as
    - some kind of sum of all m countries (the "block" to which all m countries belong), or

[40] Moreover, exchange rates are supposed to inadequately account for differences in price levels.

[41] This is exactly what is assumed away in the case of the Laspeyres (or tourist) concept as opposed to the Paasche (or migrator) concept.

- by averaging the results of all possible binary comparisons involved (or in other words by trying every country as the central country X),

2. starting with bilateral comparisons and choosing a set of transitive indices that resemble the original binary indices as closely as possible (in the "least squares" sense).

Two well known methods representing these two alternatives are

1. the *Geary-Khamis (GK)* method that treats the block as an entity (super-country) in itself with a (row) price vector $\bar{\mathbf{p}}' = [\bar{p}_1 \ldots \bar{p}_n]$ of prices of n commodities expressed in a common currency of the "block" or "community" (the common currency may be a construct only), and
2. the *Eltetö-Köves-Szulc (EKS)*-method where a set of n-1 transitive indices is found which differs from *all* $\binom{n}{2}$ possible binary *Fisher-indices* as little as possible (obviously there are many alternatives, for example the Törnqvist index instead of the Fisher index, an idea leading to the CCD method[42]).

Fig. 8.2.1 is an attempt to find a structure for the multinational methods which puts particular emphasis on the GK and the EKS method, in which, however, some methods, such as methods proposed by van Yzeren[43] or the *Minimum Spanning Tree (MST) Method*, cannot adequately be accounted for. The MST method (part d of this section) may be viewed in a way as an application of chaining. Between any two countries a number of chained (or chain-linked) comparisons via various link-countries can be defined (so called "spanning trees", each of which is characterized by the total distance between all connecting countries) and a choice can be made such that for example the total (summated) distance is minimized.

The rationale of most of the methods listed in the scheme of **fig. 8.2.1**, especially the methods called block methods, can be described as follows

1. to derive transitive inter-country comparisons of m countries with respect to price indices P_{AB} or P_{ij} (quantity indices Q_{ij} correspondingly) we have to define m positive real numbers $\lambda_1, \lambda_2, \ldots, \lambda_m$, such that[44]

$P_{ij} = \frac{\lambda_j}{\lambda_i}$ or more conveniently $P_{ij} = \frac{P_j}{P_i}$ and $Q_{ij} = \frac{Q_j}{Q_i}$

2. In order to comply with the product test the following equation should hold

$$(8.2.1) \quad P_{kj}Q_{kj} = \frac{P_j}{P_k}\frac{Q_j}{Q_k} = V_{kj} = \frac{\sum p_j q_j}{\sum p_k q_k},\ k, j = 1, \ldots, m,$$

42 Caves-Christensen-Diewert Method, see **fig. 8.2.1**.

43 or (by now becoming more common): van Ijzeren.

44 According to Rao and Banerjee 1984 and Rao 2001, the property of transitivity is satisfied if and only if such m numbers λ exist.

and the methods listed in **fig. 8.2.1** can be distinguished depending on how $P_{kj} = P_j/P_k$ and $Q_{kj} = Q_j/Q_k$ are defined, viz. by either:

Figure 8.2.1: Overview of most relevant methods, part I (esp. GK and EKS – Method)

Assume m (i = 1, ... , m) countries forming a block and n commodities (k = 1, ..., n), and

m vectors of the type	one (for the community) vector of
price vectors, $\mathbf{p}'_i = [p_{1i} ... p_{ni}]$ for country i with n prices expressed in its own (the i-th country) currency	international prices $\bar{\mathbf{p}}' = [\bar{p}_1 ... \bar{p}_n]$ expressed in the block's currency and if necessary also of international quantities $\bar{\mathbf{q}}'$
quantity vectors $\mathbf{q}'_i = [q_{1i} ... q_{ni}]$ of n quantities in country i	parities $\hat{\mathbf{p}}' = [P_1 ... P_m]$ or vector of volumes $\hat{\mathbf{q}}'$ as a result of the method

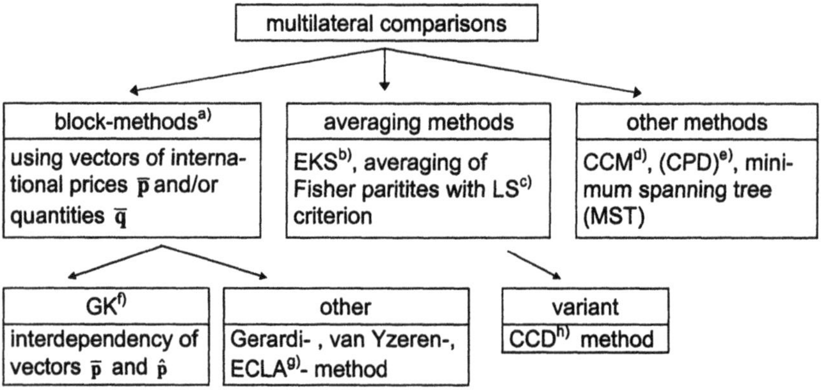

a) methods treating the block as an entity of its own

b) Eltetö - Köves - Szulc Method

c) least squares

d) Central Country Method

e) Country-Product-Dummy Method (a regression method)

f) Geary - Khamis - Method

g) Economic Commission for Latin America Method or ECLAC = Latin America and Caribean

h) Caves-Christensen-Diewert Method

- referring to an average (artificial, central, block) country, as in the case of the GK method, or
- averaging over all binary comparisons as regards prices or quantities respectively of the m countries to be compared (EKS- and related methods[45]).

Eq. 8.2.1 may be specialized as follows (X denotes the central- or block-country)

[45] which may therefore also be called "generalizations of binary comparisons" (Balk).

(8.2.2) $$\frac{P_j}{P_k}\frac{Q_j}{Q_k} = \frac{P^P_{xj}}{P^P_{xk}}\frac{Q^L_{xj}}{Q^L_{xk}} = \left(\frac{\sum p_j q_j}{\sum p_x q_j}\frac{\sum p_x q_k}{\sum p_k q_k}\right)\left(\frac{\sum p_x q_j}{\sum p_x q_k}\right)$$

such that the first factor (that is $P_j/P_k = P^P_{xj}/P^P_{xk}$) depends on prices p_x of the central country only, not on quantities q_x, and the same is true for the second factor $Q^L_{xj}/Q^L_{xk} = \sum p_x q_j / \sum p_x q_k$. We thus may rightly call methods on this basis *average price methods* (**fig. 8.2.2**), for which the GK method (Geary - Khamis) is an example. In a similar vein eq. 8.2.1 may be specialised as follows

(8.2.3) $$\frac{P_j}{P_k}\frac{Q_j}{Q_k} = \frac{P^L_{xj}}{P^L_{xk}}\frac{Q^P_{xj}}{Q^P_{xk}} = \left(\frac{\sum p_j q_x}{\sum p_k q_x}\right)\left(\frac{\sum p_j q_j}{\sum p_j q_x}\frac{\sum p_k q_x}{\sum p_k q_k}\right)$$

where the first factor $P^L_{xj}/P^L_{xk} = \sum p_j q_x / \sum p_k q_x$ as well as the second factor depends on the x-countries quantities only, such that methods based on eq. 8.2.3 such as the ECLA (or ECLAC) method may be called *average basket methods* making use of a vector of common quantites $\mathbf{q}_x$ or $\bar{\mathbf{q}}' = [\bar{q}_1 \dots \bar{q}_n]$ rather than of prices. For practical reasons this approach has much less to commend it than the *average price methods*. Finally, we may also assume

Figure 8.2.2: Overview of some of the most relevant methods, part II (with reference to R. J. Hill 1997 and B. M. Balk 2001)*

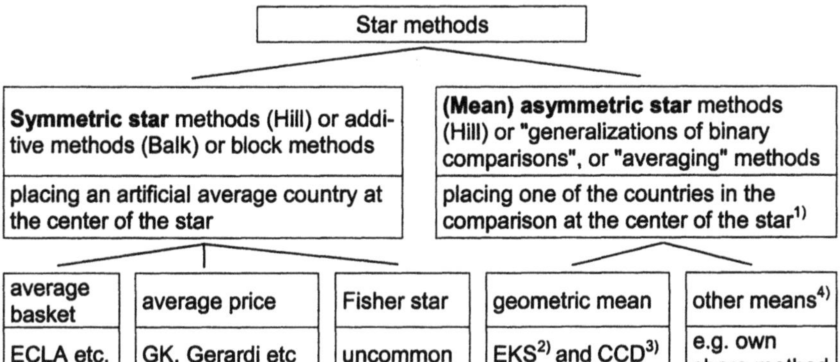

*) The scheme is again not exhaustive in that some methods like the minimum spanning tree, regression and Multivariate Generalized Törnqvist (MGT) method and some other methods are not mentioned. Furthermore, van Yzeren proposed methods belonging to different categories in this system. According to Hill, most methods can be regarded as one variant or another of star methods (i.e. making indirect comparisons over a third country, the center of a star, and often indexed X).

1) this description would of course also fit the Central Country Method (CCM)

2) using Fisher indices

3) Caves-Christensen-Diewert Method using Törnqvist indices

4) much less common variants using e.g. the arithmetic or harmonic mean

(8.2.4) $\frac{P_j}{P_k}\frac{Q_j}{Q_k} = \frac{P^F_{xj}}{P^F_{xk}}\frac{Q^F_{xj}}{Q^F_{xk}}$

giving rise to the much less known "Fisher star method" listed in **fig. 8.2.2**. Methods which may be viewed as "generalizations of binary comparisons" (Balk) conceptualize ratios like P_j/P_k (and in a similar manner Q_j/Q_k) as follows[46]

(8.2.5) $\frac{P_j}{P_k} = \underset{i}{M}\left(\frac{P^X_{ij}}{P^X_{ik}}\right) = \underset{i}{M}\left(P^X_{ij}P^X_{ki}\right)$

where $\underset{i}{M}(...)$ denotes a mean of binary comparisons (over all i = 1, ..., m countries) using index formulas of type X.

For example, the well known formula in the EKS method (in the case of m=3 countries such as A, B and C, see **sec. 8.4**)

(8.2.5a) $P^{EKS}_{AC} = \left[\left(P^F_{AC}P^F_{AA}\right)\left(P^F_{BC}P^F_{AB}\right)\left(P^F_{CC}P^F_{AC}\right)\right]^{1/3} = \sqrt[3]{\left(P^F_{AC}\right)^2 P^F_{AB}P^F_{BC}}$

represents a geometric mean of bilateral comparisons on the basis of Fisher (X = F) indices.

(8.2.5b) $\frac{P_j}{P_k} = \left(\frac{P^F_{1j}}{P^F_{1k}}\frac{P^F_{2j}}{P^F_{2k}}\cdots\frac{P^F_{mj}}{P^F_{mk}}\right)^{1/m}$ and using $P^F_{BA} = 1/P^F_{AB}$ we get $\left(\frac{P^F_{AC}}{P^F_{AA}}\frac{P^F_{BC}}{P^F_{BA}}\frac{P^F_{CC}}{P^F_{AC}}\right)^{1/3}$

in the three - countries example.

b) Evaluation of methods adequate for intra-EU comparisons

According to Diehl 1978 methods for multinational comparisons in the framework of a group of related countries, like the member countries of the European Union (EU) should be evaluated using four criteria as shown in the following **fig. 8.2.3**.

As to the list of comparable commodities (products, items) there are several possibilities:

- ideally a "*common*"-list exists that contains *all* n products being available identically (or almost identically) in all m member countries, but lack of comparability prevents this list from;
- as a compromise only a couple of so called binary lists exist or as aforesaid an "*incomplete tableau*", containing products, though not available in an (almost) identical manner in *all* countries, available at least in any two countries.

According to Diehl the following methods fail with respect to this criterion B (non existence of a complete list of items):

- the ECLA-method based on the construction of total quantities and thereby requiring a complete common list, not only partly overlapping binary lists;
- the same is true for van Yzeren's method and all methods based on averaging price relatives (p_{kA}/p_{kB} related to two countries, A and B) over *all* commodities k

[46] This applies to the EKS method or variants of it, e.g. the CCD-method using P^T instead of P^F.

= 1, ..., n, and all countries, like geometric (Walsh type) averages, as used for example in Gerardi's method.

Figure 8.2.3: Criteria to find suitable methods for inter-EU-comparisons

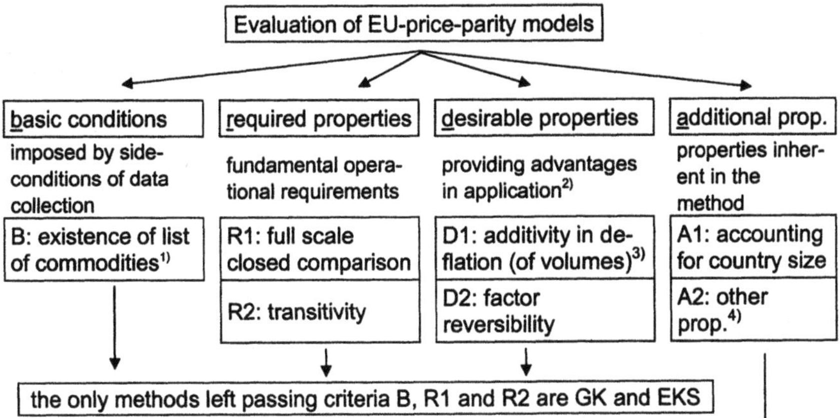

method	D1: additivity	D2: factor reversal test	A1: country size[5]
GK	not satisfied (quasi additivity in comparisons of a country i to the block as a whole)	violated (only product test), price and quantity parities differently constructed	size of country i affects results via quantities q_{ki}[7]
EKS	violated[6]	satisfied[6]	no influence

1) incomplete list of commodities can be handled

2) also theoretical elegance

3) = structural consistency (of volumes) in the sense of **sec. 5.2**

4) meaningful parameters provided as a by-product, i.e. methods permit interesting interpretations

5) or "importance" of a country

6) both results due to EKS parities being based on *geometric* means

7) however in a counterintuitive manner

It should be noticed that it makes a great difference to *average* prices rather than to compile *sums* over all quantities, as required in the ECLA method. In the latter case it is more essential to avoid gaps in the list. In the former, however, a certain amount of gaps can be handled. This is true for the GK-method, for example.[47]

The criterion B (only binary lists available) should be kept distinct from a comparison being *open or closed*. A comparison among m countries is said to be closed if

[47] Incomplete data seem to do more harm in the case of van Yzeren's method, however.

the calculation is such that adding additional countries (the country m+1, m+2, ...) or deleting one or more of the m countries entails a complete recalculation of all parities.[48] "Additivity" (D1) of price parities as deflators means here that the sum of volumes derived from deflating sub-aggregates with sub-(price) indices equals the result of deflating the total aggregate using the total price index (i.e. structural consistency in volumes, or criterion 5a in **fig. 8.1.7**).

c) Block methods: the Geary-Khamis (GK) method

Methods belonging to the group of "block methods" are those methods for multilateral international comparisons, in which the construction of a vector of "international" prices $\overline{\mathbf{p}}'$, or sometimes also a vector of "international" quantities $\overline{\mathbf{q}}' = [\overline{q}_1 ... \overline{q}_n]$, plays a decisive part. They differ in

- how an international price is defined,
- whether they also make use of international quantities $\overline{\mathbf{q}}$, and
- whether or not some kind of "weighting" of countries according to their importance (e.g. to the extent to which a country contributes to the entire quantity of the block) is involved.

The key idea of the GK-method[49] is to determine the "international" prices $\overline{p}_k$ of commodites (k = 1, 2, ..., n) and m currency converters c_i (of country i =1, 2, ..., m) *simultaneously*.[50]

The GK-method defines the common (community, or block) price $\overline{p}_k$ of commodity k with the help of the m "currency converter"[51] c_i of country i as follows:

(8.2.6) $$\overline{p}_k = \frac{\Sigma c_i p_{ki} q_{ki}}{\Sigma q_{ki}} = \frac{\Sigma c_i p_{ki} q_{ki}}{Q_k} = \sum_{i=1}^{m} c_i p_{ki} \alpha_{ki}\text{, where } \alpha_{ki} = q_{ki}/Q_k \text{ or}$$

(8.2.6a) $\overline{p}_k = \sum c_i \widetilde{p}_{ki}$	a system of n equations for n commodities in which the summation takes place over m countries

where $\widetilde{p}_{ki}$ denotes *unit values* (see **sec. 6.4**) i = 1, 2, ..., m, and Q_k is a sum of quantities (over all m countries). As known from the definition of unit-values, α_{ki} is a structural variable accounting for the size of a country i, and c_j is, as aforementioned, a currency converter (reciprocal exchange rate)[52] in order to allow for expenditures in

[48] Most of the methods to be discussed here are of the closed type.This means that parities are relative with respect to the collection of countries chosen. P_{AB} may be different depending on what the other countries are.

[49] The method was in principle devised by R. Geary (1958) from the Irish official statistics and later (1970, 1972) S. H. Khamis succeeded in proving the existence of a unique solution.

[50] The rationale of this approach will be discussed in more detail in **sec. 8.3**.

[51] or reciprocal exchange rate e_i.

[52] Unfortunately there is always some confusion because some writers use converters, c_i, as above, whereas others prefer to express the equations in terms of exchange rates $e_i = 1/c_i$.

the numerator expressed in the same common currency unit.[53] The converters c_i define the price level [or parity] of country i with respect to the whole community as follows

(8.2.7)	$c_i = \frac{\Sigma \bar{p}_k q_{ki}}{\Sigma p_{ki} q_{ki}} = \frac{\Sigma \bar{p}_k q_{ki}}{V_i}$	a system of m equations for m countries.

In eq. 8.2.7 summation takes place over k = 1, ... , n commodities, and V_i is the total value of country i). The system allows, however, only for calculation of m-1 coefficients c_i expressed in units of one of the c_i- coefficients, say $c_2, c_3, \ldots, c_m$ in units of c_1. This is sufficient as the aim is to define (purchasing power) parities between any two countries, A and B, with reference to the community as a whole. The definition of P_{ji}^{GK} (i, j = A, B) now is as follows

$$(8.2.8) \qquad P_{AB}^{GK} = \frac{c_A}{c_B} = \frac{e_B}{e_A} = \frac{\Sigma p_{kB} q_{kB} / \Sigma \bar{p}_k q_{kB}}{\Sigma p_{kA} q_{kA} / \Sigma \bar{p}_k q_{kA}}.$$

This way of expressing GK-parity between any two countries, A (base) and B, makes clear that

1. *identity* is given, that is, when all m prices for the k-th commoditiy are equal, $p_{k1} = \ldots = p_{km} = \bar{p}_k$, then all parities will be unity, or $P_{ji}^{GK} = 1$;

2. (strict) *transitivity* as well as *country reversibility* $P_{ji}^{GK} = \left(P_{ij}^{GK}\right)^{-1}$ holds since

   $$(8.2.8a) \qquad P_{AC}^{GK} = P_{AB}^{GK} P_{BC}^{GK} = \left(\frac{c_A}{c_B}\right)\left(\frac{c_B}{c_C}\right),$$

   due to the use of a constant (for all countries, A and B) vector, $\bar{\mathbf{p}}'$, all indices, P_{AB}^{GK}, V_{AB}, and Q_{AB}^{GK}, are transitive;

3. the *product test* (weak factor reversal test) is met since the factor antithesis of P_{AB}^{GK} is

   $$(8.2.8b) \qquad Q_{AB}^{GK} = \frac{\Sigma \bar{p}_k q_{kB}}{\Sigma \bar{p}_k q_{kA}} = \frac{Q_B}{Q_A},$$

   but the (strict) *factor reversal* test is *not* satisfied, because the structure of Q_{AB}^{GK} differs from the structure of P_{AB}^{GK}. This can easily be seen as follows. In order to satisfy the factor reversal test, Q_{AB}^{GK} should be constructed such that Q_{AB}^{GK} is gained from P_{AB}^{GK} by interchanging prices and quantities. Such an index would be $Q_{AB}^{*} = \frac{\Sigma \bar{q}_k p_{kA}}{\Sigma \bar{q}_k p_{kB}} \frac{\Sigma p_{kB} q_{kB}}{\Sigma p_{kA} q_{kA}} \neq Q_{AB}^{GK} = \frac{\Sigma \bar{p}_k q_{kB}}{\Sigma \bar{p}_k q_{kA}}$.

4. Both indices, P_{AB}^{GK} and Q_{AB}^{GK} are "additive" index functions, made up of sums over n commodities, in both the numerator and denominiator respectively:

[53] That is, the currency unit of the community.

$V_A = \Sigma p_{kA} q_{kA}$	$Q_A = \Sigma \bar{p}_k q_{kA}$	$c_A = Q_A/V_A$
$V_B = \Sigma p_{kB} q_{kB}$	$Q_B = \Sigma \bar{p}_k q_{kB}$	$c_B = Q_B/V_B$
$V_{AB} = \frac{V_B}{V_A}$	$Q_{AB}^{GK} = \frac{Q_B}{Q_A}$	$P_{AB}^{GK} = \frac{c_A}{c_B}$

By virtue of these relationships all GK - indices P_{AB}^{GK} and Q_{AB}^{GK} can be broken down to the commodity level and aggregated to whichever subindex is wanted. This is important for many practical purposes, but it does *not*, however, imply that structural consistency of volumes in deflation is given. Finally, it is clear that

5. what is responsible for c_A, c_B etc., and hence also for P_{AB}^{GK}, is the extent to which the prices in A or B respectively differ from the prices valid for the community of all m countries, or in other words, what matters is the extent to which
   - *values* (quantities of country i expressed in prices in their *own* currency) differ from
   - *volumes* (that is "deflated" values $\Sigma \bar{p}_k q_{ki} = Q_i$, or quantities of country i valued at *common* prices).

The continued use of the GK method is primarily owed to its structural consistency in regional deflation (or in other words its additivity in the sense of 5a in fig. 8.1.7), which is a result of using international prices (average prices $\bar{p}_k$). The method is therefore useful in analyzing the structure of the (aggregated) GDP and comparing the shares of different countries. However, the method fails a number of tests and its average prices are likely to be more reflective of the price structure of the richer countries and they therefore tend to overstate the real income of the poorer countries.

The principal shortcoming of the GK-method[54] is that it makes a considerable difference if the country resembles the community as a whole (if it is "*central*") or if it is rather dissimilar to the block as such (that is, more "*peripherical*"). Obviously it is only in the first case (both countries being central) that the GK-binary comparison between i and j resembles the direct comparison between i and j, whilst in the latter we will have to face a bias.

It will be shown in **sec. 8.3** in detail that

The GK-method tends to a price index P^{GK} which is dominated by the small (unimportant) country while the quantity index Q^{GK} is likely to be dominated by the big (important) country. Moreover, GK-indices deviate from the binary comparisons to the extent that either or both countries involved are less central (typical for the community as a whole) and hence more peripherical.

A short remark related to Gerardi's method should be added. It makes use of another way of defining the international price $\bar{p}_k$ of commodity k (in which c_i is expressed in the currency unit of the community)

[54] Another shortcoming is lack of proportionality.

(8.2.9) $\breve{p}_k = \left(\prod_{i=1}^{m} c_i p_{ki} \right)^{1/m}$,

which is of course an unweighted geometric mean. Gerardi's community prices $\breve{p}_k$ are easier to calculate than the GK-prices $\bar{p}_k$, however, it has been criticized that no country weights are applied, accounting for the different size of the m countries as measured for example by the k-th good's share $\alpha_{ki} = q_{ki}/Q_k$ of the total quantity $Q_k = \sum_i q_{ki}$ (i = 1, ..., m).[55]

d) The method of "minimum spanning trees" (MST-Method)

This method has been proposed only recently by Peter Hill.[56] It is also sometimes called a "linking method" or "spatial chaining" as opposed to block country methods and averaging methods. The basis of the method is the notion of a "distance" (dissimilarity) between any two countries, for example A and B, using the Laspeyres-Paasche Spread or "Paasche-Laspeyres-Spread" (PLS) defined as

(8.2.10) $D(A,B)) = \left| \ln\left(\frac{P^L_{AB}}{P^P_{AB}} \right) \right| = \left| \ln\left(\frac{P^P_{AB}}{P^L_{AB}} \right) \right|$.

Given that two countries are quite similar with respect to the structure of consumption, the results of a Laspeyres-type weighting (weights of the base country A) and a Paasche-type (weights of B) index would not differ much, such that D comes close to ln(1) = 0. The distance function thus serves as an indicator of structural similarity (structure of prices and quantities) of the two countries to be compared. This is an alternative to other more or less explicit measures of similarity required to find a suitable grouping of countries.[57]

In the famous ICP-project (since 1985) for example, the original attempt to compare *all* countries worldwide on the basis of the *same* list of products was given up. A subdivision into groups (or clusters) of countries was introduced instead (based on geographical distance and similarity of the political system).[58] In a way this procedure was a forerunner of the MST-method.

The novelty now is the introduction of an explicit distance measure[59] and the application of an algorithm to the matrix of distances by which clusters and sequences can be created automatically.

---

[55] Such a quantity Q_k of the community as a whole is of course a somewhat fictitious entity.

[56] On the basis of work of Kravis, as well as Heston and Summers (both authors contributing to the ICP project), and using a MST-algorithm devised by Kruskal.

[57] See below for an alternative measure τ (just an example for possibly many alternatives to D).

[58] Within each block a GK-like method or star method was used, with certain countries selected to link the groups of countries together (e.g. Turkey to provide a link between Central-European and Central-Asian countries).

[59] for example the D-measure ("PLS measure").

A spanning tree (the concept originates in graph theory) is a connection between a country (point, vertix, edge) i and each of the remaining m - 1 countries such that each country is (indirectly) linked with the others in one way only.[60] There are no two or more paths connecting any two countries (a situation which would be called a "cycle").

Hence the task consists of selecting m-1 out of $\binom{m}{2} = \frac{m(m-1)}{2}$ binary comparisons without omitting a country or producing a cycle. **Fig. 8.2.4** will give some examples of spanning trees for m = 5 countries A, ..., E along with fictitious numerical values of the selected distances[61] (ranging from 0.07 to 0.12) of the symmetric matrix of distances.

The "star" (second [central] example in **fig. 8.2.4**) is a configuration in which the central country D provides the *only* possible indirect comparison between *any* two other countries A-B, A-C etc.

Figure 8.2.4: The notion of a "spanning tree"

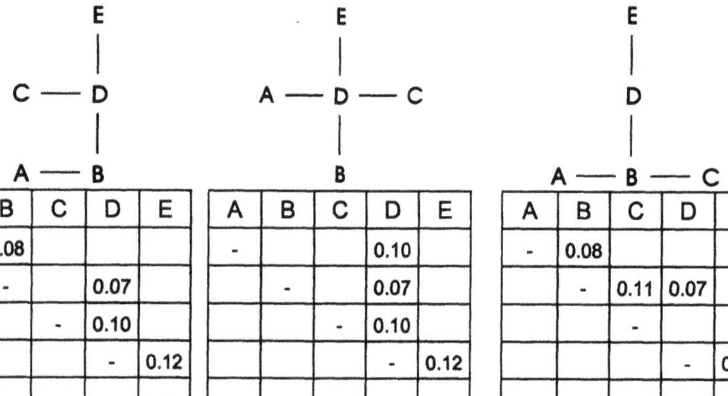

A	B	C	D	E
-	0.08			
	-		0.07	
		-	0.10	
			-	0.12
				-

sum: 0.37

A	B	C	D	E
-			0.10	
	-		0.07	
		-	0.10	
			-	0.12
				-

0.39

A	B	C	D	E
-	0.08			
	-	0.11	0.07	
		-		
			-	0.12
				-

0.38

60 It should be noticed that a "star" in all sorts of "star-methods", for example the Central Country Method (CCM), is also nothing but a special spanning tree (see **fig. 8.3.1** or central part of **fig. 8.2.4**).

61 In the left case we have for example D(A,B) = 0.08, D(B,D) = 0.07 etc.

Another spanning tree is a chain or "string" such as A-B-C-D-E with the m - 1 = 4 distances D(A,B), ..., D(D,E). The criterion for the *minimum* spanning tree (MST) is the sum of the distances, which in the examples of fig. 8.2.4 amounts to 0.37, 0.39 and 0.38. The minimum is obviously 0.37, so the left configuration is the MST for the example.[62]

The criterion of the smallest summed m-1 distances is equivalent to other reasonable criteria and it amounts to the distance of the chained spreads because

$$D(A,B)+D(B,C)=\left|\ln\left(\frac{P_{AB}^{L}}{P_{AB}^{P}}\frac{P_{BC}^{L}}{P_{BC}^{P}}\right)\right| \neq D(A,C)=\left|\ln\left(\frac{P_{AC}^{L}}{P_{AC}^{P}}\right)\right|$$

Note that transitivity would hold if and only if $P_{AC}^{L}=P_{AB}^{L}P_{BC}^{L}$ (and the same is true for the Paasche indices).

According to the rationale of the MST method we

- do not *transform* suitable binary comparisons or indirect comparisons based on them such that they comply with transitivity as closely as possible
- nor do we construct a parity as a ratio of some absolute levels like $P_{AB} = c_A/c_B$ where the c-levels are gained from a certain system of equations (which is in essence the GK way of defining a parity), but we rather
- choose a (direct or) indirect single comparison between any two countries (via one or more third countries),[63] i.e. select a spanning tree out of many such trees.

Basically, an MST defines a set of unique links which constitute unequivocally parities[64] between any two countries, such that

> the MST method does not tackle the transitivity problem by averaging or constructing artificial (fictitious) vectors of prices and/or quantities, but rather by selecting a unique set of indirect comparisons on the basis of a sort of "distance" (dissimilarity) between countries.

In practice we will already have a huge number of possible spanning trees linking countries with one another if only a small number of countries is involved.[65] A prerequisite of the method is also that the matrix of distances should not have gaps,

[62] The total distance in the case of the above mentioned string A-B-C-D-E for example amounts to 0.08 + 0.11 + 0.10 + 0.12 = 0.41, and is thus much greater than 0.37.

[63] Hence we may well choose an indirect comparison even though a direct comparison would be readily available and of course possibly also quite different from the direct one.

[64] In doing so we select one single kind of comparison between any two countries (from the set of all possible comparisons), as for example in the case of the left tree in fig. 8.2.4 the parity of relative A to D is defined via B, such that $P_{AB}P_{BD}$ constitutes the parity of A relative to D (as opposed to P_{AD} or for example $P_{AC}\,P_{CE}\,P_{ED}$).

[65] According to Hill (1998), in the comparison of m countries there are m^{m-2} spanning trees, i.e. no less than $125 = 5^3$ in the case of m = 5. There exist, however, several powerful algorithms which allow for a fast computation of the *minimum* spanning tree.

that is, all $\binom{m}{2}$ binary comparisons should be available. Finally, the distance measure - of which D is only one of several possibilities one might think of - should have the usual properties of a metric.[66]

Appearance or disappearance of a country clearly requires a completely new calculation of the MST. The method is therefore a full scale *closed* comparison.[67] Moreover, it is sensitive not only to enlargements but to all sorts of changes in the selection of countries to be compared.

To sum up:

The method permits a highly "objective", formalized solution to single out a unique set of parities and thereby to circumvent the problem of transitivity[68] by selecting one out of many ways of comparing any two countries. Hence, inconsistency between possibly many different parities is avoided

- not by averaging parities nor by making exclusively indirect comparisons of the star type,
- but rather by selecting a unique linking (spanning tree) of all countries with one another.

There is, however, little theoretical justification for the approach and it is in particular difficult to understand why a certain indirectly gained parity between two countries should be preferable to their direct parity. The economic approach would rather suggest focusing on comparability, representativity and characteristicity of the commodities and weights on which a parity is based, but these are aspects which are not (directly) accounted for in the MST method.

Further comments

The logic of the MST method can be applied to other criteria of "similarity" or "proximity". While the PLS firstly reflects the variability of prices and difference of baskets, we could also define distance matrices on the basis of overlap of price data (commodities) or other criteria. In any case, the method is based on the analysis of binary comparisons (in contrast the CPD method is to be discussed next).

An object of some criticism of the method therefore is clearly the choice of a measure of similarity of price structures. The PLS measure is said to be insufficiently sensitive to large differences and also depends on quantities and their correlation with prices. An alternative measure of the similarity of the structure of prices (taken in isolation, i.e. not affected by quantities) has been proposed by the Russian author S. *Sergueev* in the shape of his so called tau-coefficient (τ):

[66] that is D(A,A) = 0, D(A,B) = D(B,A) (symmetry) and non-negativity D(A,B) $\geq$ 0.

[67] It is possible, however, to introduce restrictions, that is, to generate a result in which the countries A and B are *not* linked, or are necessarily always linked, irrespective of whichever tree is eventually selected.

[68] i.e. inconsistency between possibly quite a number of indirect comparisons (along with the direct one).

$$\tau_{AB} = \left[\frac{\left(\sum_k \pi_k\right)\left(\sum_k \pi_k^{-1}\right)}{\sum_k \pi_k^2 \sum_k \left(\pi_k^{-1}\right)^2} \right]^{1/2} \tag{8.2.10a}$$

where $\pi_k = p_{kA} / p_{kB}$ is the (relative) price of commodity k in country A relative to country B, and of course $\pi_k^{-1} = P_{kB}/P_{kA}$. The following example may be useful to understand the formula.

Example 8.2.1

The following prices of two commodities in two countries A and B are given

k	p_{kA} p_{kB}	π_k	$(\pi_k)^2$	$(\pi_k)^{-1}$	$(\pi_k^{-1})^2$
1	5 4	1.25	1.5625	0.8	0.64
2	6 2.4	2.5	6.25	0.4	0.16
Σ		3.75	7.8125	1.2	0.8

The tau τ coefficient then is given by $\tau_{AB} = \sqrt{\frac{3.75 \cdot 1.2}{7.81 \cdot 0.8}} = 0.8485$, indicating a considerable similarity of the price structures, although $p_{1A} < p_{2A}$ while $p_{1B} > p_{2B}$. ♦

Note that τ is simply one of a great number of possible measures of similarity and that for any two countries i and j we get $\tau_{ij} = \tau_{ji}$, $0 \le \tau_{ij} \le 1$ where $\tau_{ij} = 1$ if $\pi_k = (\pi_k)^{-1} = \pi_k^2 = (\pi_k^2)^{-1}$ or $p_{kA} = p_{kB}$ for all k = 1,...,n commodities. Sergueev also proposed a modified τ-measure in which the elements π_k, $1/\pi_k$ etc. are weighted, taking the expenditure shares of the commodity k in both countries, i and j, into account.

On the basis of such τ - coefficients where the π-ratio π_k compares the price of k in country A with the *international* price of k (that is, $\bar{p}_k$ in the notation used in the introduction of the GK-method)[69] rather than with the price of k in country B, we get a vector of country-specific τ - coefficients (τ_A, τ_B, ...). Sergueev proposed a method which consists of determining the international prices such that we get in a sense "optimal" τ - values.[70]

[69] Again, making reference to international prices ensures additivity of the results (of Sergueev's method).

[70] He proposed a complicated maximin search procedure and called his method the MPCP-method (maximal possible characteristic structural international prices). As this method did not gain much importance it does not appear worthwhile going into details here.

e) Short comments on some other methods[71]

e1) Regression methods in multilateral international comparisons The CPD method

There are two applications of regression analysis in international price comparisons that should be kept distinct, viz. methods to

1. achieve transitivity by modelling price indices in a regression equation with countries as (dummy variable) regressors, like the ***Country-Product-Dummy (CPD)*** Method
2. explain individual prices p_{ki} as a function of certain characteristics of the commodity (item) k and the country i (that is the application of "hedonics" introduced in **sec. 4.2** for purposes of international instead of intertemporal comparisons).

On 1:

This method has been used in the first three phases of the ICP-project to achieve parities at the level of sector-indices (groups of homogeneous commodities), not necessarily individual commodities. Assume a matrix, **P** with prices p_{ki}, n commodities (k = 1,..., n) arranged as rows and m countries (i = 1,..., m) forming the columns of the matrix:

$$\mathbf{P} = \begin{bmatrix} p_{11} & \cdots & p_{1m} \\ p_{21} & \cdots & p_{2m} \\ \cdots & \cdots & \cdots \\ p_{n1} & \cdots & p_{nm} \end{bmatrix}.$$

The logarithm of each price p_{ki} is often assumed to be a function of the particular item dummy variable D_k (taking values 0 or 1 only) the influence of which is reflected in the coefficient α_{ki}, some other variables (dummies or measured) x_j referring to the quantities of other aspects of various goods j (among which k) and the dummy variable Δ_i of the particular country i plus a random variable U_{ki}, for example

(8.2.11) $\ln(p_{ki}) = \alpha_{ki}D_k + \eta_k x_k + \lambda_i \Delta_i + U_{ki}$ (for h = k we have $D_h = 1$ and for $h \neq k$ $D_h = 0$),

generating a system of m·n equations. The regression approach can also involve a number of "explaining" variables[72] x (not only referring to the good k) and country dummies Δ_s for basically all m countries (among which the specific country i) such as

[71] The EKS method and related methods will be discussed in greater detail in **sec. 8.4**.

[72] As referred to in no. 2 below, the method can also be applied to account for quality differences by adding dummies related to characteristics of goods rather than to goods as a whole (eq. 8.2.11a). The Country Product Dummy (CPD) Method was originally conceived as a tool to deal with missing prices in a table of unweighted means of relative prices (such as the geometric means in table 8.1.4 in order to derive "basic parities" $_AL_B$ or $_AP_B$). In general *imputed* prices may be estimated in all cases in which commodities are not (or not in the particular combination of characteristics in question) offered and purchased on the market, be it at a certain time or in a certain country.

(8.2.11a) $\ln(p_{ki}) = \alpha_{ki} + \sum_j \eta_j x_j + \sum_s \lambda_s \Delta_s + U_{ki}$,

where variables x_j denote various characteristics of commodities, among which not only good k but possibly also those goods which are close substitutes of k are included, and the Δ_s (s = 1, ..., m) refer to various countries. The method and its logic are obviously quite similar to the hedonic approach:

Conceptually this CPD approach rests on the premise that each price is composed of a limited number of parameters (or coefficients to be estimated). Notice, however, that to solve for the coefficients, α, η and λ it is necessary to dispose of many more than just m·n observations, hence many more observations than just a single one is needed for "explaining" each cell of **P**.

On the basis of eq. 8.2.11 a regression coefficient, like λ_i, is sometimes interpreted as the logarithm of a special purchasing power parity (referring to a particular commodity k) of country i with respect to the country that has been choosen as "numéraire" (that is, the country providing the [fictitious] currency in which all prices are expressed).[73]

The estimated regression coefficient $\hat{\lambda}_i$ related to the country i is regarded as measuring the price level of country i, such that $\exp(\hat{\lambda}_A)/\exp(\hat{\lambda}_B)$ is defined as purchasing power parity of country A relative to (on the basis of) country B (within the framework of unweighted means of prices on the level of basic headings).

The method can be generalized by introducing weights, e.g. expenditure shares w_{ij}, of commodity (basic heading) j in country i in the regression equation, leading to the "weighted CPD" method.

On 2:

The differences in product specification, in quantities and characteristics of products, even of those that serve the same general purpose, and the absence of certain products in certain countries is one of the basic problems in international price comparisons. It is virtually the same problem as the problem of quality change and changes in the collection of goods in intertemporal comparison,[74] which thus can be tackled with essentially the methods, e.g. the hedonic method. The basic problem this method has to face is, however, to identify a set of measurable characteristics for different commodities and to define the appropriate regression function that will give a good fit. If the multiple correlation coefficent turns out to be small, the introduction of price relatives derived from such a regression into a price index will possibly be somewhat questionable.

---

[73] Note that nothing has been said so far about how the coefficients λ_i will constitute overall-parities, that is how the individual prices p_k (k = 1, ..., n) should be combined to price indices and which part the m·n quantities q_{ki} will play in this step.

[74] It is therefore straightforward to apply the hedonic approach to these problems of international comparisons.

e2) Model based approaches

Some attempts are made to give an index such as the CCD index a motivation in terms of a regression equation. Hence there is a kind of stochastic-theory-approach- (see **sec. 2.3**) in the case of interspatial comparisons. Furthermore, we also find an economic-theory approach based on a microeconomic model of utility maximization in this area (i.e. an intertspatial variant).

The general idea of this reasoning is to aim at a (true) cost-of-living index (COLI) using the utility function and (given) utility level of the base country or the reference country respectively. However, the assumption of a household spending money according to a given cost-function (minimum cost to attain a given utility level) *simultaneously in two countries*, appears somewhat farfetched.

This applies with even greater force to some deliberations leading to an "economic" quantity index equal to the ratio of the (minimum) cost under the assumption of a common vector of prices and that all countries (!) have the same (international) preference structure.[75] It has been shown, for example, that such an index is bounded from below (above) by a weighted geometric mean of the GEKS-type using Paasche quantity indices (or Laspeyres quantity indices respectively). Here again the assumption of homotheticity, well known for its restrictive and unrealistic nature, has to be made and the result of these considerations is supposed to give support to the alleged suitability of the chained-Fisher-quantity index for volume comparisons across countries.

e3) Multilateral generalized Törnqvist method (MGT-index P^{MGT}, or Q^{MGT})

Let $v_{ik} = p_{ik}q_{ik} \big/ \sum_k p_{ik}q_{ik}$ denote the share of total expenditure in country i spent on commodity k, then the multilateral generalized Törnqvist (MGT) *quantity* index is given as

(8.2.12) $Q_{ij}^{MGT} = \prod_{k=1}^{n} \left(\frac{q_{jk}}{q_{ik}} \right)^{m_k}$ and by anlogy

(8.2.12a) $P_{ij}^{MGT} = \prod_{k=1}^{n} \left(p_{jk}/p_{ik} \right)^{m_k}$ the MGT *price* index,

where m_k is (in contrast to the "usual" Törnqvist index) a function of expenditure shares *of all m countries*, not only of just the two compared ones, countries i (base) and j. Thus, for example, a geometric mean of all m countries as regards the expenditure share of commodity k may be taken for m_k, that is,

$$m_k = \prod_{i=1}^{m} (v_{ik})^{1/m} \bigg/ \sum_{k=1}^{n} \prod_{i=1}^{m} (v_{ik})^{1/m}$$

has been proposed for a Walsh-type generalized multilateral Törnqvist index (or MGT-index). Also, an arithmetic-mean-type (instead of geometric mean Walsh-type) generalization was brought into play.

---

[75] see Balk, 2001, p.32 for more details.

Perhaps the crucial shortcoming of all methods based on the Törnqvist index is that they violate factor reversibility. They are also unable to satisfy the mean value property, let alone additivity. On the other hand, the index m_k (weight of commodity-quantity k) in Q_{ij}^{MGT} only depends on the type of commodity k as it is a function of *all* countries and not just the two compared ones (i and j). For this reason, all multilateral comparisons on the basis of the MGT-concept are transitive.[76] Furthermore, these methods (cp. Hill, 1997) are not subject to the Gerschenkron effect because no such thing as a common price vector or a common basket (quantity vector) is involved.

8.3 "Block methods" (Geary Khamis etc.) for multinational comparisons

Introductory remarks

Transitivity, though most demanding, is highly desirable.[77] In **sec. 8.1** a distinction was made between weak and strict transitivity. Only the former, less ambitious variant is ensured in the case of the Central Country Method (CCM), the first method to be discussed in what follows.

The exposition of the second method (part b of this section), i.e. the Geary Khamis (GK) method starts with the notion of strict transitivity. The GK method is an example of the so called "block methods" or "symmetric star methods", of which some other examples will be demonstrated in the remainder of this section.

The common feature of the methods of this section is that any two countries, A and B, are compared indirectly only in relation to the same third country X, where X can be one of the countries to be compared (as is the case in the CCM), or X can be some artificial or "average" country, in which case we have a true (genuine) "block method".

The Gerschenkron effect (see **sec. 8.1.g**) has received considerable attention in literature. It applies only to methods that compare all countries on the basis of either the same price vector or the same quantity vector. By virtue of the Gerschenkron effect, in the case of quantity indices Q_i and average prices (the same price vector)

- the output share of a country i (or the relative quantity index, that is, $Q_i/\sum Q_i$) tends to fall as the reference (common) price vector becomes relatively more characteristic of its (the i-th country's) own *price* vector, and conversely

---

[76] To make a weight m_k in a log-change index (symmetrically) dependent on *all* countries in order to make the index transitive comes down to the same logic as ensuring transitivity by constructing prices (or quantities) on prices (and/or quantities) of all countries under consideration. Hence, in a way, the MGT method may again be viewed as a variant of the block method. Note, however, the following sentence above.

[77] For the simple reason that it allows a one-dimensional ranking although a natural ordering is lacking in interspatial comparisons.

- the output share tends to rise as the reference quantity vector becomes relatively more characteristic of its own *quantity* vector.

By the same token, due to the Gerschenkron effect a price index such as P^P_{AB} tends to be smaller than P^L_{AB} as the common quantities in numerator and denominator resemble the reference country's quantities (Paasche concept) rather than the quantities of the base country (the Laspeyres concept).

a) The Central Country Method (CCM)

a1) Rationale of the CCM

A simple method to guarantee (weak) transitivity in the comparison of several countries, like A, B, C and so on, is to choose one land, say X, as the base land in all comparisons between any two countries. Each parity between any two countries (P_{AB}) is defined indirectly via X, expressed as a ratio of two parities, both referring to X as the base country. Thus in the case of the Laspeyres formula we get:

$$(8.3.1) \qquad \hat{P}^L_{AB} = P^L_{AB(X)} = \frac{P^L_{XB}}{P^L_{XA}} = \frac{\sum p_B q_X}{\sum p_X q_X} \Big/ \frac{\sum p_A q_X}{\sum p_X q_X} = \frac{\sum p_B q_X}{\sum p_A q_X}.$$

Note that this parity $\hat{P}^L_{AB}$ between A and B depends on the quantities q_X only (and not on q_A nor q_B). Therefore the transitive indices $P^L_{AB(X)}$ differ from the direct index P^L_{AB}, which is $\sum p_B q_A / \sum p_A q_A$, where the common budget is q_A.

Moreover, $P^L_{AB(X)}$ in general also differs from[78] $P^L_{AB(Y)} = \sum p_B q_Y / \sum p_A q_Y$.

Figure 8.3.1: Visualization of the CCM

Fig. **8.3.1** is an attempt to visualize the key idea of the CCM: any two countries A and B are compared only via X which is therefore also called the "star" country (a name intuitively explained in view of this figure). The country X is also called a "bridge" (or "link") country when the CCM is applied to two (or more) *groups* of countries, G_1 and G_2 respectively, so that all international comparisons within G_1 and those within G_2 refer to the same country X, which will have much in common with both groups.

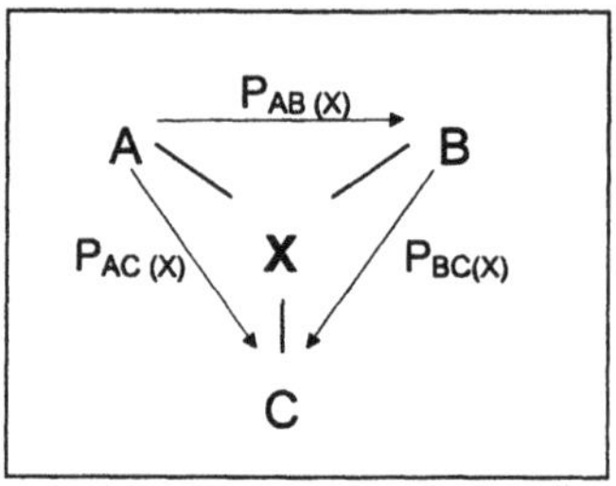

[78] This already demontrates a fact that is considered to be the main shortcoming of the CCM, in that the results are not unique.

Note that the direction of the arrows is from base country to reference country. Thus $P_{BA(X)}$ instead of $P_{AB(X)}$ means inverting the direction of the arrow. Before going into details concerning the properties of the CCM we may already conclude:

CCM is a method easy to understand, however the results are not unique but dependent on which choice has been made concerning the central country. Hence only weak transitivity is met.

a2) Properties of the CCM solution

It can easily be seen that all CCM - parities meet the country reversal test

(8.3.2) $P_{BA(X)} = \left(P_{AB(X)}\right)^{-1}$ for all countries, A and B, and also the circular test

(8.3.3) $P^L_{AC(X)} = P^L_{AB(X)} P^L_{BC(X)}$ since $\dfrac{\Sigma p_C q_X}{\Sigma p_A q_X} = \dfrac{\Sigma p_B q_X}{\Sigma p_A q_X} \dfrac{\Sigma p_C q_X}{\Sigma p_B q_X}$.

Note that $P^L_{AB(X)}$ is still a Laspeyres type index, the identity of the base country (which is A = 100) and the country from which the weights come (X) is destroyed, however. When Paasche parities are constructed in the same manner (by relating both A and B to the third country X) we get a more complicated expression

(8.3.4) $$P^P_{AB(X)} = \frac{P^P_{XB}}{P^P_{XA}} = \frac{\sum p_X q_A}{\sum p_A q_A} \frac{\sum p_B q_B}{\sum p_X q_B},$$

(8.3.4a) $$P^P_{BC(X)} = \frac{P^P_{XC}}{P^P_{XB}} = \frac{\sum p_X q_B}{\sum p_B q_B} \frac{\sum p_C q_C}{\sum p_X q_C}, \text{ and}$$

(8.3.4b) $$P^P_{AC(X)} = \frac{P^P_{XC}}{P^P_{XA}} = \frac{\sum p_C q_C}{\sum p_X q_C} \frac{\sum p_X q_A}{\sum p_A q_A}$$

Again it is easy to see that the country reversal test holds and that the indirect way to calculate P_{AC} and the direct way yield the same result so that again (weak) transitivity is satisfied since $P^P_{AC(X)} = P^P_{AB(X)} \cdot P^P_{BC(X)}$. It is interesting now to note that

1. the comparison of any two countries (say A and B) in eq. 4 depends on q_A *and* q_B but not on q_X, whilst in $P^L_{AB(X)}$ it depends on q_X only and on none of the compared countries (q_A, q_B),[79]
2. the following value decomposition equation for $P^P_{AB(X)}$ is given

$P^P_{AB(X)} = \underbrace{\dfrac{\Sigma p_B q_B}{\Sigma p_A q_A}}_{V_{AB}} : \underbrace{\dfrac{\Sigma p_X q_B}{\Sigma p_X q_A}}_{Q^L_{AB(X)}}$, and the corresponding equation for $P^L_{AB(X)}$ is

[79] See **tab. 8.3.1** for a summary of this kind of statement.

$$P^L_{AB(X)} = \frac{V_{AB}}{Q^P_{AB(X)}} = \frac{\sum p_B q_B}{\sum p_A q_A} : \underbrace{\frac{\sum q_B p_B \sum q_X p_A}{\sum q_X p_B \sum q_A p_A}}_{Q^P_{AB(X)}} = \frac{\sum p_B q_X}{\sum p_A q_X}.$$

Thus in the framework of CCM not only the Laspeyres formula but also the Paasche formula is transitive and as a consequence so is the Fisher index (the whole argument applies to price indices and to quantity indices as well). The Fisher form, however, is more complicated:

$$\text{(8.3.5)} \quad P^F_{AC(X)} = \sqrt{\underbrace{\frac{P^L_{XC}}{P^L_{XA}}}_{P^L_{AC(X)}} \underbrace{\frac{P^P_{XC}}{P^P_{XA}}}_{P^P_{AC(X)}}} = \sqrt{\underbrace{\frac{\Sigma p_C q_X}{\Sigma p_A q_X}}_{P^L_{AC(X)}} \underbrace{\frac{\Sigma p_C q_C}{\Sigma p_X q_C}}_{P^P_{XC}} \underbrace{\frac{\Sigma p_X q_A}{\Sigma p_A q_A}}_{\left[P^P_{XA}\right]^{-1}}} = P^F_{AB(X)} \cdot P^F_{BC(X)},$$

since $P^P_{AC(X)} = P^P_{XC} / P^P_{XA}$.

Table 8.3.1: Characteristicity (specificity) of the CCM solution*

Price index P			Quantity index Q
formula	quantities in P referring to	specificity	prices in Q referring to
Laspeyres $P^L_{AB(X)}$	X only	non-specific	X only
Paasche $P^P_{AB(X)}$	A and B	equi-specific	A and B
Fisher $P^F_{AB(X)}$	A, B and X	specific**	A, B and X

* all statements can easily be verified by looking at eqs. 8.3.1, 4 and 5

** but three budgets involved, so that it is of particular interest that X is representative of all countries

In multilateral comparisons the criterion of "characteristicity" or "specificity" is supposed to be desirable. This means that the weight q should be specific (typical) for the country in question (or the countries to be compared). **Tab. 8.3.1** presents a summary of the relevant relations. Note that P^F_{AB} may be called "equi-specific" because weights (quantities) of both countries, A and B, are utilized symmetrically. But this is not true for $P^F_{AB(X)} = \sqrt{\frac{P^F_{XB}}{P^F_{XA}}} \neq P^F_{AB} = \sqrt{P^L_{AB} P^P_{AB}}$. The parities are transitive in any case, P^L, P^P and P^F, but they depend on the choice of the "central country". Thus only weak[80] transitivity is given and we have $P_{AB(Y)} \neq P_{AB(X)}$ for all formulas, P^L and P^F, as demonstrated in **tab. 8.3.1**. Strict transitivity[81] requires

- not only all possible indirect comparisons being equal but also

[80] Weak transitivity means that transitivity is given but only for one specific central country at a time.

[81] It implies independence of the base (central) country; hence the name "base-land invariance".

- equal results no matter which country, X, Y or any other, has been chosen as the central country.

a3) Fixed based indices as intertemporal analogon to CCM - parities

The intertemporal counterpart to $P^L_{AB(X)}$ is $P_{0t(s)}$, the fixed weighted index with base period 0 rebased (switched[82]) to period s. If, for example, s = 1 and t = 2, the rebased Laspeyres index is

(8.3.6) $$P^L_{0t(s)} = \frac{P^L_{0t}}{P^L_{0s}} = P^L_{02(1)} = \frac{P^L_{02}}{P^L_{01}} = \frac{\sum p_2 q_0}{\sum p_1 q_0} \neq P^L_{12} = \frac{\sum p_2 q_1}{\sum p_1 q_1}.$$

In the CCM expressions like $P^L_{AB(X)} = P^L_{XB}/P^L_{XA}$ the cental country X plays the part of 0, the period to which all weights relate. Transforming a time series P_{0t} with base 0 (say 2000 = 100) to one with another (mostly more recent) base (say s = 2005) is known as "switching" (the base) or "rescaling" and is usually performed by the simple arithmetic operation

(8.3.6a) $$\hat{P}^L_{st} = \frac{P^L_{0t}}{P^L_{0s}} = P^L_{0t(s)} \neq P^L_{st},$$

which is a linear transformation since P_{0s} is constant. The roof $\wedge$ indicates that an estimation takes place, since the fixed (s) weighted index $\hat{P}^L_{0t}$, usually differs from the 0-weighted P^L_{0t}. Obviously P^L_{0t} may be regarded as the special case $P_{0t(s)} = P_{0t(0)}$ in the sense that the base period (0) and the weight period (s) coincide. Switching the base according to eq. 8.3.6a is based on the premise of proportionality of both time series P_{0t} and P_{st}. A fixed weighted *Laspeyres* index $P^L_{0t(s)}$ has a "changing cost of a fixed budget" interpretation as well as a "mean of price relatives" interpretation and is able to pass the time reversal as well as the circular test since $P^L_{13(0)} = P^L_{12(0)} P^L_{23(0)}$, whereas in general $P^L_{13} \neq P^L_{12} P^L_{23}$. The same is not true, however, for $P^P_{0t(s)}$ (or $P^P_{AB(X)}$). Eq. 8.3.6 applied to Paasche price indices leads to

(8.3.7) $$P^P_{02(1)} = \underbrace{\frac{\sum p_2 q_2}{\sum p_0 q_2}}_{B} : \underbrace{\frac{\sum p_1 q_1}{\sum p_0 q_1}}_{A} = \underbrace{\frac{\sum p_2 q_2}{\sum p_1 q_1}}_{V_{12}} : \underbrace{\frac{\sum p_0 q_2}{\sum p_0 q_1}}_{Q^L_{12(0)}} = \frac{\sum p_0 q_1}{\sum p_1 q_1} \frac{\sum p_2 q_2}{\sum p_0 q_2} \neq P^P_{12} = \frac{\sum p_2 q_2}{\sum p_1 q_2},$$

where a "changing cost of a fixed budget" interpretation is less evident than in the case of $P^L_{02(1)}$ (eq. 6) since two budgets are involved (q_1 *and* q_2). To give $P^P_{02(1)}$ or $P^P_{AB(X)}$ an economic interpretation in terms of countries we might think of comparing P^P_{XB} or

$B = \sum p_B q_B / \sum p_X q_B$ (equivalent to $P^P_{02} = \sum p_2 q_2 / \sum p_0 q_2$) to

[82] "Switching", that is, estimating P_{st} by P_{0t}/P_{0s}, should be distinguished from "revision", which implies a new definition and calculation of a market basket (budget) and which is much more demanding (requiring, among other things a new household inquiry) than switching.

$A = \sum p_A q_A / \sum p_X q_A$ (corresponding to P^P_{XA} or $P^P_{01} = \sum p_1 q_1 / \sum p_0 q_1$)

since $P^P_{AB(X)} = B/A$.[83] The interpretation of the Laspeyres approach in the framework of the CCM method is again much simpler and more direct: it is $\Sigma p_B q_X$ that is compared with $\Sigma p_A q_X$.

b) The Geary-Khamis (GK) approach to multinational comparisons

b1) Outline of the method and some properties

As mentioned in **sec. 8.2** it has been shown that[84] strict transitivity (base land invariance)[85] is given if and only if the m countries in question can be represented by a set of m positive real numbers such that $P_{ij} = \lambda_j / \lambda_i$ represents the PPP of country j relative to (in terms of) the base country i (i, j = 1,...,m).[86] It should be noted in passing that from this follows immediately that the set of λ's is not unique because any scalar multiplication of the vector of the λ's, say by θ, will yield the same parities because $P_{ij} = \theta\lambda_j / \theta\lambda_i = \lambda_j / \lambda_i$. We therefore have to choose one country (or an artificial combination of countries) as the "numeraire".

The general approach of the Geary-Khamis method[87] consists in simultaneously determining a price level of the "community" (or "block" of countries) as a whole[88] (vector $\bar{\mathbf{p}}'$ of n prices for n commodities) and a vector of parities $\hat{\mathbf{p}}' = [P^{GK}_{12} P^{GK}_{13} ...]$ with respect to a numeraire country, for example i = 1. Hence both vectors are conceived as being interrelated. Each country's (say B's) parity with respect to another country (say A) is defined as a ratio of "currency converters" denoted by c_i, such that $P^{GK}_{AB} = \frac{c_A}{c_B} = \frac{e_B}{e_A}$. The equations already introduced as eq. 8.2.6 through 8.2.8 in **sec. 8.2** are repeated in **fig. 8.3.2**.

It should be borne in mind that both systems of equations (eq. 8.2.6 an 8.2.7 in **fig. 8.3.2**) utilize national prices (p_{ki}) and national quantities (q_{ki}). This system of n and m equations can be solved iteratively for the $\bar{p}_k$ and the c_i under most general conditions. This work can be rather tedious, however, if m and n exceed 2, but the calcula-

[83] The formula thus compares the rise in costs of the most recent budget, q_2 for the whole period from 0 to 2 with the equivalent rise of prices from 0 to 1 of the budget q_1.

[84] The proof is to be credited to Rao and Banerjee 1984. See also Rao's annex to the CPI Manual in the internet (Rao 2001).

[85] to be kept distinct from bilateral symmetry (or country reversibility), which is not sufficient for achieving the multilateral property of weak transitivity, let alone strict transitivity (i.e. base land invariance).

[86] The volume ratio Q_{ij} is interpreted analogously.

[87] The method was developed by Geary in 1958 and Khamis 1972 and applied in the first three phases of the ICP project as well as 1980 in an EEC-comparison made by EUROSTAT.

[88] It is therefore a symmetric star method of the average *price* (as opposed to average *basket*) type (see **fig. 8.2.2**).

tion can be done with a simple desk calculator. Assume m = 4 countries, A, B, C and D as an example. There are only m-1, that is three, independent parities defined as follows[89]: $P_{AB}^{GK} = c_A/c_B$, $P_{BC}^{GK} = c_B/c_C$, and $P_{CD}^{GK} = c_C/c_D$, since the remaining six parities P_{AC}^{GK}, P_{CA}^{GK}, P_{AD}^{GK} etc. can be concluded due to transitivity and country reversibility.

Figure 8.3.2: The main relationships in the GK-method

determine simultaneously using national prices and quantities p_{ki}, q_{ki}

n community prices (k = 1, ..., n)	m currency converters (i = 1, ..., m)
(8.2.6) $\bar{p}_k = \sum_{i=1}^{m} c_i p_{ki} \frac{q_{ki}}{\sum_i q_{ki}} = \sum_{i=1}^{m} c_i p_{ki} \alpha_{ki}$	(8.2.7) $c_i = \sum_{k=1}^{n} \bar{p}_k q_{ki} \Big/ \sum_{k=1}^{n} p_{ki} q_{ki} = Q_i/V_i$
n equations, functions of national prices p_{ki} and national quantities q_{ki}	m equations, relating national price levels to price level of the community

m quantity indices	m price indices
(8.2.8b) $Q_{ij}^{GK} = \frac{\sum \bar{p}_k q_{kj}}{\sum \bar{p}_k q_{ki}} = \frac{Q_j}{Q_i}$	(8.2.8) $P_{ij}^{GK} = \frac{c_i}{c_j} = \frac{\sum \bar{p}_k q_{ki} / \sum p_{ki} q_{ki}}{\sum \bar{p}_k q_{kj} / \sum p_{kj} q_{kj}}$

To sum up:

1. The price level of the community, $\bar{\mathbf{p}}'$ (n commodities) is a function of (converted by c_i) national prices and quantities; division by $Q_k = \sum_i q_{ki}$ in defining $\bar{p}_k$ resembles the definition of unit values.
2. The amount by which the i-th country prices p_{ki} of the n commodities differ from those for the community as a whole in combination with national quantities q_{ki} is responsible for the c_i coefficients.
3. The indices P_{ij}^{GK} and Q_{ij}^{GK} respectively are obtained by comparing volumes (Q_i, Q_j) and values (V_i, V_j) of two countries, i and j, both with respect to the community[90] as a whole.

Thus there is a certain resemblance between the GK-method and the CCM: the community plays the part of the central country. The definition

(8.3.8) $$P_{ij}^{GK} = \frac{c_i}{c_j} = \frac{\sum \bar{p}_k q_{ki}}{\sum \bar{p}_k q_{kj}} \frac{\sum p_{kj} q_{kj}}{\sum p_{ki} q_{ki}},$$

[89] Also note that as aforesaid we can only obtain m-1 independent *relative* coefficients c_i/c_j, so that one of the c-terms has to be fixed arbitrarily if absolute figures for the c-coefficients are wanted.

[90] also known as "international price country".

not only shows that country reversibility holds since $\left(P_{ij}^{GK}\right)^{-1} = P_{ji}^{GK} = c_j/c_i$, but also that the approach to define P_{ij}^{GK} can also be viewed as a typical "indirect" (or Paasche[91]) approach to price level measurement in the sense that a change of prices has taken place to the extent to which the value ratio differs from the volume ratio:

$$\text{(8.3.8a)} \quad P_{ij}^{GK} = \frac{c_i}{c_j} = \frac{\Sigma p_{kj} q_{kj}}{\Sigma p_{ki} q_{ki}} \Bigg/ \frac{\Sigma \bar{p}_k q_{kj}}{\Sigma \bar{p}_k q_{ki}} = V_{ij}/Q_{ij}.$$

Thus the GK method is characterized by

- defining parities of m countries and a community price level and thereby also a price structure $\bar{\mathbf{p}}'$ related to n commodities as *interrelated*, that is, by a system of *simultaneous* equations, in which $\bar{\mathbf{p}}'$ is defined in a "unit-value" manner, using total quantities;
- *country weights* (thus GK-indices are *weighted* indices): the quantities q_{ki} used in defining $\bar{p}_k$ as well as c_i serve as country weights and account for the size of the i-th country,
- obtaining parities that will satisfy most of the axioms set up for international-comparisions, among which in particular strict *transitivity*. Finally
- it has been found empirically that the PPPs obtained by the GK-method are not widely different from Fisher's ideal index. But Fisher's sytem is not transitive.

The major shortcomings of the GK-method on the other hand are

- the factor reversal test is violated, the product test satisfied, however,
- equi-*characteristicity* is violated; interestingly the result is in fact more characteristic for a peripherical than for a central country (see below part b3),
- as to *consistency in aggregation* (criterion 4b in tab. 8.1.7) this axiom is violated while *structural consistency* (of volumes, criterion 4a) is satisfied by virtue of using common (community) prices for all countries,
- the GK-index system also violates the *price dimensionality* axiom in the sense of a uniform (λ-fold) change of some, not necessarily all, prices (see below),
- finally an argument of considerable practical importance concerning data requirements: the GK method assumes data on *individual* prices and quantities (referring to *each of the n identically defined commodities* in the m countries) whilst in practice very often only binary-comparision data on the level of *index* numbers (comprising all n quantities) are available.[92]

---

[91] I found the idea to compare the GK-method with both CCM and Paasche's approach in Diehl 1978, p. 157.

[92] For instance Fisher indices for all possible pairs of countries (full scale data) or for only some of the country-pairs (limited scale). The EKS-method is then applicable, whilst the GK-method is not. On the other hand, methods utilizing price-index material only (like EKS) involve a somewhat nebulous concept of quantity and there are difficulties in allowing for different sizes of the countries to be compared as well as for different quantities of commodities.

Remark as regards the lack of dimensionality

Assume only m = 2 countries and a λ-fold change of *all* prices in country 2, all quantities remain unchanged. This should result in a change of the parity c_2 of country 2 with respect to the community which is related to the amount of λ. But it can easily be seen that this will not always happen. According to eq. 8.2.6 the "new" price of the k-th commodity in the community as a whole will be

$$(8.3.9)\qquad \bar{p}_k^* = c_1^*\left(p_{k1}\frac{q_{k1}}{\Sigma q_{k1}}\right) + c_2^*\left(\lambda p_{k2}\frac{q_{k2}}{\Sigma q_{k2}}\right) \text{ for all } k = 1, \ldots, n.$$

Note that the price λp_{k2} instead of p_{k2} for commodity k in country 2 is only one out of m = 2 prices (given m = 2 countries) in determining $\bar{p}_k$. Thus the amount to which $\bar{p}_k^*$ will differ from $\bar{p}_k$ will not only depend on λ. Moreover, it has to be taken into account that the coefficients c_i^* will differ from the original coefficients c_i as well

$$(8.3.10)\qquad c_1^* = \sum_{k=1}^{n}\bar{p}_k^* q_{k1} \Big/ \sum_{k=1}^{n} p_{k1}q_{k1} \text{ and}$$

$$(8.3.10a)\qquad c_2^* = \sum_{k=1}^{n}\bar{p}_k^* q_{k2} \Big/ \sum_{k=1}^{n} \lambda p_{k2}q_{k2}$$

such that there is no reason to expect that the parity P_{12}^{GK} will reflect exactly the uniform λ-fold change of prices experienced in country 2. The exact change will be demonstrated for the two-country case in part b of this section.

It is also known that the GK-index (PPP) does not even satisfy a still somewhat weaker condition requiring that if all price-changes have a particular sign (say +, that is, all prices are rising) the change in the index number should also have the same sign. The GK-index system possesses this property only with m = 2, not with m > 2 countries.

b2) The case of m = 2 countries

It is well known that the GK-parity will simplify considerably if only two countries, A and B, are involved and that the formula P_{AB}^{GK} contains harmonic means (H) of quantities, that is

$$\left[\frac{1}{2}\left(\frac{1}{q_{k1}} + \frac{1}{q_{k2}}\right)\right]^{-1} = \frac{2q_{k1}q_{k2}}{q_{k1} + q_{k2}} = H_k$$

as weights. To verify this relationship we first examine the system eq. 8.2.6

$$(8.3.11)\qquad \bar{p}_1 = c_A p_{1A}\left(\frac{q_{1A}}{\Sigma q_{1i}}\right) + c_B p_{1B}\left(\frac{q_{1B}}{\Sigma q_{1i}}\right) = c_A p_{1A}\alpha_1 + c_B p_{1B}(1-\alpha_1)$$

$$(8.3.11a)\qquad \bar{p}_2 = c_A p_{2A}\left(\frac{q_{2A}}{\Sigma q_{2i}}\right) + c_B p_{2B}\left(\frac{q_{2B}}{\Sigma q_{2i}}\right) = c_A p_{2A}\alpha_2 + c_B p_{2B}(1-\alpha_2).$$

Upon rearranging some terms we get

$c_A p_{1A}\beta_1 + c_A p_{2A}\beta_2 = c_B p_{1B}\beta_1 + c_B p_{2B}\beta_2$ from eq. 11 and

$c_B p_{1B}\beta_1 + c_B p_{2B}\beta_2 = c_A p_{1A}\beta_1 + c_A p_{2A}\beta_2$ from eq. 11a

which turns out to be just two identical equations with coefficients defined as follows

$$\beta_k = q_{kA}(1-\alpha_k) = q_{kB}\alpha_k = \frac{q_{kA}q_{kB}}{q_{kA}+q_{kB}} = \frac{1}{2}H_k,\ k = 1, 2.$$

Hence we eventually get the famous harmonic mean expression

$$\text{(8.3.12)}\qquad P_{AB}^{GK} = \frac{c_A}{c_B} = \frac{p_{1B}\beta_1 + p_{2B}\beta_2}{p_{1A}\beta_1 + p_{2A}\beta_2} = \frac{p_{1B}\left(\frac{q_{1A}q_{1B}}{q_{1A}+q_{1B}}\right) + p_{2B}\left(\frac{q_{2A}q_{2B}}{q_{2A}+q_{2B}}\right)}{p_{1A}\left(\frac{q_{1A}q_{1B}}{q_{1A}+q_{1B}}\right) + p_{2A}\left(\frac{q_{2A}q_{2B}}{q_{2A}+q_{2B}}\right)}.$$

There is no principal diffculty in showing that in the case of n > 2 an equivalent equation for P_{AB}^{GK} will be obtained. Note that P_{AB}^{GK} as defined in eq. 12 differs from Fisher's parity which is

$$\text{(8.3.13)}\qquad P_{AB}^{F} = \sqrt{\left(\frac{p_{1B}q_{1A} + p_{2B}q_{2A}}{p_{1A}q_{1A} + p_{2A}q_{2A}}\right)\left(\frac{p_{1B}q_{1B} + p_{2B}q_{2B}}{p_{1A}q_{1B} + p_{2A}q_{2B}}\right)}.$$

Eq. 12 also shows that proportionality holds if *only two* countries are involved. Assume a change of both prices in B such that $p_{kB}^{*} = \lambda p_{kB}$ and prices in A remain unchanged. The result will be $P_{AB}^{GK} = \frac{\lambda c_A}{c_B}$, as required by proportionality.

b3) Characteristicity[93], the influence of the size of a country, and a numerical example

The size of the country has an impact on the result to be obtained for $\bar{p}_k$ by means of the quantities entering as weights in eq. 8.2.6 to determine the community's price of the k-th commodity. Consider only m = 2 countries, A and B, A being very small and B very large so that $q_{kA} < q_{kB}$ for all k (or almost all k). It is known that the GK-method now produces the somewhat awkward result that

- the price index P_{AB}^{GK} is dominated by the quantity structure of the small (or unimportant) country A and
- the quantity index Q_{AB}^{GK} is dominated by the price structure of the important country B.

This can easily be seen from eq. 8.3.12

[93] Characteristicity cannot usually be satisfied in the case of so called block methods in which prices or quantities are defined in such a way that they are influenced by the price and quantity data for all states under consideration. This applies in particular to the GK method.

(8.3.12a) $P_{AB}^{GK} = \dfrac{\sum p_{kB}\left(\dfrac{q_{kA}}{r_k+1}\right)}{\sum p_{kA}\left(\dfrac{q_{kA}}{r_k+1}\right)}$ where $r_k = \dfrac{q_{kA}}{q_{kB}}$, the summation takes place over k = 1, ..., n commodities. Since all quantities q_{kA} are small compared to q_{kB} this expression will tend to

(8.3.12b) $P_{AB}^{GK} = \dfrac{\sum p_{kB} q_{kA}}{\sum p_{kA} q_{kA}}$

as r_k tends towards zero, showing that the influence of the greater quantities in B becomes negligible and the quantities of the *smaller* country, q_{kA}, *dominate* the *price* relation.

Since Q_{AB}^{GK} is given by $Q_{AB}^{GK} = V_{AB}/P_{AB}^{GK}$ the result of eq. 12b entails

(8.3.12c) $Q_{AB}^{GK} = \dfrac{\sum p_{kB} q_{kB}}{\sum p_{kB} q_{kA}} = \sum \dfrac{q_{kB}}{q_{kA}} \dfrac{p_{kB} q_{kA}}{\sum p_{kB} q_{kA}}$.

Thus the weights attached to the *quantity* relatives *depend on* the prices p_{kB} of *the greater country* B. The following example will demonstrate this effect and also serve to verify the result obtained in part b of this section.

Example 8.3.1

Consider two commodities, 1, 2 and two countries A, B as follows

	price		quantity					values
k	p_{kA}	p_{kB}	q_{kA}	q_{kB}	Σ		i	$p_{ki}q_{ki}$
1	10	50	5	45	50		A	50 + 60 = 110
2	30	20	2	78	80		B	3810

In terms of quantities country A is small ($\Sigma q_{kA} = 7$) while B is big ($\Sigma q_{kB} = 45 + 78 = 123$) The system of equations 8.2.6 now looks as follows

$$\bar{p}_1 = c_A \cdot 10 \cdot \frac{5}{50} + c_B \cdot 50 \cdot \frac{45}{50} = c_A + 45c_B$$

$$\bar{p}_2 = c_A 30 \cdot \frac{2}{80} + c_B \cdot 20 \cdot \frac{78}{80} = 0.75c_A + 19.5c_B$$

and the system 8.2.7

$$c_A = (5\bar{p}_1 + 2\bar{p}_2)/110$$

$$c_B = (45\bar{p}_1 + 78\bar{p}_2)/3810.$$

Subtituting $\bar{p}_1$ and $\bar{p}_2$ as given by the first two equations into the second two equations gives 110c_A = 6.5c_A + 264c_B and 3810c_B = 103.5c_A + 3546c_B. Now the two equations reduce to one

$P_{AB}^{GK} = \frac{c_A}{c_B} = \frac{264}{103.5} = 2.551$, such that $P_{BA}^{GK} = 1/P_{AB}^{GK}$ = 0.392,

and the quantity index is derived from the value ratio V_{AB} = 3810/110 = 34.636 by dividing V_{AB} through P_{AB}^{GK}, which yields: $Q_{AB}^{GK} = \frac{34.636}{2.551} = 13.579$.

These results should be compared with what we would get using a Laspeyres or Paasche approach respectively

for price indices: $P_{AB}^{L} = \frac{50 \cdot 5 + 20 \cdot 2}{10 \cdot 5 + 30 \cdot 2} = 2.636$, $P_{AB}^{P} = \frac{50 \cdot 45 + 20 \cdot 78}{10 \cdot 45 + 30 \cdot 78} = 1.366$

and for quantity indices: $Q_{AB}^{L} = 25.364$, $Q_{AB}^{P} = 13.138$.

The result P_{AB}^{GK} = 2.551 is closer to P_{AB}^{L} than to P_{AB}^{P}, which shows that the influence of the smaller country A on the GK-parity is greater than the influence of the larger country B. Of course, the converse is true for the quantity indices (Q_{AB}^{GK} = 13.579 is closer to Paasche's index Q_{AB}^{P} than to Q_{AB}^{L}). ♦

To verify eq. 8.3.12 of this section the harmonic means of quantities should be calculated for this example first: $H_1 = \frac{2 \cdot 5 \cdot 45}{50} = 9$ for commodity 1, and $H_2 = \frac{2 \cdot 2 \cdot 78}{80} = 3.9$ for the second commodity. Note that the harmonic means are *markedly* smaller than both the arithmetic means of the two quantites (yielding 25 and 40) and the geometric means (15 and 12.49). This is due to assuming an extreme difference in size between A and B. Upon weighting with harmonic-mean quantities we get $P_{AB}^{GK} = \frac{50 \cdot 9 + 20 \cdot 3.9}{10 \cdot 9 + 30 \cdot 3.9} = \frac{528}{207} = \frac{264}{103.5} = 2.551$, which is exactly the result obtained from solving the system of equations, and thus equals the result we obtained above.

It may be interesting to compare now P_{AB}^{GK} with Fisher's index (which equals the EKS-result in the case of only m = 2 countries) $P_{AB}^{F} = \sqrt{P_{AB}^{L} P_{AB}^{P}}$. = 1.897.

Given a high dispersion of price and quantity relatives and a high correlation between them the results of various index formulas will differ considerably. This is true not only for intertemporal comparisons but also for international comparisons.

To sum up, the **ex. 8.3.1** yields the following results with respect to price indices: Paasche 1.366 < Marshall-Edgeworth 1.431 < Fisher 1.897 < Walsh 1.905 < Drobisch 2.001 < GK 2.551 < Laspeyres 2.636.

b4) Variants of average prices

Different alternatives of defining average (community) prices $\bar{p}_k$ have been proposed. Among other proposals (see for example Gerardi's unweighted geometric mean below) the concept of Iklé is particularly worth being mentioned. According to Iklé the impact of the diffferent size of the m countries materialized via q_{ki} is best neutralized by accounting for the different quantity shares $\gamma_i = Q_{Xi}^L \Big/ \sum_{i=1}^{m} Q_{Xi}^L$, where X is the "community" (star country) so that

(8.2.6) $$\bar{p}_k = \sum_{i=1}^{m} c_i p_{ki} \frac{q_{ki}}{\sum_i q_{ki}} = \sum_{i=1}^{m} c_i p_{ki} \alpha_{ki}$$ should be replaced by

(8.2.6*) $$\bar{p}_k = \sum_{i=1}^{m} c_i p_{ki} \frac{q_{ki}/Q_{Xi}^L}{q_{k1}/Q_{X1}^L + \ldots + q_{km}/Q_{Xm}^L} = \sum_{i=1}^{m} c_i p_{ki} \frac{q_{ki}/\gamma_i}{\sum_i \frac{q_{ki}}{\gamma_i}}$$

Divsion by Q_{Xi}^L or γ_i is a deweighting (according to country size) procedure. A consequence is that the interpretation of $\bar{p}_k$ (k = 1, ..., n) in terms of unit values no longer applies.

c) The (balanced) method of van Yzeren

The method will be briefly described in what follows despite its rather complicated nature and the fact that in the practice of price statistics it did not gain much attention.

c1) Elements of the method

The method of van Yzeren (or van Ijzeren) aims at simultaneously deriving PPP's, that is, price indices P_{ij}^Y *and* quantity indices Q_{ij}^Y, related to vectors of international prices $\bar{\mathbf{p}}$ and international quantities $\bar{\mathbf{q}}$ respectively. Thus the method combines aspects of the GK and the ECLA approach and aims at defining $\bar{\mathbf{p}}$ and $\bar{\mathbf{q}}$ in a symmetrical (balanced) way. A novel feature compared with both the GK and ECLA indices is, however, the explicit *introduction of country weights* g_i. The indices take the following forms in the case of countries, like A and B

(8.3.14) $$P_{AB}^Y = \frac{P_B}{P_A}$$ and (8.3.15) $$Q_{AB}^Y = \frac{\alpha_B}{\alpha_A} = \frac{V_{AB}}{P_{AB}^Y}$$

which entails transitivity and factor reversibility. The price levels (currency converters) P_i are determined only up to a scalar multiple (the P_i's are equivalent to the c_i coefficients in the GK method) and the coefficients α_i denote the volume shares[94] of the

[94] denoted Q_i above.

countries i (i = 1, 2, ..., m), like α_A of country A etc. They are interrelated with the price levels $P_1, ..., P_m$ as follows

(8.3.16) $\alpha_i = \frac{\mathbf{p}_i'\mathbf{q}_i/P_i}{\sum_j(\mathbf{p}_j'\mathbf{q}_j/P_j)} = \frac{K_i/P_i}{W}$ summation over j = 1, 2, ..., m countries inclusive of i such that $\sum_j \alpha_j = 1$. The "international" price and quantity vectors are defined as

(8.3.17) $\bar{\mathbf{p}} = \sum_{i=1}^{m} \frac{g_i}{P_i}\mathbf{p}_i$ and (8.3.18) $\bar{\mathbf{q}} = \sum_{i=1}^{m} \frac{g_i}{\alpha_i}\mathbf{q}_i$

where g_i are positive scalar country weights (such that $\sum_i g_i = 1$), which can be chosen arbitrarily and which also can be identically $g_i = 1/m$ for all i (unweighted case). The key idea in combining both vectors to the indices P^Y and Q^Y respectively is what van Yzeren calls his "balancing principle":

> "for each country the cost of the international basket equals the cost of the own basket at international prices" (van Yzeren 1988, p. 161).

Let i be the "own" country, the principle can be expressed in a "balancing equation" as follows

(8.3.19) $\mathbf{p}_i'\bar{\mathbf{q}} = c_i P_i$ (costs of the international basket)

(8.3.20) $\bar{\mathbf{p}}'\mathbf{q}_i = c_i \alpha_i$ (costs of own basket at international prices).

Combining both equations we obtain

(8.3.21) $\mathbf{p}_i'\bar{\mathbf{q}}/P_i = \bar{\mathbf{p}}'\mathbf{q}_i/\alpha_i$.

Substituting eq.16 for α_i in eq. 21 yields

$P_i = \left(\frac{\bar{\mathbf{q}}'\mathbf{p}_i}{\bar{\mathbf{p}}'\mathbf{q}_i}\frac{K_i}{W}\right)^{1/2}$ and we put this in eq. 19 to obtain $c_i = \bar{\mathbf{q}}'\mathbf{p}_i\left(\frac{\mathbf{q}_i'\bar{\mathbf{p}}}{\bar{\mathbf{q}}'\mathbf{p}_i}\frac{W}{\mathbf{q}_i'\mathbf{p}_i}\right)^{1/2}$.

Upon multiplication by $\sqrt{\bar{\mathbf{q}}'\bar{\mathbf{p}}}/\sqrt{\bar{\mathbf{q}}'\bar{\mathbf{p}}} = 1$ we get

(8.3.22) $c_i = \left[W(\bar{\mathbf{p}}'\bar{\mathbf{q}})(Q^L_{iX}Q^L_{Xi})\right]^{1/2} = \sqrt{W}\sqrt{\bar{\mathbf{p}}'\bar{\mathbf{q}}}\, r_i$,

where X denotes the central country to which the vectors $\bar{\mathbf{p}}$ and $\bar{\mathbf{q}}$ refer. If we define

(8.3.23) $r = (\bar{\mathbf{p}}'\bar{\mathbf{q}}/W)^{1/2}$ we have

(8.3.24) $c_i = W r r_i$.

From eq. 17 we get $\bar{\mathbf{p}}'\bar{\mathbf{q}} = \sum_i g_i \frac{\mathbf{p}_i'\bar{\mathbf{q}}}{P_i} = \sum g_i c_i$ (by eq.19). On the other hand, due to eq. 22 we have $\bar{\mathbf{p}}'\bar{\mathbf{q}} = \frac{c_i^2}{W^2 r_i^2} = \frac{r}{r_i}$, and taking the definition of r (eq. 24) into account we get

$$\bar{\mathbf{p}}'\bar{\mathbf{q}} = \sum g_i W r r_i = W r \sum g_i r_i = W r^2 .$$

Thus r is a weighted mean of the r_i-coefficients $\left(r = \sum g_i r_i\right)$ and substituting eq. 24 for c_i in eq. 19 and using 16, that is, $(\alpha_i)^{-1} = W/KP_i$, we obtain

(8.3.19a) $(\mathbf{p}_i'\bar{\mathbf{q}})/r r_i = K_i/\alpha_i$ and (8.3.20a) $(\bar{\mathbf{p}}'\mathbf{q}_i)/r r_i = K_i/P_i$

These equations link the unknown vectors $\bar{\mathbf{p}}, \bar{\mathbf{q}}$ with the constituents of the quantity index $\left(Q_{ij}^Y = \alpha_j/\alpha_i\right)$ and the price index $\left(P_{ij}^Y = P_j/P_i\right)$ together with the known expenditures $K_i = \mathbf{p}_i'\mathbf{q}_i$ (i = 1, 2, ..., m countries). Van Yzeren calls the factors r and r_i "bias correcting factors" according to which the indices are given by

(8.3.25) $$Q_{ij}^Y = \frac{\alpha_j}{\alpha_i} = \frac{r_j(\mathbf{p}_i\bar{\mathbf{q}})}{r_i(\mathbf{p}_j\bar{\mathbf{q}})} \cdot \frac{K_j}{K_i} = \frac{r_j(\mathbf{p}_i\bar{\mathbf{q}})}{r_i(\mathbf{p}_j\bar{\mathbf{q}})} V_{ij}$$

and

(8.3.26) $$P_{ij}^Y = \frac{P_j}{P_i} = \frac{r_j(\bar{\mathbf{p}}'\mathbf{q}_i)}{r_i(\bar{\mathbf{p}}'\mathbf{q}_j)} V_{ij}$$

According to eq. 23 the factor r_i is defined as $r_i^2 = Q_{Xi}^L Q_{iX}^L = \frac{\bar{\mathbf{p}}'\mathbf{q}_i}{\bar{\mathbf{p}}'\bar{\mathbf{q}}} \frac{\mathbf{p}_i'\bar{\mathbf{q}}}{\mathbf{p}_i'\mathbf{q}_i}$ and r_j^2 correspondingly.

Substitution of these expressions in eqs. 25 and 26 leads to expressions in which the r factors are eliminated and which are rarely quoted because they are based on knowledge of $\bar{\mathbf{p}}$ and $\bar{\mathbf{q}}$ such that they cannot be used for direct calculations of the two indices:

(8.3.25a) $$Q_{ij}^Y = \left(\frac{\bar{\mathbf{p}}'\mathbf{q}_j}{\bar{\mathbf{p}}'\mathbf{q}_i} \frac{\mathbf{p}_i'\bar{\mathbf{q}}}{\mathbf{p}_j'\bar{\mathbf{q}}} V_{ij}\right)^{1/2}$$

and

(8.3.26a) $$P_{ij}^Y = \left(\frac{\bar{\mathbf{p}}'\mathbf{q}_i}{\bar{\mathbf{p}}'\mathbf{q}_j} \frac{\mathbf{p}_j'\bar{\mathbf{q}}}{\mathbf{p}_i'\bar{\mathbf{q}}} V_{ij}\right)^{1/2}$$

It now easily follows that in fact eq. 15 holds according to which $P_{ij}^Y Q_{ij}^Y$ equals V_{ij} .

c2) Two countries case (m = 2)

Consider two countries only, A and B and n = 2 commodities only, then $\mathbf{p}_A' = [p_{1A} \quad p_{2A}]$, $\mathbf{q}_A' = [q_{1A} \quad q_{2A}]$, and $\mathbf{p}_B'$, $\mathbf{q}_B'$ correspondingly and

(8.3.16b) $$\alpha_A = \frac{\sum p_A q_A / P_A}{\frac{\sum p_A q_A}{P_A} + \frac{\sum p_B q_B}{P_B}} .$$

Assuming $g_A = g_B = 1/2$ we obtain the following vectors

(8.3.17a) $$\bar{\mathbf{p}} = [\bar{p}_1 \quad \bar{p}_2] = \left[\frac{1}{2P_A} p_{A1} + \frac{1}{2P_B} p_{B1} \qquad \frac{1}{2P_A} p_{A2} + \frac{1}{2P_B} p_{B2}\right]$$

(8.3.18a) $\bar{\mathbf{q}} = [\bar{q}_1 \quad \bar{q}_2] = \left[\frac{1}{2\alpha_A} q_{A1} + \frac{1}{2\alpha_B} q_{B1} \qquad \frac{1}{2\alpha_A} q_{A2} + \frac{1}{2\alpha_B} q_{B2}\right]$.

Next, the lefthand side (LHS) expression of eq. 8.3.21 should be calculated, which is in the case of the first country, A

$$\frac{1}{P_A}\left(\frac{p_{A1}q_{A1}}{2\alpha_A} + \frac{p_{A1}q_{B1}}{2\alpha_B} + \frac{p_{A2}q_{A2}}{2\alpha_A} + \frac{p_{A2}q_{B2}}{2\alpha_B}\right) = \frac{1}{P_A}\left(\frac{1}{2\alpha_A}\sum p_A q_A + \frac{1}{2\alpha_B}\sum p_A q_B\right)$$

and for country B we correspondingly get $\frac{1}{P_B}\left(\frac{1}{2\alpha_A}\sum p_B q_A + \frac{1}{2\alpha_B}\sum p_B q_B\right)$.

Now the righthand side (RHS) of eq. 8.3.21 yields in the case of country A

$$\frac{1}{\alpha_A}\left(\frac{p_{A1}q_{A1}}{2P_A} + \frac{p_{B1}q_{A1}}{2P_B} + \frac{p_{A2}q_{A2}}{2P_A} + \frac{p_{B2}q_{A2}}{2P_B}\right) = \frac{1}{\alpha_A}\left(\frac{1}{2P_A}\sum p_A q_A + \frac{1}{2P_B}\sum p_A q_B\right),$$

and for country B we get $\frac{1}{\alpha_B}\left(\frac{1}{2P_A}\sum p_A q_B + \frac{1}{2P_B}\sum p_B q_B\right)$.

Now consider both sides for countries A and B, which yields

$$\frac{1}{2\alpha_A}\sum p_A q_A + \frac{1}{2\alpha_B}\sum p_A q_B = \frac{P_A}{\alpha_A}\left(\frac{1}{2P_A}\sum p_A q_A + \frac{1}{2P_B}\sum p_B q_A\right)$$

and after rearranging and recalling eq. 8.3.16a (according to which

$\alpha_A/\alpha_B = \frac{\sum p_A q_A / P_A}{\sum p_B q_B / P_B}$) we arrive at

$$\left(\frac{P_A}{P_B}\right)^2 = \left(P_{BA}^Y\right)^2 = \frac{\sum p_A q_B}{\sum p_B q_B}\frac{\sum p_A q_A}{\sum p_B q_A} = P_{BA}^L P_{BA}^P = \left(P_{BA}^F\right)^2.$$

This proves that in the case of m = 2 countries the van Yzeren parities equal Fisher parities. Not surprisingly, the same operations performed in the case of country B will result in $\frac{P_B}{P_A} = \frac{\alpha_B}{\alpha_A}\frac{\sum p_B q_A}{\sum p_A q_B}$ and eventually in $P_{AB}^Y = P_{AB}^F$, which is the same result as obtained before for P_A / P_B.

d) Short remarks concerning further variants of the block-of-countries approach

d1) Gerardi's method

Dino Geradi's position was that the two objectives, PPP-measurement and deflation of values in an additively consistent way require different methods and he therefore proposed different variants of his method for the two purposes in question. Furthermore

1. for *deflation* with the ambitious attempt to make consistent comparisons among disaggregated deflated national aggregates of a block of countries, or aiming at aggregation of national aggregates to a block-aggregate, a *variant of the GK-method* should be chosen;
2. for PPP-measurement, on the other hand, Gerardi regarded the EKS-method preferable to the GK-method, but the EKS method treats countries of different size symmetrically, and according to Gerardi in order to provide a "balanced" comparison and to satisfy certain conditions (like mean-value test with respect to the implicit volume indices) in a block-set up requires abandoning the idea of balanced indices of the EKS type. In doing so Gerardi proposed a method utilizing weights and a vector $\tilde{\mathbf{q}}'$ of community-quantities.

The variants proposed by Gerardi are as follows:

On 1: The key element in Gerardi's method[95], like in the GK-method, is the vector $\bar{\mathbf{p}}'$ of a common price structure of the community (block) as a whole. The difference is how $\bar{\mathbf{p}}'$ is defined. The following equation

$$(8.3.27) \quad c_i^G = \frac{\sum_{k=1}^{n}\left(\prod_{i=1}^{m} p_{ki}\right)^{1/m} q_{ki}}{\sum_{k=1}^{n} p_{ki}q_{ki}} = \frac{\sum_{k=1}^{n} \tilde{p}_k q_{ki}}{\sum_{k=1}^{n} p_{ki}q_{ki}} \text{ (and correspondingly } P_{AB}^G = \frac{c_B^G}{c_A^G}\text{)}$$

plays the part of eq. 8.2.7 ($c_i = \sum_{k=1}^{n} \bar{p}_k q_{ki} \Big/ \sum_{k=1}^{n} p_{ki} q_{ki}$) in the GK-method, that is the elements of $\tilde{\mathbf{p}}'$ are now (in eq. 8.3.27) unweighted geometric means instead of weighted (the quantities q_{ki} act as country weights in determining the GK-community price $\bar{p}_k$ of the k-th commodity according to eq. 8.2.7) arithmetic means in the case of the GK method, i.e. $\bar{\mathbf{p}}'$. The argument in favor of the approach of eq. 8.3.27, i.e. geometric mean without country weights representing the size of a country are according to Gerardi:

> additivity (consistency in aggregation), sectoral independence of PPPs[96], proportionality in prices and quantities and finally applicability of this approach in the event of the negative balances.

On 2: For the comparision of the purchasing power (PPP) and comparisons of prices at aggregation levels higher than the basic headings, Gerardi recommended a formula based on weighting with constant quantities that refer to the community as a whole. These *quantities* are defined as geometric means (as *prices* were in eq. 27)

---

[95] The approach of Gerardi was used by the EEC for the 1975 volume comparison and tested in the third phase of the ICP project.

[96] By this is meant that a partial PPP or partial volume ratio relating to a certain (sub-) aggregate must only be a function of prices and quantities of that aggregate in question and of prices and quantities of the country in question. The result should not also depend on prices and quantities of other aggregates.

$$(8.3.28)\quad P_{AB}^{G}=\frac{\sum_{k=1}^{n}p_{kB}\left[\prod_{i=1}^{m}q_{k}\right]^{1/m}}{\sum_{k=1}^{n}p_{kA}\left[\prod_{i=1}^{m}q_{ki}\right]^{1/m}}=\frac{\sum_{k=1}^{n}p_{kB}\tilde{q}_{k}}{\sum_{k=1}^{n}p_{kA}\tilde{q}_{k}}$$

There is an obvious resemblace with the ECLA-method (see eq. 8.3.29) with respect to weighting with constant quantities, i.e. choosing a constant basket (as opposed to constant prices) approach. The difference is how the quantity weights are defined. Gerardi' s definition is

$$(8.3.28a)\quad \tilde{q}_{k}=\left[\prod_{i=1}^{m}q_{ki}\right]^{1/m}\quad \text{whereas the ECLA-definition is } \bar{q}_{k}=\frac{1}{m}\sum_{i=1}^{m}q_{ki}\,.$$

Gerardi's concept also resembles the index P_{0t}^{W} introduced by Walsh 1901, however, the product in eq. 28a is not calculated for two countries A, B (analogously to the geometric mean as defined for two points in time, 0 and t, in P_{0t}^{W}) but for all m countries. Hence P_{AB}^{G} may be seen as a kind of generalized (taking also the remaining m-2 countries into account) Walsh price index.

In contrast to the EKS-method the countries are no longer treated symmetrically irrespective of their size (as measured by their contribution q_{ki} to the community quantity $\bar{q}_{k}$).[97]

d2) The ECLA method

A method applied in the 1960's by the Economic Commission for Latin America (ECLA)[98] consists in utilizing the vector $\bar{\mathbf{q}}'$ of the total community quantities (and quantity-structure) of each commodity as weights for each price index

$$(8.3.29)\quad P_{AB}^{ECLA}=\frac{\sum_{k=1}^{n}p_{kB}\left[\sum_{i=1}^{m}q_{ki}\right]}{\sum_{k=1}^{n}p_{kA}\left[\sum_{i=1}^{m}q_{ki}\right]}=\frac{\sum_{k=1}^{n}p_{kB}Q_{k}}{\sum_{k=1}^{n}p_{kA}Q_{k}}=\frac{\sum_{k=1}^{n}p_{kB}\bar{q}_{k}}{\sum_{k=1}^{n}p_{kA}\bar{q}_{k}}\,.$$

Thus prices in the numerator (p_{kB}) and in the denominator (p_{kA}) are weighted with the same quantities referring to the total block of m countries (in contrast to weights, q_{iA} or q_{iB} respectively, in the Laspeyres - or Paasche approach). Obviously weights $\bar{q}_{k}$ guarantee transitivity in the same way in which a *Lowe index* P^{LW} satisfies the circular test in intertemporal comparison of any three points (0, s, t)

---

[97] The method thus guarantees "neutrality" of countries in Gerardi's understanding. Another aspect is what he calls the "similarity of countries effect", according to which not only the absolute magnitudes (where $q_{ki} > q_{kj}$ gives country i a greater weight than j) are to be considered but also the structure (distribution among the n commodities) of the total quantity.

[98] The fundamental ideas of this approach as well as the EKS-method have been proposed by Gini.

(8.3.30) $$P_{0t}^{LW} = P_{0s}^{LW} P_{st}^{LW} = \frac{\sum p_s \bar{q}}{\sum p_0 \bar{q}} \frac{\sum p_t \bar{q}}{\sum p_s \bar{q}} = \frac{\sum p_t \bar{q}}{\sum p_0 \bar{q}},$$

since weights in P^{LW} are common to all periods, not depending on either of these periods, 0, s or t.

The ECLA-method provides parities, PPP's (that is, indices P_{AB}^{ECLA}), that are transitive and able to satisfy the mean value test (for *price* indices),[99] *not*, however, structural (additive) consistency of volumes and the much weaker mean value property of *volumes* obtained from deflating with ECLA-parities. Moreover, the ECLA-type of comparison is more characteristic for a central country than for periphery countries. The more the consumption pattern of a country resembles that of the block as a whole (i.e. the vector $\bar{\mathbf{q}}$), the more the ECLA-index will be characteristic for that country. In this respect the ECLA-approach is better than the GK-approach.[100]

The ECLA-method fell in disuse because it is widely agreed that the assumption of a common quantity vector, $\bar{\mathbf{q}}$ is less acceptable than the assumption of a common price vector, $\bar{\mathbf{p}}$, being a key element in the GK-method as an example. Sums like $\sum_k \bar{p}_k q_{ik} = \bar{\mathbf{p}}'\mathbf{q}_i$ are definitely more meaningful (in fact they are "volumes") than sums, like $\sum_k \bar{q}_k p_{ik} = \bar{\mathbf{q}}'\mathbf{p}_i$.

8.4 Averaging methods (generalizations of the binary case) for multinational comparisons and related methods

This section is devoted to two representatives of the "averaging" methods. Mention is also given to some new ideas concerning indices of the "economic theory" type.

a) Generalization of country reversible index functions

This section deals with methods of making transitive multilateral comparisons based on a generalization of the index formulas of Fisher and Törnqvist. As demonstrated in **sec. 3.4** both index functions are "unbiased" in the sense that they are means of other index functions

$P_{AB}^F = \sqrt{P_{AB}^L P_{AB}^P}$ Laspeyres (P_{AB}^L) and Paasche (P_{AB}^P) in the case of Fisher where

$P_{AB}^L = \sum_{k=1}^n w_{kA} \frac{p_{kB}}{p_{kA}}$, $P_{AB}^P = \left[\sum_{k=1}^n w_{kB} \left(\frac{p_{kB}}{p_{kA}}\right)^{-1}\right]^{-1}$ expenditure shares w_{kA} and w_{kB};

[99] which applies to all variants of the constant basket method.

[100] However characteristicity (specificity, typicality) in the strict sense is failed, because the quantity vector cannot be typical for any country taken in isolation.

$P^T_{AB} = \sqrt{DP^L_{AB} DP^P_{AB}}$ log - Laspeyres DP^L_{AB} and log - Paasche DP^P_{AB} in the case of Törnqvist, that is, $DP^L_{AB} = \prod_{k=1}^{n} \left(\frac{p_{kB}}{p_{kA}} \right)^{w_{kA}}$ and $DP^P_{AB} = \prod_{k=1}^{n} \left(\frac{p_{kB}}{p_{kA}} \right)^{w_{kB}}$ respectively.

Both formulas, P^F_{0t} and P^T_{0t} (in the interspatial case P^F_{AB}, P^T_{AB}), have in common that

- they are both able to pass the country reversal test, $P^F_{BA} = \left(P^F_{AB}\right)^{-1}$, and $P^T_{BA} = \left(P^T_{AB}\right)^{-1}$ for any two countries, A and B
- they are both *not* transitive in the sense that with any three countries, A, B and C the following relation for P will generally not hold $P_{AC} = P_{AB} P_{BC}$ (for indices P^T or P^F).

By taking averages (not necessarily geometric means, an arithmetic mean would work as well for example) over all possible bilateral (two country) comparisons, however, they can be made transitive. Note that this operation (i.e. taking averages) requires country reversible index functions, such as P^F_{AB} in the case of the EKS method or P^T_{AB} in the case of the Caves-Christensen-Diewert (CCD) method.

b) The EKS-method to provide transitive PPPs

In this section some further comments on the EKS method[101] and related methods are made. Firstly, it will be demonstrated how to derive and interpret the general EKS solution for PPPs. We then show extensions in the way of country weights and alternatives to the Fisher formula

b1) The formula and some interpretations

The general equation to express EKS-parities is

$$(8.4.1) \qquad P^{EKS}_{AB} = \left[\prod_{i}^{m} P^F_{iB} P^F_{Ai} \right]^{1/m} = \left[\prod_{i}^{m} \frac{P^F_{Ai}}{P^F_{Bi}} \right]^{1/m} = \left[\left(P^F_{AB}\right)^2 \prod_{i \neq A} P^F_{Ai} \prod_{i \neq B} P^F_{iB} \right]^{1/m}.$$

Equation 8.4.1 also shows that P_{AB} depends not only on prices and quantities of the countries i = A and i = B, but also on all other m-2 countries. Thus if the set of countries is extended all price indices must be recalculated. The EKS method is therefore a closed system of parities.

To see how eq. 8.4.1 has to be read, a two and three country case will be demonstrated subsequently.

Two countries (m = 2), A and B:

A product of *two* factors has to be calculated, the first factor accounting for i = A

[101] First devised by Eletö and Köves 1964 and also independently by Szulc 1964.The very first development of this reasoning and the formula is to be credited to Gini, however.

and the second for i = B to get $P_{AB}^{EKS} = \left[\left(P_{AB}^F P_{AA}^F \right) \left(P_{BB}^F P_{AB}^F \right) \right]^{1/2} = P_{AB}^F$. Hence with only two countries involved the EKS-parity equals the Fisher-parity.

Three countries (m = 3), A, B and C:

Each parity is now derived from a product of *three* factors

$$P_{AB}^{EKS} = \left[\underbrace{P_{AB}^F P_{AA}^F}_{i=A} \underbrace{P_{BB}^F P_{AB}^F}_{i=B} \underbrace{P_{CB}^F P_{AC}^F}_{i=C} \right]^{1/3}$$ which reduces to

(8.4.1a) $$P_{AB}^{EKS} = \sqrt[3]{\left(P_{AB}^F \right)^2 P_{AC}^F P_{CB}^F}$$

if use is made of the identity and time reversal property of Fisher-indices. In the same manner the two remaining parities are obtained[102] as

$$P_{AC}^{EKS} = \left[\left(P_{AC}^F P_{AA}^F \right) \left(P_{BC}^F P_{AB}^F \right) \left(P_{CC}^F P_{AC}^F \right) \right]^{1/3} = \sqrt[3]{\left(P_{AC}^F \right)^2 P_{AB}^F P_{BC}^F}$$ and

$$P_{BC}^{EKS} = \left[\left(P_{AC}^F P_{BA}^F \right) \left(P_{BC}^F P_{BB}^F \right) \left(P_{CC}^F P_{BC}^F \right) \right]^{1/3} = \sqrt[3]{\left(P_{BC}^F \right)^2 P_{BA}^F P_{AC}^F} \, .$$

From these equations it easily follows that EKS-parities pass the time reversal test, because interchanging of A and B in eq. 8.4.1 leads to

(8.4.2) $$P_{BA}^{EKS} = \left[\prod_{i=1}^{m} P_{iA}^F P_{Bi}^F \right]^{1/m} = \left[\prod_{i=1}^{m} \frac{1}{P_{iB}^F} \frac{1}{P_{Ai}^F} \right]^{1/m} = \frac{1}{P_{AB}^{EKS}},$$

utilizing the country reversibility of the Fisher formula, that is, $P_{iA}^F = 1/P_{Ai}^F$ and $P_{Bi}^F = 1/P_{iB}^F$.

The EKS parity can also be interpreted as the geometric mean of all indirect comparisons between A and B through all possible link countries i = 1, ...,m (including A and B as link countries).

$$P_{AB}^{EKS} = \left[\underbrace{P_{AA}^F P_{AB}^F}_{\text{link A}} \underbrace{P_{AB}^F P_{BB}^F}_{\text{link B}} \underbrace{P_{AC}^F P_{CB}^F}_{\text{link C}} \right]^{1/3} .$$

A closer look at eq. 8.4.1 also reveals that transitivity holds because

(8.4.3) $$P_{AC}^{EKS} = \left[\prod_{i}^{m} P_{iC}^F P_{Ai}^F \right]^{1/m} = \left[\prod_{i}^{m} P_{iB}^F P_{Ai}^F \right]^{1/m} \left[\prod_{i}^{m} P_{iC}^F P_{Bi}^F \right]^{1/m} = P_{AB}^{EKS} P_{BC}^{EKS},$$

since $\prod P_{iB}^F \prod P_{Bi}^F = 1$. This can also be verified using the equations for the three country case $P_{AC}^{EKS} = \sqrt[3]{\left(P_{AC}^F \right)^2 P_{AB}^F P_{BC}^F} = P_{AB}^{EKS} P_{BC}^{EKS} = \sqrt[3]{\left(P_{AB}^F \right)^2 P_{AC}^F P_{CB}^F} \sqrt[3]{\left(P_{BC}^F \right)^2 P_{BA}^F P_{AC}^F}$,

the RHS of this eq. is of course $\sqrt[3]{\left(P_{AB}^F \right)^2 P_{AC}^F \frac{1}{P_{BC}^F} \left(P_{BC}^F \right)^2 \frac{1}{P_{AB}^F} P_{AC}^F}$ which is in fact P_{AC}^{EKS}.

---

[102] P_{AC} is an eq. already quoted as eq. 8.2.5a.

b2) Deriviation by generalizing the Fisher formula[103]

Consider a single parity between any two countries i and j, $P_{ij} = P_j/P_i$ instead of two measured by $P_{ij}^L = \mathbf{p}_j'\mathbf{q}_i/\mathbf{p}_i'\mathbf{q}_i$ and $P_{ij}^P = \mathbf{p}_j'\mathbf{q}_j/\mathbf{p}_i'\mathbf{q}_j$ such that[104] $P_{ij}^L = t(P_j/P_i)$ and $P_j/P_i = tP_{ij}^P$ or simply

(8.4.4) $$P_{ij}^L \frac{P_i}{P_j} = \frac{P_j}{P_i} \cdot \frac{1}{P_{ij}^P} = \frac{P_j}{P_i} P_{ji}^L = t$$

The solution for the wanted (transitive) index $P_{ij} = P_j/P_i$ is evidently $(P_j/P_i)^2 = P_{ij}^L P_{ij}^P = (P_{ij}^F)^2$, hence the Fisher formula. A "natural" generalization of eq. 4, using normalized country weighs $f_i = g_i/\sum g_i$ such that $\sum f_i = 1$ is

(8.4.5) $$\prod_i \left[P_{ij}^L (P_i/P_j)\right]^{f_i} = \prod_i \left[P_{ji}^L (P_j/P_i)\right]^{f_i}$$ (for i = 1, ...,m).

The solution of this system is

(8.4.6) $$P_j/P_i = \prod_k \left(P_{kj}^F P_{ik}^F\right)^{f_k} = P_{ij}^{GEKS}$$

which is the generalized EKS-parity (or GEKS). In the unweighted case of $f_k = 1/m$ we get the "normal" EKS solution of eq. 8.4.1. To see this, consider first the situation with j fixed and i taking on all values 1, 2, ..., m, hence

$$\left[P_{1j}^L (P_1/P_j)\right]^{f_1} \left[P_{2j}^L (P_2/P_j)\right]^{f_2} \cdots \left[P_{mj}^L (P_m/P_j)\right]^{f_m} = \left[P_{j1}^L (P_j/P_1)\right]^{f_1} \left[P_{j2}^L (P_j/P_2)\right]^{f_2} \cdots \left[P_{jm}^L (P_j/P_m)\right]^{f_m}.$$

This yields the following equation

(8.4.7) $$\left(P_{1j}^F\right)^{f_1} \left(P_{2j}^F\right)^{f_2} \cdots \left(P_{mj}^F\right)^{f_m} = P_j / P_1^{f_1} P_2^{f_2} \cdots P_m^{f_m}.$$

Consider now what eq. 5 means if i is fixed

$$\left[P_{i1}^L (P_i/P_1)\right]^{f_1} \left[P_{i2}^L (P_i/P_2)\right]^{f_2} \cdots \left[P_{im}^L (P_i/P_m)\right]^{f_m} = \left[P_{1i}^L (P_1/P_i)\right]^{f_1} \left[P_{2i}^L (P_2/P_i)\right]^{f_2} \cdots \left[P_{im}^L (P_m/P_i)\right]^{f_m}$$ giving

(8.4.8) $$\left(P_{i1}^F\right)^{f_1} \left(P_{i2}^F\right)^{f_2} \cdots \left(P_{im}^F\right)^{f_m} = P_1^{f_1} P_2^{f_2} \cdots P_m^{f_m} / P_i.$$

It is now obvious that upon multiplication of eqs. 7 and 8 we get

(8.4.6a) $$\left(P_{i1}^F P_{1j}^F\right)^{f_1} \cdots \left(P_{im}^F P_{mj}^F\right)^{f_m} = \frac{P_j}{P_i} = P_{ij}^{GEKS}.$$

As the EKS (and this applies to GEKS too) quantity indices can be obtained by interchanging prices and quantities in the price index formula (eq. 8.4.1a) we get for

[103] In what follows we decisively draw on B.M. Balk 2001 (Aggregation Methods in International Comparisons: What Have We Learned?), p. 7f (notation adapted to ours)

[104] This reasoning resembles the deriviation of the Fisher formula from a criterion of "equal distance" to P^L as well as P^P (see the uniqueness theorem UT-1 **sec. 3.1j** and also **sec. 2.6c**).

example $Q_{AB}^{EKS} = \sqrt[3]{\left(Q_{AB}^{F}\right)^{2} Q_{AC}^{F} Q_{CB}^{F}}$. And since the Fisher formula passes the factor reversal test[105] we have

$$P_{AB}^{EKS} Q_{AB}^{EKS} = \sqrt[3]{\left(\frac{\Sigma p_B q_B}{\Sigma p_A q_A}\right)^{2} \frac{\Sigma p_C q_C}{\Sigma p_A q_A} \frac{\Sigma p_B q_B}{\Sigma p_C q_C}} = \frac{\Sigma p_B q_B}{\Sigma p_A q_A},$$

or in other words: the EKS indices satisfy the factor reversal test.

b3) EKS parities derived by minimizing a distance

As mentioned repeatedly, a transitive parity should be capable of being expressed as the ratio P_k/P_i. Hence we have to find positive real numbers P_i and P_k such that the P_{ik}^{F} approximate P_k/P_i (for all i and k) in a least squares (of deviations) sense.

The general distance minimization criterion reads as follows[106]

$$\text{(8.4.9)} \qquad \min_{P_1,\dots,P_m} \sum_i \sum_k g_i g_k \left[\ln\left(P_{ik}^{F}\right) - \ln\left(P_k/P_i\right)\right]^2 .$$

The function

$$\text{(8.4.9a)} \qquad \Delta = \Delta(P_1,\dots,P_m) = \sum_i \sum_k g_i g_k \left[\ln\left(P_{ik}^{F}\right) - \ln\left(P_k/P_i\right)\right]^2$$

$$= g_1 g_2 \left\{\left(\ln P_{12}^{F}\right)^2 - 2\ln P_{12}^{F}\left(\ln P_2 - \ln P_1\right) + \left(\ln P_2 - \ln P_1\right)^2\right\}$$

$$+ \ g_1 g_3 \left\{\left(\ln P_{13}^{F}\right)^2 - 2\ln P_{13}^{F}\left(\ln P_3 - \ln P_1\right) + \left(\ln P_3 - \ln P_1\right)^2\right\} + \dots$$

$$+ \ g_2 g_1 \left\{\left(\ln P_{21}^{F}\right)^2 - 2\ln P_{21}^{F}\left(\ln P_1 - \ln P_2\right) + \left(\ln P_1 - \ln P_2\right)^2\right\} + \dots$$

has to be differentiated with respect to P_1, P_2, ..., and set equal to zero.[107] Thus $\frac{\partial \Delta}{\partial P_1} = 0$ results after division by $2g_1/P_1$ in

$$g_2\left(\ln \frac{P_{12}^{F}}{P_{21}^{F}}\right) + g_3\left(\ln \frac{P_{13}^{F}}{P_{31}^{F}}\right) + \dots 2\ln P_1\left(g_2 + g_3 + \dots\right) = 2\left(g_2 \ln P_2 + g_3 \ln P_3 + \dots\right),$$

or after some rearrangement and since $\ln\left(P_{11}^{F}\right) = \ln(1) = 0$ we get

$$\text{(8.4.10)} \qquad \ln P_1 = \sum_{i=1}^{m} g_i \ln P_i - \sum_{i=1}^{m} g_i \ln\left(P_{1i}^{F}\right).$$

---

[105] This can no longer be assumed if P^F is replaced by another index function such as Törnqvist P^T (in the CCD method, see part c of this section) or P^{ST} in Banerjee's factorial approach functions.

[106] The logarithmic distance function Δ is introduced in order to make the resulting multilateral transitive indices deviate the least from the non-transitive binary indices.

[107] We saw no proof spelled out in detail in the relevant literature. It is, however, quite useful to reproduce the reasoning step by step and we therefore demonstrate the details in what follows.

In a similar vein, examining $\frac{\partial \Delta}{\partial P_2} = 0$ leads to

(8.4.11) $$\ln P_2 = \sum_{i=1}^{m} g_i \ln P_i - \sum_{i=1}^{m} g_i \ln\left(P_{2i}^F\right).$$

Subtraction of 10 from 11 yields

(8.4.1a) $$P_{12}^{GEKS} = \frac{P_2}{P_1} = \frac{\prod_i^m \left(P_{i2}^F\right)^{g_i}}{\prod_i^m \left(P_{i1}^F\right)^{g_i}} = \prod_{i=1}^{m} \left(\frac{P_{i2}^F}{P_{i1}^F}\right)^{g_i}$$

b4) Extensions

A "natural" variant of eq. 8.4.5 is to take arithmetic (or even less frequently considered harmonic) means, that is

(8.4.5a) $\sum_i g_i P_{ij}^L \left(P_i/P_j\right) = \sum_i g_i P_{ji}^L \left(P_j/P_i\right)$ (for j = 1, ...,m)

which is related to a method proposed by ***van Yzeren*** (Balk 2001, p.10). Using harmonic means amounts to applying $\sum_i g_i P_{ji}^L \left(P_i/P_j\right) = \sum_i g_i P_{ij}^L \left(P_j/P_i\right)$ rather than eq. 5a, and it is also possible to take the Paasche formula instead of P_{ij}^L and $P_{ji}^L = 1/P_{ij}^P$.

Also, the usage of harmonic means instead of geometric means in

$P_{AB}^{GEKS} = \frac{\prod_i^m \left(P_{iB}^F\right)^{g_i}}{\prod_i^m \left(P_{iA}^F\right)^{g_i}}$ (eq. 1a) has been proposed, that is, $\frac{\left(\sum_i g_i \left(P_{iB}^F\right)^{-1}\right)^{-1}}{\left(\sum_i g_i \left(P_{iA}^F\right)^{-1}\right)^{-1}}$, leading to Diewert's "Own Share" system.

Another variant of the EKS system was provided by **Rao and Banerjee** 1986, where the factorial index approach originally devised by Banerjee for a pairwise (bilateral) comparison (see **sec. 2.6**) was extended to a multi-country-comparison. It was shown that $P_{ik}^{RB} = \prod_{j=1}^{m} \left(\frac{P_{ij}^{ST}}{P_{kj}^{ST}}\right)^{1/m}$ has the same property of base land invariance (strict transitivity) as eq, 8.4.1 using the formula of Fisher.[108] Utilization of Stuvel's formula instead of Fisher's (EKS-method!) or Törnqvist's (CCD) has the following consequences

- factor reversibility, country reversibility and symmetry of weights is given much like in the case of P^F or P^T,

[108] The corresponding Rao-Banerjee quantity index is built with Stuvel quantity indices Q^{ST} instead of price indices P^{ST}.

- P^{ST} does not, however, satisfy linear homogeneity, which is, according to Rao and Banerjee, more than offset, however, on grounds which hardly seem to be convincing, as notably
  - the choice of P^{ST} is based on the factorial approach which is (ostensibly) well established,
  - other methods do not satisfy linear homogeneity either, and
  - the results found in an empirical example were quite similar to those obtained by EKS for the same data which "prove" that their factorial approach is a viable alternative on its own right.

c) The Caves-Christensen-Diewert (CCD) - method

Like Fisher's index formula, the Törnqvist index is also country reversible but not transitive, i.e. the product

$P^T_{AB} P^T_{BC} = \prod_{k=1}^{n}\left(\frac{p_{kB}}{p_{kA}}\right)^{\overline{w}_{AB}} \prod_{k=1}^{n}\left(\frac{p_{kC}}{p_{kB}}\right)^{\overline{w}_{BC}}$ where $\overline{w}_{AB} = \frac{1}{2}(w_{kA} + w_{kB})$, and $\overline{w}_{BC}$, $\overline{w}_{AC}$ correspondingly does not necessarily equal $P^T_{AC} = \prod_{k=1}^{n}\left(\frac{p_{kC}}{p_{kA}}\right)^{\overline{w}_{AC}}$ unless $\overline{w}_{AB} = \overline{w}_{BC} = \overline{w}_{AC}$.

A simple method to guarantee transitivity is to take an average of any two-country comparison as done in the Caves-Christensen-Diewert (CCD) index, recommended for international price comparisons. In analogy to the EKS system of parities

$$(8.4.1) \qquad P^{EKS}_{AB} = \left[\prod_{i}^{m} P^F_{iB} P^F_{Ai}\right]^{1/m} = \left[\prod_{i}^{m} \frac{P^F_{Ai}}{P^F_{Bi}}\right]^{1/m} = \left[\prod_{i}^{m} \frac{P^F_{iB}}{P^F_{iA}}\right]^{1/m}$$

the CCD index is defined as

$$(8.4.12) \qquad P^{CCD}_{AB} = \left(\frac{\prod_{i=1}^{m}\left(P^T_{Ai}\right)}{\prod_{i=1}^{m}\left(P^T_{Bi}\right)}\right)^{1/m} = \left(\prod_{i=1}^{m} \frac{P^T_{Ai}}{P^T_{Bi}}\right)^{1/m} \text{ or equivalently}$$

$$(8.4.12a) \quad \ln\left(P^{CCD}_{AB}\right) = \frac{1}{m}\sum_{i=1}^{m} \ln\left(P^T_{Ai}\right) - \frac{1}{m}\sum_{i=1}^{m} \ln\left(P^T_{Bi}\right)$$

Obviously P^{CCD} is transitive since

$$P^{CCD}_{AC} = \left(\prod_{i=1}^{m} \frac{P^T_{Ai}}{P^T_{Ci}}\right)^{1/m} = \left(\prod_{i=1}^{m} \frac{P^T_{Ai}}{P^T_{Bi}}\right)^{1/m} \left(\prod_{i=1}^{m} \frac{P^T_{Bi}}{P^T_{Ci}}\right)^{1/m} = P^{CCD}_{AB} P^{CCD}_{BC}.$$

To demonstrate the calculation of the P^{CCD} index and its transitivity consider an example of three countries, A, B and C only. Since $P^T_{AA} = 1$ and therefore $\log P^T_{AA} = 0$

(identity) and $\ln\left(P_{BA}^{T}\right) = -\ln\left(P_{AB}^{T}\right)$ (country reversibility) we get the following matrix **T** of logarithms of all parities as given in **fig. 8.4.1**.

Much of what applies to the EKS method also applies to the CCD concept since in the latter, the Törnqvist index P_{0t}^{T} plays the part Fisher's index P_{0t}^{F} does in the former concept. Moreover, the CCD-index has been criticised due to the simple (unweighted) means $\psi_A = \frac{1}{m}\sum_{i=1}^{m}\ln\left(P_{Ai}^{T}\right)$ of which the formula is composed (since $\ln\left(P_{AB}^{CCD}\right)$ $= \psi_A - \psi_B$). Thus some refinements of this approach have been proposed.

Fig. 8.4.1: Demonstration of the transitivity of CCD parities (three country case)

(the formula of the EKS method can easily be verified by simply substituting P^F for P^T)

$$\mathbf{T} = \begin{bmatrix} 0 & \ln\left(P_{AB}^{T}\right) & \ln\left(P_{AC}^{T}\right) \\ -\ln\left(P_{AB}^{T}\right) & 0 & \ln\left(P_{BC}^{T}\right) \\ -\ln\left(P_{AC}^{T}\right) & -\ln\left(P_{BC}^{T}\right) & 0 \end{bmatrix}$$

(1) $\sum_{i=1}^{m}\ln\left(P_{Ai}^{T}\right) = \ln\left(P_{AB}^{T}\right) + \ln\left(P_{AC}^{T}\right)$	$3\ln\left(P_{AB}^{CCD}\right)$ = eqs. (1) - (2)
(2) $\sum_{i=1}^{m}\ln\left(P_{Bi}^{T}\right) = -\ln\left(P_{AB}^{T}\right) + \ln\left(P_{BC}^{T}\right)$	$3\ln\left(P_{BC}^{CCD}\right)$ = eqs. (2) - (3)
(3) $\sum_{i=1}^{m}\ln\left(P_{Ci}^{T}\right) = -\ln\left(P_{AC}^{T}\right) - \ln\left(P_{BC}^{T}\right)$	$3\ln\left(P_{AC}^{CCD}\right)$ = eqs. (2) - (3)

(4) $$\ln\left(P_{AB}^{CCD}\right) = \frac{2}{3}\ln\left(P_{AB}^{T}\right) - \frac{1}{3}\left[\ln\left(P_{BC}^{T}\right) - \ln\left(P_{AC}^{T}\right)\right]$$

(5) $$\ln\left(P_{BC}^{CCD}\right) = \frac{2}{3}\ln\left(P_{BC}^{T}\right) - \frac{1}{3}\left[\ln\left(P_{AB}^{T}\right) - \ln\left(P_{AC}^{T}\right)\right]$$

For the third parity we get $\ln\left(P_{AC}^{CCD}\right) = \frac{2}{3}\ln\left(P_{AC}^{T}\right) - \frac{1}{3}\left[-\ln\left(P_{BC}^{T}\right) - \ln\left(P_{AB}^{T}\right)\right]$

Obviously this is the sum of eqs. 4 and 5, that is, $\frac{2}{3}\ln\left(P_{AC}^{T}\right) + \frac{1}{3}\left[\ln\left(P_{BC}^{T}\right) + \ln\left(P_{AB}^{T}\right)\right]$ so that in fact $\ln\left(P_{AC}^{CCD}\right) = \ln\left(P_{AB}^{CCD}\right) + \ln\left(P_{BC}^{CCD}\right)$, i.e. transitivity is given, or $P_{AC}^{CCD} = \sqrt[3]{\left(P_{AC}^{T}\right)^2 P_{AB}^{T} P_{BC}^{T}}$ much like the EKS formula* $P_{AC}^{EKS} = \sqrt[3]{\left(P_{AC}^{F}\right)^2 P_{AB}^{F} P_{BC}^{F}}$.

* quoted already as eq. 8.2.5a

Selvanathan and Rao (1992) provided a generalization of the CCD approach in the following respects

1. by introducing country weights lacking in the CCD method, and
2. motivating (justifying) the CCD formula by deriving it from a regression equation and thus achieving a measure of the standard error of the generalized CCD-index (denoted GCCD index).

As in the case of the EKS method a CCD index P_{ij}^{CCD} can be interpreted as combining the direct Törnqvist index P_{ij}^{T} with all possible indirect comparisons between i and j via a third country.

There are, however, pronounced differences between the indirect ones in that for example an indirect comparison between the USA and the UK through France would be more reliable than a comparison through India.[109] The idea of introducing country weights expressing the similarity of the country as compared with the numéraire country was therefore straightforward. The resulting formula of the econometric model based GCCD index turned out to be extraordinarily complicated and also the understandability of the approach still needs to be demonstrated.

d) Model based approaches

Some attempts are made to give an index such as the CCD index a motivation in terms of a regression equation. Hence there is a kind of stochastic-theory-approach-(see **sec. 2.3**) in the case of interspatial comparisons. Furthermore, we also find an economic-theory approach based on a microeconomic model of utility maximization in this area.

The general idea of this reasoning is to aim at a (true) cost-of-living index (COLI) on the basis of the base country's or the reference country's utility function and (fixed) utility level. The assumption of a household spending money according to a given cost-function (minimum cost to attain a given utility level) simultaneously in two countries appears somewhat farfetched, however.

This applies with even greater force to some deliberations leading to an "economic" quantity index equal to the ratio of (minimum) cost, under the assumption of a common vector of prices and that all countries (!) have the same (international) preference structure.[110] It has been shown, for example, that such an index is bounded from below (above) by a weighted geometric mean of the GEKS-type using Paasche quantity indices (Laspeyres quantity indices). Here again the assumption of homotheticity, well known for its restrictive and unrealistic nature, has to be made and the result of those considerations is supposed to give support to the alleged suitability of the chained-Fisher-quantity index for volume comparisons across countries.

---

[109] Selvanathan and Rao op.cit.

[110] see Balk, 2001, p.32 for more details.

e) Multilateral generalized Törnqvist method (MGT)

Let $v_{ik} = p_{ik}q_{ik} / \sum_k p_{ik}q_{ik}$ denote the share of total expenditure in country i spent on commodity k, then the multilateral generalized Törnqvist (MGT) *quantity* index is given as $Q_{ij}^{GT} = \prod_{k=1}^{n} \left(\frac{q_{jk}}{q_{ik}} \right)^{m_k}$ and correspondingly the MGT *price* index as $P_{ij}^{GT} = \prod_{k=1}^{n} \left(\frac{p_{jk}}{p_{ik}} \right)^{m_k}$, where m_h is (in contrast to the "usual" Törnqvist index) a function of expenditure shares of all countries, not only just the two compared ones, i.e. country i (base) and j. Thus, for example, a geometric mean of all m countries as regards the expenditure share of commodity k may take the form $m_k = \prod_{i=1}^{m} (v_{ik})^{1/m} / \sum_{k=1}^{n} \prod_{i=1}^{m} (v_{ik})^{1/m}$, which has been proposed for a Walsh-type generalized Törnqvist index. Also, an arithmetic mean type (instead of geometric mean Walsh-type) generalization was brought into play.

Perhaps the crucial shortcoming of all methods based on the Törnqvist formula is that they violate factor reversibility. They are also unable to satisfy the mean value property, let alone additivity. On the other hand, the index m_k (weight of quantity k) in Q_{ij}^{GT} only depends on the type of commodity k as it is a function of all countries and not just the two compared ones (i and j). For this reason all multilateral comparisons on the basis of the MGT concept are transitive. Furthermore, these methods (cp. Hill, 1997) are not subject to the Gerschenkron effect because no such thing as a common price vector or a common basket (quantity vector) is involved.

References

Abraham, K.G., Greenlees, J.S., and Moulton, B.R. (1998): "Working to Improve the Consumer Price Index", *Journal of Economic Perspectives* 12 (1), 27-36.

Aczel, J. and Eichhorn, W. (1974): "A Note on Additive Indices", *Journal of Economic Theory*, 8, 525-529.

Adelman, Irma (1958): "A New Approach to the Construction of Index Numbers", *Review of Economics and Statistics*, 240-299.

Advisory Commission (1995): "Toward a More Accurate Measure of the Cost of Living", *Interim Report to the Senate Finance Committee (USA).*

Advisory Commission (1996): "Toward A More Accurate Measure of The Cost of Living", *Final Report to the Senate Finance Committee (USA).*

Afriat, S.N. (1977): *The Price Index*, Cambridge: Cambridge University Press.

Afriat, S.N. (1981): "On the constructability of consistent price indices between several periods simultaneously", in: Deaton, A., Essays in the theory and measurement of consumer behaviour in honour of Sir Richard Stone, Cambridge University Press, 133-161.

Al, P.G., Balk, B.M., de, Boer, S. and den Bakker, G.P. (1986): "The use of chain indices for deflating the national accounts", *Statistical Journal of the United Nations Economic Commission for Europe*, 4, 347-368.

Allen, R.G.D. (1963): "Price Index Numbers", *International Statistical Review*, 31, 281-301.

Allen, R.G.D. (1975): *Index Numbers in Theory and Practice*, London: Macmillan Press.

Allen, R. and Diewert, W.E. (1981): "Direct versus implicit superlative index number formulas", *Review of Economics and Statistics*, LXIII, 430-435.

Altermann, J., and Marimont, M. (1970): "Prices and Price Analysis in the Framework of the National Accounts", *Review of Income and Wealth*, 143-172.

Archer, C.A. (1997): "Review on the Factorial Approach Providing the True Index of Cost of Living", *Journal of the Royal Statistical Society Series A*, 142 (1), 79ff.

Astin, J. (1999): The European Union Harmonized Indices of Consumer Prices, Reykjavik.

Baker, D. (1998) (ed.): "Getting prices Right", Washington.

Balassa, Bela (1964): "The Purchasing Power Parity Doctrin: A Reappraisal", *Journal of Political Economy*, 72, 584ff.

Balassa, Bela (1973): "Just How Misleading Are Official Exchange Rate Conversions? A Comment", *The Economic Journal*, 83, 1258-1267.

Balastier, Peter (1990): "Konsumentenpreisindizes Theorie und praktische Bedeutung", Stuttgart.

Balintfy, J.L., Neter, J. and Wassermann, W. (1970): "Binary and Chain Comparisons With an Experimental Linear Programming Food Price Index", *The Review of Economics and Statistics*, 52, 324-330.

Balk, B.M. (1980): "Seasonal Commodities and the Construction of Annual and Monthly Price Indexes", *Statistische Hefte*, 21, 110-116.

Balk, B.M. (1981): "A Simple Method for Constructing Price Indices for Seasonal Commodities", *Statistische Hefte*, 22, 72-78.

Balk, B.M. (1981): "Second Thoughts on Wald's Cost-of-Living Index and Frisch's Double Expenditure Method", *Econometrica*, 49, 1553-1558.

Balk, B.M. (1983): "A Note on the True Factorial Price Index", *Statistische Hefte*, 24, 69-72.

Balk, B.M. (1985): "A Simple Characterization of Fisher's Price Index", *Statistische Hefte*, 26, 59-63.

Balk, B.M. (1989): "Changing Consumer Preferences and the Cost-of-Living Index: Theory and Nonparametric Expressions", *Zeitschrift für Nationalökonomie*, 50, 157-169.

Balk, B.M. (1989): "On van Ijzeren's approach to international comparisons and its properties", *Statistische Hefte*, 30, 295-315.

Balk, B.M. (1990): "On Calculating Cost-of-Living Index Numbers for Arbitrary Income Levels", *Econometrica*, 58, 75-92.

Balk, B.M. (1991): "Book Review of Robert A. Pollak, The Theory of the Cost-of-Living Index", *European Journal of Political Economy*, 7, 234-240.

Balk, B.M.(1993): "Malmquist Productivity Indexes and Fisher Ideal Indexes: Comment", *The Economic Journal*, 103, 680-682.

Balk, B.M. (1995) "Approximating a Cost-of-Living Index from Demand Functions", *Economic Letters*, 49, 147-155.

Balk, B.M. (1995): "Axiomatic Price Index Theory: A Survey", *International Statistical Review*, 63 (1), 69-93.

Balk, B.M. (1995): "On the use of unit value indices as consumer price subindices, Report (Dept. of Statistical Methods, Statistics Netherlands, Voorburg). Presented at the Joint ECE/ILO Meeting on Consumer Price Indices, 20-24 November, Geneva, at the International Seminar "Improving the Quality of Price Indices", 18-20 December, Florence, and at the Fourth Meeting of the International Working Group on Price Indices, 22-24 April 1998, Washington DC.

Balk, B.M. (1996): "A Comparison of Ten Methods for Multilateral International Price and Volume Comparison", *Jounal of Official Statistics*, 12, 199-222.

Balk, B.M. (2001): "Aggregation Methods in International comparisons: What have we Learned?", 1-42.

Balk, B.M. and Diewert, W.E. (2001): "A Characterization of the Törnqvist Price Index", *Economic Letters*, 73, 279-281.

Balk, B.M. and Kersten, H.M.P. (1986): "On the Precision of Consumer Price Indices caused by the Sampling Variability of Budget Surveys", *Journal of Economic and Social Measurement*, 14, 19-35.

Banerjee, K.S. (1959): "A Generalization of Stuvel's Index Number Formula", *Econometrica*, 27 (4), 676-678.

Banerjee, K.S. (1961): "A Unified Statistical Approach to Index Numbers Problems", *Econometrica*, 29, 591-601.

Banerjee, K.S. (1963): "Best Linear Unbiased Index Numbers Obtained Through Factorial Approach", *Econometrica*, 31, 712-718.

Banerjee, K.S. (1975): Cost of Living Index Numbers: Practice, Precision, and Theory, New York: Dekker.

Banerjee, K. S. (1977): On the Factorial Approach Providing the True Cost of Living, Göttingen.

Banerjee, K.S. (1979): "A Note on the Divisia Indices", *Statistische Hefte*, 20, 172-175.

Banerjee, K.S. (1979): "An Interpretation of the Factorial Indexes in the Light of Divisia's Integral Indexes", *Statistische Hefte*, 20, 261-269.

Banerjee, K.S. (1984): "Some observations on the log-change index numbers and their connection with factorial indexes", *Statistische Hefte*, 25, 143-158.

Barnett, W. A. (1980): "Economic Monetary Aggregates: An Application of Index Number and Aggregation Theory", *Journal of Econometrics*, 14 (1), 11-48.

Barnett, W.A., Offenbacher, E.K. and Spindt, P.A. (1984): "The New Divisia Monetary Aggregates", *Journal of Political Economy*, 92 (6), 1049-1085.

Barnett, W.A., Fisher, D. and Serletis, A. (1992): "Consumer Theory and the Demand for Money", *Journal of Economic Literature*, 2086-2119.

Barr, A. (1996) "Measuring growth: Is the new US approach better?", *Economic Outlook*, 18-22.

Barrow, Michael (1988): Statistics for Economics, Accounting and Business Studies, London and New York: Longmann.

Bartik, T. (1987): "The estimation and demand parameters in hedonic price models," *Journal of Political Economy*, 95, 81-88.

Basmann, R.L. and Slottje, D. (1987): "The sensitivity of the true cost-of-living to price and income induced changes of aggregate consumers' tastes", *Journal of Business and Economic Statistics*, 5, 483-498.

Blackorby, Ch. and Russel, R. (1978): "Indices and Subindices of the Cost of Living and the Standard of Living", *International Economic Review*, 229-240.

Blackorby, C. and Primont, D. (1980): "Index numbers and consistency in aggregation", *Journal of Economic Theory*, 22, 87-98.

Boskin, M., Dulberger, E.R., Gordon, R.J., Gilliches, Z. and Jorgensen, D.W. (1998): "Consumer Price, the Consumer Price Index, and the Cost of Living", *Journal of Economic Perspectives*, 12 (1), 3-26.

Bossert, W. and Pfingsten, A. (1987): "The circular and time reversal tests reconsidered in economic price index theory", *Statistische Hefte*, 28 (4), 301-315.

Brachinger, H. W. (1999): Inflationsmessung in der Schweiz: Konzeption und mögliche Verzerrungen, Frankfurt: Deutsche Bundesbank.

Braithwait, S. (1968): "Real income Levels in Latin America", *Review of Income and Wealth*, 14 (2), 113-182.

Braithwait, S. (1980): "The Substitution Bias of the Laspeyres Price Index: An Analysis Using Estimated Cost of Living Indices", *American Economic Review*, 70, 64-77.

Brown, J.N. and Rosen, S. (1982): "On estimation of structural hedonic price models", *Econometrica*, 50, 765-768.

Bundesbank, Deutsche (1998): "Probleme der Inflationsmessung", *Monatsberichte der Deutschen Bundesbank*, 53-66.

Carruthers, A. G., D J Sellwood, P W Ward (1980), Recent Developments in the Retail Prices Index, The Statistician, 29, 1-32.

Caves, D.W., Christensen, L.R. and Diewert, W.E. (1982): "Multilateral Comparisons of Output, Input adn Productivity Using Superlative Index Numbers", *Economic Journal*, 92, 73-86.

Caves, D.W., Christensen, L.R. and Diewert, W.E. (1982): "The Economic Theory of Index Numbers and the Measurement of Input, Output and Productivity", Econometrica, 50, 1393-1413.

Christensen, L., D., Jorgensen and Lau, L. (1973): "Transcendental Logarithmic Production Frontiers", *Review of Economics and Statistics*, 28-45.
Christensen, L.R., Jorgensen, D.W. and Lau, L.J. (1975): "Transcendental logarithmic utility functions", *American Economic Review*, 65, 367-383.
Clements, K. and Izan, H.Y. (1981): "A Note on Estimating Divisia Index Numbers", *International Economic Review*, 22 (3), 745-747.
Clements, K. and Izan, H.Y. (1987): "The Measurement of Inflation: A Stochastic Approach", *Journal of Business and Economic Statistics*, 5 (3), 339-350.
Combris, Pierre, Lecoq, Sebastien and Visser, Michael (1997): "Estimation of hedonic price equation for Bordeaux wine: does quality matter", *Economic Journal*, 107, 390-402.
Court, A.T. (1939): "Hedonic Price Indices with Automotive Examples", In: Horner, S.L., The Dynamics of Automobile Demand.
Craig, J. (1969): "On the Elementary Theory of Index Numbers", *Journal of the Royal Statistical Society, Series C*, 132 (18), 141-152.
Dalén, J. (1992), Computing Elementary Aggregates in the Swedish Consumer Price Index, Journal of Official Statistics, 8, pp 129-147.
Dalgaard, E. (1997): "Implementing the revised SNA, recommendations on price and volume measures", *Review of Income and Wealth*, 43 (4), 487-503.
David, P. (1962): "The Deflation of Value Added", *Review of Economics and Statistics*, 44, 148-155.
David, P. (1966): "Measuring Real Net Output: A Proposal Index", *Review of Economics and Statistics*, 48, 419-425.
Deaton, A. (1979): "The Distance Function in Consumer Behaviour with Applications to Index Numbers and Optimal Taxation", *Review of Economic Studies*, 40, 391-405.
Deaton, A. (1981): "Essays in the Theory and Measurement of Consumer Behaviour in Honour of Sir Richard Stone", *Cambridge Univ. Press*, 163-208.
Deaton, A. (1998): "Getting Prices Right: What Should Be Done?", *Journal of Economic Perspectives*, 12 (1), 37-46.
de Haan, I (2001): "Generalised Fisher Price Index", unpublished paper, Voorburg.
Diehl, H. (1978): "Examination of Purchasing-Power-Parity-Methods with a View to Choosing the Most Appropriate Method for a European-Community Purchasing-Power-Parity Model", In: *Eichhorn, W., Menn, R. Et al. Theory and Applications of Economic Indices*, Würzburg, 143-160.
Diewert, W. E. (1976): "Exact and Superlative Index Numbers", *Journal of Econometrics*, 4, 115-145.
Diewert, W. E. (1978): "Superlative Index Numbers and Consistency in Aggregation", *Econometrica*, 46, 883-900.
Diewert, W. E. (1981): "The Economic Theory of Index Numbers: A Survey", In: Deaton, A. , Essays in the theory and measurement of consumer behaviour in honour of Sir Richard Stone, Cambridge Univ Press., 163-208.
Diewert, W. and Montmarquette, C. (1983): "The Theory of the Cost-of-Living Index and the Measurement of Welfare Change", in: Diewert, W. E. and Montmarquette, C. (eds.): "Price Level Measurement", *Proceedings from a conference sponsored by Statistics Canada*, Ottawa, 163-233.
Diewert, W. (1992): "Fisher Ideal Output, Input and Productivity Indexes Revisited", *Journal of Productivity Analysis*, 3, 211-248.

Diewert, W.E. and Nakamura, A.O. (1993): Essays in Index Number Theory, Amsterdam, London, New York, North Holland.
Diewert, W. (1995): "Axiomatic and Economic Approaches to Elementary Price Indexes", *National Bureau of Economic Research Working Paper Series*, 5104.
Diewert, W.E. (1998): "Index Number Issues in the Consumer Price Index", *Journal of Economic Perspectives*, 12 (1), 47-58.
Diewert, W.E. (1999): "The Consumer Price Index and Index Number Purpose", Paper presented at the Fifth Meeting of the International Working Group on Price Indices (The Ottawa Group), Reykjavik.
Diewert, W.E. (2000): "The Consumer Price Index and Index Number Purpose", Discussion Paper 00-02 Department of Economics, The University of British Columbia Vancouver Canada.
Divisia, Francois (1925): "L'indice montaire et la theorie de la monnaie", *Revue d'Economique Politique*.
Drechsler, L. (1973): "Weighting of Indey Numbers in Multilateral International Comparisons", *Review of Income and Wealth*, 19, 17-34.
Drechsler, L. and Krzeczkowska, E. (1982): "Purchasing Power Parities in International Comparisons: Quantity vs. Price Changes", *Review of Income and Wealth*, 28, 253-259.
Drobisch, M.E.: "über Mittelgrößen und die Anwendbarkeit derselben auf die Berechnung des Steigens und Sinkens des Geldwertes," *Berichte über die Verhandlungen der königlich sächsischen Gesellschaft der Wissenschaften zu Leipzig, Mathematisch-Physikalische Classe*, 23, 25-48.
Eichhorn, W. and Voeller, J. (1976): *Theory of the Price Index. Fisher's Test Approach and Generalizations*, Berlin, Heidelberg, New York: Springer-Verlag.
Eichhorn, W., Henn, R. and Opitz, S. (1978): "Theory and Applications of Economic Index Numbers (Proceedings of an International Seminar)", Physica, Würzburg.
Eichhorn, W. and Voeller, J. (1983): "Axiomatic Foundation of Price Indexes and Purchasing Power Parities", In: W. Diewert and C. Montmarquette, eds., *Price Level Measurement*, Min. of Supply and Services Canada: Ottawa, 411-450.
Eichhorn, W. (1988): *Measurement in Economics*, Heidelberg: Physica Verlag.
Elbel, Günter (1997): "Zur Einführung Harmonisierter Verbraucherpreisindizes in Europa," *WiSta*, 3, 187.
Epple, D. (1987): "Hedonic prices and implicit markets, Estimating demand and supply functions for differentiated products", *Journal of Political Economy*, 95, 59-80.
Euler, M. (1992): "Einkommens- und Verbraucherstichprobe", *WiSta*, 7, 483.
Feenstra, Robert C. (1994): "New Product Varieties and the Measurement of International Prices," *American Economic Review*, 34, 157-179.
Feenstra, Robert C. (1995): "Exact Hedonic Price Indexes", *Review of Economics and Statistics*, 77, 634-653.
Fisher, I. (1922): *The Making of Index Numbers*, Boston: Houghton Mifflin.
Fisher, F.M and Shell, K. (1972): The Economic Theory of Price Indices: Two Essays on the Effects of Taste, Quality and Technological Change, New York.
Fisher, F. and Zvi, Grilliches (1995): "Aggregate Price Indices, New Goods, and Generics", *Quaterly Jounal of Economics*, 110, 229-244.
Flexner, W. (1959): "An Analysis of the Nature of Aggregates at Constant Prices", *Review of Economics and Statistics*, 41, 400.

Ford, J.L., Peng, W.S. and Mullineux, A.W. (1992): "Financial Innovation and Divisia Monetary Aggregates", *Oxford Bulletin of Economics and Statistics*, 54 (1), 87-101.
Forsyth, F.G. (1978): "The Practical Construction of a Chain Price Index Number", *The Journal of the Royal Statistical Society, Series A*, 141 (3), 348-358.
Forsyth, F.G. and Fowler, R.F. (1981): "The Theory and Practice of Chain Price Index Numbers", *Journal of the Royal Statistical Society, Series A*, 144 (2), 224-246.
Fowler, R.F. (1974): "An Ambiguity in the Terminology of Index Numbers Construction", *The Journal of the Royal Statistical Society, Series A*, 137 (1), 75-88.
Frisch, R. (1936): "Annual Survey of General Economic Theory: The Problem of Index Numbers", *Econometrica Vol. 4*, 1-38.
Frisch, R. (1954): "Some Basic principles of Price of Living Measurement; A Survey Article", *Econometrica*, 22 (4), 407-421.
Fuchs-Seliger, S. (1986): "Relationships between Economic and Statistical Price Indices", *Statistische Hefte*, 27.
Fuchs-Seliger, S., Pfingsten A. (1986): "Cost of Living Indices Based on Demand Functions, *Zeitschrift für Nationalökonomie*, 46.
Funke, H. and et al. (1979): "Fisher's Circular Test Reconsidered", *Schweizerische Zeitschrift für Volkswirtschaft und Statistik*, (4), 677-687.
Funke, H. (1988): "Mean Value Properties of the Weights of Linear Price Indices", *Eichhorn, W., Measurement in Economics*, 99-115.
Gandal, N. (1995): "Competing compatability standards and network externalities in the PC software market", *Review of Economics and Statistics*, 77, 599-608.
Geary, R.C. (1958): "A Note on Comparisons of Exchange Rates and Purchasing Power between Countries", *Journal of the Royal Statistical Society*, 121, 97-99.
Gerardi, D. (1982): "Selected Problems of Inter-country Comparisons on The Basis of the Experience of the EEC", *The Review of Income and Wealth*, 28, 381-405.
Gillingham, Robert (1983): "Measuring the Cost of Shelter for Homeowners: Theoretical and Empirical Considerations", *Review of Economics*, LXV, 254-266.
Gini, C. (1931): "On the Circular Test of Index Numbers", *International Review of Statistics*, 9 (2), 3-25.
Gleissner, W. (1990): "A Characterization of the Price Indices satisfying the Eichhorn-Voeller Axioms", *Statistische Hefte*, 31, 241-250.
Gleissner, W. (1992): "A parametrization of the price indices characterized by the Eichhorn-Voeller axioms", *Statistische Hefte*, 33, 75-82.
Goodhat, Charles (2001): "What Weight should be Given to Asset Prices in the Measurement of Inflation", *Economic Journal*, 111, 335-356.
Griliches, Zvi (1995): "Aggregate Price Indices, New Goods, and Generics", *Quaterly Journal of Economics*, 110, 229-244.
Gropper, M.L., Deck, L.B. and McConnell, K.E. (1988): "On the Choice of Functional Form for Hedonic Price Functions", *Review of Economics and Statistics*, 70, 668-675.
Hansen, B. and Lucas, E.F. (1984): "On the Accuracy of Index Numbers", *Review of Income and Wealth*, 30, 25-38.
Heien, D. and Dunn, J. (1985): "The True Cost-of-Living Index With Changing Preferences", *Journal of Business and Economic Statistics*, 3, 332-335.
Hershey, Robert D. (1996): "Panel Likely to Recommend New Price Index, to Better Reflect Actual Buying", *New York Times*.

Hill, P. (1988): "Index Number Theory and External Trade", Eurostat News-Special Edition, Proceedings of a Seminar held in Luxembourg.
Hill, R. (1997): "A Taxonomy of Multilateral Methods for Making International Comparisons of Prices and Quanitities," *Review of Income and Wealth*, 43 (1), 49-69.
Hill, R. (1999): "Comparing Price Levels Across Countries Using Minimum Spanning Trees", *Review of Economics and Statistics*, 81, 135-142.
Hill, T.P. (1984): "Introduction: The Special Conference on Purchasing Power Priorities," *The Review of Income and Wealth*, 30, 125-133.
Hill, T.P. (1993): "Price and Volume Measures", in System of National Accounts 1993 (Brussels/Luxembourg, New York, Paris, New York, and Washington, D.C.: Commission of the European Communities, IMF, OECD, World Bank and United Nations), 379-406.
Hillinger, C. (1970): "Comment on Invariance Axioms and Economic Indexes", *Econometrica*, 38, 773-774.
Hodgson, J.S. and Phelps, P. (1975): "The Distributed Impact of Price-Level Variations on Floating Exchange Rates", *The Review of Economics and Statistics*, 57.
Hoffmann, Johannes (1998): *Probleme der Inflationsmessung in Deutschland*, Deutsche Bundesbank.
Holland, P.W. (1986): "Statistics and Causal Inference", *Journal of the American Statistical Association*, 81.
Hoover, E.D. (1968): "Index Numbers, II, Practical Applications", *International Encyclopedia of the Social Sciences*, 7, 159-165.
Hulten, C.R. (1973): "Divisia Index Numbers", *Econometrica*, 41 (6), 1017-1025.
Ijzeren, J. Van (1958): "A Note on the useful properties of Stuvel's index numbers", *Econometrica*, 26, 429-439.
Ijzeren, J. Van (1988): "Weighting and Additivity Problems of Multilateral Comparisons", In: Eichhorn, W., *Measurement in Economics*, 157-164.
Iklé, D.M. (1972): "A New Approach to the Index Number Problem", *Quarterly Journal of Economics*, 86, 188-211.
Johnson, J.D. and Slottje, D. (1993): "Consumer Behavior and Changes in the Prices of Monetary Aggregates", *Journal of Economic and Social Measurement*, 19, 1-19.
Johnson, Christopher and Brisco, Simon (1995): *Measuring the Economy, A Guide to Understanding Official Statistics*, London: Penguin Books.
Jones, L.E. (1988): "The Characteristics model, hedonic prices and the clientele effect", *Journal of Political Economy*, 96, 551-567.
Kendall, M.G. (1969): "Studies in the History of Probability and Statistics, XXI: The early History of Index Numbers", *Review of the International Statistical Institute*, 37, 1-12.
Kersten, H.M.P. (1985): "Non response Assessment of a Consumer Price Index", *Journal of Business and Economic Statistics*, 3, 336-343.
Khamis, S.H. (1972): "A New System of Index Numbers for National and International Purposes", *Journal of the Royal Statistical Society*, 135, 96-121.
Khamis, S.H. (1984): "On Aggregation methods for International Comparisons", *Review of Income and Wealth*, 30, 185-205.
Kloek, T. and de Wit, G.M. (1961): "Best Linear and Best Linear Unbiased Index Numbers", Econometrica, 29, 602-616.
Kloek, T. and Theil, H. (1965): "International Comparisons of Prices and Quantities", *Econometrica*, 33, 535-556.

Konüs, A.A. (1936): "The Problem of the True Index of the Cost of Living", *Econometrica*, 4, 10-29.
Köves, P. (1983): *Index Theory and Economic Reality*, Budapest: Akad. Kiad.
Kravis, I.B. and Lipsey, R.E. (1969): "International Price Comparison by regression methods", *International Economic Review*, 10, 223-246.
Kravis, I.B. (1984): "Comparative Studies of National Incomes and Prices", *Journal of Economics Literature*, 22, 1-39.
Krtscha, M. (1988): "Axiomatic Characterization of Statistical Price Indices", In: Eichhorn, W., *Measurement in Economics*, 117-133.
Lau, L.J. (1979): "On Exact Index Numbers", *Review of Economics and Statistics*, 61, 73-82.
Lloyd, P.J. (1975): "Substitution Effects and Biases in Nontrue Price Indices", *American Economic Review*, 65, 301-313.
Locker, H.K. (1984): "On the Estimation of Purchasing Power Parities on the Basic Heading Level, The Review of Income and Wealth, Vol. 30, 135-152..
Lorenzen, G. (1989): "Log-Ratios and the Logarithmic Mean", *Statistische Hefte*, 30, 61-75.
Makaronidis, A. (1999): Fehlerquellen und Mindeststandards beim Harmonisierten Verbraucherpreisindex, Frankfurt: Deutsche Bundesbank, 129-148.
Martini, Marco (1992): *I numeri in un approcio assiomatico*, Milano.
Martini, Marco (1992): "A General Function of Axiomatic Index Numbers", *Journal of the Italian Statistical Society*, 1 (3), 359-376.
Martini, M. (1996): The frequency of updating of the base of consumer price index numbers (expertise for Eurostat, unpublished).
Martini, M. (2000): Index Numbers, Axiomatic Theory and Applications (unpublished translation from Italian by TES, Luxembourg).
Mauser, M.E. and Mc Donald, R.J. (1988): "An Analysis of Substitution Bias in Measuring Inflation", *Econometrica*, 56, 909-930.
Mc Carthy, P.J. (1968): "Index Numbers, III Sampling", *International Encyclopedia of the Social Sciences*, 7, 165-169.
Mc Lennan, W. (1998): Introduction of Chain Volume Measures in the Australian National Accounts.
Montgomery (1937): The Mathematical Problem of the Price Index, London.
Moulton, B. R. (1996): "Bias in the Consumer Price Index: What Is the Evidence?", *U.S. Department of Labor, Bureau of Labor Statistics, Working Paper*, 294.
Mudgett, B. D. (1951): *Index Numbers*, New York: Wiley Sons.
Muellbauer, John (1974): "Prices and Inequality: The United Kingdom Experience", *Economic Journal*, 84, 32-55.
Muellbauer, John (1974): "Household production theory, quality, and the hedonic technique", *American Economic Review*, 977-994.
Muellbauer, John (1975): "The Cost of Living and Taste and Quality Change", *Journal of Economic Theory*, 10, 269-283.
Mukherjee, M. and Prasado Rao, D.S. (1973): "On Consistent Intergroups Comparisons of Purchasing Power of Money", *Review of Income and Wealth*, 18, 35-48.
Nelson, R.A., Tanguay, T.L. and Patterson, C.D. (1994): "A Quality Adjusted Price Index for Personal Computers", *Journal of the American Statistical Association*, 12, 23-31.

Nerlove, Marc (1995): "Hedonic price functions and the measurement of preferences: The case of Swedish wine consumers", *European Economics Review*, 39, 1697-1716.

Nicholson, J.L. (1967): "The Measurement of Quality Changes", *The Economic Journal*, 77, 512-530.

Noe, N.N. and von Fürstenberg, G. (1972): "The Upward Bias in the Consumer Price Index Due to Substitution", *Journal of Political Economy*, 80, 1280-1294.

Nordhaus, William D. (1998): "Quality Change in Price Indexes", *Journal of Economic Perspectives*, 12 (1), 59-68.

Nuxoll, D.A. (1994): "Differences in Relative Prices and International Differences in Growth Rates", *American Economic Review*, 84, 1423-1436.

OECD, Statistics Directorate (1994): *Sources and Methods: Consumer Price Indices*, Paris.

Olt, B. (1996): Axiom und Struktur in der statistischen Preisindextheorie, Frankfurt/Main: Peter Lang.

Palgrave, R.H.I. (1886): "Currency and Standard of Value in England, France and India", Memorandum submitted to the Royal Commission on Depression of Trade and Industry Third Report, Appendix B., 312-390.

Pencavel, J.M. (1977): "Constant Utility Index Numbers of Real Wages", *American Economic Review*, 67 (2), 91-100.

Pfingsten, A. (1985): "A Note on the True Factorial Price Index-Addendum to a Note by Balk", *Statistische Hefte*, 26, 65-69.

Pfouts, R.W. (1966): "An Axiomatic Approach to Index Numbers", *Review of the International Statistical Institute*, 34 (2), 174-185.

Pfouts, R.W.: "Index Number Systems", *Econometrica*, 40, 931-934.

Pollack, R.E. (1975): "Subindexes of the Cost-of-Living Index", *International Economic Review*, 16 (1), 135-150.

Pollack, R. A. (1980): "Group Cost-of-Living Indexes", *The American Economic Review*, 70, 273-278.

Pollack, R.A. (1981): "The Social Cost-of-Living Index", *Journal of Public Economics*, 15, 311-336.

Pollack, R.A. (1983): "The Theory of the Cost-of-Living Index", in: *Diewert, W.E. and Montmarquette, C.*, Price Level Measurement, (eds.), Ottawa: Statistics Canada, 87-161.

Pollack, R.A. (1989): "The Theory of the Cost-of-Living Index", *New York, Oxford.*

Pollak, Robert A. (1998): "The Consumer Price Index: A Research Agenda and Three Proposals", *Journal of Economic Perspectives*, 12 (1), 69-78.

Porter-Hudak, S. and Hayes, K. (1991): "A numerical methods approach to calculating cost-of-living indices", *Journal of Econometrics*, 50, 91-105.

Prasada Rao, D.S. and Banerjee, K.S. (1986): "A Multilateral Index Number System Based on the Factorial Approach", *Statistische Hefte*, 27, 297-313.

Richter, M.K., "Invariance Axioms and Economic Indexes", *Econometrica*, 34, 739-755.

Rosen, S. (1974): "Hedonic Prices and Implicit Markets: Product Differentiation in Pure Competition", *Journal of Political Economy*, 82, 34-55.

Ruggles, R. (1983): "The United State, National Income Accounts 1947-1977: Their conceptual Basis and Evolution", In: Foss, M.F., *The U.S. National Income and Product Acconts, University of Chicago Press.*

Ruist, E. (1968): "Index Numbers, International Theoretical Aspects", *International Encyclopedia of the Social Sciences*, 7, 154-159.
Sato, K. (1974): "Ideal index numbers that almost satisfy the factor reversal test", *The Review of Economics and Statistics*, 56, 549-552.
Sato, K. (1976): "The Ideal log. change Index Number", *The Review of Economics and Statistics*, 58, 223-228.
Sato, K., (1976): "The Meaning and Measurement of the Value Added Index", *Review of Economics and Statistics*, 58, 434-442.
Sato, K. (1980): "The Price Reversal Test and Economic Indexes", *Statistische Hefte*, 21, 127-130.
Schimmler, K. (1973): "On National Accounts at Constant Prices", *Review of Income and Wealth*, 19, 457-459.
Schoch, H. and Wagener, M. (1984): "Price Index Paradoxa", Methods of Operation Research, 48, 113ff.
Schultz, M. (1939): "A Misunderstanding in Index-Number Theory: The True Konüs Condition on Cost-of-Living Index Numbers and ist Limitations", *Econometrica*, 7, 1-9.
Schultze, C.L. (2003): "The Consumer Price Index: Conceptual Issues and Practical Suggestions", *Journal of Economic Perspectives*, 17 (1), 3-22.
Selvanathan, E.A., "A note on stochastic approach to index numbers", *Journal of Business and Economic Statistics*, 19, 471-474.
Selvanathan, E.A. and Prasada Rao, D.S. (1992): "An Econometric Approach to the Construction of Generalized Theil-Törnquist Indices for Multilateral Comparisons", *Journal of Econometrics*, 54, 335-346.
Selvanathan, E.A. and Prasada Rao, D.S. (1994): *Index Numbers, A Stochastic Approach*, Basingstoke: Macmillan.
Sen, A. (1979): "The Welfare Basis of Real Income Comparisons: A Survey", *Journal of Economic Literature*, 17, 1-45.
Shapiro, M.D. and Wilcox, D.W. (1996): "Mismeasurement in the Consumer Price Index: An Evaluation", *Working Paper 5590, National Bureau of Economic Research*.
Silver, M. and Heravi, S. (2001): "Scanner Data nad the Measurement of Inflation", *Economic Journal*, 111, 383-404.
SNA (1993): *The System of National Accounts*, Washington.
Stanley, L. R. and Tschirhart, J. (1991): "Hedonic Prices for a Nondurable Good: The Case of Breakfast Cereals", *Review of Economics and Statistics*, 73, 537-541.
Staehle, H. (1935): "A Development of the Economic Theory of Price Index Numbers", *Review of Economic Studies*, 2 (3), 163-188.
Steiger, H.H. (1970): "Definition und Messung von Preisstrukturen", *Statistische Hefte*, 11, 57-63.
Stigler, G.I. and Becker, G.S. (1977): "De gustibus non est disputandum", *American Economic Review*, 67, 76-90.
Strecker, H. (1963): "Zur Genauigkeit wirtschaftsstatistischer Grössen", *Metrika*, 6, 133-142.
Stuvel, G. (1957): "A New Index Number Formula", *Econometrica*, 25, 123-131.
Stuvel, G. (1989): *Index-number Problem and its Solution*, Basingstoke: Macmillan.
Subramanian, S. (1965): "Consistency of Fisher's Tests", *Econometrica*, 33 (3), 619-623.

Summers, R. (1973): "International Price Comparisons Based Upon Incomplete Data", *Review of Income and Wealth*, 19, 1-16.
Swamy, S. (1965): "Consistency of Fisher's Tests", *Econometrica*, 33 (3), 619-623.
Szulc, B.J. (1983): "Linking Price Index Numbers", In Diewert, W.E. and C. Montmarquette, eds., *Price Level Measurement*, Min. of Supply and Services Canada: Ottawa, 537-566.
Teekens, R. (1989): "Consumer Price Indices in the European Community, Similarities, Differences and Proposals for Harmonization", *Eurostat*.
Theil, H. (1960): "Best Linear Index Numbers of Prices and Quantities", *Econometrica*, 28 (2), 464-480.
Theil, H. (1965): "The Information Approach to Demand Analysis", Econometrica, 33 (1), 67-87.
Theil, H. (1973): "A New Index Number Formula," *Review of Economics and Statistics*, 55, 498-502.
Theil, H. (1974): "More on Log-Change Index Numbers", *Review of Economics and Statistics*, 56, 552-554.
Triplett, J. and Mc.Donlad, R. (1977): "Assessing the quality error in output measures: the case of refrigerators", *Review of Income and Wealth*, 23, 137-156.
Turvey, R. (1989): *Consumer Price Indices: An ILO Manual*, Geneva: International Labour Office.
Törnqvist, L., Vartia P. and Vartia, V.O. (1985): "How Should Relative Changes be Measured?", *The American Statistician*, 39, 43-46.
Usher, D. (1974): "The Suitability of the Divisia Index for the Measurement of Economic Aggregates", *Review of Income and Wealth*, 20, 273-288.
Vartia, Y.O. (1976): "Ideal Log-Change Index Numbers", *Scandinavian Journal of Statistics*, 3, 121-126.
Vartia, Y.O. (1978): "Fisher's FiveTines Fork and Other Quantum Theories of Index Numbers", In: Eichhorn, W., Henn, R., Opitz, O., Sheppard, R.W., *Theory and Applications of Economic Indices*, 271-296.
Vartia, Y.O. and Vartia, P.L.I.: "Descriptive Index Number Theory and the Bank of Finland Currency Index", *Scandinavian Journal of Economics*, 86 (3), 352-364.
Vogt, A. (1977): "Zum Indexproblem, Geometrische Darstellung sowie eine neue Formel", *Schweizerische Zeitschrift für Volkswirtschaft und Statistik*, 113, 73-80.
Vogt, A. (1978): "Der Wertindextreue-Test und eine Vereinfachung des Indexproblems", *Statistische Hefte*, 19, 131-139.
Vogt, A. (1978): "Divisia Indices on Different Paths", In Eichhorn, W., Henn, R. and S. Opitz, Hg., *Theory and Applications of Economic Index Numbers*, Physica: Würzburg, 297-305.
Vogt, A. (1979): Das statistische Indexproblem im zwei Situationen Fall, Zürich.
Vogt, A. (1981): "Characterizations of indices, especially of the Stuvel and the Banerjee index," *Statistische Hefte*, 22, 241-245.
von Auer, L. (2001): "Consistency in Aggregation", *Jahrbücher für Nationalökonomie und Statistik*, 224(4), 383-398.
von Bortkiewicz, L. (1923): „Zweck und Struktur einer Preisindexzahl", Nordisk Statistik Tidsskrift, 2, 369-408.

von der Lippe, P. (1999): "Kritik internationaler Empfehlungen zur Indexformel für Preisindizes in der amtlichen Statistik: Einige Bemerkungen zur ökonomischen Theorie der Indexzahlen und zu Kettenindizes", *Jahrbücher für Nationalökonomie und Statistik*, 218 (3+4), 385-414.
von der Lippe, P. (1999): "The Political Role of Official Statistics in the Former GDR (East Germany)", *Historical Social Research*, 24 (4), 3-28.
von der Lippe, P. (2000): "Der Unsinn von Kettenindizes", *Allgemeines Statistisches Archiv*, 84, 67-82.
von der Lippe, P. (2000): "Preisindizes mit konstanten Wertgewichten: Bemerkungen zu einem Vorschlag von Werner Neubauer", *Jahrbücher für Nationalökonomie und Statistik*, 220 (6), 742-758.
von der Lippe, P. (2001): *Chain Indices, A Study in Price Index Theory*, Stuttgart: Federal Statistics Office of Germany, Vol. 16 (Spectrum of Federal Statistics); for more information see http://www.von-der-lippe.de/werbeseite1.php.
von der Lippe, P. (2005): "Das Ideal des "reinen Preisvergleichs" (The Principle of Pure Price Comparison), *Jahrbücher für Nationalökonomie und Statistik*, 225/4, 499-509.
Wald, A. (1937): "Zur Theorie der Preisindexziffern", *Zeitschrift für Nationalökonomie*, 8, 179-219.
Westergaard, H. (1890): *Die Grundzüge der Theorie der Statistik*, Jena: Gustav Fischer.
Witte, A.D., Sumka, H.I and Erekson, O. (1979): "An estimate of a structural hedonic price model of the housing market, an application of Rosen's theory of implicit markets", *Econometrica*, 47, 1151-1173.
Young, A.H. (1812): "An Inquiry into the Progressive Value of Money in England as marked by the Price of Agricultural Products (Picadilly: Harchard).
Young, A.H. (1992): "Alternative Measures of Change in Real Output and Prices", *Survey of Current Business*, 72, 32-48.